World Geomorphological Landscapes

Series Editor

Piotr Migoń, Institute of Geography and Regional Development, University of Wrocław, Wrocław, Poland

Geomorphology – 'the Science of Scenery' – is a part of Earth Sciences that focuses on the scientific study of landforms, their assemblages, and surface and subsurface processes that moulded them in the past and that change them today. Shapes of landforms and regularities of their spatial distribution, their origin, evolution, and ages are the subject of geomorphology. Geomorphology is also a science of considerable practical importance since many geomorphic processes occur so suddenly and unexpectedly, and with such a force, that they pose significant hazards to human populations. Landforms and landscapes vary enormously across the Earth, from high mountains to endless plains. At a smaller scale, Nature often surprises us creating shapes which look improbable. Many geomorphological landscapes are so immensely beautiful that they received the highest possible recognition – they hold the status of World Heritage properties. Apart from often being immensely scenic, landscapes tell stories which not uncommonly can be traced back in time for millions of years and include unique events. This international book series will be a scientific library of monographs that present and explain physical landscapes across the globe, focusing on both representative and uniquely spectacular examples. Each book contains details on geomorphology of a particular country (i.e. The Geomorphological Landscapes of France, The Geomorphological Landscapes of Italy, The Geomorphological Landscapes of India) or a geographically coherent region. The content is divided into two parts. Part one contains the necessary background about geology and tectonic framework, past and present climate, geographical regions, and long-term geomorphological history. The core of each book is however succinct presentation of key geomorphological localities (landscapes) and it is envisaged that the number of such studies will generally vary from 20 to 30. There is additional scope for discussing issues of geomorphological heritage and suggesting itineraries to visit the most important sites. The series provides a unique reference source not only for geomorphologists, but all Earth scientists, geographers, and conservationists. It complements the existing reference books in geomorphology which focus on specific themes rather than regions or localities and fills a growing gap between poorly accessible regional studies, often in national languages, and papers in international journals which put major emphasis on understanding processes rather than particular landscapes. The World Geomorphological Landscapes series is a peer-reviewed series which contains single and multi-authored books as well as edited volumes.

World Geomorphological Landscapes – now indexed in Scopus® !

More information about this series at https://link.springer.com/bookseries/10852

Frank D. Eckardt
Editor

Landscapes and Landforms of Botswana

Springer

Editor
Frank D. Eckardt
Department of Environment and Geographical Sciences
University of Cape Town
Rondebosch, South Africa

ISSN 2213-2090 ISSN 2213-2104 (electronic)
World Geomorphological Landscapes
ISBN 978-3-030-86101-8 ISBN 978-3-030-86102-5 (eBook)
https://doi.org/10.1007/978-3-030-86102-5

This Springer imprint is published by the registered company Springer Nature Switzerland AG
The registered company address is: Gewerbestrasse 11, 6330 Cham, Switzerland

Fig. 1 Botswana is home to approximately 2 million people, most of whom reside in the east. The country, which is 581,730 km^2 in size, hosts a wide range of largely subtle but also distinct and unique landforms in the continental interior of Southern Africa. *Source* Google Earth

This book is dedicated to Marty McFarlane

Foreword

It is with great pleasure that I write this foreword for *Landscapes and Landforms of Botswana.* As a soil scientist, I appreciate the role of pedogenesis, the location of the different soil types and linkages to geomorphological formations. Pedogeomorphology has over the years helped national planners and decision-makers on numerous projects, given the rapid development of the country. While at the University of Botswana, where I grew as an academic for almost 30 years, geomorphology was an essential course in Environmental Science. Students graduated from UB to work among other things on road and bridge construction, mining and the building of dams. I have had the privilege of getting feedback from their employers who confirmed that geomorphology graduates played a valuable role in their operations.

As a former Head of Environmental Science and Dean of Science at the University of Botswana, I also noted that the study of landscapes and landforms impacted other subjects like geology, biology, chemistry, physics and climatology. It gave students an opportunity to develop interdisciplinary and multidisciplinary perspectives and graduate with an open mind, in preparation for the world of work.

Many of the contributors here have been active in practical, national applications and research. I would like to mention in particular Prof. Susan Ringrose, who has spent many years teaching at the University of Botswana, in the Department of Environmental Science and thereafter was a researcher and director of the Okavango Research Institute. Professor Ringrose has made a great impact on the lives of many Batswana, including the author of this foreword. Professor Julius Atlhopheng, the Dean of Science at UB, is a geomorphologist par excellence. I have read his publications, which are of the highest international standard in the subject. He has mentored junior colleagues and some of them have reached the pinnacle of academia, which is full professorship. Former colleagues at UB, Prof. Marek Wendorff and Prof. Read Mapeo bring the additional geological perspective to this work, while Associate Professor Jeremy Perkins adds the biogeographical dimension. Other local contributors of note are Dr. Peter Eze, Prof. Parida and Prof. Elisha Shemang, fellow colleagues at BIUST and Dr. Senwelo Isaacs, in the Directorate of Intelligence and Security and Dr. Mark Stephens who is now at the University of The Bahamas.

Clearly given the small population size of Botswana, the total number of active geomorphic practitioners is relatively large. This speaks strongly to the landscape of the country, which also continues to attract scientists from all corners of the world. Many authors here have had a long history of geomorphological research in Botswana. This includes Prof. David Thomas, who is an authority on dryland geomorphology and published extensively on southern Africa. I congratulate all contributors for a job well done and conclude by saying that this book will be of value to universities as well as to planners and decision-makers in Botswana and beyond.

I also applaud Associate Professor Frank D. Eckardt for mobilising distinguished academics and researchers. I have no doubt that with their combined wealth of experience in teaching, research, service and leadership, the readers will find rich insights and opportunities for future geomorphic pursuits.

Prof. Otlogetswse Totolo
Vice-Chancellor
Botswana International University
of Science and Technology
Palapye, Botswana

Dedication to Marty McFarlane

February 1st, 1941–Sept 14th, 2018

A kimberlite erupts in the Kalahari, a rising column of hot ash and quartz rises, which generates lightning discharge. Suspended quartz beads melt at high temperatures in the vicinity of lightning bolts, which in an instant turn quartz into molten glass, prior to free fall and accumulation in the sandy regolith overburden. Kalahari kimberlites are hard to locate, given the limited surface expression. McFarlane et al. (2018) suggest that these glassy lightning-induced volcanic beads are a much-needed predictor in pursuit of kimberlites and diamonds. Much of Marty's work had a way of getting your attention. In many instances it would question what has come before but would be based on the detailed examination of sediment grains, cores or air photos.

Marty and her twin sister Hilde were born in London in 1941 at the height of the Blitz. The Patz parents were refugees and had just arrived from Sudeten Czechoslovakia and soon settled in Bearsden, a lush suburb of Glasgow, where Marty attended school and later earned a geography degree from Glasgow University. After the graduation in 1963 she joined the Baker Centenary Expedition to Lake Mwitanzige, Uganda (Former Lake Albert) (Bishop 1967). Upon her return to England, clearly having been influenced by her 6 weeks in East Africa, she attended a course on the Tropical Environments at London University, Institute of Education. Marty thereafter embarked on a PhD at University College London, studying laterites, while also working as a demonstrator in the Geography Department at Makerere University in Uganda and lecturing at the Department of Geography, Nairobi University, Kenya. This period culminated in a well-received and much cited book (McFarlane 1976).

Following the career path of her husband, the family moved to Java, Indonesia, and then back to the England. Remarkably, through these moves, whilst continuing to be the linchpin of the family, Marty constructed a successful career of her own, building on her early research. Apart from raising a family, she worked on hydrogeological aspects of weathering profiles in Malawi and Zimbabwe for the British Geological Survey, Hydrogeology Group, Wallingford. She also consulted for Du Pont on Titanium mobilisation on tropical weathering profiles as well as British Titan Products, who were interested in microbial activity in soil profiles from Sri Lanka. The Brazilian National Research Council sponsored a visiting fellowship at the University of Sao Paulo and University of Salvador, where she gave talks on soils, weathering, hydrogeology and lateritic mineral dissolution by micro-organisms. She became an Honorary Research Fellow in the Department of Microbiology at University of Reading, and Research Associate at the School of Geography at Oxford University.

In 1994 she joined the Department of Environmental Science of the University of Botswana in Gaborone, teaching Physical Geography, Tropical Geomorphology and Terrain Evaluation. She became a private consultant in 1997, based in Maun, where she completed a half-built house on the banks of the Thamalakane River. Marty brought her tropical experience to fruition by providing much insight into groundwater exploration and associated drill cores. She particularly contributed to Government Water Development Projects in Northern Botswana as well as Environmental Impact Assessments. Among other topics, she also published on Kalahari weathering profiles (McFarlane et al. 2010) and identified Makgadikgadi spring mounds (McFarlane and Long 2015).

Marty has published on a range of topics including karstification associated with pans in northern Australia, the degradation of dunes in Northwest Ngamiland and African weathering profiles. Arguably her biggest contribution was drawing attention to the vertical extent of deep weathering and associated palimpsest underlying the Kalahari surface. While most geomorphologists would generally focus on the recent or ongoing physical movement of sediment at the surface by either wind or water, Marty's work would highlight the vertical hydro-biological processes, which promote denudation and accumulation of weathering products. This seminal work was directly informed by her tropical experience in East Africa, Brazil and Asia.

Margaret Joan McFarlane, geomorphologist, hydrogeologist, paleo-climatologist and environmental consultant was born on the 1st of February 1941 and passed away peacefully at her home in Maun on the 14th of September 2018, aged 77, following a prolonged illness. She leaves behind her life partner, two sons and two grandchildren.

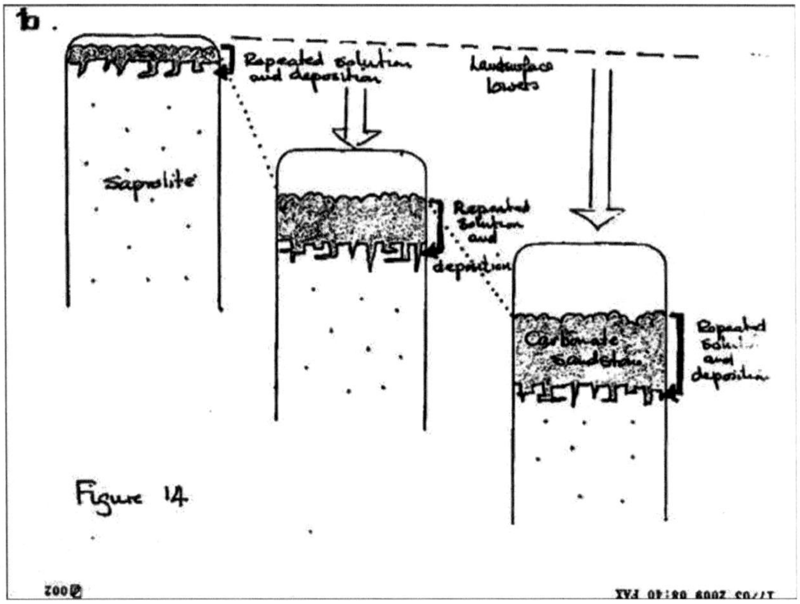

Fig. 2 Vertical weathering and denudation profile for the Kalahari, drawn and faxed by McFarlane

Fig. 3 Marty McFarlane in her Hilux while conducting fieldwork

References

Bishop, W.W., 1967. The Lake Albert Basin. The Geographical Journal, 133(4), pp.469-480.)

McFarlane, M. J. Laterite and landscape. Academic Press, 1976.

McFarlane, M.J., Eckardt, F.D., Coetzee, S.H. and Ringrose, S., 2010. An African surface weathering profile in the Kalahari of North West Ngamiland, Botswana: processes and products. Zeitschrift für Geomorphologie, 54(3), pp.273-303

McFarlane, M.J. and Long, C.W., 2015. Pan floor 'barchan'mounds, Ntwetwe Pan, Makgadikgadi, Botswana: Their origin and palaeoclimatic implications. Quaternary International, 372, pp.108-119.

McFarlane, M.J., Long, C.W. and Coetzee, S.H., 2018. Lightning-induced beads,'fulguroids', associated with kimberlite eruptions in the Kalahari, Botswana. International Journal of Earth Sciences, 107(7), pp.2627-2633.

Series Editor Preface

Landforms and landscapes vary enormously across the Earth, from high mountains to endless plains. On a smaller scale, nature often surprises us, creating shapes that look improbable. Many physical landscapes are so immensely beautiful that they received the highest possible recognition—they hold the status of World Heritage properties. Apart from often being immensely scenic, landscapes tell stories which not uncommonly can be traced back in time for tens of million years and include unique events. In addition, many landscapes owe their appearance and harmony not solely to the natural forces. For centuries, and even millennia, they have been shaped by humans who have modified hillslopes, river courses, and coastlines, and erected structures that often blend with the natural landforms to form inseparable entities.

These landscapes are studied by geomorphology—'the science of scenery'—a part of Earth Sciences that focuses on landforms, their assemblages, surface and subsurface processes that moulded them in the past and that change them today. To show the importance of geomorphology in understanding the landscape, and to present the beauty and diversity of the geomorphological sceneries across the world, we have launched a book series *World Geomorphological Landscapes*. It aims to be a scientific library of monographs that present and explain physical landscapes, focusing on both representative and uniquely spectacular examples. Each book will contain details on the geomorphology of a particular country or a geographically coherent region. This volume presents the geomorphology of Botswana—a country whose geomorphological landscapes are somehow overshadowed by the grand sceneries of neighbouring Namibia and South Africa. But this book clearly demonstrates that the rather subdued topography of Botswana contains vital clues to decipher the history of quaternary environmental change in the southern part of Africa and tells fascinating stories of shifting dunes, disappearing and reappearing lakes, changing river patterns of vast inland deltas and many others. In addition, the unrestricted accessibility of satellite imagery helps to appreciate the beauty of these seldom visited landscapes. It is also worth highlighting how convincingly the individual chapters demonstrate linkages in the physical world between landforms, hydrology, soils and the biota.

The World Geomorphological Landscapes series is produced under the scientific patronage of the International Association of Geomorphologists (IAG)—a society that brings together geomorphologists from all around the world. The IAG was established in 1989 and is an independent scientific association affiliated with the International Geographical Union (IGU) and the International Union of Geological Sciences (IUGS). Among its main aims are to promote geomorphology and to foster the dissemination of geomorphological knowledge. I believe that this lavishly illustrated series, which sticks to the scientific rigour, is the most appropriate means to fulfil these aims and to serve the geoscientific community. To this end, my great thanks go to Prof. Frank D. Eckardt who allowed himself to be persuaded to add this publication project to his busy agenda, managed to recruit the best specialists in Botswanan geomorphology and delivered such a high-quality final project. I am also grateful to all individual contributors who agreed to contribute to the volume and shared their enormous expertise with the global geomorphological community. I have no doubt that the book will long serve as the primary reference source about the physical geography of Botswana.

Series Editor
Piotr Migoń

Wrocław, Poland

Preface

Following South Africa and Namibian releases, this is the third book in the World Geomorphological Landscapes Series covering the southern African region. Despite being neighbouring countries, each differs remarkably. The preceding books did not examine many of the landforms that dominate the landscape of Botswana. For example, the subcontinental Kalahari sands cover a part of all of Botswana's neighbours, but it did not feature in either of the previous editions and neither did the Chobe or Limpopo rivers. Nevertheless, Namibia and South Africa feature dramatic landscapes, exemplified in the Namib Sand Seas, high energy coastlines and a skirting great escarpment, including the Drakensberg. Botswana's geomorphology, in contrast, is a much more subtle affair, with few notable modern-day processes.

Post-Cretaceous Kalahari Group sediments mantle nearly three-quarters of Botswana; the older exposed metamorphic and intrusive rocks of Archaean affinities, dominate in the southeast, exposed across the Limpopo's catchment. In suitable topography, the relatively higher rainfall of this region allows small dams, and a variety of soil types to support much of the country's agriculture. Here, most of the human population is concentrated among the vegetated hills and inselbergs. Elsewhere, Kalahari sediments obscure much of the country's geology, along with its mineral wealth: notably the kimberlite pipes, hydrocarbons, the so-called Kalahari Copperbelt which extends from Namibia into Zambia, regional-scale duricrusts and groundwater resources.

Despite it being readily accessible, the vast Kalahari surface fails to offer limited lines of sight. I have on occasions referred to it as "geomonotony". This is perhaps unfair, considering how the Kalahari has attracted and engaged many scholars since the late nineteenth century. They have been intrigued by fossil dune ridges and rivers, which extend over hundreds of kilometres, along with countless pans and curvilinear shorelines of ancient mega lakes. Collectively, these fossil landforms testify to aeolian and fluvial landscapes of impressive extent. The subdued sediments of the Kalahari surface muffle a network of recent fault lines. This zone of recurrent seismic activity represents the active southwest extension of the East African rift, which traverses northern Botswana. Initiated in the Late Neogene, this rift maintains dominant control over the geography of the endorheic Okavango and Makgadikgadi basins, where modern-day flooding events control intricate vegetation patterns and sustain the renowned wildlife of Botswana. In recent years, the country has been subject to notable earthquakes, attributed to extensional reactivation of ancient faults in the deep crust.

The geomorphology has also acted as a home and conduit of our deep collective ancestry and today serves the modern, semi-arid, landlocked nation of Botswana in the pursuit of mineral and water resources. Botswana authorities even used fluvial geomorphological evidence to resolve a dispute over their international border with neighbouring Namibia.

Notwithstanding Botswana's subtle topography, we can study this uplifted continental basin as a complex palimpsest; subject to remarkable endogenic and exogenic histories, its landscapes have been reshaped by ancient denudation processes, sedimentation, incipient rifting and subtle epeirogenic buckling. Many aspects of the Botswana landscape are challenging, and most of these landforms defy classification into "textbook" categories. This is likely why the country has attracted so many researchers over the years. While some have been affiliates of the University of Botswana or other academic establishments in the region, the country stands out as an international attraction. In returning from abroad, many researchers have made the region their lifelong endeavour. Therefore, it has not proved difficult to populate chapter topics, nor find appropriate reviewers. My deep appreciation goes out to dozens of authors, who have given their time in the pursuit of this book, as well as Piotr Migon, who has lent much support and steady commitment in this effort with Springer. The authors are particularly indebted to The Office of the President for providing research permits, without which the primary research underpinning many of the chapters would not have been possible. This work, some of it going back many decades, would also not have been feasible without the support and collaborations of The National Museum, Geological Survey, University of Botswana and Okavango Research Institute. We thank the communities and other landowners for permission to access land during field-based research activities. A special word of thanks also goes to Greg Botha, Rob Bryant, Woody Cotterill, Stefan Grab, Murray Gray, Suzanne Grenfell, Thomas Gumbricht, Paul Hesse, Roger Key, Mike Meadows, Dave Nash, Tony Palmer, Susan Ringrose, Francis Sefe, Paul Shaw, Matt Telfer, Benny van der Waal, Heike Wanke and Mike de Wit who assisted in the review process.

The book starts with the active and past drainage basins of northern Botswana, which support much of the nation's tourist sector: including the Okavango Delta, a World Heritage site since 2014. It then expands into the wider Kalahari, covering fluvial and aeolian landforms along with concealed aquifers, duricrusts, caves and kimberlites. From here the book transitions into the eastern Limpopo catchment, gorges, dams and soils and concludes with a look at the Tsodilo Hills, a cultural world heritage site since 2001, geoconservation in Botswana and zoogeomorphology.

Cape Town, South Africa Frank D. Eckardt

Contents

Editor and Contributors

About the Editor

Frank D. Eckardt is a dryland geomorphologist at the Environmental and Geographical Science Department at the University of Cape Town. He holds an undergraduate degree from Kings College London and a Masters in Remote Sensing from Silsoe College. His DPhil from Oxford on gypsum crusts in Namib, provided an introduction to the southern African environment. He was a lecturer at the University of Botswana between 1998 and 2005. During his time in Botswana, his work mostly entailed the study of contemporary surface processes at the Makgadikgadi as well as applied remote sensing in the pursuit of groundwater.

Contributors

Kathleen A. Alexander Chobe Research Institute, Centre for African Resources: Animals, Communities and Land use (CARACAL), Kasane, Botswana;
Department of Fish and Wildlife Conservation, Virginia Tech, USA

Julius Atlhopheng Department of Environmental Science, The University of Botswana, Gaborone, Botswana

Lukas Belz Institute for Chemistry and Biology of the Marine Environment (ICBM), Carl Von Ossietzky University, Oldenburg, Germany

Read Brown Mthanganyika Mapeo Department of Geology, The University of Botswana, Gaborone, Botswana

Sallie L. Burrough Trapnell Fellow of African Environments, Environmental Change Institute, School of Geography and the Environment, University of Oxford, Oxford, UK

Mike de Wit Earth Sciences Department, University of Stellenbosch, Matieland, South Africa

Nathalie Diaz Service Culture et Médiation scientifique (SCMS), University of Lausanne, Lausanne, Switzerland

Frank D. Eckardt Department of Environmental and Geographical Science, University of Cape Town, Rondebosch, South Africa

Karen Ellery Science Extended Studies Unit, CHERTL, Rhodes University, Grahamstown, South Africa

William N. Ellery Geography Department, Rhodes University, Grahamstown, South Africa

Peter N. Eze Department of Earth and Environmental Science, Botswana International University of Science & Technology, Palapye, Botswana

Douglas G. Goodin Department of Geography and Geospatial Sciences, Kansas State University, Manhattan, USA

Senwelo M. Isaacs Department of Archaeology, Faculty of Humanities, University of Botswana, Gaborone, Botswana;
Department of Environmental Science, Faculty of Science, University of Botswana, Gaborone, Botswana

Jasper Knight School of Geography, Archeology and Environmental Studies, University of the Witwatersrand, Johannesburg, South Africa

John Mendelsohn Ongava Research Centre, Outjo district, Namibia

Thuto Mokatse Institute of Earth Surface Dynamics, FGSE, University of Lausanne, Lausanne, Switzerland

Andy Moore Department of Geology, Rhodes University, Grahamstown, South Africa

David J. Nash School of Geography, Archaeology and Environmental Studies, University of the Witwatersrand, Johannesburg, South Africa;
University of Brighton, School of Applied Sciences, Brighton, United Kingdom

Bhagabat P. Parida Department of Civil & Environmental Engineering, Botswana International University of Science & Technology, Palapye, Botswana

Jeremy S. Perkins Department of Environmental Science, University of Botswana, Gaborone, UB, Botswana

Susan Ringrose School of Interdisciplinary Studies, University of Glasgow (Dumfries Campus), Dumfries, Scotland, UK;
Okavango Research Institute, University of Botswana, Maun, Botswana

Mike Roberts De Beers Exploration, Gaborone, Botswana

Irka Schüller Marine Research Department, Senckenberg am Meer, Wilhelmshaven, Germany;
Institute for Chemistry and Biology of the Marine Environment (ICBM), Carl Von Ossietzky University, Oldenburg, Germany

Elisha Shemang Earth and Environmental Sciences, Botswana International University of Science and Technology, Botswana, South Africa

Mark Stephens School of Chemistry, Environmental and Life Sciences, Faculty of Pure and Applied Sciences, University of the Bahamas, Nassau, Bahamas;
Department of Environmental Science, Faculty of Science, University of Botswana, Gaborone, Botswana

Abi Stone School of Environment, Manchester Environmental Research Institute, FSE Research Institutes, Manchester, UK

David S. G. Thomas School of Geography and the Environment, Oxford University Centre for the Environment, University of Oxford, Oxford, UK

Stephen Tooth Department of Geography and Earth Sciences, Aberystwyth University, Wales, UK

John Van Thuyne Institute of Earth Surface Dynamics, FGSE, University of Lausanne, Lausanne, Switzerland;
Van Thuyne Ridge Research Centre, Satau, Botswana, South Africa

Mark Vandewalle Chobe Research Institute, Centre for African Resources: Animals, Communities and Land use (CARACAL), Kasane, Botswana

Torsten Vennemann Institute of Earth Surface Dynamics, FGSE, University of Lausanne, Lausanne, Switzerland

Eric P. Verrecchia Institute of Earth Surface Dynamics, FGSE, University of Lausanne, Lausanne, Switzerland

Pascal Vittoz Institute of Earth Surface Dynamics, FGSE, University of Lausanne, Lausanne, Switzerland

Achim Wehrmann Institute for Chemistry and Biology of the Marine Environment (ICBM), Carl Von Ossietzky University, Oldenburg, Germany

Marek Wendorff Faculty of Geology, Geophysics and Environmental Protection, AGH University of Science and Technology, Kraków, Poland

Giles F. S. Wiggs School of Geography and the Environment, Oxford University Centre for the Environment, University of Oxford, Oxford, UK

Heinz Wilkes Institute for Chemistry and Biology of the Marine Environment (ICBM), Carl Von Ossietzky University, Oldenburg, Germany

Check for updates

1 Introduction to the Landscapes and Landforms of Botswana

Julius Atlhopheng, Read Brown Mthanganyika Mapeo, and Frank D. Eckardt

Abstract

This chapter provides a short overview of geological units that have surface expression along with the dominant landforms and landscapes of Botswana. It introduces the western portion of the country which features young regional arenosols and is home to few outcrops. This is often simply referred to as the Kalahari. The exception is the Ghanzi Ridge, a highly denuded Precambrian Orogen, sitting between two regional, but largely concealed cratons. The lower-lying east is home to Archean hills and the countries major drainage network. The second part of the introduction covers the geomorphic settings associated with the various book chapters with much focus on the landscapes of the Kalahari, northern Botswana and eastern lowveld. Especially the north and northeast are home to a variety of fluvial systems, which include the Okavango, Zambezi and Makgadikgadi dry lake. In conclusion, this chapter draws attention to the tertiary education of geomorphology in Botswana.

Keywords

Botswana • Geomorphology • Geology

J. Atlhopheng (✉)
Department of Environmental Science Private Bag UB00704, The University of Botswana, Gaborone, Botswana
e-mail: atlhophe@mopipi.ub.bw

R. Brown Mthanganyika Mapeo
Department of Geology Private Bag UB00704, The University of Botswana, Gaborone, Botswana

F. D. Eckardt
Department of Environmental and Geographical Science, University of Cape Town, Private Bag X3, Rondebosch, 7701, South Africa
e-mail: frank.eckardt@uct.ac.za

1.1 The Regional Landscape of Botswana

This chapter will briefly overview Botswana's geology and geomorphology and introduce the subsequent chapters. Botswana's regional mega-geomorphology can be described as an elevated, gentle basin, which straddles two cratons and an orogenic belt, which are largely covered by Karoo age volcanics and associated sediments ($\sim$300–250 Ma) as well as younger duricusts and arenosols. These accumulations are the result of the raised rim represented by the southern African's subcontinental escarpment (Figs. 1.1 and 1.2). The eastern portion of the country is devoid of recent cover and is dominated by outcrops of Precambrian age. The east and south of Botswana provide the headwaters to major rivers of South Africa, such as the Limpopo and Orange catchment. The Molopo and Nossop rivers are ephemeral tributaries to the Orange and, on human time scales, along with other rivers of the Kalahari, can be described as fossil channels/valleys. A short section of the country also houses the banks of the Chobe-Zambezi floodplain. The rivers collectively demarcate the borders to the east and north of the country. The sandy Kalahari cover, which extends from Botswana to all neighbouring nations, dominates much of the country's west and centre. Despite such contrasting landforms, much of the country's topography is subtle. Although Botswana approximates France or Texas in size, its relief varies by less than 1,000 m from the promontory of Otse Hill at 1,491 m, down to an elevation of 513 m at the confluence between the Limpopo and Shashe rivers in the far east. Most of the country's relief is associated with exposed basement terrain (Fig. 1.3).

This book follows in the footsteps of a number of national post-independence "geo-milestones" for Botswana, which include the systematic mapping of Botswana's landforms (Mallick et al. 1981), the publication of the Kalahari Environment book by Thomas and Shaw (1991), the release of an annotated bibliography on the Geomorphology of Botswana (Shaw and Nash 1998), the release of a digital geological map of Botswana (Key and Ayres 2000) and the publication

F. D. Eckardt (ed.), *Landscapes and Landforms of Botswana*, World Geomorphological Landscapes,
https://doi.org/10.1007/978-3-030-86102-5_1

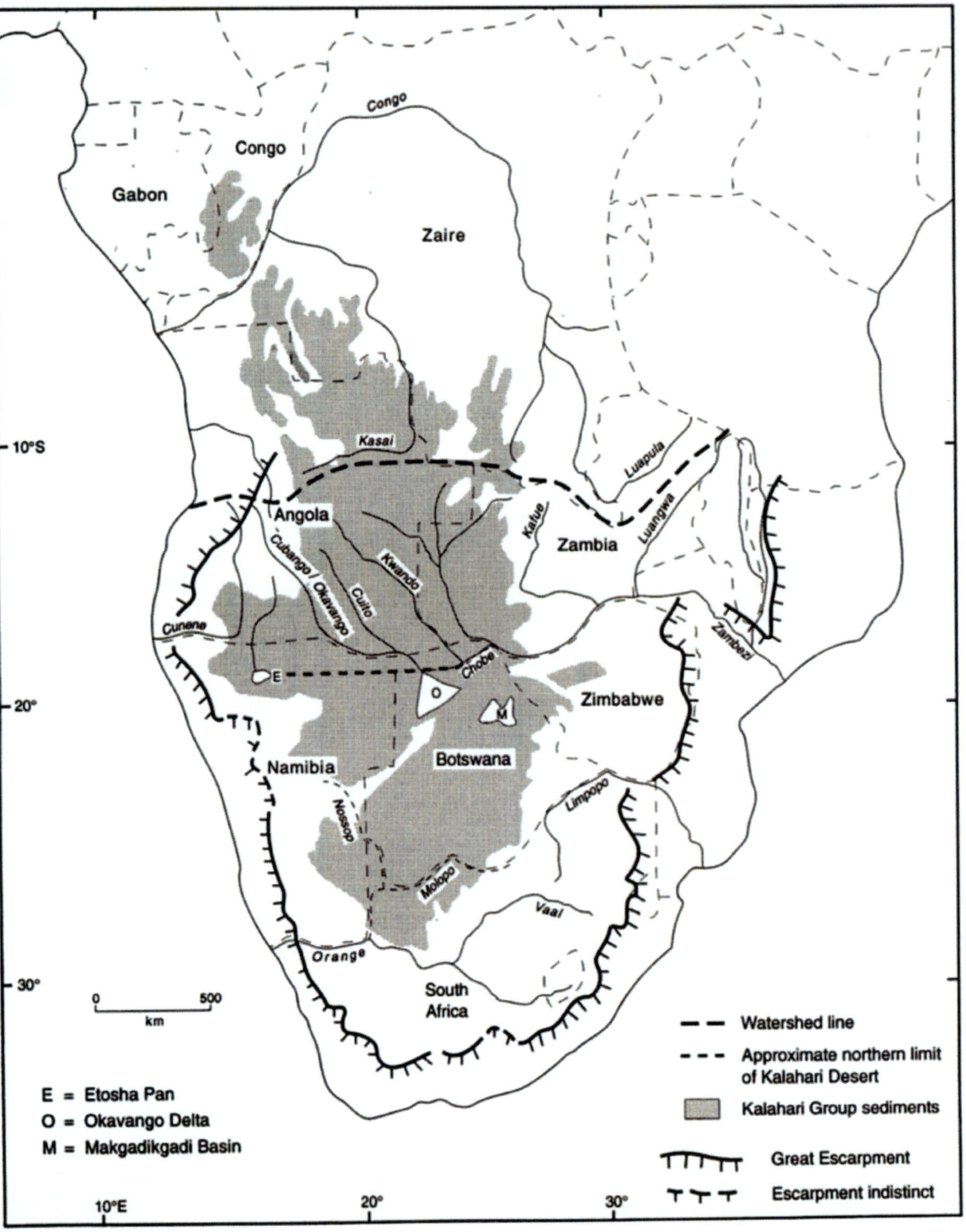

Fig. 1.1 Botswana, major drainage and Kalahari cover in their regional context (*Source* Thomas and Shaw 1990)

Fig. 1.2 Botswana elevated and flat topography in relation to the rest of Africa (*Source* Frank D. Eckardt, NASA Blue Marble draped over vertically exaggerated ETOPO5)

of the National Atlas featuring a chapter on Geology and Geomorphology of Botswana (Mapeo et al. 2001).

1.2 The Geology of Botswana

Botswana's geology consists of three main units: the Archean cratons, the Proterozoic units and the Phanerozoic cover (Carney et al. 1994) (Fig. 1.4). Most rock units in Botswana are exposed in the country's eastern quarter and along the Ghanzi-Chobe Ridge in the northwestern region. The rest is largely covered by Cenozoic sediments. The country's geology offers many economic resources such as coal, diamonds, base and precious metals, aggregates and soils, which evolved from basement geology (Atlhopheng 2004) and contribute to national wealth and livelihoods.

The Precambrian units in Botswana include the Archaean Zimbabwe craton to the east and the Kaapvaal craton in the

Fig. 1.3 A fibreglass model of Botswana's topography on display at the Botswana National Museum. Note the flat Kalahari environment in the western portion of the country and the hilly topography to its east (Photo: Frank D. Eckardt)

south. These formed between 3.7 and 2.6 billion years ago and consist of granitoid gneisses and narrow greenstone belts. The Archaean Limpopo Belt was reworked in the Proterozoic era and formed during the Zimbabwe and Kaapvaal cratons collision 2.7–2.6 billion years ago (Begg et al. 2009). The basement geology is also known as the Kalahari craton, which defines several major tectonic terranes, upon which younger volcano-sedimentary sequences were deposited. The major Proterozoic terranes of Botswana formed between 2.1 and 1.2 billion years ago and consist of the Okwa Block, exposed along the Okwa Valley and orogenic belts that wrap around Archean terranes such as the Kheis Belt in southern Botswana and the Magondi Belt in the northeast. In the northwest Proterozoic terranes consist of the Kgwebe rhyolite which in turn are overlain by the Ghanzi Group sediments. These were altered during the Neoproterozoic Pan-African deformation, creating the Damaran Orogenic Belt. This extends from the coastal

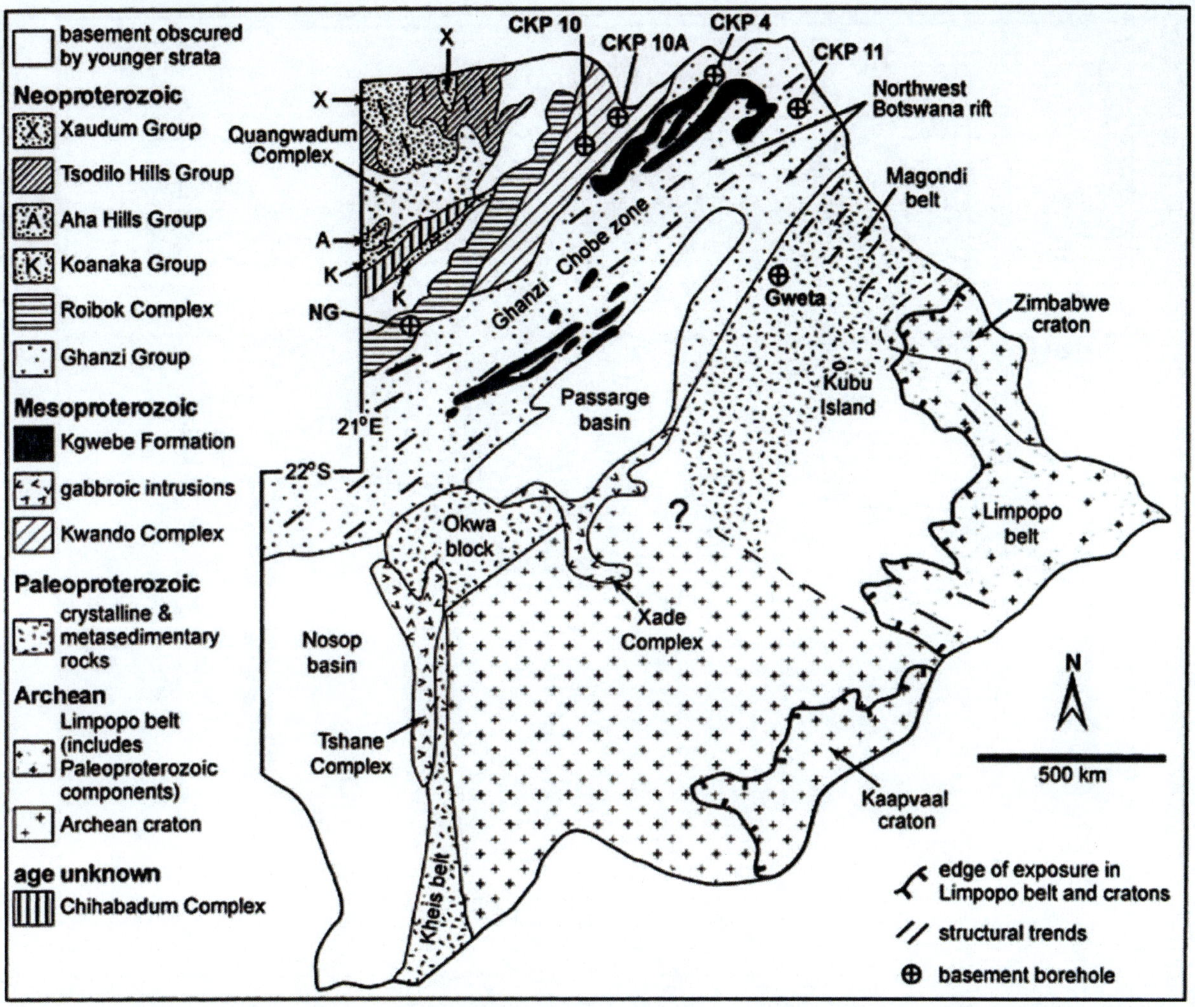

Fig. 1.4 The subsurface Precambrian geology of Botswana (*Source* Singletary et al. 2003)

Namib to northern Zambia (Key and Mapeo 1999) and is bound by the Congo Craton to the northwest of Botswana (Figs. 20.3 and 20.4).

The Transvaal Supergroup overlies the Archean basement to the southeast. Other sedimentary basins in Botswana include the Passarge Basin in the country's centre and the Nossop Basin in the southwest. The basement is mostly exposed in the country's eastern part (Carney et al. 1994; Key and Ayres 2000). The Precambrian terranes of Botswana are also overlain by Phanerozoic Karoo Supergroup (Smith, 1984; Carney et al. 1994) (Fig. 1.3), which occupies an intracratonic sag basin, made up of a succession of sedimentary and volcanic rocks (Figs. 1.4 and 1.5). The Karoo Supergroup is an important geologic unit in Botswana that underlies a significant portion of the country and contains coal and gas resources (Smith 1984). It also hosts the Ntane Sandstone Formation, a reliable groundwater aquifer. The last major episode of geological activity in Botswana coincides with the deposition of Cenozoic to recent Kalahari sediments, related to the evolution of surficial deposits, which limit the exposure of the Karoo Supergroup. Much of these sediments entail conglomerates, gravel, marl, sandstone, alluvium and lacustrine deposits and include Kalahari Sands, arenosols and duricrusts (Haddon and McCarthy 2005).

Despite the intracontinental position, Botswana is home to incipient rifting which has created the Okavango (Kinabo et al. 2007) and Makgadikdai Rift Zones (Eckardt et al. 2016). The country has also seen a magnitude 6.5 intraplate earthquake in April 2017 (Gardonio et al. 2018).

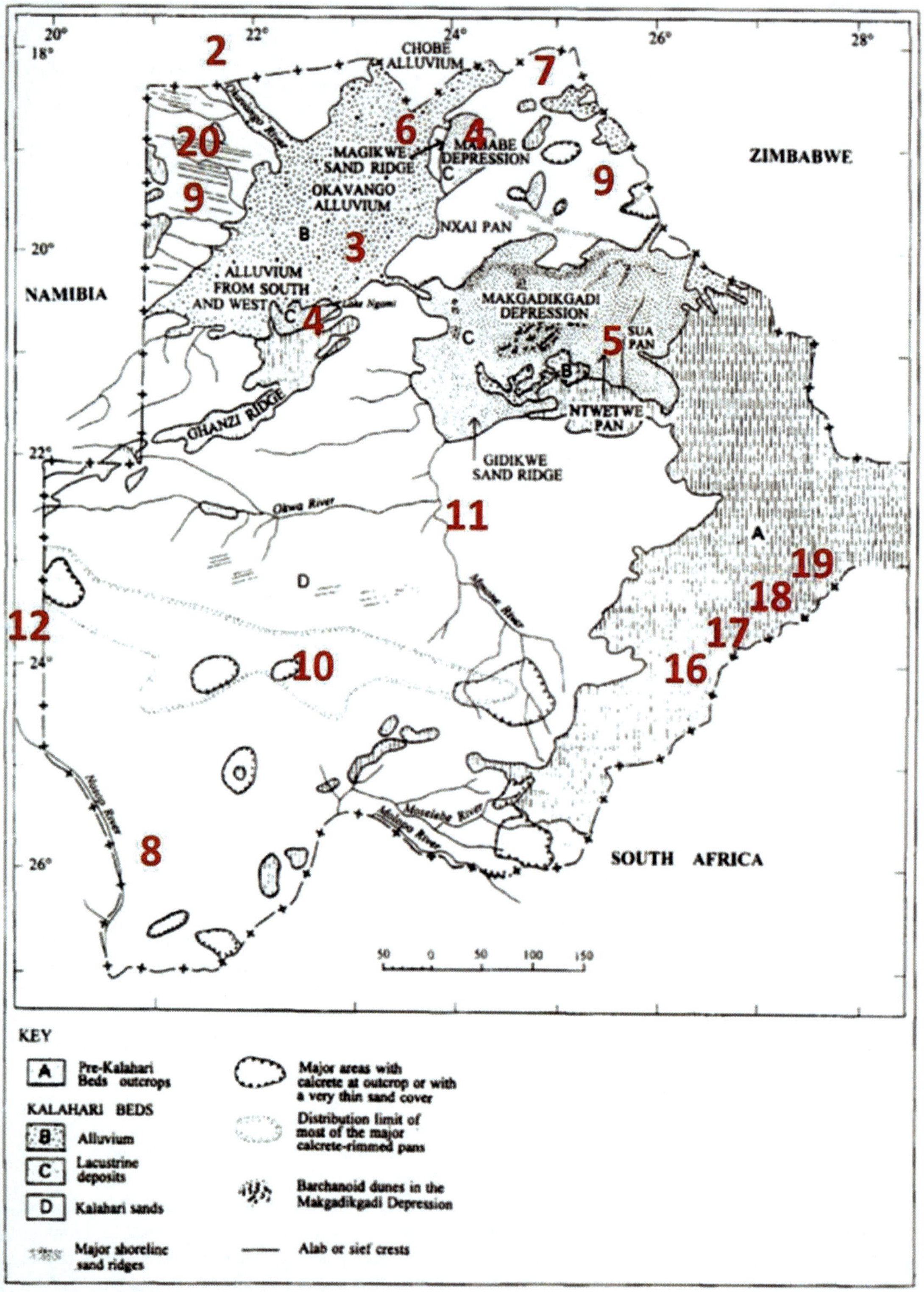

Fig. 1.5 The post Karoo, Kalahari deposits of western Botswana. Numbers indicate some of the site-specific chapters in the book. Chapters 16–19 cover various aspects of the eastern lowveld (*Source* Key and Ayres 2000)

1.3 The Geomorphology of Botswana

Notwithstanding the geological complexity, much of which remains invisible at the surface, differences in the surficial landscape remain relatively subtle. However, two major geomorphic units can be identified, which are the sandveld (Figs. 1.5 and 1.6) that coincides with the Kgalagadi (mostly referred to as the Kalahari Desert) and the hardveld to the east (De Wit and Bekker 1990) (Fig. 1.7).

Grove (1969) provided the seminal introduction to the geomorphology of northern Botswana, which forms part of the sandveld. This region is home to the Cubango, Cuito (tributaries to the Okavango), Cuando and Zambezi catchment (Chap. 2), associated wetlands, delta and buried peat (Chap. 3). Lacustrine extensions include Lake Ngami and the Mababe (Chap. 4) and the expansive Makgadikgadi palaeo-lake depressions (Chap. 5). The northern wetlands also include the Chobe enclave (Chap. 6), which merges with the floodplains of the Chobe-Zambezi rivers (Chap. 7). These relatively open and undisturbed settings, along with the provision of surface water, both perennial and ephemeral, sustain much of the wildlife (Chap. 22) and in turn, the country's tourism.

The remainder of the dry surfaces of Kalahari is associated with vegetated E-W trending linear dunes that are considered largely fossil in nature (Chap 8). In the south, where they are SW-NE trending, dunes may potentially be prone to aeolian reactivation (Chap. 9), given future climatic drought scenarios (Bhattachan et al. 2013). The central and southern Kalahari is drier than the north, both in terms of rainfall and surface water. Here pans are small and remain dry (Chap. 10), while dry valleys are mostly fossil in nature (Chap. 11). The sandy soils and ephemeral pans still play a role in recharging aquifers of varying salinity (Chap. 12), which provide pumped surface water to Kalahari rangelands and associated livestock. The regional Kalahari sands (Haddon and McCarthy 2005) conceal widespread subsurface duricrusts (Chap. 13) and saprolites (McFarlane et al. 2010), the result of groundwater processes and chemical denudation. These cover materials have generally obscured kimberlite pipes (Chap. 15) until their discovery at Orapa, Letlhakane and Jwaneng, which led to Botswana's economic transformation. No alluvial diamonds are known in Botswana.

The eastern hardveld, in contrast, is home to Precambrian hills and ridges and is characterized by gently sloping plains as part of the Limpopo catchment (Chap. 16), which also demarcates the extent of the Kalahari sands. Seasonal surface waters fill recently constructed dams (Chap. 17), which form the backbone to the growing urban and economic centres of eastern Botswana and sustain diverse land use, agriculture being amongst the most prominent as well as human settlements and service infrastructure. The evolved soils are loamy and clayey and generally ferric in nature (Chap. 19). Surficial processes in the hardvelt may lead to pronounced erosion and the origin of isolated gullies. The rocky outcrops also favour some notable landforms, including caves and rock shelters (Chap. 14) and gorges (Chap. 18) and much of the countries' geoheritage (Chap. 22).

1.4 Geomorphology Taught in Botswana

In conclusion, it would be apt to reflect on the history of the Department of Environmental Science at the Gaborone campus of the University of Botswana which made a significant contribution to the appreciation and study of Botswana's geomorphology. In 1975, John Cooke was the department's first staff member. He appointed Paul Shaw in 1982 to teach geomorphology. The department was housed in a small single-story office block, with laboratories for geology and physical geography. The storeroom held army surplus tents, jerrycans and a two-way radio, which, with luck, could link into the Department of Lands and Surveys network. The annual field trip took second-year students to the surrounding villages such as Kanye and Ranaka. The first long-distance field trip in 1987, specifically for geomorphology and remote sensing, took students to the Chobe via the Makgadikgadi. By the late 1980s, the geomorphology class had around 30 students, drawn from Environmental Science and Geology, some of which would later join the Geological Survey and other Government Departments. Many early junior staff members are now professors at the University of Botswana, having been educated at UB and abroad. The current multi-story building was completed in 1993, just as university student numbers increased. It now houses more than 30 academic staff, serving Environmental Science and the Department of Geology (Fig. 1.8). The Okavango Research Institute of the University of Botswana was established in 1994 to be a leading wetlands and dryland research institute. Many of the authors and some of the reviewers of the various book chapters here served at the University and other institutions of Botswana. All of this points to a longstanding and ongoing national commitment to the study of Botswana's landscapes and associated environment (Fig. 1.9).

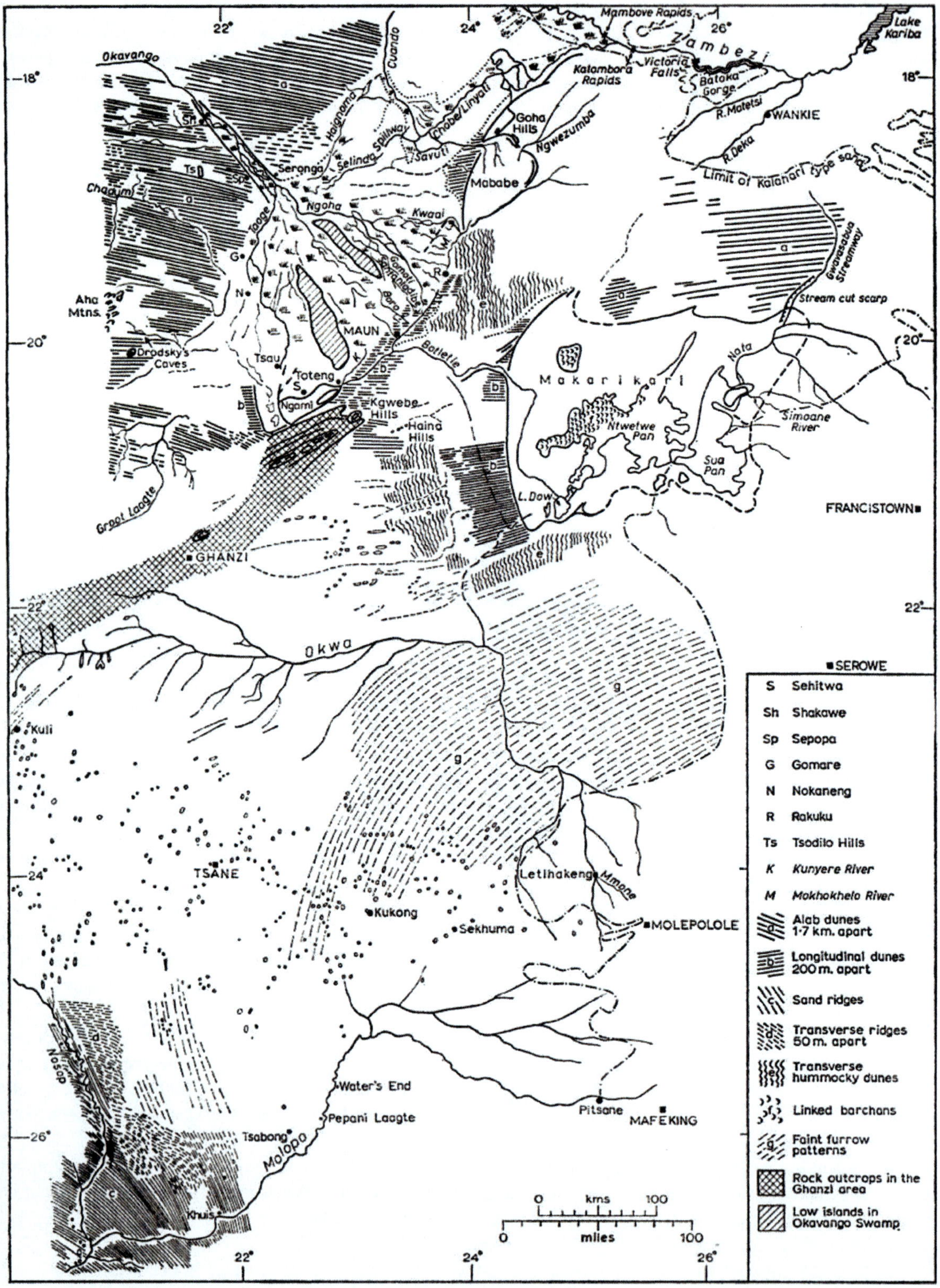

Fig. 1.6 Detailed surface geomorphology for the Kalahari in Botswana (Grove 1969)

Fig. 1.7 (following double-page spread) Scanned and georectified paper map of the photogeology of Botswana (Mallick et al. 1981) draped over vertically exaggerated elevation data from SRTM30 captures the mega geomorphology of Botswana (*Source* Frank D. Eckardt)

Fig. 1.7 (continued)

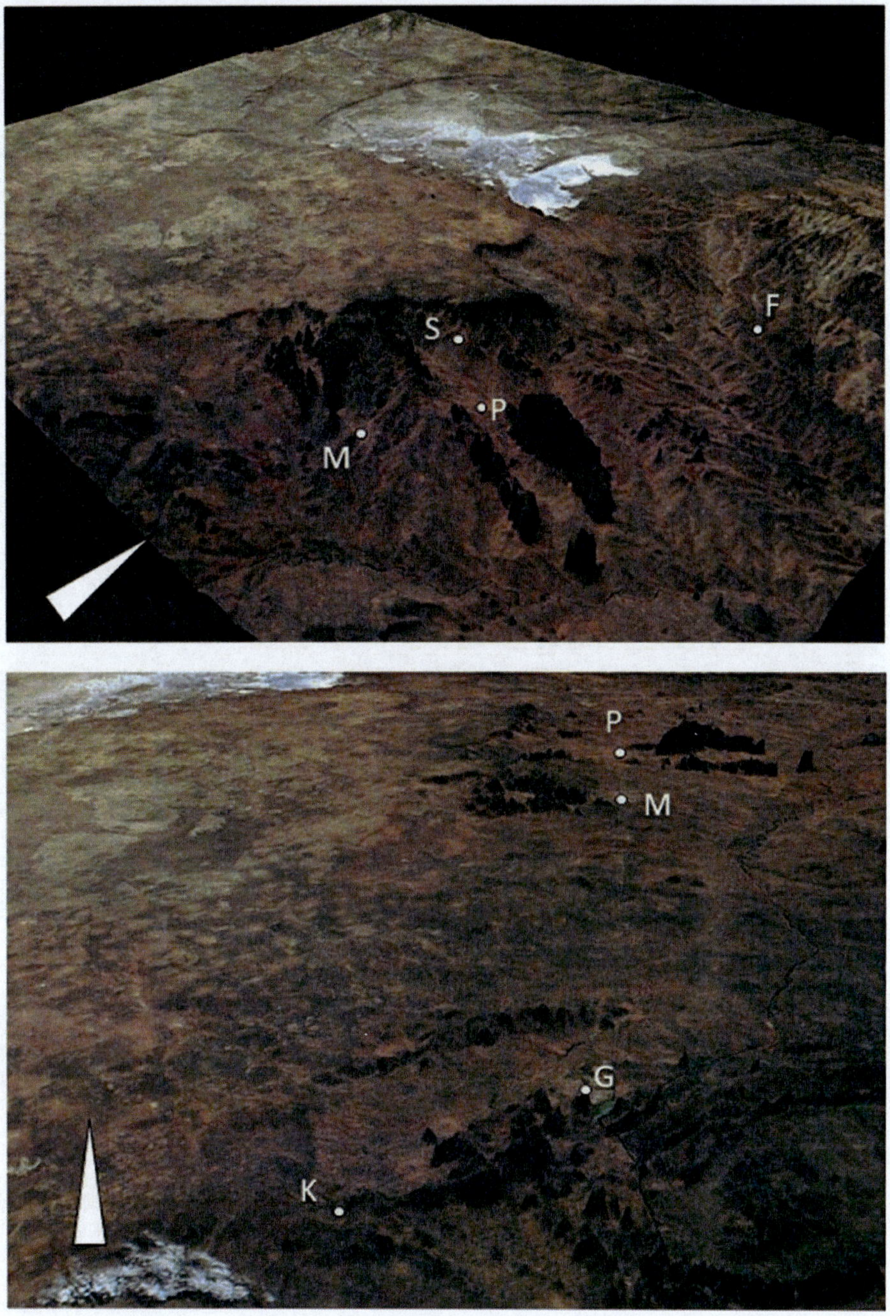

Fig. 1.8 Eastern Botswana's low-lying hardveld contrasted against the Kalahari to the west. Major settlements are Kanye (K), the capital Gaborone (G), Mahalapye (M), Palapye (P), Serowe (S) and Francistown (F). The east is home to Precambrian hills, caves (Chap. 14), the Limpopo headwaters (Chap. 16), numerous dams (Chap. 17) gorges (Chap. 18), evolved soils (Chap. 19) and much geoheritage (Chap. 21) (*Source* Frank D. Eckardt, MODIS image draped over vertically exaggerated SRTM30, with lines of sight of approximately 500 km)

Fig. 1.9 Aerial view of University of Botswana, Science Block, Environmental Science and Geology Departments on the southside of campus (Photo: Frank D. Eckardt)

References

Atlhopheng JR (2004) Terrain evaluation and the Botswana environment. Botsw Notes Rec 36(1):37–47

Begg GC, Griffin WL, Natapov LM, O'Reilly SY, Grand SP, O'Neill CJ, Hronsky JMA, Djomani YP, Swain CJ, Deen T, Bowden P (2009) The lithospheric architecture of Africa: seismic tomography, mantle petrology, and tectonic evolution. Geosphere 5(1):23–50

Bhattachan A, D'Odorico P, Okin GS, Dintwe K (2013) Potential dust emissions from the southern Kalahari's dunelands. J Geophys Res Earth Surf 118(1):307–314

Carney JN, Aldiss DT, Lock NP (1994) The geology of Botswana, vol 37. Geological Survey Department

Eckardt FD, Cotterill FP, Flügel TJ, Kahle B, McFarlane M, Rowe C (2016) Mapping the surface geomorphology of the Makgadikgadi Rift Zone (MRZ). Quatern Int 404:115–120

Gardonio B, Jolivet R, Calais E, Leclère H (2018) The April 2017 Mw6. 5 Botswana earthquake: an intraplate event triggered by deep fluids. Geophys Res Lett 45(17):8886–8896

Grove A (1969) Landforms and climatic change in the Kalahari and Ngamiland. Geogr J 135(2):191–212

Haddon IG, McCarthy TS (2005) The Mesozoic-Cenozoic interior sag basins of Central Africa: the late-cretaceous–Cenozoic Kalahari and Okavango basins. J Afr Earth Sc 43(1–3):316–333

Key RM, Ayres N (2000) The 1998 edition of the national geological map of Botswana. J Afr Earth Sci 30:427–451

Key RM, Mapeo R (1999) The Mesoproterozoic history of Botswana and the relationship of the NW Botswana Rift to Rodinia. Episodes 22(2):118–122

Kinabo BD, Atekwana EA, Hogan JP, Modisi MP, Wheaton DD, Kampunzu AB (2007) Early structural development of the Okavango rift zone, NW Botswana. J Afr Earth Sc 48(2–3):125–136

Mallick DIJ, Habgood F, Skinner AC (1981) Geological interpretation of Landsat imagery and air photography of Botswana Overseas Geology and Mineral Resources, vol 56. Institute of Geological Sciences, London

Mapeo R, Atlhopheng J, Ntsimanyana M, Chatupa J, Tsimako J, Siamisang T, Ramonyane P (2001) Geology and geomorphology of

Botswana. In: Botswana National Atlas. Department of Surveys and Mapping, pp 13–29
McFarlane MJ, Eckardt FD, Coetzee SH, Ringrose S (2010) An African surface weathering profile in the Kalahari of Northwest Ngamiland, Botswana: processes and products. Z Geomorphol 54(3):273–303
Shaw PA, Nash DJ (1998) Geomorphology of Botswana an annotated bibliography, Botswana Society
Singletary SJ, Hanson RE, Martin M, Bowring SA, Key RM, Ramokate LV, Direng BB, Krol MA (2003) Geochronology of basement rocks in the Kalahari Desert, Botswana, and implications for regional tectonics. Precambrian Res 121:47–71
Smith RA (1984) The lithostratigraphy of the Karoo Supergroup in Botswana. Bull Geol Survey Botswana 26
Thomas DSG, Shaw PA (1990) The deposition and development of the Kalahari Group sediments, Central Southern Africa. J Afr Earth Sci (and the Middle East). 10(1–2):187–197
Thomas DSG, Shaw PA (1991) The Kalahari environment. Cambridge University Press.
De Wit PV, Bekker RP (1990) Explanatory note on the land systems map of Botswana. Food and agriculture organization of the United Nations

Julius Atlhopheng has a PhD in regolith dating using stable isotopes with palaeomagnetism applications, from Australia. As a biophysical scientist, at the University of Botswana, and Professor in the Department of Environmental Science, his main emphasis is on applied geomorphology. His academic and scientific activities are mainly related to dryland geomorphology, climate change and environmental management issues, including Rio conventions (e.g.UNCCD) and solutions/interventions at different spatial-temporal scales

Read Brown Mthanganyika Mapeo is Associate Professor of Geology in the Department of Geology at the University of Botswana. He holds a PhD in structural Geology from the Southampton University. He began his carrier with the Botswana Geological Survey in 1985, where he was involved in geological mapping projects. He specializes in structural geology, regional geology and Remote Sensing. and teaches courses in Structural Geology, Global tectonics, African Geology and Remote Sensing. His research activities encompass the regional, geochronological, and tectonic geology of geologic units in southern Africa.

Frank D. Eckardt is a dryland geomorphologist at the Environmental and Geographical Science Department at the University of Cape Town. He holds an undergraduate degree from Kings College London and a Masters in Remote Sensing from Silsoe College. His DPhil from Oxford on gypsum crusts in the Namib, provided and introduction to Southern Africa environment. He was a lecturer at the University off Botswana between 1998-2005. During his time in Botswana his work most entailed the study of contemporary surface processes at the Makgadikgadi as well as applied remote sensing in the pursuit of groundwater.

The Angolan Catchments of Northern Botswana's Major Rivers: The Cubango, Cuito, Cuando and Zambezi Rivers

2

John Mendelsohn

Abstract

Other than modest amounts of local rainfall, all other surface water in northern Botswana comes from river catchments in Angola, and to a lesser degree from north-western Zambia. The four major river systems that supply water to Botswana are the Cubango and Cuito (which jointly supply the Okavango Delta), the Cuando (which provides the water of the Linyanti Swamps and Savuti Channel) and the Zambezi (which fills the Chobe river). The catchments lie either on shallow soils (much of the Cubango catchment and the eastern upper catchment of the Zambezi), or on deep sands (all of the Cuito and Cuando, part of the Cubango, and much of the Western Zambezi catchment). River discharges following rain from the former are relatively rapid, reaching northern Botswana towards the end of the rains. Those from catchments of deep sand are extremely slow, arriving in Botswana later during the dry months.

Keywords

Okavango • Cubango • Cuito • Cuando • Zambezi

2.1 Introduction

Much of the surface water, wildlife and economic production in northern Botswana is a direct or indirect product of river water from catchments that are spread across central and eastern Angola. The water is delivered by rivers from four sub-catchments (Fig. 2.1): the Cubango (also called the Okavango or Kavango) and Cuito rivers into the Okavango Delta; the Cuando river (also known as Kwando or Mashi) into the Linyanti Swamps; and the Zambezi river into the Chobe river. The catchments of the Cubango, Cuito and Cuando lie entirely in Angola, while the catchment of the Zambezi is split between Angola and Zambia. Strong flows into the Okavango Delta may continue further down the Boteti river and into the Lake Xau. Likewise, high flows reaching the Linyanti Swamps may flow into the Savuti Channel and Mababe Depression.

Much of the highland catchment consists of thick layers of aeolian Kalahari sands which hold and then gradually release water, rather like a gigantic sponge. This controls the quality, volume and seasonality of water that flows to Botswana. Parts of the upper catchments also lie on the Bié or Central Angolan Plateau.

Topographically, south-eastern Angola drops gradually from the highest elevations in the north (above 1,500 masl) to the lowest areas along the Namibian border, which are about 1,000 masl. The terrain is hilliest in the northern areas, particularly in the headwaters of the Cuito, Cuando and Lungue-Bungo of the Western Zambezi, while the southern parts of the catchments are extremely flat. The origin of the high topography (1,500–1,850 masl) of the Bié or Central Angolan Plateau is not clear (R. Swart pers. comm.).

The four rivers run south off the Great Equatorial Divide, a watershed running across Angola from east to west (Fig. 2.2). Major rivers flowing northwards off the Divide include the Casai (or Kasai), the Cuanza and the Queve.

The upper and lower catchments of the Cubango, Cuito and Cuando separate northern areas that are steeper and more elevated (see Fig. 2.1) with surface flows that are more permanent, from the lower areas that are much flatter, with more ephemeral tributaries. The Western Zambezi catchment is likewise a hilly upland, while the Bulozi floodplain is flat.

Annual average rainfall in the upper catchments ranges between about 800 mm in the south and 1,200 mm in the north of the Zambezi catchment and 1,300 mm in the north of the Cubango catchment (Mendelsohn and Martins 2018). Almost all rain falls between early October and late April. Rainfall along the Namibian border in the south of the lower

J. Mendelsohn (✉)
Ongava Research Centre, Outjo district, Namibia
e-mail: jm@orc.eco

F. D. Eckardt (ed.), *Landscapes and Landforms of Botswana*, World Geomorphological Landscapes,
https://doi.org/10.1007/978-3-030-86102-5_2

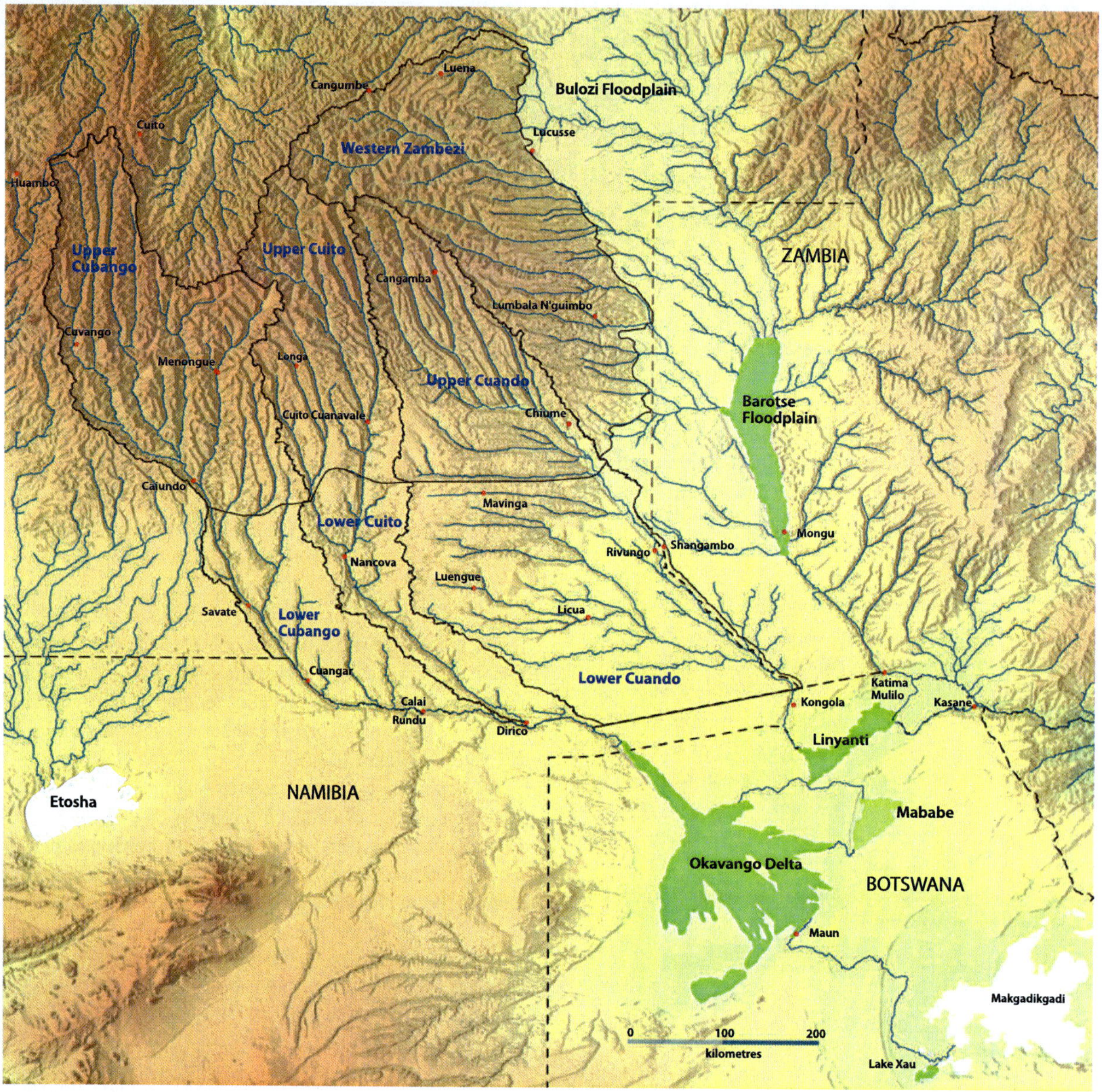

Fig. 2.1 Elevations and major catchment areas in Angola, and the major wetlands fed by rivers flowing off the central and eastern highlands. Elevation data from NASA SRTM (90 m)

catchments varies between 500 and 650 mm/year. Most flows of water into Botswana therefore come from the upper catchments, while limited supplementary flows from the lower catchments occur only during extremely wet periods.

The complementary functioning of these rivers and catchments is considerable. For example, the Western Zambezi in Angola produces a steady flow of water, while river discharges from the Bulozi and Zambian zone to the east are more episodic and variable. How these complementary flows support downstream wetlands—such as the Barotse floodplains—remains to be understood. The complementary nature of the Cubango and Cuito rivers in supplying water to the Okavango Delta is described below.

2.2 Gradients

The profiles of the Cuito and Cuando rivers are similar (Fig. 2.3). They and their respective upstream sister rivers—the Cuanavale and Kembo—flow in parallel from comparable elevations and over similar gradients until they merge. Within approximately 300 kms of their sources, the Cuito

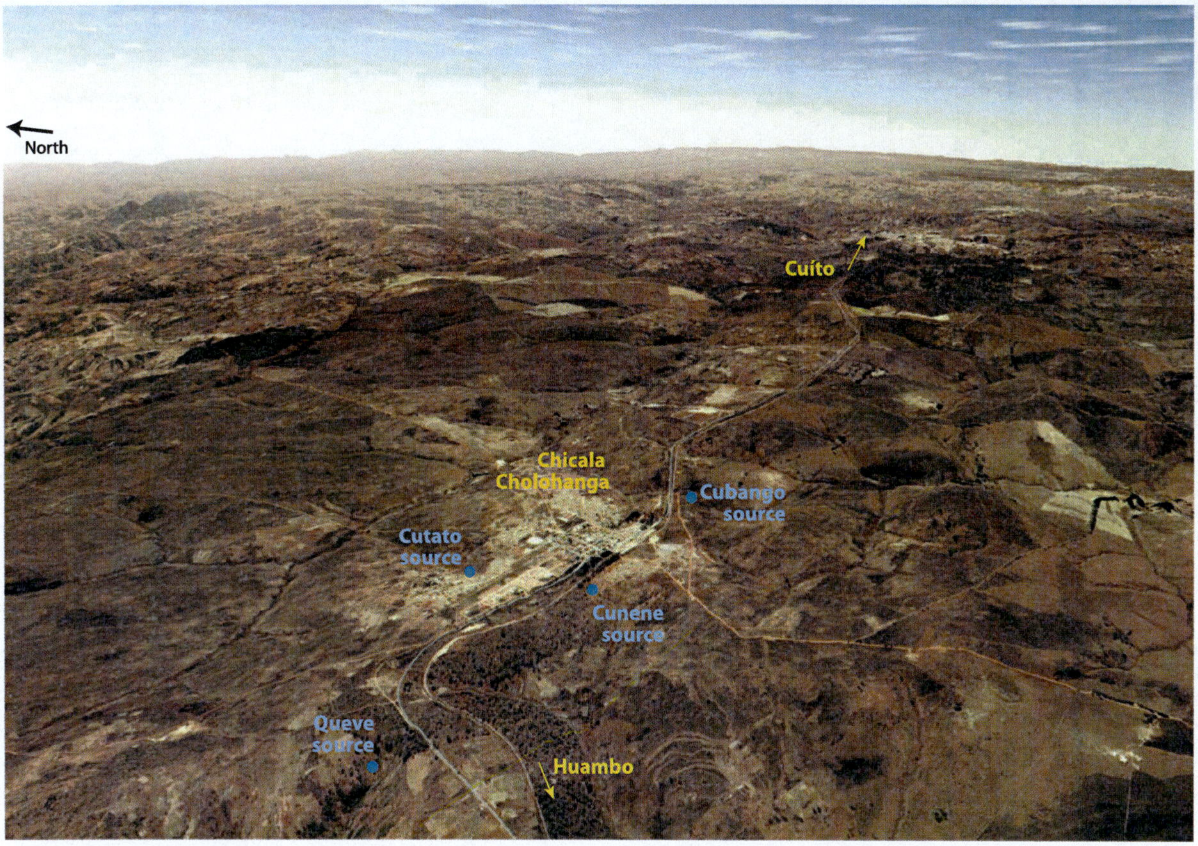

Fig. 2.2 A major road and the railway line (the Caminho de Ferro de Benguela) run across Angola from Benguela on the Atlantic to Luau in the far east. Both the road and the railway line follow the Great Equatorial Divide. This watershed separates rivers flowing north and south in Angola, including those that are the focus of this review. The small town of Chicala Cholohanga is on a gentle hill on this watershed. Within a few hundred metres of the town centre are the sources of two rivers flowing south—the Cunene and Cubango—and two that flow north—the Queve and Cutato (Microsoft Bing Maps Image rendered by RAISON)

and Cuando drop about 210 and 250 m, respectively. In contrast, the Cubango and its associated parallel rivers (the Cuchi and Cutato Nganguela) drop about 400 m over their *first* 300 kms.

The Cubango, Cuito and Cuando then flatten out, and their flows slow over the lower gradients. Thus, in their *last* 300 kms before reaching the southern border of Angola, the Cubango falls about 100 m, the Cuito 90 m and the Cuando only 70 m. As the most sluggish of the three, the Cuando, therefore, drops just 23 cm per kilometre, or 0.23 mm per metre.

2.3 Soils

The highlands are one fundamental feature of the catchments. Another is the expanse of unconsolidated Arenosol sands and Ferralsols which make up 73% and 24% of the area, respectively (Fig. 2.4). These soils have several impacts on the geography of the catchments. First, the broad ridge—or water tower—of sandy, permeable and deep soils functions as a sponge, slowly releasing water to provide steady flows into the major rivers. Perhaps it also stores vast volumes of water in deep aquifers, such as in the newly discovered aquifer in the Cuvelai (www.bgr.bund.de/EN/Themen/Wasser/Projekte/laufend/TZ/Namibia/ceb_fb_en.html).

Second, the soils have extremely few plant nutrients and little water-holding capacity (Asanzi et al. 2006; Ucuassapi & Dias 2006; Wallenfang et al. 2015). Thus, most catchment areas have poor agricultural potential, yields of staple cereals and manioc being among the lowest in Africa (https://datamarket.com). As a result, relatively few people live in this part of Angola. Of those living here, the majority reside in towns that provide other economic opportunities.

Third, soils in areas with the highest rainfall are generally less fertile than the same soils in drier areas, a probable result of most nutrients being leached by rain or absorbed and bound up in luxuriant vegetation growing in high rainfall areas.

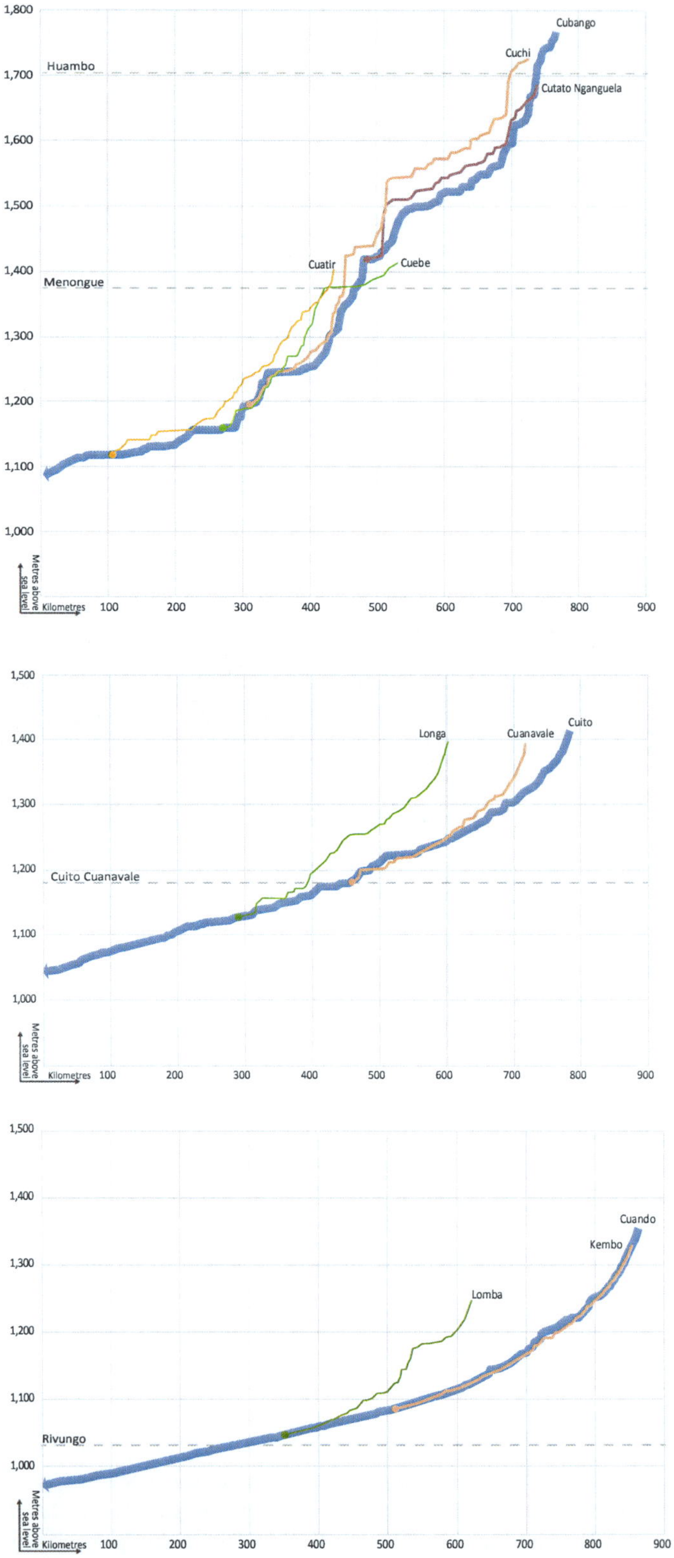

Fig. 2.3 Elevational profiles of rivers of the Cubango, Cuito and Cuando systems running from their sources on the right to the southern border of Angola. The x-axis is the distance from the Angola border to the sources, while the y-axis is the elevation in metres above sea level

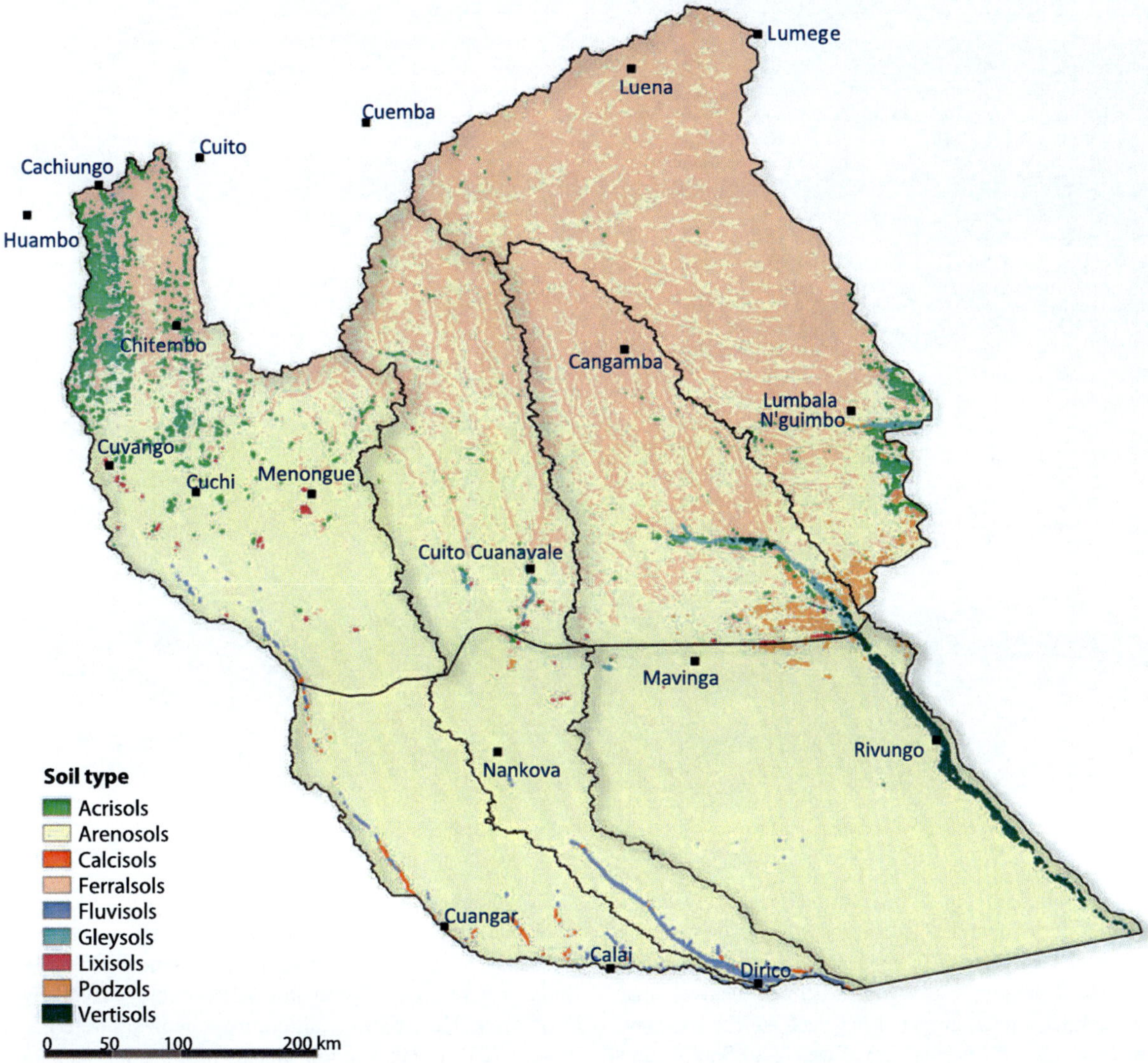

Fig. 2.4 The distribution of soils in the catchments of south-eastern Angola. Although soils in the upper catchments of the Cuito, Cuando and Western Zambezi are classified in the ISRIC data set as Ferralsols, they are more likely to be Ferralic Arenosols. (*Source* International Soil Reference and Information Centre (ISRIC) https://soilgrids.org)

Fourth, there is little erosion, particularly over the vast, flatter areas of sandy arenosols.

Fifth, as a result of the limited erosion and seepage of groundwater filtered by the sandy soils, most river water is extremely clear and free of dissolved minerals and suspended solids (NGOWP 2020a, b; von Brandis and Boyes 2017; Mendelsohn and Martins 2018). This is especially true of the relatively acidic water seeping out of arenosol sands.

Sixth, due to winds prevailing from the east, soils on the western sides of rivers are generally more fertile than those to the east. The easterly winds carry fine alluvial silts from the rivers and their floodplains, when these are dry, to deposit the fine sediments in the west. Here they mix with sand, forming soil that is more fertile and loamy. Most people therefore live and farm to the west. This is also where most roads run, and where the natural vegetation is normally more luxuriant than east of the rivers.

Substantial beds of peat fill the low-lying centres of many river valleys, especially those that meander across broad, sandy landscapes, and particularly those in the broad Cuando river valley. The depth and value of the peat remain unknown (the Vertisols shown in Fig. 2.4 are most likely peat).

A characteristic feature of the sandy upper catchments of the Cuito, Cuando and Western Zambezi is the presence of numerous headless valleys shaped as amphitheatres. These range in size from a few hundreds of metres to over 10 kms wide

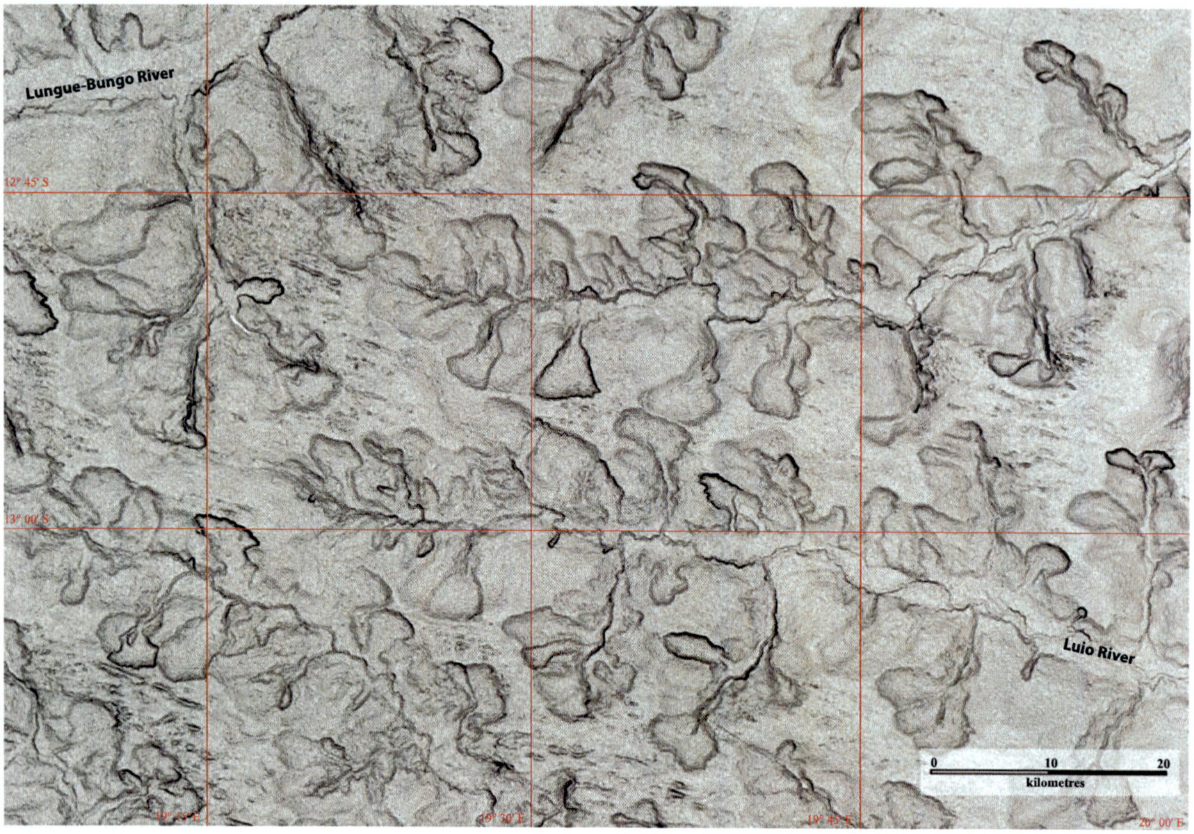

Fig. 2.5 Shaded SRTM elevations showing amphitheatre valleys along the Lungue-Bungo and Luio rivers

and up to 100 m deep (Fig. 2.5). In any one area, many of these features may be of a similar size and shape suggesting that they formed at the same time and by similar processes (R. Swart pers. comm.). Most are now vegetated and perhaps relatively inactive, but a small number are active (e.g., at Lake Tchanssengwe (18.64 East, 12.42 South); at 19.32 East, 11.80 South; and at 18.99 East, 12.22 South; and in the city of Luena (Fig. 2.6)). The amphitheatres probably form where impermeable layers force water to flow horizontally and out of the ground, which may cause unconsolidated sands on slopes to slump.

2.4 Water Chemistry

A fundamental feature of water in the four catchments is its extreme clarity (Fig. 2.7) due to low concentrations of suspended solids. Water draining the catchments of deep sands have the greatest clarity, while water in the upper Cubango is often more opaque. All the rivers become somewhat murky once the first summer rains carry ash from the many bush fires into the river water. The fires are widespread and frequent across south-eastern Angola (Stellmes et al. 2013; Mendelsohn and Martins 2018).

Expeditions of the National Geographic Okavango Wilderness Project (NGOWP) collected water quality measurements at daily stops from May to June 2017 on the Cubango river, and May to July 2018 on the Cuando river (NGOWP 2020a, b). In addition, measurements were taken in May 2018 in lakes at the sources of the Cuito and Cuanavale rivers (von Brandis and Boyes 2017).

Total dissolved solids (TDS), salinity and conductivity are similar parameters of water quality, largely measuring the dissolved salt or ion content of water. Unsurprisingly, the three measures varied in similar ways along the Cuando. The lowest levels were close to the sources of the Cuando and Kembo (a major tributary nearby (Fig. 2.16)), while the highest values were downstream in the lower catchment area of the Cuando and close to Zambia. Similar trends were observed along the Cubango, but with two exceptions. First, levels of salts and dissolved solids between the uppermost and middle reaches of the river rose more rapidly than along the Cuando. Second, levels also declined somewhat in the middle reaches south of Cuvango town, perhaps because of the diluting effects of water added by the Cacuchi, Cuelei, Cuebe and Cuatir rivers. These four rivers drain large areas of sandy substrates (like those of the Cuando and Cuito) and

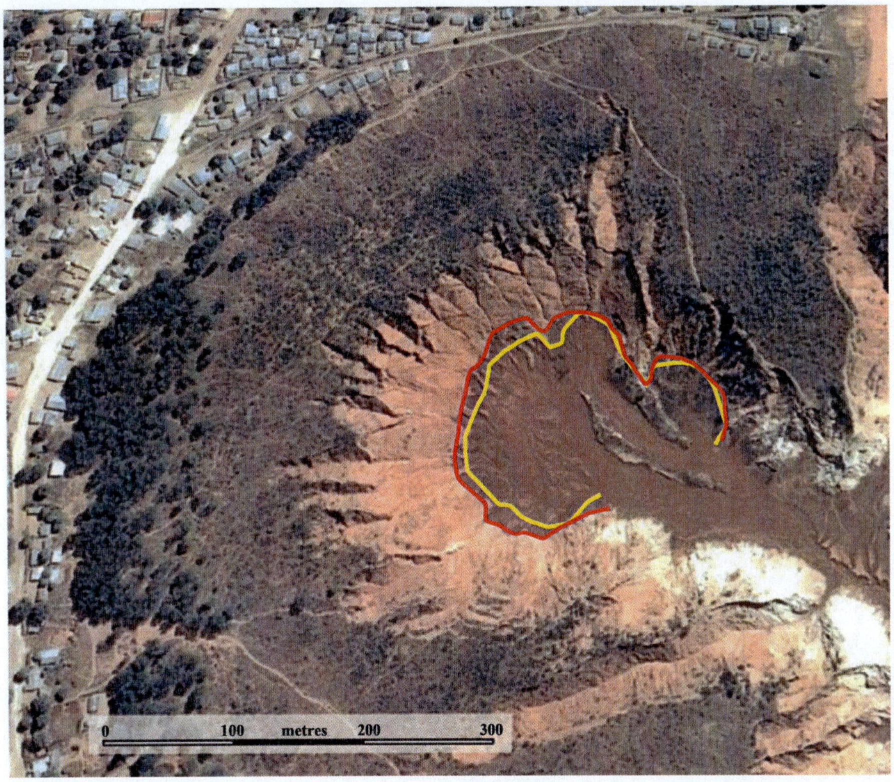

Fig. 2.6 Amphitheatre valley at Luena (19.93 East, 11.78 South). The image was taken in July 2013. The red line marks the top of the dark, saturated zone of sand in July 2013, while the yellow line is the top of the saturated zone 4 m lower in September 2011. The saturated zone lies about 60 m below the surrounding plateau. Image from Google Earth

Fig. 2.7 Meander along the Cuito river near the town of Rito. Note the clear water, white sands and dry woodland above the under-cut river bank (Photograph: Helge Denker)

their waters are thus also likely to carry fewer salts and dissolved solids.

Even the higher values of salinity recorded in these waters are low. For example, they are about 1,500 times lower than in sea water, much lower than those in most fresh waters, and indeed comparable to those found in purified water.

Oxygen saturation levels were found to vary little along the Cubango, generally exceeding 90%. In contrast, there were major fluctuations along the Cuando and Kembo. Low saturation levels were measured close to the sources, but down the river, percentages rose rapidly to above 90% before progressively dropping again to less than 70% along parts of the river dominated by broad floodplains of phragmites reeds. Why oxygen levels should rise and fall so much along the Cuando's rivers is not clear.

Soils in the headwaters of the Cuando (and doubtless the Cuito, Zambezi and even the Cuebe, Cuelei and Cuatir) are acidic with low base cation contents, as is true for rivers in most parts of south-eastern Angola. In contrast, water in the upper Cubango (at least north of Cuvango town) was only marginally acidic (NGOWP 2020a).

2.5 Rivers of Northern Botswana

2.5.1 Cubango River

The Cubango catchment is divided between two groups of rivers (Fig. 2.8). The first is to the west, and farthest north, from where the Cubango, Cutato Nganguela, Cuchi and Cacuchi rivers flow south in parallel. They originate from the highest elevations at about 1,700 masl (Fig. 2.3) from areas underlain by granitic, gneissic and meta-sedimentary sequences which were subsequently weathered into gently rolling hills (de Carvalho 1981; De Araújo and Guimarães 1992; McCourt et al. 2013). Most soils are derived either from these rocks—and are thus old, weathered and leached—or are more recent sandy sediments, likely of aeolian origin.

The four rivers have comparatively steep gradients, with flows that accelerate in places where the rivers tumble down rapids or small waterfalls (Fig. 2.9). Of these, the Cubango's flow is most even, generally following a straight course. However, the rivers meander through sizeable beds of phragmites reeds in places (Fig. 2.10) which are most extensive along the Cutato Nganguela and Cuchi. Observations suggest that the reeds may filter the water since both rivers appear murkier in their upper than their lower reaches. The extent to which they may reduce pollutants and nutrient levels in these rivers requires study.

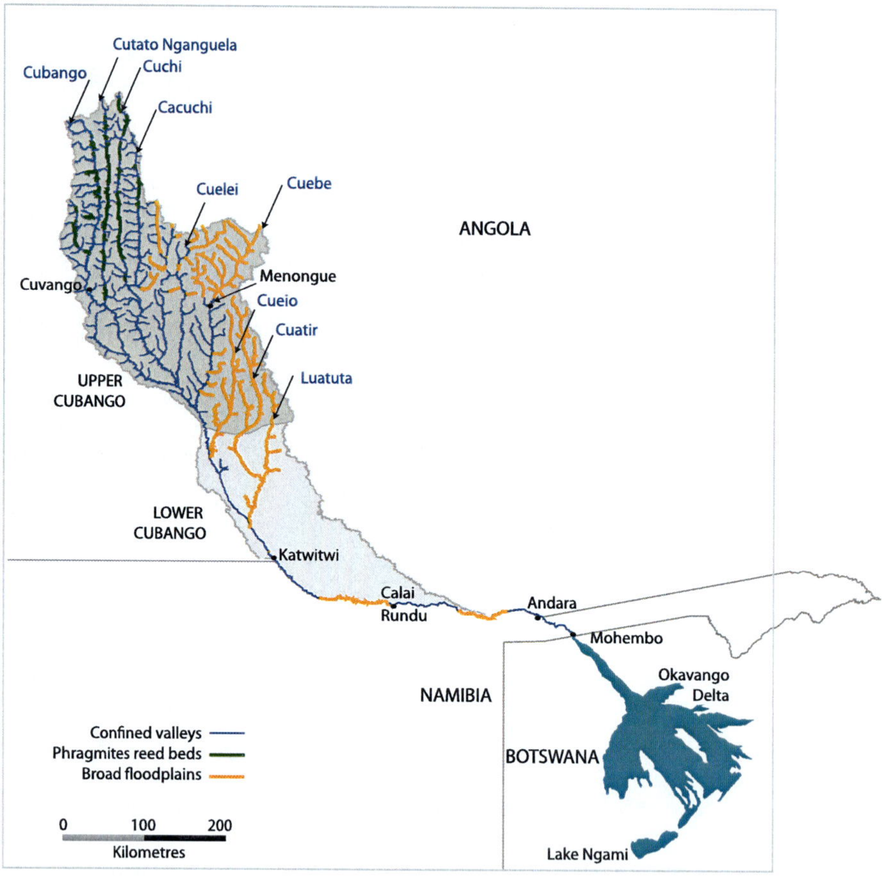

Fig. 2.8 Rivers of the Cubango catchment

Fig. 2.9 Waterfall and rapids on the Cutato Nganguela river just downstream of the town of Cutato (Photograph: John Mendelsohn)

Fig. 2.10 Phragmites marshes, meanders and oxbows on the Cutato Nganguela river (Photograph: Helge Denker)

Fig. 2.11 Broad floodplains along the Cuebe river, with its very clear water. Brown areas close to the river are moist and peaty, while grey areas further away are drier grasslands. Miombo forest flanks the floodplains. Black fire scars and a cloud shadow are also visible (Photograph: John Mendelsohn)

The second group of rivers are to the east: the Cuelei, Cuebe, Cueio, Cuatir and Luatuta, shown in orange in Fig. 2.8. All five largely drain sandy substrates, and their courses are often flanked by broad floodplains of grasslands and some beds of peat close to the actual river lines (Fig. 2.11). Water flows are slower and seasonally more stable than those of the north-western group.

The two groups of rivers also provide complementary supplies of water. From the north-west, fast, episodic flows, probably with more minerals; and from the sandier eastern areas, cleaner, slower and more stable flows. Water leaving Angola in the Cubango in the rainy, summer months is likely dominated by discharges of the western group, while slower, later flows from the eastern group of rivers probably make up much of the water in the dry season.

On a larger scale, flows from the upper Cubango complement those from the Cuito in ways that add much value to the Okavango Delta, which broadly consists of three wetland environments: permanent swamps, seasonal floodplains and occasional floodplains. The permanent swamps provide stable habitats for plants and animals, which live there from year to year with little new production. The steady flows of the Cuito river do much to maintain these permanent waters and the life they support, especially in the dry season when the Cubango's levels are low. But it is following good rains in its upper catchment that the Cubango delivers pulses of floodwaters which trigger flourishes of production in the Delta's seasonal and occasional floodplains. Plants germinate and grow, the dormant eggs of aquatic animals hatch, frogs emerge from beneath the ground, fish swim in to spawn, birds arrive to feed and nest, and herbivorous mammals move into a feast, fatten and breed. It is indeed during these periodic floods that much of the Delta's rich biodiversity is produced (Cronberg et al. 1996; McCarthy et al. 2000; Hogberg et al. 2002; Krah et al. 2006; Lindholm et al. 2007; Mendelsohn et al. 2010). This abundance of life would not be generated in the absence of floodwaters coming from the western upper Cubango catchment. Arguably, this catchment is the most important supplier of productive water to the Okavango Delta.

2.5.2 Cuito River

The Cuito and Cuanavale's sources are around 1,400 masl (Fig. 2.3). The two rivers flow south in parallel to their confluence at the town of Cuito Cuanavale, which lies at 1,250 masl. From there, the Cuito continues southwards for 140 kms to Nankova where it is joined by the Longa, carrying water from the upstream Luassinga and Cuiriri rivers (Fig. 2.12). After Nankova, no regular or substantial tributaries flow into the Cuito. Close to Dirico, the Cuito river joins the Cubango on the Namibian border (Fig. 2.13).

The entire Cuito catchment is underlain by aeolian, highly permeable sands. Except for a few small waterfalls in the northern reaches of the Cuito and Cuanavale and close to the Namibian border, all its rivers meander across flat, grassy floodplains (Fig. 2.14). These are often underlain by peat close to the drainage courses. As elevations rise, swathes of grassland that extend away from the rivers become increasingly shrubby and woody, mainly with the so-called underground trees or geoxylic suffrutices (Zigelski et al. 2019). These grasslands are located mainly in the upper catchment and are probably maintained largely by frequent fires and hardpans or aquitards which keep the upper soil layers saturated during wet periods.

Fig. 2.12 Confluence of the Longa and Cuito Rivers near the town of Nankova (Photograph: Helge Denker)

Fig. 2.13 The upper and lower catchments of the Cuito river and its major tributaries

Six major source lakes have been identified in the uppermost catchment areas: the Cuito (Fig. 2.15), Cuanavale, Limpulo, Mbambi, Cunde and Calua lakes (Von Brandis and Boyes 2017) (Fig. 2.16).

2.5.3 Cuando River

The Cuando is the most undisturbed and enigmatic river in southern Africa, perhaps in all of Africa. Its lowest reaches straddle the Namibia/Botswana border where they form an inland delta (Nkasa Ruparo) and terminal swamp (Linyanti) which are major attractions to wildlife and tourists. Its active catchment lies entirely within Angola (because the Zambian border runs along the eastern edge of Cuando's floodplain). There are only four significant towns on the Cuando river: Cangamba, Cangombe and Rivungo in Angola, and Shangombo on the Zambian side opposite Rivungo. Lake Saliakembo (on the Kembo river) is the only major lake known in the upper Cuando. A substantial and recent compendium of information on the Cuando was compiled by the National Geographic Okavango Wilderness Project (2020b).

Perhaps the most unusual aspect of the Cuando is its structure and hydrological functioning. Average flow rates at Kongola (the only place where water flows are measured) are 33.9 cubic metres/second, equivalent to just over one billion cubic metres per year. No flows of any significance enter the Cuando below Kongola; thus all the water in the Cuando represented by these figures comes from Angola (Fig. 2.17).

Of all the rivers in Angola, the Cuando has the broadest floodplain, which is 5–15 km wide for much of its length, an expanse easily seen in satellite images (Fig. 2.18). The floodplain extends over about 3,450 km^2 from where it first widens in the north, approximately at the confluence of the Kembo and Cuando, to where it narrows sharply on Angola's southern border. Apart from its narrow channels of open water, the entire floodplain is covered in macrophytes of tall grasses, phragmites reeds and papyrus (which are seen only in limited patches on the Cubango and Cuito and the Barotse floodplain of the Zambezi). Where the substantial nutrient supplies needed to support thousands of square kilometres of macrophytes come from, is a puzzle, given that soils throughout the Cuando's catchment are so infertile. One possibility is that the nutrients were derived from basaltic rocks upstream along the Cuando, Kembo, Cubangui and Cussivi rivers where there are outcrops of Calondo metasediments. In northern Angola, these rocks are associated with rivers rich in alluvial diamonds which probably accompany nutrients from kimberlites. In addition, other rich sources of nutrients may lie west of the main Cuando, for example along the Cueio and Lomba rivers where there are alluvial diamonds which might also be derived from local kimberlites.

2.5.4 Zambezi River

Although the upper Zambezi is often assumed to be in Zambia, its western upper catchment in Angola is large and probably has significant impacts on the downstream functioning of the Zambezi. First, the sandy highlands of the Western Zambezi operate like a giant sponge that stores and gradually releases steady flows into its major rivers, such as the Luena, Lungue-Bungo and Luanguinga

Fig. 2.14 The confluence of the Cuito and Cubango rivers on the border between Angola and Namibia. The sluggish, meandering Cuito is flanked by multiple scroll bars and oxbows, unlike the Cubango with its faster, more direct flow (image from Microsoft Bing Maps Imagery Services)

(Figs. 2.19 and 2.20) that later feed the Zambezi. It is not known how much water these rivers contribute to the upper Zambezi, but the flows may be significant, especially in the dry season when water from Angola may do much to support perennial wetlands along the Zambezi, such as the Barotse floodplains. Three source lakes have been identified in the Western Zambezi catchment: Lakes Tchanssengwe, Dala and Sapua.

The second feature is the contribution of the Western Zambezi catchment to the seasonal inundation of the Bulozi floodplain (Fig. 2.21) when it becomes an enormous wetland and breeding ground for fish in the upper Zambezi. The

Fig. 2.15 The lake and source of the Cuito river. Most lakes in the catchments appear to form behind narrowed sections of the river valleys (Photograph: John Mendelsohn)

western rivers may also maintain year-round water levels in refugia for adult fish within this flat nutrient-rich expanse. The Bulozi floodplain is effectively the lower catchment of the Western Zambezi (Mendelsohn and Weber 2015; Zigelski et al. 2018).

2.6 Recharge and Discharge

Comparatively little is known about mechanisms of recharge and discharge in the catchments of all the rivers in south-east Angola. Water levels and flows have only been measured for any reasonable time at Rundu (for the Cubango), at Andara and Mohembo (for the Okavango/Cubango as it enters its Delta) and at Kongola (for the Cuando) in recent times. Because no tributaries join the Cubango downstream of Rundu or after its confluence with the Cuito, the Cuito's discharges can be estimated by subtracting measures at Andara/Mohembo from those at Rundu. To my knowledge, no flow measurements are available for the Zambezi's tributaries in Angola.

Rivers in the upper western Cubango catchment are recharged comparatively rapidly by rainfall because rock surfaces are exposed in places, and many soils are shallow and less permeable than in other sandier areas. These surges of water are also apparent from the large seasonal changes in discharges on the Cubango. Recharge is slower elsewhere where almost all rain sinks into the ground. How much rainwater emerges later as seepage into tributaries is unknown. Likewise, how long it takes rainwater to emerge as seepage is unknown, but this is certain to vary in relation to the depth and nature of sediments.

Fig. 2.16 Rivers and sub-catchments of the Cuando

There are major differences in the volumes and timing of discharge between the Cubango, Cuito and the Cuando. For example, the Cubango supplies about 55% of all the water flowing into the Okavango Delta, with the remaining 45% coming from the Cuito. However, these proportions vary during the year. Thus, the Cubango supplies about double the Cuito's volume water between January and June, whereas the reverse holds from July to December when the Cubango discharges are lowest and about half those of the Cuito (Mendelsohn et al. 2010).

The Cuito's annual discharge is almost four times higher than that of the Cuando's total of 1.068 billion cubic metres (Fig. 2.22). This is surprising because their catchments are similar in size and lie side by side, sharing apparently the same upper topography, soils, rainfall and geomorphology. However, they differ greatly in their lower catchment areas: the Cuito meanders across short grassy floodplains that are seldom wider than one or two kilometres, while the Cuando's meandering path is across a floodplain that is generally ten or more kilometres wide (Figs. 2.17 and 2.18). Moreover, the Cuando's floodplain supports vast areas of phragmites reeds, papyrus and tall grasses. Losses of water due to evaporation and transpiration from the Cuando must

Fig. 2.17 The Cuando river and its 10-km-wide floodplain near its confluence with the Lomba river (Photograph: John Mendelsohn)

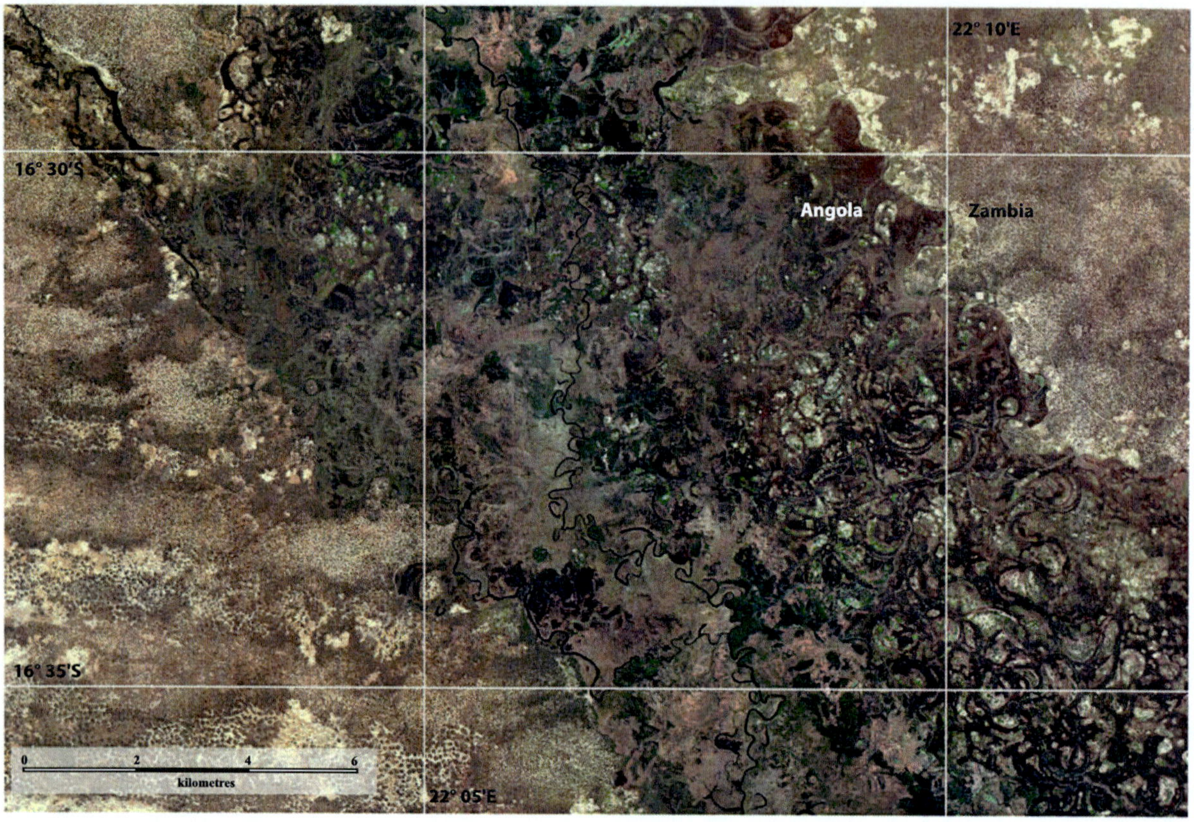

Fig. 2.18 The Cuando river and its broad floodplain south of Rivungo and Shangombo. The main channel is 10–20 m wide in most places (image from Microsoft Bing Maps Imagery Services)

therefore be greater than from the Cuito, and this may explain the great differences in discharge between the two rivers.

The Cuando's flows are also slower than those of the Cuito, and very much slower than those of the Cubango. Peak discharges along the Cubango reach the southern border of Angola in April and May, one or two months after high rainfall in the upper catchment of the Cubango. Those of the Cuito reach the border between May and July, and thus two to five months after rain in the upper catchment of the Cuito (Fig. 2.23).

The highest flows on the Cuando at Kongola are usually measured in July and August, but discharges vary considerably from year to year, as shown over 22 years when its flows were measured consistently (Fig. 2.23). Discharges sometimes peaked at the end of the year, suggesting that rainwater that fell in the upper catchment during the previous summer may have taken much of the year to reach the gauging station at Kongola. In other years, flows peaked towards the end of summer, perhaps as a result of local rains adding to river flows in the Kongola area.

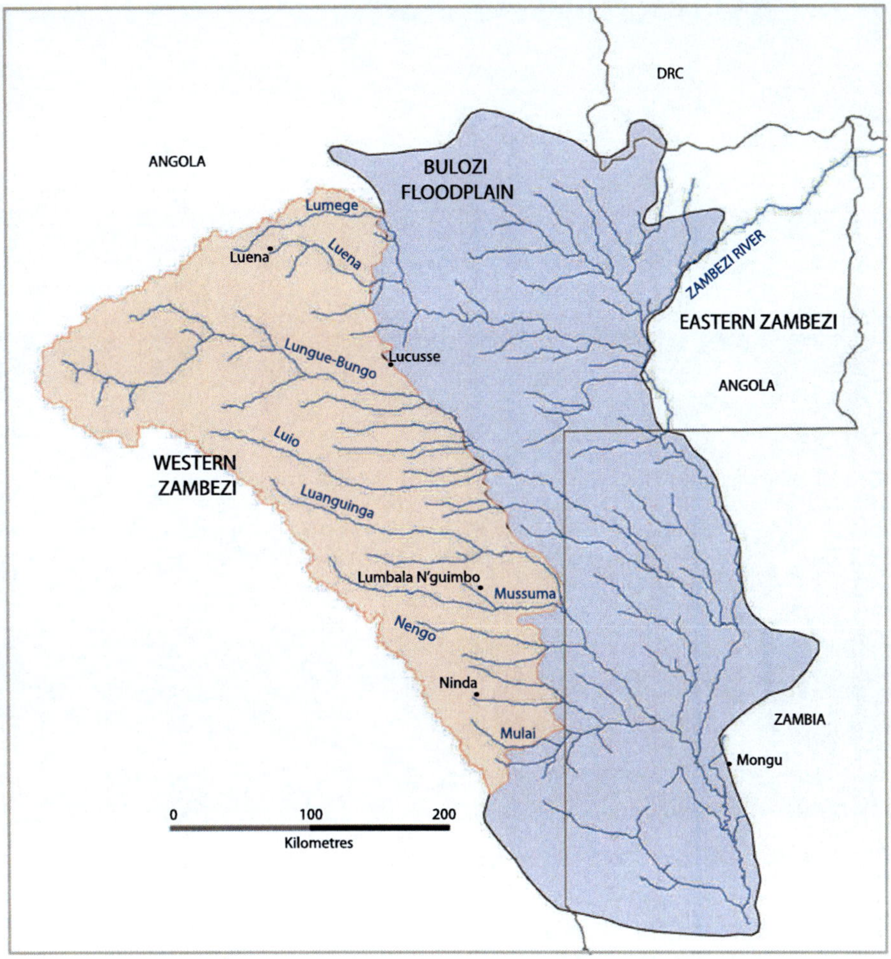

Fig. 2.19 Rivers and the catchment of the Western Zambezi in Angola, as well as the Bulozi floodplain

Although the main Cubango, Cuito and Cuando rivers have many tributaries in their lower catchments, none of these watercourses delivers much, or any water. While rainfall is somewhat lower than in the upper catchments, the watercourses are dry mainly because the lower catchments are extremely flat with soils of highly permeable sands (Fig. 2.4). Most rain, therefore, seeps away, leaving no water to flow on the surface.

2.7 Mixing Water

The Cubango/Cuito, Cuando and Zambezi are discrete bodies of water in their Angolan catchments. But they can connect with each other when water levels are high, in north-eastern Namibia and north-eastern Botswana (Fig. 2.24). Thus, when water levels are high in the

Fig. 2.20 The Lungue-Bungo is the largest river meandering off the Western Zambezi catchment (Photograph: John Mendelsohn)

Zambezi, some of its water flows westwards along the Chobe river and via the Bukalo Channel into Lake Liambezi. The Linyanti Swamps which are fed by the Cuando may also merge with Lake Liambezi when water levels are high. Water from the Cuando and Zambezi thus mix from time to time in these wetlands, and some of the wetland's mixed water can later flow into the Zambezi when it drops, causing the Chobe to reverse and flow eastwards.

Furthermore, Cubango/Cuito water in the Okavango Delta may reach the Cuando and Linyanti Swamps via the Selinda Spillway. When high, water from both the Cuando and Okavango Delta can flow into the Mababe Depression, the former via the Savuti Channel and the latter along the Khwai river. Thus, water collected from across the broad Angolan catchments can later be spread across other broad areas in northern Botswana, including into Lake Xau via the Boteti river. High water conditions which permit these connections have been recorded sporadically in recent years, such as in the early and late 1960s and again between 2008 and 2012. But river flows have also been low, indeed so low that Lake Liambezi and Lake Ngami were dry from the mid-1980s until 2004.

Fig. 2.21 Flows from the Western Zambezi in Angola and the Eastern Zambezi flowing out of Zambia's north-western province merge in the massive expanse of the Bulozi floodplains which straddle the border of Zambia and Angola. The floodplains cover approximately 400 kms north to south and up to 200 kms west to east (Photograph: Rich Beilfuss)

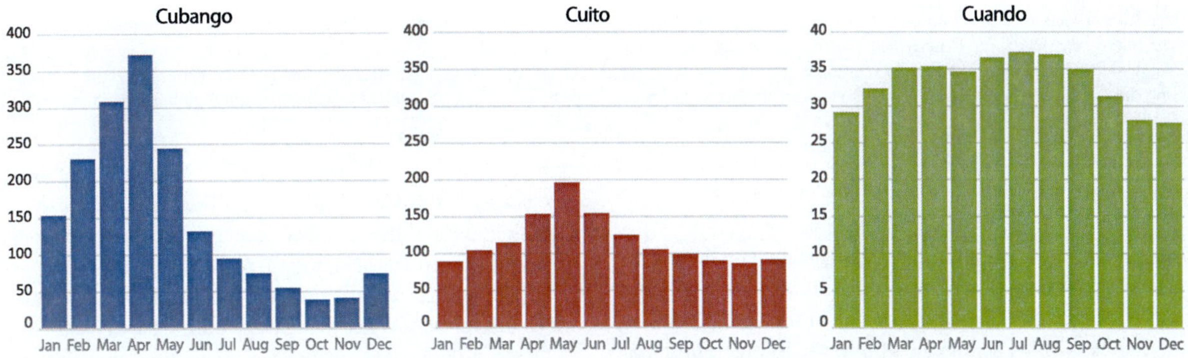

Fig. 2.22 Average monthly discharges in cubic metres per second along the Cubango (at Rundu), the Cuito (at Dirico) and Cuando (at Kongola). Note how Cubango flows vary much more than those of the Cuito, which, in turn, are much more variable than flows along the Cuando. Note the different scale for Cuando's y-axis. Data kindly supplied by the Hydrology Division, Ministry of Agriculture, Water and Forestry, Namibia. Averages for the Cubango and Cuito cover the period 1972–2007; those for the Cuando from 1972 to 2016

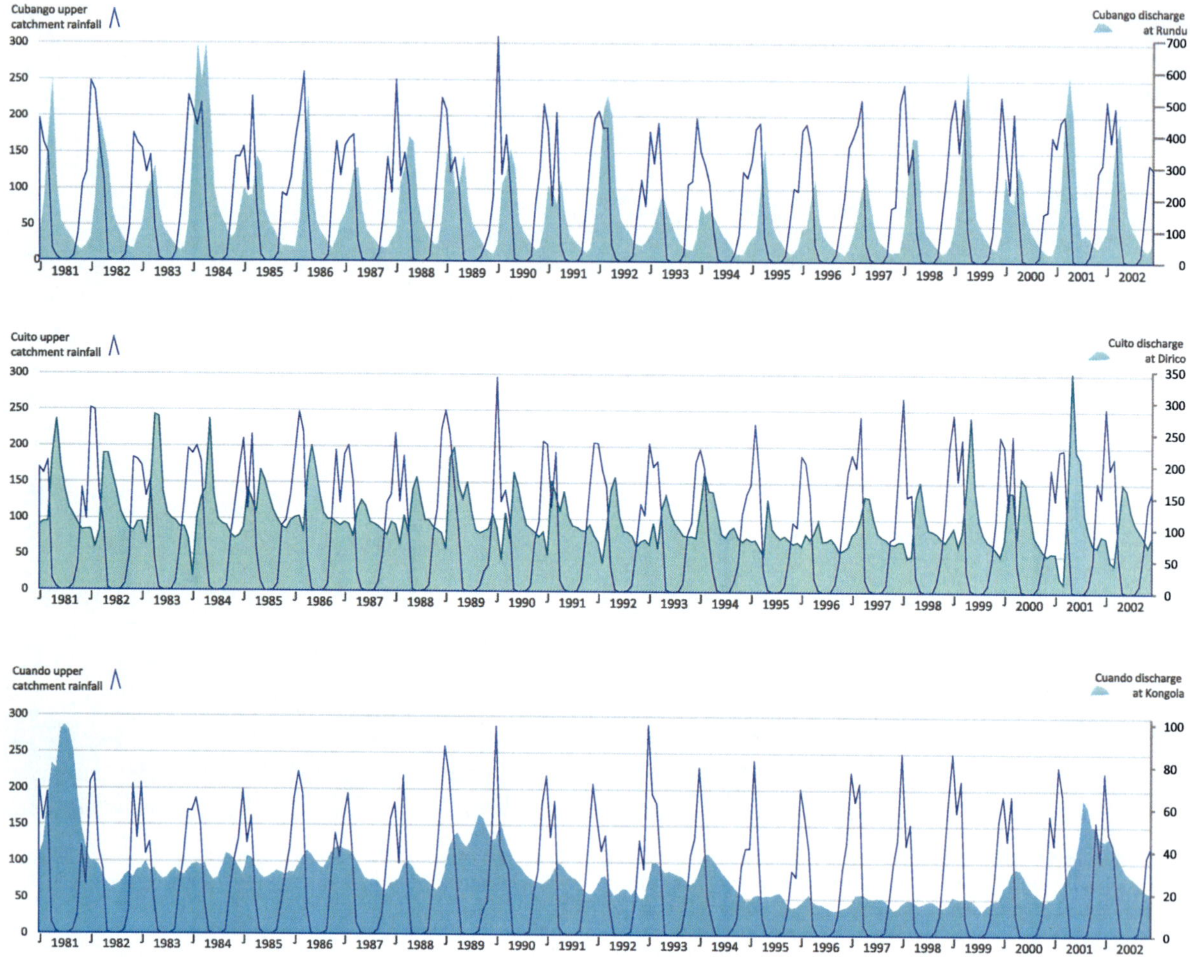

Fig. 2.23 Discharges in cubic metres per second along the Cubango (at Rundu), the Cuito (at Dirico) and Cuando (at Kongola) between 1981 and 2002, and average monthly rainfall in their upper catchments. Note the different scales of the y-axis. Each year runs from January to December. Discharge data kindly supplied by the Hydrology Division, Ministry of Agriculture, Water and Forestry, Namibia. Rainfall estimates from CHIRPS data (https://iridl.ldeo.columbia.edu/SOURCES/.UCSB/.CHIRPS/.v2p0/.monthly/.global/.precipitation/)

2.8 Conclusion

Most river flows are from the upper catchments because of their elevated and dissected topography and relatively high rainfall. Seepage into the upper catchment rivers and their tributaries is slow, acidic and almost totally free of minerals or suspended solids. In contrast, most rainwater seeps away in the flat lower, southern catchment areas, probably to levels beneath any rivers, and deep into the Kalahari Basin. The lower catchments thus only provide limited supplementary flows during extremely wet periods.

The north-western catchment of the Cubango differs from the other sandy upper catchments due to its geological foundations of granitic, gneissic and meta-sedimentary sequences, its higher elevation and consolidated soils. These conditions together with substantial falls of rain lead to rapid pulses of discharge during the summer months that bring periodic floodwaters and bouts of biological production to the Okavango Delta. These pulses complement the steadier flows of the Cuito which deliver most water to the Delta during the dry season.

Other complementary flows come from the Western Zambezi catchment and the eastern catchments in Zambia; and from the north-western and eastern rivers within the upper Cubango catchments.

The slowest flows are along the Cuando, an enigmatic river which seemingly loses much of its discharge to evapotranspiration from reeds and papyrus covering the very broad floodplains that flank the main channel along much of its journey south.

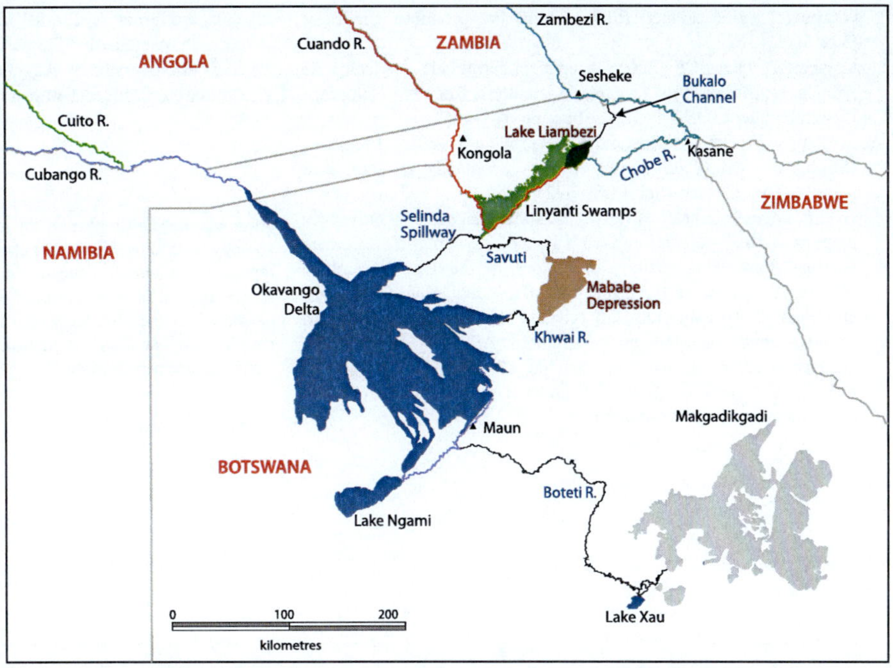

Fig. 2.24 The complex of interconnected wetlands formed from flows down the Okavango, Cuito, Cuando and Zambezi rivers in north-eastern Namibia and northern Botswana and include the Okavango Delta (Chap. 3), Lake Ngami and Mababe Depression (Chap. 4), Makgadikgadi (Chap. 5), Chobe enclave (Chap. 6) and Chobe-Zambezi floodplain (Chap. 7)

Acknowledgements The author is grateful to Rich Beilfuss, Helge Denker, Frank D. Eckardt, Marijn Goud, Antonio Martins, Celia Mendelsohn, Brent Mudde, members of the National Geographic Okavango Wilderness Project, Roger Swart, John Ward and two referees for their help and advice.

References

De Araújo AG, Guimarães F (1992) Notícia explicativa da carta geológica à escala 1:1: 1.000.000. Serviço Geológico de Angola, Luanda (and accompanying map)

Asanzi C, Kiala D, Cesar J, Lyvers K, Querido A, Smith C, Yost RS (2006) Food production in the Planalto of southern Angola. Soil Sci 171:810–820

von Brandis RG, Boyes CA (2017) Final report on the National Geographic Okavango Wilderness Project 2017 baseline scientific surveys of three river source lakes in the south-eastern highlands of Angola

de Carvalho H (1981) Geologia de Angola. Serviços de Geologia e Minas, Instituto de Investigação Cientifica, Companhia de Diamantes de Angola

Cronberg G, Gieske A, Martins E, Prince Nengu J, Stenström L-M (1996) Major ion chemistry, plankton, and bacterial assemblages of the Jao/Boro River, Okavango Delta, Botswana: the swamps and flood plains. Arch Hydrobiol Suppl 107:335–407

Hogberg P, Lindholm M, Ramberg L, Hessen DO (2002) Aquatic food web dynamics on a floodplain in the Okavango Delta, Botswana. Hydrobiologia 470:23–30

Krah M, McCarthy TS, Huntsman-Mapila P, Wolski P, Sethebe K (2006) Nutrient budget in the seasonal wetland of the Okavango Delta, Botswana. Wetl Ecol Manag 14: 253–267

Lindholm M, Hessen DO, Mosepele K, Wolski P (2007) Food webs and energy fluxes on a seasonal floodplain: the influence of flood size. Wetlands 27:775–784

McCarthy TS, Cooper GRJ, Tyson PD, Ellery WN (2000) Seasonal flooding in the Okavango Delta, Botswana – recent history and future prospects. S Afr J Sci 96:25–33

McCourt S, Armstrong RA, Jelsma H, Mapeo RBM (2013) New U-Pb SHRIMP ages from the Lubango region, SW Angola: insights into the Palaeoproterozoic evolution of the Angolan Shield, southern Congo Craton. Afr J Geol Soc Lond 170:353–363

Mendelsohn JM, Martins A (2018) River catchments and development prospects in south-eastern Angola. Report for World Wildlife Fund (WWF), The Nature Conservancy (TNC) and National Geographic Okavango Wilderness Project (NGOWP)

Mendelsohn JM, Weber B (2015). Atlas e perfil do Moxico, Angola/An atlas and profile of Moxico, Angola. Mendelsohn J, & Weber B. RAISON, Windhoek.

Mendelsohn JM, vanderPost C, Ramsberg L, Murray-Hudson M, Wolski P, Mosepele K (2010) Okavango delta: floods of life. RAISON, Windhoek

NGOWP (2020a) Final report: scientific exploration in Angola during 2017. National Geographic Okavango Wilderness Project. http://www.wildbirdtrust.com/wp-content/uploads/2020/05/Report-3_Annual-2017_Cubango.pdf.

NGOWP (2020b) Final report: scientific exploration in Angola during 2018. National Geographic Okavango Wilderness Project

Stellmes M, Frantz D, Finckh M, Revermann R, Röder A, Hill J (2013) Fire frequency, fire seasonality and fire intensity within the

Okavango region derived from MODIS fire products. Biodivers Ecol 5:351–362
Ucuassapi AP, Dias JCS (2006) Acerca da fertilidade dos solos de Angola. In: Moreira I (ed) Angola: Agricultura, Recursos Naturais e Desenvolvimento. ISA Press, Lisboa, pp 477–495
Wallenfang J, Finckh M, Oldeland J, Revermann R (2015) Impact of shifting cultivation on dense tropical woodlands in south-east Angola. Trop Conserv Sci 8:863–892
Zigelski P, Lages F, Finckh M (2018) Seasonal changes of biodiversity patterns and habitat conditions in a flooded savanna - The Cameia National Park Biodiversity Observatory in the Upper Zambezi catchment, Angola In: Revermann, R, Krewenka, KM, Schmiedel, U, Olwoch JM, Helmschrot J, Jürgens N (eds) Climate change and adaptive land management in southern Africa – assessments, changes, challenges, and solutions, pp 438–447; Biodiversity & ecology, vol 6. Klaus Hess Publishers, Göttingen & Windhoek. https://doi.org/10.7809/b-e.00356
Zigelski P, Gomes A, Finckh M (2019) Suffrutex dominated ecosystems in Angola. In: Huntley BJ, Russo V, Lages F, Ferrand N (eds) Biodiversity of Angola. Science & conservation: a modern synthesis. Springer Nature

John Mendelsohn's passion for research has driven much of his work in zoology, education, household economies, land rights and uses, and geography. The nature and functioning of rivers in the Kalahari Basin are a special interest, particularly the Cunene, Cuvelai, Okavango, Zambezi and Cuando river systems. Most of his recent research has been as a freelance consultant in Namibia and Angola. He now works at the Ongava Research Centre in northern Namibia.

The Okavango Delta Peatlands

3

William N. Ellery and Karen Ellery

Abstract

The presence of a large (approximately 2000 km^2) peatland in a semi-arid climatic setting such as the Kalahari is unusual. Peat forms in permanently flooded areas in the Okavango Delta primarily due to the perennial input of large volumes of water from a distant catchment in the highlands of Angola, into a valley formed by rifting. Peat deposits form in three distinct settings in the Okavango: backswamp settings where open water is converted into homogeneous emergent peatlands, lake and channel margins where the peatland is patchy, and the inlets to lakes that connect to the primary distributary channel, which presently is the Okavango-Nqoga-Maunachira River system. An unusual feature of peat formation in backswamp areas, as well as in lake and channel margin settings, is that frequently mats of fine organic detritus on the bed rise to the water surface and are colonised by emergent plants. Once thus colonised, peat production is accelerated due to the higher productivity and less easily decomposed tissue of emergent plants compared to submerged and floating-leaved plants. In backswamp settings, the floating mats are extensive (hundreds of square metres to hectares) and lead to the formation of homogeneous plant communities that cover large areas. In the case of lake and channel margins, floating mats form small isolated features (up to a few square metres) that are blown to the lake or channel margin by wind. Their accumulation on the leeward sides of lakes and broad streams gives rise to a patchy and heterogeneous plant community. Where lakes are connected to the primary distributary channel, papyrus debris collects as large floating rafts along channel margins, ultimately to be deposited in the lake inlet. Thus, large lakes are converted to papyrus swamp over periods of decades. Channel switching of primary channels leads to radical changes in the flow such that formerly flooded areas dry out and peat deposits are destroyed over periods of decades due to desiccation. However, a new cycle of peat formation takes place in the newly flooded area. Peat deposits in the Okavango are thus not permanent features but have a lifespan of about one or two centuries. Given increasing recognition that peat formation in "dryland" wetlands requires an elevated base level, the hypothesis proposed here is that chemical sedimentation in the lower reaches of the Okavango elevates the base level, and peat formation is an inevitable and passive consequence. This leads to the formation of an alluvial fan with a remarkably uniform slope from the fan apex to the toe of the system.

Keywords

Wetlands in Drylands • Vegetation succession • Peat fires • Clastic • Organic and chemical sedimentation

3.1 Introduction

Although much of southern Africa is classified as "dryland", it is characterised by several freshwater wetlands greater than 2000 km^2 in extent (Lidzhegu et al. 2019), all of which occur along the courses of large rivers such as the Zambezi and its tributaries. One of the largest freshwater wetlands in the region is the Okavango Delta in northern Botswana with an estimated area of approximately 12,000 km^2 (McCarthy and Ellery 1998). It is remarkable that this large peatland landscape occurs in a region with a negative water balance where potential evapotranspiration exceeds rainfall annually by a factor of three.

W. N. Ellery (✉)
Geography Department, Rhodes University, Grahamstown, South Africa
e-mail: f.ellery@ru.ac.za

K. Ellery
Science Extended Studies Unit, CHERTL, Rhodes University, Grahamstown, South Africa

F. D. Eckardt (ed.), *Landscapes and Landforms of Botswana*, World Geomorphological Landscapes,
https://doi.org/10.1007/978-3-030-86102-5_3

This chapter describes the three distinct settings that host peat formation in the Okavango Delta: backswamp, channel margin and lake inlet. Compared with many peat swamp forests, bogs and fens that develop over millennia, the Okavango Delta peatlands exist for much shorter periods because of dramatic switches in the flow that take place over timescales of centuries. It is only through understanding the interplay between processes of peat formation, peat destruction, clastic and solute sedimentation, and changing water distribution, that we can appreciate the significance of peatlands in the morphology of the Okavango Delta ecosystem as a whole. This chapter is therefore wide-ranging, not only describing the peatlands themselves but also the broader landscape-level processes that characterise the system.

3.2 The Inputs and Distribution of Water in the Okavango Delta

The Okavango River rises in the highlands of eastern Angola, which receives rainfall greater than 1500 mm a^{-1}. Runoff from the catchment drains south-eastwards, entering Botswana at the town of Mohembo (Fig. 3.1). Downstream of Mohembo the Okavango River is confined within a fault-bounded depression in a region known as the Panhandle. At the toe of the Panhandle, and perpendicular to it, lies a half-graben between the Gomare Fault in the north and the Kunyere and Thamalakane Faults in the south. When the Okavango River reaches this half-graben it divides into a number of distributary channels to form the Okavango Delta, or, more correctly, Fan.[1] Connected directly to the Okavango River is the main distributary Nqoga River, which flows eastwards to the north of Chief's Island. The remaining distributary channels receive the bulk of their water supply as overbank flow and seepage of water through peat deposits flanking the primary Okavango-Nqoga river system. The Thaoge River flows southwards down the western margin of the Delta and the Maunachira River flows eastwards along the northern margin of the Delta. Both of these river systems eventually disappear as they lose water. The Jao and the Boro Rivers flow south-eastwards to the south of Chief's Island and the Mboroga River arises from the Maunachira River via diffuse flow and drains south-eastwards to the north of Chief's Island. During wet periods the Mboroga and Boro Rivers flow into the Thamalakane River, which flows south-westwards along the Thamalakane Fault, mainly into the Boteti River, which is the only river to flow out of the Delta. The Boteti River terminates in Ntwetwe Pan of the Makgadikgadi Pans, 200 km to the east.

The upper reaches of the Okavango Delta are permanently flooded in a region known as the permanent swamps and the lower reaches are seasonally flooded in a region known as the seasonal swamps. Extending into the alluvial fan are a number of sandveld tongues dominated by species of *Acacia*, while areas with clay-rich soils are dominated by *Colophospermum mopane* (Ellery and Ellery 1997).

The water balance of the Okavango Delta is characterised by a significant surface inflow of water from the catchment in eastern Angola via the Okavango River, which contributes 9.2×10^9 m^3 a^{-1} (McCarthy and Ellery 1998; McCarthy et al. 2003). Local summer rainfall (mainly November to March) contributes a further 6.0×10^9 m^3 a^{-1}, which amounts to about 35% of the water input. Despite January being the month of the highest rainfall in the Delta region, the peak flood wave that has travelled over 1000 km from central Angola, reaches Mohembo at the top of the Panhandle in April, with an average monthly flow in April of about 15×10^9 m^3. Flow in the Okavango River at Mohembo is lowest in the month of November, with an average monthly flow of about 4×10^8 m^3. This variation in inflow contributes to considerable seasonal variation in the extent of inundation in the Okavango Delta, with the greatest extent being in the dry season in July to August while the minimum extent of inundation is generally in the wet season in January to February (McCarthy et al. 2003).

Flow in the Okavango Delta has varied spatially for millennia. The first written records of flow in 1854 describe flow down the Thaoge River into Lake Ngami (Fig. 3.1, Andersson 1856). Today, the Thaoge River no longer flows into Lake Ngami, with its course restricted to about a quarter of its original length. According to oral accounts provided by Indigenous people, the Thaoge River dried progressively during the late 1800s, but during this time a series of hippo trails leading eastwards from the Thaoge River were enlarged by erosion to divert flow into the Nqoga and Mboroga Rivers to the north and east of Chief's Island, respectively (Wilson and Dincer 1976). In the early twentieth century, the Mboroga River via the Thamalakane River was the main supplier of water to the town of Maun (Stigand 1923). In the 1930s the lower reaches of the Nqoga River started failing and this was associated with an increase in flow along the more northerly Maunachira River (Wilson and Dincer 1976). At first glance, such a dynamic environment appears an unlikely setting for the development of extensive peatlands.

[1] Although the Okavango has a triangular shape, deltas form where rivers enter standing bodies of water, whereas alluvial fans are subaerial features. The Okavango is an alluvial fan and is more correctly referred to as the Okavango Fan.

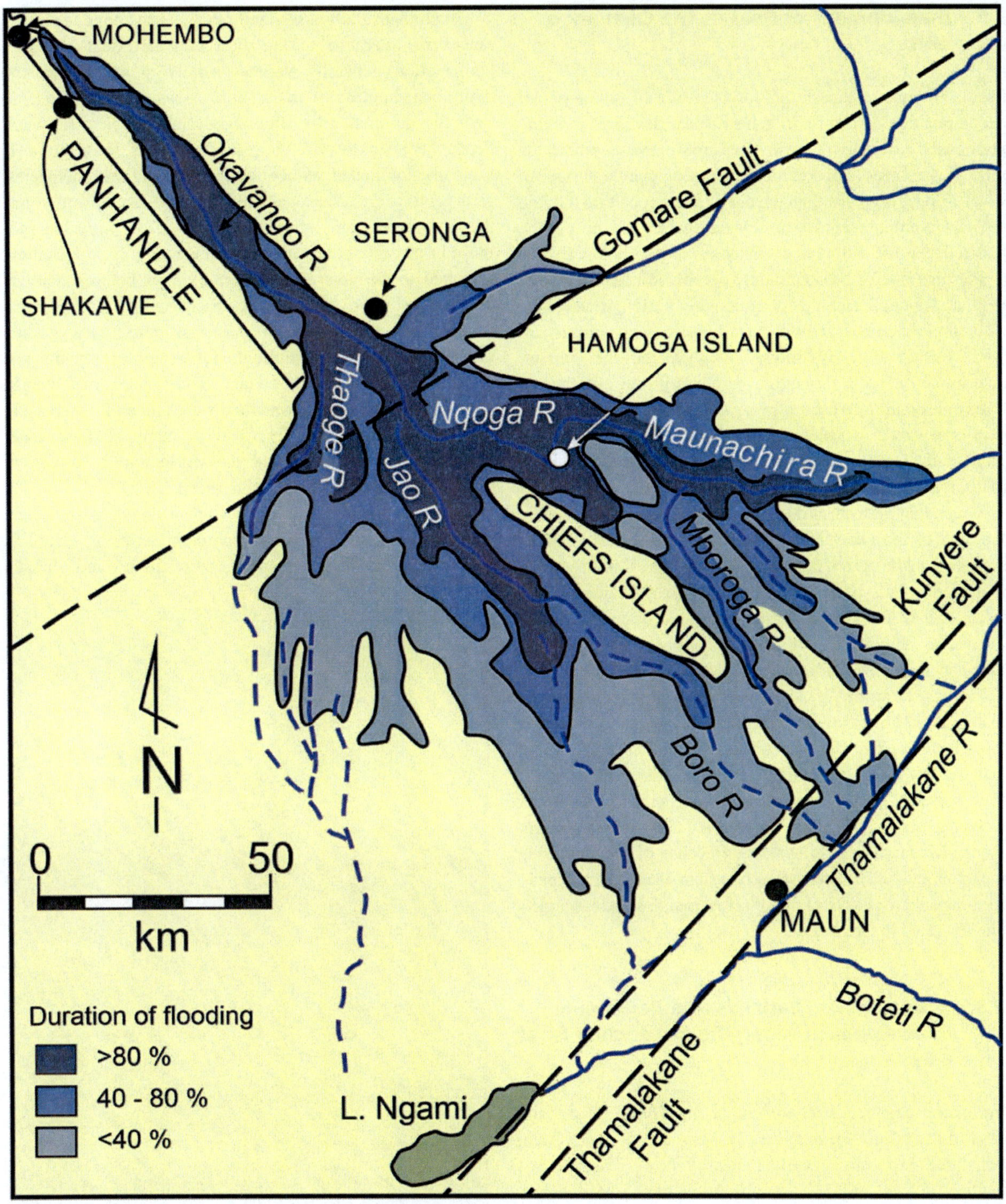

Fig. 3.1 Physiographic features and duration of flooding of the Okavango Delta (modified after McCarthy et al. 2003; Gumbricht et al. 2004)

3.3 Development of Peat in the Okavango Delta

Peat is an organic deposit that forms as a result of the partial decomposition of plants. In general, peat develops in calm, permanently inundated, freshwater environments where the rate of accumulation of organic matter, at or near the surface of a water body, is greater than the rate of decomposition (Clymo 1984). Organic material at or near the surface of a peat body, referred to as the acrotelm, decomposes relatively rapidly due to the presence of predominantly aerobic conditions. However, as organic material is progressively buried, it enters the catotelm, which is anaerobic such that decomposition is slowed considerably. Hydrology is considered the primary determinant of peat formation, particularly the presence of anaerobic conditions caused by permanent flooding, but other factors reported to influence peat formation are many and include low temperature, limited nutrient availability and low pH (Clymo 1984; Craft 2016).

Given that permanent flooding is required for peat to form, the area of the Okavango Delta where peat exists is much smaller than the area of the wetland as a whole. If it is assumed that flooding must be present more than 80% of the time for peat to form, it is likely that the area of peat accumulation in the Okavango Delta matches the area of near-permanent flooding as mapped by J McCarthy et al. (2003) and Gumbricht et al. (2004), as shown in Fig. 3.1. It is estimated to cover approximately 2000 km^2. However, the channel switching processes that alter flow patterns over relatively short time periods of centuries would to some degree serve to reduce the extent of such deposits. Nonetheless, they are widespread in the upper Delta reaches and have been well studied along the Okavango-Nqoga-Maunachira river systems. Because of their differing locations and functioning in the ecosystem, the three main peat-forming areas of backswamps, channel margins and lake inlets are described in turn below.

3.3.1 Backswamp Peatlands and Vegetation Development Along the Maunachira River System

Following channel abandonment of the Thaoge River and later of the lower Nqoga River, the Maunachira river system benefitted from significantly increased flows over the last century (Smith 1976). A small portion of the Maunachira river system consists of lakes, channels and islands, and the rest consists of the permanently inundated backswamps that experience very low water flow rates of generally much less than 0.03 m.s^{-1}. These backswamps have experienced peat accumulation associated with a number of vegetation successional processes (Ellery 1986).

Given that different wetland plant species have particular ranges of tolerance to variation in water depth, species composition generally varies spatially depending on the depth and duration of inundation. Areas of deep water are typically dominated by submerged plant species. As water depth declines towards terrestrial habitats, vegetation is typically dominated sequentially by floating-leaved species, emergent species, seasonally flooded species, and then terrestrial species. This zonation of vegetation is visible spatially in places where water depth decreases from a shallow lake bed to the lake edge (Fig. 3.2). This heterogeneity in plant distribution can also represent temporal vegetation successional changes that occur in response to a gradual decline in water depth through, for example, organic peat accumulation on the bed of a lake. Such successional change in species composition over time with gradually decreasing water depth has been described in many shallow water bodies and wetlands around the world and is known as "hydrarch succession" (Gleason 1926).

Classification of the Maunachira River wetland plant communities resulted in eight communities being identified that vary in their species composition and habitat

Fig. 3.2 Typical spatial zonation of wetland vegetation in the seasonal swamps of the Okavango Delta. In the foreground are floating-leaved species in the deepest water, which give way to emergent wetland species as the depth declines, and ultimately to grasses (not visible) and trees on dry land

characteristics. One of the communities is restricted to channel and lake margins, being dominated by *Miscanthus junceus* or *Cyperus papyrus* (Community G; Fig. 3.3a). A second community is dominated by the floating-leaved plant *Trapa natans*, which occurs in monospecific stands in localised areas where channels enter lakes (Community H; Fig. 3.3a). Neither of these communities play a significant role in backswamp community development, but the channel margin community is key in influencing channel dynamics and is considered in detail in the following section.

The remaining communities are distributed in relation to variation in water depth and the thickness of the substrate in which plants are rooted (Communities A to F; Fig. 3.3a and b). In areas where water depth is greater than about 1.8 m, the plant community is submerged (generally as monospecific stands of *Najas pectinata*; Community A). Where water depth is between 1.6 and 1.8 m deep, a floating-leaved community (Community B) is present (Fig. 3.3b), usually rooted in a benthic mat of organic detritus 0.3–0.4 m thick resting on the sandy bed of an otherwise open water body. Areas dominated by floating-leaved species include *Brasenia schreberi*, *Nymphaea nouchali*, *N. lotus* and *Nymphoides indica*.

Typically, floating-leaved communities would eventually become emergent communities with the gradual accumulation of benthic detritus and peat—a process that can take millennia. However, an interesting and somewhat unusual phenomenon occurs commonly in the Maunachira floating-leaved communities which considerably speeds up the transition from floating-leaved to emergent backswamp communities: the formation of floating mats or rafts of detritus. The fine organic detritus that is resting on the sandy bed of the floating-leaved communities becomes buoyant, likely through gases being produced in decomposition processes in the benthic mat, which rises to the water surface. Because of the different ways in which such floating mats form, two different plant communities arise. The first is the floating-leaved and sudd community (Community C, Fig. 3.4a and b) which results from the formation of small, isolated mats or sudds of detached organic detritus held together by the root mass of the formerly floating-leaved plants (Fig. 3.4a). The second is the floating *Pycreus nitidus* sedge community (Community D; Figs. 3.3a and b, 3.4b) which forms on extensive floating detrital rafts that cover areas from twenty to hundreds or even thousands of square metres, and therefore remain in a position where they originally form. In contrast, the small isolated floating mats are moved by wind and typically accumulate on the leeward side of open water bodies. The formation of the two "floating" communities is summarised diagrammatically in Fig. 3.5a–c.

Once they occur as floating mats plant community succession is dramatically altered from being driven by gradual rates of benthic detrital accumulation to almost instantaneously being dominated by emergent vegetation. Both the isolated sudds as well as the extensive rafts rapidly become dominated by *Pycreus nitidus* which is able to grow clonally via submerged stolons. Lateral expansion via clonal growth occurs at a rapid rate of between 1 and 1.5 m per annum (Ellery et al. 1990). The succession then also proceeds with colonisation by a diverse array of later successional species such as sedges *Cyperus pectinatus*, *Fuirena pubescens*, and *F. stricta*, the insectivorous *Drosera madagascariensis*, the broad-leafed woody shrub *Ficus verruculosa* and the fern *Thelypteris confluens*, in a small-scale patchy manner (Community E, Fig. 3.3a and b). Over time the substrate becomes more homogeneous and the vegetation is dominated primarily by grasses, sedges and the shrub *Ficus verrucIosa* (Community F, Figs. Fig. 3.3a and b, 3.6). As mentioned, the development of such climax backswamp communities from floating communities occurs much more rapidly than would take place from gradual infilling of deeper water communities due to the higher growth rates and biomass production of emergent species, as well as their much coarser nature, which slows decomposition rates.

As succession progresses towards the shallow emergent homogeneous community, the thickness of the floating mat increases, as does the thickness of the organic benthic sediment, due to the gradual raining down of organic material from the floating mat (Fig. 3.3b). As these processes continue, the floating and benthic layers meet forming a relatively uniform layer of peat. The peat deposits associated with each plant community varies from the floating *Pycreus nitidus* community and its precursors, where the organic matter is very fine and unconsolidated (sludgy), held together by fine root material associated with submerged species, to the climax community where the peat is consolidated and contains considerable coarse root material that makes up a large proportion of the volume of the deposit (Fig. 3.7).

Whilst Fig. 3.8 indicates the spatial relationship between the different backswamp plant communities in the Maunachira River landscape, similar communities are found throughout the permanently flooded upper reaches of the Okavango Delta. The submerged and floating-leaved communities are located in the few open water bodies that exist. However, relatively rapid accumulation of peat in this system gives rise to large expanses of relatively stable climax, diverse, emergent backswamp communities. At a landscape level, in a dynamic system such as the Okavango Delta, such stability is not permanent and when this river system loses its permanent supply of water due to channel avulsion further upstream, the drying and burning of peat deposits would then play a key role in the renewal of the landscape. These processes are described later in this chapter.

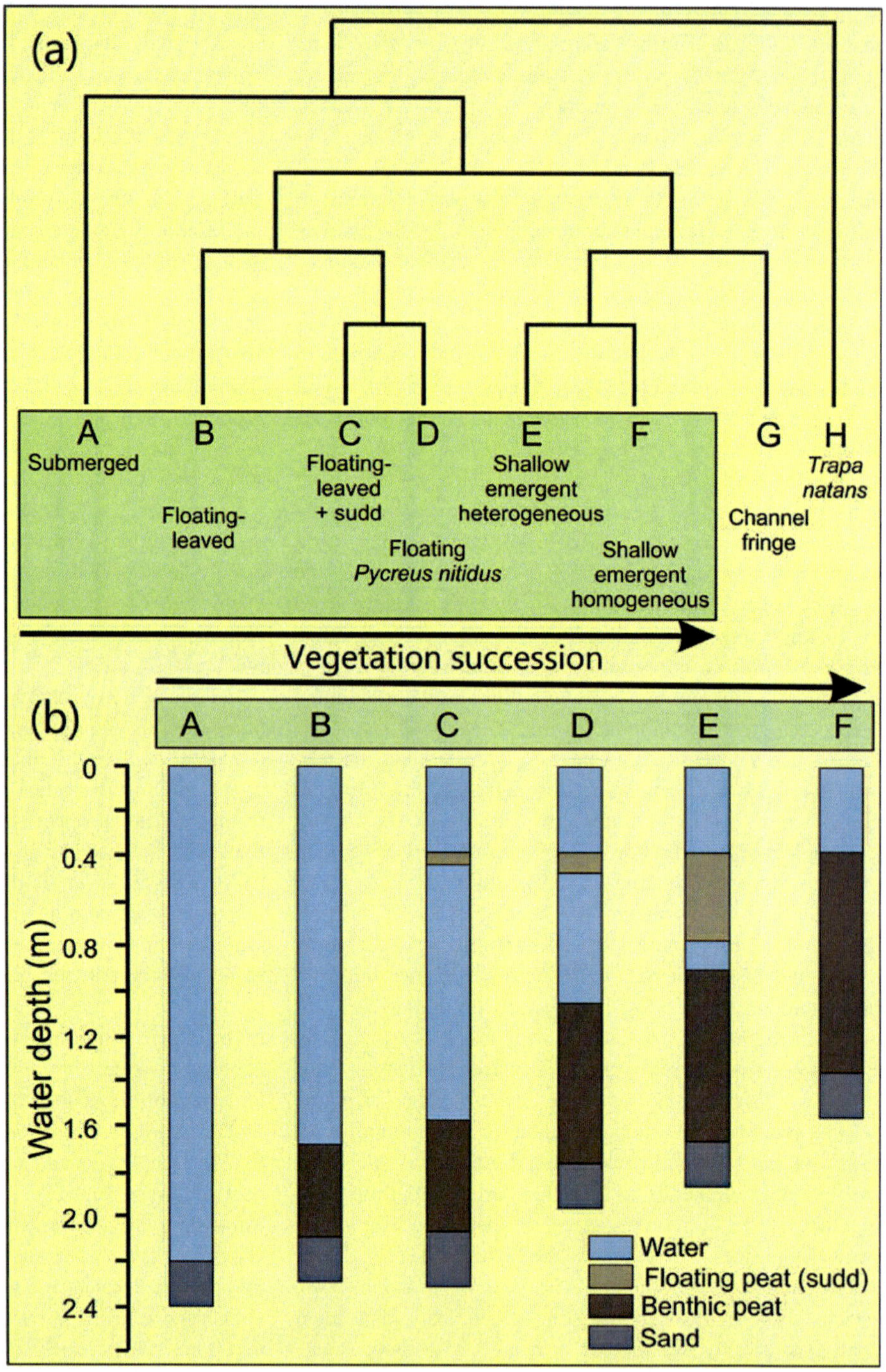

Fig. 3.3 Vegetation community classification (**a**) and depth to rooting substrate in relation to successional processes (**b**) in the permanent swamps of the Maunachira River wetlands of the Okavango Delta (modified after Ellery et al. 1991)

3.3.2 Channel-margin Peat Deposits and Associated Vegetation

Whilst the backswamp peatlands are an important part of the Okavango Delta landscape due to their extent, the relatively localised channel margin peat deposits and associated vegetation communities (Community G, Fig. 3.3a) play a key role in the landscape-level water delivery and channel-switching events previously alluded to. Channels form in the wetland largely due to the network of hippo trails that are used repeatedly in backswamp areas to get from daytime resting grounds to feeding areas on islands (McCarthy et al. 1998). These channels are generally oriented parallel to the hydraulic slope and promote effective

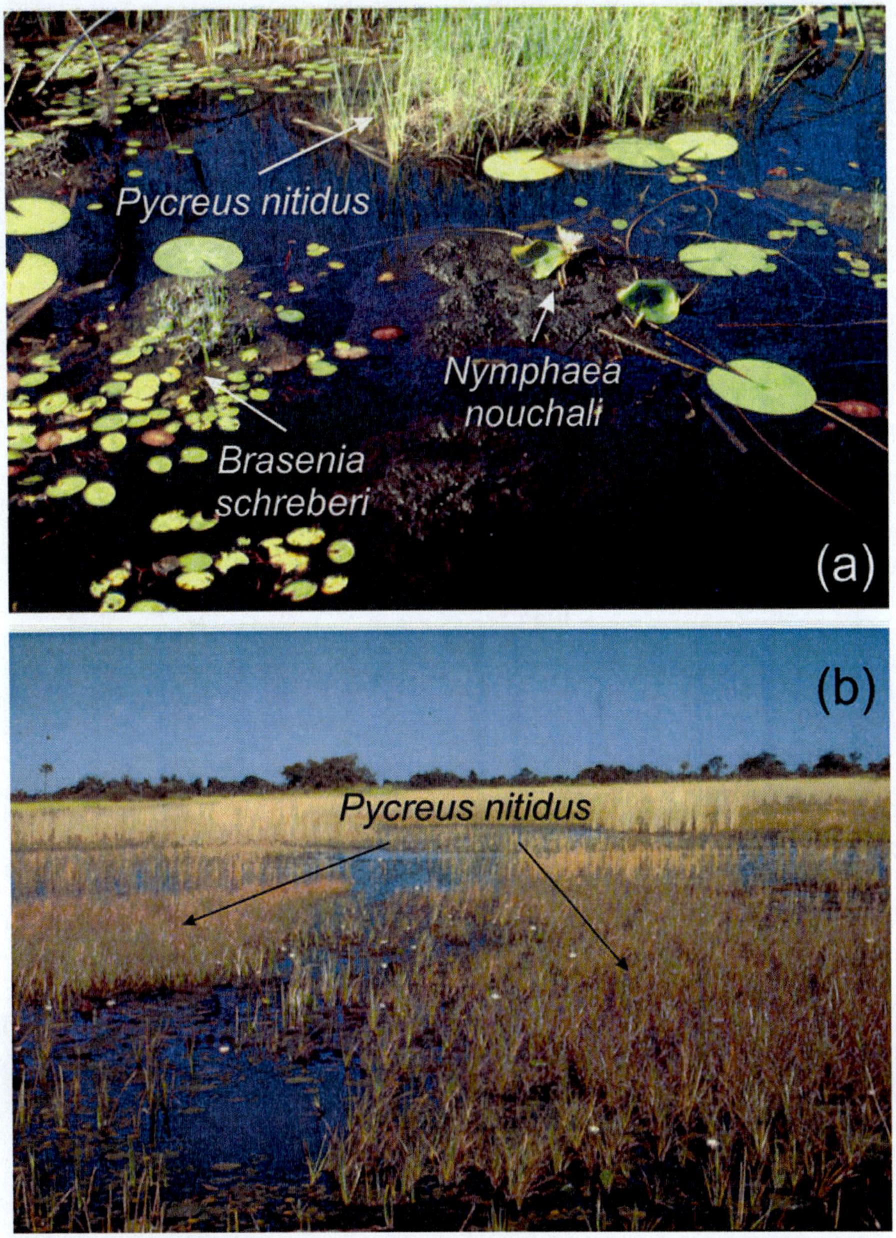

Fig. 3.4 Detached, isolated rafts of floating detritus held together by the root masses of formerly bottom-rooted floating-leaved plants of *Nymphaea nouchali* and *Brasenia schreberi* (**a**), and extensive rafts of detritus floating near the water surface of an open water body, colonised by *Pycreus nitidus* (**b**)

transmission of water through the permanent swamp to sustain water delivery to the distal reaches of the system. However, it is the interaction between the accumulation of clastic and organic sediments within and alongside any channel that is linked directly to the incoming Okavango River, that drives avulsion events that lead to radical changes in flow over timescales of decades to centuries (McCarthy et al. 1986).

Clastic sediment entering the Okavango Delta from the catchment amounts to about 200 000 t a^{-1}, of which 170 000 t is sand transported as bedload, which accumulates on the beds of primary channels (Okavango-Nqoga; McCarthy et al. 1991). Suspended clastic sediment accumulates largely in the peat deposits of the channel margins (McCarthy et al. 1991). This results in a downstream decline in clastic sediments within the channel margin peats, with between 25 and 40% organic matter in the Panhandle and Nqoga River margins, and between 40 and 90% organic matter along the Maunachira River margins (McCarthy et al. 1989). The grass *Pennisetum glaucocladum* dominates where the organic content is lowest, followed by species such as the reed *Phragmites mauritianus* and the sedge

Fig. 3.5 Change in vegetation from an open water body with submerged and floating-leaved communities distributed in relation to water depth (**a**), to an emergent plant community rooted in shallow water (<0.1 m depth) growing on an extensive detrital floating raft (**b**) or small, isolated floating detrital sudds that accumulate on the leeward side of open water bodies (**c**)

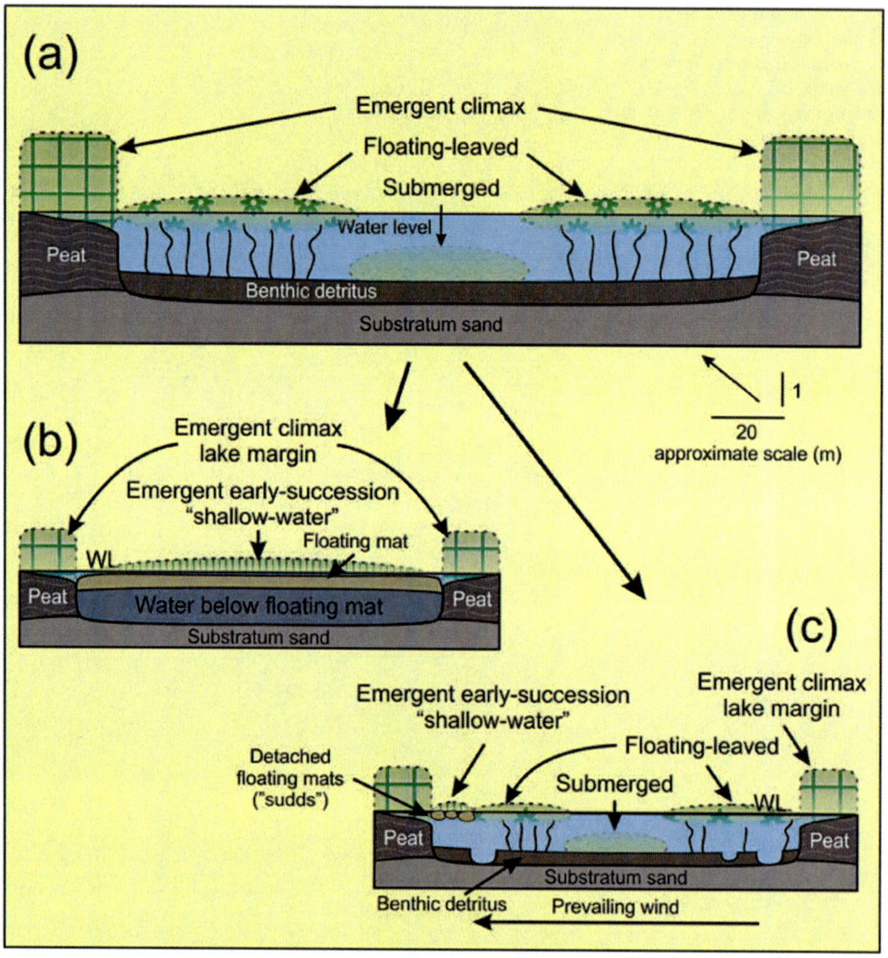

Fig. 3.6 The climax community (Community F) that covers large areas of permanently flooded peat (thickness is approximately 2 m), with a diverse range of species present including *Ficus verruculosa* and *Miscanthus junceus*. Isolated patches of woody plants are on termite mounds in the peatland

Fig. 3.7 The character of peat deposits of early successional stages dominated by *Brasenia schreberi* and *Pycreus nitidus* (Communities C and D; top) and by climax species such as *Miscanthus junceus* (Community F; bottom)

Cyperus papyrus with increasing organic contents (Ellery et al. 2003). Channel margin communities dominated by *Miscanthus junceus* are characterised by peat deposits that have very high organic contents (Ellery et al. 2003).

There is a sharp transition between the channel and the channel margins, which are made of peat such that channels are roughly rectangular in cross-section (Fig. 3.9). A key feature of peat channel margins is that the channels lose water by overbank flow and flow through the acrotelm of the peat deposits, which means that channels get progressively smaller downstream in respect of their width and discharge such that water entering the ecosystem is dispersed widely across the wetland. The width of the Okavango-Nqoga-Maunachira channel system declines from almost 150 m to 20 m. Equally important is the role of peat banks in confining the bedload sediments to within channels as the sediments are rolled and bounced on the bed of the channel and unable to escape. As channels lose water their ability to transport sediment declines and bedload sediment is deposited on the channel bed, leading to channel bed aggradation, which in the region of the lower Nqoga River near Hamoga Island (see Fig. 3.1) is approximately 0.05 $m.a^{-1}$ (Fig. 3.10). In places where channel bed aggradation is as high as this, the channel margins are dominated by luxuriant, dense stands of papyrus rooted in peat (Ellery et al. 2003). Aggradation of the channel margin takes place largely through extremely rapid growth of papyrus that ultimately forms a coarse and robust peat of partially decomposed and entangled rhizomes, roots and shoots, and can match these high rates of bedload aggradation. Such rapid rates of peat formation have not been reported elsewhere in the region, with reported rates of peat formation elsewhere of 0.002–0.004 $m.a^{-1}$ (Grundling et al. 2013), although rates may be as high as 0.02 $m.a^{-1}$ (Thamm et al. 1996). With bedload and channel margin accumulation rates being similar, channel width and depth are maintained (Ellery et al. 2003). Over time the primary channels are gradually elevated relative to the surrounding backswamp, resulting in substantial loss of water from the channel to the surrounding wetland and ultimately to channel avulsion once sufficient relief has been created.

This set of processes has happened progressively along the lower Nqoga River since the late 1930s. Aggradation of the sandy channel bed and peat margins of the

Fig. 3.8 Plant communities of the backswamp environments of the Maunachira River, Okavango Delta

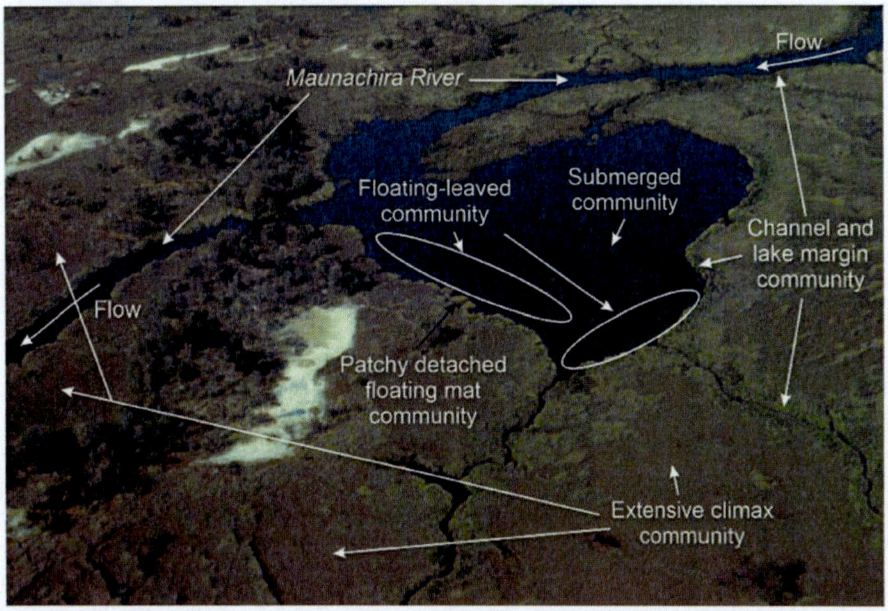

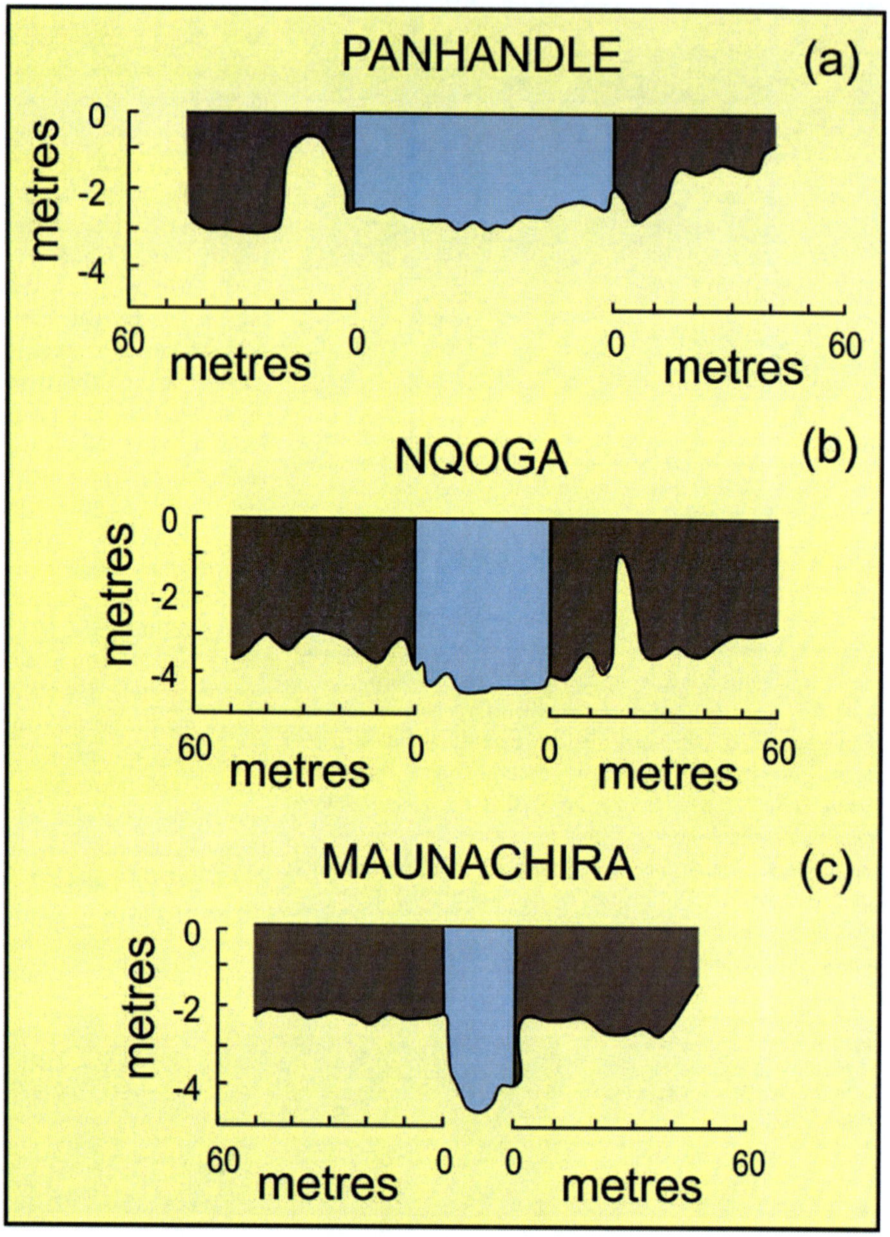

Fig. 3.9 Relationship of channels and channel margins, showing variation in depth and width of channels of the Okavango, Nqoga and Maunachira Rivers over a distance of nearly 250 km arranged sequentially from the town of Shakawe at the top of the Panhandle to the lower Maunachira River

eastward-flowing lower Nqoga River was accompanied by erosion and enlargement of a small hippo trail linking the Nqoga River to Bokoro Lediba (Tswana word for "lake") to the north (Fig. 3.11a). As flow was increasingly diverted along the hippo trail it became navigable and was given a name: Letenetso Channel (Fig. 3.11b). During this time, the lower Nqoga River started blocking by vegetation encroachment (Ellery et al. 1995). By 1983 all flow from the lower Nqoga River was redirected along the Letenetso Channel into the Maunachira River system (Fig. 3.11c). This was accompanied by, first, lake infilling in Bokoro Lediba and Dxergha Lediba, which is described in the next section, and secondly, peat fires along the abandoned Nqoga River, which is described in the following section.

3.3.3 Lake Inlet Peat Accumulation and Closure

It is inevitable that the first lake situated along a papyrus-lined river course linked to the main Okavango River will ultimately close due to clastic and inorganic sedimentation processes that occur at the lake inlet. *Cyperus*

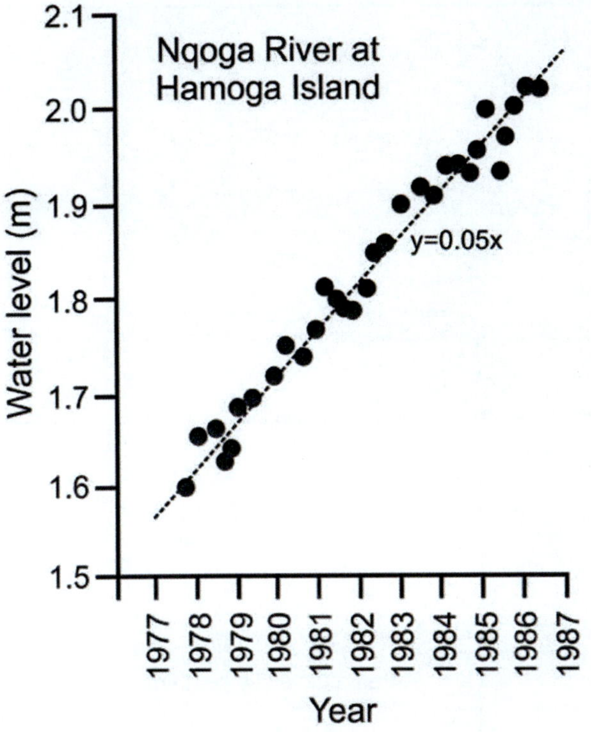

Fig. 3.10 Aggradation rate as reflected by the measured water level at Hamoga Island on the lower Nqoga River

papyrus along channel margins is rooted in peat but it extends from the peat bank into the channel via rhizomes that produce successive shoots. Both older shoots, as well as entire rhizomes, become detached from the parent plant, float downstream on the water surface. These individual floating fragments accumulate in large floating mats on the inner (convex) banks of the sinuous streams, where current velocity is very low compared to the outer (concave) banks. At some point, these floating mats become too large to remain in place and move downstream as large floating rafts. As channels lose water to the surrounding swamp they become progressively narrower downstream such that these rafts eventually are larger than the channel is wide, and a surface blockage forms (Fig. 3.12).

Such blockages tend to undergo a process of consolidation through vegetative growth of papyrus, which tends to strengthen the mat, as well as decomposition where no growth is occurring, which weakens the integrity of the blockage. Ultimately, all of the organic material that temporarily blocks channels ends up in the first lake encountered along the river, which leads to its infilling and the establishment of papyrus swamp.

Such lake infilling processes have been tracked for Dxerega Lediba for over 60 years (Fig. 3.13; see McCarthy et al. 1993) and involve a combination of clastic and organic sedimentation. Lake closure initially involves the formation of a linear sandy bar at the mouth, which propagates out into the lake. The toe end of the delta initially becomes colonised by very dense stands of *Eichhornia natans* which can be submerged or have leaves floating on the water surface (Fig. 3.14). *Eichhornia natans* effectively traps floating rafts of *Vossia cuspidata* debris generated from the channels upstream such that in areas close to the mouth, this species forms dense monospecific stands rooted in the lake bed (Fig. 3.14). This robust emergent species traps rafts of papyrus debris, which rapidly colonise the mouth adjacent to the mouth bar. Once papyrus has been established, peat formation is relatively rapid compared with other species, and the lake is converted to a papyrus marsh with a central channel. In this way, the lake mouth progresses further into the original lake until such time that a single channel passes through papyrus-dominated peatland where the lake used to exist. In the case of Dxerega Lediba, even the channel has almost disappeared (see 2018, Fig. 3.13), as a secondary inlet on the south side of the lake has become the main point of inflow. The small areas of open water at this inlet are rapidly being colonised by emergent plants.

3.4 Channel Abandonment and the Occurrence of Peat Fires

The emphasis up to this point has been on peat formation and the consequences of this at a landscape scale. The focus now changes to processes associated with a failing channel, such as the abandoned lower Nqoga River in Fig. 3.11, where peatlands are destroyed because of desiccation and burning in peat fires.

Channel margin and backswamp peat deposits in the Okavango Delta generally reach thicknesses of between 2 and 5 m. As channel abandonment gradually moves in an upstream direction, peat deposits dry progressively and become susceptible to combustion in peat fires. Peat fires have been observed along the abandoned section of the lower Nqoga River from before the 1970s (probably since the 1950s; Ellery et al. 1989) and were visible in remotely sensed images in 2000, showing that burning of these deposits takes place over periods of several decades. Remarkably, peat fires along the upper Thaoge River were also observed in the same remotely sensed imagery, a channel that was abandoned in the late 1800s, illustrating the prolonged desiccation that takes place in association with

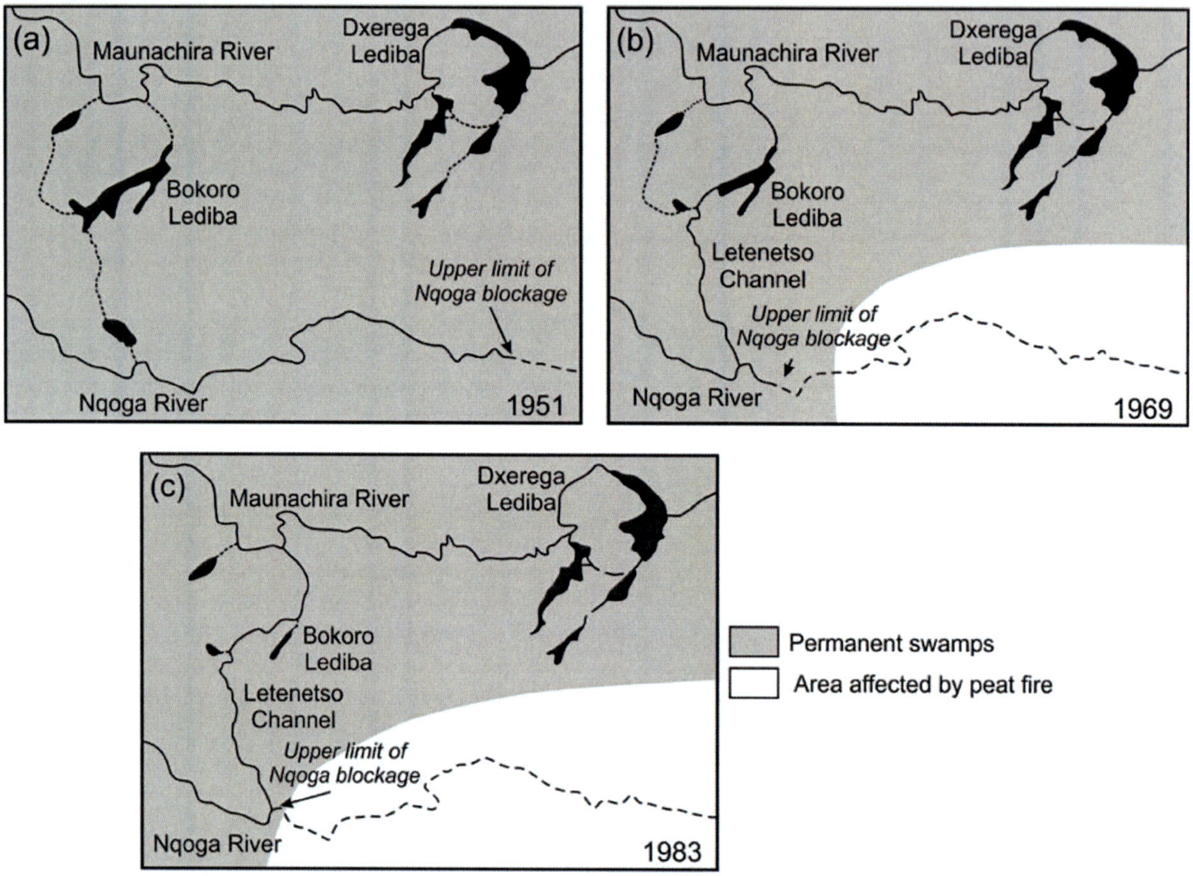

Fig. 3.11 Abandonment of the lower Nqoga River and channel enlargement of the Letenetso Channel based on aerial photography over the period 1951–1983 (modified after McCarthy et al. 1993)

channel abandonment and the equally lengthy time over which peat fires persist following abandonment.

Peat fires on the abandoned lower Nqoga River were observed by the authors in the 1980s (Fig. 3.15a). Generally, the peat fires burn to a depth of less than 1 m at a time due to both oxygen limitations and increasing moisture with increasing depth. But as desiccation proceeds, the peat deposit is burnt to increasing depth in successive fires. In the lower reaches of the abandoned Nqoga River that have been drying for the longest period, burning has destroyed all the peat, thereby exposing the original sandy surface that was beneath the peat (Fig. 3.15b). The former course of the lower Nqoga River is clearly visible from the air as a slightly raised feature (as a result of channel-bed aggradation as described earlier) and is surrounded by a flat and featureless ash-covered plain interrupted occasionally by former hippo trails entering the former watercourse from the adjacent former backswamp (Fig. 3.15b). Since peatlands have the ability to retain nutrients (Ellery et al. 1989), their burning results in the release of these nutrients, which allows these areas to become rapidly colonised by dry land plant species, particularly grasses. As such, these areas are more productive than other areas of the Okavango Delta and support large populations of herbivores including buffalo (*Syncerus*

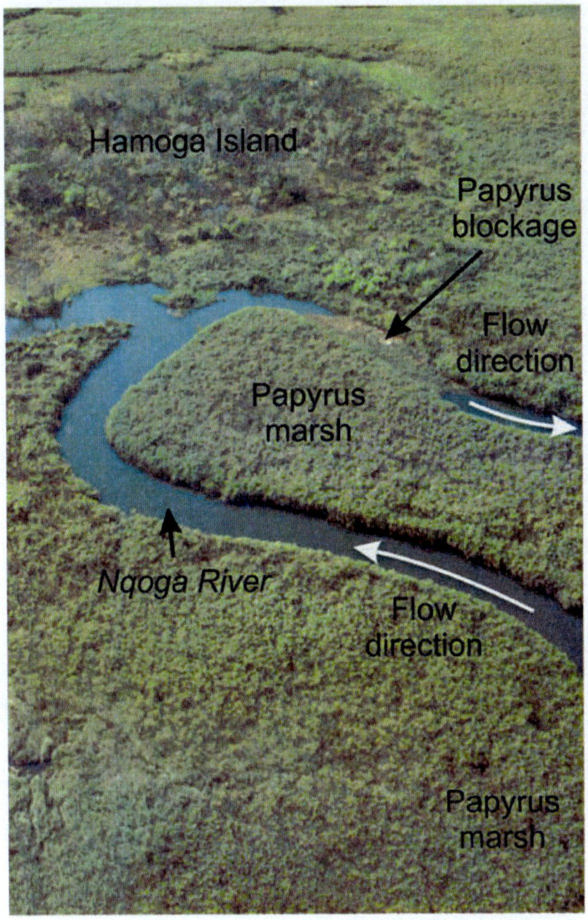

Fig. 3.12 Surface papyrus blockage on the Nqoga River at Hamoga Island

caffer) and red lechwe (*Kobus leche*), which in turn attract large predators. These areas also support a variety of bird species. Should flooding occur at some time in the future, plant and animal diversity would likely decline, and, in time, the cycle is completed by the development of a vegetated peatland dominated by emergent plants.

In summary, peat deposits of up to 5 m thick can be converted to a very thin sedimentary layer of ash, peat and soil with a total thickness of less than 0.5 m, which can be colonised by dryland species (Fig. 3.16). A further feature is the presence of a channel bed that is elevated relative to the adjacent burnt-out plain. The lowering of the basin level through the destruction of peat deposits makes it inevitable that at some point in the future, the area surrounding the abandoned channel will be reflooded. Depending upon the depth of inundation it is, therefore, possible that the former channel will form an island surrounded by permanent swamp (Fig. 3.17). Channel abandonment and the subsequent peat fires are therefore suggested as important processes in the maintenance of habitat diversity in the ecosystem as a whole.

3.5 A Conceptual Model of Peat Formation at a System-wide Scale

We end this chapter on a slightly conjectural note. The localised, relatively short-term, dynamic channel avulsion events that take place as a result of interactions between incoming clastic sediments from the catchment and organic sediments produced in situ, are well documented and understood. However, we suggest that another key process, occurring at a much wider spatial scale and over much longer time spans, is also influencing peatland formation in the Delta. This is the process of chemical sedimentation that influences the longitudinal slope of the system, creating conditions that favour the formation of peat deposits.

As much as 460,000 t of dissolved sediments enter the Okavango Delta system each year (McCarthy and Metcalfe 1990), which is more than twice the clastic sediment input. Of this amount, only about 30,000 t exits via the Boteti River, with possible minor loss of solutes from the system via groundwater outflow (McCarthy and Metcalfe 1990). It is therefore likely that a minimum of 250 000–350 000 t a^{-1} of dissolved sediment accumulates in the Delta each year, with the dominant solutes being SiO_2 and $CaCO_3$ (McCarthy and Metcalfe 1990). The accumulation of these solutes has an appreciable effect on the Okavango Delta landscape and on the morphology of the system as a whole, as these are concentrated in the seasonal floodplains (McCarthy and Ellery 1994). Solute accumulation takes place mainly in the seasonally inundated floodplains (mainly as opalline silica in the form of silcrete) and islands as magnesian calcite and trona beneath and on the surface of islands, leading to aggradation (McCarthy and Ellery 1994).

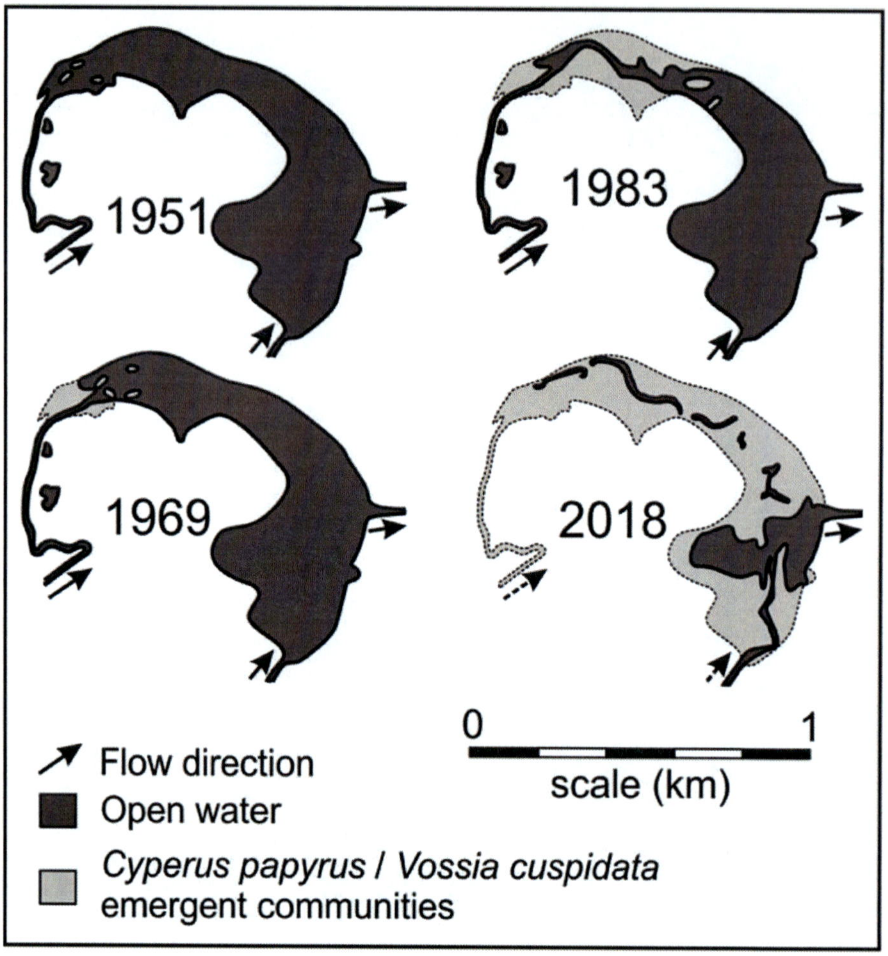

Fig. 3.13 Closure of Dxerega Lediba over the period 1951–2018

With most of the clastic sediment entering the Okavango Delta from the catchment in Angola being deposited in the Panhandle (McCarthy et al. 1991), and most of the dissolved sediment entering the system being deposited on the floodplains and islands of the seasonal swamps (McCarthy and Ellery 1994), accommodation space is created in the mid-reaches of the Delta in which deposition of organic sediment can occur (Fig. 3.18a). Large-scale geomorphic controls such as this, and their influence on peatland formation, are poorly understood. The creation of accommodation space for peat accumulation has been described in the Stillerust, Mfolozi and Mkuze floodplains in KwaZulu-Natal, South Africa as a consequence of aggradation of trunk stream floodplains, which blocks tributary streams (Grenfell et al. 2008, 2010; Ellery et al. 2012). Similarly, the elevation of the base level along trunk streams by tributary sediment input can create accommodation space along trunk streams, which has been documented for the Wakkerstroom and Krom River valley-bottom wetlands in South Africa (Joubert and Ellery 2013; Pulley et al. 2018). We believe these geomorphic processes may by key in many dryland peatlands and are worthy of

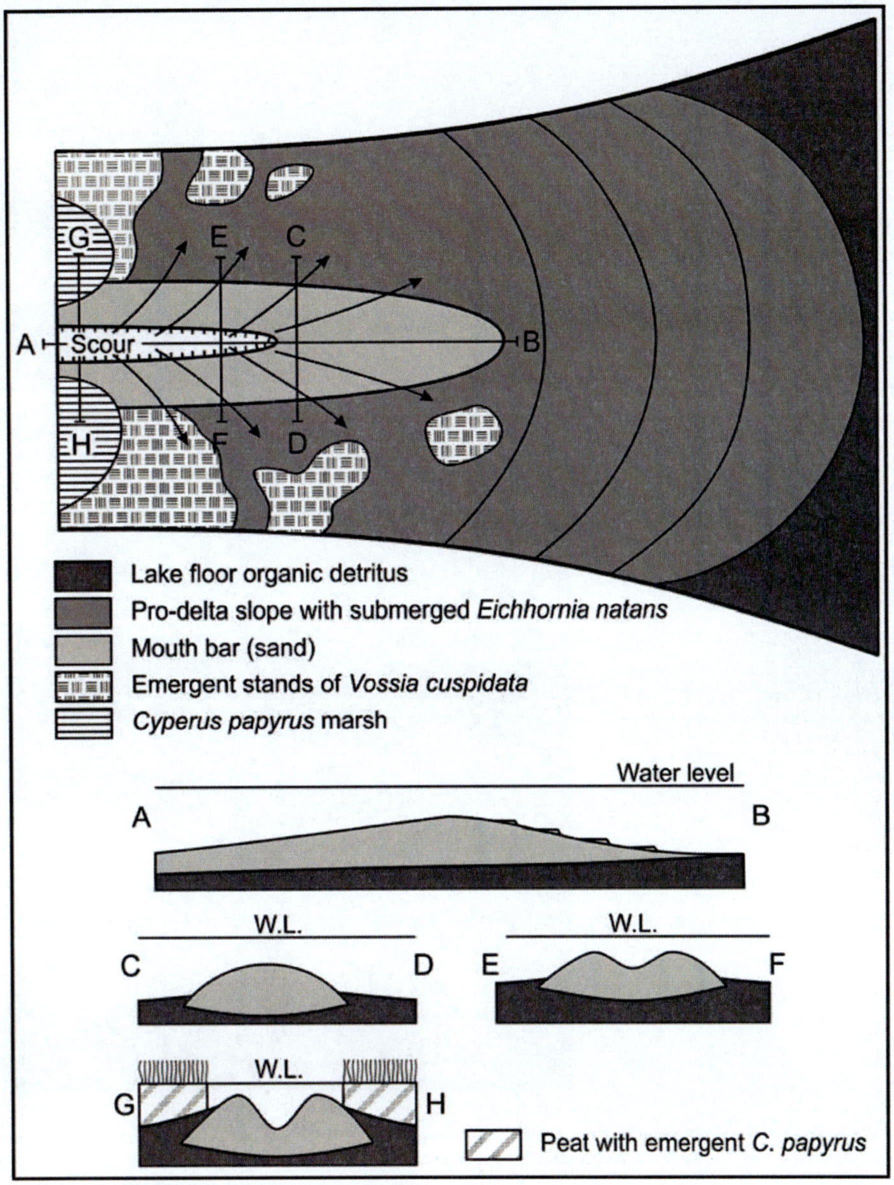

Fig. 3.14 Schematic illustration of the features of the mouth bar and channel at the inlet to Dxerega Lediba (modified after McCarthy et al. 1993)

further study, and in the case of the Okavango, chemical sedimentation in the lower reaches creates accommodation space in an upstream direction that leads to peat formation in the mid-reaches of the system. For this reason, peat formation occurs as a consequence of system-wide landscape processes, and its formation leads to the formation of a system with a remarkably uniform longitudinal slope (Fig. 3.18b, McCarthy et al. 1997).

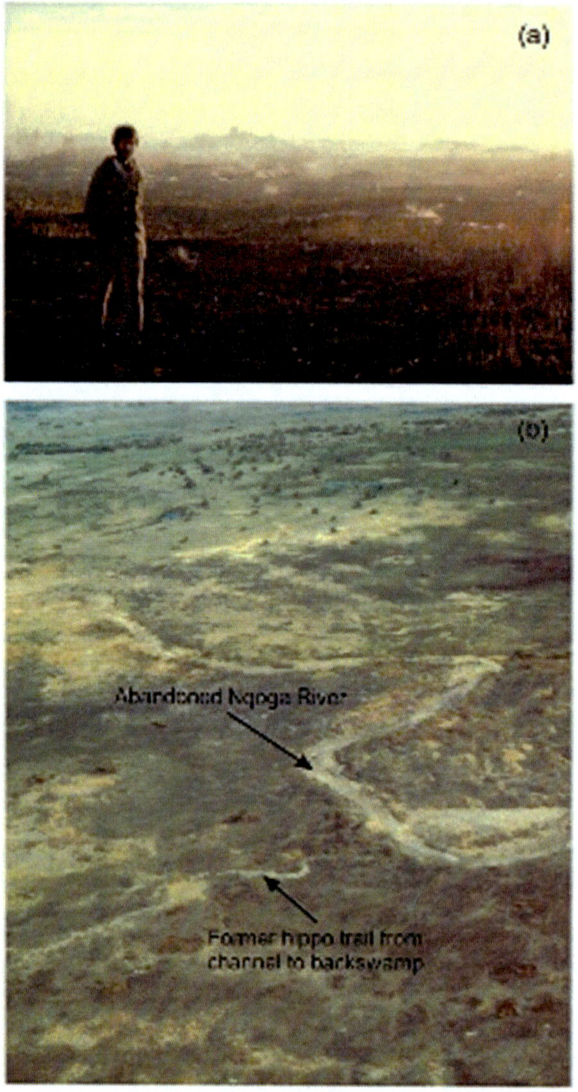

Fig. 3.15 Actively burning peat fires (**a**) and the landscape following decades of combustion of peat deposits along the abandoned lower Nqoga River, revealing a flat and featureless plain surrounding the former channel (**b**)

3.6 Conclusion

Organic sedimentation in the Okavango Delta is characteristic of the permanent swamps. It happens largely due to the presence of extensive organic detritus that rises and floats at or just below the water surface, leading to accelerated succession by emergent wetland plants that have very high rates of primary production compared to submerged and floating-leaved plants, and therefore greatly accelerate organic sedimentation. The closure of lakes by the accumulation of rafts of papyrus is also an important process leading to the formation of large areas with homogeneous surface topography and plant community distribution.

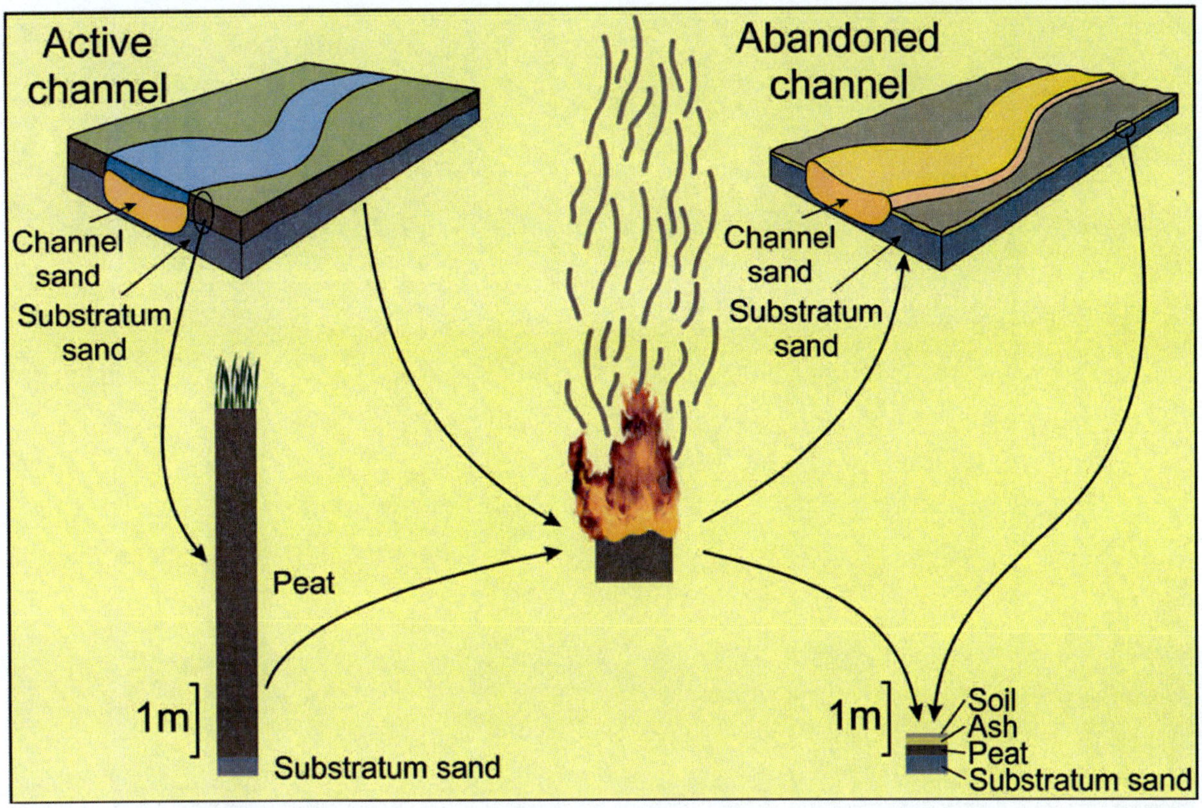

Fig. 3.16 Conceptual model of the destruction of peat deposits following channel abandonment. A 2–5 m peat deposit is converted to a 0.2–0.5 m thick ash/soil deposit and the former channel bed is elevated above the surrounding burnt-out plain

However, aggradation on the beds of primary channels due to the confinement of bedload sediment to in-channel areas, and the simultaneous aggradation of the peat banks, leads to elevation of the channel relative to the surrounding swamps and to channel avulsion. Following avulsion, peat deposits are destroyed such that extant peat deposits in the Okavango Delta are generally only hundreds of years old.

If the average peat thickness in the permanent swamps is between 2 and 4 m, then peat accumulation amounts to 20–40 mm per annum, which is far higher than rates measured elsewhere in the world, including Africa, which are typically two to ten times lower than this.

A key factor in the Okavango Delta is that peat formation may reflect interactions between clastic, dissolved and organic sediment at a system-wide scale such that peat formation is likely to be largely responsible for the overall structure of the alluvial fan.

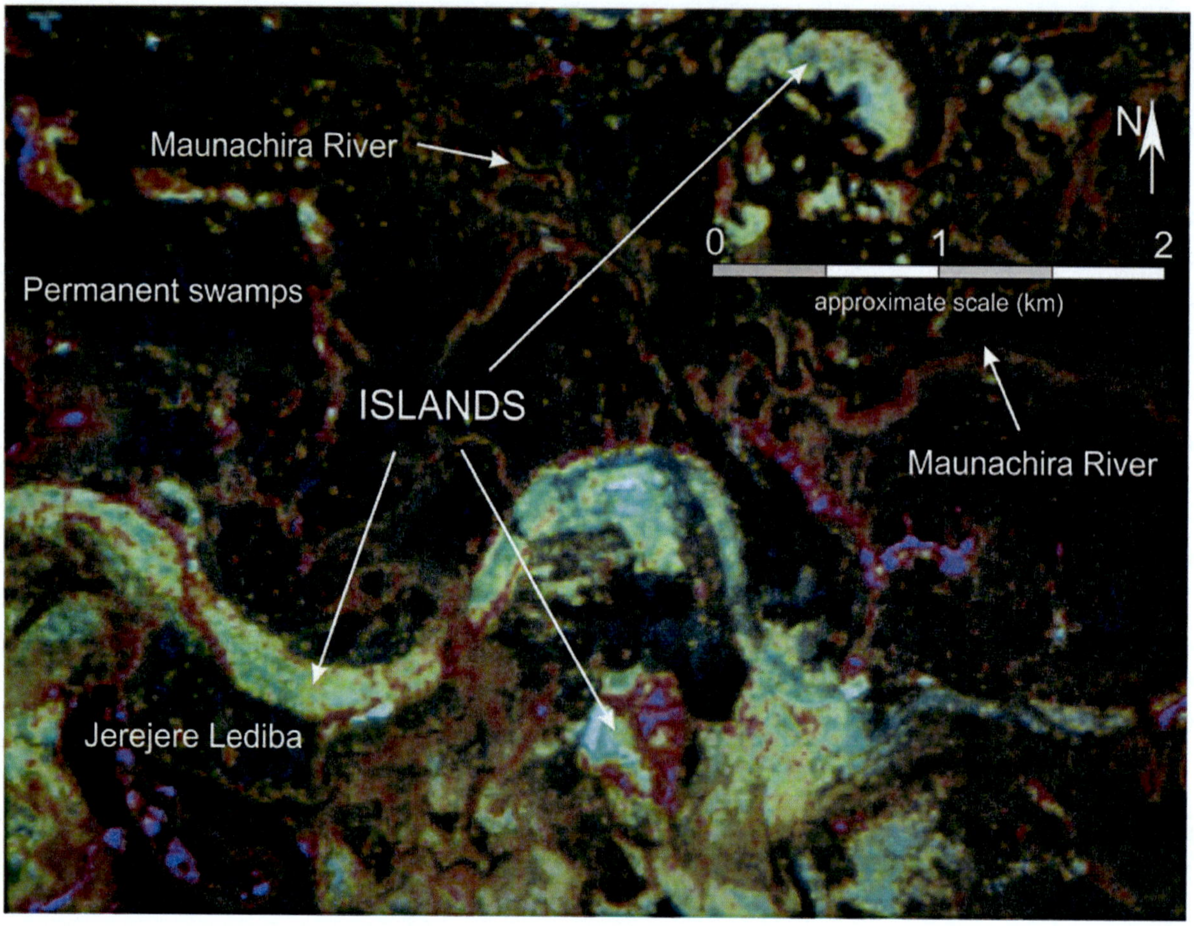

Fig. 3.17 SPOT2 satellite imagery on 8 October 1991 (scene reference number 118.388 and 119.388) of a region of permanent swamp between the abandoned lower Nqoga River and the Maunachira River. The image has been processed to reveal standing phytomass based on a transformed vegetation index (McCarthy et al. 1993). Lakes appear black, permanent backswamp communities are dark colours (charcoal and dark blue), with channel margin communities golden to red, while islands appear yellow to light green. The sinuous island feature must have been formed by a meandering channel in a previous (much larger than at present) flood cycle

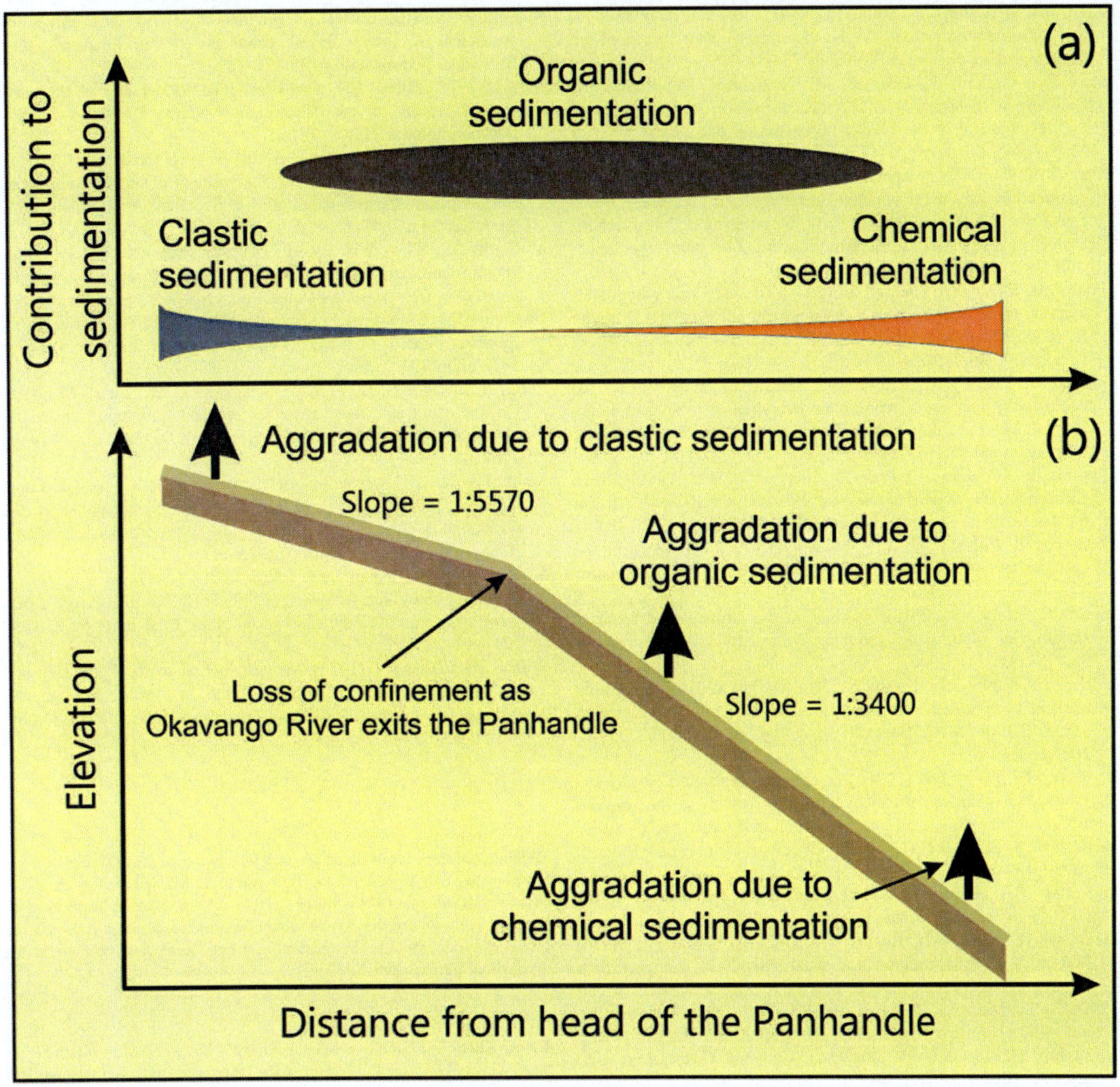

Fig. 3.18 A conceptual model of organic sedimentation in the Okavango Delta as a consequence of lowering the longitudinal slope in an upstream direction of the locus of chemical sedimentation in the floodplains and islands of the seasonal swamps

References

Andersson CJ (1856) Lake Ngami: explorations and discovery during the four years in the wilds of South-Western Africa. Hurst and Blackett, London

Clymo RS (1984) The limits to peat bog growth. Philos Trans R Soc Lond B 303:605–654

Craft C (2016) Creating and restoring Wetlands: from theory to practice. Elsevier, Amsterdam

Ellery K, Ellery WN (1997) Plants of the Okavango delta: a field guide. Tsaro Publishers, Durban

Ellery WN, Ellery K, McCarthy TS, Cairncross B, Oelofse R (1989) A peat fire in the Okavango Delta, Botswana and its importance as an ecosystem process. Afr J Ecol 27:7–21

Ellery K, Ellery WN, Rogers KH (1990) Formation, colonisation and fate of floating sudds in the Maunachira river system of the Okavango Delta. Aquat Bot 38:315–329

Ellery K, Ellery WN, Rogers KH, Walker BH (1991) Water depth and biotic insulation: major determinants of back-swamp plant community composition. Wetlands Ecol Manag 1:149–162

Ellery WN, Ellery K, Rogers KH, McCarthy TS (1995) The role of *Cyperus papyrus* in channel blockage and abandonment in the north-eastern Okavango Delta. Afr J Ecol 33:25–49

Ellery WN, McCarthy TS, Smith ND (2003) Vegetation, hydrology, and sedimentation patterns on the major distributary system of the Okavango Fan, Botswana. Wetlands 23:357–375

Ellery W, Grenfell S, Grenfell M, Humphries M, Barnes K, Dahlberg A, Kindness A (2012) Peat formation in the context of the development of the Mkuze floodplain on the coastal plain of Maputaland, South Africa. Geomorphology 141–142:11–20

Ellery K (1986) The composition and dynamics of wetland plant communities in the Maunachira River system, Okavango Delta, Botswana. MSc thesis, University of Witwatersrand, Johannesburg

Gleason HA (1926) The individualistic concept of the plant association. Bull Torrey Bot Club 53:7–26

Grenfell M, Ellery WN, Grenfell SE (2008) Tributary valley impoundment by trunk river floodplain development: a case study from the KwaZulu-Natal Drakensberg foothills, Eastern South Africa. Earth Surf Proc Land 33:2029–2044

Grenfell SE, Ellery WN, Grenfell MC, Ramsay LF, Flugel TJ (2010) Sedimentary facies and geomorphic evolution of a blocked-valley lake: Lake Futululu, northern KwaZulu-Natal, South Africa. Sedimentology 57:1159–1174

Grundling P, Grootjans AP, Price JS, Ellery WN (2013) Development and persistence of an African mire: how the oldest South African fen has survived in a marginal climate. CATENA 110:176–183

Gumbricht T, Wolski P, Frost P, McCarthy TS (2004) Forecasting the spatial extent of the annual flood in the Okavango delta, Botswana. J Hydrol 290:178–191

Joubert R, Ellery WN (2013) Controls on the formation of Wakkerstroom Vlei, Mpumalanga province, South Africa. Afr J Aquat Sci 38:135–151

Lidzhegu Z, Ellery WN, Mantel SK, Hughes DA (2019) Delineating wetland areas from the cut-and-fill method using a Digital Elevation Model (DEM). S Afr Geogr J. https://doi.org/10.1080/03736245.2019.1638825

McCarthy TS, Ellery WN (1994) The effect of vegetation on soil and ground water chemistry and hydrology of islands in the seasonal swamps of the Okavango Fan, Botswana. J Hydrol 154:169–193

McCarthy TS, Ellery WN (1998) The Okavango Delta. Trans R Soc S Afr 53:157–182

McCarthy TS, Metcalfe J (1990) Chemical sedimentation in the Okavango Delta, Botswana. Chem Geology 89:157–178

McCarthy TS, Ellery WN, Rogers K, Cairncross B, Ellery K (1986) The roles of sedimentation and plant growth in changing flow patterns in the Okavango Delta, Botswana. S Afr J Sci 82:579–584

McCarthy TS, McIver J, Cairncross B, Ellery WN, Ellery K (1989) The inorganic chemistry of peat from the Maunachira channel-swamp system, Okavango Delta, Botswana. Geochim Cosmochim Acta 53:1077–1089

McCarthy TS, Stanistreet IG, Cairncross B (1991) The sedimentary dynamics of active fluvial channels on the Okavango fan, Botswana. Sedimentology 38:471–487

McCarthy TS, Ellery WN, Stanistreet IG (1993) Lakes of the north eastern region of the Okavango swamps, Botswana. Z Für Geomorphologie 37:273–294

McCarthy TS, Barry M, Bloem A, Ellery WN, Heister H, Merry C, Ruther H, Sternberg H (1997) The gradient of the Okavango fan, Botswana, and its sedimentological and tectonic implications. Afr J Earth Sci 24:65–78

McCarthy TS, Ellery WN, Bloem A (1998) Some observations on the geomorphological impact of hippopotamus (*Hippopotamus amphibius* L.) in the Okavango Delta, Botswana. Afr J Ecol 36:44–56

McCarthy JM, Gumbricht T, McCarthy TS, Frost P, Wessels K (2003) Flooding Patterns of the Okavango Wetland in Botswana between 1972 and 2000. J Hum Environ 32:453–457

Pulley S, Ellery WN, Lagesse JV, Schlegel PK, McNamara SJ (2018) Gully erosion as a mechanism for wetland formation: An examination of two contrasting landscapes. Land Degrad Dev 29:1756–1767

Smith PA (1976) An outline of the vegetation of the Okavango drainage system. In: Proceedings of the symposium on the Okavango Delta and Its Future Utilisation. National Museum, Botswana Society, Gaborone, pp 93–112

Stigand AG (1923) Ngamiland. Geogr J 62:401–419

Thamm AG, Grundling P, Mazus H (1996) Holocene and recent peat growth rates on the Zululand coastal plain. J Afr Earth Sc 23:119–124

Wilson BH, Dincer T (1976) An introduction to the hydrology and hydrography of the Okavango Delta. In: Proceedings of the symposium on the Okavango Delta and Its Future Utilisation. National Museum, Botswana Society, Gaborone, pp 33–48

William N. Ellery is a wetland scientist who worked in the Okavango Delta for over 20 years as part of a team trying to understand the structure and dynamics of this remarkable wetland. His work focussed on interactions of hydrology, clastic and dissolved sediment deposition, plants and animals, as factors that contribute to the creation of a large freshwater peatland in the drylands of the Kalahari.

Karen Ellery is currently a science educator but worked for a number of years with a multidisciplinary team in the Okavango Delta. Her own work in the area focussed on plant successional processes in the permanent swamps, which included aspects of peat development.

4 Landscape Evolution of the Lake Ngami and Mababe Depressions Within the Okavango Rift Zone, North-Central Botswana

Susan Ringrose

Abstract

Lake Ngami and the Mababe Depression form elongated troughs peripheral to the Okavango Delta from which they currently receive inflow. The two basins originated as structural depressions resulting from East African Rift (EAR) propagation along the Kunyere and Thamalakane (Mababe) fault lines and are also embedded in older Okavango Delta fans. Both basins are partially ringed by palaeo-shorelines at heights that vary from 945 to 920 m with a dominant 936 m level. Dates of shoreline formation are currently under debate. (Moore et al. 2012) suggest that the Ngami and Mababe basins were mostly submerged as a result of major inflowing river captures during the Early-Mid Pleistocene. Major shorelines in the basins developed during the Palaeo-Lake Thamalakane (PLT-936 m) period of the Mid Pleistocene at 200–500 ka. This contrasts with interpretations in (Burrough and Thomas 2008), who suggest that palaeo-shorelines at ca. 936 m were formed on numerous occasions within the last 100 ka due to climatic conditions and feedback factors. Hence controversies revolve around possible dates and palaeo-climatic conditions for shoreline formation. Preliminary work on basin sediments suggests that palaeo-lakes in both basins operated as separate mostly closed system alkaline lakes for the last 65 ka, inferring an absence of numerous large high-level palaeo-lakes (at 936 m) during this interval. As further information is needed, a comparative deep drilling programme is recommended for both basins with full sedimentological analysis and sample dating to resolve issues regarding past climates and the possible extent of both early (Mid Pleistocene) and later (Late Glacial to Holocene) palaeo-lakes in the region.

Keywords

Palaeo-Lake Thamalakane • Reworked shorelines • Basin sediments

4.1 Introduction

Lake Ngami and the Mababe Depression are structural basins that are periodically filled by overspill from the Okavango Delta in northern Botswana. Drainage into these basins, both now and in the Pleistocene, has varied commensurately with flow variations into the Delta, which itself varies with rainfall over the central African interior (Fig. 4.1). The volume of drainage into the Okavango Rift Zone (ORZ) has also been subject to periodic and sequential reduction resulting mainly from tectonic factors related to the East African Rift (EAR) (Modisi et al. 2000). Climatic variability has doubtless also played a key role. Significantly, both the Ngami and Mababe basins occur at opposite ends of the north-west to south-east trending fault system, which controls the distal terminus of the Okavango Delta. The basins themselves form structurally controlled depressions, which are currently partially ringed by former shorelines as a result of occupancy or over-topping by numerous palaeo-lakes (Shaw 1985b). Palaeo-lake chronology is, therefore, key to understanding the geomorphological evolution of the Ngami and Mababe basins. This chapter looks into the geological and hydrological origins of the basins through time by way of explaining the characteristic features of the geomorphological landscape. The terminology of 'Lake Ngami' and the 'Mababe Depression' is entrenched in the literature and stems from the fact that the Ngami basin has been recognised as long containing lake-water while much of the Mababe basin has remained dry in historic and recent times.

S. Ringrose (✉)
School of Interdisciplinary Studies, University of Glasgow (Dumfries Campus), Dumfries, DG1 42L, Scotland, UK
e-mail: sringrose66@gmail.com

Okavango Research Institute, University of Botswana, Shorobe Road, Maun, Botswana

F. D. Eckardt (ed.), *Landscapes and Landforms of Botswana*, World Geomorphological Landscapes,
https://doi.org/10.1007/978-3-030-86102-5_4

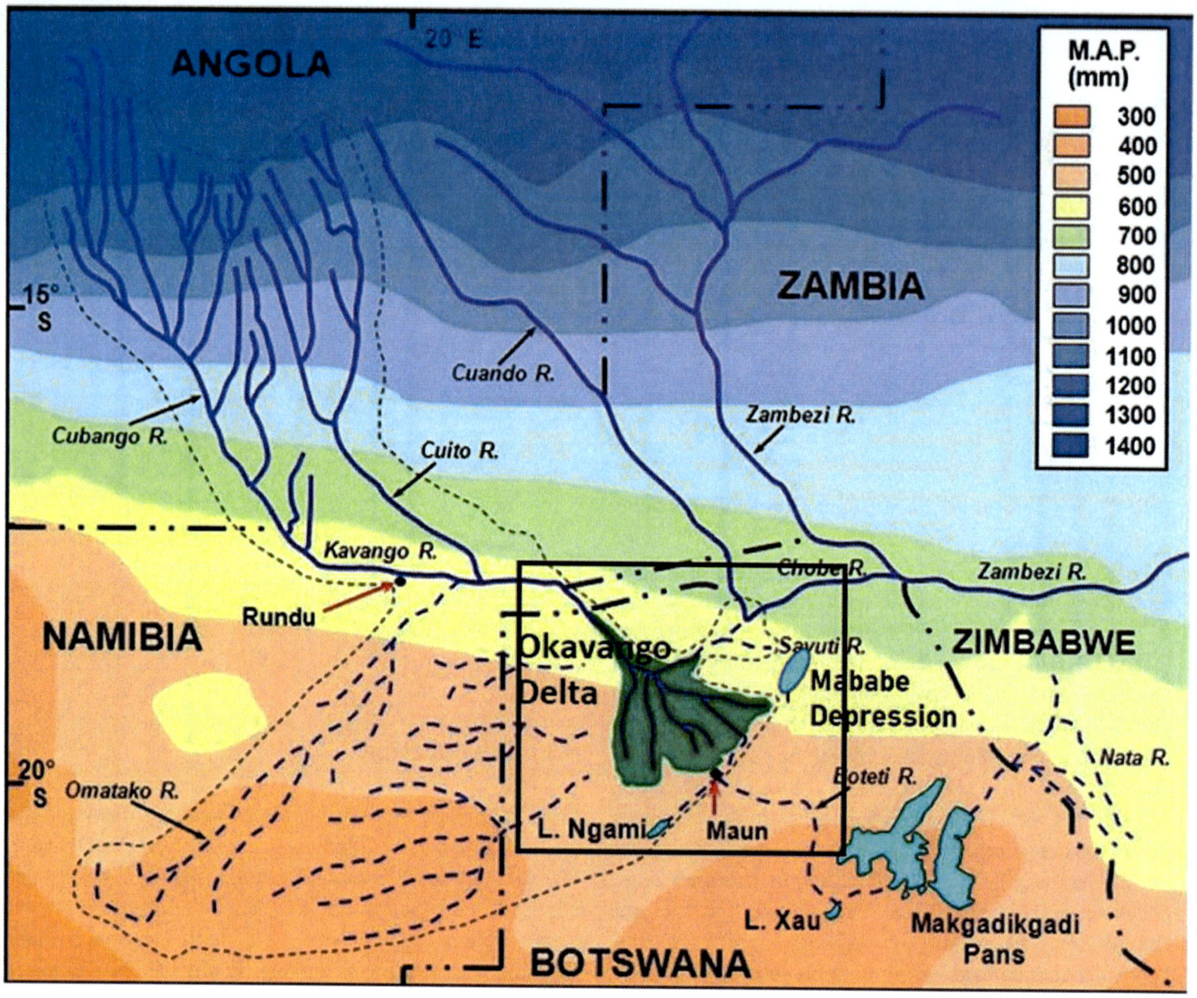

Fig. 4.1 Drainage map showing the extent of Zambezi, Cuando-Chobe and Okavango (Cuito, Cubango and Kavango) headstreams over the Angolan plateau. MAP = Mean Annual Precipitation. *Source* Okavango Research Institute

This geomorphological review of Lake Ngami and the Mababe Depression is tied up with the palaeo-lakes, which have developed in them despite the inherent semi-aridity and overall recurrence of arid events in the region (e.g. Stokes et al. 1997). Palaeo-lake incursions are controlled by the basins' locations whereby they receive inflow from the endoreic Okavango Delta distributary system and to a lesser extent from the Cuando (aka Kwando) River (Figs. 4.1 and 4.2). The extent and duration of flow through the Okavango system are controlled by climatic factors influencing rainfall over the Angolan and Zambezian Highlands. In terms of overall climate, the entire area lies within the southern African summer rainfall zone where under the influence of the Inter-Tropical Convergence Zone (ITCZ) migration, 80% of precipitation occurs between October and March (McCarthy et al. 2000) and is interspersed by a long dry season. The distance from the upper Cubango River to the terminus of the Okavango Delta extends over 1860 km with a drop in altitude from 1780 to 900 m, providing extremely low gradients (Mendelsohn and El Obeid 2004). These low gradients result in a delay of 5–6 months between rainfall events over the Angolan Plateau and flooding downstream such that the Okavango Delta and the hydrologically linked Lake Ngami and Mababe Depression receive irregular inflow during the dry winter months, notably in June and July. When operative, this causes a greening effect and water availability downstream for numerous cattle and abundant wildlife including fishing, cattle and tourism industries in Lake Ngami (Fig. 4.3a, b). Floodwater inflow is currently less in the Mababe Depression where infrequent partial filling has led to widespread re-vegetation. Nonetheless, the depression is also a magnet for herbivores (Fig. 4.3c) and predators. In terms of relative human impacts, the total population around Lake Ngami is 6589 (Kurugundla et al. 2018) whereas the total population around the Mababe

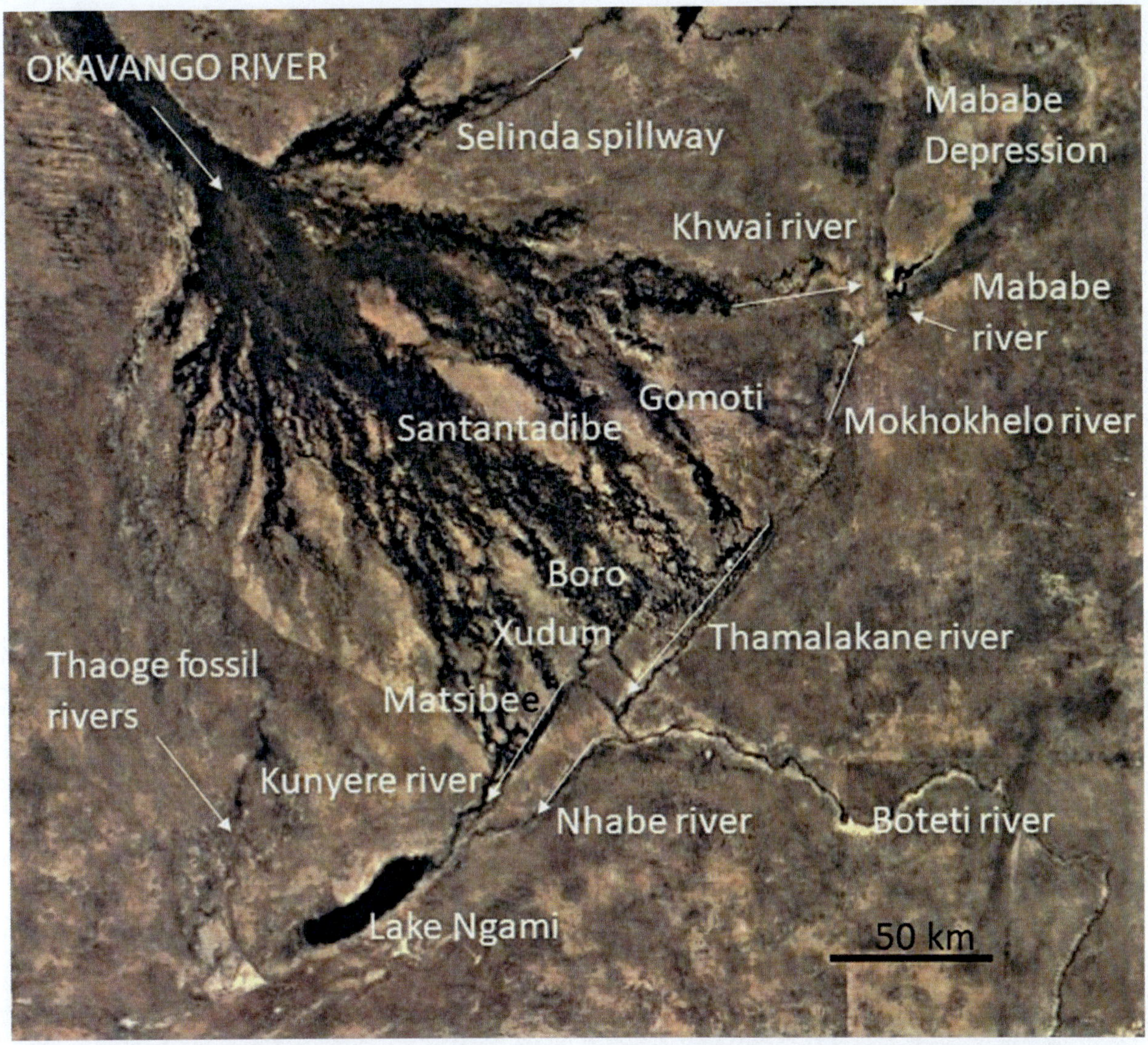

Fig. 4.2 Relationships between Lake Ngami, the Mababe Depression and inflow rivers derived mainly from the Okavango Delta. *Source* Google Earth, image date 31 December 2014

Depression is negligible and confined mainly to Mababe village in the south. Hence, human impacts and resource exploitation around the two basins are variable. More active farming and fishing activities occur around Lake Ngami while tourism and limited hunting are prevalent around the Mababe Depression, which lies mostly within Chobe National Park.

4.2 Geology, Tectonics and Hydrological Change

Much of the earlier geological history of the area continues to be significant as the present-day ORZ is still tectonically active with landscape features resulting from the re-activation of older bedrock trends (e.g. Kinabo et al. 2008). The ORZ overlies much of the Damaran mobile belt, which mainly comprises folded and faulted Neoproterozoic and Mesoproterozoic lithologies on north-east trends (Key and Ayres 2000). These and overlying Karoo sediments and volcanics underlie the elongate basins comprising Lake Ngami and the Mababe Depression. The normal faults defining the elongate basins form sub-graben, which are smaller segments of the larger ORZ structural network. Much of this network was reactivated through earth movements along with a south-west extension of the East African Rift (EAR) (Modisi et al. 2000). Specifically, Lake Ngami lies within a downfaulted sub-graben between a relatively steep horst block to the south (separated by the Kunyere

a
b
c
d
e
f
g
h
i

Fig. 4.3 Photos depicting aspects of Lake Ngami and the Mababe Depression. **a** Livelihoods around Lake Ngami include cattle raising—note Lake Ngami in background (Photo December 2018, W. Matheson) **b** Birdlife around Lake Ngami, which attracts tourists to the area (Photo December 2018, W. Matheson) **c** Savuti marsh in the Mababe Depression is frequently dry and attracts herbivores (Photo January 2017, W. Matheson) **d** The Magikwe Ridge south of Savuti camp showing exposed aeolian sand towards the ridge surface (Photo January 2017, W. Matheson) **e** Flooding along the Mababe River and recent road bridge re-construction (Photo December 2018, W. Matheson) **f** One of the Gubatsa hills that mark the periphery of the Mababe Depression in the vicinity of Sauvti **g**The Goha hills occur immediately north of the Mababe Depression **h** Part of lower Ngami trench showing sandy silt overlain by white, diatomaceous silt at the 350 cm hiatus. Upper bed comprises organic-rich silt, hammer 35 cm (Photo P. Huntsman, 2003) **i** Part of the upper Ngami trench showing burnt peat over diatomaceous silt. Lower bed comprises silty sand with calcareous nodules, hammer 35 cm (Photo P. Huntsman, 2003)

fault) and the less well-developed Lecha fault to the north. Structurally, the Kunyere fault continues along the southern Okavango axis to the north-east (Fig. 4.4). A north–north-east extension referred to as the Mababe fault defines the eastern margin of the Mababe Depression. This normal fault displaced the eastern margin downwards and partially parallels a second displaced fault scarp on the western margin forming the Mababe Depression (Kinabo et al. 2008). The Depression is also bounded to the north by the more frequently active Chobe fault and related horst (Kinabo et al. 2008). The Chobe fault also forms the southern escarpment of the elongate Linyanti swamp-trough, through which a link is provided via the Savuti River between the Mababe Depression to the south and Cuando and Zambezi

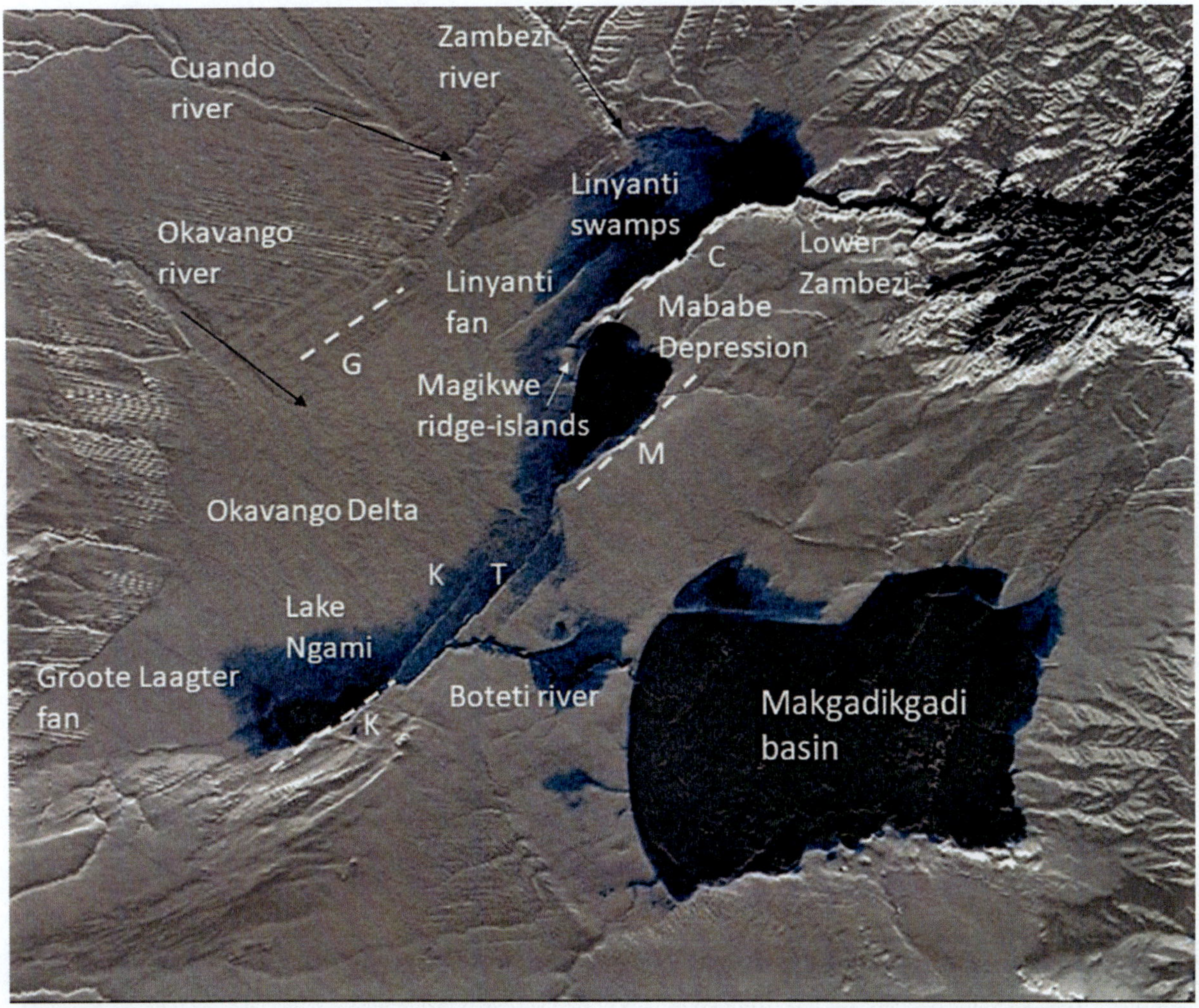

Fig. 4.4 SRTM imagery showing key structural features within the Okavango Rift Zone, fans of the early Okavango and inundation extent of palaeo-lake Makgadikgadi (PLM). PLM shown as blue gradations ca. 945–950 m. G = Gumare fault, K = Kunyere fault, T = Thamalakane fault, M = Mababe fault, C = Chobe fault (SRTM via C. Hartnady)

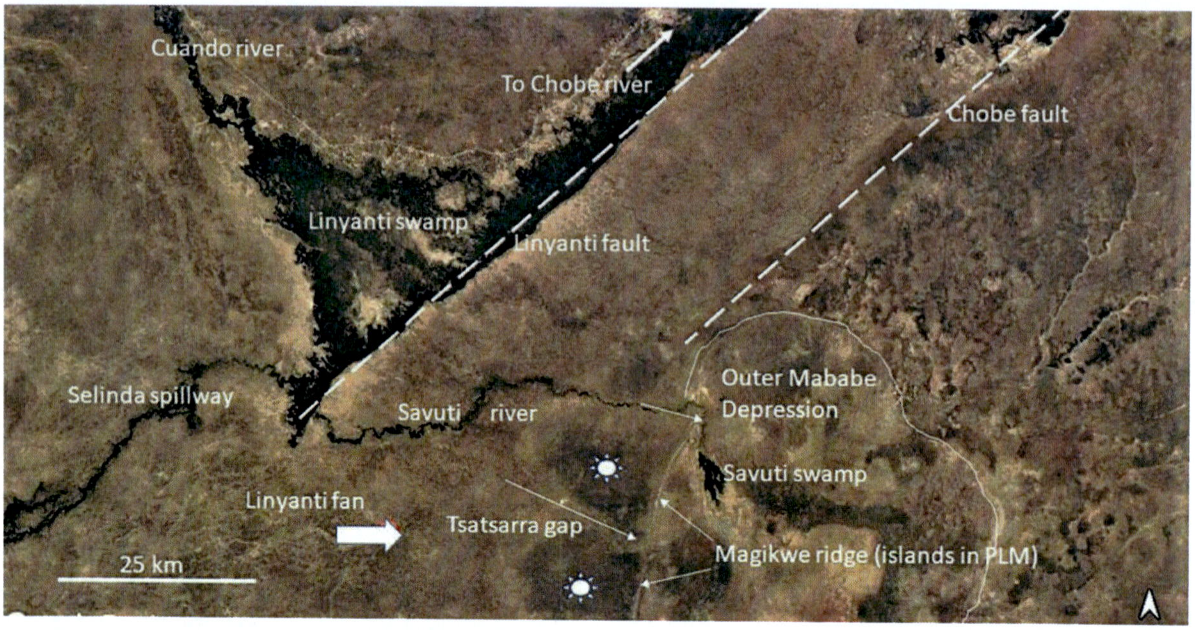

Fig. 4.5 The north-western extent of the Mababe Depression and the Savuti River inflow from the Linyanti swamps. The Linyanti swamps receive inflow from the Cuando river (see Fig. 4.1) but also from the Okavango River via the Selinda spillway. Bold arrow indicates the direction of flow from early Okavango Linyanti fan. Stars depict areas of reworked sediment to the west of the Magikwe ridge. *Source* Google Earth image date 31 December 2014

drainage further north (Fig. 4.4). The land surface of the Linyanti trough tilts towards the northeast, thereby facilitating mainly Cuando flow towards the Chobe River (Fig. 4.5). Major displacements have also taken place along the southern boundary of the Okavango graben, for instance along the Kunyere and Thamalakane/Mababe faults forming in places minor fault-scarps along the southern margin of the Okavango Delta. As indicated above, further reactivation on any of these faults could lead to radical landscape change, which is considered possible as the area is prone to medium–high earthquake activity (Scholz et al. 1976). Movement along the Kunyere, Mababe and Chobe faults (Fig. 4.4), which are associated with the Ngami and Mababe basins, has likely played an important role in influencing both the volume and direction of inflow over time.

The geomorphological development of the Okavango Delta and related basins, therefore, hinges on the volume and duration of drainage via rivers heading south-east off the Angolan-Zambian highlands into the ORZ (Fig. 4.1). Moore et al. (2012) and McCarthy et al. (2013) describe tectonic events leading to the long-term evolution of this drainage network across the Kalahari Plateau (Table 4.1). Drainage evolution occurred since the late Palaeogene/Miocene and has involved major rivers such as the Cuando and Zambezi and their tributaries in mid-stream lake formation and sequential river capture. Over time, these events have caused drainage diversions into and away from the ORZ, instigating and sustaining palaeo-lake development, augmented by climatic feedback mechanisms. The dating of these diversions has been achieved in terms of minimum dates derived from a combination of archaeological and geomorphological evidence and the phylogenetic records of aquatic vertebrates (Moore et al. 2012, their Fig. 7 ABCD). The authors have dated the occurrence of palaeo-lakes in the ORZ basin from Early Stone Age and early Middle Stone Age artefacts as indicated in, for instance, Robbins and Murphy (1998) and McFarlane and Segadika (2001). These results have led to a controversial re-evaluation of the antiquity of the major palaeo-lakes (see below). The re-evaluated early inundation events are directly linked to upstream drainage diversions involving upstream lake formation. The downstream consequences in Lake Ngami and the Mababe Depression are typified by sequential palaeo-lakes, which were initially extensive and then progressively smaller in size as discharge from the input rivers and lakes diminished (Moore et al. 2012).

As indicated, the development of the early ORZ palaeo-lakes coincides with drainage development in the Zambezi-Cuando-Okavango rivers as these became re-oriented as a result of tectonic events. Uplift and river capture including north-east to south-west faulting initiated basin development along with Lake Ngami and Mababe Depression trends during the Early Pleistocene, alongside the formation of the Okavango graben (Table 4.1). Early

Table 4.1 Drainage evolution and the development of ORZ palaeo-lakes (modified after Moore et al. 2012 and McCarthy 2013 and references therein)

Approximate age	Causal Event	Consequence	End result	*LN and MD
Miocene/Late Palaeogene	Uplift along the Ovambo-Kalahari-Zimbabwe Axis	Link severed between the Palaeo-Chambeshi/Upper Zambezi drainage system and the Limpopo River	Endoreic drainage and fluvio-lacustrine deposition within the resultant mid-Kalahari basin	
Late Pliocene	Mid-Zambezi captured the Palaeo-Chambeshi/Proto-Kafue/Upper Zambezi	Major source of water/ sediment to the mid-Kalahari basin cut-off,	Cuito-Cubango Rivers relics of the earlier endoreic drainage system	
Early Pleistocene	Uplift across the course of the Upper Zambezi along the NE-SW Chobe Fault associated with the SW propagation of the EARS	Diversion of considerable flow of the Palaeo-Chambeshi/Upper Zambezi drainage system into northern Botswana	Palaeo-Lake Deception initiated at 990–1000 m asl	Initiation of early NE-SW faulting Early Okavango and LN and MD depressions submerged
Early Pleistocene	Deepening of ORZ leads to inflow and sedimentation	Development of early Okavango	Cuito-Cubango Rivers form Okavango River which flows south over the Gumare fault	Early inflow-deposition into MD from Linyanti fan and some deposition into LN from Groote Laager fan
Early to Mid-Pleistocene	Link between the Upper Chambeshi and Kafue severed by uplift of Congo-Zambezi watershed	Inflow into Kalahari basin diminished	Later contraction to lower 945 m shoreline Palaeo-Lake Makgadikgadi (PLM)	Extensive PLM submerged LN. MD to Linyanti with Magikwe ridge forming islands
Later in Mid Pleistocene	Link between the Kafue and Upper Zambezi was severed	Palaeo-Lake Patrick initiated in the Kafue graben	Contraction of former PLM to the 936 m level or Palaeo-Lake Thamalakane (PLT)	PLT dated using ESA-early MSA tools to 200–500 ka. Formation of palaeo-shorelines s around LN and MD
Later in Mid-Pleistocene	Palaeo-lake Bulozi impounded in the vicinity of N'gonye Falls	Further reduction of inflow into the Kalahari basin	Contraction of the 936 m (PLT) level to the 920 m shoreline	Lowest recorded palaeo-lake levels on LN and MD at ca. 920 m
Later in Mid-Pleistocene	Link between the Upper Cuando and Zambezi severed and Upper and Mid-Zambezi reconnected	Further reduction of inflow into the Kalahari basin	Contraction of the 920 m level to the 912 m shoreline	

*LN = Lake Ngami MD = Mababe Depression

river capture diverted water into the southern basins, initially causing extensive flooding in the ORZ referred to as Palaeo-Lake Deception (PLD) (McFarlane and Eckardt 2008). Around this time, the early Okavango River disgorged sediment into the graben, which extended further north-east and south-west than the present delta. This early flow formed extensive fans referred to as the Groote Laagter fan and Linyanti fan (Fig. 4.4). These early fans provided a backdrop into which the faulted Ngami and Mababe basins became nested (McCarthy 2013).

Meanwhile, continuing Early to Mid-Pleistocene drainage diversions involving the Cuando and Zambezi rivers and their tributaries led to continuing but diminished inundation within the ORZ (Moore et al. 2012). This led to the contraction of the earlier PLD to a later, slightly smaller Palaeo-Lake Makgadikgadi (PLM) (Table 4.1 and Fig. 4.4). Traces of PLM shorelines at ca. 945 m are found intermittently along the margins of the Ngami and Mababe basins (cf. Moore et al. 2012, their Fig. 3B). SRTM data imply that at the maximum PLM extent, the Mababe Depression was almost completely submerged at the 945 m level leaving the central Magikwe ridge (ca. 945 m) exposed as a series of islands (Fig. 4.3d and 4). During PLM time, the Lake Ngami area was also practically submerged. The north-west margin of the PLM extended into the present-day Linyanti trough to the north of the Chobe fault (cf. Thomas and Shaw 1991).

Later in the Early-Mid Pleistocene, the link between the Kafue and Upper Zambezi was severed leading to further declining inflow to the ORZ and the contraction of the former PLM to the 936 m level (Table 4.1). The 936 m level is referred as Palaeo-Lake Thamalakane (PLT), which has now been dated using a range of evidence including ESA and early MSA tools to at least 200–500 ka (e.g. Moore et al.

2012; McFarlane and Segadika 2001). This ca. 936 m level is prevalent as a series of dominant palaeo-shorelines, which characterise both the Ngami and Mababe basins. PLT was an extensive, mid-level palaeo-lake that ran north-east to south-west along the low-lying line of the Thamalakane and Kunyere faults and extended southwards into the Makgadikgadi basin (Ringrose et al. 2005). The northern extent of the PLT is unclear but based on mapped contours (in McCarthy 2006), land heights to the west of the Mababe Depression appear too high (940 m) to facilitate a northern extension of the PLT into the Linyanti trough, but this requires further work.

Later in the Mid-Pleistocene, further diversions reduced inflow into the ORZ, which led to the contraction of the PLT to later 920 m and 912 m palaeo-shorelines. This is attributed to the link between the Upper Cuando and Zambezi being severed, leading to the re-connection of the Upper and Mid-Zambezi (Moore et al. 2012). The re-connection of the Upper and Mid-Zambezi had the effect of directing more flow away from the ORZ. Hence as the present-day drainage evolved (Fig. 4.1), a number of extensive palaeo-lakes are believed to have developed in the ORZ mostly during the Early-Mid Pleistocene (Table 4.1). However, other younger dates for ostensibly the same palaeo-lakes (PLM, PLT, etc.), but resulting from more direct climatic and feedback factors, have been indicated in Thomas and Shaw (1991) and later by authors such as Burrough and Thomas (2008). Hence, both perspectives are considered later in this chapter.

4.3 Landforms and Geomorphological Evolution

4.3.1 Lake Ngami

Lake Ngami in 2018–2019 formed a picturesque water-body lying at the heart of fishing and farming communities (Figs. 4.3a, b and 4.6) but was desiccated in 2020. A useable lake was probably intermittently prevalent on the landscape at least since the Lower Stone Age (Robbins et al. 2009) and was first noted and mapped by early European explorers such as Passage and Livingstone in the 1850s (Shaw 1985a). From the various descriptions, the lake level has been calculated at this time as ca. 930 m (Shaw et al. 2003). The nineteenth century lake was characterised by a series of inflow channels including the anastomosing branches of a former Okavango distributary called the Thaoge River (Fig. 4.6), which is thought to have been blocked ca. 1880 as a result of excessive *Papyrus* growth (Shaw 1985a). Currently, much of the Lake Ngami inflow comes from west and central Okavango distributaries (Fig. 4.2), which are highly variable in terms of their proportional flow volume (Wolski

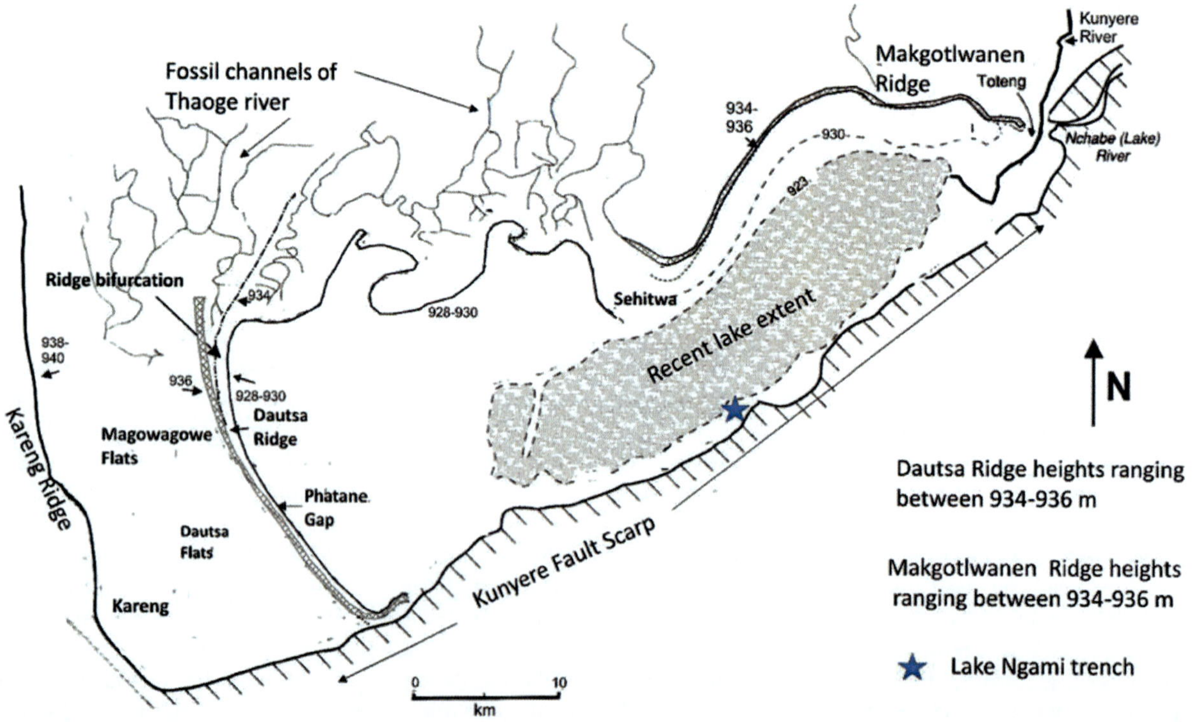

Fig. 4.6 Main geomorphological features of the Ngami basin, described in text. (Map modified after Thomas and Shaw 1991)

and Murray-Hudson 2008). The western distributaries include the Xudum and Matsibe rivers, which coalesce north of the Kunyere River forming a channel, which flows south-westwards along the fault line towards Lake Ngami. Inflow also originates from the more central Okavango distributaries including the Gomoti, Santantadibe and Boro rivers, which flow south-eastwards until reaching the Thamalakane fault, where they converge into the Thamalakane River (Fig. 4.2). Much of the Thamalakane flow is diverted south-eastwards down through the Boteti River towards the Makgadikgadi Pans (Shaw 1984). The remaining flow continues south-west as the Nchabe (or Lake) River, which continues along with the trend of the Thamalakane fault to Lake Ngami. The Kunyere and Nchabe rivers converge near Toteng and flow into the lake.

A recent history of lake levels in Lake Ngami and their variability, using a combination of historic records and inflow data from the upper Okavango, has been compiled by Kurugundla et al. (2018). Their work indicates that the lake derives more than 90% of its water from Okavango drainage and the remaining ca. 10% from local precipitation. The floodwaters typically arrive between June and July, peaking in late August. The water then recedes steadily between October and May of the following year. Multiple observations indicate that the behaviour of the lake does not directly reflect the impact of historic climate changes in the local area, but variation in the Okavango inflow, which is dependent on periodic rainfall fluctuations in the headwater catchments in Angola (Kurugundla et al. 2018). After documenting over a century of observations and measurements, the authors concluded that Lake Ngami has mainly functioned since historical times (1850s) as a shallow wet basin with intermediate (and sometimes extensive) dry periods (Table 4.2). Water depths measured in September 2012 showed that the average lake depth was ca. 3.5 m. In September 2012, the volume of water determined for the lake was 974 Mm^3 occupying 277.7 km^2, with an average water level contour of 923.3 m (cf. 923.5 m, Shaw et al. 2003). At this level, the lake functions as a closed basin, whereas lake levels > 930 m infer a through-flow system to the Boteti River (Shaw 1985b).

The physical landscape of much of the present-day Ngami basin is determined by the Kunyere fault scarp with its north-west facing down-slip and the termination of former inflow channels of the Thaoge River, giving the basin its west-southwest by east-south east alignment (Fig. 4.6). The outer basin is approximately 70 km long and 38 km wide, representing a surface area of about 2 660 km^2, while the flat inner lake bed is 45 km long and 8 km wide (or about 360 km^2). Hence ca. 88% of the entire Ngami basin comprises outer stepped palaeo-shorelines and offshore bars, in addition to pans, back swamps and lacustrine foreshore areas. Most geomorphological descriptions emphasise the multiple palaeo-shorelines, first described by Grove (1969) and later by Cooke (1980), Mallick et al. (1981) and Shaw (1985b). Palaeo-shorelines from ca. 928 m to 940 m are also discussed in Thomas and Shaw (1991) and Burrough et al. (2007, 2009b). The predominant palaeo-shorelines comprise the western Magotlawanen ridge (934–936 m) and the Dautsa ridge (928–936 m) complexes, which fringe the inner basin. The Dautsa ridge is incised at the Phatane Gap (Fig. 4.6) and is regarded as an offshore barrier beach, capped by summit dunes formed of reworked aeolian sand (Shaw 1985b). The offshore bars were formed by wind-generated wave action on the leeward side of the palaeo-lakes that once occupied the basin (Grove 1969; Shaw 1988). In addition, towards the western periphery of the basin, the possibly older Kareng ridge stands between 938 and 940 m. A series of former back-swamps such as the Dautsa flats occur between the Kareng and Dautsa ridges (Fig. 4.6). Measured average heights from prominent ridges indicate that the Lake Ngami basin has experienced minor downward tectonic displacement (<5 m) with no tilting since the Mid-Pleistocene (Gumbricht et al. 2001; McCarthy 2013), hence bifurcation on the northern Dautsa ridge may be related to tectonic controls (Huntsman-Mapila et al. 2006).

Despite the ESA-early MSA, Mid-Pleistocene ages (500–200 ka) discussed earlier for the PLT ridges (e.g. Moore et al. 2012) OSL dates from the Kareng, Dautsa and Magotlwanen palaeo-shorelines (936–934 m) suggest that these ridges only date back to ca. 90–100 ka with younger

Table 4.2 Conditions reflecting lake levels in Lake Ngami between 1850 and 2016 (after Kurugundla et al. 2018)

Date	Wet/Dry	Date	Wet/Dry
1850s	Wet	1959–1972	Longest wet
1881–1883	Probable dry	1972–1973	Seasonal wet/dry
1898–1899	Wet	1974–1979	Wet
1924–1926	Very wet	1980–2003	Longest dry
1934–1941	Dry	2004	Seasonal wet/dry
1948–1949	Seasonal wet/dry	2008	Seasonal wet/dry
1956–1958	Seasonal wet/dry	2009–2015	Wet*

*2012 = water in lake to 923.3 m level

ridges dated at 3–4 ka (Shaw et al. 2003). A perspective that has been endorsed by later MSA dates on lithic tool sites (Nash et al. 2013). The PLT palaeo-shorelines are regarded in Shaw et al. (2003) as being older landforms subsequently modified by Holocene fluctuations in lake level and subsequent aeolian reworking. Holocene reworking included water levels peaking at 936 m between 4 and 3 ka and returning to 934 m at 2.8–2 ka, before falling rapidly to 932 m and later historical levels. According to Shaw et al. (2003), the 936 m level, in particular, is correlated with increased through flow in the Okavango Delta, hence increased rainfall in the catchment. The subsequent decline is attributed to episodic closure and rerouting of Okavango distributaries, in particular the blocking of the Thaoge River.

Later Burrough et al. (2007) obtained 74 OSL dates from Lake Ngami palaeo-shoreline samples from depths of 3 to 6 m. The OSL results are interpreted as being derived from sediments deposited as sequential stacked palaeo-shorelines resulting from 'multiple lake highstands' at ca. 935 m over a period from ca. 140 ka to 0.3 ka. Burrough et al. (2007) indicate that the highest lake levels were recorded from the 938 m Kareng ridge (Fig. 4.6), which resulted in palaeo-lake events recurring at 140 ± 11 ka, 124 ± 8 ka, 115 ± 8 ka, 87 ± 7 ka, 59 ± 4 ka, 32 ± 2 ka and 12 ± 1 ka. Other dates were obtained from the lakeward flanks of the Dautsa ridge (936–934 m), which record palaeo-lake deposition at 15.2 ± 1.8 ka, 30 ± 3 ka, 32 ± 3 ka, 33 ± 3 ka, 38 ± 3 ka, 47 ± 4 ka and 56 ± 5 ka. Burrough et al. (2007 and references therein) discuss these highstand features in terms of the regional climatic controls, increased local water availability, the feedback effects of a large body of water on local rainfall and the relationships between dated Holocene wet periods, which show local precipitation increases and flow through the upper Okavango. Burrough et al. (2007) conclude that the Dautsa 934–936 m ridge complex began to be constructed during the Late Pleistocene (ca.18 ka), with further palaeo-shoreline activity until ca. 13.5 ka. This work emphasises the frequency of extensive later PLT palaeo-lakes at the 934–936 m level in the Ngami basin and beyond during the Late Pleistocene (Burrough et al. 2007). Furthermore, Shaw et al. (2003) suggest that equally extensive later PLT palaeo-lakes continued in the basin up to 3–4 ka during Holocene times.

4.3.2 The Mababe Depression

There is a relative absence of historical accounts pertaining to water bodies in the relatively featureless Mababe Depression. Travellers like Passarge (1896–1898) who encountered Lake Ngami mapped it as the 'Ngami See' as it was full of water at the time (Table 4.2), whereas a general area to the east of the Okavango is entitled 'Mababe Sumpt' on 1896–98 maps due to marshes in the Depression (Vanderpost and Hancock 2018). The present-day lake bed is dry with minor ephemeral wetlands, specifically the Savuti marsh to the north-west (Fig. 4.3c) and the Mababe marsh to the south of the Depression (Fig. 4.7). Currently, drainage into the Mababe basin mainly comprises overflow from the Okavango Delta in addition to some flow from the Cuando River (Fig. 4.1). South-eastern Okavango drainage enters the Mababe Depression at its southern tip and is derived mostly from the Khwai distributary through the Machaba and Kudumane rivers (Shaw 1984). Historical evidence suggests that these rivers (Fig. 4.7) fed a substantial swamp around 1910. Since then the southern basin has contained water on seven occasions derived from Khwai inflow (Shaw 1985b). The Khwai River also flowed variably between 2009 and 2015, causing flooding in and around Mababe village. The floods damaged the Mababe road bridge (Fig. 4.3e) and filled the southernmost part of the Depression. The second southern inflow is currently a fossil backswamp, which runs north-east along the Thamalakane fault line called the Mokhokhelo River (Fig. 4.7). Furthermore, north inflow to the Mababe Depression takes place through the misfit Savuti River, which crosses over the Linyanti fault before flowing eastwards along with the Linyanti fan. The river cuts through the Magikwe ridge into the northern section of the Mababe depression, forming the Savuti swamp (Fig. 4.5). Historically, the Savuti River fed the Savuti marsh regularly until the late 1880s, after which it dried up completely until flow resumed in 1958 (Shaw 1985b). Following a long dry spell, flow again resumed in 2009–2015.

Interestingly, the Savuti River inflow arises from both the Okavango Delta and the Linyanti swamps, which are fed by the Cuando River. Some Cuando-Linyanti outflow in turn drains circuitously towards the Zambezi River via the Chobe River backflow. The upper Savuti River drains from the Linyanti swamp across from where the Selinda spillway (an Okavango distributary) accesses the Linyanti swamp. This co-incidence of inflow and outflow drainage suggests that when Okavango flooding takes place through the Selinda spillway, this initially fills the southern Linyanti swamp, which increases flow down through the Savuti River (Fig. 4.5). Hence, it is currently difficult to trace the origins of northern Mababe drainage as this may originate as Cuando inflow through the Linyanti swamp (which is mainly tilted NE) or as Okavango inflow diverted through the Selinda spillway (Fig. 4.2). Furthermore, movement on the relatively active Linyanti or Chobe faults (Kinabo et al. 2008) could readily divert inflow to the Mababe Depression in favour of either the Cuando River or Selinda spillway (Fig. 4.5).

In terms of dimensions, using the outer circumscribing boundaries, the Mababe Depression is approximately 92 km long and 64 km wide (or 5 888 km^2), while the inner featureless lake bed is 51 km long and 20 km wide (or 1 020 km^2). Hence, the depression is significantly larger than in the Ngami basin. About 85% of the Mababe Depression is made up of palaeo-shorelines, foreshore plains, estuaries and deltas (Fig. 4.7). The basin is subject to relatively high levels of seismic activity, with a higher rate of earthquake incidence than Lake Ngami. This occurs mainly along the southern margin of the Cuando-Linyanti trough and the west-central part of the Mababe Depression (Scholz et al. 1976). Gumbricht et al. (2001) indicated that the main Mababe palaeo-shoreline, the Magikwe ridge (Figs. 4.3d and 4.7), varies in elevation from 930 m in the south to 945 m in the north, suggesting that the entire basin is tilting southwards. Given the tilting, the minimal elevation (sump) level occurs towards the southern basin at ca. 920 m.

The geomorphology of the Mababe basin is determined mainly by the uplifted Mababe fault scarp to the south-east, with its north-west facing down-slip and the Magikwe ridge which stands 15–20 m above the lake bed along the western margin (Figs. 4.3f and 4.7). Due to structural controls, the Mababe basin is aligned north–north-east to south–south-west. The Mababe Depression also comprises a number of unusual rocky outcrops towards the north of the main Magikwe ridge. These mainly comprise six small hills, which occur around the Savuti River inflow to the Savuti marsh including the Gubatsa Hills (Fig. 4.3f). The larger Goha Hill (Fig. 4.3g) arises beyond the northern edge of the basin. Extensive palaeo-shorelines, estuarine deposits and deltas in the Depression have been described by Cooke (1980), Mallick et al. (1981), Shaw (1985b), Thomas and Shaw (1991) and Burrough and Thomas (2008). The main Magikwe ridge runs along the entire western side of the Depression as an extensive palaeo-shoreline rising between 936 and 950 m (Fig. 4.3d). The ridge height has led to considerations of its formation by wind-generated wave action since it occurs on the leeward side of the Mababe palaeo-lakes (Grove 1969; Shaw 1988). The Magikwe ridge is capped by aeolian deposits (to 950 m), with sand blow-outs interspersed along the ridge crest (Fig. 4.3d) (e.g. Thomas and Shaw 1991). The ridge bifurcates northwards near the Savuti channel where 'tombolo like features' (Shaw 1985) anchor segments of the ridge to the Gubatsa Hills. The gentle west-facing back-slope of the ridge consists of 'reworked Kalahari sand' (Thomas and Shaw 1991), which may reflect foreshore reworking during the waning phase of PLM during the Early to Mid-Pleistocene (Figs. 4.5 and 4.7). The reworked sand area comprises linear 'lagoonal' ridges not unlike those identified on the backslope of the Gidikwe ridge of the Makgadikgadi system (e.g. Cooke and Verstappen 1984; McFarlane and Eckhardt 2008). Further fossil

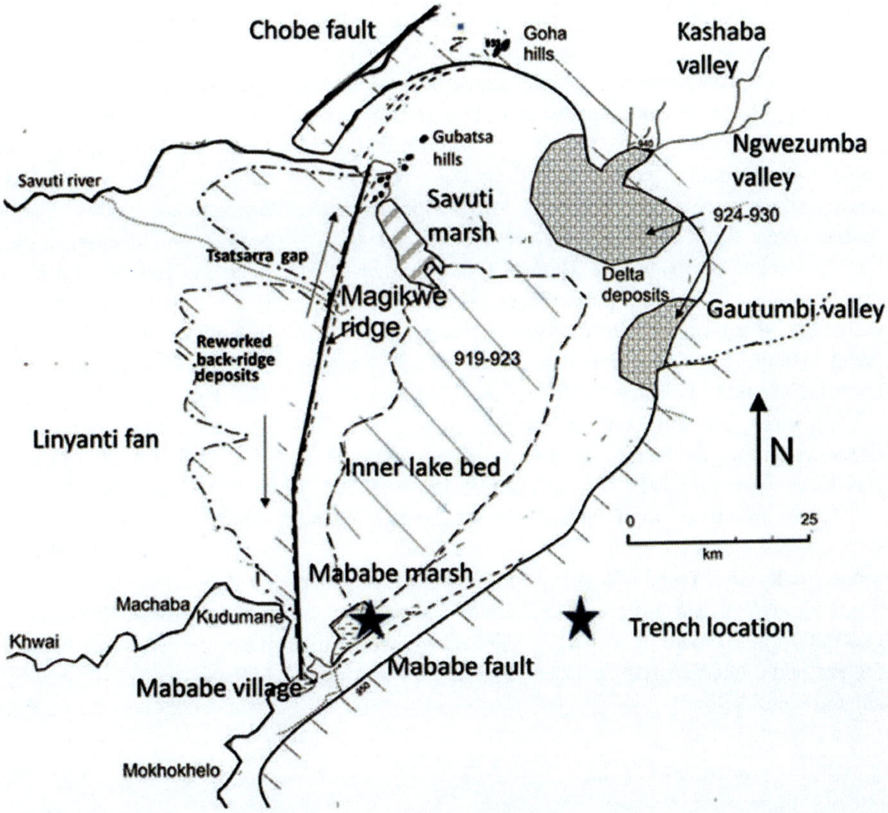

Fig. 4.7 Main geomorphological features of the Mababe Depression, described in text. (Map modified after Thomas and Shaw 1991)

drainage on the west side of the Mababe Depression includes a south-east trending branch of the Savuti River, which cuts through the Magikwe ridge at the Tsatsarra Gap (Fig. 4.7). The area north of the eastern fault scarp is breached at two locations by the misfit Ngwezumba River and its tributary the Kashaba River, which drain intermittently off the eastern plateau. The Ghautumbi River to the south is less distinctive with no known flow (Shaw 1985b). The entry of these rivers into the Mababe basin has led to the formation of deltas and offshore bars attributed to palaeo-lakes standing between 924 and 936 m (Fig. 4.7) (Thomas and Shaw 1991).

Comparing the maps of Lake Ngami (Fig. 4.6) and the Mababe Depression (Fig. 4.7), it becomes apparent that palaeo-lake levels in the two basins show a measure of consistency, although the altitudinal rise on the Magikwe ridge (from ca. 936 to 950 m) detracts from this. While Moore et al. (2012) describe a single-event 936 m palaeo-shoreline (PLT) in both basins dating back to ca. 200–500 ka, Burrough and Thomas (2008) had previously identified and OSL dated apparently much younger palaeo-lake levels along the same Magikwe ridge at the PLT 936 m level. Their first palaeo-lake event is regarded as having taken place during Late Pleistocene time between 38 ± 6 ka and 35–36 ± 5 ka. This event is regarded as being consistent with the formation of the Dautsa Ridge in Lake Ngami between 38 ± 3 ka and 30 ± 3 ka and the Kareng ridge at ca. 32 ± 2 ka (Burrough and Thomas 2008). The second palaeo-lake event to attain the 936 m height is regarded as having occurred in both basins between 18 and 13 ka. This appears to be augmented by evidence of delta and terrace formation associated within the Ngwezumba and Savuti rivers (Fig. 4.7). The third 936 m palaeo-lake event that occurred in both basins is regarded as having taken place during the Early-Mid Holocene in the Mababe Depression at 8.4 ± 1.1 ka, 6.6–6.3 ± 1.1 ka, 5.5 –5.2 ± 0.6 ka. Burrough and Thomas (2008) suggest that these extensive inter-basin palaeo-lakes (as the PLT extended from Ngami to the Mababe basins, along the Thamalakane fault and southwards) resulted from positive hydrological budgets in the catchment area and a considerable variability in inflow through the Okavango Delta. The debate regarding the relative feasibility of one single Mid-Pleistocene palaeo-lake PLT at 936 m (Moore et al. 2012) vis a vis the three or four Late Pleistocene-Holocene palaeo-lakes at ca. PLT levels (Burrough and Thomas, 2008) hinges partly on a reasonable source for the volume of water required to feed and sustain such extensive water bodies (>40 000 km^2, Moore et al. 2012). The occurrence in the Ngami and Mababe basins of such diverse events is considered again below.

4.3.3 Basin Lake Floor Sediments

Work on lake floor sediments in both the Ngami and Mababe basins began in 2002 and has resulted in trench data being independently reported by Huntsman-Mapila et al. (2006) and Gamrod (2009). Here a comparison of the two trench data sets is presented for the first time to invoke further discussion on Late Pleistocene-Holocene events in the basins. The data sets were derived from open trenches dug by mechanical excavators such that 3D bedding relationships and hiatuses in the sediment pile were readily identifiable. Trench-derived stratigraphic records from Lake Ngami were excavated at 922 m asl, close to the local palaeo-lake sump at 921 m, while the Mababe Depression excavation was taken from 920 m, close to the 919 m sump (Figs. 4.6 and 4.7). As basin depths may be as great as 800 m in Mababe and 400 m in Ngami (Kinabo et al. 2008), the trench data are evidently restricted to a fraction of the entire sequence. Sediment descriptions from the 460 cm deep Ngami trench and the 560 cm deep Mababe trench are shown in Fig. 4.8. These are augmented by comparative dates from both profiles (Table 4.3), diatom habitat interpretations from the Lake Ngami profile (Table 4.4) and particle size data that are used to help characterise the Mababe Depression profile (Fig. 4.9).

Dates for the two profiles were obtained using OSL (Mababe Depression) and TL with ^{14}C (Lake Ngami) methods (Table 4.3). These are plotted on the profiles (Fig. 4.8) and indicate how the Mababe deposits appear generally older than those in the Ngami trench. To illustrate the relative antiquity of the two profiles, lines were drawn to identify similar dates from each. The area between the semi-isochronous lines has led to the identification of four depositional phases. For instance, the first line is drawn between the location of the sample used for the 40.5 ka date in Ngami and the 41.3 ka date in the Mababe profile. Hence, Phase 1 extends from an unknown time at depth to ca. 40 ka in both profiles. A second line is drawn to represent ca. 27 ka in both profiles, whereas the third line represents ca. 11.5 ka. Interestingly, depositional hiatuses representing non-deposition and/or drying events were also identified (Figs. 4.8 and 4.9). In most cases, the hiatuses occur at or near the dates described above. Hence, the profile data are subdivided into what becomes four depositional phases mostly separated by near-contemporary hiatuses and based on dates.

The earliest Phase 1 depositional period comprises initially deeper water lacustrine sediment at the base of the Mababe trench followed by organic clayey silt, indicative of swamp conditions at ca. 65.8 ka. This changes vertically into

Table 4.3 Comparison of age dates from different dating techniques in the Ngami and Mababe sub-basins (after Huntsman-Mapila et al. 2006 and Gamrod 2008)

Site	Depth cm	Method	Cosmic Dose*	Dose Rate**	Age Estimate
Ngami-PHM	90	^{14}C			3.675 ± 0.030 ka BP
Ngami-PHM	220	TL	170.00 ± 17.00	2.672 ± 277	12.197 ± 1.277 ka
Ngami-PHM	220	^{14}C			14.740 ± 0.080 ka BP
Ngami-PHM	320	^{14}C			17.890 ± 0.080 ka BP
Ngami-PHM	380	TL	150.00 ± 15.00	943 ± 91	35.602 ± 3.851 ka
Ngami-PHM	440	TL	140 ± 14	1824 ± 132	40.499 ± 3.187 ka
Mababe-JG	45	OSL	21.6 ± 1.3	1.847 ± 0.087	11.70 ± 0.90
Mababe-JG	135	OSL	35.9 ± 2.0	1.332 ± 0.061	27.0 ± 2.0
Mababe-JG	380	OSL	60.7 ± 2.6	1.469 ± 0.070	41.3 ± 2.7
Mababe-JG	500	OSL	147.3 ± 7.7 lePara>	2.27 ±	64.8 ± 4.7

*In the case of the TL measurements, this is the cosmic dose rate in uGya-1 whereas in the case of the OSL measurements, this is regarded as the equivalent dose in Gy.
**In the case of the TL measurements, this is the dose rate in uGya-1, whereas in the case of the OSL measurements, this is regarded as the dose rate in Gy/ka.

Table 4.4 Interpretation of diatom genera, Ngami profile with examples of dominant taxa (original data in Huntsman-Mapila et al. 2006)

Depth cm	Dominant genera/taxa	Sediment	Interpretation of environment (inferred)
120–100	*Achnanthes minutissima* *Cymbella silesiaca* *Denticular kuetzingii* *Nitzschia amphibia* *Rhopolodia gibba* *Surirella engleri*	Laminar diatomaceous silt–dark grey -abundant diatoms	Shallow lake waters—mainly epiphytic diatoms living in freshwater may be slightly alkaline ^{14}C dated 3.7 ka BP
160–240	*Denticular kuetzingii* *Nitzschia amphibia* *Fragilaria construens* *Fragilaria pinnata* *Achnanthes minutissima* *Cymbella gracilis* *Cyclotella meneghiniana*	Massive diatomaceous silt—abundant diatoms	Shallow lake waters—fairly alkaline basic pH. Associated with significant macrophyte growth in the extended littoral zone. Fluctuating lake levels TL and ^{14}C dated 12–15 ka
330–300	*Fragilaria construens* *Denticular kuetzingii* *Fragilaria brevistriata*	Organic rich clayey-silty—few broken diatoms*	Includes planktonic fragments indicative of relatively deeper waters TL dated at ca.18 ka
440–350	*Denticular kuetzingii*	Pale yellow sandy-silt—very few diatoms	Shallow lake waters—associated with macrophyte growth in littoral zone TL dated at ca.36 ka
460–440	*Aulacoseira ambigua,* *Denticular kuetzingii,* *Fragilaria construens* *Rhopolodia gibba*	Silty-sand with minor calcrete—abundant diatoms	Lake relatively deep but brackish** as high electrolytic composition indicated TL dated at ca.40.5 ka

*330 fragments—disturbance could be tectonic or storms
**Brackish—could be waters originally saline but mixed with rainwater
Taxa were identified using a range of diatom flora including those from African lakes (Gasse 1986) and oligotrophic temperate lakes (e.g. Krammer and Lange-Bertalot, 1968–1991)

more shallow water sediments, which terminate at the second hiatus dated at ca. 41.3 ka. This compares with sediment at the base of the Lake Ngami profile, which comprises silty sand with calcrete nodules, which may also be indicative of an increasingly shallow lacustrine environment terminating with possible sub-aerial exposure around 40 ka. This non-depositional interval appears to be in some agreement with a drying event (41–46 ka) indicated in Stokes et al. (1997), but is at variance with the high palaeo-lake levels at 38 ± 6 ka from both basins in Burrough and Thomas (2008). A ca. 40 ka date is also at variance with data from Ringrose et al. (2008) in terms of floodplain formation in the

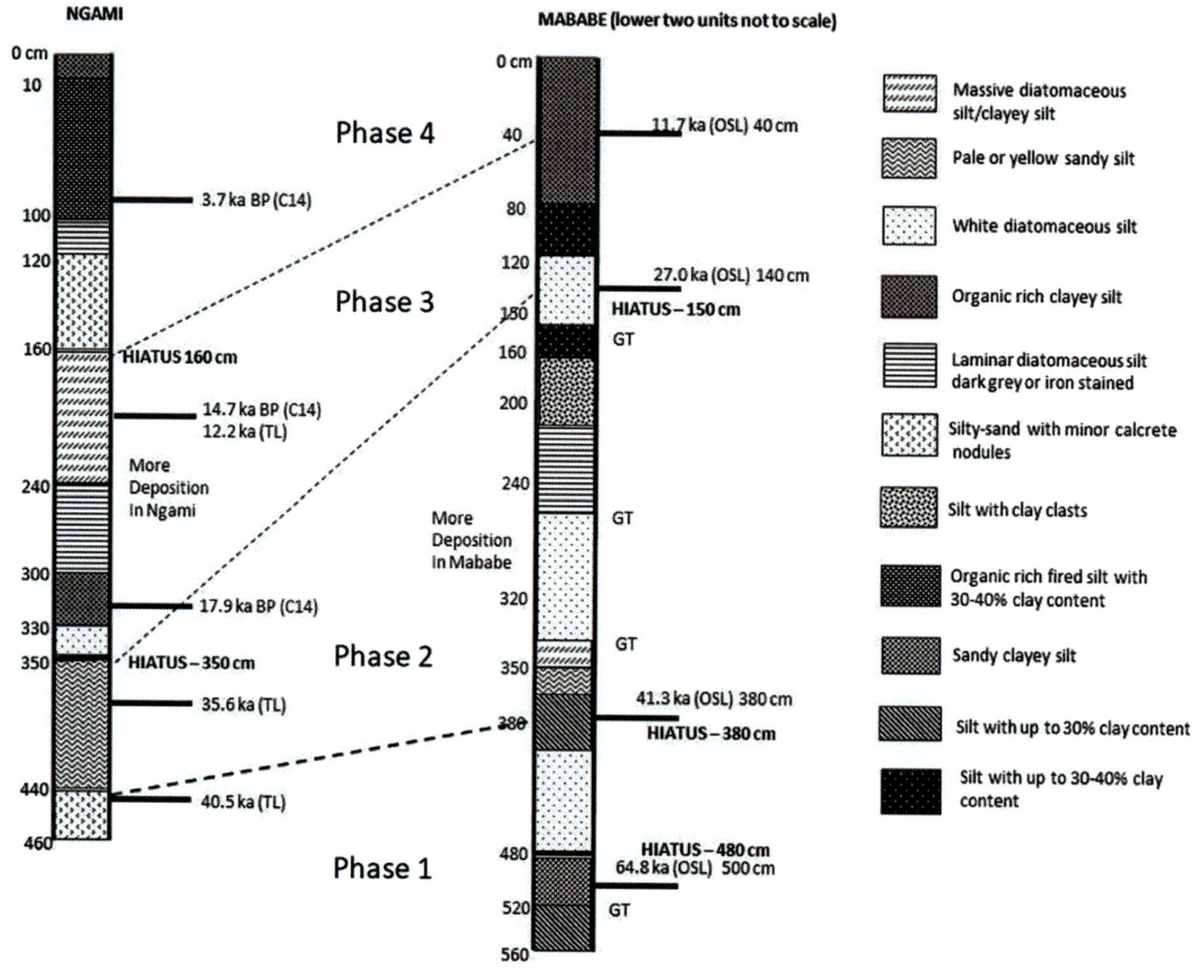

Fig. 4.8 Stratigraphic profiles from the Ngami and Mababe trenches indicating four depositional phases described in text. Late Pleistocene-Holocene dates and depositional hiatuses are shown for each profile. (Individual profiles modified after Huntsman-Mapila et al. 2006 and Gamrod 2009)

Okavango Delta at this time. Phase 2 extends over ca. 13 000 years, (Fig. 4.8) when 230 cm of sediment was deposited in the Mababe Depression contrasting with only 90 cm in Lake Ngami. Phase 2 deposits in the Ngami profile comprise a relatively thin pale yellow sandy-silt possibly attributable to fluvial or through-flow deposition (Fig. 4.3h). The thicker sediment in the Mababe trench commences with deeper water silts and a sandy-silt, similar to that found in the Ngami trench (Fig. 4.8). However, most of the profile in Phase 2 comprises thick white diatomaceous silts indicative of shallow lacustrine conditions, capped by deeper water silt and clay. Hence, a contrast exists between thin, fluvial sediments in Lake Ngami and the thicker lacustrine sediments of the Mababe Depression between ca. 40 ka and 27 ka. The 27 ka marker is again represented by a hiatus, which is coincident with a drying interval between 20 and 26 ka (Stokes et al. 1997) and a non-palaeo-lake period in Burrough and Thomas (2008).

Following the dry period, Phase 3 starts after 27 ka and continues to ca. 11.7 ka (Fig. 4.8). During this time, the Mababe sequence initially depicts shallow water deposition followed by a deeper water interval, which later changed to an intermittently wet, swampy environment. Over the same time period, Lake Ngami deposits show an initial period of shallow, followed by deeper water sediments. This is followed by two totally different laminar and massive diatomaceous beds up to 140 cm in thickness, which are indicative of shallow, alkaline waters suggesting a closed basin environment (Table 4.4). The diatomaceous beds terminate in a further hiatus suggesting drying at ca. 11.7 towards the beginning of the Holocene. Hence, a contrast exists between the thicker lacustrine diatomaceous sediments

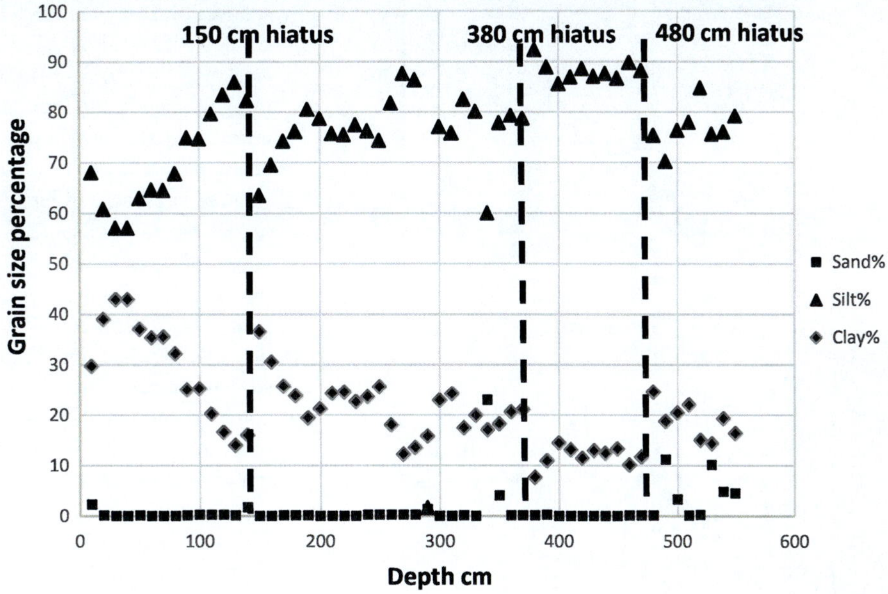

Fig. 4.9 Particle size data from Mababe trench showing dated depositional discontinuities at three hiatus level. 480 cm hiatus dated ca. 64.8 ka (OSL), 380 cm hiatus dated ca.41.3 ka and 150 cm hiatus ca. 27.0 ka. *Source* Gamrod (2009)

of the closed Ngami basin and the thinner lacustrine-swamp sediments of the also closed Mababe basin between ca. 27 ka and 11 ka. This is co-incident in terms of timing with overall drying between 9 and 16 ka (Stokes et al. 1997) and a non-palaeo-lake period in Burrough and Thomas (2008). During Phase 4, drying swamp conditions continued to prevail in the Mababe Depression to the point of desiccation. However, the Phase 4 Lake Ngami deposits represent an extended period of shallow water depicted by calcareous sandy-silt and diatomaceous beds (Fig. 4.8). These were later capped with peat beds inferring swamp conditions (Fig. 4.3i), here dated at 3.7 ka. The swamp vegetation was later burnt (cf. Shaw et al. 2003).

Cordova et al. (2017) describe the interval ca. 16 600–12 500 cal BP as being one of fluctuating but high summer rainfall in the Ngami basin, which augmented Okavango inflow and probably contributed to the deposition of shallow, diatomaceous sediment (Fig. 4.8). This was followed by reduced rainfall at ca.11.4 ka, reflecting the dry interval implied by the hiatus at ca. 11.0 ka in the Ngami profile (Fig. 4.8). Interestingly, a date of 11.7 ka was also obtained from the middle of the upper organic-rich silty-clays in the Mababe profile and so recognising the early part of Phase 4 as a very Early Holocene drying event. In contrast, the upper beds in the Ngami profile may initially reflect increasing rainfall with shallow water conditions throughout the last 10 000 years (cf. Cordova et al. 2017).

Additional geochemical evidence was obtained to shed light on possible reasons for different conditions being prevalent in both lake basins in terms of the timing of inflow from different sources (cf. Huntsman-Mapila et al. 2005; Ringrose et al. 2009). For instance, in Table 4.5, the

Table 4.5 Summary of Ngami (LN) and Mababe (MD) selected major and trace metals (original data from Huntsman-Mapila et al. (2006; Gamrod 2008)

	Max—LN	Min—LN	Average LN	Max—MD	Min—MD	Average MD
Al_2O_3wt%	16.3	2.04	8.78	12.66	5.67	9.31
Fe_2O_3wt%	6.44	0.99	3.78	4.72	2	3.54
MgO wt%	1.03	0.14	0.69	0.54	0.23	0.41
CaO wt%	2.66	0.11	0.88	1.4	0.42	0.76
Sr ppm	110	42.4	72.45	387	117	293.57
Sr/Ca (%)	0.042	0.004	0.018	0.09	0.01	0.05
Pb ppm	20	5	12.36	55	15	37.43
Co ppm	21	2	11.24	73	61	66.21
Cr ppm	250	80	145.71	107	74	92.36
Ni ppm	41	8	24.93	97	27	52.36
V ppm	164	14	93.86	221	128	163.71

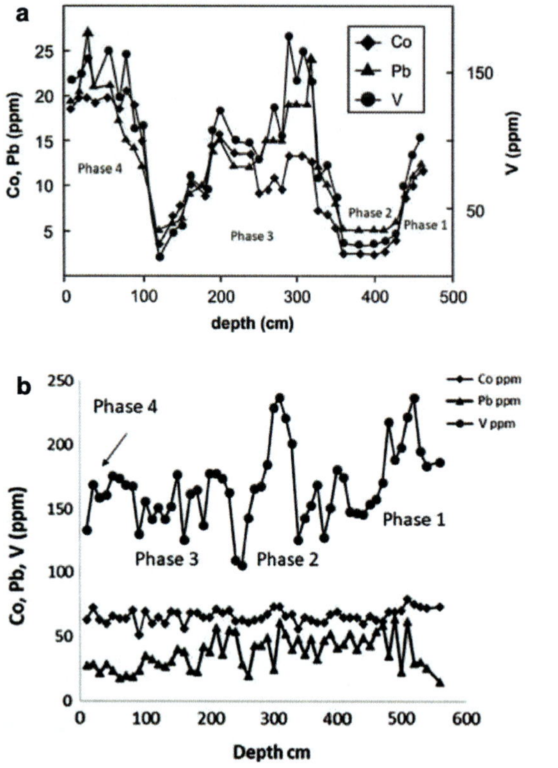

Fig. 4.10 a Transition metal plot from Ngami trench showing trends through four depositional phases. Note differences between the V scale on right and Co, Pb scale on left of graph. *Source* P. Huntsman, 2006. b Transition metal plot from Mababe trench showing trends through four depositional phases. *Source* Gamrod 2009

transition metals, V, Co, Pb and Ni, show noticeably higher abundances in Mababe sediments, whereas Cr values are higher in the Ngami deposits. The transition metals are likely derived from headstream areas characterised by the Palaeoproterozoic Congo craton of the Angolan Highlands (Fig. 4.1). V is relatively common in the schists and related meta-sediments of the Angolan highlands forming vanadiferous (with chromite and magnetite) mafic to ultra-mafic intrusions (e.g. Jelsma et al. 2018). Comparative plots were constructed of V, Co and Pb abundances within Lake Ngami and Mababe Depression profile sediments (Fig. 4.10a, b). Most data variability is depicted by the V trends, which generally decline through Phase 1 in both the Ngami and Mababe data sets. During this early time span (ca. 65 ka to 40 ka), the range of values in the Lake Ngami profile lies between 75 and 25 ppm, whereas the range in the Mababe Depression lies between 240 and 100 ppm. Since these ranges of values are so dissimilar, different inflow sources are inferred for both basins. In Phase 2, the Ngami V data in Fig. 4.10a essentially occupy a trough suggesting low sediment inputs at around 25 ppm, corresponding to sandy-silt deposition (Fig. 4.8). In contrast, Phase 2 in the Mababe data (Fig. 4.10b) is represented by fluctuating peaks and troughs ranging through 240–110 ppm, suggesting variable inflow and sediment inputs commensurate with a diatomaceous palaeo-lake. The overall much higher V content in Mababe suggests a different inflow source from that received in the Ngami basin. This changes in Phase 3 where the Ngami data (post 27 ka) now show a range of fluctuating inputs from 175 to 52 ppm whereas the Mababe data are relatively flat between 175 and 125 ppm. This may reflect a new regime when river inflow was essentially derived from a similar source. The similar though fluctuating V values continue through Phase 4 when the Ngami data show a wide range of V peaks and troughs (between 155 and 22 ppm) throughout the Mid-Late Holocene.

The relatively high abundance of V in addition to Co and Pb, in the Mababe Depression during Phases 1 and 2, may be explained by the closer proximity of Mababe to vanadiferous deposits in the headstream source area, which are more directly accessible through the Cuando River (Fig. 4.1). Conversely, Lake Ngami appears to have long obtained inflow via a longer route through the Okavango system. In addition to receiving inflow from the Okavango River, the Delta is known to be a sink for a wide range of metals (Masamba and Muzila 2005), which are retained within the sand and peat rich sediments. Hence, the Okavango Delta distributaries that flow into Lake Ngami likely became relatively depleted in transition metals (Fig. 4.2). The abrupt transition from higher V-enriched Cuando inflow into the Mababe basin between ca. 65 and 27 ka (depositional Phases 1 and 2) to Okavango inflow after 27 ka (during depositional Phases 3 and 4) may be related to local tectonic shifts (cf. McCarthy et al. 1993) particularly within the earthquake active zone towards the southern tip of the Linyanti swamp-trough (Fig. 4.5) (Scholz et al. 1976; Cruse et al. 2009; Gamrod 2009) after which both basins received inflow via Okavango Delta sources. Hence, tectonic movement along the Linyanti or Chobe faults is invoked to explain the transition metal change, which occurred only in the Mababe Depression sediments. This, in turn, suggests that the two basins operated separately in terms of inflow at least until ca. 27 ka. After this date and throughout the Holocene, the predominant separate inflow appears to have been derived from mainly decreasing Okavango River and Delta sources.

The general picture provided by basin sediments in both Lake Ngami and the Mababe Depression is of two separately functioning lake basins often characterised by shallow but often fluctuating palaeo-lake levels with earlier sediments derived from different sources. Hence the overall results of palaeo-lake sediment analyses fail to endorse the existence of sequential, deep (for example ca. 15 m deep at the 936 m) and persistent palaeo-lakes within the Lake Ngami and the Mababe Depression, between ca. 65 ka and 4 ka (cf. Burrough and Thomas 2008).

4.4 Issues and Future Work

Two main controversial issues have been described during this review of the geomorphological histories of two relatively small lake basins, Lake Ngami and the Mababe Depression, which are peripheral to the Okavango Delta. The first involves the age differential indicated by different literature sources for the palaeo-shorelines, especially those of Palaeo-Lake Thamalakane (PLT). The second issue centres around the frequency or persistence of the palaeo-lakes hinging on whether these were single sequential occurrences or multiple occurring events.

Table 4.1 depicts incremental palaeo-lake incursions into the ORZ, which formed shoreline sequences in both Lake Ngami and the Mababe Depression. Moore et al. (2012) and McCarthy (2013) suggest that the palaeo-shorelines (from PLM and PLT in particular) formed from relatively ancient palaeo-lakes, which developed from ca. 1.4 Ma to 200 ka. The existence of the large palaeo-lakes is based on inflow (and later diversions) of the Cuando, Zambezi and Okavango rivers. These palaeo-lakes have been allocated minimum ages using ESA and early MSA stone tools (Robbins and Murphy 1998; McFarlane and Segadika 2001) associated with the two basins and the adjacent Makgadikgadi. Work by Moore et al. (2012) demonstrates that the existence of the early mega-lakes effectively terminated after the reconnection of the Upper and Mid-Zambezi (Table 4.1). Hence the palaeo-lakes were single ever diminishing, sequential occurrences. On the other hand, Burrough et al. (2009a) have inferred several later mega-lake phases (covering both the Ngami and Mababe basins) using numerous OSL dates, at the PLM/PLT levels during the last ca. 100 ka. Burrough et al. (2009b) argue that an augmented Okavango River provided most of the water for the later mega palaeo-lakes and suggest that cooler temperatures led to reduced evapotranspiration which in turn drove mega-lake expansion from ca. 100 to 20 ka. For the period from 19 to 15 ka, however, Burrough et al. (2009b) propose a southward shift of the ITCZ resulting in higher precipitation in the catchment area of the Okavango River which thus drove later high-level palaeo-lake events. Hence, mainly climatic factors drove these multiple occurring events to the same palaeo-lake level. While it may be argued that possible errors in these younger dates result from aeolian sediment reworking on the palaeo-shorelines (Moore et al. 2012) or bioturbation possibly via termites (McFarlane et al. 2005; Bateman et al. 2007), the discrepancies remain. Since the dates 200–500 ka versus 100–15 ka pertain to the exact same palaeo-shorelines (945 m PLM and 936 m PLT), then additional work is required to help resolve the differences presented including the palaeo-climatic implications.

Obtaining an unequivocal answer may lie in a different approach to the problem such as the initiation of an interdisciplinary deep drilling (to near bedrock) project. This could provide a series of cores to several 100 m (Kinabo et al. 2008) through both the Ngami and Mababe basins and involve the analysis and dating of as many samples as possible. There is abundant evidence in the literature to suggest that the interpretation of deep drill cores is increasingly useful (e.g. Johnson et al. 2016). The location of the Mababe-Ngami structural depressions between the Lake Malawi drill site(s) and the Pretoria Salt Pan (Partridge et al. 1997) should provide ideal locations from which to obtain conclusive palaeo-lacustrine evidence and useful additions to the palaeo-climatic record of southern Africa (A. Moore, pers.com).

4.5 Conclusions

1. Lake Ngami and the Mababe Depression currently receive inflow from the Okavango River via the Okavango Delta and as such form overflow basins of temporary storage from the Okavango system. Both the Ngami and Mababe basins form deep, geomorphologically diverse elongated troughs with maximum extents ca. 2 660 km^2 and 5 888 km^2, respectively.
2. Whereas the timing of their origin is largely uncertain, the basins were formed as structural depressions originating as a result of East African Rift propagation along the Kunyere and Thamalakane (Mababe) fault lines within the Okavango Rift Zone. The two basins are embedded in older Okavango Delta fans, namely the Groote Laagater and Linyanti fans (McCarthy 2013).
3. Recent evidence from Moore et al. (2012) suggests that the Ngami and Mababe basins were entirely submerged during the Palaeo-Lake Deception period of the Early Pleistocene, with higher points emerging as islands during the Palaeo-Lake Makgadigadi period (Early-Mid Pleistocene). Palaeo-shorelines mainly developed along the basins' peripheries during the Palaeo-Lake Thamalakane (PLT-936 m) period possibly ca. 200–500 ka. However, Shaw et al. (2003), Burrough et al. (2007) and Burrough and Thomas (2008) suggest that the 936 m palaeo-shorelines within the ORZ were formed on numerous occasions within the last 100 ka due to mainly climatic and feedback factors.

4. Limited trench data from Ngami and Mababe sediments suggest that after 65 ka the two basins operated as mostly separate, often closed system alkaline lakes, with early inflow rivers from different sources.
5. Controversies prevail in terms of the dating of the palaeo-shorelines (500–200 ka vs. 100–4 ka) and in terms of the palaeo-climatic implications of the same large palaeo-lakes being prevalent at different times. Such divergent results may be resolved by a comparative comprehensive drilling programme with full sedimentological analyses and the dating of samples in both basins.

Acknowledgements I would like to acknowledge the significant contributions provided by Dr Philippa Huntsman to this chapter which draws heavily on her work especially in Lake Ngami. Collaboration with Philippa throughout the formulation of the chapter and resulting conclusions has also added invaluably to this work. The author also wishes to thank Dr Andy Moore for his useful contributions. Additional thanks are extended to Dr Eliot Atekwana for facilitating the use of Mababe trench data and Wilma Matheson for photographic and editorial assistance.

References

Bateman MD, Boulter CH, A.S. Carr AS C.D. et al (2007) Preserving the palaeoenvironmental record in Drylands: Bioturbation and its significance for luminescence-derived chronologies. Sed Geol 195:5–19

Burrough SL, Thomas DSG (2008) Late Quaternary lake-level fluctuations in the Mababe Depression: Middle Kalahari palaeolakes and the role of Zambezi inflows. Quat Res 69:388–403

Burrough SL, Thomas DSG, Shaw PA et al (2007) Multiphase Quaternary highstands at Lake Ngami, Kalahari, northern Botswana. Palaeogeogr. Palaecol. 253:280–299

Burrough SL, Thomas DSG, Bailey RM (2009a) Mega-lake in the Kalahari: A Late Pleistocene record of the Palaeolake Makgadikgadi system. Quat Sci Rev 28:1392–1411

Burrough SL, Thomas DSG, Singarayer JS (2009b) Late Quaternary hydrological dynamics in the Middle Kalahari: forcing and feedbacks. Earth-Sci Rev 96:313–326

Cooke HJ (1980) Landform evolution in the context of climatic change and neotectonics the middle Kalahari of north-central Botswana. Trans Inst Br Geogr 5:80–99

Cooke HJ, Verstappen TH (1984) The landforms of the western Makgadikgadi basin in northern Botswana, with a consideration of the chronology of the evolution of Lake Palaeo-Makgadikgadi. Z Geomorphol 28:1–19

Cordova CE, Scott L, Chase BM et al (2017) Late Pleistocene-Holocene vegetation and climate change in the Middle Kalahari, Lake Ngami. Botswana. Quat Sci Rev. 171:199–215

Cruse AM, Atekwana EA, Gamrod J et al. (2009) Environmental change driven by tectonic processes and climate shifts as recorded in the sedimentary record of paleolake Mababe, northern Botswana. Portland GSA Annual Meeting, Paper 134–10

Gamrod JL (2009) Paleolimnological Records of Environmental Change Preserved in Paleo-lake Mababe. Thesis Oklahoma State University, Northwest Botswana M. S

Gasse F (1986) East African Diatoms: Taxonomy, ecological distribution. Bibliotheca Diatomologica J. Cramer, Berlin

Grove AT (1969) Landforms and climatic change in the Kalahari and Ngamiland. Geogr J 135:191–212

Gumbricht T, McCarthy TS, Merry CL (2001) The topography of the Okavango Delta, Botswana and its tectonic and sedimentological implications. S Afr J Geol 104(3):243–264

Huntsman-Mapila P, Kampunzu AB, Vink B et al (2005) Cryptic indicators of provenance from the geochemistry of the Okavango Delta sediments. Botswana Sed. Geol. 174(1):123–148

Huntsman-Mapila P, Ringrose S, Mackay AW et al (2006) Use of the geochemical and biological sedimentary record in establishing palaeo-environments and climate change in the Lake Ngami basin. NW Botswana. Quat Intl. 148:51–64

Jelsma HA, McCourt S, Perritt SA et al (2018) The Geology and Evolution of the Angolan Shield, Congo Craton, Geology of Southwest Gondwana. In: Siegesmund S, Basej MAS, Oyhantcabal P, Oriolo S (eds) Ch 9 in Geology of Southwest Gondwana, Springer Nature p 217–239

Johnson TC, Werne JP, Brown TP et al. (2016) A progressively wetter climate in southern East Africa over the past 1.3 million years. Nature Research Letter, doi:https://doi.org/10.1038/nature19065. 16 pp

Key R, Ayres N (2000) Geology of Botswana, A new geological dataset of Botswana. J. Afr. Earth Sci. 30(3):CD1–25

Kinabo BD, Hogan JP, Atekwana EA et al (2008) Fault growth and propagation during incipient continental rifting: Insights from a combined aeromagnetic and Shuttle Radar Topography Mission digital elevation model investigation of the Okavango Rift Zone, northwest Botswana. Tectonics 27: TC3013, doi:10.1029/2007TC002154

Krammer K, Lange-Bertalot H (1986–1991) Bacillariophyceae, Gustav Fisher Verlag Stuttgart

Kurugundla CN, Parida BP, Buru JC (2018) Revisiting Hydrology of Lake Ngami in Botswana. Hydrol. Current Res. 9:301. https://doi.org/10.4172/2157-7587

Mallick DIJ, Hapgood F, Skinner AC (1981) A geological interpretation of Landsat Imagery and Air Photography of Botswana. Inst. Geol. Sci, London, p 56

Masamba WRL, Muzila A (2005) Spatial and Seasonal Variation of Major Cation and Selected Trace Metal Ion Concentrations in the Okavango-Maunachira-Khwai Channels of the Okavango Delta. Botsw. Notes Rec 37:218–226

McCarthy TS (2006) Groundwater in the wetlands of the Okavango Delta, Botswana, and its contribution to the structure and function of the ecosystem. J Hydrol 320:264–282

McCarthy TS (2013) The Okavango Delta and its place in the geomorphological evolution of southern Africa 30th Alex L. du Toit Memorial Lecture, S. Afri. J. Geol. 116.1:1–54doi:https://doi.org/10.2113/gssajg.116.1.1

McCarthy TS, Green RW, Franey NJ (1993) The influence of neo-tectonics on water dispersal in the northeastern regions of the Okavango swamps. Botswana. J. Afr. Earth Sci. 17:23–32

McCarthy TS, Cooper GRJ, Tyson PD et al (2000) Seasonal flooding in the Okavango Delta, Botswana – recent history and future prospects. S. Afri. J. Sci. 96:25–33

McFarlane MJ, Eckardt FD (2008) Lake Deception: A New Makgadikgadi Palaeolake. Bots Notes Rec 38:195–201

McFarlane MJ, Segadika P (2001) Archaeological evidence for the reassessment of the ages of the Makgadikgadi palaeolakes. Botsw Notes Rec. 33:83–89

McFarlane MJ, Eckardt FD, Ringrose S et al (2005) Degradation of linear dunes in Northwest Ngamiland, Botswana and the implications for luminescence dating of periods of aridity. Quat Int 135:83–90

Mendelsohn J, El Obeid S (2004) Okavango River. Struik Publishers, Cape Town, The flow of a lifeline, p 176

Modisi MP, Atekwana EA, Kampunzu AB et al (2000) Rift kinematics during the incipient stages of continental expansion: Evidence from the nascent Okavango rift basin, northwest Botswana. Geol 28:939–942

Moore AE, Cotterill FPD, Eckhardt FD (2012) The evolution and ages of Makgadikgadi palaeo-lakes: consilient evidence from Kalahari drainage evolution south-central Africa. S. Afri. J. Geol. 115 (3):385–413

Nash DJ, Coulson S, Staurset S et al (2013) Provenancing of silcrete raw materials indicates long-distance transport to Tsodilo Hills, Botswana, during the Middle Stone Age. J Hum Evol 64:280–288

Partridge TC, Demenocal B, Lorentz SA et al (1997) Orbital forcing of climate over South Africa: a 200,000-year rainfall record from the Pretoria saltpan. Quat Sci 16:1125–1133

Ringrose S, Huntsman-Mapila P, Kampunzu AB (2005) Geomorphological and geochemical evidence for palaeo feature formation in the northern Makgadikgadi sub-basin. Botswana. Palaeogeogr. Palaeocl 217:265–287

Ringrose S, Huntsman-Mapila P, Downey W et al (2008) Diagenesis in Okavango fan and adjacent dune deposits with implications for the record of palaeo-environmental change in Makgadikgadi-Okavango-Zambezi basin, northern Botswana. Geomorphology 101:544–557

Ringrose S, Harris C, Huntsman-Mapila P et al (2009) Origins of strandline duricrusts around the Makgadikgadi Pans (Botswana Kalahari) as deduced from their chemical and isotope composition. Sediment Geol 219(1–4):262–279

Robbins LH, Campbell AC, Murphy ML et al (2009) Mogapelwa: Archaeology, Palaeoenvironment and Oral Traditions at Lake Ngami. Botswana. S. Afr. Archaeol. Bull. 64(189):13–32

Robbins LH, Murphy ML (1998) The Early-Middle Stone Ages of Botswana (P.Lane, A.Reid and A Segebye, eds) Pula Press and Botswana Society, Gaborone, Botswana. 50–64

Scholz CH, Koczynski TA, Hutchins DG (1976) Evidence for Incipient Rifting in Southern Africa Geophys. J. R. asfr.SOC: 44:135–144

Shaw PA (1984) A historical note on the outflows of the Okavango system. Bots. Notes Rec. 16:127–130

Shaw P (1985a) The desiccation of Lake Ngami: An historical perspective. Geogr J 151:318–326

Shaw P (1985b) Late Quaternary Landforms and Environmental Change in Northwest Botswana: The Evidence of Lake Ngami and the Mababe Depression. Trans Inst Br Geogr 10(3):333–346. https://doi.org/10.2307/622182

Shaw PA (1988) After the flood: The fluvio-lacustrine landforms of Northern Botswana. Earth-Sci Rev 25:449–456

Shaw PA, Bateman MD, Thomas DSG et al (2003) Holocene fluctuations of Lake Ngami, Middle Kalahari; chronology and responses to climate change. Quat Int 111:23–35

Stokes S, Thomas DSG, Washington R (1997) Multiple episodes of aridity in southern Africa since the last interglacial period. Letters to Nature 388:154–158

Thomas DSG, Shaw PA (1991) The Kalahari Environment. Cambridge University Press, Cambridge

Vanderpost C, Hancock P (2018) Makgadikgadi: Botswana's dusty wetlands. Sandor Books Pty Ltd., Maun, Botswana, p 135

Wolski P, Murray-Hudson M (2008) An investigation of permanent and transient changes in flood distribution and outflows in the Okavango Delta, Botswana. Phys Chem Earth 33:157–164

Susan Ringrose Susan Ringrose obtained a master's degree in Soil Science from the University of Manitoba and a PhD in Quaternary Geology from London University. After several years working as a Quaternary Geologist with the Manitoba Department of Mineral Resources, she accepted lecturing positions first in Sierra Leone and later with the University of Botswana. Thirty years research in Botswana was initially remote sensing oriented involving range degradation mapping evolving later into geomorphological applications around the Okavango Delta. She retired as Director of the Okavango Research Institute in 2012 and later became an affiliate (Honorary Professor) with the School of Multidisciplinary Studies, at the University of Glasgow, Dumfries Campus while continuing Botswana oriented research.

The Makgadikgadi Basin

5

Sallie L. Burrough

Abstract

Visible from space, the Makgadikgadi pans are one of the most distinctive geomorphic features of southern Africa. Now seasonally dry, this collection of pans is the sump of what was once one of Africa's largest inland lakes. Formed by uplift along the Kalahari-Zimbabwe axis in the early Miocene and re-sculpted during the Pleistocene by both climatic events and neo-tectonism, the geomorphology of this basin tells a story of significant hydrological dynamism within northern Botswana, a story that has arguably influenced the evolution and dispersal of our own species (Chan et al. 2019). The boundary of the 37,000 km^2 basin withholds a suite of landforms including saltpans, dunes, relict shorelines, lacustrine spits and fluvial deltas, each with its own inter-linked geomorphic history. Today, its landscape remains geomorphically active, providing one of the largest sources of dust in southern Africa with knock-on effects for global climate.

Keywords

Saltpan • Palaeolake • Shoreline • Diatoms • Spit • Islands • Dust • Quaternary

5.1 Introduction

Viewed from space, the landscape of northern Botswana is dominated by the endorheic drainage basins that lie at the end of the Okavango/Kwando rivers and in particular, the Makgadikgadi pans, a complex of salt flats and shallow seasonal lakes that lie beyond the Boteti river at the system terminus. The basin is dominated by Ntwetwe Pan to the west (4700 km^2) and Sua Pan to the east (3400 km^2), but beyond these, there are many smaller pans, including Nxai and Kudiakam to the north, and Makarikari, Dzibui, Mopipi, Rysana, Tsokotsa, Guguaga, Nkokwane, Tshitsane and Ntsokotsa to the south (Cooke 1980; White and Eckardt 2006). Now ephemerally dry, this collection of saltpans which covers a total area of 37,000 km^2 has long been recognised as the remnants of an extensive former lacustrine system, known variously as palaeolake Makgadikgadi, megalake Makgadikgadi in its late Quaternary form and palaeolake Deception at its greatest and oldest configuration (McFarlane and Eckardt 2006). Made up of three principal sub-basins: Lake Ngami, the Mababe Depression (Chap. 4) and the Makgadikgadi Basin, it occupies a rift depression (the Makgadikgadi–Okavango–Zambezi or MOZ depression) at the end of the Okavango delta, part of the south-western extension of the East African Rift System (EARS) which propagates into northern Botswana. Its relict shorelines are today clearly visible in remote imagery and elevation data (Fig. 5.1a, b), but its former existence as a lake has been recognised and documented since the mid-nineteenth century (e.g. Livingstone 1858) and has become a source of fascination to geomorphologists, geologists and Quaternarists particularly within the last 50 years (e.g. Grove 1969; Ebert and Hitchcock 1978; Cooke 1984; Cooke and Verstappen 1984; Helgren 1984; Shaw and Cooke 1986; Shaw 1988; White and Eckardt 2006; Burrough et al. 2009a; Burrough et al. 2012; Riedel et al. 2014; Nield et al. 2015; Eckardt et al. 2016).

The contemporary landscape of the Makgadikgadi (formerly known as the Makarikari) basin is a palimpsest of geomorphic features moulded by lacustrine, tectonic, fluvial and aeolian processes reflecting a long morphodynamic history stretching back into the Miocene. The most prominent of these features includes the broad, gently curved 250 km long lacustrine strandline known as the Gidikwe

S. L. Burrough (✉)
Trapnell Fellow of African Environments, Environmental Change Institute, School of Geography and the Environment, University of Oxford, South Parks Road, Oxford, OX1 3QY, UK
e-mail: sallie.burrough@ouce.ox.ac.uk

F. D. Eckardt (ed.), *Landscapes and Landforms of Botswana*, World Geomorphological Landscapes,
https://doi.org/10.1007/978-3-030-86102-5_5

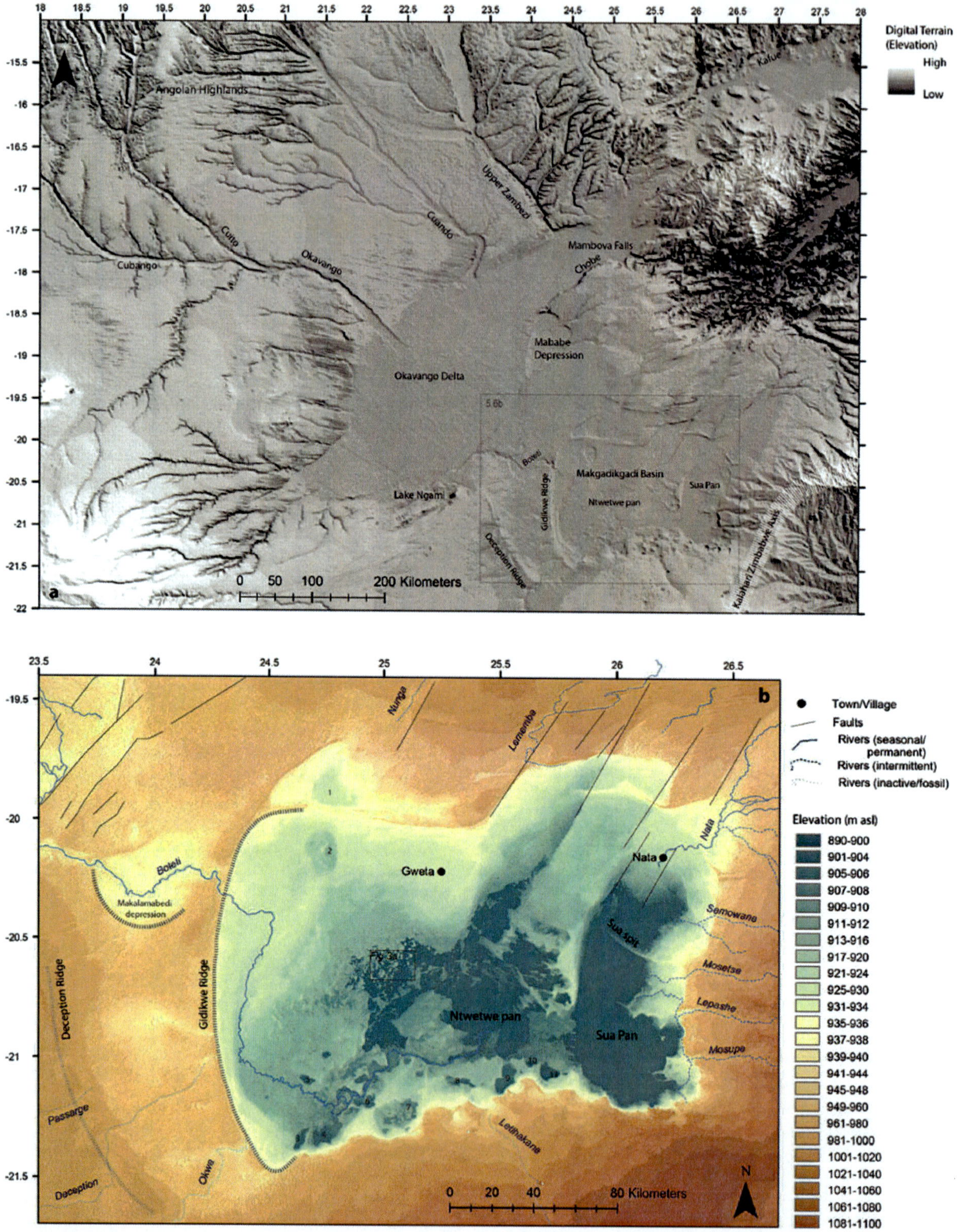

Fig. 5.1 Digital terrain model of **a** the wider Makgadikgadi region, fluvial systems and major features referred to in text; **b** the Makgadikgadi basin. Pans referred to in text are numbered: (1) Nxai; (2) Kudiakam; (3) Makarikari; (4) Lake Xau; (5) Dzibui; (6) Tsokotsa; (7) Rysana; (8) Guguaga; (9) Nkokwane; (10) Tshitsane; and (11) Ntsokotsa. The location of Fig. 5.6a is also shown

ridge (Grey and Cooke 1977) on the western margin of the basin, first identified and mapped in Grove's (1969) seminal work on the geomorphology of the Kalahari. Associated with diatomaceous and ostracod deposits (Cooke 1979), this conspicuous feature is clearly lacustrine in origin, though both its precise nature as an offshore bar or bounding coastline and the timing of its formation are more contentious. Active fluvial systems entering the basin include the Boteti from the northwest, which intermittently carries seasonal floodwater from the Okavango Delta towards the basin and, following channelisation works (SMEC 1987) terminates in Lake Xau. Previously however, the main Boteti channel entered Ntwetwe pan through a 'palaeo-estuary' (Ringrose et al. 2014) in the south-west of the basin. Very clearly a 'misfit' channel (Cooke 1980), its conveyance capacity is at least equal to the full Okavango flow (UNDP/FAO 1977) suggesting a hydrological regime in the past that was very different to today. On the other side of the basin, the Nata, Semowane, Mosetse, Lepashe and Mosupe rivers flow seasonally from the eastern Botswana/western Zimbabwe into Sua pan. Other drainage lines now remain dry all year round including the Passage, Deception and Okwa-Mmone systems which enter the basin from the southwest and the Letlhakane system from the south. Drainage into the pans is also possible via fault controlled channelled groundwater flow and has been observed in the Nunga and Lememba channels entering the basin from the north (Eckardt 2010).

Over the last 30 years, advances in scientific technology have enabled the geomorphic evolution of the suite of sediments and landforms which lie within these pan systems to be partially pieced back together, a process that, until recently, remained largely speculative. The primary aim of this chapter is to unpick the complex landscape of the basin into its component landforms and to outline what we know (and don't' know) of how they have been shaped by long-term geological processes, high amplitude climate change, and contemporary morphodynamics.

5.2 The Morphology of Makgadikgadi's Geological History

The origins of the Makgadikgadi lacustrine system can be traced back to the epeirogenic uplift of the Kalahari-Zimbabwe axis (Moore 1999) (Fig. 5.1a) which divided the Limpopo drainage from the Okavango and Zambezi systems (Du Toit 1933). There is scant evidence for the timing of this event, with initiation of the endoreic fluvio-lacustrine system suggested as the late Palaeogene (Moore et al. 2008); Miocene or Pliocene (Partridge and Maud 2000). Ultimately however, uplift resulted in the separation of the Kalahari basin from the exoreic Limpopo drainage and led to the accumulation of both water and sediment in the region of the Makgadikgadi basin as rivers flowing from the north terminated in the interior. Contributing to the effects of fluvial impoundment, both the Cuito and Cubango rivers were captured by the headwaters of the Okavango sometime during the Pliocene, increasing the flow of internally draining rivers towards proto-Makgadikgadi (Haddon and McCarthy 2005). Evidence in the form of Pliocene-age pollen in fluvio-lacustrine deposits overlying the Makgadikgadi basement lends some support to the otherwise uncertain timing of the initiation of lacustrine sedimentation (Moore et al. 2012).

Horst and graben structures associated with N-NE striking faults have long been observed within the basin (Fig. 5.1b) (Baillieul 1979; Smith 1984) and are testament to the southwesterly propagation of the EARS. In particular, large-scale faulting between Sua and Ntwetwe pans explains the presence of isolated blocks of granite that make up the well-known Khubu 'island' on the western side of Sua Pan (Baillieul 1979) and the long arm of the Ntwetwe pan that parallels a north-easterly trending fault (Fig. 5.1b). The influence of faulting in this system continues to the present day (see Sect. 5.4), but during the Pliocene and Pleistocene the impact of the extending rift was a gradual and progressive reorganisation of drainage systems with profound geomorphological consequences for Makgadikgadi. Uplift along the Chobe Fault across the Upper Zambezi eventually diverted the Palaeo-Chambeshi/Upper Zambezi drainage system into northern Botswana (Moore et al. 2012) and it is this configuration of river systems that, it is hypothesised, is associated with Makgadikgadi's largest lacustrine phase, Palaeo-Lake Deception. Geomorphological evidence for this huge lake system comes in the form of an offshore bar or shoreline standing at ~ 985 m asl in its western margins, the degraded remnants of which were identified from remote imagery by both Cooke (1979) and McFarlane and Eckardt (2006) (Fig. 5.1b). This westernmost margin separates a belt of transverse dunes to the west from a 50 km wide flat zone between the 985 m strandline and the Gidikwe ridge (Cooke 1979), an area suggested to represent a former lagoon. East–west hummocky lineations within this lagoon zone have been variously ascribed to be the remnants of 'wave-induced current formed ridges' (Grey and Cooke 1977) or giant desiccation stripes formed on the gently sloping mud floor as the lake level retreated (Cooke 1979). The areal extent of the lake at this level is not entirely known because of the degraded nature of the Deception shoreline but it is likely the Palaeolake covered a region of approximately 185,000 km^2, its level ultimately controlled by an outlet point at the Mambova rapids to the north (McFarlane and Eckardt 2006).

High resolution Helicopter time-domain electromagnetic (HTEM) data in conjunction with deep borehole records have been used to identify low resistivity silt/clay sediments

Fig. 5.2 Visualisation of Palaeolake Makgadikgadi. Vertically exaggerated SRTM digital elevation data draped over Landsat (7, 4, 2) mosaic with added water level depicting the 945 m level in the basin. *Source* Frank D. Eckardt

up to 200 m thick that lie expansively below the Okavango Delta (Podgorski et al. 2013). These have been interpreted as the deposits of a very extensive lake system (>90,000 km^2) relating to lake high stands in Palaeolake Makgadikgadi (Fig. 5.2). Whether these deposits relate to Palaeolake Deception, or to a later configuration of Palaeolake high-stand in the Makgadikgadi basin, is not known but they provide strong evidence for a palaeo-lacustrine system that would have been one of Africa's largest.

Inflow to Palaeolake Deception was subsequently reduced in the early Pleistocene by further geological processes, namely, the uplift of the Congo-Zambezi watershed which severed the link between the Chambeshi and Kafue sometime between 2.2 and 1.4 Ma (Cotterill and De Wit 2011) significantly reducing the lake's catchment size. Sometime after this, continued propagation of the EARS led to uplift along the Linyanti Fault, diverting the Cuando river into the Zambezi via the Chobe (Moore et al. 2012) and depriving the palaeolake of a significant contributor to inflow. This was followed by the capture of the upper Zambezi by the middle Zambezi. These geological events, the timing of which is generally poorly constrained, form the backdrop to a progressively shrinking Palaeolake in the lake Quaternary. Several researchers (e.g. McFarlane and Eckardt 2006; Moore et al. 2012) not unreasonably suggest the total disappearance of the lake in the mid-Pleistocene leaving an exposed basin floor that was, by comparison, geomorphically and hydrologically inactive during the late Quaternary and Holocene. This idea is evidenced largely on the existence of Early Stone Age (ESA) archaeology on the basin floor below the 936 m palaeo-lake shoreline (McFarlane and Segadika 2001), which, they suggest, indicates the basin must have been dry since the ESA which probably ended at least 500 kyrs ago. In a similar vein, Grove (1969) whose work first identified and mapped the shorelines of the Palaeolake system, suggested the lake must have pre-dated the Middle Stone Age (MSA) deposits identified by (Bond and Summers 1954) on the Nata River. Since then, other MSA sites have subsequently been identified right across the Makgadikgadi basin floor (e.g. Ebert 1979; Ebert and Hitchcock 1978; Robbins 1987). More recently, however, there is a growing body of data that suggests lake levels continued to rise and fall during the late Quaternary, with a similar timing and amplitude to transgressions and regressions that occurred in many lakes across eastern and northern Africa during this time (e.g. Armitage et al. 2015; Drake et al. 2018; Moernaut et al. 2010; Scholz et al. 2011). At least some part of this late Quaternary hydrological dynamism in the Makgadikgadi basin seems to have occurred in response to high amplitude climate changes that radically modified the landscape of much of Africa (Burrough et al. 2009a,b). The seeming discordance with apparently undisturbed archaeological records in the basin can now be explained through detailed landscape reconstructions at lake floor archaeological sites (Burrough et al. (in prep)), which suggest cyclical lacustrine burial and deflation exposure of archaeological material at several locations within the basin.

5.3 Makgadikgadi in the Late Quaternary, the Morphological Record of Climate Change

The idea that much of the landscape of the Makgadikgadi basin owes its existence to geomorphological processes that were operating in the late Quaternary, dates back to the work of Dick Grove, John Cooke and Paul Shaw who meticulously mapped, measured and, where possible, dated many of the landforms associated with the basin. Issues of landscape preservation and limitations of dating techniques bias our understanding of these landforms to the late Quaternary but here we examine the insights into landscape evolution these analyses have provided.

5.3.1 Beach Ridges

Detailed mapping of the 945 m Gidikwe Ridge and lower beach ridges (relict shorelines) within the basin at 936, 920 and 912 m asl (Cooke 1979, 1980; Grove 1969; Shaw and Cooke 1986), suggest stable lake phases at several levels within the Makgadikgadi. These shorelines are largely composed of unconsolidated fine to medium-sized sand particles with little detectable sediment structure (Fig. 5.3). Burrough and Thomas (2009) suggest depositional shorelines such as these largely record the regressive phase of lake high-stands and can be thought of in a similar way to coastal zones with a probable mix of both wave-deposited and, in the backshore zone, wind-deposited sediment. The controls on these features relate to both the rate of water level change and the rate of sediment supply so that shoreline deposits accumulate when the latter keeps pace, or exceeds water level rise and are preserved when water level declines again (Burrough and Thomas 2009; Thompson and Baedke 1996) assuming some stabilisation effects from vegetation colonisation or duricrust formation. Buffering and amplifying effects of existing shorelines, through dissipating energy and provisioning sediment, mean there is some persistence of these features when formed, a morphodynamic lag-time, which is the time taken for a shoreline such as the Gidikwe ridge to adjust to changes in the sediment or water budget (Burrough and Thomas 2009). There is also the possibility of periods of erosive transgression when water level rise exceeds sediment supply (for example, at a time when rainfall is higher than or less-seasonal than present and sediment erodibility is lower within the landscape due to stronger stabilisation by vegetation). In these cases, the shoreline would, to some extent, migrate landwards and some of the shoreline record may be lost. For this reason, these shorelines do not provide a complete record of lake level change and only preserve high stand events, telling us nothing of the timing or duration of lake absence (Burrough et al. 2009a, 2009b). They do, however, definitively provide a means to date lake presence. Early attempts to place a chronology on the formation of Makgadikgadi beach ridges were undertaken using radiocarbon dating (Cooke and Verstappen 1984; Shaw et al. 1997, 1992). In a system where organic preservation is poor however, ^{14}C measurements were necessarily limited to surface bone, shell or calcrete deposits. The latter, in particular, are post-depositional and so place only a minimum age on the geomorphic feature being dated. Nevertheless, it seems unlikely that these calcretised surfaces form sub-aqueously so they provide, at the very least, a good estimate of times when lake levels were lower than, or fluctuating around, the dated level.

Over the last 30 years, the development of Optically Stimulated Luminescence (OSL) dating which directly dates the last deposition event of a sediment, has enabled researchers to add significantly to this body of dates, facilitating the development of an approximate timeframe for the formation and evolution of a range of depositional landforms within the Makgadikgadi (Fig. 5.4). Importantly, what has emerged from this research is that, as predicted by the shoreline formation model described above, the prominent sandy shorelines around the basin, including the Gidikwe ridge, are composite accumulations of sediment, recording multiple phases of beach ridge deposition during the late Quaternary. As the relative precision on OSL dates is at best ~8–10% and the dates provide no clear information on duration of high-stand, these data are crude. Nonetheless, they provide the best estimate of the timing of Makgadikgadi lake high-stands to date. Age-groups for high lake stands within Makgadikgadi itself, identified using a cluster analysis and tested for significance using a weighted t-test, fall at 105 ± 3 ka, 92 ± 2 ka, 64 ± 2 ka, 39 ± 2 ka, 27 ± 1 ka, 17 ± 2 ka and 8.5 ± 0.2 ka, the latter phase being

Fig. 5.3 Shoreline ridges on the western (top) and eastern (bottom) margins of Sua Pan

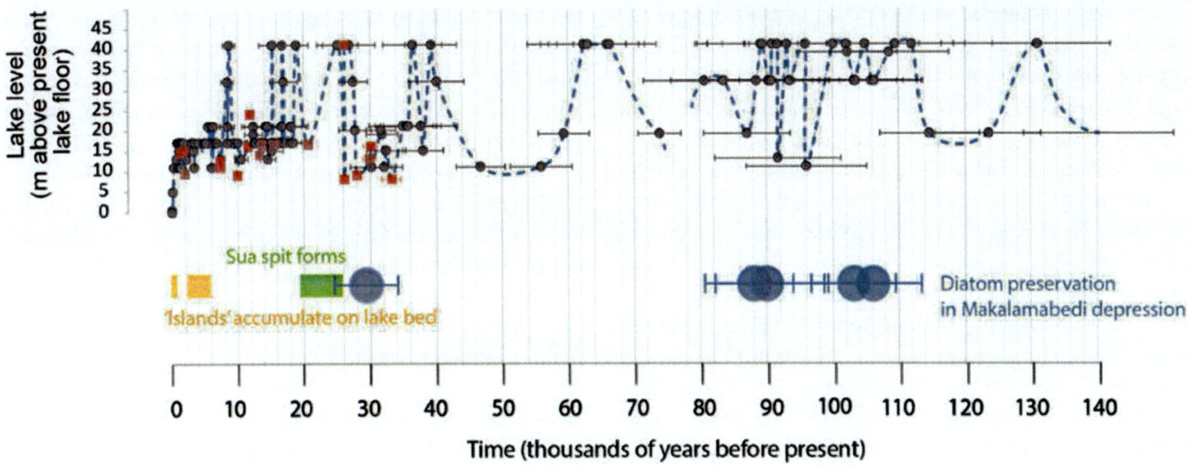

Fig. 5.4 Lake level curve for the Makgadikgadi system (including Ngami and Mababe) as reconstructed from OSL ages on shorelines (grey dots, uncertainties shown as grey bars) and calibrated radiocarbon ages (red squares, uncertainties shown as red bars). The chronology of other sedimentary deposits and landforms within the basin are shown below

represented by only two dates at the top of two shoreline sequences (Burrough et al. 2009a, 2009b). Older shoreline building events are present but represented by only single dates with uncertainties too large to provide meaningful significance. More generally, these dates can be placed within the greater body of data from across the Makgadikgadi system (Fig. 5.4). There is some agreement between dates across the sub-basins of the larger Makgadikgadi system but also a significant degree of inconsistency. Ngami appears to be the most sensitive basin in the system, perhaps too sensitive for its geomorphology to be used as a palaeo-hydrological tool. Both Ngami and Mababe have undergone independent localised filling that has not, apparently, left a strong geomorphic signal in the Makgadikgadi. This is not surprising given that the largest, and most active part of the catchment lies 1000 km to the north-west of the entire lake system and the Boteti, at 930 m asl, acts as an 'overspill valve' into the Makgadikgadi basin when the Thamalakane is in full flood. It is therefore, difficult to envisage a hydrological configuration where Makgadikgadi could be full of water without both the Ngami and Mababe basins also supporting high lake stands. Little is known however, about former contributions from fluvial systems that drain directly into the Makgadikgadi, including the Okwa (Fig. 5.1b).

The existence of beach ridges is strong evidence for very different palaeoenvironmental conditions than present. At its greatest extent, this megalake system likely covered an area of at least 66,000 km^2 (White and Eckardt 2006) and, modelling experiments suggest, may have recycled enough water to substantially alter the regional climate and ecosystem (Burrough et al. 2009b). The regional origin of hydrological or climatological changes with the potential to sustain palaeo-lake Makgadikgadi is difficult to derive since the hydrological system for which the entire palaeo-lake forms the terminal sump, covers over 12° of latitude. Estimates suggest a substantial 50 km^3 inflow per year under present day conditions would be required to balance evaporative loss (Grove 1969). It is most likely that this inflow was provided by additional contributions from the Zambezi via ponding up of water at the Mambova rapids when discharge was extremely high and back-flow into the palaeo-lake system via the extremely low gradient Chobe (Shaw and Thomas 1996, 1988; Thomas and Shaw 1992). At least within historical records, the Okavango and Zambezi Rivers follow a similar pattern of flow variability (Mazvimavi and Wolski 2006) and there is some geomorphological and geochronological evidence for ponding back of floodwaters in the form of beach ridges in the Chobe enclave (Burrough and Thomas 2008). Shoreline construction in the Makgadikgadi basin itself however, requires not only a substantial water-body but sufficient sandy material and wave action. This implicates both sediment supply and wind energy as important additional factors that should not be neglected.

5.3.2 Diatom Deposits

In addition to shorelines, pockets of diatom beds particularly along the Boteti have also been identified and in some cases dated (Schmidt et al. 2017; Shaw et al. 1997). Originally described by Passarge (1904), these 2–3 m thick, diatom-rich exposures extend along the banks of the Boteti at 924 m asl (Shaw et al. 1997) and 935–940 m asl (Schmidt et al. 2017). In the Makalamabedi depression (Fig. 5.1b), the composition of the diatom taxa suggests standing or slow-moving waters of shallow depth with little wave disturbance, conditions interpreted as being caused by the ponding back of Boteti flow due to the presence of a high-lake stand to the east. OSL dates placed the age of this deposit at between 32 and 27 ka (Fig. 5.4) (Shaw et al. 1997), consistent with 14C ages on local terrace calcretes. Further downstream, where the Boteti cuts a gorge through the Gidikwe Ridge, a second exposure at a higher level reveals a 30 cm thick unit consisting of consolidated, fine-laminated lacustrine sediments rich in diatoms which again suggest alkaline, shallow water conditions (Schmidt et al. 2017). This deposit however, has been OSL dated to between 89 and 106 ka (Schmidt et al. 2017), correlating with both shoreline records (Burrough et al. 2009a, 2009b) and climate model predictions for substantially wetter climate within the southern African interior at this time (Singarayer and Burrough 2015). Diatoms and other terrestrial proxies, including leaf wax hydrogen isotopes and branched glycerol dialkyl glycerol tetraether (br GDGT) membrane lipids, have recently become important targets for ongoing lake-bed drilling programmes in the Makgadikgadi which seek to extract information from lakebed muds on both the hydrological status of the lake and local climatic conditions (Fig. 5.5).

5.3.3 Cuspate Spits

In Sua pan, a series of protruding ridges extend out west into the basin in a southeast-northwest orientation (Fig. 5.1b), the largest of which is known as Sua Spit and is currently home to a soda ash (Na_2CO_3) and salt (principally NaCl) mine (Botswana Ash (Pty) Ltd.). Each is bounded on either side by fluvial systems with associated delta-like deposits. Moore and Larkin (2001) have speculated these landforms are relict fluvial channels in a similar vein to comparable processes operating in the Okavango delta where a dense reed fringe along the active channel develops into peat beds and the channel then aggrades between raised peats until it is abandoned. The organic peats are eventually destroyed leaving a raised sandy bar marking the place of the abandoned channel (Moore and Larkin 2001). They envisage this process occurring prior to uplift along the Kalahari-Zimbabwe axis which cut off the Kalahari rivers from the Limpopo (Sect. 5.2) with a former link between the Cuando to the northwest of the basin with the Motloutse, and displacement progressively northwards as represented by the three distinct spits. A recent suite of OSL dates on both drill cores and pits dug within and through these spits, however, indicates a much younger age for their deposition, within the last 30 kyrs (Burrough et al. (in prep), Fig. 5.4). This would favour a simpler explanation that they are indeed true cuspate 'spits' and likely formed when lake levels were moderate, wind-driven wave-action was strong and sediment loads were high in the rivers flowing into the basin from the east. The formation of such spits, sometimes known as 'Azov-type' spits has been shown to occur in shallow basins when waves are obliquely incident at an angle larger than 45° (Uda et al. 2018; Zenkovitch 1967). Digital elevation models pick out the form of former positions of spit features to the north and associated shoreline concavities along the eastern margin of the lake (Fig. 5.1b).

5.3.4 Makgadikgadi 'Islands'

Across the western and northwestern shores of Ntwetwe pan, and to a lesser extent, the northwest corner of Sua pan, a large number of vegetated crescentic shaped islands are visible, transverse to the dominant easterly wind (Fig. 5.6a). These landforms appear to be coalesced into ridges in the west of Ntwetwe, becoming less well-defined as sediment banks up towards the Gidikwe ridge. Their vegetated surfaces, which can sustain a dense covering of halotrophic grasses (pre-dominantly *Odyssea paucinervis* and *Sporobolus ioclados*), make these features visibly distinct in remotely sensed imagery, in contrast to the generally de-vegetated pan floor (Burrough et al. 2012). Many individual forms also possess subtle ridging which parallels the landform's concentric shapes. Grove (1969) described these landforms as 'barchans' and inferred they are aeolian origin, formed when the basin was drier than present. Cooke (1980) suggested if this were the case, then the dunes must be severely degraded, and instead inferred the 'islands' were formed sub-aqueously at times when the lake level was high. In a systematic study of the landforms that used sedimentological and morphological analyses in conjunction with known deposition contexts and analysis of the optically stimulated signal from luminescence measurements, Burrough et al. (2012) concluded the landforms were likely aeolian in origin but that neither hypothesis could be clearly ruled out. McFarlane and Long (2015) used Google Earth imagery to infer these 'islands' were spring mounds formed by ongoing discharge of potable groundwater and sediment, initiated during lacustrine periods and tentatively inferred to be pre-Quaternary in age. Most recently, Franchi et al. (2020) suggested the islands were remnants of former shorelines (Fig. 5.5).

Field investigations and full-profile dating of several of these 'island' landforms, however, revealed no evidence of former or contemporary groundwater discharge from the mounds but did provide a clear late Holocene age for sediment deposition c. 2–3 ka (Burrough and Thomas 2013) post-dating any recorded lacustrine phases. Ages at the westerly (downwind) end of the pan where 'island' density, and therefore sediment volume, are greater, yielded the older ages (c. 4.9–4.4 ka), than further east where forms are more

Fig. 5.5 A team from the Geoscience Institute of Botswana extract lake-bed cores from Ntwetwe pan for terrestrial proxy analysis to reconstruct lake-level fluctuation and past climate change

discrete. Burrough and Thomas (2013) suggested that the most likely formation was through aeolian processes and that the earlier phase of stabilisation in the west was a function of greater net sediment accumulation in this (end) part of the system. Richards et al. (2021) convincingly suggest the islands may be formed due to differential drying of the pan floor creating 'sticky' surfaces on which sediment becomes trapped and vegetation can accumulate.

5.4 Contemporary Morphodynamics

Today, the Makgadikgadi remains far from being geomorphically inactive with regular flood inundations, episodes of huge aeolian dust emissions and ongoing tectonic processes. Faulting mapped using SRTM and geomagnetic data, has clearly demonstrated the surface expression of active neotectonic processes in and around the Makgadikgadi, some of which control the flow of rivers into the basin including the Nunga, Lememba and Nata Rivers to the north (Eckardt et al. 2016) and Mmone, Deception and Letlhakeng valleys in the south (Nash and Eckardt 2016). Modification of the highest shoreline levels in the north of the basin in particular (Fig. 5.1b) suggests activity in tectonically controlled fault zones has occurred since the last time significant shorelines were built at the 945 m level confirming the impact of tectonism on the basin geomorphology and groundwater flows remains significant.

To what extent groundwater controls seasonal inundation of the pan today is poorly understood and likely to vary variably across the basin. Under present day conditions, Makgadikgadi can generally be described as a clay-rich discharge pan with shallow groundwater. The high clay content of the sediment facilitates rapid surface ponding under a seasonal rainfall regime. Moderate salt crust accumulations at the surface include halite, thenadite and trona (Eckardt et al 2008; Eckardt 2010). Some of these salts, together with soda-ash (sodium carbonate) are currently

Fig. 5.6 **a** CNES/Airbus Image 2016 (accessible via Google Earth 2018) of 'island' landforms in Ntwetwe pan (location of image shown in Fig. 5.1b); **b** Drone image of 'island' showing concentric ridging within 'island'

mined in the northern part of Sua pan via pumping of groundwater from a depth of ~40 m.

The basin is presently affected by a clear rainfall gradient ranging from 500 mm yr^{-1} in the northeastern part (Sua, Nata) to less than 400 mm yr^{-1} in the west (Rakops and Orapa) (Bryant et al. 2007) with most rainfall being delivered between December and February. The reliability of wet season rains is, however, subject to strong inter-annual and decadal variability linked to the El Nino Southern Oscillation (ENSO) with wetter periods (1971–1985) (Eckardt 2010) associated with enlarged bodies of standing water in the basin (Bryant et al. 2007). Seasonal standing water is additionally derived from a combination of both local inflows and groundwater though the proportional contribution from these inputs is poorly delineated. Sua pan, for example, is host to the most regular inundations with seasonal flows from the Nata and Semowane contributing to up to 2 m of standing water in the north of the basin and the Mosetse river to shallower extents in the central Sua pan (Eckardt 2010). Strontium isotope studies in the north of the basin suggest these surface waters are disconnected from the underlying brines (Eckardt et al. 2008). The implication of this is that inflows from the eastern rivers do not, under present conditions, contribute significantly to subsurface recharge. In the south of Sua pan however, regular seasonal standing water seems to be unrelated to fluvial surface flow and instead has a likely larger contribution from groundwater (Eckardt 2010). Numerous other 'wetspots' where frequent flooding occurs have been identified from MODIS data across the Makgadikgadi basin (Bryant, pers comm) but little is known about the source of the water that these areas seasonally host. Typically, the pan surface within the entire basin dries out between April and November when evaporation greatly exceeds inputs from rainfall and groundwater.

During the dry season, Makgadikgadi provides a significant contribution to atmospheric dust, being ranked within the top 10 of the world's dustiest places (Bryant et al. 2007; Vickery et al. 2013; Washington et al. 2003). Dust production within the pan (Fig. 5.7) is related both to the accumulation of sediment and salts through fluvial and groundwater inputs and to the specific surface conditions including the nature and structure of surface crusts (Nield et al. 2015). Following 'resetting' events at the surface by rainfall, fluvial input or groundwater, there is an observable

Fig. 5.7 Common dust devils on Sua Pan during the dry season carrying aerosols high into the atmosphere (top). On occasion, major dust plumes may extend well beyond the pan margin (bottom photo: Frank D. Eckardt)

development of polygonal ridge thrusting during the dry season as the surface dries out and efflorescence from shallow groundwater drives the development of ridges. Nield et al. (2015) monitored this process on Sua Pan using high resolution Terrestrial Laser Scanner (TLS) and suggested a geomorphic cycle whereby higher, exposed ridges increase evaporation, salt phase change and ridge growth. As thrusting continues however, ridge growth is reduced via negative feedback as ridges collapsed, either because of structural weaknesses or mechanical and chemical breakdown by wind. It is these degraded crusts that expose low density 'fluffy' sediment leading to high rates of dust emission from the surface.

The landscape of the pan has not gone without impact from human processes in recent times. Groundwater pumping by BotAsh mine, which extracts 2400 m^3 of brine/h from a depth of 38 m in north of Sua basin may well be a contributing factor to changing surface conditions, in particular increased grass cover (Eckardt 2010) observed in the area over the last decade. Pumping is not however limited to the Sua region and shallower borehole pumping for settlements, livestock and more recently, in Ntwetwe Pan, for wildlife all have poorly known consequences for surface processes and their expression within the landscape and ecosystem of the basin.

5.5 Conclusions

The Makgadikgadi basin is a complex geomorphic system preserving a dynamic evolutionary history that has taken it from one of African's largest megalake systems to a dusty saltpan with many stable and unstable states in between. The processes that shaped its landforms span a period from the Pliocene to the present day and hold huge importance for Botswana's landscape, climate and ecosystem, not least our own species (Burrough 2016; Chan et al. 2019). At its geomorphic extremes in the past, its shorelines likely held enough water to significantly alter regional climate (Burrough et al. 2009b). Today, that same landscape pumps enough dust into the atmosphere to potentially alter global climate (Schepanski 2018; Vickery et al. 2013). Our knowledge of the processes that shaped the palimpsest of landforms in the Makgadikgadi remains incomplete and often contentious. This endeavour, together with its visibility and significance in the African landscape, has ensured it has long remained a source of fascination to many, both inside and outside the academic realm.

References

Armitage SJ, Bristow CS, Drake NA (2015) West African monsoon dynamics inferred from abrupt fluctuations of Lake Mega-Chad. Proc Natl Acad Sci USA 112:8543–8548. https://doi.org/10.1073/pnas.1417655112

Baillieul TA (1979) The Makgadikgadi pans complex of Botswana. J Sediment Petrol 45:494–503

Bond G, Summers R (1954) A late Stillbay hunting camp on the Nata River Bechuanaland protectorate. S Afr Archaeol Bull 9:89–95

Bryant RG, Bigg GR, Mahowald NM, Eckardt FD, Ross SG (2007) Dust emission response to climate in southern Africa. J Geophys Res. D Atmos 112

Burrough SL (2016) Late Quaternary environmental change and human occupation of the southern African interior. In: Jones S, Stewart BA (eds) Africa from MIS 6–2: population dynamics and paleoenvironments. Springer, Dordrecht, pp 161–174

Burrough SL, Thomas DSG (2008) Late Quaternary lake-level fluctuations in the Mababe depression: middle Kalahari palaeolakes and the role of Zambezi inflows. Quat Res 69:388–403. https://doi.org/10.1016/j.yqres.2008.02.003

Burrough SL, Thomas DSG (2009) Geomorphological contributions to palaeolimnology on the African continent. Geomorphology 103:285–298. https://doi.org/10.1016/j.geomorph.2008.07.015

Burrough SL, Thomas DSG (2013) Central southern Africa at the time of the African humid period: a new analysis of Holocene palaeoenvironmental and palaeoclimate data. Quat Sci Rev 80:29–46. https://doi.org/10.1016/j.quascirev.2013.08.001

Burrough SL, Thomas DSG, Bailey RM (2009a) Mega-lake in the Kalahari: a late Pleistocene record of the palaeolake Makgadikgadi system. Quat Sci Rev 28:1392–1411. https://doi.org/10.1016/j.quascirev.2009.02.007

Burrough SL, Thomas DSG, Singarayer JS (2009b) Late Quaternary hydrological dynamics in the middle Kalahari: forcing and feedbacks. Earth-Sci Rev 96:313–326. https://doi.org/10.1016/j.earscirev.2009.07.001

Burrough SL, Thomas DSG, Bailey RM, Davies L (2012) From landform to process: morphology and formation of lake-bed barchan dunes, Makgadikgadi, Botswana. Geomorphology 161–162:1–14. https://doi.org/10.1016/j.geomorph.2012.03.027

Burrough S, Thomas D, Allin J, Nash D, Mothulatshipi S, Coulsen S, Staurset S (2021) Lessons from a lake bed: unpicking hydrological change and early human landscape use in the southern African interior.

Chan EKF, Timmermann A, Baldi BF, Moore AE, Lyons RJ, Lee SS, Kalsbeek AMF, Petersen DC, Rautenbach H, Förtsch HEA, Bornman MSR, Hayes VM (2019) Human origins in a southern African palaeo-wetland and first migrations. Nature. https://doi.org/10.1038/s41586-019-1714-1

Cooke HJ (1979) The origin of the Makgadikgadi pans. Botsw Notes Rec 11:37–42

Cooke HJ (1980) Landform evolution in the context of climatic change and neo- tectonism in the middle Kalahari of north-central Botswana. Trans Inst Br Geogr 5:80–99

Cooke HJ (1984) The evidence from northern Botswana of late Quaternary climatic change. Late Cainozoic palaeoclimates of the southern hemisphere. In: Proceedings of SASQUA symposium, Swaziland, 1983, Balkema, Rotterdam

Cooke HJ, Verstappen TH (1984) The landforms of the western Makgadikgadi basin in northern Botswana, with a consideration of the chronology of the evolution of Lake Palaeo-Makgadikgadi. Z Geomorphol 28:1–19

Cotterill FPD, De Wit MJ (2011) Geoecodynamics and the Kalahari epeirogeny: linking its genomic record, tree of life and palimpsest into a unified narrative of landscape evolution. S Afr J Geol 114:489–514. https://doi.org/10.2113/gssajg.114.3-4.489

Drake NA, Lem RE, Armitage SJ, Breeze P, Francke J, El-Hawat AS, Salem MJ, Hounslow MW, White K (2018) Reconstructing palaeoclimate and hydrological fluctuations in the Fezzan Basin (southern Libya) since 130 ka: a catchment-based approach. Quat Sci Rev 200:376–394. https://doi.org/10.1016/j.quascirev.2018.09.042

Du Toit AL (1933) Crustal movements as a factor in the evolution of South Africa. S Afr J Sci 24:88–101

Ebert JI (1979) The significance of archaeological sites located near or in association with ancient strandlines of Lake Makgadikgadi, Botswana. Nyame Akuma 15:2–9

Ebert JI, Hitchcock RK (1978) Ancient Lake Makgadikgadi, Botswana: mapping, measurement and palaeoclimatic significance. Palaeoecol Afr 10–11:47–56

Eckardt F (2010) The Makgadikgadi wetland system framework management plan

Eckardt FD, Bryant RG, McCulloch G, Spiro B, Wood WW (2008) The hydrochemistry of a semi-arid pan basin case study: Sua Pan, Makgadikgadi Botswana. Appl Geochem 23:1563–1580. https://doi.org/10.1016/j.apgeochem.2007.12.033

Eckardt FD, Cotterill FPD, Flügel TJ, Kahle B, McFarlane M, Rowe C (2016) Mapping the surface geomorphology of the Makgadikgadi rift zone (MRZ). Quat Int 404:115–120. https://doi.org/10.1016/j.quaint.2015.09.002

Franchi F, MacKay R, Selepeng AT, Barbieri R (2020) Layered mound, inverted channels and polygonal fractures from the Makgadikgadi Pan (Botswana): possible analogues for Martian aqueous morphologies. Planet Space Sci 192. https://doi.org/10.1016/j.pss.2020.105048

Grey DRC, Cooke HJ (1977) Some problems in the quaternary evolution of the landforms of northern Botswana. CATENA 4:123–133

Grove AT (1969) Landforms and climatic change in the Kalahari and Ngamiland. Geogr J 135:191–212

Haddon IG, McCarthy TS (2005) The Mesozoic-Cenozoic interior sag basins of central Africa: the Late-Cretaceous-Cenozoic Kalahari and Okavango basins. J Afr Earth Sci 43:316–333

Helgren DM (1984) Historical Geomorphology and Geoarchaeology in the southwestern Makgadikgadi basin Botswana. Ann Assoc Am Geogr 74:298–307. https://doi.org/10.1111/j.1467-8306.1984.tb01454.x

Livingstone D (1858) Missionary travels and researches in South Africa. Harper Bros, New York

Mazvimavi D, Wolski P (2006) Long-term variations of annual flows of the Okavango and Zambezi Rivers. Phys Chem Earth 31:944–951

McFarlane MJ, Eckardt FD (2006) Lake deception: a new Makgadikgadi palaeolake. Botsw Notes Rec 38:195–201

McFarlane MJ, Long CW (2015) Pan floor "barchan" mounds, Ntwetwe Pan, Makgadikgadi, Botswana: their origin and palaeoclimatic implications. Quat Int 372:108–119. https://doi.org/10.1016/j.quaint.2014.10.008

McFarlane MJ, Segadika P (2001) Archaeological evidence for the reassessment of the ages of the Makgadikgadi palaeolakes. Botsw Notes Rec 33:83–92

Moernaut J, Verschuren D, Charlet F, Kristen I, Fagot M, De Batist M (2010) The seismic-stratigraphic record of lake-level fluctuations in Lake Challa: hydrological stability and change in equatorial East Africa over the last 140 kyr. Earth Planet Sci Lett 290:214–223. https://doi.org/10.1016/j.epsl.2009.12.023

Moore AE (1999) A reappraisal of epeirogenic flexure axes in southern Africa. S Afr J Geol 102:363–376

Moore AE, Larkin PA (2001) Drainage evolution in south-central Africa since the breakup of Gondwana. S Afr J Geol 104:47–68

Moore AE, Cotterill FPD, Main MPL, Williams HB (2008) The Zambezi River. In: Large rivers: geomorphology and management, pp 311–332. https://doi.org/10.1002/9780470723722.ch15

Moore AE, Cotterill FPD, Eckardt FD (2012) The evolution and ages of Makgadikgadi palaeo-lakes: consilient evidence from Kalahari drainage evolution south-central Africa. S Afr J Geol 115:385–413. https://doi.org/10.2113/gssajg.115.3.385

Nash DJ, Eckardt FD (2016) Drainage development, neotectonics and base-level change in the Kalahari Desert, southern Africa. S Afr Geogr J 98:308–320. https://doi.org/10.1080/03736245.2015.1028987

Nield JM, Bryant RG, Wiggs GFSS, King J, Thomas DSGG, Eckardt FD, Washington R (2015) The dynamism of salt crust patterns on playas. Geology 43:31–34. https://doi.org/10.1130/G36175.1

Partridge TC, Maud RR (2000) The Cenozoic of southern Africa. In: Oxford monographs on geology and geophysics, vol 40

Passarge S (1904) Die Kalahari. Dietrich Riemer, Berlin

Podgorski JE, Green AG, Kgotlhang L, Kinzelbach WKH, Kalscheuer T, Auken E, Ngwisanyi T (2013) Paleo-megalake and paleo-megafan in southern Africa. Geology 41:1155–1158

Richards J, Burrough S, Wiggs G, Hills T, Thomas D, Moseki M (2021) Uneven surface moisture as a driver of dune formation on ephemeral lake beds under conditions similar to the present day: a model-based assessment from the Makgadikgadi basin, northern Botswana. Earth Surf Process Landf 1–18. https://doi.org/10.1002/esp.5215

Riedel F, Henderson ACG, Heußner KU, Kaufmann G, Kossler A, Leipe C, Shemang E, Taft L (2014) Dynamics of a Kalahari long-lived mega-lake system: hydromorphological and limnological changes in the Makgadikgadi basin (Botswana) during the terminal 50 ka. Hydrobiologia 739:25–53. https://doi.org/10.1007/s10750-013-1647-x

Ringrose S, Cassidy L, Diskin S, Coetzee S, Matheson W, Mackay AW, Harris C (2014) Diagenetic transformations and silcrete-calcrete intergrade duricrust formation in palaeo-estuary sediments. Earth Surf. Process. Landforms 39:1167–1187. https://doi.org/10.1002/esp.3516

Robbins LH (1987) Stone age archaeology in the Northern Kalahari, Botswana: Savuti and Kudiakam Pan. Curr Anthropol 28:567–569

Schepanski K (2018) Transport of mineral dust and its impact on climate. Geosciences 8. https://doi.org/10.3390/geosciences8050151

Schmidt M, Fuchs M, Henderson ACG, Kossler A, Leng MJ, Mackay AW, Shemang E, Riedel F (2017) Paleolimnological features of a mega-lake phase in the Makgadikgadi basin (Kalahari, Botswana) during marine isotope stage 5 inferred from diatoms. J Paleolimnol 58:373–390. https://doi.org/10.1007/s10933-017-9984-9

Scholz CA, Cohen AS, Johnson TC, King J, Talbot MR, Brown ET (2011) Scientific drilling in the Great Rift Valley: the 2005 Lake Malawi scientific drilling project - an overview of the past 145,000 years of climate variability in southern hemisphere East Africa.

Palaeogeogr Palaeoclim Palaeoecol 303:3–19. https://doi.org/10.1016/j.palaeo.2010.10.030
Shaw P (1988) After the flood: the fluvio-lacustrine landforms of northern Botswana. Earth Sci Rev 25:449–456. https://doi.org/10.1016/0012-8252(88)90011-6
Shaw PA, Cooke HJ (1986) Geomorphic evidence for the late Quaternary paleoclimates of the middle Kalahari of northern Botswana. CATENA 13:349–359
Shaw PA, Thomas DSG (1988) Lake Caprivi: a late Quaternary link between the Zambezi and middle Kalahari drainage systems. Z Geomorphol 32:329–337
Shaw PA, Thomas DSG (1996) The Quaternary palaeoenvironmental history of the Kalahari, southern Africa. J Arid Environ
Shaw PA, Thomas DSG, Nash DJ (1992) Late Quaternary fluvial activity in the dry valleys (mekgacha) of the middle and southern Kalahari, southern Africa. J Quat Sci 7:273–281
Shaw PA, Stokes S, Thomas DSG, Davies FBM, Holmgren K (1997) Palaeoecology and age of a Quaternary high lake level in the Makgadikgadi basin of the middle Kalahari, Botswana. S Afr J Sci
Singarayer JS, Burrough SL (2015) Interhemispheric dynamics of the African rainbelt during the late Quaternary. Quat Sci Rev 124:48–67. https://doi.org/10.1016/j.quascirev.2015.06.021
SMEC (Snowy Mountains Engineering Corporation) (1987) Final report, technical study, southern Okavango integrated water development, phase 1. Government of Botswana Ministry of Mineral Resources and Water Affairs, Department of Water Affairs
Smith RA (1984) Lithostratigraphy of the Karoo stratigraphy in Botswana. Botsw Geol Surv Bull 26:34
Thomas DSG, Shaw PA (1992) The Zambezi River: tectonism, climatic change and drainage evolution: is there really evidence for a catastrophic flood? a discussion. Palaeogeogr Palaeoclim Palaeoecol 91:175–178
Thompson TA, Baedke SJ (1996) Beach-ridge development in Lake Michigan: shoreline behavior in response to quasi-periodic lake-level events. Mar Geol 129:163–174
Uda T, Serizawa M, Miyahara S (2018) Formation of cuspate foreland. In: Uda T, Serizawa M, Miyahara S (eds) Morphodynamic model for predicting beach changes based on Bagnold's concept and its applications. Intech, London, pp 188–177. https://doi.org/10.5772/intechopen.81417
UNDP/FAO (1977) Investigation of the Okavango Delta as a primary water resource for Botswana, DP/Bot/71/506 Technical report. 3 vols, Gaborone
Vickery KJ, Eckardt FD, Bryant RG (2013) A sub-basin scale dust plume source frequency inventory for southern Africa, 2005–2008. Geophys Res Lett 40:5274–5279. https://doi.org/10.1002/grl.50968
Washington R, Todd M, Middleton NJ, Goudie AS (2003) Dust-storm source areas determined by the total ozone monitoring spectrometer and surface observations. Ann Assoc Am Geogr
White KH, Eckardt F (2006) Geochemical mapping of carbonate sediments in the Makgadikgadi basin, Botswana using moderate resolution remote sensing data. Earth Surf Process Landf 31(6)
Zenkovitch VP (1967) Processes of coastal development. Wiley, New York

Sallie L. Burrough is Trapnell Fellow of African Environments at the University of Oxford in the United Kingdom. Originally a Geographer, she has an MSc in Quaternary Science from the University of London and a DPhil in African Hydroclimate Change from the University of Oxford. A geochronologist by training, her research interests include long-term hydro-climate variability, landscape dynamics, palaeoecological change and ancient human occupation of African dryland regions. She has carried out research in Botswana for more than 15 years working with colleagues from the National Museum, the Okavango Research Institute and the University of Botswana.

6 Landscapes and Landforms of the Chobe Enclave, Northern Botswana

Thuto Mokatse, Nathalie Diaz, Elisha Shemang, John Van Thuyne, Pascal Vittoz, Torsten Vennemann, and Eric P. Verrecchia

Abstract

The northern part of the Chobe Enclave (an administrative district of northern Botswana) is an agricultural area situated between relatively pristine national parks situated in the Middle Kalahari Basin. It belongs to the Linyanti-Chobe structural basin and constitutes a syntectonic depocenter formed within a large structural depression, known as the Okavango Graben, a tectonic structure of a likely trans-tensional nature. The landscape includes fossil landforms, such as sand dunes, pans, sand ridges, and carbonate islands resulting from palaeo-environmental and palaeo-drainage changes through the Quaternary and associated to (neo)tectonic processes. In addition to river- and wind-reworked Kalahari sands, the sediments include diatomites and carbonate deposits, forming inverted reliefs and originating from palustrine palaeo-environments. The Linyanti-Chobe basin is at the convergence of several ecoregions from tropical and subtropical grasslands to savannas and shrubland biomes. The hydrological cycle in the northern Chobe Enclave is governed by a complex interplay between the Okavango, Kwando, and Upper Zambezi drainage basins, which originate from tropical watersheds of the Angolan highlands. Finally, the widespread development of termite mounds impacts the diversity of soils and sediments of the northern Chobe Enclave, which is also reflected in the vegetation.

Keywords

Trans-tensional basin • Palustrine and floodplain environments • Calcrete • Savana vegetation and soils • Termites • Linyanti-Chobe basin • Environmental change • Quaternary

T. Mokatse · J. Van Thuyne · P. Vittoz · T. Vennemann · E. P. Verrecchia (✉)
Institute of Earth Surface Dynamics, FGSE, University of Lausanne, Lausanne, Switzerland
e-mail: eric.verrecchia@unil.ch

N. Diaz
Service Culture et Médiation scientifique (SCMS), University of Lausanne, Lausanne, Switzerland

E. Shemang
Earth and Environmental Sciences, Botswana International University of Science and Technology, Botswana, South Africa

J. Van Thuyne
Van Thuyne Ridge Research Centre, Satau, Botswana, South Africa

6.1 Introduction

6.1.1 Where Is the Chobe Enclave and Why Is It so Interesting?

Landscape evolution in the Chobe-Linyanti basin is closely associated with active tectonics and surface processes. Sedimentary basins forming under extensional tectonic regimes are important records of geological history, among other types of depositional environments. The architectures of these basins and the basin-fill are influenced by the rheological structure of the lithosphere, the availability of crustal discontinuities that can be tensionally reactivated, the mode and amount of extension, and the lithological composition of pre- and syn-rift sediments (Ziegler and Cloetingh 2004). An important factor in controlling geomorphic processes during landscape evolution can be the topography of a region affected by recent and/or active processes such as faulting (Mayer 1986; Cox 1994; Bishop 2007; Giaconia et al. 2012a, 2012b). Therefore, it is crucial to report the underlying geology in order to understand the way landforms evolve.

The study site, as a part of the Middle Kalahari Basin of northern Botswana, is situated in the northern Chobe Enclave, an administrative district, close to the Chobe-Linyanti depocenter (Fig. 6.1). It is bounded by the Linyanti

F. D. Eckardt (ed.), *Landscapes and Landforms of Botswana*, World Geomorphological Landscapes,
https://doi.org/10.1007/978-3-030-86102-5_6

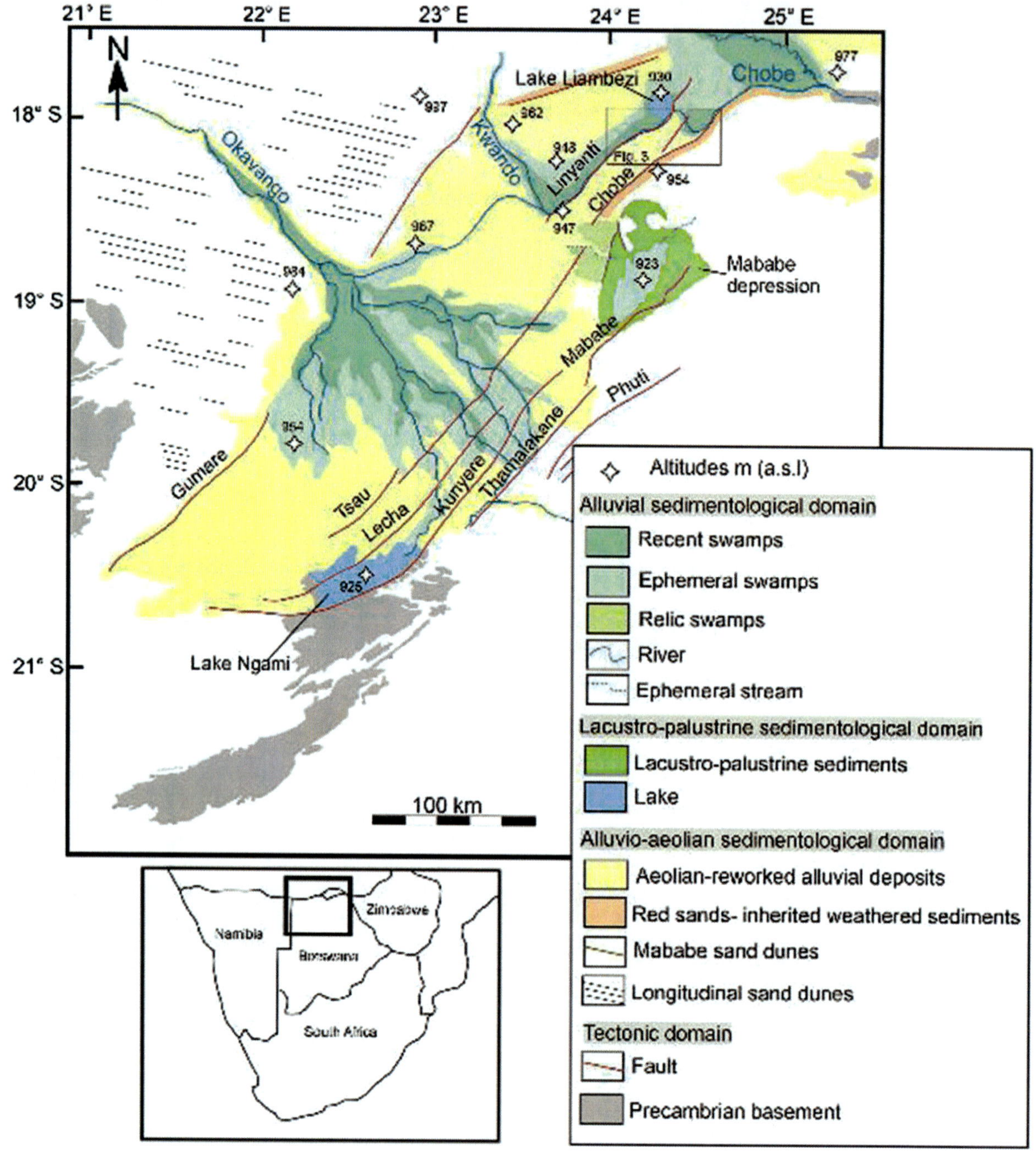

Fig. 6.1 Map of the sedimentary domains, Okavango region

river and Lake Liambezi in the north, bordering the eastern part of the Caprivi Strip, and the Savuti and Chobe rivers in the west and south, respectively. It is an agricultural area between relatively pristine national parks covered mostly by the Kalahari sands with unexpected occurrences of carbonate beds (Diaz et al. 2019). Equally so, the various types of sedimentary deposits (Fig. 6.1), including diatomites, fluvial, and aeolian sands, span the site. Palaeo-shorelines within the Middle Kalahari have been investigated as Quaternary landform relicts, which formed under different hydrological conditions (e.g., Burrough and Thomas 2008; Burrough et al. 2009a, 2009b; Shaw et al. 1988). Some investigations focused on understanding geomorphological features, such as sand ridges associated with lacustrine systems, namely, the Mababe Depression, Lake Ngami, and the Makgadikgadi Pan (Grey and Cooke 1977; Grove 1969; Shaw 1985). Understanding the timing of formation of these sand ridges has been helped by thermoluminescence (TL) and optically stimulated luminescence (OSL) dating techniques (e.g., Cooke and Verstappen 1984; Shaw 1985; Burrough et al. 2007; Burrough and Thomas 2008; Ringrose et al. 2008; Burrough et al. 2009a, 2009b). Burrough and Thomas (2008) reported OSL ages of sand ridges from the Mababe Depression and Lake Ngami consistent in ages with sand ridges in the northern Chobe Enclave. However, given the nature of emplacement of these sand ridges and associated sedimentary deposits, such as carbonate beds (Diaz et al. 2019), and the complexity of interpreting the emplacement of geomorphological features, some questions remain unanswered, such as:

- What is the origin of the sediments forming the sandy landforms?
- What is the origin of the carbonate deposits?
- What depositional environments and landscapes are recognized?
- What could be the main processes at work in this landscape?

The area is also characterized by a large diversity of soils and vegetation (Romanens et al. 2019; Vittoz et al. 2020). The region covered in this study is unique as it mirrors some aspects of the inland alluvial fans of the Okavango and Kwando rivers, as geologic and tectonic processes extend into this arid to semiarid region, hence establishing a regional-scale and even local-scale zoning of the same large-scale vegetation zones adjacent to the subtropical swamps/flooded areas. The Linyanti-Chobe basin, in particular, is at the convergence of several ecoregions that are all part of the "tropical and subtropical grasslands, savannas and shrubland" biome (www.worldwildlife.org/biomes/tropical-and-subtropical-grasslands-savannas-and-shrublands). It includes the Zambezian flooded grasslands, the Zambezian and mopane woodlands, the Kalahari *Acacia-Baikiaea* woodlands, and the Kalahari xeric savannah. The Linyanti-Chobe basin flora thus consists of a mix of species from these contrasting, neighboring ecosystems. Finally, the hydrological cycle in the Chobe Enclave is governed by a complex interplay between the Okavango, Kwando, and Upper Zambezi drainage basins. In addition to local rainfall, it receives waters from the Okavango, Kwando-Linyanti, and Zambezi-Chobe River systems, which drain tropical watersheds in the Angolan highlands.

6.1.2 Topography and Climate of the Chobe Enclave

Tectonics influenced both the topography and hydrologic conditions of the study region, possibly altering palustrine/shallow lacustrine and sedimentary budgets. The Middle Kalahari is a semiarid environment with a record of wetter climate in the past (Cooke 1975; Shaw and Cooke 1986), and depositional environments that preserve a complex history of Quaternary environmental change (Thomas and Shaw 1991, 2002). Today, northern Botswana is classified as a hot semiarid (steppe) climate (BSh) (Peel et al. 2007). This type of climate is characterized by a mean annual temperature above 22 °C, low mean annual precipitation of which 90% falls between October and April (Skarpe et al. 2014). The northern Chobe Enclave district (average precipitation of 650 mm; Jones 2002) has a wet season from November to March with average minimum and maximum temperatures of 20 °C and 32 °C, respectively, and a dry season from April to October with average minimum and maximum temperatures of 13 °C and 30 °C, respectively.

6.1.3 Hydrological Setting of the Chobe-Linyanti Basin

As noted above, this region is bounded by the Linyanti river and Lake Liambezi in the north and the Savuti and Chobe rivers in the west and south, respectively. These bounding rivers are part of a complex drainage system that includes all three of the regionally important rivers of the Okavango, the Kwando, and the Zambezi (McCarthy et al. 2000).

The headwaters of the catchments to the Okavango, Kwando, and Zambezi are characterized by a humid, subtropical climate with a mean annual precipitation of up to 1300 mm, decreasing down to about 570 mm in the region of the Chobe Enclave district. The rainy season in the headwater region, as well as the study region, is commonly between November and March (Mizlow et al. 2009) and the intensity depends critically on fluctuations in the positioning of the Intertropical Convergence Zone (e.g., McCarthy et al. 2000). The discharge of the three river systems, along their course which is individually >1800 km, provides a maximum flooding to their respective distal wetlands, towards the south, with a delay of about 6 months. The Okavango's and Kwando's maximum floods occur from June to August (e.g., Pricope 2013), while the Zambezi has an earlier annual flood pulse that may spill over into the Linyanti-Chobe wetlands through the Bukalo Channel, as well as the Chobe River. The flooding in the Linyanti-Chobe basin thus peaks between March and May, a few weeks after the Zambezi reaches its maximum discharge (Tweddle and Hay 2011; Pricope 2013). The Okavango waters penetrate the Linyanti-Chobe basin only during exceptional floods via the Selinda spillway. The relative exchange from the Zambezi and/or the Okavango/Kwando into the Linyanti-Chobe area may well have been more important during the past humid periods, and the existence of a larger palaeo-lake during the Quaternary is postulated (Burrough and Thomas 2008), as supported by inherited geomorphological features in the landscape.

The complexity of the drainage system in the Chobe Enclave is related to uplift along the Okavango-Kalahari-Zimbabwe axis and extension of the East African Rift system during the Cretaceous (Moore and Larkin 2001; Tooth et al. this book). The Kwando River has been deviated from the Zambezi due to faulting along the Linyanti fault, creating the endoreic systems and resultant wetlands of the Linyanti river and Lake Liambezi. The southern bounding fault of the Chobe, in turn, separated the Kwando from the Chobe River, together with movements along the Linyanti fault, also limiting the Linyanti overflow into the Chobe River. Instead, the Chobe River and its wetlands may now receive backflow waters from the Zambezi (Moore and Larkin 2001; Tooth et al. this book). However, depending on the relative amounts of precipitation in the Okavango, the Kwando, and finally also the Zambezi catchments, additional water in the Linyanti-Liambezi basin may enter via the Selinda spillway from the Okavango drainage basin and/or via the Bukalo Channel from the Zambezi drainage basin (McCarthy et al. 2000). Depending on the water levels in the Linyanti and Lake Liambezi drainage system, the Chobe River serves as the major surface outflow (Seaman et al. 1978; Burrough and Thomas 2008; Kurugundla et al. 2010; Peel et al. 2015) and finally returns the waters through a confluence with the Zambezi River further south.

6.2 Geological Settings

6.2.1 Structural Context

The northern Chobe Enclave district forms part of the Linyanti-Chobe sub-basin, bounded by the Linyanti and Chobe faults (Fig. 6.1). The Linyanti-Chobe sub-basin together with the Ngami and Mababe basins are the three characteristic syntectonic depocenters (Kinabo et al. 2007) within the large structural depression known as the Okavango Graben. This structural depression is made up of a series of NE-SW trending normal to dextral strike-slip faults (Modisi et al. 2000; Campbell et al. 2006; Bufford et al. 2012). The bounding faults of the Okavango Graben in the southern part form an *en-echelon* pattern with a direction following the strike of the Precambrian basement structures (Mallick et al. 1981; Moorkamp et al. 2019). The Okavango Graben has been associated with the formation of the southwestern branch of the East African Rift system (Modisi et al. 2000; Alvarez and Hogan 2013); however, Pastier et al. (2017) argued that the tectonic structure of the Okavango Graben better fits a trans-tensional basin model. Trans-tensional or pull-apart basins are topographic depressions that form at releasing bends or steps in a basement strike-slip fault system, with basin margin characterized by the developments of *en-echelon* oblique—extensional faults that soft- or hard-link with increasing displacement in the principal displacement zones (Wu et al. 2009). Kinabo et al. (2008) coupled analysis of Shuttle Radar Topography Mission (SRTM), Digital Elevation Model (DEM), and aeromagnetic data in the Chobe region and revealed the development of soft linkage on segments of the Linyanti fault and evidence of a hard linkage forming between two *en-echelon* right-stepping segments of the Chobe fault.

6.2.2 Bedrock Geology

The Ghanzi-Chobe basin in northwestern Botswana consists of a linear belt of volcano-sedimentary sequences that were deposited during Meso-Neoproterozoic times following extension tectonics associated with the Namaqua orogeny (Modisi et al. 2000). This basin was subsequently deformed as part of the continent-wide tectonic event, the Pan-African Damara orogeny, resulting in the inversion of the Proterozoic volcano-sedimentary basins and formation of a fold belt along the northern margin of the Kalahari Craton (Modie 1996). Sediments of the Mesozoic Karoo Supergroup then became part of this fault-bounded graben system (Bordy et al. 2010). The giant Okavango dyke swarm cut across the basin during this time (Le Gall et al. 2002). Significantly, this sequence in northwestern Botswana was further affected by the reactivation of ancient faults in Pan-African belts, resulting in the Proterozoic NE-SW strike (Pastier et al. 2017). Thus, the Ghanzi-Chobe basin in northwestern Botswana preserves an important record of subsequent tectonic and depositional events that brought together two prominent southern African cratons—the Kalahari and Congo cratons.

6.2.3 Surficial Sedimentary Geology

The Kalahari sands mainly dominate the sediments observed at the ground surface of the area. They outcrop as the upper part of a sedimentary body of more than 200 m in thickness

(Thomas and Shaw 1991). Two other sedimentary units, not documented in the region, have also been identified: continental carbonate landmasses (described as "islands" in this chapter) and diatomites.

First, X-ray diffraction (XRD) analyses confirm the almost pure quartz nature of the sands in the region. They are usually extremely well sorted and are present in all types of sediments and soils, whatever their nature (Fig. 6.2). They obviously belong to the large body of the Kalahari sands (Table 6.1). They are also found associated with finer fractions, likely originating from (i) a fine loess fraction and/or dust (Crouvi et al. 2010), (ii) diatoms, or (iii) pedogenic redistribution and neoformation, the latter including the role of termites (see below; Jouquet and Lepage 2002; Jouquet et al. 2011). Their ubiquity, and their mostly aeolian and fluvial reworked nature at the surface of the landscape (Thomas and Shaw 1991, 2002), make these sands suitable for OSL dating (see below).

Second, carbonate formations in the region appear as limited landmasses identified as "islands" in the sand sea (Fig. 6.3). They have variable thicknesses but are frequently thicker than 1 m (Figs. 6.4 and 6.5e–g). In the literature, these carbonate layers are usually defined as "calcrete". Although this term refers to a large variety of processes explaining their formation (and consequently palaeo-environments; Wright, 2007), their genesis has been attributed in Botswana to a combination of pedogenic, groundwater, and geochemical-diagenetic processes (Nash et al. 1994; Nash and Shaw 1998; Ringrose et al. 2002, 2014). The beds, hardened to various degrees, include all the features observed in palustrine limestones at the macroscale (e.g., traces of roots, burrows and/or galleries, shells,

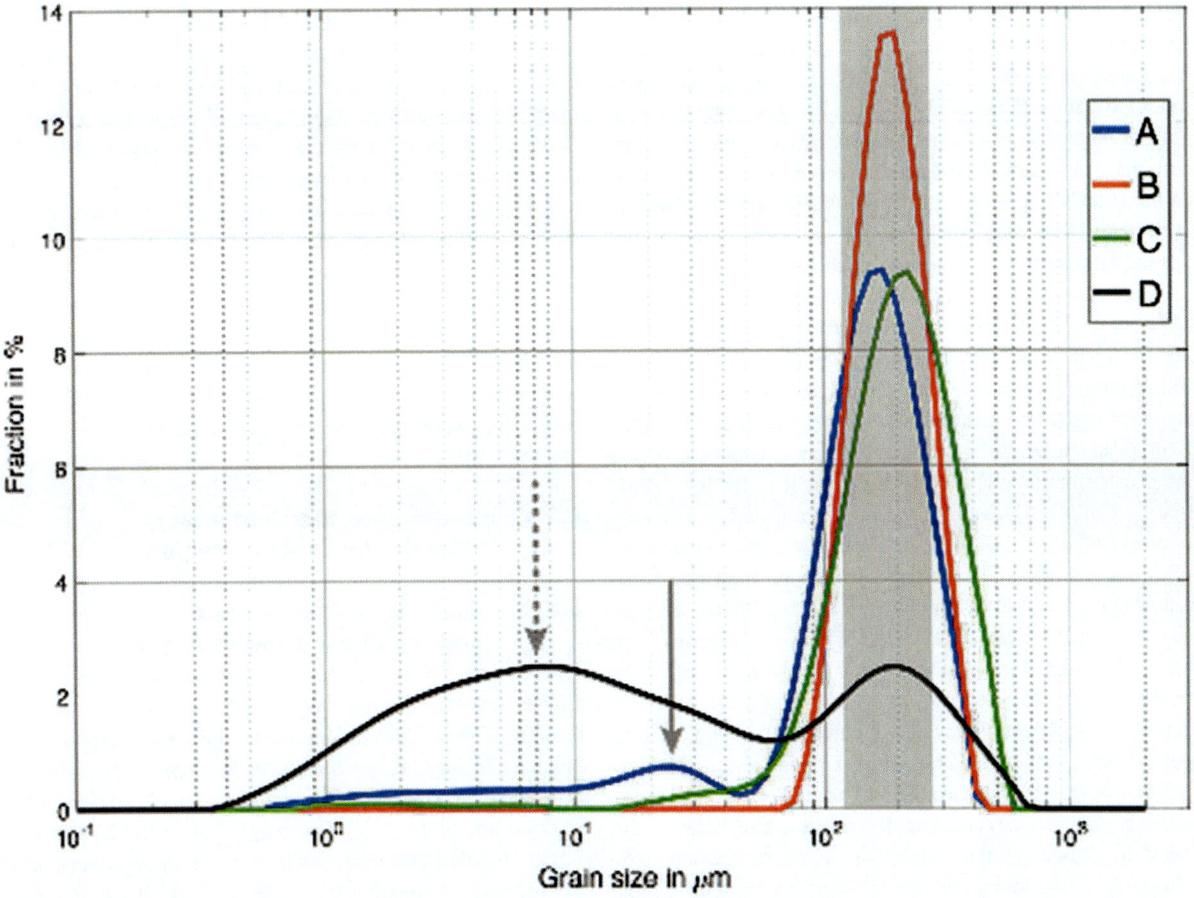

Fig. 6.2 Four examples of grain-size distributions from the Chobe Enclave: the Kalahari sand mode is always displayed (shaded area centered on 200 µm). In addition, two modes of fine particles are detected and likely correspond to fine peri-desertic loess (solid arrow at 30 µm) and/or dust/very fine silt due to termite activity and/or diatomite (dashed arrow at 8 µm; see Crouvi et al. 2010 and Sect. 4.3). **A** Sandy and homogeneous layer (5.40 m deep), Baobab pit (18°06′10″ S, 24°18′30″ E, 935 m asl). **B** Recent fluvial sand from the Linyanti (18°04′30″ S, 24°08′43″ E, 937 m asl). **C** Recent aeolian sand, east plateau of the Linyanti river (18°10′16″ S, 24°14′54″ E, 936 m asl). **D** Baobab pit, upper layer (soil)

Table 6.1 Average grain sizes of the Kalahari sands from various locations

Locations	Average sizes (μm)	Authors
Kalahari dune field	125–220	Cook (1980)
Namib sand sea	200	Lancaster (1981)
South West Kalahari	190–240	Lancaster (1986)
Kalahari dune crest	175	Watson (1986)
Northern Botswana	200	Thomas and Shaw (1991)
Central Kalahari (Ghanzi)	170–225	Wang et al. (2007)
Central Kalahari (Tshane)	210–230	Wang et al. (2007)
Namib sand sea	170–340	Crouvi et al. (2010)

paleosol imprints) and at the microscale (e.g., desiccation features, redistributions in the micromass and nodulization, multiple vadose and phreatic phases in cements; see Freytet and Verrecchia 2002; Verrecchia 2007). Consequently, although designated as "calcretes" in the literature (sensu Wright and Tucker 1991), the carbonate beds from the northern Chobe Enclave district could also be described as palustrine limestones. Their position as low relief features in the landscape are due to an inverted relief, emphasizing their inherited nature (Diaz et al. 2019). This raises the question "Is the palustrine origin of these limestones likely?". Fine carbonate deposits have been observed in the Okavango Delta (McCarthy et al. 2012) in island-mounds forming within the Delta. However, these tree-covered islands do not accumulate significant amounts of calcium carbonate, only traces as subsurface crystals. Their formation is related to the chemistry at the interfaces between fresh groundwater flow and trapped saline groundwater through tree uptake and transpiration (McCarthy et al. 2012), with dust contributing substantially to the material found on the surface of islands (Humphries et al. 2014). If such a process cannot generate large calcium carbonate accumulations, it is for the simple reason that the Okavango waters are presently extremely poor in Ca^{2+}, i.e., 3–7 mg/L on average, with a conductivity of 33–40 μS cm^{-1} (Dr. Mogobe, Okavango Research Institute, Maun; personal communication, spring 2016). But one can imagine a palaeo-river system significantly enriched in Ca^{2+}, in which the same processes at work during large floods could have led to some of the fossil carbonate deposits outcropping today. They could correspond to large ponds and marshes and would constitute precious palaeo-climate and environmental indicators (Diaz et al. 2019; see below) to reconstruct the relationships between the Zambezi-Chobe system and the southern palaeo-lakes observed by Burrough and Thomas (2008). Presently, most of these outcropping carbonate deposits are karstified and eroded under the present-day climate and weathering conditions.

Third, diatomites are easily recognized in the sediments: they form a fine, white material, with a very low density and are sometimes so soft that they can be crushed to a powder in the field. X-ray fluorescence (XRF) analyses confirmed their pure siliceous nature (SiO_2). The presence of quartz, probably of aeolian origin, is also observed using scanning electron microscope (SEM). However, in the study region, diatomites can also be associated to calcite as vein, crack, fracture, and tubular infillings (Fig. 6.6). Such pure diatomites can only form in ponds or shallow lakes, in a range of water salinities, from freshwater to slightly brackish. They must correspond to periods during which wetter conditions prevailed. Their presence can correspond to periods similar to what has been suggested by Burrough and Thomas (2008), i.e., "that increased flow in the Chobe and Zambezi system significantly contributed to the Middle Kalahari lake phases" for the palaeo-lakes observed in the south and east of the Chobe. It can be hypothesized that their alternation with the carbonate deposits indicates that the Chobe underwent fluctuations in the hydrochemistry of water sources, as well as in climatic conditions. But this theory needs to be properly assessed and documented by further studies.

6.2.4 Insights into the Quaternary in the Chobe-Lynianti Region

6.2.4.1 Quaternary Palaeo-Lakes and the Chobe-Lynianti Region

The endorheic nature of the basin has led to aeolian, fluvial, and/or lacustrine sedimentary accumulations since the Cretaceous (Grove 1969). Regarding the Quaternary palaeodrainage history, it has mostly been reconstructed through geological and geomorphological archive studies (e.g., Thomas and Shaw 2002; Burrough et al. 2007; Moore et al. 2012). Fossil landforms, such as sand dunes, pans, sand ridges, and carbonate islands (Diaz et al. 2019; Fig. 6.3), testify to changes in hydrological conditions. However, the present-day semiarid climate leads to the deflation of sedimentary archival deposits and does not favor the preservation of organic proxies (Thomas and Burrough 2012). Regional palaeo-environmental records are thus fragmentary

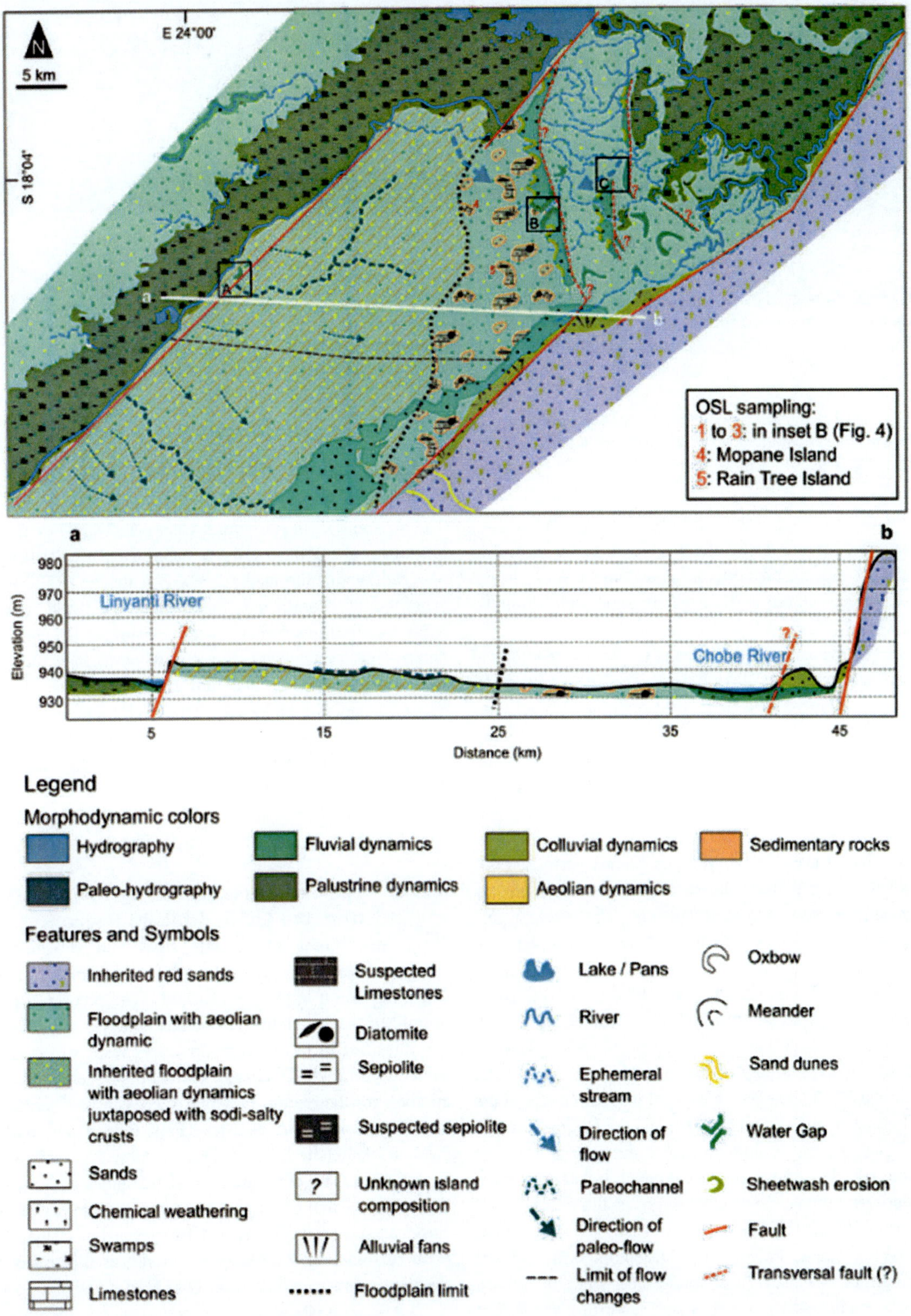

Fig. 6.3 Geomorphological map from the Chobe Enclave and elevation profile through the Chobe Enclave

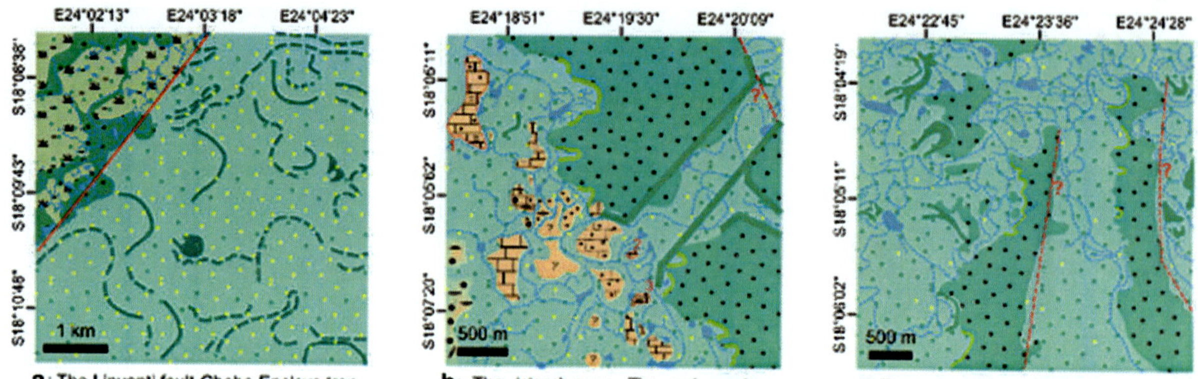

a: The Linyanti fault-Chobe Enclave transition.

b: The island area. The red numbers represent the sites sampled for OSL dating: 1. Baobab Island, 2. Nata Island, 3. Acacia Island.

c: The eastern floodplain area.

Fig. 6.4 Detailed geomorphological maps in three key areas from west to east (see Fig. 6.3 for location). **a** Palaeo-floodplain, **b** the island area, **c** the sand ridge area. The red numbers correspond to the sites where samples were taken for OSL measurements. Red question marks refer to potential transversal faults. See legend in Fig. 6.3

and difficult to assess (e.g., Cordova et al. 2017), making interpretations sometimes invalidated, particularly regarding influences of climate change and/or tectonic events on the palaeo-drainage (Moore et al. 2012; Bäumle and Himmelsbach 2018).

In the Okavango region, extensive geomorphological evidence for large palaeo-lake exist, referring to wetter climate periods (Grove 1969). Most previous research has focused on palaeo-lakes in this region, notably Lake Ngami (Huntsman-Mapila et al. 2006), Lake Mababe (Gamrod 2009), and Lake Makgadikgadi (Thomas and Shaw 1991). Burrough and Thomas (2008) proposed that these palaeo-lakes might have coalesced into a single large Mega-Kalahari Lake during previous humid periods. OSL dating of the main palaeo-lake shorelines (beach ridges) in the Okavango region suggested that high lake levels occurred during the Holocene and Late Pleistocene at 8.5 ± 0.2 ka, 17.1 ± 1.6 ka, 26.8 ± 1.2 ka, 38.7 ± 1.8 ka, 64.2 ± 2.0 ka, 92.2 ± 1.5 ka, and 104.6 ± 3.1 ka (Burrough and Thomas 2008; Burrough et al. 2009a, 2009b).

However, it seems that tectonic processes also played a critical role in the drainage evolution. Moore et al. (2012) reviewed the Plio-Pleistocene history of the main rivers in the region belonging to the Congo Basin and the Upper Zambezi Basin. They proposed a model in which the general drainage conditions resulted in smaller lakes, whereas larger lakes were controlled by climate feedbacks, such as the rainfall and evaporation balance. The authors recognized different palaeo-lake systems: Palaeo-Lake Deception with a fossil sand ridge at ~995 m (McFarlane and Eckardt 2008), Palaeo-Lake Makgadikgadi with fossil sand ridges at 945 m (Thomas and Shaw 1991; Burrough and Thomas 2008), and Palaeo-Lake Thamalakane with fossil sand ridges at 936 m, 920 m, and 912 m (Grey and Cooke 1977). They proposed one mega-lake (Palaeo-Lake Deception and after Palaeo-Lake Makgadikgadi) during the Early Pleistocene (<500 ka) that progressively contracted, rather than oscillated, due to a reduction of inflow from the tributaries of the Upper Zambezi palaeo-river system that were cut off by tectonic subsidence and lifting, inducing headward erosion and river capturing (Bäumle and Himmelsbach 2018). The decrease of water inputs from the Boteti River controlled by the fault system at the foot of the Okavango Delta caused the Palaeo-Lake Thamalakane to shrink, indicated by a shoreline decreasing from 920 to 912 m between 300 and 100 ka. The 912 m lake would have desiccated in the last 100 ka (Moore et al. 2012).

6.2.4.2 The Carbonate Islands from the Chobe-Lynianti Region

Carbonate islands of various sizes and surrounded by ephemeral waterbodies are a common feature of the landscape (Figs. 6.3 and 6.4; Diaz et al. 2019). The precise origin and timing of such landforms remain unclear, but they are hypothesized to have been formed in response to regional palaeo-hydrological changes, making them interesting archives for the region. Different successions of sedimentary beds are exposed in quarries (Figs. 6.5e–g and 6.6). The Baobab Island (site 1, Fig. 6.4) is one of these quarries opened in an island. The other studied islands (sites 2, 3, 4, and 5, Figs. 6.3, 6.4 and 6.7) refer to outcrops; the sands below the carbonate layers are described from pits dug in the alluvial plain surrounding the carbonate island, as illustrated in the supplementary materials from Diaz et al. (2019).

The Baobab Island quarry is 5 m deep and the only place where the Kalahari sands can be reached directly below the carbonate beds. These sands have a thickness of 1.30 m

Fig. 6.5 Different landscapes and landforms from the Chobe Enclave. **a** View from the Chobe fault near Katchikawe on the Chobe backflow, **b** view from the Linyanti swamps toward the Linyanti fault during the dry season, **c** view on the western side of the N-S sand ridge, slightly elevated with tree vegetation and surrounded by ephemeral streams covered with grass, **d** Sandy island covered with trees and surrounded by ephemeral streams covered with grass, **e** view of a carbonate island (Diaz et al. 2019) slightly elevated, covered with trees and surrounded by ephemeral streams covered with grass, **f** Baobab quarry view with exposure of the carbonate beds (see location on Fig. 6.4b), **g** Carbonate layer from another quarry, Acacia pit (see location on Fig. 6.4b), where a diatomitic layer was also observed (Fig. 6.6), **h** view on the eastern floodplain with small N-S sand ridges covered by trees and surrounded by ephemeral streams covered with grass during the dry season

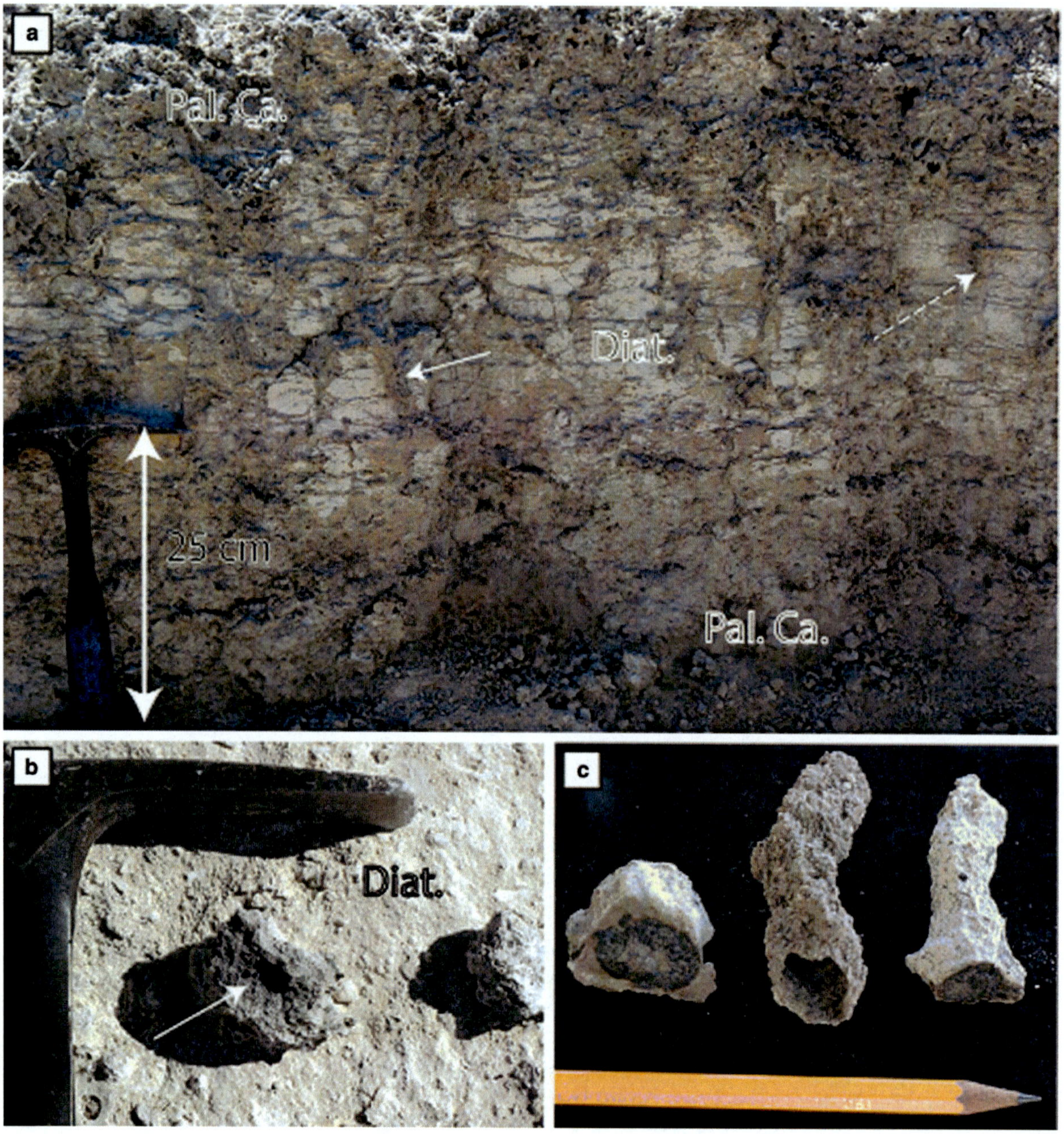

Fig. 6.6 Field photographs showing **a** the outcrop of a diatomite layer (Diat.) between two palustrine carbonate deposits (Pal.Ca.) at Acacia quarry (18°07′03″ S, 24°19′31″ E, 933 m asl; see Fig. 6.5g). The age of such deposits is still unknown. There are some diagenetic features, such as burrows or galleries (solid white arrow) and craze planes and fractures (dashed arrow), both infilled with $CaCO_3$. **b** Close-up of a carbonate-rich burrow or gallery inside a diatomite layer. **c** Comparison between present-day termite gallery (sample in the center) and fossil features found in diatomites. All these hand specimens are cemented by Ca-carbonate

(Fig. 6.5f). They are enriched in secondary silica concretions, as nodules or rhizoliths. The sands become carbonate-rich just below, and at the contact, with the overlying carbonate layer with a well-marked boundary between the sand-rich layers and the carbonate bed. The bottom part of the carbonate bed includes few small carbonate nodules, increasing in size and abundance toward the surface. Siliceous nodules display the same distribution as the carbonate ones. The uppermost 1.70 m of the pit comprises a hard carbonate bed without any apparent nodules.

The pit from site 2 (3 m deep) was observed next to Nata Island (Fig. 6.4b) located between the northern and the

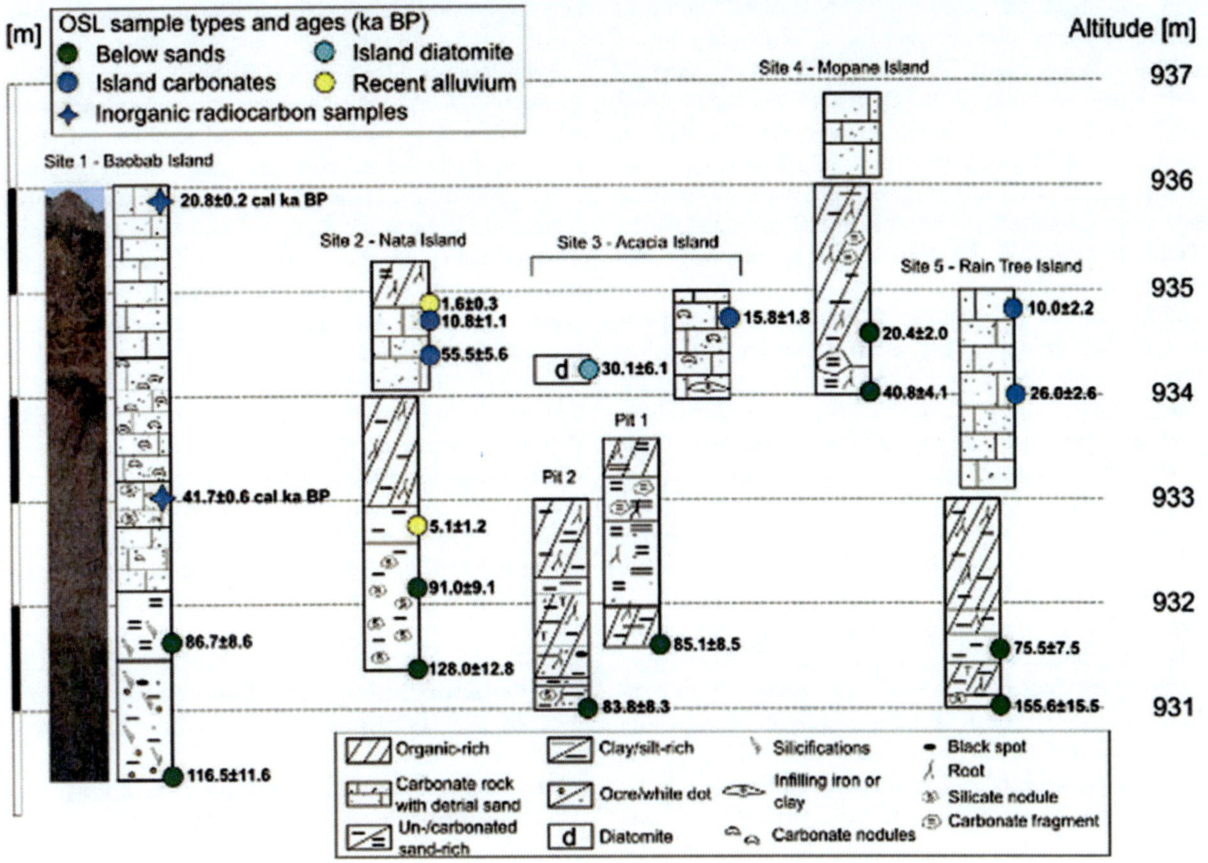

Fig. 6.7 Descriptions of the pits and/or soil profiles in each site represented in Figs. 6.3 and 6.4b, with OSL (circles) and radiocarbon (four-branches stars) ages of the different key sedimentary layers

southern N-S sand ridges. The elevation of the island compared to the surrounding sediments is about 2.5 m. A sandy layer, about 1.60 m thick and enriched in siliceous nodules, forms the bottom part of the pit. It is overlaid by a 40 cm-thick grey sandy layer without any siliceous nodules. An organic-rich sandy layer composes the upper part of the outcrop. The carbonate rock from the bed shows root and vegetation tissue debris and dendritic Fe–Mn features.

Two 2 m-deep pits were dug at site 3 located next to the most southern N-S sand ridge (Fig. 6.4). The first pit is just next to Acacia Island. Its bottom part is mostly an organic-rich and carbonate-free sand 40 cm in thickness. It is overlaid by a carbonate-rich sandy layer 1.10 m thick, without any organic matter. The uppermost part, 30 cm thick, is organic- and carbonate-rich. The second pit is located within the alluvial plain surrounding Acacia Island. The bottom part is a grey sand-rich layer 20 cm thick with abundant siliceous nodules. It is overlain by a thin sandy layer, 10 cm thick, in turn overlain by a 80 cm organic-rich layer and then, a thin sandy layer (15 cm), the top of the outcrop capped with an organic-rich layer, 75 cm thick. The elevation of the carbonate island compared to the surrounding alluvial plain is about 2 m. The carbonate bed studied was 1 m thick. It is a nodular carbonate rock rich in iron/clay infillings. It seems to overlay a diatomite layer (30 cm thick), showing some dark and carbonate veins forming a grid pattern (Fig. 6.6).

Another 2 m deep pit was dug next to Mopane Island located 10 km from the N-S sand ridge in the W (site 4, Fig. 6.3). The elevation of this island compared to the surroundings is about 0.5 m. The pit is characterized by a carbonate-rich layer, 20 cm thick, with small roots and a dominance of a sand size fraction ($\sim$70%). The layer above is 1.80 m thick and is organic-rich, without any carbonate in the fine fraction, but includes carbonate blocks and gravels. The carbonate rock fragments from the bed are very similar to the Nata Island carbonate layer and comprise tissue debris of roots and plants as well as dendritic Fe–Mn features.

A 2 m deep pit was dug next to Rain Tree Island located 10 km from the N-S sand ridge in the W (site 5, Fig. 6.3) in the surrounding alluvial floodplain. The elevation of the carbonate bed compared to the floodplain is about 2.5 to

3 m. The lowest part is a white sandy layer (10 cm thick) having abundant siliceous nodules. It is overlain by an organic-rich sandy layer (30 cm thick), in turn overlain by a grey sandy layer (20 cm thick). Above this light-colored sandy layer, there is a second organic-rich layer (20 cm thick) with an organic content increasing toward the upper part of the pit (1.10 m thick). The carbonate rock from the bed is very similar to those of Nata and Mopane Islands. Tissue debris of roots and plants, as well as dendritic Fe–Mn features, are present.

Thin section observations of carbonate rock samples suggest that the carbonate precipitation occurred in palustrine environments. A petrographic study showed that the quartz deposition in the palustrine environment would have been contemporaneous to the carbonate precipitation (Diaz et al. 2019). Consequently, to assess the timing of the carbonate island formation, optically stimulated luminescence dating (OSL) was used on quartz from key sedimentary layers (Fig. 6.5); (i) quartz from sandy layers situated below the carbonate islands (green circles in Fig. 6.5), (ii) quartz from sands trapped within the carbonate rock composing the island carbonate beds (blue circles), and (iii) quartz from sandy layers situated above the island carbonate beds (yellow circles). At Baobab Island, the ages of the carbonate rocks were determined using radiocarbon dating on the inorganic carbon fraction (four-branches blue stars). Finally, quartz extracted from the diatomite described at Acacia Island have been used for OSL dating (light blue circle). It can be hypothesized that OSL and radiocarbon ages from the key sedimentary layers may give an indication of the ages of the respective phases (i) the pre-palustrine carbonate, (ii) the palustrine carbonate, and (iii) the post-palustrine carbonate formations, respectively (Fig. 6.7).

Resulting ages show that the pre-palustrine carbonate formation phase would have occurred before 75 ka BP (before MIS 4). There are two younger samples from site 2 that deposited during MIS 2 (20.4 ka BP) and MIS 3 (40.8 ka BP). Were they already deposited during a different phase? The difference between the carbonate island and the top of the sand pit was only 0.5 m instead of 2 m for the other sites. Could this suggest that erosion was less intense in site 4? And that we are measuring alluvium sands possibly deposited during the palustrine carbonate formation phase? Indeed, the palustrine carbonate formation phase could have spanned from 55.5 to 10.0 ka BP (MIS 3 and MIS 2). Interestingly, quartz trapped in the diatomite are also from this phase (30.1 ka BP). It seems that the chemistry of waterbodies was not homogeneous. Finally, the post-palustrine carbonate phase would have occurred after 10 ka BP (MIS 1; Fig. 6.7, Diaz et al. 2019). Thus, it seems that, while lakes were forming in the Makgadikgadi Basin, palustrine environments were developing in the Chobe Enclave (Fig. 6.8). There are still questions pending regarding the origin of waters, e.g., were these palustrine areas forming before the onset of the activity of the Linyanti fault, allowing waters from the Linyanti to enter directly in the Chobe? Bäumle and Himmelsbach's (2018) work on the Linyanti aquifer proposes that palaeo-lakes along the Linyanti could have formed at the beginning of the extension of the graben system occurring during the Early and Middle Pleistocene, contemporaneously to Palaeo-Lake Makgadikgadi. The continuous development of the graben system during the Late Pleistocene led to the present-day Linyanti swamps. Consequently, the Chobe islands seem to be contemporaneous to the development of the graben system, cutting off more and more the inflow from the Linyanti through time, until a tipping point after which only occasional inflows prevailed, such as today (Fig. 6.8). Moreover, it seems that the water inflow would have been at least sufficient to sustain a palustrine environment and the formation of carbonate until 10 ka BP. After that, the carbonate-rich palustrine system dried out and started to be differentially eroded and solely recent alluvial sands were deposited (MIS 1; Fig. 6.8).

6.3 Geomorphology and Landforms of the Northern Chobe Enclave District

6.3.1 Importance of Hydrology in Shaping Geomorphological Features

The fault systems of the Linyanti and the Chobe control the hydrological settings in the study area. These settings are the main factors controlling the landforms in the enclave. Due to the uplifts linked to the faults, the general topographic slope in the Chobe Basin is <0.02% on average over long distances. The slope is decreasing from SW to NE, as well as from NW to SE. It increases again close to the Chobe strike-slip fault. This slope configuration induces the development of anastomosed river systems that are completely dry in the western part and become ephemeral to continuous going to the east. This hydrological gradient from W to E induces landforms associated with different morphodynamic processes found in three typical areas (Fig. 6.4) as described below: the western palaeo-floodplain, the island area, and the eastern floodplain area.

6.3.1.1 The Western Palaeo-Floodplain

The southwestern part of the northern Chobe Enclave district is characterized by an inherited floodplain composed by old fluvial sediments reworked by aeolian processes (Fig. 6.3). Numerous palaeo-channels are observable, having flow directions to the E in the northeastern part and to the SSW in the southwestern part. A soil and vegetation study of the area (Romanens et al. 2019) showed that there is a sodium-rich layer at the soil surface or under a layer of reworked aeolian

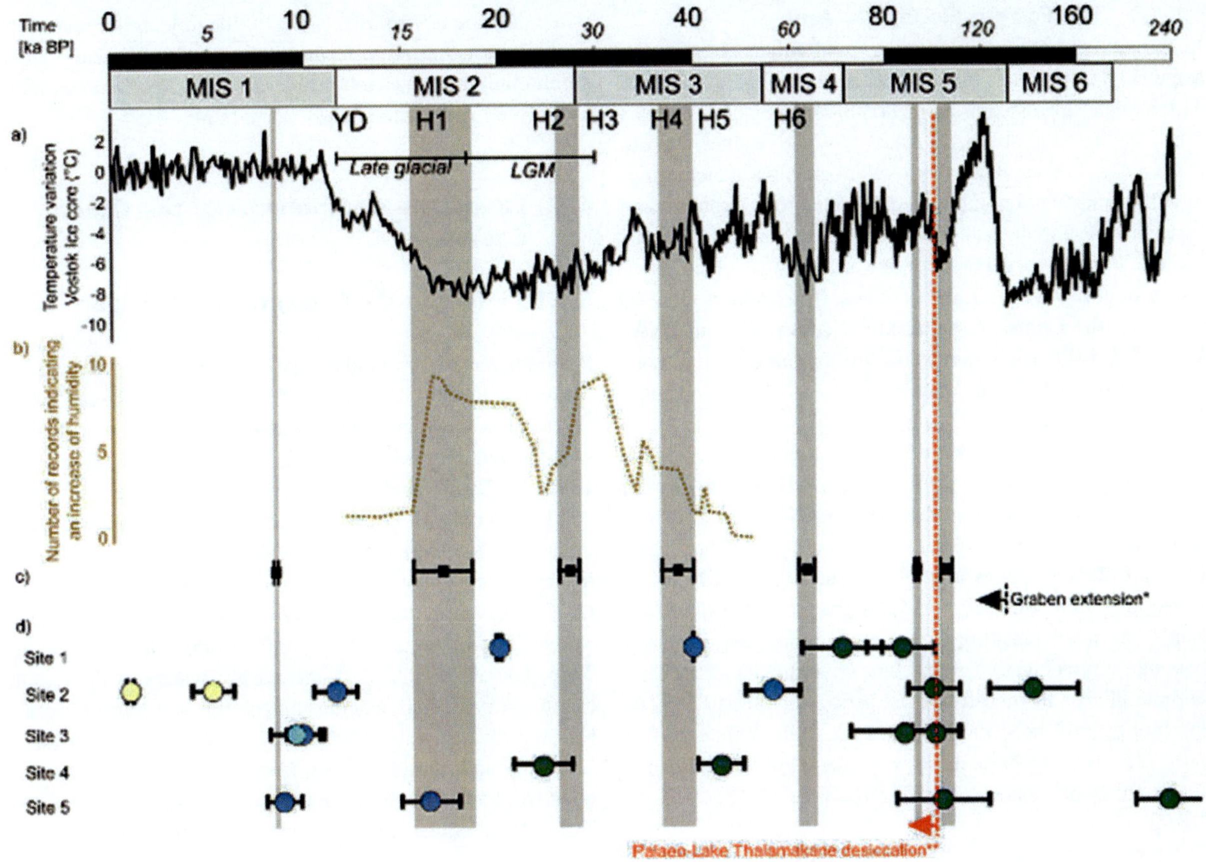

Fig. 6.8 Comparison of the carbonate island ages (circles and four-branches stars) with other types of archives. **a** Temperature variations assessed from ^{2}H in the Vostok Ice Core (Petit et al. 1999); data are available online (https://cdiac.ess-dive.lbl.gov/ftp/trends/temp/vostok/vostok.1999.temp.dat); **b** Numerous records in the Kalahari area are interpreted as an increase in humidity (Chase and Meadows 2007); the grey bars refer to lake highstands of the Mega lake Makgadikgadi (Burrough and Thomas 2008). *Graben extension in the Linyanti fault area (Bäumle and Himmelsbach 2018), **Thamalakane fault activation inducing the formation of the Okavango Delta and the desiccation of the Palaeo-Lake Thamalakane (Moore et al. 2012)

sands and between the palaeo-channels. Such a formation is probably inherited but its extent under the whole palaeo-floodplain remains unclear. Its origin can be attributed to a palaeo-hydrological system similar to what is observed today in the Okavango Delta. A detailed geomorphological map (Fig. 6.4a) shows the Linyanti swamps, and the inherited floodplain delimited by the Linyanti fault (Fig. 6.5b). The palaeo-channel configurations at this boundary suggest an older connection with the Linyanti river.

6.3.1.2 The Island Area

Lying between the palaeo-floodplain in the west and the floodplain in the east, the landscape of the island area (Figs. 6.3 and 6.5d–g) is dominated by the presence of islands the size of several hundred squared meters and a height of a few meters. Ephemeral river channels occasionally surround them and, when the water input is sufficient, small pans form (Fig. 6.4b). Sedimentary rocks of different natures form the islands, i.e., carbonate rocks, diatomite, sepiolite layers, amorphous silica-rich sands, and quartz sands (see above). Some of the islands have been documented (Diaz et al. 2019) but the composition of other islands can only be suggested (dashed lines in Fig. 6.4b) on the basis of field observations; finally, some have an unknown composition (marked with "?" in Fig. 6.4b). The succession of islands and channels forms an undulating landscape, which attenuates and finally disappears completely to the west (black dashed line in Fig. 6.3). In the east, a NNE-SSW sand ridge, around 20 km long and 2 km large, marks the limit with the eastern floodplain (Fig. 6.5c). It is composed by fluvial sands undergoing erosion due to sheet wash and forming triangular morphologies along their western side. On the eastern side, the sand ridge has a straight border possibly associated to an NNW-SSE transferal fault. The sand ridge is split into two halves by a river coming from the eastern floodplain, forming a water gap (Fig. 6.4b).

6.3.1.3 The Eastern Floodplain Area

In the eastern floodplain, smaller sand ridges, having the same characteristics as the one presented above, are present. These ridges are also crosscut by rivers forming sand patches that might have been attached to the ridges before (Fig. 6.4c). The number of ephemeral streams is increasing, and their anastomosed nature creates numerous oxbows. The water can accumulate in multiple pans, which can be connected. Further to the east of the sand ridges, permanent rivers are flowing from Lake Liambezi (from WNW to ESE) and from the Chobe River backflow (from NNE to SSW; Fig. 6.5 a, h). Large swamps are forming around their confluence and sand ridges counter the extent of these swamps in the west (Fig. 6.3).

Finally, two different landscapes are found at the borders of the area. In the north, the Linyanti swamps form around the Linyanti river, flowing along the fault from the Linyanti delta, towards Lake Liambezi. In the south along the Chobe fault, the slope increases abruptly (about 60 m) as illustrated by the elevation profile (Fig. 6.3). At the top of the slope, red sands, a color resulting from aluminum- and iron-rich coatings around quartz grains, are covering the landscape and are likely inherited from past pedogenic weathering. At the foot of the Chobe fault towards the study area, colluvial fans are forming in contact with the alluvial sediments deposited by the Chobe river (Fig. 6.3).

6.3.1.4 The Complexity of the Chobe's Geomorphology

The geomorphology of the drier Linyanti-Chobe basin is thus complex. A relative chronology of events can be proposed, as also supported by the regional Quaternary chronology (see above): first, red sands formed on a palaeo-surface, observed today at the top of the Chobe fault. They can be considered as paleosols, possibly very old and inherited from a period before the Chobe fault activation by the extension of the graben system (Pliocene? before mega-lakes?). The Chobe extension started and led to possible water accumulation along the Chobe fault. While the graben system was still active, the Linyanti fault formed. This event led to palustrine-lacustrine-fluvial environments from which the palaeo-floodplain and the island landforms are inherited. According to OSL dating, carbonate deposition occurred from 55 to 10 ka BP, contemporaneous to Palaeo-Lake Thalamakane retraction in the SSW. Tectonic events might have affected the drainage, inducing drier conditions through time, cutting off the Upper Zambezi and the Linyanti inputs. From 10 to 5 ka BP, the Linyanti inputs should have been stopped, due to continuous lifting, leading to the fossilization of the palaeo floodplain and carbonate-rich palustrine environments. Since this time, relief inversion could have occurred with the preferential erosion of sands around the carbonate-rich formations that are observed today as elevated carbonate islands (Diaz et al. 2019).

6.4 Landscape Components of the Chobe Enclave

6.4.1 Water in the Linyanti-Chobe Region

Both surface and groundwaters of the study region can be compared to the Okavango and the Zambezi (Dyer 2017). While the Okavango Delta hydrology has been particularly well studied over the last few decades (e.g., McCarthy et al. 1991; McCarthy and Ellery 1998; McCarthy et al. 1998; 2000; McCarthy 2006; Atekwana et al. 2016; Akondi et al. 2019), the Kwando and Zambezi rivers, which may also be of importance to the Chobe region, have been less studied so far (Tooth et al. this book). By analogy to the previous studies in the region (McCarthy, 2006; Atekwana et al. 2016; Akondi et al. 2019), the surface waters of the region can be classified as calcium-magnesium-(sodium)-bicarbonate type waters for the major rivers (Okavango, Kwando/Linyanti, and Zambezi; Fig. 6.9). The Okavango waters, however, have a larger proportion of potassium, whereas the Zambezi waters have a higher electrical conductivity. While the dissolved elemental proportions in the groundwaters do not change markedly relative to the surface waters, they have a higher overall conductivity (ion concentrations and total dissolved ions) compared to the large river entries. Consequently, a more marked change is noted for the Makgadikgadi pan surface waters and extremely evaporated terminal swamp waters of the Linyanti/Chobe as these become sodium-(potassium)-chloride type waters with locally also important sulfate concentrations, notably in the Linyanti/Chobe area. These highly evaporated terminal waters also reach calcite and/or anhydrite saturation (Dyer 2017; Atekwana et al. 2016). Zambezi river waters have a higher Ca/Mg compared to the Okavango River waters, which may indicate geologically slightly different catchments. As the pH of the Zambezi River is about 7.5, while that of the Okavango is about 6.9, this could indicate the presence of carbonate-bearing rocks in the Zambezi basin. The Okavango waters, in contrast, are more concentrated in sodium and potassium compared to the Zambezi River, suggesting that sodium and potassium as well as Ca are likely derived from silicate (feldspar) weathering. These trends in the major ion compositions are clearly represented in a modified Gibbs diagram (Fig. 6.9).

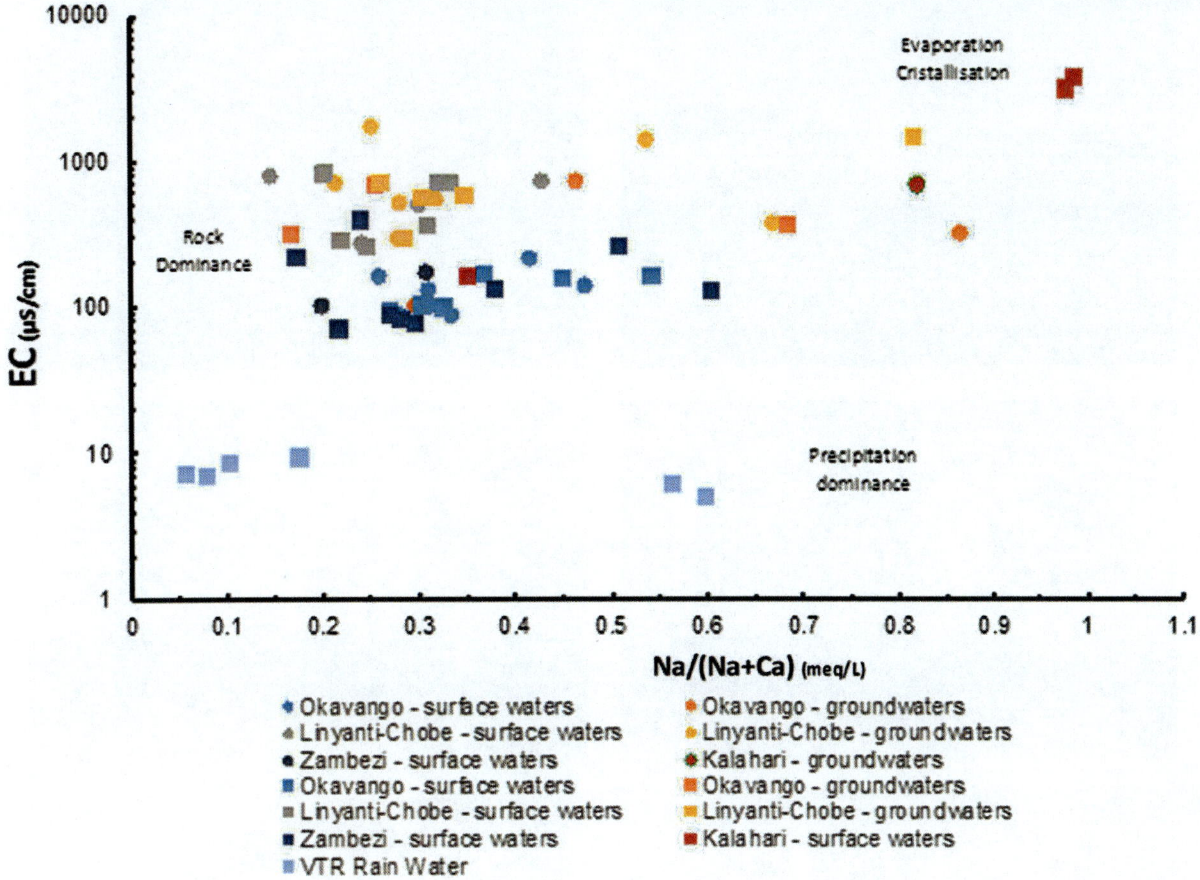

Fig. 6.9 Modified Gibbs diagram illustrating the changes in Na and Ca concentrations as a function of source terrain, evaporation-crystallization processes in the studied region

6.4.2 Vegetation and Soils

Eight plant communities can be distinguished in the Chobe Enclave (Vittoz et al. 2020): three grassland units, three woodland units, and two forest units (Fig. 6.10; Table 6.2). Their spatial distribution depends on the fine microtopography of the northern Chobe region. Based on the vegetation map (Fig. 6.11) and the geomorphology, the study area can be divided into four areas: (1) the floodplains along the Linyanti river, characterized by a gradient of flood duration and frequency; (2) a mosaic of *Combretum* woodlands and dambo grasslands in the island area, with contrasted topography positions and substrates; (3) a mosaic of mopane woodlands and sandveld in the western palaeo-floodplain, similarly related to topography; (4) and the *Baikiaea* forests on red sands, located on the eastern side of the Chobe Fault, extending to the Chobe Forest Reserve.

In a parallel study, five soil types were recognized in the study region (Romanens et al. 2019). The following soil types were found (Figs. 6.12 and 6.13), named according to the WRB (IUSS Working Group WRB, 2014):

- Arenosols are largely dominated by sand along the whole profile, generally poor in nutrients and organic matter (Skarpe et al. 2014).
- Kastanozems are characterized by a dark, organic topsoil, saturated with bases, and by a calcium-rich horizon.
- Chernozems and Phaeozems have a high organic content in most of their profile, with a black topsoil horizon; compared to Phaeozems, the Chernozems are characterized by a calcium-rich horizon.
- Solonchaks and Solonetz are characterized by a salt- or sodium-rich horizon in their profile, the second being richer in clays.

Fig. 6.10 Illustrations of the vegetation units. **a** Wet floodplain; **b** Dry floodplain grassland; **c** Riverine forest; **d** *Combretum* woodland; **e** Dambo grassland; **f** Mopane woodland; **g** Sandveld; **h** *Baikiaea* forest (Photos: F. Pellacani, a, b, c, d, g; P. Vittoz, e, f, h)

Table 6.2 Scientific names used in Vittoz et al. (in press) for the different plant communities

Vegetation unit	Scientific name (Vittoz et al. 2020)
Dry floodplain grassland	*Aristida junciformis–Aristida meridionalis* Grassland
Riverine forest	*Croton megalobotrys–Setaria verticillata* Forest
Combretum woodland	*Eragrostis superba–Combretum hereroense* Woodland
Dambo grassland	*Geigeria schinzii–Setaria sphacelata* Grassland
Mopane woodland	*Colophospermum mopane–Jasminum stenolobum* Woodland
Sandveld	*Ipomoea chloroneura–Oxygonum alatum* Woodland
Baikiaea forest	*Baikiaea plurijuga–Baphia massaiensis* Forest

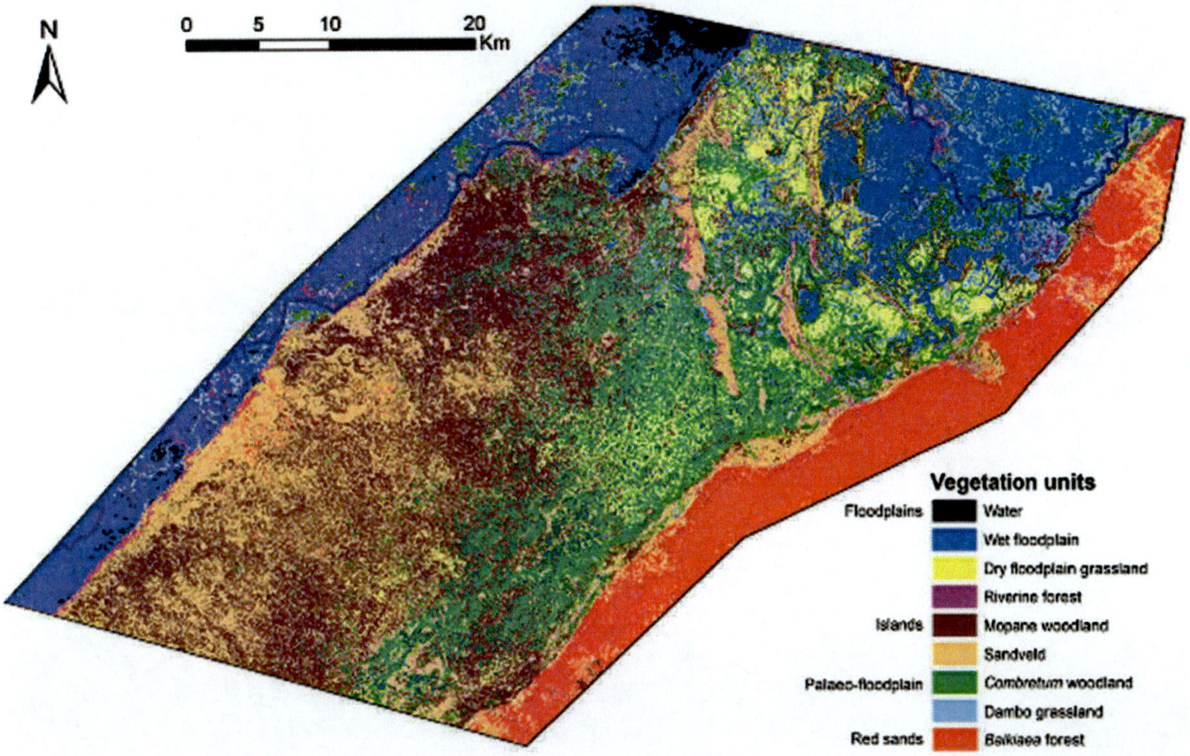

Fig. 6.11 Vegetation map of the Chobe Enclave, with vegetation unit groups according to landscape areas. The Linyanti River is delineated by the dark blue line

- Calcisols contain calcium-rich horizons but compared to Kastanozems, are not particularly rich in organic carbon.

6.4.2.1 Floodplains

The wettest parts of the floodplains were not considered in detail. However, a high diversity of plant communities is expected, close to the one observed in the Okavango Delta (Sianga and Fynn 2017). These plant communities are related to the water depth in permanently flooded areas (Ellery and Ellery this book) and the importance and recurrence of flood events by the Linyanti river in the seasonal swamps (Murray-Hudson et al. 2011; Pricope 2013).

The dry floodplain grasslands are present on higher topographic positions, which are flooded at a low frequency, possibly not each year or not at all, but are still under the seasonal influence of a high water table. This plant community has a low cover of grasses (~40%), established on former alluvial deposits, with a very high sand proportion (~98%; soils classified as Arenosols) related to a very low Cation Exchange Capacity (CEC). The dominant species are the grasses *Aristida meridionalis*, *A. junciformis*, and *Hyperthelia dissoluta*, and the forbs *Dicerocaryum eriocarpum* and *Meeremia tridentata* are some of the typical species encountered. These dry floodplain grasslands are

Fig. 6.12 Soil map of the northeastern part of the Chobe Enclave. This map was produced by conducting parallel soil (Romanens et al. 2019) and vegetation studies (Vittoz et al. 2020) and the soil map was partly based on the vegetation map

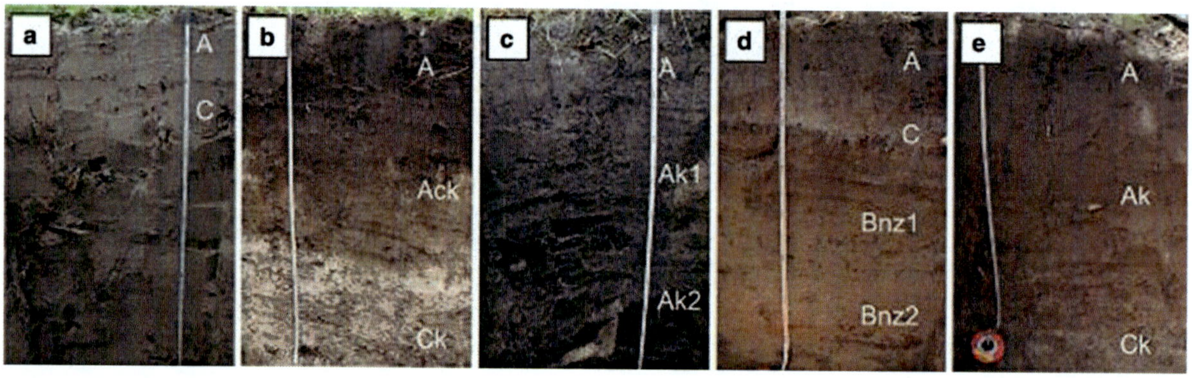

Fig. 6.13 Examples of soils found in the Chobe Enclave. **a** Arenosol: soil of floodplains and wind-reworked areas. **b** Kastanozem: soil found on sandy areas and diatomitic/carbonate islands. **c** Chernozem: soil of the humid depressions (dambos) with some hydromorphic properties. **d** Solonetz: soil often developed under Mopane woodlands. **e** Calcisol: soil easily formed on carbonate islands. Letters on pictures refer to the nomenclature of horizons (for more details, see IUSS Working Group WRB 2014 and Romanens et al. 2019)

characterized by numerous termite mounds (Mujinya et al. 2010, 2014), often occupied by trees (McCarthy 1998; McCarthy et al. 1998), some of them found almost exclusively in such situations (e.g., *Phoenix reclinata*, *Imperata cylindrica;* Roodt 1998).

Riverine forests form a species-rich community, encountered as small fragments on prominences in or along the floodplain of the Linyanti river. It has the highest mean tree cover (~50%) and the tallest trees (~8.5 m) of all vegetation units, their density and size related to their possibility to reach a permanent water table with their roots. The alkaline soils (Kastanozems) are dominated by sand but with 15–33% of silt, probably due to the high density of termite mounds. *Berchemia discolor, Vachellia tortilis, Croton megalobotrys, Combretum mossambicense,* and *Dichrostachys cinerea* are some of the typical trees and shrubs, often climbed by *Capparis tomentosa*. *Kigelia africana, Terminalia prunioides,* and *Ficus thonningii* are less frequent woody species but encountered only in riverine forests. Some typical species in the understory are the grasses *Setaria verticillata, Panicum maximum,* and *Urochloa trichopus* and the forbs *Ipomoea dichroa, Acalypha fimbriata, Justicia heterocarpa, Achyranthes aspera, Dicliptera paniculata, Commelina petersii, Cyathula orthacantha,* and *Leonotis nepetifolia.*

6.4.2.2 Islands

Two plant communities are distributed in this area according to the topography. The *Combretum* woodlands is found on small islands underlain by carbonate beds. The soils (Kastanozems) are characterized by the highest clay content of the study area (14–20%) that favors a high CEC dominated by calcium, promoting an alkaline pH (7.8–8.8). The typical trees are *Combretum hereroense, Sclerocarya birrea* subsp. *caffra, Senegalia nigrescens,* and *Vachellia hebeclada*. The grasses *Eragrostis superba, Heteropogon contortus, Digitaria eriantha,* and *Cenchrus ciliaris*, and the forbs *Rhynchosia minima, Hoslundia opposita* and *Hibiscus caesius,* and *Solanum panduriforme* are frequently encountered.

The dambo grasslands are found in pans. This community occupies shallow, depressions that are waterlogged during the wet season (Acres et al. 1985). Without a river draining them, fine mineral materials tend to accumulate, and the waterlogging restrains organic matter mineralization. Hence, the soils (Chernozems or Phaeozems) show the lowest sand content (19–29%) but the highest silt (59–68%) and organic carbon (3–5%) contents of all the plant communities in the Chobe Enclave. The high CEC is dominated by calcium, resulting in a neutral to alkaline soil pH. Grasses largely dominate the community, with *Setaria sphacelata, Digitaria milanjiana, Hyparrhenia rufa, Bothriochloa bladhii, Cymbopogon caesius, Panicum repens,* and *Trachypogon spicatus* as frequent or dominant species.

6.4.2.3 Western Palaeo-Floodplain

The western palaeo-floodplain is similarly characterized by a mosaic of two vegetation units according to topography. The area is a fossil delta with islands separated by channels inherited from ancient hydrological dynamics, before being separated from the Linyanti river by tectonic activity (Kinabo et al. 2008; Fynn et al. 2014).

The islands are occupied by the mopane woodlands. Some areas are characterized by tall *Colophospermum mopane* trees, while in other places the species is limited to stunted specimens, with a low tree cover. It seems that these sizes are related to different soil conditions: stunted mopanes grow on silty soils characterized by a high CEC, saturated by sodium, and with high alkaline pH (9.5–10), classified as Solonchaks. In contrast, tall trees occupy soils with a sandy upper horizon, with moderate alkaline pH (7.0–8.2), but with the presence of a sodic horizon below 30 cm (Solonetz or other soil types, like Calcisols, with a *salic* horizons). These high sodium concentrations suggest that these areas have an origin comparable to the present islands in the Okavango Delta (Ramberg and Wolski 2008; McCarthy et al. 2012). Besides mopane, the typical species are the grasses *Schmidtia pappophoroides, Aristida adscensionis,* and *Digitaria eriantha*; the sedge *Kyllinga buchananii*; and the forbs *Ipomoea coptica* and *Chamaecrista absus.*

The palaeo-channels are filled with leached, river-washed sands (Arenosols) and occupied by sandvelds. Compared to dry floodplain grasslands, this plant community has a higher topographic position, with a complete absence of visible water table during the wet season. Tree and shrub layers are characterized by the presence or dominance of *Senegalia cinerea, Terminalia sericea, Grewia retinervis,* and *Dicrostachys cinerea*, and the herb layer by the grasses *Panicum maximum, Urochloa trichopus,* and *Digitaria eriantha*, and the forbs *Acanthosicyos naudinianus, Spermacoce senensis, Indigofera flavicans, Gisekia africana, Cleome hirta,* and *Pavonia senegalensis.*

6.4.2.4 Baikiaea Forest

Southeast of the Chobe Fault, the *Baikiaea* forests have a dense woody stratum. This vegetation unit, which covers a large area in Chobe National Park (*Baikiaea plurijuga–Combretum apiculatum* woodland) according to Skarpe et al. (2014), grows on thick, reddish Kalahari sands. The soils are classified as Arenosols and their red color is due to an iron and aluminum-rich oxyhydroxide coating, which is responsible for acidic conditions (pH 5.7–6.0). This area is completely disconnected from the rivers and the plants have to rely only on rainwater. Trees and shrubs show high diversity, the most frequent species being *Combretum elaeagnoides, Baikiaea plurijuga, Baphia massaiensis, Croton gratissimus,* and *Bauhinia petersiana*. The grasses *Dactyloctenium giganteum, Panicum maximum, Urochloa trichopus,* and *Digitaria eriantha,* and the forbs *Jacquemontia tamnifolia, Ipomoea pes-tigridis, Vigna unguiculata, Chamaecrista absus,* and *Harpagophytum zeyheri* are constant or locally important species.

6.4.3 The Role of Termites

The most notable species of termites encountered in the northern Chobe Enclave district are the fungus-growing termites (FGT) (subfamily Macrotermitinae, Isoptera). They are important for two reasons: (i) their large iconus epigeal mounds can be seen from afar and (ii) their ability to modify the distribution of sediments and soils in order to build their nest (Jouquet and Lepage 2002; Mujinya et al. 2011). Fungus-growing termites are a subfamily composed of 12 genera and 345 species. One of these genera, *Odontotermes* (142 species) present in this region, tends to build cryptic subterranean fungus-chambers 20–50 cm below ground. An important species found in the area is the *Trinervitermes trinervoides*, also commonly known as snouted harverster termite for the protruding portion of their face. These are non fungus-growing termites and do not build large fungus-chamber but only small scattered domed mound of only a few centimeters high. The most common species observed in the study area is *Macrotermes michaelseni*, a FGT species part of he *Macrotermes* genus (65 species). It has long been considered as an ecosystem engineer (Jones 1990) for the modifications it brings to the soil, its ability to concentrate nutrients, and its capacity to create patches of fertile land in tropical and subtropical savanna ecosystems (Jouquet and Tavernier 2005; Corenblit et al. 2016).

6.4.3.1 The Potential Impact of Termites in Savanna Ecosystems of the Chobe Enclave

Fungus-growing termites have shared an exosymbiosis since 30 Ma (Roberts et al. 2016) with a fungi belonging to the *Termitomyces* genus. In order to maintain the symbiosis, *Macrotermitinae* must maintain specific hydric and thermic conditions in their nests. Therefore, fungus-growing termites build large biogenic structures in which they are able to increase the alkalinity of soils by an order of magnitude of 3–4. They also increase their carbonate content, the C/N ratio, and concentrate nutrients such as potassium and phosphorus. FGT modify the chemical compositions and mineralogical propreties of clays through the process of selection and transportation of sand grains in their buccal cavity, where they are mixed with saliva (Jouquet et al. 2011). They also act as accelerating agents of clay alteration and chemical weathering in tropical ecosystems. The activities of FGT tend to slightly raise the land surface locally, providing some recolonization advantages. They form a pattern of fertile lands by conentrating nutrients, and enhance the growth of vegetation by creating islands of fertility (see above; Dangerfield et al. 1998). The selection of very fine sands, in order to meet the construction requirement for their mounds, create patches of clayey sands that have the property to retain water for long periods of time, producing scattered pockets of water in semiarid regions such as the Chobe-Linyanti region (Pennisi et al. 2015).

Different stages of FGT mounds can be observed in the northern Chobe Enclave district (i) active termite mounds (Fig. 6.14a), generally between 50 cm and a few meters high; they are homogenous and hard consolidated forms; (ii) mounds that have been recently abandoned by the termite colony, due to a predator or a natural hazard; their shape appears as similar to a flattened promontoir with broken pieces laying on the adjacent ground; they can be recolonized by the same species of termites; (iii) relic termite mounds (Fig. 6.14b); these mounds have seen many stages of occupation over a relatively long period of time, from a few dozen years up to hundreds of years and their size can reach 18 m in diameter and be 7 m high. These relic mounds can probably be considered as geomorphological objects by their size, their impact on the environment, and their role in the ecosystem. They are distinctive features in the landscape of which many animals take advantage, such as snakes, monkeys, leopoards, lions, elephants, as well as plants (including trees).

Like in many parts of Africa, people from the the study region are using mounds for different purposes, e.g., soil amendment, geophagy (Mills 2008), or building material. Their volume is considerable: a single mound in the Chobe had an above ground volume of 1,350 m^3. They also can remain for long periods of time: FGT relic mound ages have been estimated by optically stimulated luminescence (Kristensen et al. 2015) to be 4,000 yrs old in Ghana and to be 2,200 yrs old using ^{14}C in Congo (Erens et al. 2015).

6.4.3.2 Hydric Gradients Influencing FGT Activities

In regard to termite activities, two gradients are observed in the Chobe Enclave: (i) an EW gradient assigned to the topographic flooding pattern and (ii) a NS gradient attributed to the distance to the Linyanti river in the north. Distribution of *Macrotermes* mounds in the landscape represent these gradients fairly well. In the eastern part of the enclave, large active mounds, similar to the one encountered in the center of the Okavango Delta today, dominate the topography. Moving further west, these big mounds become less common and after a short distance toward the center of the enclave, they totally disappear. In reference to the NS gradient, within the riverine woodland of the Linyanti river, all mound stages are observed (active, abandoned, or relic). Moving south, the active termite mounds are rarer, most of them being abandoned; finally, in the middle part of the enclave, there are no active *Macrotermes* mounds.

6.4.3.3 Termites Influence the Evolution of the Landscape

Regarding the formation of islands, Dangerfield et al. (1998) were the first to describe the role of FGT in the context of the Okavango Delta. A similar process may have existed for most of the northern Chobe Enclave's past. FGT are probably still involved in the geomorphological shaping of the region through this specific island formation process found in its eastern part. The sediment composition of the islands being an important factor, a recent study quantified the way in which FGT modify local soil grain-size distributions to make the building material for their nest (Van Thuyne et al. 2021). When built on a coarse sand soil, FGT tend to

Fig. 6.14 **a** Young active termite mound of the *Macrotermes michaelseni* species, situated on a diatomaceous soil in an open grassland on the eastern side of the Chobe Enclave. **b** A non-re-colonized relic termite mound of the *Macrotermes michaelseni* species. Its size is impressive

moderately increase the upper part with finer material and tend to increase substantially the lower part in finer material, whereas, when built on a silt soil, the opposite is observed. Once abandoned and destroyed by erosion, the altered soil material consequently modifies the grain-size distribution of the local sediments.

The above-mentioned modifications brought by FGT have considerable importance and impact on this semiarid region. For example, the soil pH is often low and the soil remains deficient in many nutritive elements: the mounds are able to provide better soil conditions with a more alkaline pH and a higher nutritive content. The mounds have also the capacity to retain water for long period of time, (i) favoring a de-aridification process (Pennisi et al. 2015), (ii) offering a fire protection in these savanna-prone fire zones, and (iii) contributing to higher stability and resilience of the ecosystem (Pringle et al. 2010).

6.5 Challenges and Conclusions

The northern Chobe Enclave district constitutes a fascinating location in order to understand the complex relationships between the development of landforms in a high African plateau under the influence of trans-tensional pull-apart basins and the various external agents at work during the Quaternary. At the crossroads of various rivers and influenced by neotectonics, large parts of the landscape are not yet explained: issues such as (i) how did the syntectonic depocenters evolve through the Quaternary? Or (ii) what are the ages of the various faults and water gaps affecting the region? The presence of large palustrine carbonate deposits is not totally explained: how were such deposits possible in the past? And what would have been the environmental conditions that enabled such calcium carbonate accumulations in an essentially siliceous basin? In addition, if the source of carbon can easily be identified (CO_2 and organic matter), where is the source of calcium?

This chapter presented a limited state of the art of the science conducted in the area. It is clear that the present-day composition of the various water bodies cannot totally explain the diversity of the geochemical nature of the sediments. Nevertheless, it seems likely that the large presence of sodium in soils and sediments could be related to some water bodies, associated with efficient evaporation processes. The vegetation is perfectly adapted to the landscape constraints. Another important point is the role of termites: they redistribute the particle sizes of sediments at the surface of the geomorphological features and partly shape the landscape. Today, the main exogenous parameters affecting the landscape seem to be wind, water balance (rain and the prominent role of rivers and floodplains), as well as the biological impact of the living realm. But the conditions prevailing today were probably different during the past, making the reconstruction of the landscape history challenging.

Acknowledgements The authors would like to thank Prof. Frank D. Eckardt for his confidence and his help in writing this Chapter, the Ministry of Environment, Natural Resources Conservation, and Tourism of the Republic of Botswana as well as the Chobe Enclave Conservation Trust for the research permits, without which these studies would not have been possible. L. Ballif, S. Dyer, F. Pellacani, and R. Romanens are thanked for providing documents and data. The authors would like to acknowledge all the staff from the Van Thuyne-Ridge Research Center in the Chobe Enclave for their support in the field and base camp. K. Verrecchia kindly edited the manuscript. Prof. Susan Ringrose and an anonymous reviewer kindly improved the first version of this Chapter. This research has been funded by the University of Lausanne (International Relations), the Herbette Foundation, and grants from the Swiss National Foundation nos P2LAP2_174595 to ND and 200021_172944 to EPV.

References

Acres BD, Rains AB, King RB, Lawton RM, Mitchell AJB, Rackham LJ (1985) African dambos: their distribution, characteristics and use. Z Geomorphol 52:62–83

Akondi RN, Atekwana EA, Molwalefhe L (2019) Origin and chemical and isotopic evolution of dissolved inorganic carbon (DIC) in groundwater of the Okavango Delta Botswana. Hydrol Sci J 64:105–120

Alvarez NA, Hogan JP (2013) The role of tectonic inheritance in the geometry and location of continental rifts-an example from the Okavango rift zone, Botswana. Geol Soc Am Abstr Programs 45 (7):832

Atekwana EA, Molwalefhe L, Kgaodi O, Cruse AM (2016) Effect of evapotranspiration on dissolved inorganic carbon and stable carbon isotopic evolution in rivers in semi-arid climates: the Okavango Delta in north west Botswana. J Hydrol 7:1–13

Bäumle R, Himmelsbach T (2018) Exploration of deep, previously unknown semi-fossil aquifers of the Kalahari Basin (southern Africa). Grundwasser 23:29–45

Bishop P (2007) Long-term landscape evolution: linking tectonics and surface processes. Earth Surf Process Landf 32:329–365

Bordy EM, Segwabe T, Makuke B (2010) Sedimentology of the upper Triassic-lower Jurassic (?) Mosolotsane formation (karoo supergroup) Kalahari Karoo Basin, Botswana. J Afr Earth Sci 58:127–140

Bufford KM, Atekwana EA, Abdelsalam MG, Shemang E, Atekwana EA, Mickus K, Moidaki M, Modisi MP, Molwalefhe L (2012) Geometry and faults tectonic activity of the Okavango rift zone, Botswana: evidence from magnetotelluric and electrical resistivity tomography imaging. J Afr Earth Sci 65:61–71

Burrough SL, Thomas DSG (2008) Late Quaternary lake-level fluctuations in the Mababe depression: middle Kalahari palaeolakes and the role of Zambezi inflows. Quat Res 69:388–403

Burrough SL, Thomas DSG, Shaw PA, Bailey RM (2007) Multiphase Quaternary highstands at Lake Ngami, Kalahari, northern Botswana. Palaeogeogr Palaeoclim Palaeoecol 253(3–4):280–299

Burrough SL, Thomas DSG, Bailey M (2009a) Mega-lake in the Kalahari: a late Pleistocene record of the Palaeolake Makgadikgadi system. Quat Res 28:1392–1411

Burrough SL, Thomas DSG, Singarayer JS (2009b) Late Quaternary hydrological dynamics in the middle Kalahari: forcing and feedbacks. Earth-Sci Rev 96:313–326

Campbell G, Johnson S, Bakaya T, Kumer H, Nsatsi J (2006) Airborne geophysical mapping of aquifer water quality and structural controls in the lower Okavango Delta, Botswana. S Afr J Geol 109:475–494

Chase BM, Meadows ME (2007) Late Quaternary dynamics of southern Africa's winter rainfall zone. Earth Sci Rev 84:103–138

Cooke HJ (1975) The palaeo-climatic significance of caves and adjacent landforms in western Ngamiland, Botswana. Geogr J 141:430–444

Cook HJ (1980) Landform evolution in the context of climatic change and neo-tectonism in the middle Kalahari of north-central Botswana. Trans Inst Br Geogr 5:80–99

Cooke HJ, Verstappen TH (1984) The landforms of the western Makgadikgadi basin in northern Botswana, with a consideration of the chronology of the evolution of Lake Palaeo-Makgadikgadi. Z Geomorphol 28:1–19

Cordova CE, Scott L, Chase BM, Chevalier M (2017) Late Pleistocene-Holocene vegetation and climate change in the middle Kalahari, Lake Ngami, Botswana. Quat Sci Rev 171:199–215

Corenblit D, Corbara B, Steiger J (2016) Biogeomorphic ecosystems within stressful and disturbed environments: a focus on termites and pioneer plants. In: Johnson EA, Martin YE (eds) A biogeoscience approach to ecosystems. Cambridge University Press, Cambridge, pp 456–474

Cox RT (1994) Analysis of drainage-basin symmetry as a rapid technique to identify areas of possible Quaternary tilt-block tectonics: an example from the Mississippi embayment. Geol Soc Am Bull 106:571–581

Crouvi O, Amit R, Enzel Y, Gillespie AR (2010) Active sand seas and the formation of desert loess. Quat Sci Rev 29:2087–2098

Dangerfield JM, McCarthy TS, Ellery WN (1998) The mound-building termite Macrotermes michaelseni as an ecosystem engineer. J Trop Ecol 14:507–520

Diaz N, Armitage SJ, Verrecchia EP, Herman F (2019) OSL dating of a carbonate island in the Chobe Enclave, NW Botswana. Quat Geochronol 49:172–176

Dyer S (2017) Water cycle in the northern Kalahari. Unpublished Master of Science in Biogeosciences, University of Lausanne

Ellery WN, Ellery K (2020) The Okavango Delta as a peatland landscape. In: Eckardt FD (ed) Landscapes and landforms of Botswana. Springer, Berlin

Erens H, Boudin M, Mees F, Mujinya BB, Baert G, Van Strydonk M, Boeckx P, Van Ranst E (2015) The age of large termite mounds—radiocarbon dating of Macrotermes falciger mounds of the Miombo woodland of Katanga, DR Congo. Palaeogeogr Palaeoclim Palaeoecol 435:265–271

Freytet P, Verrecchia EP (2002) Lacustrine and palustrine carbonate petrography: an overview. J Paleolimnol 27:221–237

Fynn RW, Chase M, Röder A (2014) Functional habitat heterogeneity and large herbivore seasonal habitat selection in northern Botswana. S Afr J Wildl Res 44:1–15

Gamrod JL (2009) Paleolimnological records of environmental change preserved in Paleo-Lake Mababe, northwest Botswana. Doctoral dissertation, Oklahoma State University

Giaconia F, Booth-Rea G, Martinez-Martinez JM, Azañon JM, Peréz-Peña JV (2012a) Geomorphic analysis of the Sierra Cabrera, an active pop-up in the constrictional domain of conjugate strike-slip faults: the Palomares and Polopos fault zones (eastern Betics, Spain). Tectonophysics 580:27–42

Giaconia F, Booth-Rea G, Martínez-Martínez JM, Azañón JM, Pérez-Peña JV, Pérez-Romero J, Villegas I (2012b) Geomorphic evidence of active tectonics in the Sierra Alhamilla (eastern Betics, SE Spain). Geomorphology 145–146:90–106

Grey DRC, Cooke HJ (1977) Some problems in the Quaternary evolution of the landforms of northern Botswana. CATENA 4:123–133

Grove AT (1969) Landforms and climatic change in the Kalahari and Ngamiland. Geogr J 135:191–212

Humphries MS, McCarthy TS, Cooper GRJ, Stewart RA, Stewart RD (2014) The role of airborne dust in the growth of tree islands in the Okavango Delta, Botswana. Geomorphology 206:307–317

Huntsman-Mapila P, Ringrose S, Mackay AW, Downey WS, Modisi M, Coetzee SH, Vanderpost C (2006) Use of the geochemical and biological sedimentary record in establishing palaeo-environments and climate change in the Lake Ngami basin, NW Botswana. Quat Int 148:51–64

IUSS Working Group WRB (2014) World reference base for soil resources 2014: international soil classification system for naming soils and creating legends for soil maps. FAO, Rome

Jones JA (1990) Termites, soil fertility and carbon cycling in dry tropical Africa: a hypothesis. J Trop Ecol 6:291–305

Jones BTB (2002) Chobe Enclave, Botswana - lessons learnt from a CBNRM project 1993–2002. CBNRM Support Project, Gaborone

Jouquet P, Lepage M (2002) Termite soil preferences and particle selections: strategies related to ecological requirements. Insectes Soc 49:1–7

Jouquet P, Tavernier V (2005) Nests of subterranean fungus-growing termites (isoptera, macrotermitinae) as nutrient patches for grasses in savanna ecosystems. Afr J Ecol 43:191–196

Jouquet P, Traoré S, Choosai C, Hartmann C, Bignell D (2011) Influence of termites on ecosystem functioning. ecosystem services provided by termites. Eur J Soil Biol 47:215–222

Kinabo BD, Atekwana EA, Hogan JP, Modisi MP, Wheaton DD, Kampunzu AB (2007) Early structural development of the Okavango rift zone NW Botswana. J Afr Sci 48:125–136

Kinabo BD, Hogan JP, Atekwana EA, Abdelsalam MG, Modisi MP (2008) Fault growth and propagation during incipient continental rifting: insights from a combined aeromagnetic and shuttle radar topography mission digital elevation model investigation of the Okavango rift zone, northwest Botswana. Tectonics 27:TC3013. https://doi.org/10.1029/2007TC002154

Kristensen JA, Thomsen KJ, Murray AS, Buylaert J-P, Jain M, Breuning-Madsen H (2015) Quantification of termite bioturbation in a savannah ecosystem: application of OSL dating. Quat Geochronol 30:334–341

Kurugundla CN, Dikgola K, Kalaote K (2010) Restoration and rehabilitation of Zibadianja Lagoon in Kwando-Linyanti river system in Botswana. Botsw Notes Rec 2010:79–89

Lancaster N (1981) Grain size characteristics of Namib Desert linear Dunes ». Sedimentology 28(1):115–122

Lancaster N (1986) Grain size characteristics of linear dunes in the southwestern Kalahari. J Sediment Res 56:395–400

Le Gall B, Tshoso G, Jourdan F, Feraud G, Bertrand H, Tiercelin JJ, Kampunzu AB, Modisi MP, Dyment J, Maia M (2002) $^{40}Ar/^{39}Ar$ geochronology and structural data from the giant Okavango and related mafic dyke swarms, Karoo igneous province, northern Botswana. Earth Planet Sci Lett 202:595–606

Mallick DIJ, Habgood F, Skinner AC (1981) A geological interpretation of Landsat imagery and air photography of Botswana. Overseas geology and mineral resources, vol 56. Institute of Geological Sciences, London, 36 pp

Mayer L (1986) Tectonic geomorphology of escarpments and mountain fronts. In: Wallace RE (ed) Active tectonics. Studies in geophysics. National Academy Press, Washington, DC, pp 125–135

McCarthy TS (1998) The role of biota in the initiation and growth of islands on the floodplain of the Okavango alluvial fan, Botswana. Earth Surf Process Landf 23:291–316

McCarthy TS (2006) Groundwater in the wetlands of the Okavango Delta, Botswana, and its contribution to the structure and function of the ecosystem. J Hydrol 320:264–282

McCarthy TS, Ellery WN (1998) The Okavango Delta. Trans R Soc S Afr 53:157–182

McCarthy TS, McIver JR, Verhagen B (1991) Groundwater evolution, chemical sedimentation and carbonate brine formation on an island in the Okavango Delta swamp, Botswana. Appl Geochem 6:577–596

McCarthy TS, Bloem A, Larkin PA (1998) Observations on the hydrology and geohydrology of the Okavango Delta. S Afr J Geol 101:101–117

McCarthy TS, Cooper GJR, Tyson PD, Ellery WN (2000) Seasonal flooding in the Okavango Delta - recent history and future prospects. S Afr J Sci 96:25–33

McCarthy TS, Humphries MS, Mahomed I, Le Roux P, Verhagen BT (2012) Island forming processes in the Okavango Delta, Botswana. Geomorphology 179:249–257

McFarlane MJ, Eckardt FD (2008) Lake deception: a new Makgadikgadi palaeolake. Botsw Notes Rec 38:195–201

Mills AJ (2008) Fungus culturing, nutrient mining and geophagy. a geochemical investigation of Macrotermes and Trinervitermes mounds in southern Africa. J Zool 278:24–35

Milzow C, Kgotlhang L, Bauer-Gottwein P, Meier P, Kinzelbach W (2009) Regional review: the hydrology of the Okavango Delta, Botswana—processes, data and modelling. Hydrogeol J 17:1297–1328

Modie BN (1996) Depositional environments of the Meso- to Neoproterozoic Ghanzi-Chobe belt, northwest Botswana. J Afr Earth Sc 22:255–268

Modisi MP, Atekwana EA, Kampunzu AB, Ngwisanyi TH (2000) Rift kinematics during the incipient stages of continental extension: evidence from the nascent Okavango rift basin, northwest Botswana. Geology 28:939–942

Moore AE, Larkin PA (2001) Drainage evolution in south-central Africa since the break-up of Gondwana. S Afr J Geol 104:47–68

Moore AE, Cotterill FPD, Eckardt FD (2012) The evolution and ages of Makgadikgadi palaeo-lakes: consilient evidence from Kalahari drainage evolution south-central Africa. S Afr J Geol 115:385–413

Moorkamp M, Fishwick S, Walker RJ, Jones AG (2019) Geophysical evidence for crustal and mantle weak zones controlling intra-plate seismicity–the 2017 Botswana earthquake sequence. Earth Planet Sci Lett 506:175–183

Mujinya BB, Van Ranst E, Verdoodt A, Baert G, Ngongo LM (2010) Termite bioturbation effects on electro-chemical properties of Ferralsols in the upper Katanga (D.R. Congo). Geoderma 158:233–241

Mujinya BB, Mees F, Boeckx P, Bodé S, Baert G, Erens H, Delefortrie S, Verdoodt A, Ngongo M, Van Ranst E (2011) The origin of carbonates in termite mounds of the Lubumbashi area, D.R Congo. Geoderma 165:95–105

Mujinya BB, Adam M, Mees F, Bogaert J, Vranken I, Erens H, Baert G, Ngongo M, Vanranst E (2014) Spatial patterns and morphology of termite (Macrotermes falciger) mounds in the upper Katanga, DR Congo. CATENA 114:97–106

Murray-Hudson M, Combs F, Wolski P, Brown MT (2011) A vegetation-based hierarchical classification for seasonally pulsed floodplains in the Okavango Delta, Botswana. Afr J Aquat Sci 36:223–234

Nash DJ, Shaw PA (1998) Silica and carbonate relationships in silcrete-calcrete intergrade duricrusts from the Kalahari of Botswana and Namibia. J Afr Earth Sc 27:11–25

Nash DJ, Thomas DSG, Shaw PA (1994) Siliceous duricrusts as palaeoclimatic indicators: evidence from the Kalahari desert of Botswana. Palaeogeogr Palaeoclim Palaeoecol 112:279–295

Pastier A-M, Dauteuil O, Murray-Hudson M, Moreau F, Walpersdorf A, Makati K (2017) Is the Okavango Delta the terminus of the east African rift system? towards a new geodynamic model: geodetic study and geophysical review. Tectonophysics 712–713:469–481

Peel MC, Finlayson BL, McMahon TA (2007) Updated world map of the Köppen-Geiger climate classification. Hydrol Earth Syst Sci 11:1633–1644

Peel RA, Tweddle D, Simasiku EK, Martin GD, Lubanda J, Hay CJ, Weyl OLF (2015) Ecology, fish and fishery of Lake Liambezi, a recently refilled floodplain lake in the Zambezi region, Namibia. Afr J Aquat Sci 40:417–424

Pennisi E, Doak D, Tarnita C (2015) Africa's soil engineers: termites. Science 347(6222):596–597

Petit JR, Jouzel J, Raynaud D, Barkov NI, Barnola JM, Basile I, Delmotte M (1999) Climate and atmospheric history of the past 420,000 years from the Vostok ice core, Antarctica. Nature 399:429–436

Pozo M, Calvo JP (2018) An overview of authigenic magnesian clays. Minerals 8:520–542

Pricope NG (2013) Variable-source flood pulsing in a semi-arid transboundary watershed: the Chobe River, Botswana and Namibia. Environ Monit Assess 185:1883–1906

Pringle RM, Doak DF, Brody AK, Jocqué R, Palmer TM (2010) Spatial pattern enhances ecosystem functioning in an African savanna. PLoS Biol 8:e1000377

Ramberg L, Wolski P (2008) Growing islands and sinking solutes: processes maintaining the endorheic Okavango Delta as a freshwater system. Plant Ecol 196:215–231

Ringrose S, Kampunzu AB, Vink BW, Matheson W, Downey WS (2002) Origin and palaeo-environments of calcareous sediments in the Moshaweng dry valley, southeast Botswana. Earth Surf Process Landf 27:591–611

Ringrose S, Huntsman-Mapila P, Downey W, Coetzee S, Fey M, Vanderpost C, Vink B, Kemosidile T, Kolokose D (2008) Diagenesis in Okavango fan and adjacent dune deposits with implications for the record of palaeo-environmental change in Makgadikgadi–Okavango–Zambezi basin, northern Botswana. Geomorphology 101:544–557

Ringrose S, Cassidy L, Diskin S, Coetzee S, Matheson W, Mackay AW, Harris C (2014) Diagenetic transformations and silcrete–calcrete intergrade duricrust formation in palaeo-estuary sediments. Earth Surf Process Landf 39:1167–1187

Roberts M, Tod C, Aanen DK, Norbe T (2016) Oligocene termite nests with in situ fungus gardens from the Rukwa rift basin, Tanzania, support a Paleogene African origin for insect agriculture. Plos One 11(16):e0156847

Romanens R, Pellacani F, Mainga A, Fynn R, Vittoz P, Verrecchia EP (2019) Soil diversity and major soil processes in the Kalahari basin, Botswana. Geoderma Reg 19:e00236

Roodt V (1998) Trees and shrubs of the Okavango Delta: medicinal uses and nutritional value. Shell Oil Botswana (Pty) Ltd, Maun

Seaman MT, Scott WE, Walmsley RD, van der Waal BCW, Toerien DF (1978) A limnological investigation of Lake Liambezi, Caprivi. J Limnol Soc South Afr 4:129–144

Shaw PA (1985) The desiccation of lake Ngami: an historical perspective. Geography J 151:318–326

Shaw PA, Cooke HJ (1986) Geomorphic evidence for the late Quaternary paleoclimates of the middle Kalahari of northern Botswana. CATENA 13:349–359

Shaw PA, Cooke HJ, Thomas DSG (1988) Recent advances in the study of Quaternary landforms in Botswana. Palaeoecol Afr 19:15–26

Sianga K, Fynn R (2017) The vegetation and wildlife habitats of the Savuti-Mababe-Linyanti ecosystem, northern Botswana. Koedoe 59:1–16

Skarpe C, du Toit JT, Moe SR (2014) Elephants and savanna woodland ecosystems: a study from Chobe national park, Botswana. Wiley Ltd., Chichester
Thomas DSG, Burrough SL (2012) Interpreting geoproxies of late Quaternary climate change in Africa drylands: implications for understanding environmental change and early human behavior. Quat Int 253:5–17
Thomas DSG, Shaw PA (1991) The Kalahari environment. Cambridge University Press, Cambridge
Thomas DSG, Shaw PA (2002) Late Quaternary environmental change in central southern Africa: new data, synthesis, issues and prospects. Quat Sci Rev 21:783–797
Tooth S, Vandewalle M, Goodin D, Alexander KA (2022) Anatomy of a wetland in a dryland: the Chobe-Zambezi channel-floodplain system. In: Eckardt FD (ed) Landscapes and landforms of Botswana. Springer, Berlin
Tweedle D, Hay CJ (2011) Fish protection areas: documentation for their establishment in Sikunga and Impalila conservancies. Report for the WWF 2011:1–19
Van Thuyne J, Darini I, Mainga A, Verrecchia EP (2021) Are fungus-growing termites super sediment-sorting insects of subtropical environments? J Arid Env 193:104566
Verrecchia EP (2007) Lacustrine and palustrine geochemical sediments. In: Nash DJ, McLaren S (eds) Geochemical sediments and landscapes. Blackwell Publishing, Oxford, pp 298–329
Vittoz P, Pellacani F, Romanens R, Mainga A, Verrecchia EP, Fynn RWS (2020) Plant community diversity in the Chobe Enclave, Botswana: insights for functional habitat heterogeneity for herbivores. Koedoe 62:a1604
Wang L, D'Odorico PD, Ringrose S, Coetzee S, Macko SA (2007) Biogeochemistry of Kalahari sands. J Arid Environ 71:259–279
Watson A (1986) Grain size variations on a longitudinal dune and a barchan dune. Sediment Geol 46(1–2):49–66
Wright VP (2007) Calcrete. In: Nash DJ, McLaren SJ (eds) Geochemical sediments and landscapes. Balckwell Publishing, Oxford, pp 10–45
Wright VP, Tucker ME (1991) Calcretes: an introduction. In: Wright VP, Tucker ME (eds) Calcretes. IAS reprint series, vol 2. Blackwell, Oxford, pp 1–22
Wu JE, McClay K, Whitehouse P, Dooley T (2009) 4d analogue modelling of transtensional pull-apart basins. Mar Pet Geol 26:1608–1623
Ziegler PA, Cloetingh S (2004) Dynamic processes controlling evolution of rifted basins. Earth Sci Rev 64:1–50

Thuto Mokatse is a Ph.D. student working with Prof. Eric P. Verrecchia in the Institute of Earth Surface Dynamics at University of Lausanne (Unil, Switzerland). After having studied geology at the University of Botswana (UB), he obtained an M.Sc. degree in geology from Botswana International University of Science & Technology (BIUST). His doctoral research incorporates a multidisciplinary approach that encompasses geophysics, geochemistry, and sedimentology to investigate the Late Quaternary evolution of the Chobe Enclave landscape in northern Botswana.

Nathalie Diaz is a soil scientist working at the University of Lausanne in the Culture and Scientific Outreach Service. She is a Master in Biogeosciences (Universities of Lausanne and Neuchâtel, Switzerland) and a Ph.D. in Earth Sciences (University of Lausanne, Switzerland). She worked on pedo-sedimentary processes related to African landscape evolution throughout the Quaternary. She has developed a project in collaboration with the Royal Holloway University of London (UK) with the aim of assessing the geochronology of carbonate bed islands in the Chobe Enclave (Botswana) in order to relate them to the evolution of the Quaternary hydrological system.

Elisha Shemang is a Professor at the Department of Earth and Environmental Sciences, Botswana International University of Science and Technology (BIUST). He holds a Ph.D. in Applied Geophysics from Ahmadu Bello University. His research interests include the application of geophysics to the study of geomorphological features and geothermal resources.

John Van Thuyne is a Ph.D. researcher at the Institute of Earth Surface Dynamics at the University of Lausanne (Unil, Switzerland). He is a soil scientist doing his Ph.D. on fungus-growing termites. He studies the role and the impact these insects have on semi-arid sub-tropical savannas and in the ways in which termites transform these landscapes. His study area is the Okavango Delta and the Chobe Enclave in northern Botswana. A few years ago, he built an inter-disciplinary research centre in this unique environment in order to welcome researchers from all over the world to do fieldwork there.

Pascal Vittoz is senior lecturer at the Institute of Earth Surface Dynamics, University of Lausanne (Unil, Switzerland), and is a Ph.D. in Biology (Unil). His research interests include the composition, ecology and distribution of plant communities and the vegetation dynamics under climate change and other anthropic or natural processes. His projects are mainly based on field observations, by relocating and repeating historical plant inventories or with the help of permanent plots.

Torsten Vennemann is currently a full Professor of Stable Isotope Geochemistry at the Institute for Earth Surface Dynamics, University of Lausanne (Unil, Switzerland). He received his Ph.D. degree (1989) in Geochemistry from the University of Cape Town, South Africa. His research work has focused on developments of analytical methods and applications of stable isotope geochemistry to a broad range of questions in geology, ore deposits, (paleo)environments, and climate change.

Eric P. Verrecchia is full professor at the Institute of Earth Surface Dynamics, University of Lausanne (Unil, Switzerland). He is a Ph.D. in sedimentary petrology and biogeochemistry from Sorbonne University (1992, UPMC, Paris, France) and started his career as a senior researcher at CNRS (France). He is a recognized specialist of the coupled cycle of carbon and calcium in arid and semiarid environments. A few years ago, he started conducting research in the Chobe Enclave under the auspice of the Swiss National Foundation in partnership with BIUST (Botswana).

Check for updates

The Chobe-Zambezi Channel-Floodplain System: Anatomy of a Wetland in a Dryland

7

Stephen Tooth, Mark Vandewalle, Douglas G. Goodin, and Kathleen A. Alexander

Abstract

In this chapter, the Chobe-Zambezi channel-floodplain system is defined as the fluvially influenced area that is located around and between the Chobe and Zambezi rivers approaching their confluence. This area is located in the 'Four Corners' region, the informal term given to the region where the Botswana, Namibia, Zambia and Zimbabwe borders meet. The large-scale structure and medium-term (10^2–10^5 years) development of the channel-floodplain system is related to a combination of tectonic activity and climatically-driven changes to flow and sediment supply. The system has developed in a region of subsidence that is related to the East African Rift System. Upstream of the Mambova Rapids, the modern sinuous, alluvial channels are flanked by extensive floodplain wetlands, with crevasse splays, gullies, oxbows, scroll plains, abandoned channels and backwaters (stagnant or slow-flowing, channel-like depressions) all widespread. Collectively, these fluvial landforms create the physical template for shorter-term water, sediment and ecosystem dynamics. A strong flood and drying season dynamic is evident; river stages typically rise from January and peak around April, before subsequently falling again. The Zambezi provides the largest flow volumes, with flow spreading gradually from north to south through a complex system of active and partially active channels and floodplain wetlands towards the Chobe. Along the two rivers, lateral channel migration and extension of splays, gullies and backwaters has been negligible over at least the last 40–50 years, with few new oxbows forming. To the east, both rivers cross the uplifting Chobe fault, with each river forming complexes of steeper, bedrock anabranching channels in the region of the Mambova Rapids. The two rivers ultimately coalesce ~10 km farther downvalley, and continue as the Zambezi River. A longer term (>10^6 years) developmental model is outlined, which posits that headward retreat of the Victoria Falls, at present located ~80 km downstream of the Chobe fault, will initiate a phase of erosion that will cross the fault in ~1–2 million years' time. This phase of erosion will initiate deep channel incision, river network reorganisation and wider landscape denudation.

Keywords

Alluvial channel • Bedrock channel • Chobe River • Drylands • Floodplain wetlands • Zambezi River

7.1 Introduction

The Chobe River is located in northeastern Botswana, and flows from Lake Liambezi in the west to Kazungulu in the east, whereupon it forms a confluence with the larger Zambezi River (Fig. 7.1). The Chobe and Zambezi rivers are key landscape elements of the so-called 'Four Corners' region (Moore et al. 2007), with their typically sinuous, locally dividing, courses forming parts of the borders between Botswana, Namibia, Zambia and Zimbabwe (Figs. 7.1 and 7.2a). This complex river geomorphology has led to past conflict, with the ~3–4 km^2 Sedudu (Kasikili) Island, located between a northern and southern channel of the Chobe (Fig. 7.2a, b), being the subject of a past border

S. Tooth (✉)
Department of Geography and Earth Sciences, Aberystwyth University, Wales, UK
e-mail: set@aber.ac.uk

M. Vandewalle · K. A. Alexander
Chobe Research Institute, Centre for African Resources: Animals, Communities and Land use (CARACAL), Kasane, Botswana

D. G. Goodin
Department of Geography and Geospatial Sciences, Kansas State University, Manhattan, USA

K. A. Alexander
Department of Fish and Wildlife Conservation, Virginia Tech, USA

F. D. Eckardt (ed.), *Landscapes and Landforms of Botswana*, World Geomorphological Landscapes,
https://doi.org/10.1007/978-3-030-86102-5_7

dispute between Namibia and Botswana (see Information Box).

Beyond its role as an international border, the hydrology and geomorphology of the Chobe River have wider significance. Along with the Limpopo River (Chap. 16), and the Okavango (Chap. 2) and Kwando (Cuando) rivers (Chap. 2), the Chobe River represents one of the few permanently flowing ('perennial') rivers in Botswana territory. The spectacular fluvial landscape approaching the confluence of the Chobe and Zambezi rivers (Fig. 7.2a) is characterised by a complex of alluvial channels and extensive floodplain wetlands, with bedrock-influenced channels more prominent in the vicinity of the features known as the Mambova Rapids (Clark 1950; Moore and Cotterill 2010) or the Mambova Falls (Shaw and Thomas 1988; Thomas and Shaw 1991). In previous publications, the extensive area ($\sim$3000–4000 km^2) of alluvial channels and floodplain wetlands upstream of the Mambova Rapids (the preferred name herein) has been referred to by various names, including the Chobe swamps (Nugent 1990; Moore and Cotterill 2010; McCarthy 2013), the Chobe-Zambezi or Chobe floodplain (Shaw and Thomas 1988; Moore et al. 2007), and the Zambezi wetlands (Pricope 2013; Burke et al. 2016). Hereafter, the broader term Chobe-Zambezi channel-floodplain system is adopted to also encompass the more bedrock-influenced river reaches downvalley of the main area of alluvium and wetlands (Fig. 7.2a).

With the exception of the studies of Sedudu (Kasikili) Island that formed part of the aforementioned border dispute (see Information Box), until now there have been no detailed investigations of the fluvial landscape approaching the Chobe-Zambezi confluence, with the channel-floodplain system having received only passing mention in studies of long-term river development (e.g. Nugent 1990; Moore et al. 2007; McCarthy 2013) or reconstructions of Quaternary hydrological and lacustrine fluctuations (e.g. Shaw and Thomas 1988; Thomas and Shaw 1991). Furthermore, although part of the Chobe River lies within the Chobe National Park (located on the Botswana side of the Botswana-Namibia border—Fig. 7.2b) and is a focal point for wildlife populations and associated tourism activities (Fig. 7.2c), there is little specific visitor information regarding the channel-floodplain system and its importance for ecosystem service delivery. By comparison with the considerable research attention that has been devoted to the fluvial and fluvio-lacustrine landscapes elsewhere in

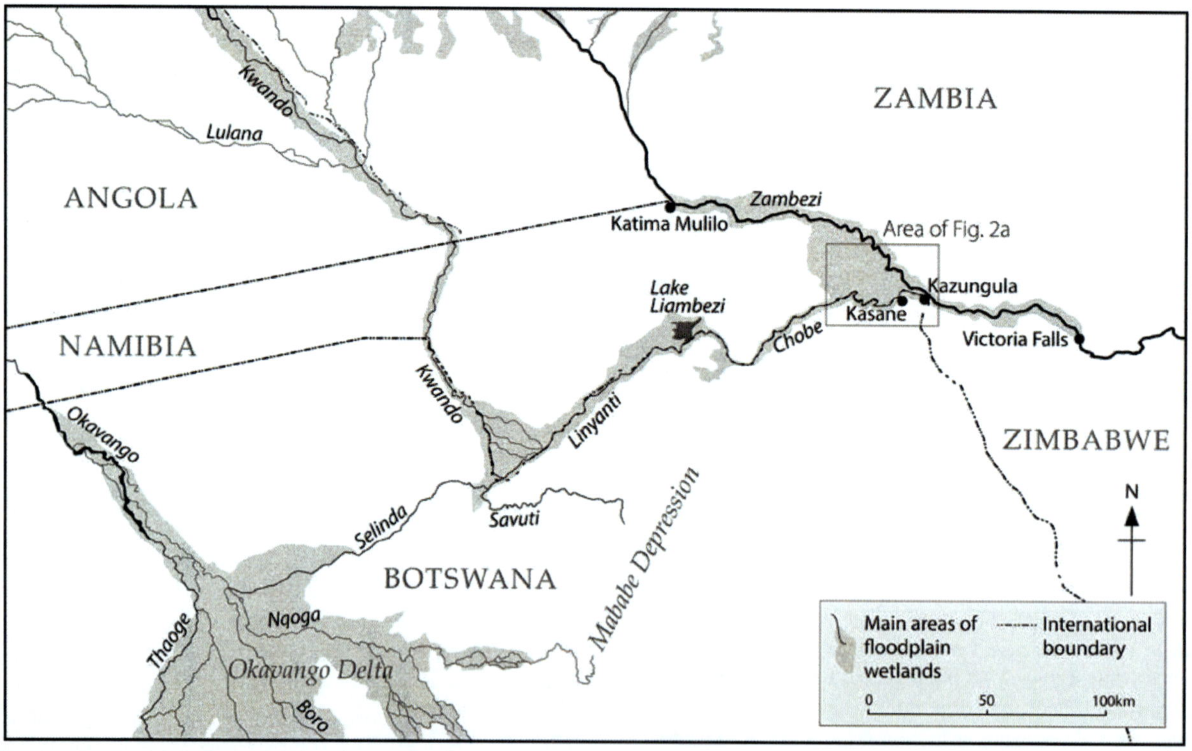

Fig. 7.1 Map encompassing the Four Corners region, showing the main landscape features referred to in this chapter, key locations, and the area covered by Fig. 7.2a. Note the northwest-southeast orientation of the three main rivers (Okavango, Kwando (Cuando), Zambezi) and the dog-legged but overall southwest-northeast orientation of shorter rivers that link these three main rivers, including the Linyanti and Chobe rivers

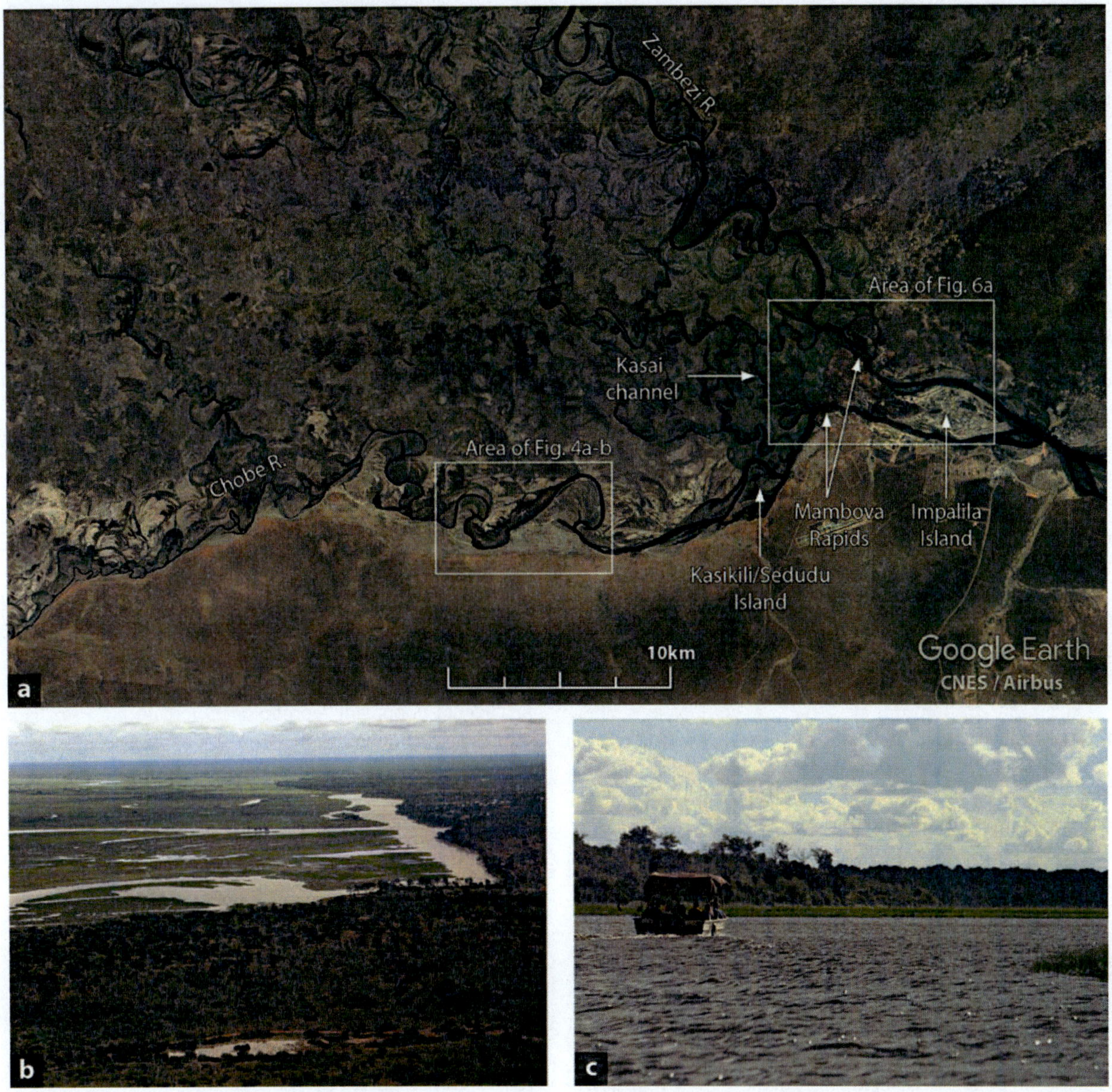

Fig. 7.2 **a** Satellite image (from Google Earth) of the area approaching the Chobe-Zambezi confluence, illustrating the complex of alluvial and bedrock fluvial landforms, other landscape features referred to in this chapter, and the areas covered by some other figures. **b** Oblique aerial view (looking downstream towards Kasane) of the Chobe River and escarpment taken during a time of rising stage (February 2016). The area in view forms part of the Chobe National Park and includes the downstream end of the disputed Sedudu (Kasikili) Island (see Information Box). **c** Ground level view of the Chobe River and escarpment in the Chobe National Park (view looking upstream)

northern and central Botswana, particularly the Okavango delta (Chap. 2) and Makgadikgadi (palaeo) lakes complex (Chap. 5), the Chobe-Zambezi channel-floodplain system appears to be a neglected cousin. This chapter describes the main fluvial landforms in the region, and outlines how they owe their origin and ongoing development to a complex interplay of tectonic activity and climate changes. We hope that this chapter and other ongoing studies will provide the basis for enhanced appreciation of this aspect of Botswana's geoheritage.

7.2 Geographical and Environmental Setting

The large-scale physiography of the Chobe-Zambezi region is related to a series of major southwest-northeast trending faults and subsidiary northwest-southeast trending faults that have resulted in regions of relative uplift (horsts) and subsidence (grabens). The more alluvial part of the Chobe-Zambezi channel-floodplain system has developed upstream of the Mambova Rapids in a low-gradient region of subsidence (Figs. 7.2a and 7.3a), essentially forming the northeastern part of what has been variably and loosely termed the Caprivi or Ngami Depression (e.g. du Toit 1927; Wellington 1955), Chobe graben (Nugent 1990), upper Zambezi Trough (Shaw and Thomas 1988), Kalahari Rift (Shaw and Thomas 1992), the Okavango-Linyanti Trough, Graben or Depression (McCarthy 2013) and the Okavango Graben (Pastier et al. 2017). These faults are related to the southwest propagation of the southwestern branch of the East African Rift System (EARS) but the limited geodetic and geophysical data mean that local crustal and lithospheric structures, mantle activity, and fault displacement patterns and rates are poorly known. Consequently, various deformation models for the region have been proposed, including an incipient rifting zone (e.g. Scholz et al. 1976; Modisi et al. 2000; Bufford et al. 2012), a composite graben (graben-within-graben) structure (e.g. McCarthy 2013) and a transtensional basin (Pastier et al. 2017).

The present-day hydrology of the Chobe-Zambezi channel-floodplain system is dominated by a strong flood and drying season dynamic. Whilst this marked seasonal regime is common to this part of central southern Africa, the location of the Chobe-Zambezi channel-floodplain within a depression amplifies the significance of floods. Seasonal movement of the Intertropical Convergence Zone and Congo Air Boundary are associated with austral summer rains. A strong north-south rainfall gradient means that the highest totals occur in the catchment headwaters in Angola and Zambia, so in the study region the Zambezi and Chobe river stages typically rise from January and peak around April, before subsequently falling again during the austral winter (Fig. 7.3b). Overbank flooding occurs with rising stage (Fig. 7.2b) but annual floodplain inundation patterns and extents are highly variable, and depend on the relative volumes and timing of peak flows along the two rivers, as well as the contributions from local rainfall and groundwater (Pricope 2013; Burke et al. 2016). The larger Zambezi provides the largest flow volumes, with overbank flow typically spreading gradually from north to south through a complex system of channels and floodplain wetlands (Fig. 7.2a) towards the Chobe River (Pricope 2013; Burke et al. 2016). Decadal-scale flow variations are also associated with El Niño-Southern Oscillation (ENSO) dynamics, with La Niña conditions tending to be associated with higher rainfall and higher flooding (Pricope 2013; Alexander et al. 2018; Heaney et al. 2019). In particularly wet years, such as occurred around the end of the first decade of the new millennium, some flow may also come from the Okavango Delta via the Selinda River (also termed the Selinda Spillway or Makgwekwana), to the Linyanti River and Lake Liambezi (Fig. 7.1; see also 'Fluvial Landforms' below), and thence to the Chobe River, with this smaller flood pulse typically peaking around June, July or August (Pricope 2013; Burke et al. 2016). Upstream of the Mambova Rapids, the characteristically low channel gradients (typically < 0.0001 m/m) mean that the Chobe has been reported to occasionally reverse its flow (i.e. flow east–west) over short distances (see Shaw and Thomas 1988; Pricope 2013; Burke et al. 2016). Near to Kasane, unambiguous measurements of this phenomenon are lacking, but 30–40 km farther west the phenomenon has been observed, and is driven by local rainfall patterns, the relative timing of flow from the Zambezi and Linyanti, and the associated relative flood levels in different parts of the system.

Importantly, the rivers in the study region are still largely unaffected by large dams, large-scale flow abstraction or significant channel engineering and so the flood dynamics and channel-floodplain morphologies remain close to their natural state. Although beyond the scope of this chapter, it is nevertheless worth noting that there is growing concern over declining river and groundwater quality, with widespread antibiotic resistance reported in river and sediment microbial populations as well as faecal isolates from many water-associated wildlife populations (Sanderson et al. 2018). In addition, flood dynamics and ENSO phenomena have been shown to be strongly correlated with increased under-5 diarrheal case reports (e.g. Fox and Alexander 2015; Alexander et al. 2018; Heaney et al. 2019).

7.3 Fluvial Landforms

Despite the ongoing debates over the most appropriate deformation model for the Chobe-Zambezi region, it is clear that fault activity has been a key factor in the development of many large-scale landscape elements, and also a profound influence on river drainage organisation and flood flow patterns (Fig. 7.1). To the west, the Okavango and Kwando (Cuando) rivers (Chap. 2) essentially terminate as large fan-shaped features (the 'inland deltas' of McCarthy 1993), with only relatively minor channels crossing the southwest-northeast oriented bounding faults at the toe of these features (Fig. 7.1). At the toe of the Okavango Delta, the Boteti River has breached the Kunyere and Thamalakane faults and follows a southeasterly course towards the

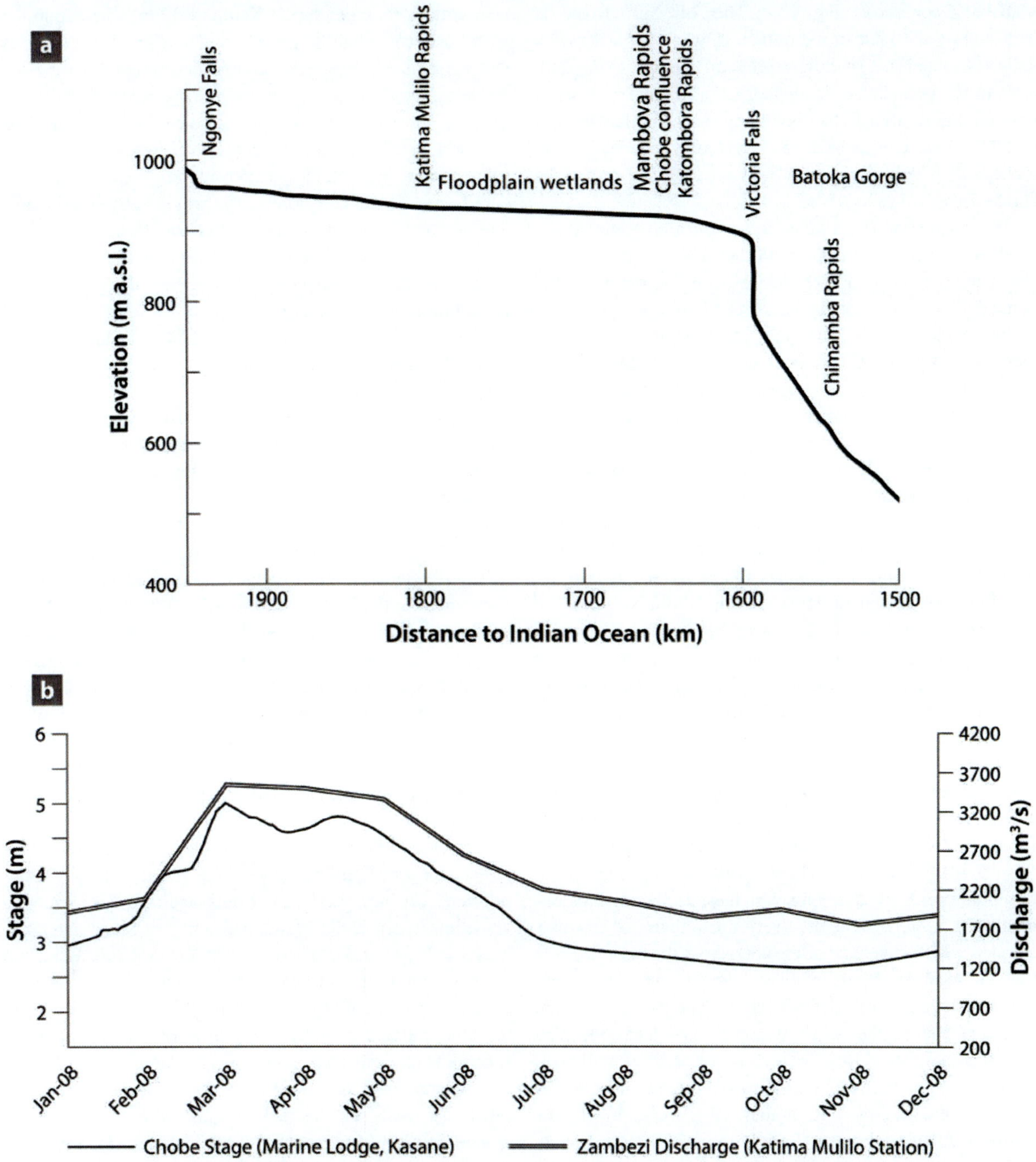

Fig. 7.3 **a** Long profile of part of the upper and middle sections of the Zambezi River, showing the location of the Chobe-Zambezi channel-floodplain system on a low-gradient reach upstream of the Mambova Rapids (modified after Nugent 1990 and Moore et al. 2007). **b** Pattern of mean monthly discharge (Zambezi River) and stage height (Chobe River) in an average year (2008) (after Pricope 2013 and Burke et al. 2016). The Zambezi data are from Katima Mulilo station (Department of Water Affairs, Namibia) and the Chobe stage data are from upstream of Kasane (derived from the ENVISAT Altimeter)

Makgadikgadi Pans complex (Chap. 5). At the foot of the Kwando Delta (also sometimes termed the 'Linyanti Swamps' or 'Mamili wetland'), the Linyanti fault has disrupted the flow, but the Savuti channel breaches the fault in the southwest and is joined by the Selinda from the west, and thereupon follows an easterly course towards the

Mababe Depression (Fig. 7.1). The Linyanti River is deflected by the fault to the northeast, whereupon it essentially terminates in a shallow depression adjacent to the fault to form the periodically inundated Lake Liambezi. A small channel has breached the bounding fault on the eastern margin of Lake Liambezi and commonly conveys flows towards the Chobe fault, essentially marking the start of the Chobe River (Fig. 7.1).

The course of the Chobe River is influenced by the southwest-northeast orientation of the Chobe fault. The river gradually increases in size downstream, becoming better defined as it approaches the main area of floodplain wetlands in the Chobe National Park (Fig. 7.2a). Although displacement rates along the Chobe fault are poorly known, uplift of some tens of metres is suggested by the fact that the prominent (up to ~30 m high) Chobe escarpment bounds the river (Fig. 7.2b, c), providing rare local exposure of sandstone and basalt. The escarpment rim is dissected by numerous small tributaries that deliver clastic sediment (dominantly clay, silt and fine-medium sand) to the Chobe (e.g. Sedudu River). There are few data on river sediment loads, but over time, the sinuous Chobe River has laterally migrated and reworked much of this sediment, forming well-defined scroll plains (Fig. 7.4a, b), commonly with 2–3 m of relief between the sandier ridges and the adjacent muddier, organic-rich swales (Fig. 7.4c). Banklines are formed mainly in fine-grained sediment (clay through fine-medium sand), with grasses and sedges widespread and shrubs and trees only present locally. Levees up to ~1.5 m high and some tens of metres wide are prominent features along some sections of bankline, and have been breached locally by crevasse splays that range up to ~0.1 km^2 in area (Fig. 7.4d). Numerous oxbows with various degrees of infilling provide evidence for past meander bend abandonment (Fig. 7.4a, b). The relatively open nature of oxbows, coupled with the patterns of erosion across the neck of some modern meander bends, indicates that many bend abandonment events result from chute cutoffs; across the upstream part of the neck, overbank flow leads to splay channels that extend downvalley and redistribute sediment, whilst on the downstream part, flow returning to the channel initiates headward-eroding gullies (Fig. 7.4a, b). If the upstream splays and downstream gullies ultimately connect, then increasing volumes of flow and sediment may be diverted from the main channel, ultimately leading to bend abandonment. Some of the stagnant or slow-flowing backwaters that parallel the main channel (Fig. 7.4a, b), including those in the vicinity of the disputed Sedudu (Kasikili) Island (Fig. 7.2a, b), may have formed as a consequence of gullies that have deepened rapidly relative to their rate of headward extension, and have since been inundated by river water.

The course of the upper Zambezi is also fault controlled. Initially, the river follows a south-southeasterly course, but then crosses rock outcrop at Katima Mulilo, whereupon the river steepens slightly across rapids (Fig. 7.3a), and is deflected more to the east-southeast, essentially skirting the northern margin of the main region of subsidence (Fig. 7.2 a). Like the Chobe River, the Zambezi follows a sinuous course with alluvial, partly vegetated banklines (Fig. 7.2a). Evidence for past lateral channel migration and local chute cutoff-driven bend abandonment is provided by scroll plains and oxbows that are preserved adjacent to the modern channel (Fig. 7.2a), but in comparison to the Chobe, the limited number of tributaries in this river reach indicates a more limited sediment supply.

The low gradient region between the Chobe and Zambezi rivers is also marked by numerous fluvial landforms (Fig. 7.2a), including many scroll plains, oxbows and partially active and abandoned channels. As the courses of the two rivers start to converge towards the southeast, several secondary channels diverge from the right-bank of the Zambezi and follow sinuous courses southward and southeastward towards the Chobe (e.g. the Kasai channel—Figs. 7.2a and 7.5). By contrast with the main Chobe and Zambezi channels, but similar to many channels in the Okavango Delta (Tooth and McCarthy 2004a), the banklines of these secondary channels have relatively little clastic sediment and are largely formed by sedges and grasses (principally papyrus and phragmites) that are rooted in peat (Fig. 7.5). As such, the main function of these channels appears to be for flow rather than sediment conveyance, although observations of subaqueous bedforms (ripples, 2D and 3D dunes) nonetheless indicate some active transport of bedload sand.

Towards their confluence, both the Chobe and Zambezi rivers cross an uplifting horst block formed along a more northerly oriented part of the Chobe fault (in this area, also sometimes termed the Mambova fault–see Shaw and Thomas 1988; Fox and Alexander 2015) (Fig. 7.6a). Here, the rivers encounter basalt outcrop and initially follow separate courses separated by Impalila Island (Fig. 7.2a). Both rivers are characterised by interlacing networks of steeper, sediment-deficient channels that typically have incised ~5 m (Shaw and Thomas 1988) into the m-scale jointed but otherwise resistant bedrock (Fig. 7.6a–c). These river reaches closely correspond to the bedrock anabranching river style that is characteristic of some other southern African river reaches, such as along the Orange River approaching Augrabies Falls in western South Africa, and the Sabie and Olifants rivers in eastern South Africa (e.g. Heritage et al. 1999; Tooth and McCarthy 2004b; Tooth 2015; Milan et al. 2018, 2020). River gradients steepen across the Mambova Rapids (Figs. 7.3a and 7.6b), with areas of hydraulically abraded and plucked (quarried) bedrock and local boulder bars indicating active rock erosion and deposition during floods (Fig. 7.6c). On the lower

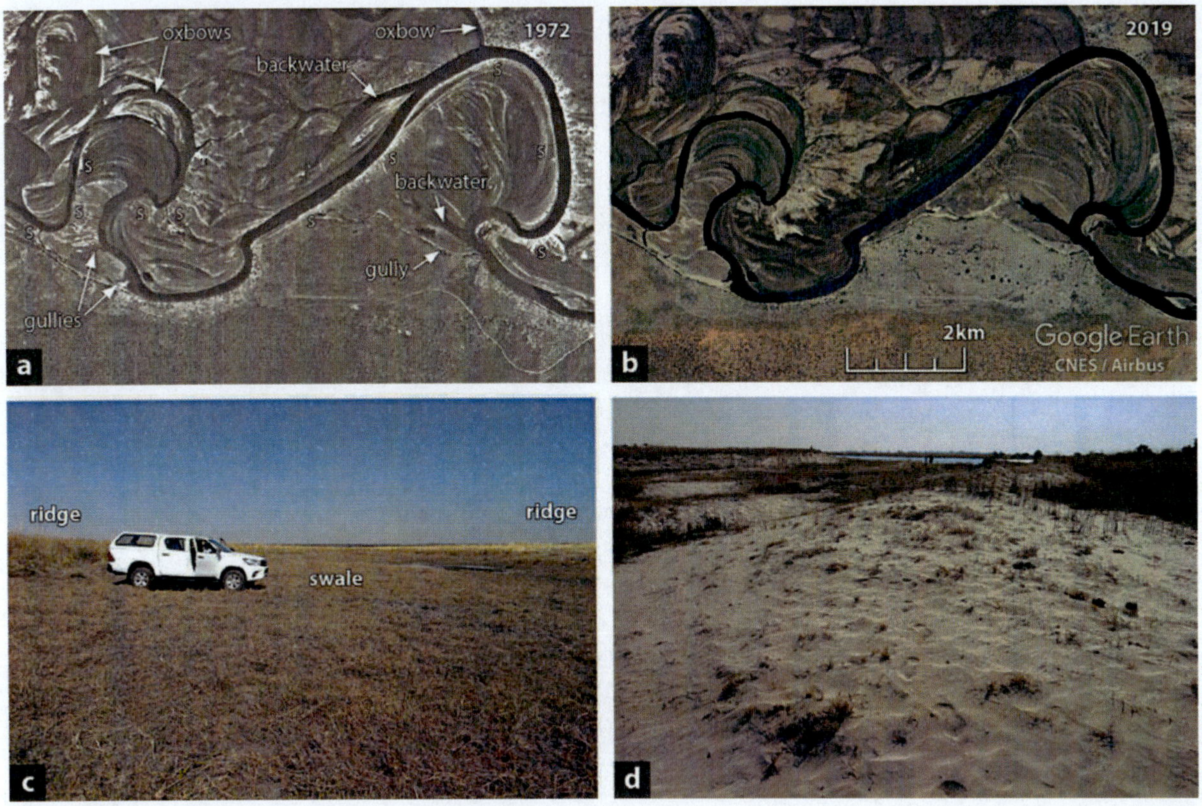

Fig. 7.4 The Chobe River and floodplain in the Chobe National Park illustrating alluvial landforms: **a** aerial image from 1972; **b** aerial image (from Google Earth) from 2019 (S indicates crevasse splay locations); **c** scroll plain on the inside of a meander bend, showing the topographic relief typically developed across the series of ridges and swales; **d** crevasse splay on the inside of a meander bend, showing sandy sediment emanating from a well-defined breach in the channel bank (view looking up the splay towards the main channel, located in the centre far distance)

Fig. 7.5 The Kasai channel, a secondary channel linking the Zambezi and Chobe rivers upstream of the Mambova Rapids (view looking downstream)

gradient reaches downstream of the falls, bedrock crops out locally and confines the channels and floodplains, but scroll plains provide evidence of some lateral channel migration and sediment reworking (Fig. 7.2a). The confluence of the two rivers occurs ~10 km downstream of the rapids near the town of Kazungula, whereupon the river continues westward as the Zambezi towards Victoria Falls, located on the Zimbabwe-Zambia border (Fig. 7.1).

7.4 Development of the Fluvial Landscape

The large-scale structure and medium-term (10^3–10^5 years) development of the Chobe-Zambezi channel-floodplain system is related to a combination of tectonic activity and climatically-driven changes to flow and sediment supply. The floodplain wetlands have developed in a region of long-term subsidence but a key control on fluvial landform

Fig. 7.6 **a** Satellite image (from Google Earth) of the Chobe and Zambezi rivers in the region of the Chobe fault and Mambova Rapids, illustrating the juxtaposition of alluvial fluvial landforms (left of image) and more bedrock-influenced fluvial landforms (middle and right of image). **b** Ground view of one of the shorter, steeper, permanently inundated, bedrock reaches at the Mambova Rapids (view looking upstream). **c** Ground view of one of bedrock reaches at the Mambova Rapids that is inundated during higher stages (view looking downstream towards the main channel, located in the far distance)

development is movement on the Chobe fault where it crosses the two rivers. The uplifting fault block provides the local base level for the rivers upstream, and so the critical factor is the relative uplift rate compared to the upstream channel-floodplain aggradation rate and/or the bedrock channel incision rates across the uplifting block. No data exist to constrain these rates, but four basic (end member) scenarios can be envisioned (Fig. 7.7):

(1) the fault uplift rate is greater than both the upstream channel-floodplain aggradation rate and the bedrock channel incision rates. This scenario results in partial or complete blockage (ponding) of channel flow and sediment transport in some upstream reaches, and abandonment of some bedrock channel reaches;
(2) the fault uplift rate is less than the channel-floodplain aggradation rates in reaches upstream. This scenario results in greater longitudinal sediment transfer, with downstream channels becoming more mixed bedrock-alluvial and parts of the outcrop becoming covered by sediment;
(3) the fault uplift rate is roughly in balance with the channel-floodplain aggradation rate and/or bedrock channel incision rates. This scenario results in equilibrium conditions, with rivers maintaining well-defined bedrock channels across the fault;
(4) the fault uplift rate is significantly less than the bedrock channel incision rates. This scenario results in the upstream passage of knickpoints, channel incision, gully formation and progressive evacuation of sediment from upstream reaches.

Aerial imagery (Fig. 7.6a) shows that whilst a few minor alluvial channels appear to be partially blocked by the uplifting fault block and some minor bedrock channels appear to be active only during the highest flows, the main channels of the two rivers are maintaining well-defined, dominantly bedrock courses. These observations suggest that the present-day channel-floodplain aggradation rate and/or bedrock channel incision rates are generally able to keep pace with uplift (scenario 3). In the past, however, scenario 1 may have applied for periods of time. A previous period (or periods) of more widespread blockage is suggested by evidence for the existence of an extensive former lake in the region upstream of the Mambova Rapids (Shaw and Thomas 1988; Thomas and Shaw 1991). Based on aerial image interpretation and fieldwork in the area south of Lake Liambezi, Shaw and Thomas (1988) described parallel series of curvilinear sandy ridges, which were interpreted as offshore bars formed during high lake phases, and adjacent diatomaceous, silty sediments, which were interpreted as lagoonal sediments. At several locations along the Chobe escarpment upstream of Kasane, Shaw and Thomas (1988) also described 'alluvial terraces' (or 'lacustrine terraces') of grey silty sand overlying calcrete hardpans. The calcrete contains localised concentrations of freshwater gastropod shells, some species of which were interpreted as indicating shallow, low-energy lacustrine conditions at the time of deposition. Based on these landforms and sediments, Shaw and Thomas (1988) posited the former existence of a 2000 km^2 palaeolake at ~936 m elevation that they named Lake Caprivi. Two radiocarbon ages from the shells indicated deposition of the shells at ~15 000 years BP, followed by subsequent formation of calcrete with falling water levels, with one other radiocarbon age from a shell indicating a return to high water levels between 2000 and 3000 years BP. The sites, sample materials and radiocarbon ages led Shaw and Thomas (1988) to conclude that Lake Caprivi was coeval with, and linked to, the 936 m Lake Thamalakane stage identified farther to the west. The relative importance of tectonic activity and climatic changes in promoting these and other, older high lake level phases in northern Botswana is unclear but Shaw and Thomas (1988) suggested that that the rock outcrop at the Mambova Rapids would have been a factor, as it would have formed a major impedance to the overflow of the 936 m lake.

By comparison with these lacustrine features, as yet there are no geochronological data for the channel-floodplain landforms and sediments in the Chobe-Zambezi channel-floodplain system. Nevertheless, the channel-floodplain landforms of the Chobe River that have been described in this chapter (e.g. scroll plains, oxbows, splays—Fig. 7.4) have all developed at elevations ~5–10 m lower than the surfaces of the terraces described by Shaw and Thomas (1988). This suggests that the disappearance of Lake Caprivi was followed by 5–10 m of channel incision, and then renewed lateral migration and floodplain formation. If this interpretation is correct, then the channel-floodplain landforms of the Chobe River—and by implication, many of the similar landforms along the Zambezi River—are probably of Holocene age. Based on comparison with ongoing optically stimulated luminescence (OSL) dating of similar fluvial landforms (e.g. scroll plains) in the Okavango delta (Larkin 2019; Tooth et al. 2022 submitted), a mid to late Holocene age for much of the Chobe River scroll plains is most likely, but this remains to be confirmed by ongoing studies.

Regardless of the channel and floodplain landform ages, analyses of aerial photographs and satellite images show that modern rates of change are very slow; along both the Chobe and Zambezi rivers, channel lateral migration, extension of crevasse splays and headward retreat of gullies and backwaters has been negligible over at least the last 40–50 years,

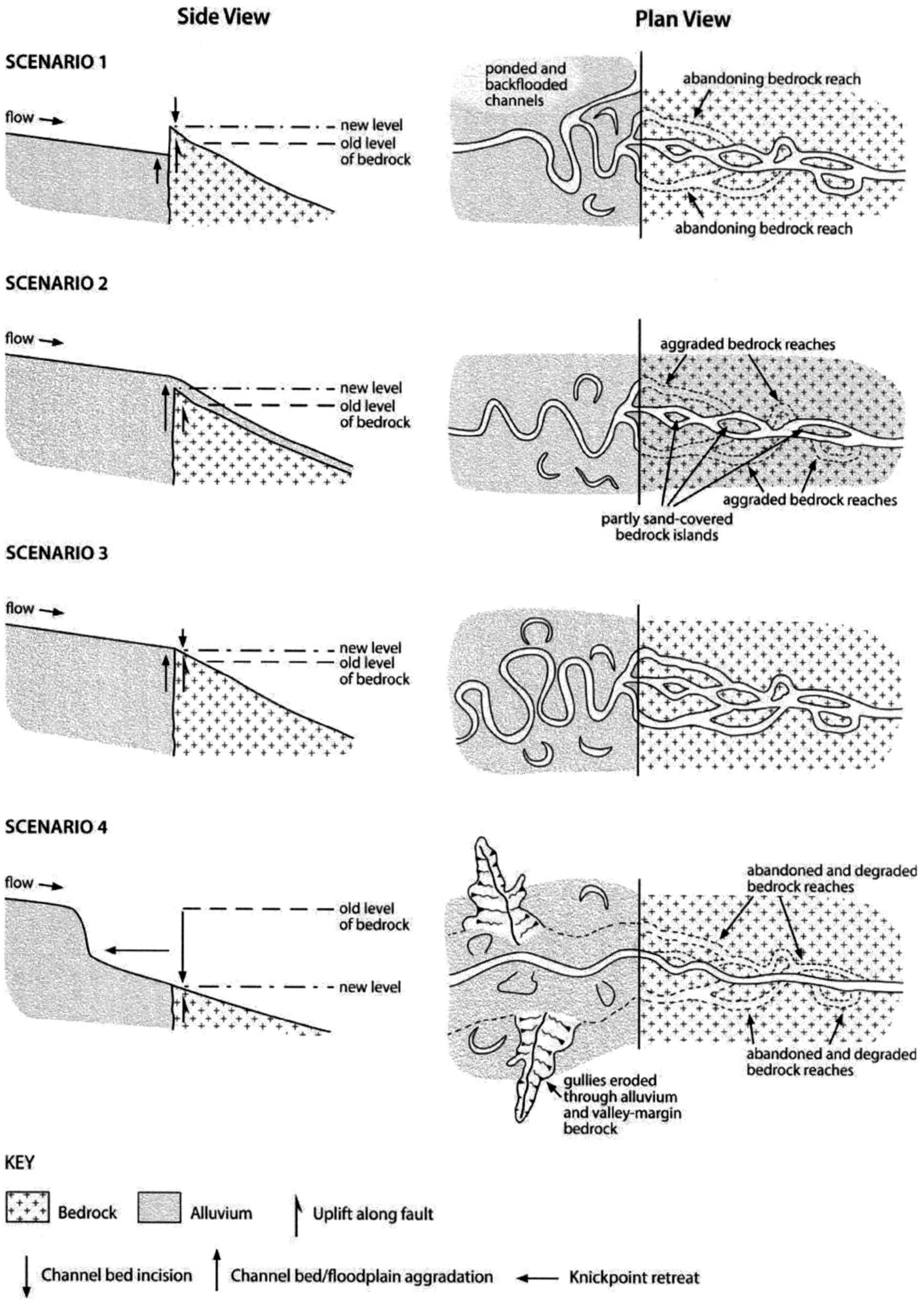

Fig. 7.7 Schematic illustrations of different scenarios of channel and floodplain change in the area around the Chobe fault

with few oxbows forming (e.g. Fig. 7.4a, b). Hence, although the observable evidence for former meandering channels, oxbows, scroll plains and splays might give the impression of a highly dynamic channel-floodplain system, on a timescale of a typical individual human lifetime (10^1–10^2 years), this is not the case. In essence, the modern Chobe-Zambezi channel-floodplain system provides a largely static—at best, very slowly changing—physical template that is nested within a complex of older, relic fluvial and lacustrine landforms.

7.5 Possible Future Developmental Trajectories

Along with the Okavango and Kwando rivers, the long-term behaviour of the Chobe-Zambezi channel-floodplain system is a key driver in the geomorphological development of northern Botswana and the surrounding areas of central southern Africa (Fig. 7.1). As this chapter has outlined, geological factors (e.g. tectonic activity, lithological outcrop) and long-term climate change have combined to drive various river responses, leading to a range of alluvial and bedrock fluvial landforms. Subsequent to the disappearance of Lake Caprivi, the modern bipartite character of the system was (re-)established, with alluvial channels and extensive floodplain wetlands developing upstream of the uplifting Chobe fault block, and more bedrock-influenced channels developing across and downstream of the fault block (Fig. 7.2a). Although small changes in the relative rates of fault uplift, channel-floodplain aggradation and bedrock channel incision are likely to control the details of reach-scale dynamics (Fig. 7.7–scenarios 1–3), this bipartite character is likely to persist on timescales of 10^4–10^5 years.

On longer timescales (>10^6) years, however, profound changes are likely to occur. Victoria Falls is located on the Zambezi River ~80 km downriver of the Mambova Rapids (Fig. 7.1), and represents a major knickpoint in the river long profile (Fig. 7.3a). In full flood, the falls have a maximum vertical drop of 108 m and width of 1700 m (Moore and Cotterill 2010). Like other great African waterfalls (e.g. Tooth 2015), the inevitable upstream retreat of Victoria Falls and the associated deep bedrock incision is progressively delivering significant base level change into the continental interior, and ultimately will influence the pattern and tempo of wider landscape denudation in the Chobe-Zambezi region. Based on the archaeological record, Moore and Cotterill (2010) have estimated that Victoria Falls is retreating at rates between ~0.040–0.080 m/yr. Assuming that these rates are maintained far into the future, then the falls will reach the Chobe-Zambezi confluence area in ~1 million years (with retreat rate at upper end of the range) to ~2 million years (with retreat rates at lower end of the range), and ultimately cross the Chobe fault. On the same timescale, the upstream retreat of the Mambova Rapids will likely already have delivered many metres of base level fall into the main region of alluvial channel and floodplain wetlands, but the arrival of the significantly larger falls will initiate a renewed phase of deep channel incision and wider landscape denudation (Fig. 7.7—scenario 4) that will pass northwest along the Zambezi River to its upper reaches and west-southwest along the Chobe and the Linyanti towards the Okavango Delta. Irrespective of future climate change scenarios and possible patterns and rates of EARS development, a likely outcome of this phase of deep incision is significant loss of alluvial river reaches and associated floodplain wetlands, with wider development of steeper, more confined, bedrock or mixed bedrock-alluvial reaches. Along the main region of subsidence, the present configuration of only partly connected alluvial river segments—each with different names, such as the Selinda, Linyanti, and Chobe (Fig. 7.1)—will progressively link more strongly. These linkages will give rise to a major river that erodes deep into the heart of central southern Africa as an increasingly integrated part of an enlarging Zambezi drainage network (cf. Wellington 1955; Moore and Larkin 2001; Moore et al. 2007, 2012; McCarthy 2013).

7.6 Conclusion

This chapter has provided the first comprehensive overview of the fluvial landforms of the Chobe-Zambezi channel-floodplain system, a neglected cousin of the better studied Okavango and Kwando systems. The descriptions of the different alluvial and bedrock fluvial landforms, and the outline of the key tectonic and climatic drivers involved in their development, hopefully will provide the basis for more detailed work that can improve understanding of the timing and rates of channel-floodplain changes. Collectively, the fluvial landforms in the region create the physical template within which occur shorter-term water, sediment and ecosystem dynamics, all of which have strong links to a range of contemporary environmental management concerns such as water quality and human and wildlife health (Fox and Alexander 2015; Alexander et al. 2018; Heaney et al. 2019). In addition, improved knowledge of past, present and possible future channel-floodplain dynamics provides opportunities for enhancing awareness of Botswana's rich geoheritage, with potential knock-on benefits for geoscience education, geotourism promotion and geoconservation efforts. These opportunities and benefits have yet to be exploited. In the Chobe National Park in particular, knowledge and insights regarding the past, present and possible future changes to the channel-floodplain system and associated ecosystem services need to be conveyed to visitors in simplified, widely accessible forms. Targeted use of improved signboarding, educational dioramas or computer animations could be used to help enhance the visitor experience provided by other aspects of the landscape and biota.

Information Box: The Role of Geomorphology in the Sedudu (Kasikili) Island Dispute Sedudu Island (known as Kasikili Island in Namibia) is a 3–4 km^2 island in the middle of the Chobe River, and is located ~5–10 km upstream of Kasane (Fig. 7.2a, b). The island is surrounded by two channels of the river, namely a 'northern channel' and a 'southern channel', and is frequently inundated during the seasonal floods. The island has no permanent residents but assumes significance because it was the subject of a long-running territorial dispute between Botswana and Namibia that culminated towards the end of the twentieth century. The dispute arose because of the imprecise wording of the agreement regarding the northern border in the 1890 Heligoland-Zanzibar Treaty. At that time, Southwest Africa (now Namibia) was under German control and Bechuanaland Protectorate (now Botswana) was under British control. Following their gaining of independence, the imprecise wording enabled both Namibia and Botswana to claim ownership of the island. The key issue was over which channel was the "main channel", as this would determine where the border between the two countries should be placed. Neither the German nor English versions of the 1890 Treaty: (i) clearly or consistently defined the notion of "main channel"; (ii) provided any criteria or guidance for identifying such a channel; or (iii) provided an unambiguous description or map of such a channel in the vicinity of the island. Consequently, the dispute centred on whether the border should run down the northern channel, thus ceding the island to Botswana, or whether it should run down the southern channel and so cede the island to Namibia.

In 1996, the two countries reached a Special Agreement to resolve the dispute in the International Court of Justice (ICJ) in The Hague. As reported by various commentators (Alexander 1999; Salman 2000; Johnson 2000), documentation associated with both the Botswana and Namibian cases included data and information regarding the hydrology, topography, sedimentology, and dynamics of the Chobe River channels and Sedudu (Kasikili) Island. Sequential aerial photographs from 1925 onwards show that the island has essentially been stable for many decades and has undergone no detectable changes in size and shape, so the northern and southern channels are effectively fixed in position. Hence, in its deliberations, and alongside other arguments concerning historical use, jurisdiction and navigability, the ICJ took into consideration the channel depths and widths, flow volumes, bed profile configurations, and bed and bank sediment characteristics (see https://www.icj-cij.org/en/case/98).

The 1999 ruling of the ICJ was that the border should run down the thalweg of the river (the line of deepest soundings) in the northern channel. This established the entirety of Sedudu (Kasikili) Island within Botswana's territory. The court recalled, however, that under the terms of a 1992 agreement (the Kasane Communiqué), the two countries had accepted that there should be unimpeded navigation for craft of their nationals and flags in the channels around the island, including free movement of tourists. Both countries were commended for their commitment to the peaceful resolution of the dispute, for respecting freedom of navigation, and for injecting environmental considerations into the navigable uses of the river (Salman 2000).

Sedudu Island hosts a significant wildlife population and remains one of the major tourist attractions in the Chobe National Park but in addition is a significant part of Botswana's geoheritage. In particular, the history outlined above illustrates the value of colonial-era historical aerial photographs in determining channel/island dynamics on decadal-centennial timescales, and also provides a rare example of where geomorphology has featured prominently in an international court of law to settle a border dispute.

Acknowledgements This study was conducted under permit from the Ministry of Environment, Natural Resources Conservation and Tourism (EWT8/36/4). Support for this work was provided by the National Science Foundation Dynamics of Coupled Natural and Human Systems (Award #1518486 to Alexander, https://www.nsf.gov) and by CAR-ACAL. We would like to thank Lipa Nkwalale for his assistance in obtaining historical aerial photographs, the Department of Wildlife and National Parks for facilitation of fieldwork and Gareth Edwin for cartographic support. Aberystwyth University's Centre for International Development Research at Aberystwyth (CIDRA) is providing support for ongoing geomorphological investigations by Tooth and colleagues in the Four Corners region. We also appreciate the review and editorial comments of Fenton (Woody) Cotterill, Frank D. Eckardt, and Piotr Migoń, as these helped to shape the final version of the chapter.

References

Alexander KA, Heaney AK, Sharman J (2018) Hydrometeorology and flood pulse dynamics drive diarrheal disease outbreaks and increase vulnerability to climate change in surface-water-dependent populations: a retrospective analysis. PLoS Med 15(11):e1002688

Alexander WRJ (1999) Science, history and the Kasikili Island dispute. S Afr J Sci 95:321–324

Bufford KM, Atekwana EA, Abdelsalam MG, Shemang E, Atekwana EA, Mickus K, Moidaki M, Modisi MP, Molwalefhe L (2012) Geometry and faults tectonic activity of the Okavango Rift Zone, Botswana: evidence from magnetotelluric and electrical resistivity tomography imaging. J Afr Earth Sc 65:61–71

Burke JJ, Pricope NG, Blum J (2016) Thermal imagery-derived surface inundation modeling to assess flood risk in a flood-pulsed savannah watershed in Botswana and Namibia. Rem Sens 8:676

Clark JD (1950) The stone age cultures of Northern Rhodesia: with particular reference to the cultural and climatic succession in the upper Zambesi Valley and its tributaries. South African Archaeological Society, Cape Town, p 157

du Toit A (1927) The Kalahari and some of its problems. S Afr J Sci 24:88–101

Fox JT, Alexander KA (2015) Spatiotemporal variation and the role of wildlife in seasonal water quality declines in the Chobe River, Botswana. PLoS One 10:e0139936

Heaney AK, Shaman J, Alexander KA (2019) El Niño-Southern oscillation and under-5 diarrhea in Botswana. Nat Commun 10:5798, 9 pp

Heritage GL, van Niekerk AW, Moon BP (1999) Geomorphology of the Sabie River, South Africa: an incised bedrock-influenced channel. In: Miller AJ, Gupta A (eds) Varieties of fluvial form: Chichester. Wiley, UK, pp 53–79

Johnson C (2000) Case concerning Kasikili/Sedudu Island (Botswana/Namibia). Int J Mar Coast Law 15:581–599

Larkin Z (2019) Dryland rivers and hydroclimatic change: past, present and future, unpublished PhD thesis, Macquarie University

McCarthy TS (1993) The great inland deltas of Africa. J Afr Earth Sc 17:275–291

McCarthy TS (2013) The Okavango Delta and its place in the geomorphological evolution of Southern Africa. S Afr J Geol 116:1–54

Milan DM, Heritage G, Tooth S, Entwistle N (2018) Morphodynamics of bedrock-influenced dryland rivers during extreme floods: insights from the Kruger National Park South Africa. Geol Soc Am Bull 130:1825–1841

Milan DM, Tooth S, Heritage G (2020) Topographic, hydraulic, and vegetative controls on bar and island development in mixed bedrock-alluvial, multichanneled, dryland rivers. Water Resour Res 56:23 pp

Modisi M, Atekwana E, Kampunzu A, Ngwisanyi T (2000) Rift kinematics during the incipient stages of continental extension: evidence from the nascent Okavango rift basin, northwest Botswana. Geology 28:939–942

Moore AE, Cotterill FPD (2010) Victoria falls: Mosi-oa-Tunya—the smoke that thunders. In: Migoń P (ed) Geomorphological landscapes of the world. Springer Science+Business Media, pp 143–153

Moore AE, Larkin PA (2001) Drainage evolution in south-central Africa since the breakup of Gondwana. S Afr J Geol 104:47–68

Moore AE, Cotterill FPD, Main MPL, Williams HB (2007) The Zambezi River. In: Gupta A (ed) Large rivers: geomorphology and management. John Wiley and Sons, pp 311–332

Moore AE, Cotterill FPD, Eckardt FD (2012) The evolution and ages of Makgadikgadi palaeo-lakes: consilient evidence from Kalahari drainage evolution, south-central Africa. S Afr J Geol 115:385–413

Nugent C (1990) The Zambezi River: tectonism, climatic change and drainage evolution. Palaeogeogr Palaeoclimatol Palaeoecol 78:55–69

Pastier A-M, Dauteuil O, Murray-Hudson M, Moreau F, Walpersdorf A, Makati K (2017) Is the Okavango Delta the terminus of the East African Rift System? Towards a new geodynamic model: geodetic study and geophysical review. Tectonophysics 712–713:469–481

Pricope NG (2013) Variable-source flood pulsing in a semi-arid transboundary watershed: the Chobe River, Botswana and Namibia. Environ Monit Assess 185:1883–1906

Salman MAS (2000) International rivers as boundaries: the dispute over Kasikili/Sedudu Island and the decision of the International Court of Justice. Water Int 25:580–585

Sanderson CE, Fox JT, Dougherty ER, Cameron ADS, Alexander KA (2018) The changing face of water: a dynamic reflection of antibiotic resistance across landscapes. Front Microbiol 9: 1894, 13 pp

Scholz CH, Koczynski TA, Hutchins DG (1976) Evidence for incipient rifting in southern Africa. Geophys J Int 44:135–144

Shaw PA, Thomas DSG (1988) Lake Caprivi: a late Quaternary link between the Zambezi and middle Kalahari drainage systems. Z Geomorphol 32:329–337

Shaw PA, Thomas DSG (1992) Geomorphology, sedimentation, and tectonics in the Kalahari Rift. In: Schick AP (ed) Surficial processes and landscape evolution: rift valleys and arid terrains. Isr J Earth Sci 41:87–94

Thomas DSG, Shaw PA (1991) The Kalahari Environment. Cambridge University Press, 284 pp

Tooth S (2015) The Augrabies falls region: a fluvial landscape divided in flow but magnificent in spectacle. In: Grab S, Knight J (eds) Landscapes and landforms of South Africa. World geomorphological landscapes. Springer-Verlag, Berlin-Heidelberg, pp 65–73

Tooth S, McCarthy TS (2004a) Controls on the transition from meandering to straight channels in the wetlands of the Okavango Delta Botswana. Earth Surf Process Landforms 29:1627–1649

Tooth S, McCarthy TS (2004b) Anabranching in mixed bedrock-alluvial rivers: the example of the Orange River above Augrabies Falls Northern Cape Province, South Africa. Geomorphology 57:235–262

Tooth S, McCarthy TS, Duller GAT, Assine ML, Wolski P, Coetzee G (2022) Significantly enhanced mid Holocene fluvial activity in a globally-important, arid-zone wetland: the Okavango Delta, Botswana, Earth Surf Process Landforms, in press

Wellington JH (1955) Southern Africa: a geographical study. Physical geography, vol 1. Cambridge University Press, Cambridge

Stephen Tooth Graduated from the University of Southampton, UK, and completed a Ph.D. at the University of Wollongong, Australia. He undertook postdoctoral work at the University of the Witwatersrand, South Africa, before joining Aberystwyth University, UK. His research focuses on geomorphology, sedimentology, and environmental change, especially with respect to rivers and wetlands in the drylands of southern Africa, Australia, South America and southern Europe.

Mark Vandewalle Completed his B.Sc. degree at the University of the Witwatersrand, South Africa, majoring in Zoology, Botany and Ecology. He completed a Ph.D. at the same university in Wildlife Ecology, having conducted his fieldwork in northern Botswana. He subsequently joined the Department of Wildlife and National Parks on contract with the Government of Botswana and was the Principal Wildlife Biologist for the northern

Districts of the country. He co-founded the NGO "Centre for the Conservation of African Resources; Animals, Communities and Land use" (CARACAL) and managed the organisation as the Chief Executive Officer. His main research interests are wildlife ecology and conservation, and environmental and human health, and he is also involved in community outreach and conservation education

Douglas G. Goodin Received his Ph.D. from the University of Nebraska–Lincoln, USA, where he emphasized the use of multi- and hyper-spectral remote sensing for the analysis of surface energy budgets and mass-energy exchanges. His current research focuses on use of remote sensing for analysing the biophysical effects of land use and land cover change, and how these land cover dynamics affect human-environment interactions.

Kathleen A. Alexander Received her Ph.D. and veterinary degree from the University of California, Davis, and has been conducting research in east and southern Africa for over 20 years. She has worked for the Government of Botswana as the Chief of the Wildlife Veterinary Unit in the Department of Wildlife and National Parks, and later as the Ecological Advisor to the Office of the President of Botswana and the Attorney Generals Chambers. Most recently, she served on the Botswana Presidential Covid Task Force as a scientific advisor. She has spent most of her professional life working with local communities, integrating scientific approaches with traditional understanding in order to identify interventions for improved rural livelihoods. She is a member of the World Conservation Union's Wildlife Health Specialist Group and the Commission for Ecosystem Management. She moved to the Department of Fisheries and Wildlife Conservation at Virginia Tech, USA, in 2007 where she continues to conduct research in her long-term Botswana study site on the dynamics of emerging infectious disease at the human-animal interface.

Dunes of the Southern Kalahari

8

David S. G. Thomas and Giles F. S. Wiggs

Abstract

The dunes of the southern Kalahari are amongst the most studied in the world. Occurring in a region embracing arid and semiarid conditions, and dunes that are both marginally dynamic as well as more stabilised, research has focussed both on the processes operating on partially and variably vegetated surfaces, and the palaeoenvironmental histories of dunefield accumulation. Whilst commonly classified as linear dunes, this chapter examines the range and variability of forms in southern Botswana and contiguous areas, their temporally variable dynamics, and the developing chronometric histories of their development during the late Quaternary period.

Keywords

Linear dunes • Partial vegetation cover • Disturbance • Holocene

8.1 Introduction

The Kalahari is a predominantly sandy dryland region, and dunes are an important part of southern African landscapes as a whole (Thomas and Wiggs 2012). Despite this, the Kalahari does not totally fulfil common conceptions of deserts in terms of the degree of aridity or the occurrence of sand dunes, which are absent from large tracts, especially in Botswana. It is in the southern and northern-most regions of the country that dunes dominate the landscape, but in very contrasting ways. In the driest southwestern part of Botswana, westwards from Tshabong, an extensive region of relatively low but pronounced linear dunes dominates a landscape that also includes distinctive pans (Chaps. 5 and 10) and river valleys (Chap. 11). North of around latitude 23°S, more degraded dune forms are found in a number of contrasting systems juxtaposed with the major lacustrine and fluvial systems of Makgadikgadi, Okavango and Zambezi. The northern dunes are analysed in the next Chap. 9; here we focus on the dunes of southern Botswana and neighbouring territories.

The southern Kalahari dunefield extends into the Northern Cape of South Africa and southeastern Namibia (Fig. 1.6), with the dunefield as a whole spanning latitudes 23°S (where the northern limit is met, in Namibia) to 28°20′S in the Northern Cape, with the Orange River providing the southerly limit to the main system, though some pockets of dunes are found south of the river. The system may have its origins at the Plio-Pleistocene transition (Miller 2014) with subsequent reworking of sediments represented by today's dune bodies, which record both evidence of episodic accumulation (Thomas and Burrough 2016) and contemporary activation (Wiggs et al. 1995).

From west to east the system extends from 18°E to 22°30′E longitude (Thomas and Shaw 1991). Most of this dunefield, which covers an area of around 100,000 km^2 (Fryberger and Goudie 1981), is in fact found outside Botswana, although the driest area, which has the most active dune surfaces under modern climate conditions, is in the southwest of the country, centred on Bokspits in the area traversed by the Nossop and Molopo river valleys, which mark the border between Botswana and South Africa (Fig. 8.1). On average, mean annual rainfall in the part of the summer precipitation zone of southern Africa that the southern dunefield occupies is 150–300 mm, but in any year can be significantly less or significantly more, with inter-annual rainfall variability as high as 50%. There is enough moisture in the system to support, under natural conditions in non-drought periods, a partial vegetation cover on the dunes which are dominated by, but not confined to, linear forms (Fig. 8.1), with vegetation including grasses, shrubs and

D. S. G. Thomas (✉) · G. F. S. Wiggs
School of Geography and the Environment, Oxford University Centre for the Environment, University of Oxford, South Parks Road, Oxford, OX1 3QY, UK
e-mail: david.thomas@ouce.ox.ac.uk

F. D. Eckardt (ed.), *Landscapes and Landforms of Botswana*, World Geomorphological Landscapes,
https://doi.org/10.1007/978-3-030-86102-5_8

trees (Fig. 8.2). Vegetation limits the ability of the wind to mobilise sand, with the operation of aeolian processes today commonly restricted to the upper slopes and crests of dunes (Wiggs et al. 1994), but both drought periods and grazing pressures (Perkins 2018) can result in more extensive sand mobilisation in the southern Kalahari (Thomas and Leason 2005, Fig. 8.3).

8.1.1 Early Work

Early twentieth century descriptions of the southern Kalahari dune landscape were limited to observations made by travelers. Arnold Hodson, the first British policeman in Bechuanaland, perceptively described the southwest of the country as 'not quite an ordinary desert… it is made up of a sea of sand hills' (Hodson 1912:21) going on to note that it 'constitutes by far the most dreary and depressing part of the desert' (ibid). In Namibia, Korn and Martin (1937) noted how the dunes that form the northwestern part of the system are cut by river valleys, suggesting formation prior to valley incision. Lewis (1936) provided the first scientific descriptions focussed on the dunes, with his field observations confined to the South African part of the dune system and conducted as part of a systematic topographic survey of the Union (Fig. 8.4) that also made use of discontinuous air photograph coverage flown in 1934.

It is worth considering some of the information in these first scientific reports, as they frame and provide context for more recent research. Amongst Lewis's pertinent observations was the parallelism and generally uniform size and spacing of the dune ridges, which at the large scale arced from c. 20° to 40° from north. He also noted that in some areas, such as north of the confluence of the Nossob and Auob rivers, dune patterns were less regular, whilst in other areas dunes branched one from another, with branches often terminating abruptly. Lewis also noted that the dunes were on average 27 feet (8.2 m) high with a mean crest to crest wavelength of 250 yards (228.6 m), had broad flat crests up to 30 feet (9.1 m) wide, with the crests being the only parts

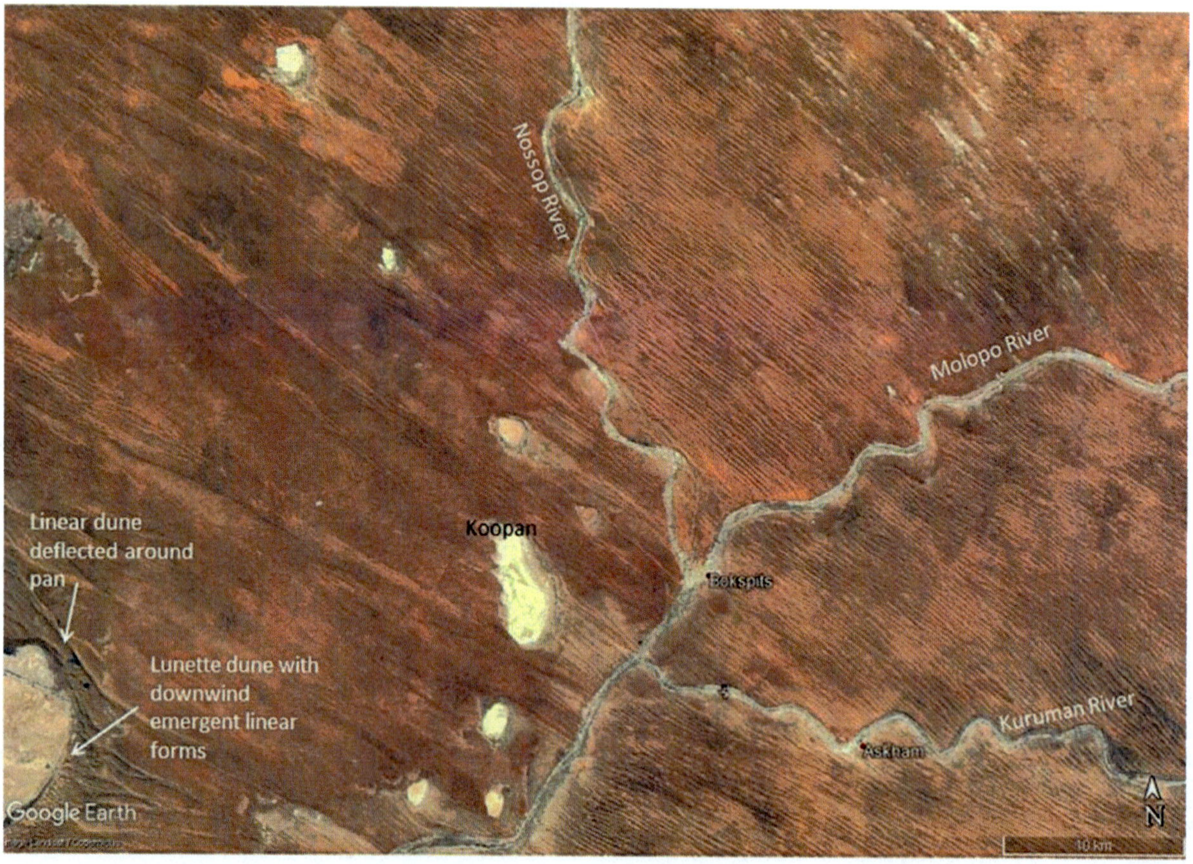

Fig. 8.1 Google Earth image of part of the southern Kalahari dune system at the intersection of Botswana and South Africa. Three (usually) dry valleys cut through the linear dunes, in some places leading to dune-end deflections on the upwind (NW) side of valleys and in other places sourcing sediment for smaller downwind dunes. Pan depressions interact with dunes in various ways: sourcing downwind lunette dunes, which in turn may source linear forms, or seemingly deflecting airflow and dunes around their margins

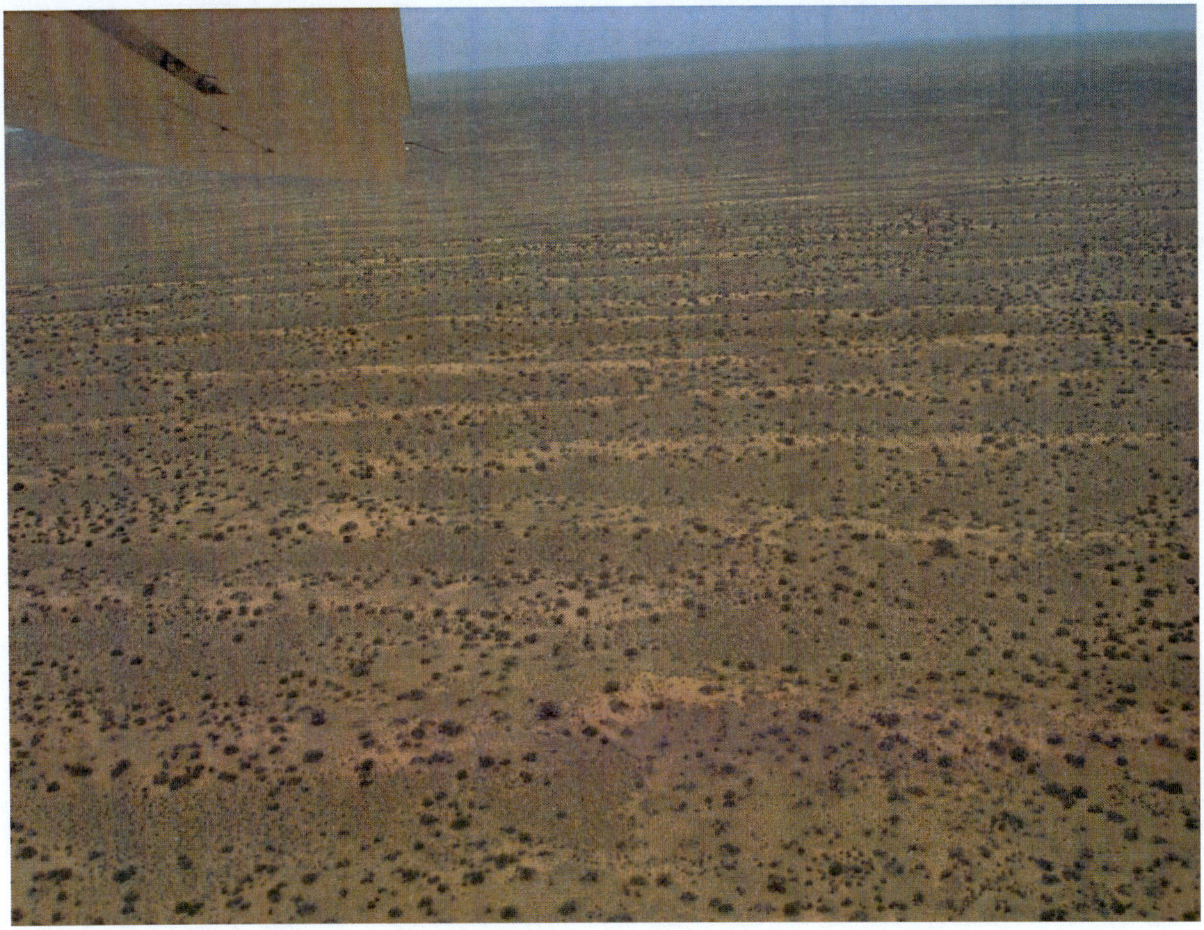

Fig. 8.2 Partially vegetated linear dunes near the Aoub valley, supporting a mixed vegetation cover. Inter-dunes are commonly well vegetated with grass species. Photograph taken in 2011. *Credit* David Thomas

of the dunes where the wind appeared to move sand. Photographs in Lewis's (1936) paper actually illustrate dune sites with very different vegetation covers, with some showing dunes and interdune areas devoid of vegetation except for widely spaced small trees. Based on the estimated age of one tree stump, 'a kameel-doorn tree said to be 400 years old' (p. 29) Lewis considered the dunes to be older than this and younger than the underlying but unspecified Kalahari limestone. He also regarded the features as transverse dunes, formed under the influence of easterly winds.

It was the aerial and field-based descriptions and analyses of Grove (1969) and Goudie (1969, 1970) that marked the beginning of a more systematic approach to dune science in the Kalahari. Grove's study drew on complete air photograph coverage of Botswana, with limited extension into South Africa, and six weeks of field reconnaissance to provide detailed morphological data and descriptions of landforms (Fig. 1.6). Building on Lewis's (1936) paper, much of Grove's description of the southern dunes mirrored these earlier observations but with some key differences and additions. First, Grove recorded the spatially systematic nature of dune system vegetation, with sparse grasses on crests, grasses, shrubs and small trees on flanks and taller (up to 5 m high) acacia trees in the grassed interdune areas. East–west variations in vegetation densities reflected the regional rainfall gradient. Second, Grove (1969) correctly ascribed the dominant forms as linear (rather than transverse) dunes, with formative sand transporting winds from the north-northwest over much of the curved dune system. He also reported that most dune 'Y junctions' occurred with the stems pointing downwind. Rather than branching—that is dunes splitting into two ridges—convergence has dominated, with adjacent ridges merging downwind, a point picked up in a statistical analyses of the ridge patterns by Goudie (1969) and Thomas (1985).

Third, over and above the occurrence of Y junctions within the dunefield morphometry, Grove recognised the

Fig. 8.3 Grazing pressures, especially close to water points, and droughts, can contribute to dune surface de-vegetation that in turn facilitates surface sediment mobilization by the wind. Photograph near Bokspits, 2008. *Credit* David Thomas

considerable complexity of dune patterns and forms in the southern Kalahari. This was reflected in his map of the system that included transverse ridges between 25°S and 25.8°S, particularly on the Botswana side of the border with South Africa (Fig. 1.6), as well as faint ridges ('furrow patterns') over an extensive area of central and eastern southern Botswana. Other than in Mallick et al. (1981), there has been little reference to these features since, though they might represent remnants of a southwest-northeast curvature of the main southern dune system (Thomas and Shaw 1991). Fourth, Grove also reflected on the ways in which the linear ridges interacted with other landscape features in the region. This included the downwind development of linear ridges from localised transverse (lunette) dunes on the margins of pan depressions (see Chap. 5) and interactions with dry river valleys that were subsequently analysed in-depth by Bullard and Nash (1998, 2000). Figure 8.1 shows examples of these contexts.

8.2 Dune Sediments

Aeolian processes are size selective, so that the sands that form desert dunes tend to be comprised of grains that have an upper size (relating to grain diameter) of c. 2 mm (2000 μm). Grains with a diameter less than 0.63 μm are classed as silt and clay, the silt fraction forming the main component of aeolian dust.

The sands of southern Kalahari linear dunes generally comprise over 90% quartz (Lewis 1936), the remainder comprising feldspars, zircon, garnet and other accessory minerals. Reddish-yellow hues are common (Thomas and Martin 1987), though sands may be paler and greyer downwind of valleys and pans (Stone and Thomas 2008). The redness of the dune sand is a result of thin iron coatings on grain surfaces (Lewis 1936). Sands are relatively coarse compared to those in many desert dunefields (Thomas and

Fig. 8.4 Lewis's (1936) map of southern Kalahari dunes, largely confined to the south African sector

Shaw 1991) although sediments generally become finer in a south-easterly direction in the system as a whole (Lancaster 1986). From the analysis of almost 50 samples from dune sediment cores from eastern Namibia, well-sorted medium sands dominated with the mean grain size commonly in the 250–300 μm range (Stone and Thomas 2008). Around 350 km further east in the Northern Cape, sands fall in the fine-medium size range. Cross-profile sediment sorting trends have also been reported, with dune crests generally being better sorted and comprising coarser sand than that found on the lower flanks (Lancaster 1986; Livingstone et al. 1999).

Silt particles are relatively rare in the sands of active dunes, because the saltation process that mobilises sand grains commonly liberates finer particles into the atmosphere where they can often be transported over longer distances by suspension. In most of the dune sand samples analysed by Stone and Thomas (2008) and Telfer (2011), less than 5% comprised silt or finer clay-sized material. An analysis of sediments from bare and partially vegetated dunes at Struizendam and Bokspits, on the Botswana side of the Nossop valley, reported less than 2% silt and clay in dune surface sediments (Bhattachan et al. 2013).

The interdune areas (or 'straats') between dune ridges are commonly sand covered, with sands generally having a higher silt and clay content than the dunes and also a greyer or whiter colouration (Thomas and Martin 1987). The thickness of these deposits is highly variable, being thinnest in western and southern parts of the system (Bullard and Nash 1998). The system as a whole is located on what was described as a limestone plateau by Mabbutt (1957) but is actually a partially silicified calcrete duricrust (Chap. 13) that is exposed in the flanks of the valleys that dissect the dunefield (Bullard and Nash 1998) or in pan margins (Lancaster 1986).

8.3 Technological Developments: The Advent of Systematic Research on Southern Kalahari Dunes

Despite only a small number of studies or descriptions of the southern Kalahari dunes being published in the first 70 years of the twentieth century, those considered above set the scene for the in-depth analyses of dune forms, histories and processes that have followed. From the 1970s onwards significant technical advances in the geosciences have also facilitated both observation of and research in the southern dunefield. Satellite imagery (e.g. Fig. 8.4), first available with global coverage from 1972 when the Landsat programme commenced, generated the ability for high resolution analysis of dunefield-wide patterns, morphometric analyses, and change analysis over time. For the southern Kalahari, this commenced with the work of Breed et al. (1979) within E. D. McKee's seminal edited volume *A study of global sand seas*. Luminescence dating was first applied to dune sediments in the Thar Desert, India, by Singhvi et al. (1982) with subsequent applications leading to the Kalahari becoming one of the best-dated dune systems on earth (Lancaster et al. 2016). This has allowed dune formation and accumulation histories to be produced, whilst questions surrounding aeolian sediment movement on partially vegetated dunes has benefited from the evolution and application of a range of field measurement and monitoring methodologies that allow sediment transport and surface changes to be recorded (Thomas and Wiggs 2008).

It is easy to ignore one further simple development that has changed our ability to investigate Kalahari dunes. Until the twenty-first century, access to the dunefield was largely restricted to sand tracks and dirt roads. Travel by horse, camel (a police camel breeding station was situated in the dunes near Bokspits and Askham on the South African side of the Molopo valley) or early motor vehicles at best made reconnaissance and research a slow process. Grove's (1969) paper, for example, was contingent on a slow six-week traverse of Botswana that barely entered the southern dunefield, getting only as far as Khuis on the northern side of the Molopo valley. Now tar roads allow the heart of the dunefield to be reached in a matter of hours from Upington in South Africa or Tsabong in Botswana, whilst Google Earth allows the full dune system to be explored and analysed from a desk.

The following sections pursue in detail the three themes considered above—dunefield morphology, dunefield accumulation and dune surface processes and activity in the southern Kalahari, focussing primarily on the principal linear and linear-like dune forms, before consideration of other, often under analysed forms found within the system.

8.4 Southern Kalahari Linear Dune Morphology and Morphometry

Linear dunes are the most common desert dune type on earth, and broadly speaking linear forms develop in wind environments dominated by a bimodal sand transport direction (Lancaster 1982; Thomas 2013), with dunes forming largely by extension in the resultant downwind direction (Thomas 1988; Telfer 2011). When the southern Kalahari dune system is considered as a whole, the pattern forms a 'wheelround' (Shaw 1997), or semi-circular arc (Lancaster 1981) that reflects the winds that blow anticlockwise around the semi-stable southern African anticyclone (Tyson 1986). Seasonal movements in the position of this atmospheric circulation feature mean that fluctuations in mean wind direction can occur throughout the year. Breed and Grow (1979), Lancaster (1981) and Thomas (1984) all used the 'Fryberger method' (Fryberger 1979) to calculate modern potential (not absolute as other factors such as ground cover and soil moisture determine whether sand actually moves) sand transport in the Kalahari dune systems. Lancaster (1981) noted that sand movement potentials were generally simple (unimodal), and only in the west of the dune system, around Keetmanshoop in Namibia, are potential sand transporting wind regimes truly directionally bimodal.

Imbalances between components of the wind regime that shape the ridges can lead to asymmetry in dune cross-profiles, which is a characteristic of these landforms recorded for example in Australia (Mabbutt and Wooding 1983) as well as in the Kalahari (Lewis 1936; Bullard et al. 1995). It can also lead to some lateral migration of ridges over time, recorded in linear ridge studies from Namibia and China (Bristow et al. 2005; Rubin et al. 2008). The occurrence of Y junctions, which have fascinated researchers in the southern Kalahari from Lewis (1936) through to Grove (1969), Goudie (1969) and Thomas (1988), may relate to complexities in the wind regime and sand transport over time leading to the deflection of ridge ends and their merging with adjacent forms (Breed and Grow 1979). Where ridges are spaced further apart, the opportunities for merging are fewer compared to situations where dunes are more closely spaced. The controls on dune patterns were subsequently investigated in modelling studies during the 1990s by for example Werner (1995) and Werner and Kucurek (1999).

Lancaster (1987) estimated that around a third to a half of the dunes in the southern Kalahari were simple linear ridges, simple dunes being those that do not have contact with neighbouring dunes (McKee 1979). These equate to the parallel/near-parallel forms recognised by Lewis (1936) in the South African sector and the extremely straight and continuous ridges that are particularly prevalent in the northern parts of the system, especially to the east of Mariental and around Stampriet in Namibia (Stone and Thomas 2008; White et al. 2015). Here Y junctions are rare and dunes are respectively c. 9 m high and 500–1000 m apart, trending NNW-SSE, and 16 m and 1–2 km apart, trending NW–SE. Individual ridges may be tens of kilometres long without converging on adjacent dunes. Through the system as a whole, the ridges are generally higher where ridge spacing becomes greater (Thomas 1988).

Factors affecting dune spacing and dune height are potentially complex and controversial, but certainly the volume of available sediment is one factor that limits or facilitates the number of ridges that can form in a given area and the heights that they can attain (Wasson and Hyde 1983). Wind regime, and changes therein over time, and the frequency of dune building episodes, both potentially linked to long term climate changes (Livingstone and Thomas 1993; Stokes et al. 1997; Bailey and Thomas 2014) are further elements that contribute to the complexities that have influenced dunefield patterns and morphologies in the southern Kalahari.

The mostly parallel and continuous forms apart, the remaining half- to two-thirds of the system possesses more complex dune patterns, where interactions between individual forms are more common and the identification of specific continuous ridges is more difficult. According to McKee (1979), these types of dunes are either compound, when dunes of the same type merge (i.e. in the Kalahari at Y junctions) or complex, where dunes of different forms interact, for example where linear forms emerge from pan margin dunes, as shown in Fig. 8.4. In parts of the southern Kalahari, the complexity of the patterns of dunes generates forms that are more akin to networks of dunes rather than strict linear forms.

Bullard et al. (1995) used detailed aerial photograph analysis and digital elevation models to investigate the variations in dune morphology and morphometry in a 4000 km^2 area of the dune field regarded as containing representative examples of the forms found throughout the system. This investigation led to a statistically rigorous five-fold classification of dune plan forms and morphologies that was then applied to an analysis of the system as a whole (Fig. 8.5). Dune classes embrace discontinuous ridges (class 1), through to simple continuous ridges (class 2) and network-like forms (class 5). With the formation of the dunefield associated with winds broadly from the northwest to westerly sector, and therefore with sediment transport occurring from a similar direction, it was shown that dune patterns become more complicated in a south-easterly direction (Lancaster 1988; Bullard et al. 1995) with an associated increase in Y junction occurrence. A possible secondary trend may relate to the regional aridity gradient (Bullard et al. 1995). With conditions becoming wetter towards the northeast (Thomas and Shaw 1991), the overall temporal opportunities for dune building and organisation, either in terms of long term periods of aridity or shorter aridity/drought phases, may have been less than in drier parts of the system.

The morphology and pattern of dunes can be affected by interactions with other landforms in the system, particularly river valleys and pans (Fig. 8.4), with changes in airflow (Eitel and Blummel 1997; Bullard and Nash 2000) and sediment availability (Thomas et al. 1993) being causal factors. Bullard and Nash (1997) recognised four types of dune valley contexts in the southern Kalahari: where dune patterns and orientations are the same on both up- and downwind sides of a valley (for example at the lower Nossop valley at 26°39'S 20°38'E); where dunes adjoin a valley on the upwind side and the pattern adjacent to the valley is altered (for example on the Nossop valley at 26°24'S 20°43′

	General Description	Example Location
1	Dunes parallel/sub-parallel, discontinuous occuring as short lengths (< 2km). Y-junctions uncommon and there are no transverse elements.	26° 49'S 21° 00'E
2	Dunes parallel/sub-parallel and continuous for several km, few Y-junctions and no (or very rare) transverse elements. Those Y-junctions which do occur tend to form at the junction of two long dunes rather than as short spurs at the side of a dune. Occasional slip faces on crests but < $2m^2$.	26° 42.57'S 20° 42.15'E
3	Dunes parallel/sub-parallel and continuous for several km. Y-junctions common,both as parallel dunes merge and as short spurs < 600m on either side of dune. No slip faces on undisturbed dunes but may be common where grazing occurs.	26° 33.01'S 20° 31.42'E
4	Linear dune network comprising large steep dunes and smaller gently sloping dunes. Larger dunes have broadly linear trend but are very sinuous. Small dunes tend to be orientated perpendicular to this trend. Small pans occur in deep interdune areas. Both types of Y-junction occur, Y-junctions and termini are common.	26° 26.84'S 20° 48.35'E
5	No obvious linear trend and a chaotic hummocky appearance. Dune slopes shallow with very rounded crests. Dunes have low relief but occasional dunes up to 7 - 8m high. Very little interaction between dunes	26° 08.74'S 20° 36.37'E

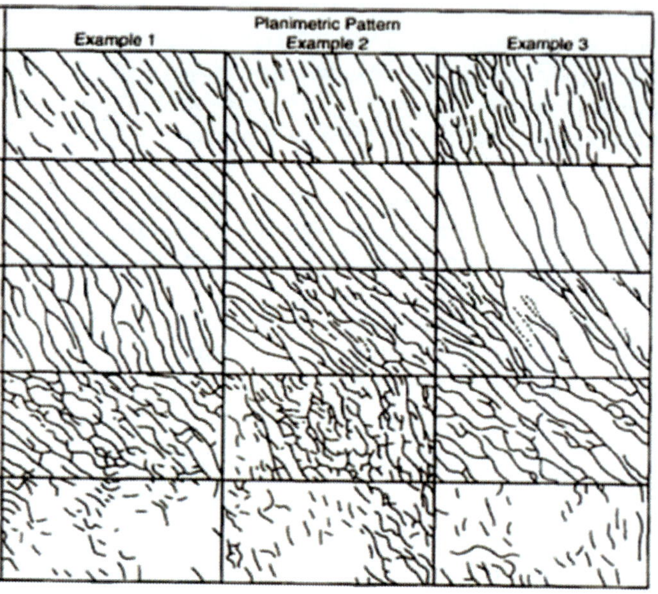

Fig. 8.5 Dune pattern and morphology variations in the southern Kalahari derived from aerial photograph and statistical analyses (Bullard et al. 1995). Reproduced with the permission of John Wiley and Sons

E); where dunes adjoin a valley on the upwind side but are absent on the downwind flank (i.e. the dunes terminate at the valley, or recommence several kilometres downwind of the valley, for example at 26°52'S 20°47'E on the Molopo valley); and where dune-free zones occur on both valley sides (for example along the Aoub valley at 24°46'S 18° 44'E). The type of association appears to be dependent on the topography and relative orientation of a valley at the point of contact with the dunes (Bullard and Nash 1998). Where dune patterns appear unaltered despite being traversed by a valley, a further explanation is that a once-continuous dune pattern has been disrupted by later channel flow (Thomas et al. 1998).

Some of the interactions between linear dunes and pan depressions are not dissimilar to those found between dunes and valleys, though other relationships also occur. In some contexts linear dune orientations clearly deflect around the margins of pans, which may be incised over 50 m into underlying sediments, presumably resulting in localised airflow deflection that has influenced net sediment transport direction on the linear ridges. In other contexts pans occur in broad, duneless corridors, but linear dunes extend downwind from the pan margin lunette. This is very evident, for example, at Koopan and other pans near the confluence of the Kuruman and Molopo valleys (Fig. 8.4), at Witpan (Thomas et al. 1993) and at numerous pans in southwest Botswana (Lancaster 1986).

8.4.1 Other Dune Forms in the Southern Kalahari and Their Significance

Aside from the patterns of linear and linear-like dune forms that dominate the southern Kalahari, other dune forms are an important if more localised part of the landscape too. Surface vegetation disturbances, often caused by fire, drought or grazing, can cause localised sand movement at sites often called 'blow outs'. In places, significant patches of superimposed secondary dune development can result where sand is moved downwind from disturbed patches, resulting in a parabolic-shaped dune front with trailing arms. Nested patches of parabolic dunes, up to 1 km long, are found in several locations in the dunefield (Eriksson et al. 1989; Thomas and Shaw 1991; Fig. 8.6). Other localised forms also occur, such as small barchan dunes developed from valley floor sediments in the Molopo (Lewis 1936; Bullard and Nash 1998), and narrow arcuate forms associated with valleys that Bullard and Nash (2000) term valley-marginal dunes.

Over 110 pan depressions in the southern Kalahari (Mallick et al. 1981), both within (Grove 1969) and beyond (Lancaster 1978a) the main dunefield, possess fringing dunes on their downwind margins (Fig. 8.7). Often described as lunettes (Hills 1940) due to their distinctive planform, these dunes (Telfer and Thomas 2006) are commonly regarded as developing from sediment deflated from the pan

Fig. 8.6 Patches of parabolic dunes are found at several locations in the dunefield, including here centred on 25.973°S 20.631°E in the Kalahari Transfrontier Park. Google Earth image, July 2013

floor, either directly during dry conditions when pans are excavated (Lancaster 1978b), from sediments initially transported to the downwind margin by wave action during wetter conditions (Bowler 1986), or from sediment that has washed or blown into the basin from the margins (Thomas et al. 1993).

Most lunette dunes in the southern Kalahari are vegetated and do not appear to be accumulating significant sediment under present conditions (Goudie and Thomas 1986). A notable exception is at Witpan (20.17°E 26.67°S) in the Northern Cape, where the western sector of a large fringing lunette has a distinctive bare and active lunette crest with a slip face (Thomas et al. 1993). The pan-facing plinth of the lunette is indurated with clay facilitating runoff and gullying during rainfall events that return sediment to the pan floor that is subsequently recycled back to the dune by the wind during drier conditions.

Grove (1969) noted that pans may have more than one downwind lunette dune, with differences in orientation, which can be particularly distinct in central southern Botswana reflecting changes in net sediment transport direction (Lancaster 1978b). Lawson and Thomas (2002) applied luminescence dating to a series of lunette sequences from pans within the main dunefield to the west of the Molopo valley, whilst the large 9 m high lunette at Witpan was subject to intensive full-profile luminescence dating by Telfer and Thomas (2007). Both studies found lunette age records dominated by Holocene ages, including accumulation in the last few centuries on inner lunettes and the upper sediments at Witpan.

8.5 Late Quaternary Linear Dune Accumulation in the Southern Kalahari

The partial vegetation cover that southern Kalahari dunes support is a significant characteristic that might relate to dune development and behaviour in two ways. It likely indicates that the main dune bodies formed in the past, and are now largely stabilised, and it may represent the generally static, and extending, nature of linear dunes being conductive to vegetation colonisation (Thomas 1992). These two

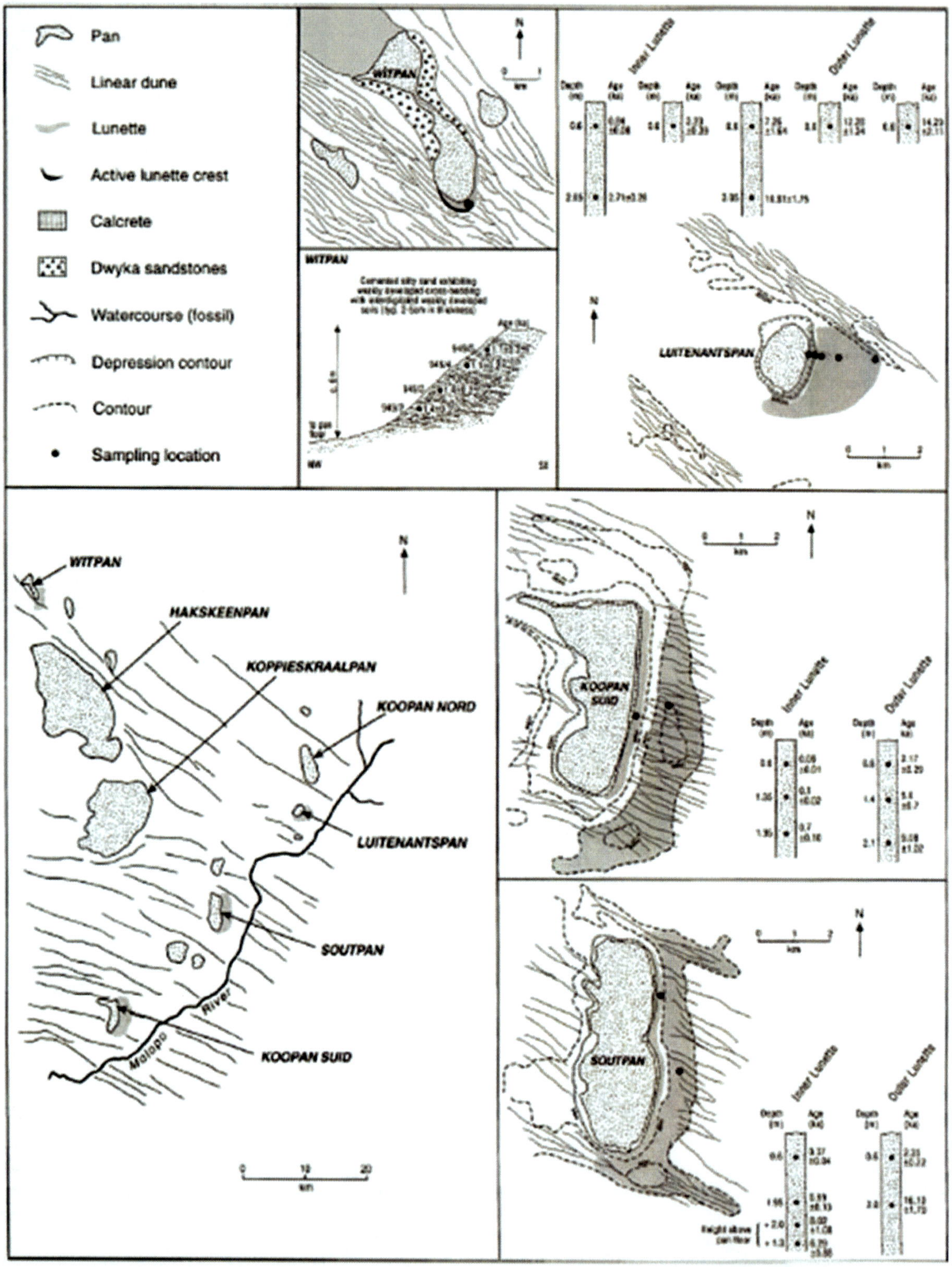

Fig. 8.7 Variations in lunette dune forms and associated dune ages from pans west of the Molopo valley (Lawson and Thomas 2002). Reproduced with the permission of Elsevier publishers

explanations are not, of course, mutually exclusive, especially in environments such as the southern Kalahari where rainfall is both seasonal and highly variable from year to year. Vegetation cover, and the sand on dune surfaces, is not immutable to change even at the relatively short timescales of years to decades (Livingstone and Thomas 1993). Knowing when the dunes were formed is therefore an important facet of understanding their status in the modern landscape.

Deacon and Lancaster (1988: 62) regarded the extensive Kalahari dune systems as 'one of the most impressive pieces of evidence for the nature of late Quaternary climatic change in the Kalahari'. Lancaster (1981) compared the orientation of dunes with the modern sand transport vectors from wind data analysis, to infer subtle changes in the position of the southern African anticyclone between the time when dunes formed and the present day. He went on to develop a dune mobility index (*MI*) integrating both wind data and effective moisture data (annual precipitation minus potential evapotranspiration) (Lancaster 1988; Fig. 8.8), that was used to infer how climate conditions had changed in the region since the dunes were constructed. The interpretation of *MI* values depended on calibrating the index to a range of modern dune states, and the assumption that when mean annual precipitation exceeded 150 mm, there is too much moisture and vegetation in the environment for sufficient aeolian sand transport for dune bodies to accumulate.

From the 1990s onwards, luminescence dating, which measures the time that has elapsed since sediment was buried (i.e. since it was last transported on the surface of a dune and exposed to sunlight), and studies of sediment transport and environmental conditions affecting these dunes, considered in the next section, have gone a long way to explaining the history and behaviour of these features, and understanding the potential relationships, and differences, between dune building and surface sediment movement.

Luminescence dating was first applied to Kalahari linear dune sands by Stokes et al. (1997), since which 270 ages have been produced from sediments in the southern Kalahari dunefield out of 600 ages from dune features in southern Africa as a whole (Thomas and Burrough 2016, Fig. 8.9a). In the southern Kalahari 194 ages are from the linear ridges and associated features of the main dune system with most of the remainder from the lunette dunes on the margins of pan depressions. Thomas and Burrough (2016) divide the southern Kalahari linear dunes as a whole into the western system, which comprises the long, quasi-straight ridges of Namibia, and the southern system, which includes the more complex forms in the Northern Cape and in Botswana.

In some respects, the chronologies of dune emplacement and accumulation have proved difficult to interpret, especially as more age data have been published (Fig. 8.9b). With the notable exception of one full dune profile that was exposed and sampled, above underlying calcrete, in a road pit near Twee Rivieren on the South Africa – Botswana border (at 26°31'S, 20°36'E), the earliest ages were derived from a small number of samples primarily collected from hand-dug pits in the crests of ridges (Stokes et al. 1997). From this, a relatively simple picture of discrete episodes of late Quaternary dune building emerged for the southern Kalahari. Stokes et al. (1997) suggested that the linear dunefield was primarily constructed in phases at 6–10 ka and 22–16 ka ago, with the deepest basal age from the pit of 28 ± 8 ka.

As the sampling of dune ridge sediments progressed through the use of augering methods that allowed samples suitable for luminescence dating to be extracted from the full depth of dune sediments, new insights occurred. These included the identification of dune basal ages as old as 104 ± 8.3 ka near Witpan (Telfer and Thomas 2007) and in excess of 180 ka from the long straight ridges near Stampriet (Stone and Thomas 2008).

New questions and challenges also emerged regarding the interpretation of dune age records, reviewed by Telfer and Hesse (2013), and Thomas and Burrough (2016). Amongst these, emerging from work in the southern Kalahari, Stone and Thomas (2008) showed how the depth of sampling, and the sampling intervals used, affected interpretation of age records. Other imperfections in dune luminescence age chronologies from the region (Chase 2009), including the gaps in records and the relatively high statistical errors on ages, have also made interpretation challenging, compounded by uncertainties surrounding the climatic drivers of linear dune dynamics. Further, when age statistical errors are considered, the identification of distinct periods of dune building often becomes blurred through the overlapping of the one sigma statistical errors that age calculations produce (Thomas and Burrough 2016, Fig. 8.9b).

In some respects, these issues limit the contribution that dune age studies have been able to make to understanding the landscape history of the southern Kalahari region during the Quaternary period (Chase 2009; Chase and Brewer 2009). Part of the difficulty has derived from the ages themselves being used as the proxy record of past environmental conditions, without evidence of an association with a particular environmental process. To address this problem, Thomas and Bailey (2017) developed a methodology to quantify dune sediment accumulation associated with sequences of luminescence ages derived from dune bodies. This approach was built on Bailey and Thomas's (2014) numerical model of linear dune accumulation that incorporated the climatic factors that drive dune surface activity and sediment accumulation. The model also recognised, and factored in, the potential effects of post-depositional reworking of older dune sediments on preserved records and the impacts of sampling and analytical processes on the

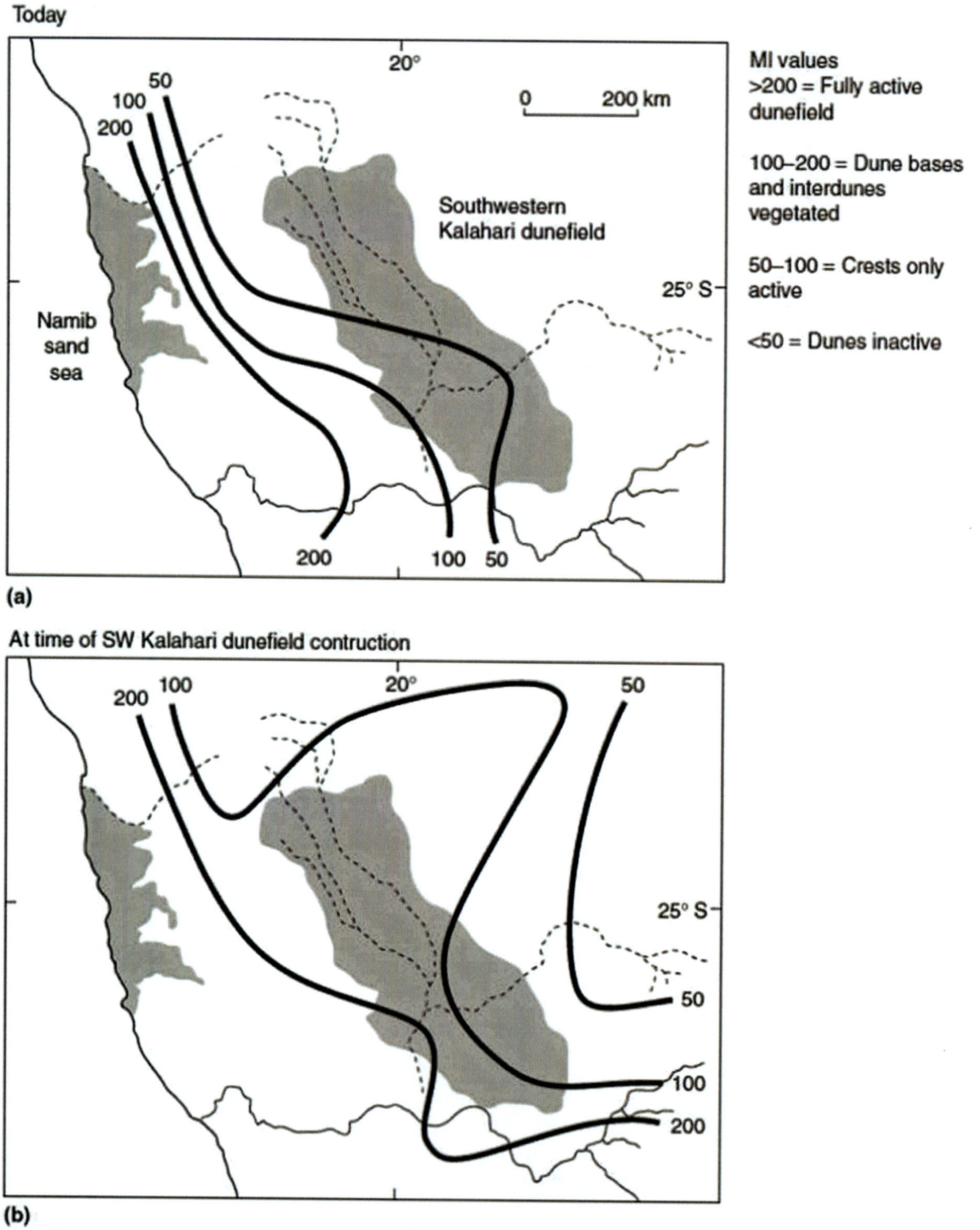

Fig. 8.8 Mobility Index values for the southern dunefield **a** today and **b** inferred for the time of dune accumulation. Index values are: 200 = fully active dunes; 100–200 = interdunes and lower dune slopes vegetated; 50–100 = only dune crests active and unvegetated; < 50 = dunes wholly inactive and vegetated. From Lancaster (1988). Reproduced with the permission of Elsevier publishers

(a)

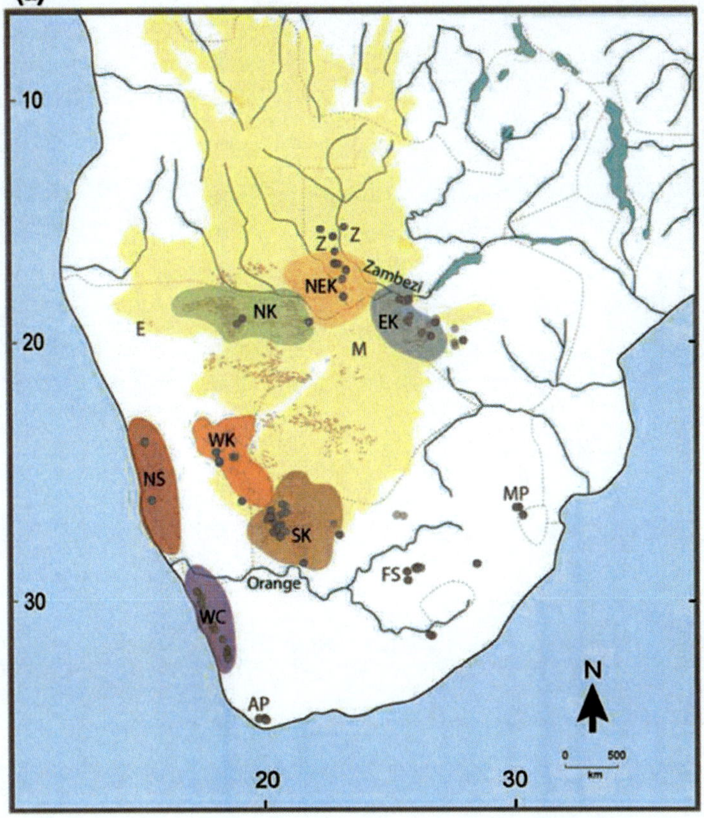

Fig. 8.9 **a** Locations of luminescence dated dune systems in southern Africa as a whole, showing also the distinction between the western and southern components of the wider Southern Kalahari dune system (after Thomas and Burrough 2016). Major dunefield areas are shaded: SK = southern Kalahari; WK = western Kalahari; EK = eastern Kalahari; NK = northern Kalahari; NEK = northeastern Kalahari; NS = Namib Sand Sea; WC = West Coast dunefield. Smaller dune areas or regions where dunes occur but are scattered rather than landscape-covering features are shown by letters: AP = Agulhas Plain; FS = Free State; MP = Mpumalanga; Z = Zambia east of the Zambezi River. E = Etosha Pan, M = the Makgadikgadi basin. Some of these systems are considered further in Chap. 9. Reproduced with the permission of Elsevier publishers. **b** Plots of southern Kalahari linear dunefield age data, showing how the overall record has evolved as new data have appeared. Data are shown for three time periods relating t the publication of information: 1997–2002; 2003–2007; and 2008–2013. The black histogram bars show the central ages plotted in 5 ka groups, while individual ages plotted with the central ages and statistical errors (horizontal lines associated with each age show an apparent record with indistinguishable phases of accumulation due to the age errors overlapping. The broad pale horizontal bars show the 9–16 ka and 20–26 ka periods of dune building that was proposed in the initial age study in the region by Stokes et al. (1997). Reproduced with the permission of Elsevier publishers

resultant data. The model was then developed to interpret dune luminescence age datasets (Thomas and Bailey 2017), quantifying the records of sediment accumulation that exists within stacked dune age records, examining the varying thicknesses of accumulation intervals between successive dated sediment units, and integrating the records for whole dunefields.

This approach was first applied to the total dunefield age and sediment dataset from the Kalahari, producing continuous records of variations in dune building (called accumulation intensity, or *AI*) through time (Thomas and Bailey 2017). Figure 8.10 shows the *AI* curves for the linear dunes of the southern and western Kalahari for the last 50 ka, alongside the record of water accumulation in the Stampriet aquifer (Stute and Talma 1998). This substantial aquifer underlies the southern dunefield (Chap. 12) and its recharge record, derived from ^{13}C dating of waters, is regarded as a proxy for regional rainfall changes. It can be seen that the linear dunes presently found in the southern and western parts of the system accumulated most intensively during the early Holocene period, around 13–9 ka, prior to significant aquifer recharge in the mid-Holocene. This represents a period of increased rainfall compared to today's conditions that may have generated sufficient vegetation on dune

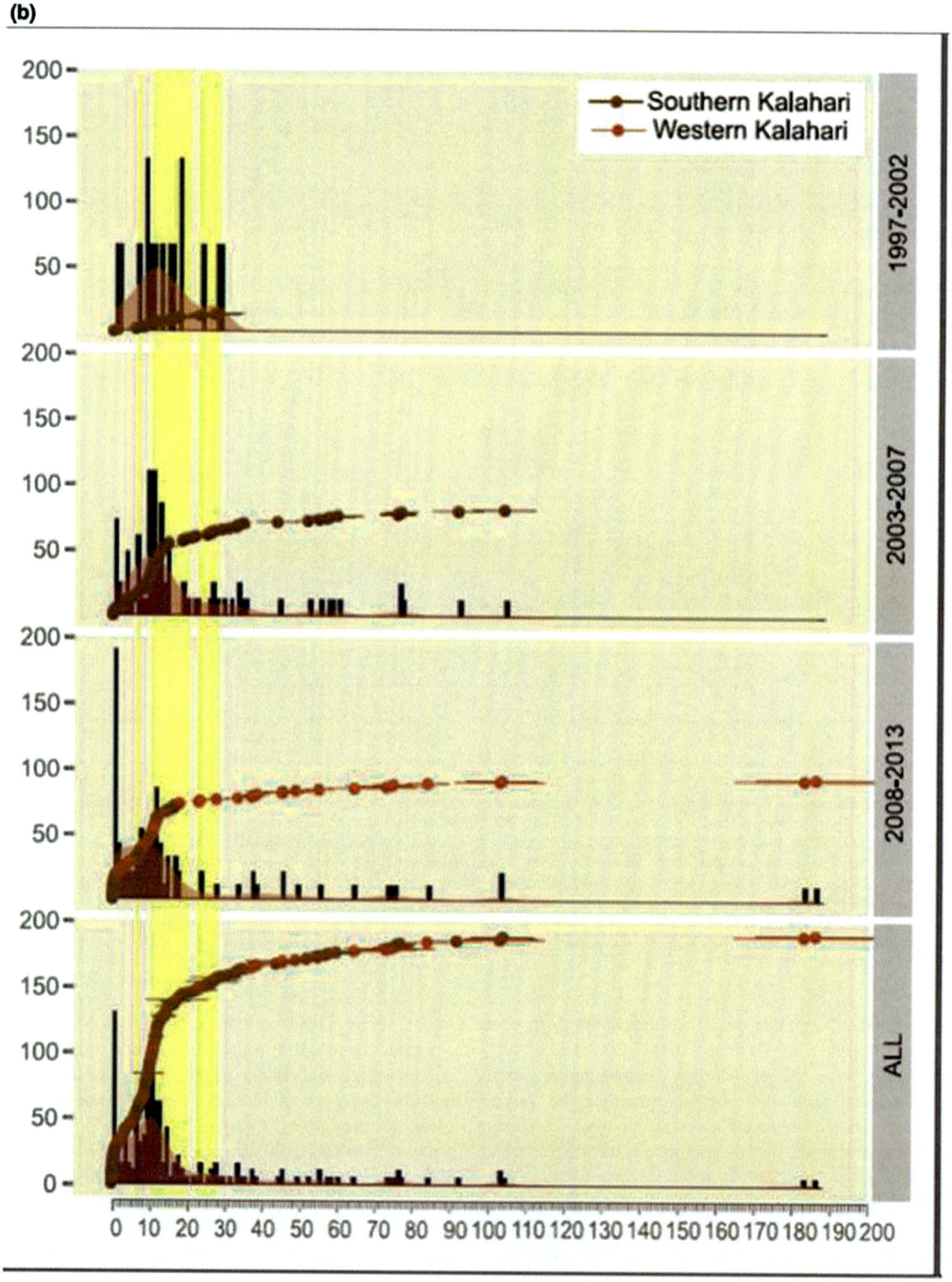

Fig. 8.9 (continued)

surfaces to inhibit dune activity. The last ~2 ka have also seen dune accumulation in the drier southern part of the system, as noted by Telfer and Thomas (2007).

In parts of the southern Kalahari, linear dunes encroach onto the upwind surface of pan depressions (e.g. Eitel and Blumel 1997; Telfer and Thomas 2007). The systematic and extensive application of luminescence dating to one of these, at Bettenstadtpan (27°25'S, 20°33'E) was used by Telfer (2011) to examine the assumption that linear dunes develop through downwind extension. 42 luminescence ages, derived from seven augered full-dune profiles at 100 m intervals along the downwind end of the dune, showed that basal ages increased from 7.5 ± 0.6 ka at the southeastern-most tip of the dune on the pan floor to 17.6 ± 1.3 ka 600 m upwind in a northwesterly direction. Whilst ages from the upper 2–3 m of the 8 m high ridge preserved evidence of reworking in recent centuries, sediments from the core of the dune also recorded a general trend of ages getting younger in a downwind direction, evidencing the role of dune extension in the accumulation histories of these forms.

In summary, it can be seen that there have been significant advances in the interpretation of longer-term southern Kalahari linear dunefield development, facilitated through the application of luminescence dating. It is likely that environmental conditions during the major early Holocene period of dune accumulation recorded in the sedimentary record were generally drier than those prevailing today and certainly drier than the wetter mid-Holocene period recorded in the Stampriet aquifer record. Luminescence dating also records significant sediment reworking and accumulation within the dune system in the late Holocene up to the present day, such that a full understanding of dune behaviour and dynamics requires consideration of surface activity under the partially vegetated conditions that prevail today.

8.6 Dune Surface Processes and Dynamics

The partial vegetation cover (Figs. 8.1 and 8.2) evident on the dunes of the Kalahari has led to them being described as fixed, stable or relic features (Grove 1969; Goudie 1970; Lancaster 1981) in contrast to the far-more active and dynamic dunes of the Namib Desert. This suggests that the dunes formed and developed in drier and/or windier conditions in the past partly stimulated the use of OSL dating on the dunes (as described above, Sect. 8.5), with the expectation that dated episodes of Kalahari dune-building could be utilised for palaeo-environmental interpretation in a relatively straightforward manner (e.g. Stokes et al. 1997). Whilst we now have a far better understanding of the sophistication required for collecting and interpreting OSL data (Telfer and Hesse 2013; Thomas and Burrough 2016), we also have a more highly-developed appreciation of the non-binary (i.e. not active vs inactive) nature of sand dune mobility and dynamics.

Livingstone and Thomas (1993) first discussed the extent to which the dunes of the southern Kalahari could be considered geomorphologically inactive features on account of their vegetation cover. Noting that there was no direct or indirect evidence for the Kalahari dunes having ever existed in a state of complete denudation, in contrast to the Namib dunes, they recognised that the relationship between dune

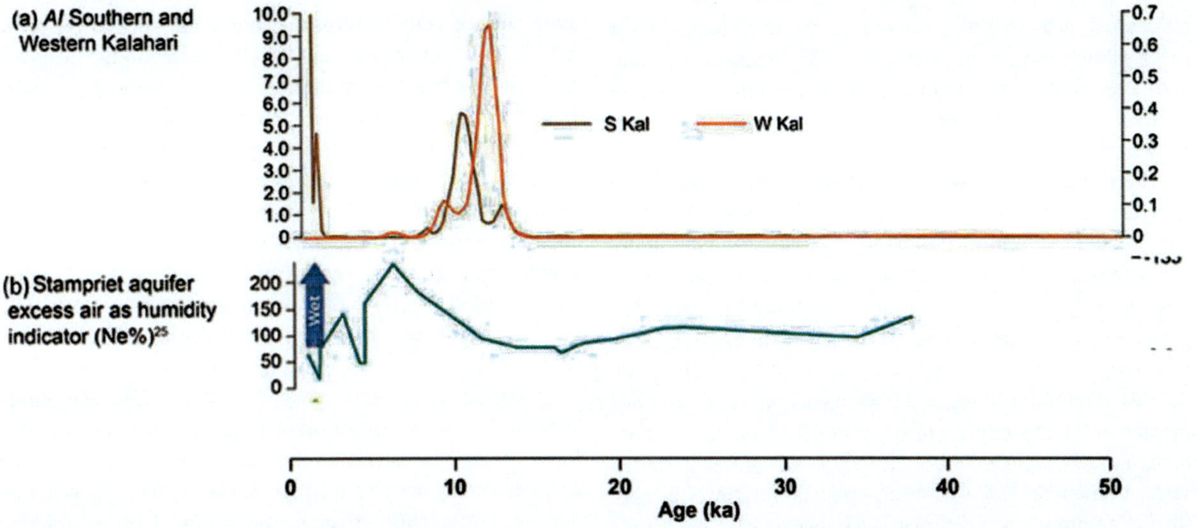

Fig. 8.10 Dune accumulation intensity (*AI*) curves for the southern and western linear dunes over the last 50 ka (Thomas and Bailey 2016), compared to the precipitation record derived from the Stampriet Aquifer (Stuut and Talma 1998). Reproduced with the permission of John Wiley and Sons

activity, vegetation cover, and environmental variability might be more nuanced and subtle than a simple dry/active and wet/inactive narrative. Rather, they considered that the presence of vegetation on a dune might not necessarily indicate dune inactivity, but rather represent a mobility dynamic existing as part of an environmental continuum. In this way, they considered that dune activity in the Kalahari could be both episodic, responding to changes in vegetation cover driven by the considerable rainfall variability in the semi-arid environment, and also as a more subtle and continual gradation in landscape change, rather than a hard threshold delineating dune activity.

Taking these ideas as a starting point, Bullard et al. (1996, 1997) undertook a meteorological analysis of the southern Kalahari dunefield between the years 1960 and 1992. In an attempt to identify the potential activity (*PA*) of the dunes in response to changing rainfall, wind, and vegetation cover, they applied Lancaster's (1988) Mobility Index, *MI*:

$$MI = W/(P/PE)$$

Where W is the amount of time wind blows above a threshold for sand entrainment, and *P/PE* encompasses the effectiveness of rainfall (*P*) accounting for potential evapotranspiration (*PE*). In this way, the index balances wind energy available for erosion against the resistance to erosion provided by a vegetation cover responding to moisture availability. As discussed above Lancaster (1988) had designed the Mobility Index to investigate the paleoenvironmental development of the Kalahari dunes. Extending the analysis to more recent times and at shorter temporal scales, Bullard et al. (1997) were able to identify high variability in the potential activity of the dunes on a year-to-year basis. As shown in Fig. 8.11 their analysis recognised the potential for partial sediment transport activity on the dune surfaces for much of the study period, and specifically identified the mid- and late- 1980s, a period of significant drought in southern Africa, as having the potential for active dune crests and fully active dunes.

The research of Bullard et al. (1996, 1997) showed that the key drivers for enhanced dune activity varied at sub-decadal timescales, intimating that some dune activity was possible in the Kalahari today despite a partial vegetation cover. Importantly, this study also showed the potential for a strong dynamic in dune mobility in response to periods of drought, where they found a correlation between decreased rainfall and increased wind energy. These findings pointed to the capacity for the dunes of the Kalahari to show both a muted, gradual mobility dynamic under a partial vegetation cover, and a stronger, episodic response to more intense environmental extremes, as indicated by Livingstone and Thomas (1993).

Bullard et al. (1996, 1997) showed the *potential* for landscape-scale mobility in the dunes of the southern Kalahari in response to changing environmental variables over annual temporal scales. Smaller, dune-scale studies of *actual* dune activity as a result of variability in vegetation cover were provided by Wiggs et al. (1994, 1995, 1996). Field-based measurements of airflow over de-vegetated and partially vegetated dunes in the south-west Kalahari allowed Wiggs et al. (1994, 1996) to quantify the acceleration of near-surface wind velocity evident with the removal of the vegetation canopy, and the consequent increase in erosion and deposition (total surface change) manifest on the respective dune surfaces (Fig. 8.12).

From Fig. 8.12 Wiggs et al. (1994) recognised that the impact of a reduction in the surface vegetation cover on the Kalahari dunes, provided by a natural burn event in this case, was most acutely experienced on the upper slopes and crests of the dunes. In these upper portions of the dunes sediment transport activity leading to erosion and deposition was enhanced in comparison to the far less active lower dune slopes and interdune areas. This was explained as a response of the windflow to the topography of the dunes, with airflow acceleration on the windward slopes leading to enhanced erosive stresses on the upper slopes and crests in the absence of vegetation (Wiggs et al. 1996). In this way, the much more dynamic crests of the non-vegetated dunes display a sharper crestal shape, as a result of the development of an avalanche slope, in contrast to the more rounded and flat crestal shapes of the partially vegetated dunes typical of the Kalahari environment (see Sect. 2.1.1).

Field measurements led Wiggs et al. (1994) to conclude that changes in airflow patterns as a result of vegetation removal on the Kalahari dunes resulted in an increase in dune surface activity (erosion and/or deposition) by up to 200%. This suggested that the dynamics of the Kalahari dunes were a function of variability in wind velocity but also vegetation dynamics (i.e. the frequency, scale, and duration of vegetation cover decline and growth). This sensitivity of Kalahari dune activity to vegetation cover change was explored more widely by Wiggs et al. (1995). Monitoring erosion and deposition activity on seven dunes with varying degrees of vegetation cover (from 10 to 29% lateral cover) over a period of 10 weeks, they noted the strong influence that vegetation cover has on dynamics at the dune-scale (Fig. 8.13).

Wiggs et al. (1995) recognised that whilst the dunes supporting a relatively substantial vegetation cover (ranging from 15 to 29% lateral cover) demonstrated some erosion and deposition activity on their surfaces, those dunes supporting a vegetation cover below around 14% exhibited a substantial increase in geomorphic activity, by up to an order

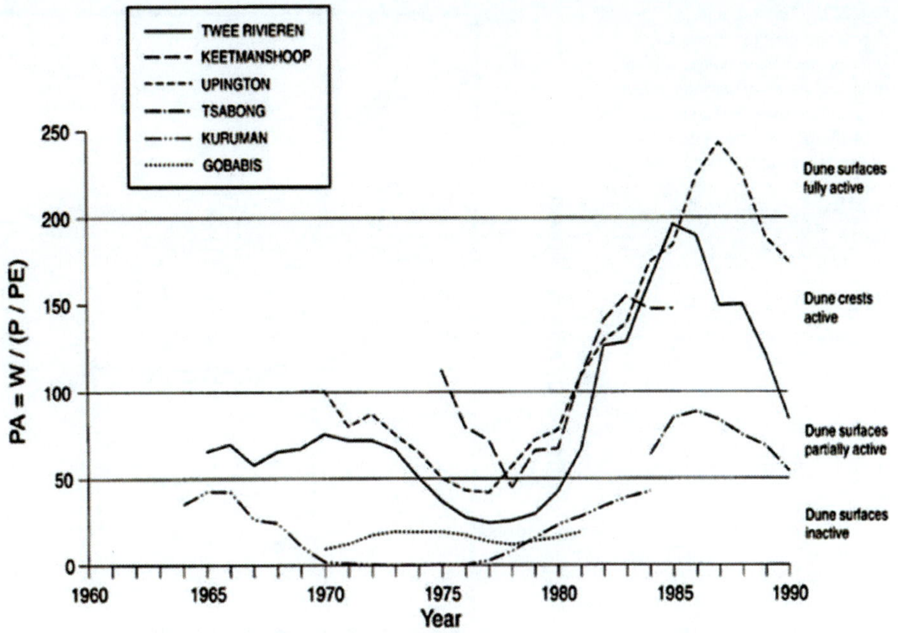

Fig. 8.11 Potential Activity (*PA*, 5-year running mean) of the Kalahari dunes as calculated from meteorological analysis by Bullard et al. (1997) using data from six weather stations in the southwest Kalahari. The thresholds for different stages of dune activity were determined by Lancaster (1988). Reproduced with the permission of John Wiley and Sons

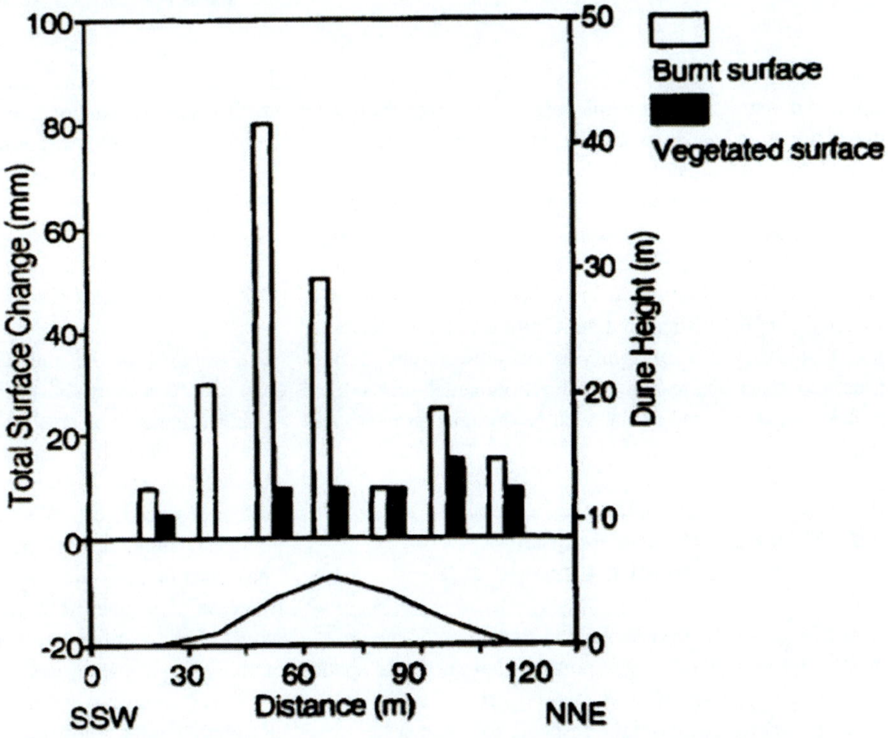

Fig. 8.12 Comparison of surface erosion/deposition activity (total surface change) between a naturally vegetated dune (vegetated surface) and a dune which was denuded of vegetation due to fire (burnt). Measurements taken over a period of 10 weeks by Wiggs et al. (1994). Bars represent surface change (left axis), line represents dune topography (right axis). Reproduced with the permission of John Wiley and Sons

of magnitude. Using the data shown in Fig. 8.12 Wiggs et al. (1995) tentatively identified a threshold lateral vegetation cover of around 14% which, in the Kalahari environment, appeared to differentiate between enhanced and subdued dune mobility.

This field-based evidence offered support to the conceptual arguments of Livingstone and Thomas (1993) against the notion of the partially vegetated Kalahari dune system as an inactive relic. Rather, the dunefield analysis of Bullard et al. (1995, 1996, 1997) and dune-scale field data of Wiggs

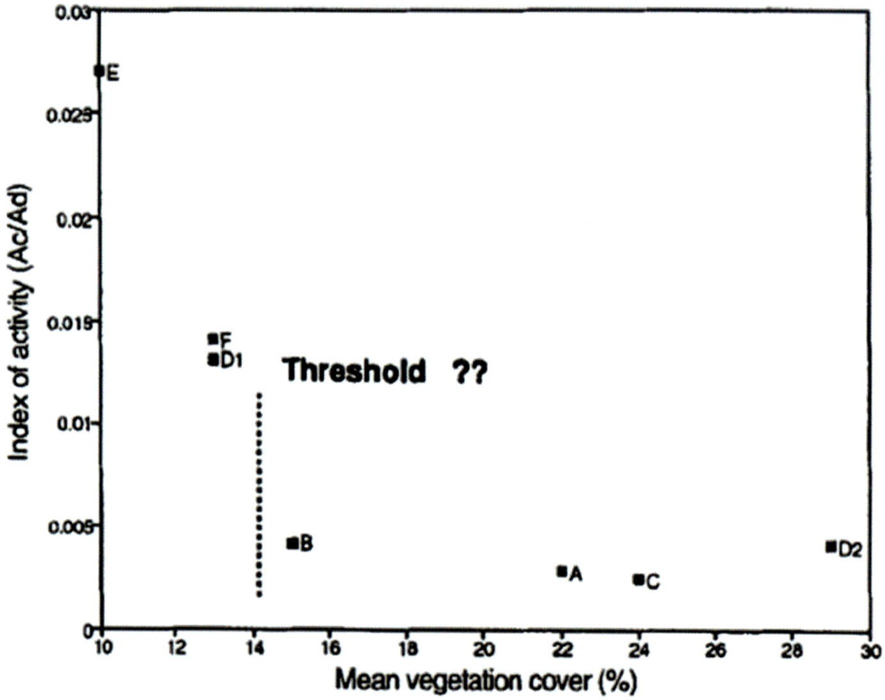

Fig. 8.13 Erosion and deposition activity (index of activity) measured on the surfaces of seven dunes of varying mean vegetation cover. Reproduced with the permission of John Wiley and Sons

et al. (1994, 1995, 1996) suggested that the dunes displayed a muted dynamic under a partial vegetation cover, with their upper slopes and crests exhibiting surface activity that varied in response to climatologically driven cycles of moisture availability and vegetation growth. However, the dunes demonstrated a much more intense geomorphological dynamic, with increased activity across whole dunes, when and where the vegetation cover dropped below about 14%. Hence the partially vegetated dunes of the Kalahari have the potential to display very significant episodic activity, in both time and space, in response to environmental drivers that impose vegetation reduction, such as drought, grazing, and fire.

8.7 Dune Landscape Response to Environmental Change

Research on the geomorphological dynamics of the south-west Kalahari dunefield in response to temporal and spatial variability in vegetation cover raises questions as to how the Kalahari environment might respond to large-scale pressures, such as predicted future global warming. Dintwe et al. (2015) report that in the twenty-first-century soil organic carbon in the Kalahari could reduce by as much as 14% due to declining values in mean annual precipitation and a drying of the soil (IPCC, 2013). This would have severe deleterious impacts on the ecological system. Research on the likely influence of such environmental changes on the Kalahari dune landscape has focussed on two issues; the potential for enhanced dune mobilisation, and the prospect of the dunes of the Kalahari becoming a significant emitter of aeolian dust.

8.7.1 Enhanced Sand Dune Mobilisation

The response of the dunefield vegetation to significant climatic shifts was established by Thomas and Leason (2005). Using Landsat TM data from a drought year (1984) and a wet year (1993) they monitored the changing extent of the dunefield comprising a vegetation cover <14%, the threshold recognised by Wiggs et al. (1995) below which enhanced dune activity might be expected. They analysed their data to determine maps for aeolian hazards, noting an increase of around 300% in the area with < 14% vegetation cover, and so potentially at risk of enhanced sand dune activity, in drought years (Fig. 8.14).

This recognition that large-scale climatic variability in the Kalahari Desert could lead to rapid and intense sand dune activity at the dunefield and sub-dunefield scales led Knight et al. (2004) to develop strategies by which outputs from Global Circulation Models (GCMs) could be best applied to driving indices of sand dune mobility. This allowed for the integration of climate model outputs relevant to vegetation growth and dune movement (such as precipitation, potential

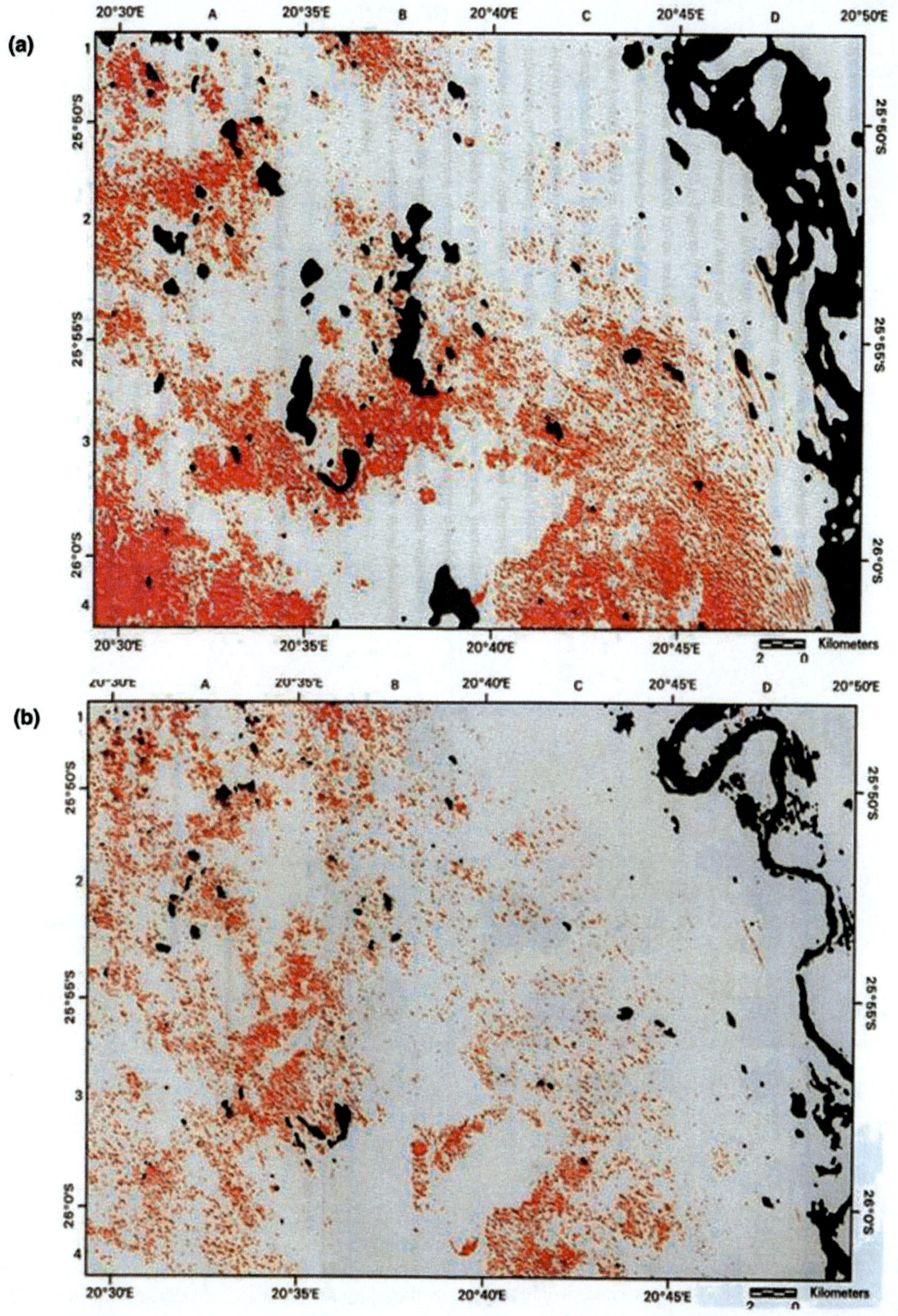

Fig. 8.14 Aeolian hazard maps determined by Thomas and Leason (2005) using Landsat TM imagery for **a** a dry year (1984) and **b** a wet year (1993). Red areas denote those with < 14% vegetation cover and are considered hazardous for increased aeolian activity. White areas have > 14% vegetation cover. Black areas are cloud cover, pan and river valley areas. Reproduced with the permission of Elsevier publishers

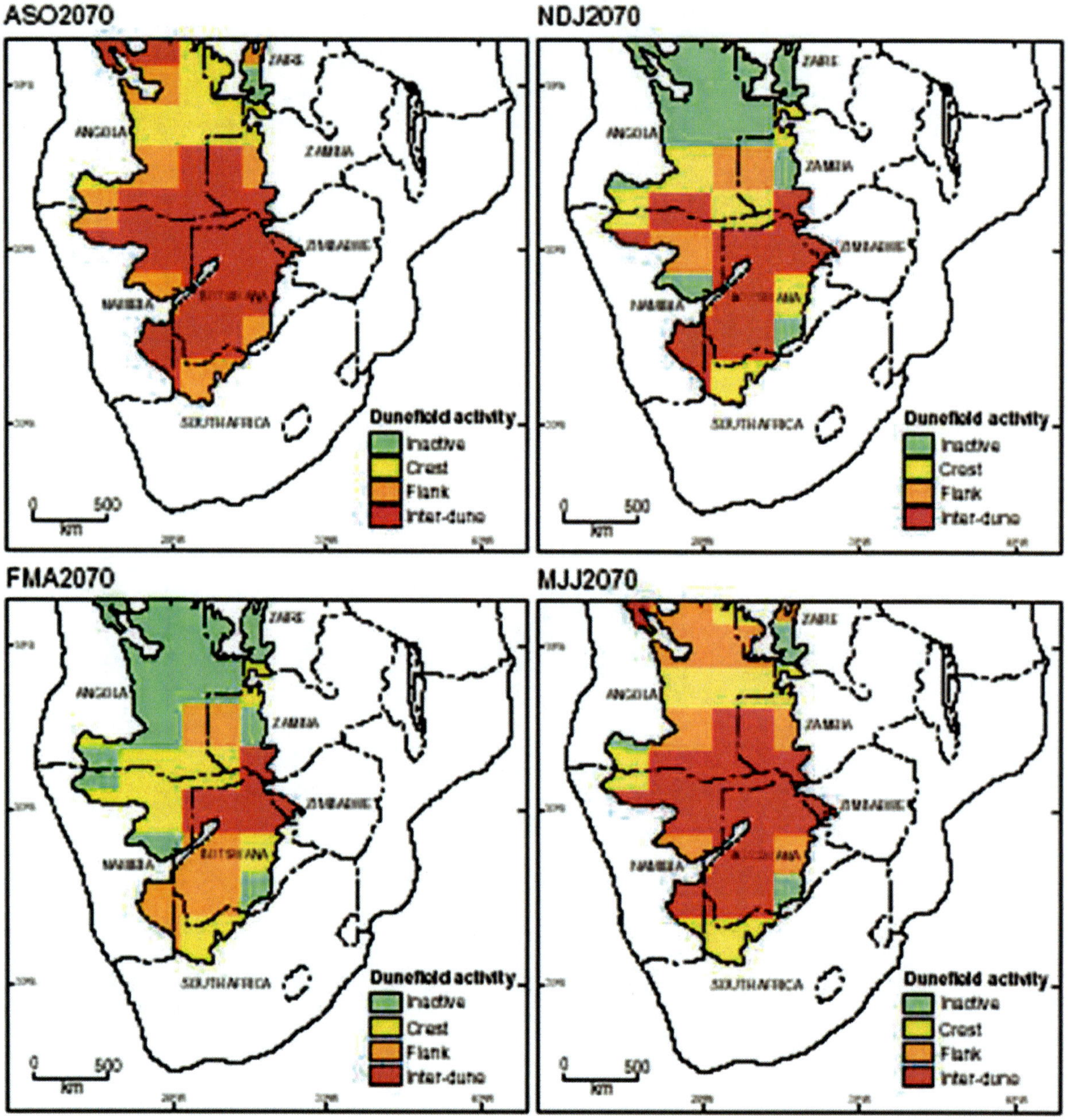

Fig. 8.15 Predicted dunefield activity in different 3-month blocks after 2070 (Thomas et al. 2005). Reproduced with permission of Springer Nature BV

evaporation, and wind velocity) with dune mobility indices to calculate sand dune activity with predicted changing future climate. There have since been a number of attempts to accomplish such predictions of dune mobility in the southwest Kalahari. Thomas et al. (2005) used an adaptation of the Lancaster (1988) Mobility Index incorporating mean windspeed and a weighted measure for net rainfall to account for the lag in response of vegetation to changing moisture status. Applying outputs from three GCMs and a range of emissions scenarios Thomas et al. (2005) found that all modelled outputs predicted marked increases in dune activity throughout the twenty-first century. Particularly intense activity was predicted for the dunes in the southwest Kalahari with a fully active dunefield anticipated by the year 2070 (Fig. 8.15), resulting in likely considerable impacts on both the landscape and agricultural potential of the dunefield.

The method used by Thomas et al. (2005) stimulated discussion as to the difficulties and challenges for predicting sand dune mobility under future climate scenarios in the Kalahari and elsewhere (Ashkenazy et al. 2012) and this led Mayaud et al. (2017) to attempt an alternative approach. Mayaud et al. (2017a, b) employed a cellular automaton model (ViSTA, Mayaud et al. 2017b) to simulate known and quantitative feedbacks between vegetation growth, wind flow dynamics, and sediment flux (Mayaud et al. 2016). The model was driven using projected climatic changes resulting from two principal emissions scenarios from the IPCC with outputs modelled at three sites in the Kalahari. Similar to the study of Ashkenazy et al. (2012), the Mayaud et al. (2017a) model did not predict the same degree of dunefield activity in response to twenty-first-century climate change as the study of Thomas et al. (2005). However, an appealing element of the Mayaud et al. (2017a) study was the inclusion of the potential impacts of changes in fire frequency and grazing dynamics. Here, Mayaud et al. (2017a) found that fire frequency (both anthropogenic and natural in origin) had a primary impact on vegetation cover in the Kalahari and, together with grazing pressure, dictated the intensity of sediment mobility. Mayaud et al. (2017a) therefore concluded that human activities could shape future Kalahari aeolian and dune landscapes to an extent at least equivalent to climate change impacts.

The potential degradational impact of human activity through increased grazing pressure was also the focus of the field observations and seedbank experiments of Bhattachan et al. (2014). Concentrating on the resilience and potential recovery of the vegetated dune surfaces after periods of intense grazing, Bhattachan et al. (2014) noted that whilst the seedbank of perennial grasses on the Kalahari dunes was depleted by grazing pressure, the impact of grazing on soil nutrient concentrations was relatively small. This meant that, in the absence of larger scale environmental pressures, vegetation could re-establish on the dunes fairly rapidly once grazing pressure had been removed. The implication from this finding, together with the findings of Mayaud et al. (2017a, b) and Thomas et al. (2005), is that the stability of the Kalahari dune system, provided by the existence of a partial surface vegetation cover, is at risk from both climatic variability and from any increasing intensity in human activity, including grazing pressure and burning. These changing environmental and human pressures may combine and result in severe degradation to the vegetation on dunes, resulting in considerable risk of dune activity and mobilisation.

8.7.2 Dunes as Emitters of Aeolian Dust

Emissions of aeolian dust in southern Africa are generally focused around topographic depressions (Thomas and Wiggs 2012). Here, silt-sized material is concentrated by the action of water, dried-out in the semi-arid climate, and subsequently eroded by winds. Source areas in southern Africa consisting of dry lake beds, inland pans, and dry river valleys can readily be identified (Vickery et al. 2013). Globally, dune systems are thought to have low potential as dust sources because of the absence of fine particles. However, evidence from the semi-vegetated Australian dune systems (e.g. Bullard et al. 2008) suggests that stable dunefields can accrue fine material primarily through weathering and abrasion. Where the surface vegetation becomes degraded due to human or environmental stresses, winds may then have the capacity to erode the fine material. There is therefore the potential for semi-vegetated dune systems to become sources of aeolian dust. In the Kalahari dunefield this potential has been explored by Bhattachan et al. (2012, 2013).

Using a dust impact generator to evaluate the relative potential of Kalahari dune sands to produce dust-sized material, Bhattachan et al. (2012) found that sediments in currently vegetated interdunes contained a ready supply of fine dust material that is also rich in nutrients (Bhattachan et al. 2013). The authors therefore suggest that the removal of vegetation in these areas, due to a more arid climate and/or increased grazing pressure, may allow the erosion of this fine material into the atmosphere. Bhattachan et al. (2012) note that any potential dust emissions from the dune sands would likely be unsustainable in the long term because, once de-stabilised and exhausted of dust-sized material, there would be no mechanism to re-supply fine material into the dune surface sediments. However, whilst the loss of nutrients from the dune sands due to erosion would be severe (Bhattachan et al. 2013), the authors also note that the emittance of dust from the Kalahari could potentially result in a very significant pulse of iron-enriched (and nutrifying) sediment into the Southern Ocean.

8.8 Conclusion

The dunes of southwestern Botswana and contiguous areas of Namibia and South Africa have been the subject of scientific investigations since the 1930s, with earlier non-scientific accounts providing important insights to this still-often remote landscape. We now know, from chronometric analyses, that the current dunes have formed over both a long period of time and in multiple periods of construction.

The partial vegetation cover on the Kalahari dune system today results in a muted response of the dunes to erosive forces, with activity generally limited to the upper flanks and crests of the dunes. However, periods of drought and intense human activity, reducing vegetation cover below a stability threshold, can result in episodic dune activity limited in spatial extent (with respect to human activity) and/or temporal extent (with regards to the longevity of periods of drought). Further, the dynamics of the Kalahari dune system is sensitive to these destabilising pressures such that there is evidence for increasing climatic and human pressures through the twenty-first century to induce a widespread increase in dune activity and mobility, and also in dust emission potential.

The southern Kalahari dune system represents a landscape oscillating around the threshold between stability and dynamism. The response of the dunes to changing environmental and climatic drivers is beginning to be understood and documented. Without careful management, the enduring susceptibility and sensitivity of the dune landscape to climate change and human activity is likely to become apparent in the century ahead.

References

Ashkenazy Y, Yizhaq H, Tsoar H (2012) Sand dune mobility under climate change in the Kalahari and Australian deserts. Clim Change 112:901–923

Bailey RM, Thomas DSG (2014) A quantitative approach to understanding dated dune stratigraphies. Earth Surf Proc Land 39:614–631

Bhattachan A, D'Odorico P, Okin GS, Dintwe K (2013) Potential dust emissions from the southern Kalahari's dunelands. J Geophys Res Earth Surf 118:307–314

Bhattachan A, D'Odorico P, Dintwe K, Okin GS, Collins SL (2014) Resilience and recovery potential of duneland vegetation in the southern Kalahari. Ecosphere 5(1):2. https://doi.org/10.1890/ES13-00268.1

Bhattachan A, D'Odorico P, Baddock MC, Zobeck TM, Okin GS, Cassar N (2012) The southern Kalahari: a potential new dust source in the southern hemisphere. Environ Res Lett 7. https://doi.org/10.1088/1748-9326/7/2/024001

Bowler JM (1986) Spatial variability and hydraulic evolution of Australian lake basins: analogue for Pleistocene hydraulic change and evaporate formation. Palaeogeogr Palaeoclim Palaeoecol 54:21–41

Breed CS, Grow T (1979) Morphology and distribution of dunes in sand seas observed by remote sensing. In: McKee ED (ed) A study of global sand seas, pp 253–303. USGS Professional Paper 1052. 429 pp

Breed CS, Fryberger SG, Andrew S, McCauley C, Lennartz F, Gebel D, Horstman K (1979) Regional studies of sand seas using Landsat (ERTS) imagery. In: McKee ED (ed) A study of global sand seas, pp 253–30305–397. USGS Professional Paper 1052. 429 pp

Bristow CS, Lancaster N, Duller GAT (2005) Combining ground penetrating radar surveys and optical dating to determine dune migration in Namibia. J Geol Soc Lond 162:315–321

Bullard JE, Nash DJ (1998) Linear dune pattern variability in the vicinity of dry valleys in the southwest Kalahari. Geomorphology 23:35–54

Bullard JE, Nash DJ (2000) Valley-marginal sand dunes in the south-west Kalahari: their nature, classification and possible origins. J Arid Env 45:369–383

Bullard JE, Baddock MC, McTainsh G, Leys J (2008) Sub-basin scale dust source geomorphology detected using MODIS. Geophys Res Lett 35:L15404

Bullard JE, Thomas DSG, Livingstone I, Wiggs GFS (1995) Analysis of linear sand dune morphological variability, southwestern Kalahari Desert. Geomorphology 11:189-203

Bullard JE, Thomas DSG, Livingstone I, Wiggs GFS (1996) Wind energy variations in the southwestern Kalahari desert and implications for linear dunefield activity. Earth Surf Proc Land 21:263–278

Bullard JE, Thomas DSG, Livingstone I, Wiggs GFS (1997) Dunefield activity and interactions with climatic variability in the southwest Kalahari Desert. Earth Surf Proc Land 22:165–174

Chase B (2009) Evaluating the use of dune sediments as a proxy for palaeoaridity: a southern African case study. Earth Sci Rev 93:31–45

Chase BM, Brewer S (2009) Last Glacial Maximum dune activity in the Kalahari Desert of southern Africa: observations and simulations. Quatern Sci Rev 28:301–307

Deacon J, Lancaster N (1988) Late Quaternary palaeoenvironments of southern Africa. Oxford Science Publications, Oxford, p 225

Dintwe K, Okin GS, D'Odorico P, Hrast T, Mladenov N, Handorean A, Bhattachan A, Caylor KK (2015) Soil organic C and total N pools in the Kalahari: potential impacts of climate change on C sequestration in savannas. Plant Soil 396:27–44

Eitel B, Blumel WD (1997) Pans and dunes in the southwestern Kalahari (Namibia): geomorphology and evidence for Quaternary paleoclimates. Z Fur Geomorphol Suppl 111:73–95

Eriksson PG, Nixon N, Snyman CP, DuP BJ (1989) Ellipsoidal parabolic dune patches in the southern Kalahari desert. J Arid Environ 16:111–124

Fryberger SG (1979) Dune forms and wind regime. In: McKee ED (ed) A study of global sand seas, pp 305-397. USGS Professional Paper 1052

Fryberger SG, Goudie AS (1981) Arid geomorphology. Prog Phys Geogr 5:420–428

Goudie AS (1969) Statistical laws and dune ridges in southern Africa. Geogr J 135:404–406

Goudie AS (1970) Notes on some major dune types in southern Africa. S Afr Geogr J 52:93–101

Goudie AS, Thomas DSG (1986) Lunette dunes in southern Africa. J Arid Environ 10:1–12

Grove AT (1969) Landforms and climatic change in the Kalahari and Ngamiland. Geogr J 135:191–212

Hills ES (1940) The lunette, a new landform of aeolian origin. Aust Geogr 3:15–21

Korn H, Martin H (1937) Die jüngere geologische und klimatische Geschichte Südwestafrikas. Zentalblad Für Mineralogie, Geologie Und Paläontologie B11:456–473

Knight M, Thomas DSG, Wiggs GFS (2004) Challenges of calculating dunefield mobility over the 21st century. Geomorphology 59:197–213

Lancaster IN (1978a) The pans of the southern Kalahari, Botswana. Geogr J 144:81–98

Lancaster IN (1978b). Composition and formation of southern Kalahari pan margin dunes. *Zeitschrift fur Geomorphologie* NF22: 148–169

Lancaster N (1981) Palaeoenvironmental implications of fixed dune systems in southern Africa. Palaeogeogr Palaeoclimatol Palaeoecol 33:327–346

Lancaster N (1982) Linear dunes. Prog Phys Geogr 6:475–504

Lancaster N (1986) Grain-size characteristics of linear dunes in the south-western Kalahari. J Sediment Petrol 56:395–400

Lancaster N (1987) Grain-size characteristics of linear dunes in the south-western Kalahari. J Sedim Petrol 57:573–574

Lancaster N (1988) Development of linear dunes in the southwestern Kalahari, southern Africa. J Arid Environ 14:233–244

Lancaster N, Wolfe S, Thomas D, Bristow C, Bubenzer O, Burrough S, Duller G, Halfen A, Hesse P, Roskin J, Singhvi A, Tsoar H, Tripaldi A, Yang X, Zarate M (2016) The INQUA Dunes Atlas chronologic database. Quatern Int 410:3–10

Lawson MP, Thomas DSG (2002) Lunette dune development chronologies in the SW Kalahari. Quatern Sci Rev 21:825–836

Lewis AD (1936) Sand dunes of the Kalahari within the borders of the Union. S Afr Geogr J 19:22–32

Livingstone I, Bullard JE, Wiggs GFS, Thomas DSG (1999) Grain-size variation on dunes in the southwest Kalahari, southern Africa. J Sediment Res 69:546–552

Livingstone IP, Thomas DSG (1993) Modes of linear dune activity and their palaeoenvironmental significance: an evaluation with reference to southern African examples. In: Pye K (ed) The dynamics and context of aeolian sedimentary systems. Geological Society Special Publication, vol 72, pp 91–101

Mabbutt JA (1957) Physiographic evidence for the age of the Kalahari Sands of the south west Kalahari. In: Clark JD (ed) The 3rd Pan-African Congress on Prehistory, Livingstone 1955. Livingstone Museum, Livingstone

Mabbutt JA, Wooding RA (1983) Analysis of longitudinal dune patterns in the north western Simpson Desert, central Australia. Zeitschrift Fur Geomorphologie Supplementbund 45:51–70

McKee ED (ed) (1979) A study of global sand seas, USGS Professional Paper 1052. 429pp.

Mallick DIJ, Habgood F, Skinner AC (1981) A geological interpretation of Landsat imagery and air photography of Botswana. In: Overseas geology and mineral resources, vol 56. Institute of geological Sciences, London 36pp.

Mayaud JR, Wiggs GFS, Bailey RM (2016) Characterising turbulent windflow around dryland vegetation. Earth Surf Proc Land 41 (10):1421–1436

Mayaud JR, Bailey RM, Wiggs GFS (2017a) Modelled responses of the Kalahari Desert to 21st century climate and land use change. Sci Rep 7:3887. https://doi.org/10.1038/s41598-017-04341-0

Mayaud JR, Bailey RM, Wiggs GFS (2017b) A coupled vegetation/sediment transport model for dryland environments. J Geophys Res Earth Surf 122(4):875–900

Miller R (2014) Evidence for the evolution of the Kalahari dunes from the Auob River, southeastern Namibia. Trans R Soc S Afr 69:195–204

Perkins JS (2018) Southern Kalahari piospheres: looking beyond the sacrifice zone. Land Degradation Dev 29:2778–2784

Rubin DM, Tsoar H, Blumberg DG (2008) A second look at western Sinai seif dunes and their lateral migration. Geomorphology 93:335–342

Shaw PA (1997) Africa and Europe. In: Thomas, DSG (ed) Arid zone geomorphology. Wiley: Chichester 467–486

Singhvi AK, Sharma YP, Agrawal DP (1982) Thermoluminescence dating of sand dunes in Rajasthan, India. Nature 295:313–315

Stokes S, Thomas DSG, Washington R (1997) Multiple episodes of aridity in Africa south of the equator since the last interglacial cycle. Nature 388:154–158

Stone AEC, Thomas DSG (2008) Linear dune accumulation chronologies from the southwest Kalahari, Namibia: challenges of reconstructing late Quaternary palaeoenvironments from aeolian landforms. Quatern Sci Rev 27:1667–1681

Stute M, Talma AS (1998) Glacial temperatures and moisture transport regimes reconstructed from noble gas and $\dot{O}^{18}O$, Stampriet aquifer, Namibia. Isotope techniques in the study of past and current environmental changes in the hydrosphere and the atmosphere. In: Proceedings, IAEA Vienna symposium *1997,* Vienna, pp 307–328

Telfer M (2011) Growth by extension, and reworking, of a southwestern Kalahari linear dune. Earth Surf Proc Land 36:1125–1135

Telfer M, Hesse P (2013) Palaeoenvironmental reconstructions from linear dunefield: recent progress, current challenges and future directions. Quatern Sci Rev 78:1–21

Telfer M, Thomas DSG (2006) Spatial and temporal complexity of lunette dune development, Witpan, South Africa: implications for palaeoclimate and models of pan development in arid regions. Geology 34:853–856

Telfer M, Thomas DSG (2007) Late Quaternary linear dune accumulation and chronostratigraphy of the southwestern Kalahari: implications for aeolian palaeoclimatic reconstructions and predictions of future dynamics. Quatern Sci Rev 26:2617–2630

Thomas DSG (1984) Ancient ergs of the former arid zones of Zimbabwe, Zambia and Angola. Inst Br Geogr Trans NS 9:75–88

Thomas DSG (1988) Analysis of linear dune sediment-form relationships in the southern Kalahari Dune Desert. Earth Surf Proc Land 13:545–553

Thomas DSG (2013) Reconstructing paleoenvironments and palaeoclimates in drylands: what can landform analysis contribute? Earth Surf Proc Land 38:3–16

Thomas DSG, Bailey RM (2017) Is there evidence for global-scale forcing of Southern Hemisphere Quaternary desert dune accumulation? A quantitative method for testing hypotheses of dune system development. Earth Surf Proc Land 42:2280–2294

Thomas DSG, Burrough SL (2016) Luminescence-based dune chronologies in southern Africa: analysis and interpretation of dune database records across the subcontinent. Quatern Int 410:30–45

Thomas DSG, Leason HE (2005) Dunefield activity response to climate variability in the southwest Kalahari. Geomorphology 64:117–132

Thomas DSG, Martin HE (1987) Grain size characteristics of linear dunes in the southwestern Kalahari — a discussion. J Sediment Petrol 57:572–573

Thomas DSG, Nash DJ, Shaw PA, Van der Post C (1993) Present day sediment cycling in Witpan in the arid southwestern Kalahari Desert. CATENA 20:515–527

Thomas DSG, Shaw PA (1991) The Kalahari environment. Cambridge University Press, 284p

Thomas DSG, Stokes S, O'Connor PA (1998) Late Quaternary aridity in the southwestern Kalahari Desert: new contributions from OSL dating of aeolian deposits, northern Cape Province, South Africa. In: Alsharan AS, Glennie KW, Whittle GL, Kendall CG (eds) Quaternary deserts and climate change. Balkema, Rotterdam, pp 213–224

Thomas DSG, Wiggs GFS (2008) Aeolian system responses to global change: challenges of scale, process and temporal integration. Earth Surf Proc Land 33:1396–1418

Thomas DSG, Wiggs GFS (2012) Aeolian systems. In: Holmes PJ, Meadows M (eds) Geomorphology of Southern Africa. Stuik, Cape Town

Thomas DSG, Knight M, Wiggs GFS (2005) Remobilization of southern African desert due systems by twenty-first century global warming. Nature 435:1218–1221. https://doi.org/10.1038/nature03717

Tyson PD (1986) Climatic change and variability in southern Africa. Oxford University Press, Oxford, p 220

Vickery KJ, Eckardt FD, Bryant RG (2013) A sub-basin scale dust plume source frequency inventory for southern Africa, 2005–2008. Geophys Res Lett 40(19):5274–5279

Wasson RJ, Hyde R (1983) Factors determining desert dune type. Nature 304:337–339

Werner BT (1995) Eolian dunes: computer simulations and attractor interpretation. Geology 23:1107–1110

Werner BT, Kocurek G (1999) Bedform spacing from defect dynamics. Geology 27:727–730

White K, Bullard J, Livingstone I, Moran L (2015) A morphometric comparison of the Namib and southwest Kalahari dunefields using ASTER GDEM data. Aeol Res 19:87–95

Wiggs GFS, Livingstone I, Thomas DSG, Bullard JE (1994) Effect of vegetation removal on airflow patterns and dune dynamics in the southwest Kalahari Desert. Land Degrad Rehabil 5:13–24

Wiggs GFS, Livingstone I, Thomas DSG, Bullard JE (1996) Airflow and roughness characteristics over partially vegetated linear dunes in the southwest Kalahari Desert. Earth Surf Proc Land 21:19–34

Wiggs GFS, Thomas DSG, Livingstone BJE, I, (1995) Dune mobility and vegetation cover in the southwest Kalahari Desert. Earth Surf Proc Land 20:515–529

David S. G. Thomas is Professor of Geography and former Head at the School of Geography and Environment, University of Oxford, and has worked in the Kalahari and surrounding areas since 1981 on geomorphology, long-term Quaternary environmental change and human-environment interactions. He has published over 200 research papers and 10 books, including The Kalahari Environment with Paul Shaw. He is Honorary Professor at the University of the Witwatersrand and has held a similar position at the University of Cape Town. He has supervised almost 50 doctoral students, many now in their own academic posts and even researching in Botswana, with four contributing to this volume.

Giles F. S. Wiggs is Professor of Aeolian Geomorphology at the School of Geography and the Environment at the University of Oxford. He undertook his first work in Botswana in 1993 as a postdoctoral researcher investigating sand dune mobility and vegetation cover in the SW Kalahari. This focus of research led to frequent visits to the region over the following 20 years and his work has extended more recently to investigations of wind-blown dust emissions from the Makgadikgadi pans. He has published over 50 research papers on aeolian geomorphology and processes in southern Africa. His current research interests focus on the origins of sand dune formation and the controls on dust emissions from dry lake beds and ephemeral river valleys.

9 Dunes of the Northern Kalahari

David S. G. Thomas

Abstract

Dune ridges in northern Botswana and adjacent territories are well beyond the limit of the occurrence of aeolian sand transport in southern Africa today. Striking for their distinct patterning in the landscape evident from the air and in satellite imagery, on the ground these are subdued features that in places can be barely discernible apart from through the patterning of vegetation. Overall, these features have characteristics that attest to their inactivity today and their degradation by non-aeolian processes since their formation. Analysis of ridge ages, orientations and sediments sheds light, and some remaining uncertainty, on the controls on their formation during the Quaternary Period.

Keywords

Linear ridges • Vegetated dunes • Degradation • Late quaternary

9.1 Introduction: When is a Dune Not a Dune?

A year before A T Grove's seminal paper on the Botswanan Kalahari was published (Grove 1969), Flint and Bond (1968) reported previously undescribed 'long, wide but extremely low' sand ridges from northwestern Zimbabwe (then Rhodesia). They described them as the remnants of longitudinal dunes; Grove (1969) used the term 'alab dune' for these features, which extend into northeastern Botswana and abut the northeastern margins of the Makgadikgadi basin (Fig. 1.6). These features have a wavelength of approximately 1.7 km from ridge top to ridge top (Grove 1969), low relief ('a few feet': Flint and Bond 1968) and a wooded vegetation cover. These characteristics led Flint and Bond (1968) and Grove (1969) to conclude that the landforms are dune remnants, rather than dunes per se. Significantly, their presence was taken as an indication of arid-humid climate changes in the region during the Quaternary period, since the areas where these forms are found today have modern mean annual rainfall values in the range of 500–800 mm and sand dunes are thought to form only when rainfall is below around 100–150 mm p.a.

These reports were in fact not the earliest considerations of degraded dune ridges in the northern parts of the Kalahari and its margins. Over 60 years earlier, Passarge (1904) had considered these forms to be evidence of drier climates in the past, and Lewis (1936) made passing mention of the dunes even further to the north, in southern Angola between the Linyanti and Okavango rivers. King (1952) reported 'peculiar streaks of bush and grass running in parallel lines' in Zimbabwe. It was however the overview provided by the detailed aerial photograph analysis by Grove (1969) that allowed the full extent and morphometry of the ridges to be established, and some earlier misinterpretations, including their mapping on 1:250,000 Rhodesian topographic map sheets as a dense network of parallel channels between sandy interfluves (described in Thomas and Shaw 1991) to be dispelled.

Grove's (1969) study also demonstrated that the suite of northern Kalahari dunes comprised not only parallel ridges, whether 'great rolling dunes mainly on the west and north sides of the Okavango delta' and 'low linear ridges of the kind described by Flint and Bond', but other forms too. These include barchan-like (crescentic) dunes on the floor of Ntwetwe Pan in Makgadikgadi (see Chap. 5), small longitudinal ridges to the west of the shorelines of palaeolake Makgadikgadi, and transverse hummocky dunes in the area that is now the Central Kalahari Game Reserve (see Figs. 1.6, 9.1 and 22.3). The transverse ridges were described by

D. S. G. Thomas (✉)
School of Geography and the Environment, Oxford University Centre for the Environment, University of Oxford, South Parks Road, Oxford, OX1 3QY, UK
e-mail: david.thomas@ouce.ox.ac.uk

F. D. Eckardt (ed.), *Landscapes and Landforms of Botswana*, World Geomorphological Landscapes,
https://doi.org/10.1007/978-3-030-86102-5_9

Cooke (1984: 267) as 'confusing masses of sand hills up to 25 m high'. The analysis of dune forms has also extended further to the north as far as latitude 16.5° S, towards the northeastern extremities of the occurrence of Kalahari sediments in the vicinity of the Upper Zambezi River. Both linear ridges and pan-margin lunettes have been described, dated and interpreted in terms of Quaternary environmental changes (Williams 1986; O'Connor and Thomas 1999; Burrough et al. 2015).

For the system as a whole, subsequent studies have detailed morphologies and dunefield patterns (e.g. Lancaster 1981; Cooke 1984; Burrough et al. 2012), relationships to wind regimes (Lancaster 1981; Thomas 1984), Quaternary histories and chronologies (Stokes et al. 1998; Thomas et al. 2000, 2003; Thomas and Burrough 2016), and post-formation tectonic modifications (Jones 1982; McFarlane and Eckardt 2007). These themes are considered further in the following sections.

9.2 Dunefield Patterns and Morphologies

Broadly speaking, the linear dunes to the north and west of the Okavango Delta occur in a shallow arc whereby ridges are orientated ESE-WNW in the east (Fig. 1.6), E-W to the west of the delta, arcing to ENE-WSW at their westerly extent, around 16° E (O'Connor and Thomas 1999). The eastern-most ridges, that extend westwards from Zimbabwe into Botswana, are orientated ENE-WSW through the system. Ridges are commonly parallel to sub-parallel throughout the region, though there is some evidence of ridge merging (Y junctions, (see Chap. 8), including in western Zimbabwe (Thomas 1984). Y junctions are regarded as evidence of ridge deflection during formation. This may potentially be due to wind regime shifts or variability during formation, though they may also relate to processes of dunefield self-organisation in relation to sediment availability (Telfer et al. 2017). Dunes are generally spaced between 1–2.5 km from crest to crest, though in most instances the term 'crest' refers to a broad, relatively flat zone that may be up to 1 km wide, rather than to a distinctive upper part reminiscent of active linear dunes. Ridge sediments are commonly yellow-buff in colour, contrasting with the paler, often finer-grained, sediments of dry interdunes or the darker peats that occur in places where the interdunes are today occupied by wetland deposits.

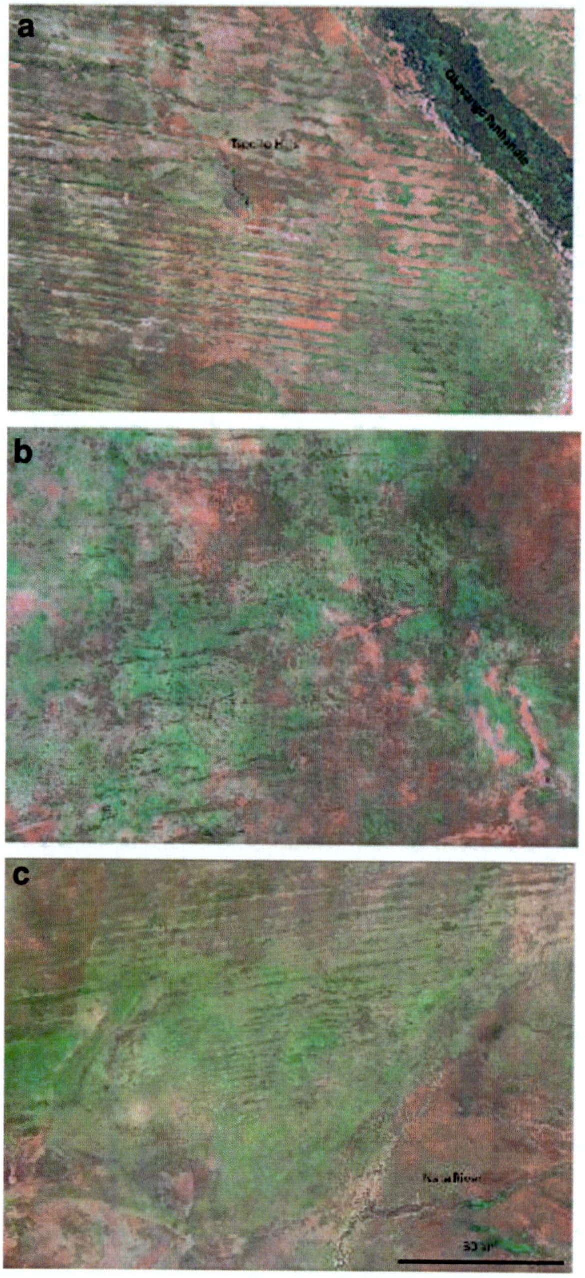

Fig. 9.1 a Dune ridges west of the Okavango panhandle with sand build up on the east of Tsodilo. The absence of dune sand to the west and uninterrupted ridges to the south (Also refer to Chap. 20). b Transverse ridges west of the Makgadikgadi basin. c Ridges north of Sua Pan (Fig. 9.7). (All Images Landsat 7 false Colour 742)

Lancaster (1981) and Thomas (1984) provided the first overall analyses of the Kalahari's northern linear dune patterns across the region as a whole, including their northernmost extent in Angola and Zambia. Over the years the spatial distribution of these extensive features has been described in different ways, and terminology can be confusing. Lancaster (1981) referred to the total system as the northern dunes, but with a Group A and Group B. Group A referred to the ridges north and west of the Okavango Delta, occurring in northern Botswana, southern Angola, northern Namibia including the Caprivi Strip, and into western Zambia; with Group B broadly corresponding to the dunes first described by Flint and Bond (1968) in western Zimbabwe and into northeast Botswana, but restricted to areas north of Makgadikgadi. Thomas (1984) subsequently referred to Group A as the northern dunes and group B as the eastern dunes. As more field data has been collected from various locations through the region as a whole, and more information gathered on the age of formation of the ridges, a further revision has occurred (Fig. 8.10a in previous dune chapter) into three areas: the Northern Kalahari dunefield to the west of the Okavango and east of Etosha pan in Namibia, with some expression in southern Angola; the Northeastern Kalahari dunefield to the east of the Okavango Delta, extending north to the Caprivi Strip, southeast Angola and Barotseland in Zambia, west of the Zambezi river, and the Eastern Kalahari dunefield, in western Zimbabwe and the eastern part of Botswana, between Nata and Kasane (Thomas and Burrough 2016).

All the linear ridges in northern Botswana and adjacent countries have been noted as relatively low forms and have experienced post-depositional degradation (Flint and Bond 1968; Thomas 1984; McFarlane and Eckardt 2007), potentially by sheet flow under conditions that were wetter than when the dunes formed (Flint and Bond 1968). There are marked variations within the system, however. The rolling ridges of the Caprivi Strip for example are up to 25 m in height above the interdune areas (Lancaster 1981) and thus have a distinct surface expression. Some ridges on the Botswana-Zimbabwe border are also this high, but many of the ridges in western Zimbabwe, and the majority in western Zambia, are extremely low to the point of barely having any morphological expression (Flint and Bond 1968) or at most an elevation of 5 m above interdune areas, with sometimes imperceptibly low gradients on the ridge flanks (O'Connor and Thomas 1999). In places, linear patterns have been disrupted by degradation over time by processes such as sheet wash (Thomas 1984; McFarlane and Eckardt 2007). In these contexts, both on the ground and in aerial imagery, ridges can be most diagnostically present due to striking linear vegetation differences. The ridge sands themselves tend to support woody species, for example dense teak (*Baikiaea plurijuga*) woodland in western Zimbabwe (Thomas 1984) or mukwa (*Pterocarpous angolensis*) woodland in, for example, western Zambia (Burrough et al. 2015). Other important ridge species include *Guibourtia coleosperma, Burkea africana, Ochna pulchra, Croton zambesiam* and *Terminalea sericea.* Interdune sediments tend to have a higher silt and clay component and may be seasonally wet, and consequently are more grassed, usually with *Erogrostis* and *Aristida* species. Figure 9.2 illustrates variations in ridge morphologies and vegetation in the region.

9.3 Controls of Northern Dune Development in the Past

The vegetated and degraded state of the northern Kalahari dune systems is clear evidence that these features are palimpsests of past environmental conditions that were significantly different from those of today. The changes in environmental state that occurred since accumulation are considered below, along with a range of issues and uncertainties that remain unresolved.

9.3.1 Airflow

Atmospheric systems of interior southern Africa are dominated by the quasi-stable southern African anticyclone, around which winds blow in an anticlockwise direction (Tyson 1986). The arcuate curvature of the dune ridge system planforms reflects the generally easterly airflow that relates to the anticyclone, but with subtle discrepancies between modern mean wind directions and ridge orientations used to infer equally subtle shifts in airflow when dunes were actively forming (Lancaster 1981; Thomas 1984; Fig. 9.3). The generally westward movement of sand in the region is further demonstrated in the few contexts was dunes encounter topographic obstacles in the region. Both the Aha and Tsodilo Hills in northwestern Botswana display significant sand build-up on their easterly flanks (Fig. 9.1 and Chap. 20). To the west of the Zambezi River in Zambia, large pan depressions (locally called plains) often also display distinctive lunette dunes on their east-northeastern margins (O'Connor 1997; Burrough et al. 2015).

9.3.2 Sediment Supply

Although the linear ridges in the northern Kalahari occur within the extensive Kalahari sand zone of interior southern Africa (Thomas and Shaw 1991), the location of dunefield ridges may hint towards regional factors also influencing sediment supply at times of dune building. Ridges commonly extend westward from the margins of the major northwest-southeast river valleys of the Okavango, Cuando and Zambezi Rivers (Fig. 9.4). They also occur predominantly to the west of the Gwaai River in Zimbabwe, a south bank tributary of the Middle Zambezi. Given many ridges are tens, if not hundreds, of kilometres long, it is probably inappropriate to consider them to be true 'source bordering' forms in the sense of Nanson et al. (1995), where dunes are restricted to the immediate environs of the fluvial system they

Fig. 9.2 Vegetation cover and subtle surface morphology of linear ridges in the northern Kalahari. Top photo taken in the Makgadikgadi Rift Zone (Fig. 9.7) and bottom photo take south of the Aha Hills in Namibia (Fig. 9.6)

emanate from. Nevertheless, the river systems could have been important factors in influencing the sustainability of sand availability dune periods of dune construction (Thomas and Bailey 2017). A mineralogical analysis by McFarlane et al. (2010) showed affinities of the ridge sands in northwest Ngamiland with underlying basement weathering products (saprolite), suggesting that local, rather than fluvially transported, sediment contributed to ridge sediments.

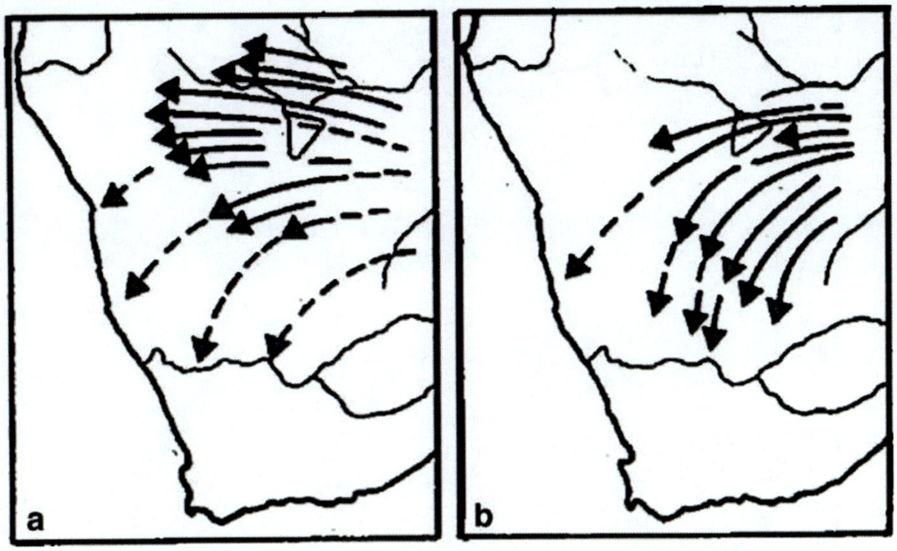

Fig 9.3 Proposed palaeoairflow and sand transport when **a** northern/northeastern dunes and **b** eastern dunes were formed (Lancaster 1981)

9.3.3 Aridity

The occurrence of aeolian processes in the northern parts of the Kalahari today is limited to the deflation of dust from the floors of salt pans, notably Makgadikgadi in Botswana (Chap. 5) and Etosha in Namibia. Elsewhere, except for locations where human activity may have disturbed surfaces, vegetation cover inhibits sediment transport by the wind. Consequently, from Grove (1969) onwards, significant reductions in precipitation from today's levels have been inferred at the time of dune formation, with the dunes of the northern Kalahari as marked evidence of past enhanced aridity in the region.

However, relationships between climate and dune activity are not straightforward (Chase and Brewer 2009), such that indices of dune activity integrate data that embrace not only precipitation change but changes in temperature, that affect evaporation and thus *effective* precipitation, and windiness (Chase 2009), which can be difficult to access from many proxy records of past climatic change. What is clear however, notwithstanding difficulties in inferring precise changes in climatic elements, is that the dune ridges of the northern Kalahari are very heavily vegetated, as well as being over 500 km distant from the dunes in the drier southern Kalahari that can experience surface sediment movement today (see Chap. 8). Though determining the degree of difference between modern climates and those occurring when ridges were constructed, it is clear that they are a palimpsest of environmental conditions that must have been markedly different from those prevailing today, including a significant degree of rainfall reduction (Grove 1969; Thomas and Burrough 2016). The regional rainfall gradient is relatively steep, such that relatively small shifts in circulation patterns might have brought about significant shifts in rainfall, and aridity, in the sub-continent.

9.4 Dunefield Accumulation

Early interpretations of dune building episodes were based on comparison with a limited suite of other palaeoenvironmental proxy records (Lancaster 1981) and on the relative degree of degradation of the different ridge systems (Thomas 1984). Luminescence dating has subsequently been applied to sediment samples from the three northern ridge systems, but to a lesser degree compared to the dunefields in the southern Kalahari (see Chap. 8). To date 33 ages have been produced from the eastern Kalahari dune system, commencing with the study of Stokes et al. (1997, 1998), and 19 ages from each of the northern and northeastern systems (see Thomas and Burrough (2016) for a discussion of the available data). These data have in turn been subject to analysis by the accumulation intensity (*AI*) modelling approach of Bailey and Thomas (2014), producing the accumulation curves in Fig. 9.5 (Thomas and Bailey 2017).

These suggest that the sediment preserved in the ridges today accumulated predominantly around c 35 ka ago in the northern system, west of the Okavango Delta and including ridges adjacent to the Tsodilo Hills (Thomas et al. 2003). Tfv(Zimbabwe: Stokes et al. 1997) and northeastern (Caprivi and western Zambia: O'Connor 1997, O'Connor and Thomas 1999) dunes appear to preserve sediment from after the last glacial maximum, peaking respectively at c 15 ka and c 13 ka ago. Interestingly, this corresponds to the regional drying phase identified in the leaf wax data by Collins et al. (2014). A lesser peak in ridge accumulation, c 30–24 ka, is also evident in the northeastern dune

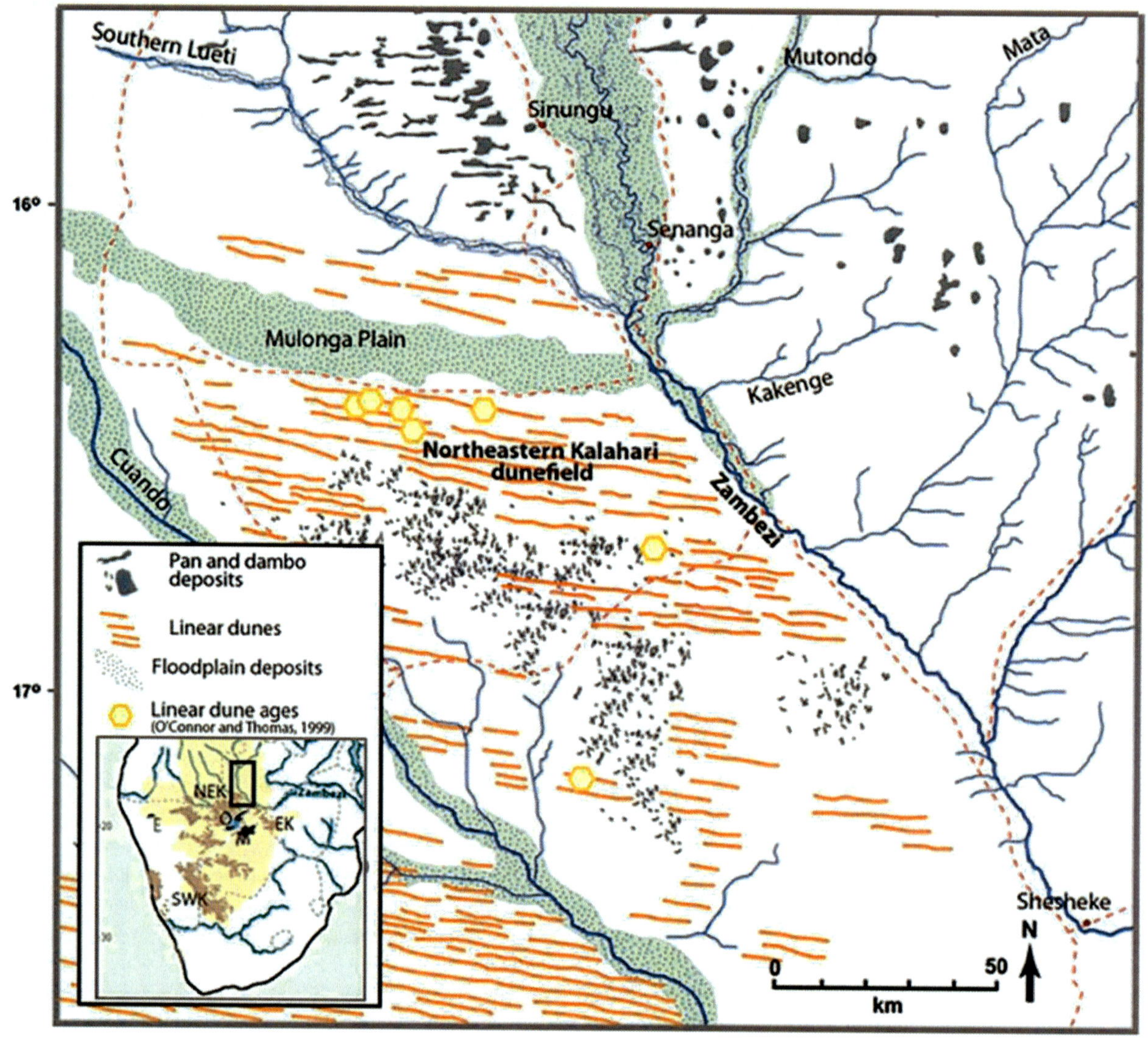

Fig 9.4 Linear ridges in western Zambia, and relationships with the Upper Zambezi valley and tributaries. Ridges in Zambia are part of the wider system that includes northern Botswana. Dunes ridges are predominantly located to the west (downwind) of the main Zambezi valley, suggesting a potential sediment supply relationship with the fluvial system. Ridges south of the Mulonga alluvial plain appear to have had their ends truncated by river flow. North of the Southern Lueti River, dambo orientations may preserve the pattern of former linear ridges. East of the Zambezi, large pans in the Mongu region commonly have east-side lunette dunes (Burrough et al. 2015)

accumulation intensity curve and is derived mainly from data for samples near the Tsodilo Hills (Thomas et al. 2003).

It is important to emphasise that these phases of ridge accumulation are not indicative of the origins/initial emplacement of the features, the timing of which is currently unknown and may extend back beyond the normal time range of luminescence dating ($\sim$ 150–200 ka). Though initial studies, before direct dating was feasible, considered each system to have been formed in a single climatic event (e.g. Lancaster 1981), subsequent analyses of dune systems and sediments around the world (e.g. Stokes et al. 1997; Fitzsimmons et al. 2007; Thomas and Wiggs 2008; Roskin et al. 2011; Thomas and Bailey 2019) clearly demonstrate that many dunefields have accumulated sediments over multiple periods of activity and reactivation that may be separated by periods of quiescence or degradation.

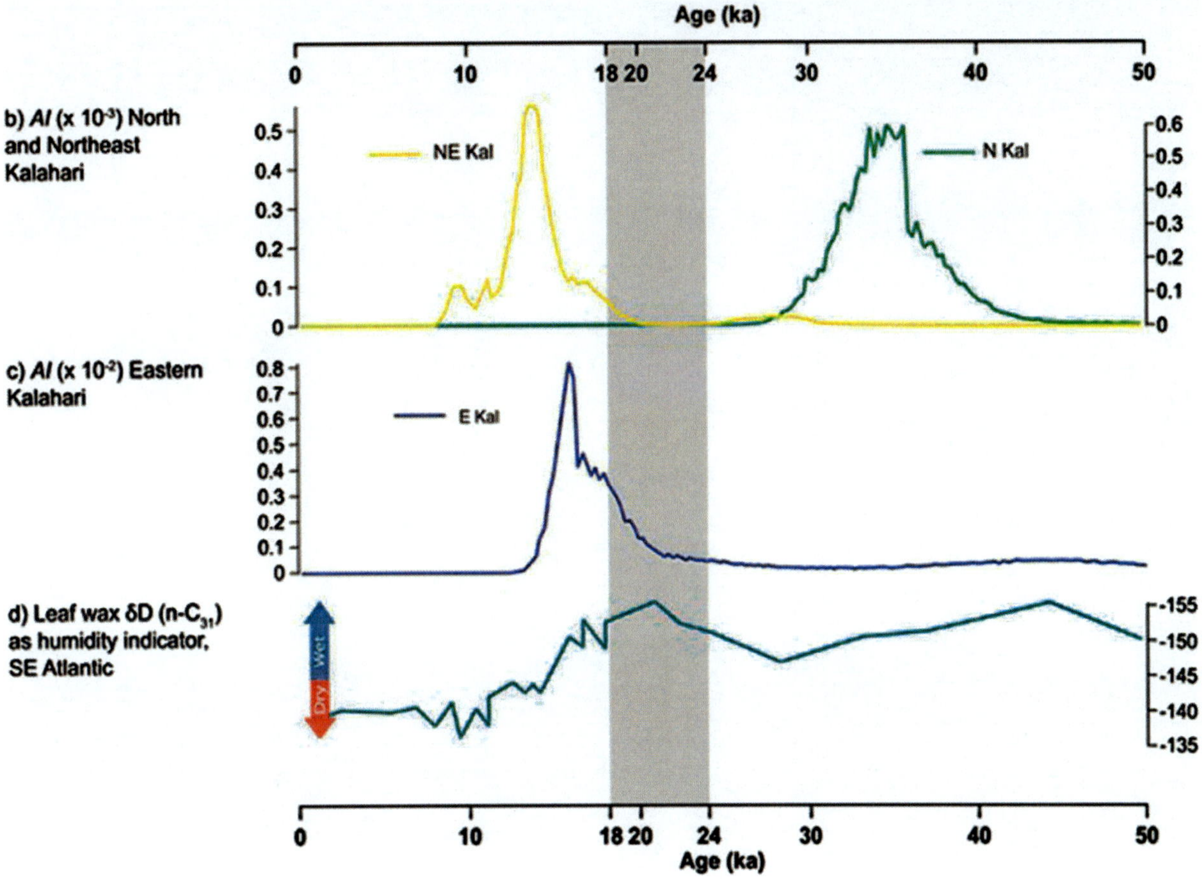

Fig. 9.5 Dune accumulation intensity (*AI*) curves for the three groups of northern Kalahari linear dune ridges over the last 50 ka (Thomas and Bailey 2017). Also shown is Collins et al's (2014) regional humidity curve derived from analysis of leaf wax isotope data from an offshore Atlantic marine core

9.5 Dunefield Degradation

There is no doubt that periods of degradation have impacted on the morphologies and sediments of northern Kalahari linear dune ridges. The are several factors that indicate the occurrence of degradation, relating to ridge cross-sectional profiles and plan forms, and also various processes through which degradation has occurred, through slope erosion, flooding and tectonic activities.

Generally speaking, the further apart linear dune ridges are, the higher they become (Lancaster 1982), so that the relative low relief of northern Kalahari ridges relative to their spacing, which is considerably greater than that of the ridges in the southern Kalahari discussed in Chap. 8, has often been regarded as evidence of significant post-depositional erosion. For example, based on the wavelength of the ridges, and comparison with forms in the active Namib Sand Sea, Grove (1969) considered that the dune features to the west of the Okavango Delta must have been 60 m high when active. Caution however needs to be exercised in this regard on two counts. First, sediment availability impacts greatly on the height and spacing of dunes: for example, linear dunes of comparable height in the southern Kalahari, Simpson-Strzelecki Desert in central Australia and Great Sandy Desert in northwestern Australia display considerable differences in crest-to-crest wavelengths (Thomas 1984). Second, individual linear dune ridges in active dunefields can show marked variations in height along their lengths, in part a function of the controls on surface activity (Tsoar 1985). The widths and heights of ridges in the northern Kalahari have already been noted above as showing marked variability in height (from <5 to 25 m), such that height alone cannot be regarded as a good diagnostic of post-depositional lowering, with other factors providing better evidence of the ridges being degraded over time.

Flint and Bond (1968) noted how gentle the slopes of the ridges in western Zimbabwe are, compared to those of active dunes today. This they regarded as evidence of sheet wash, under wetter climate conditions, moving sediment over time

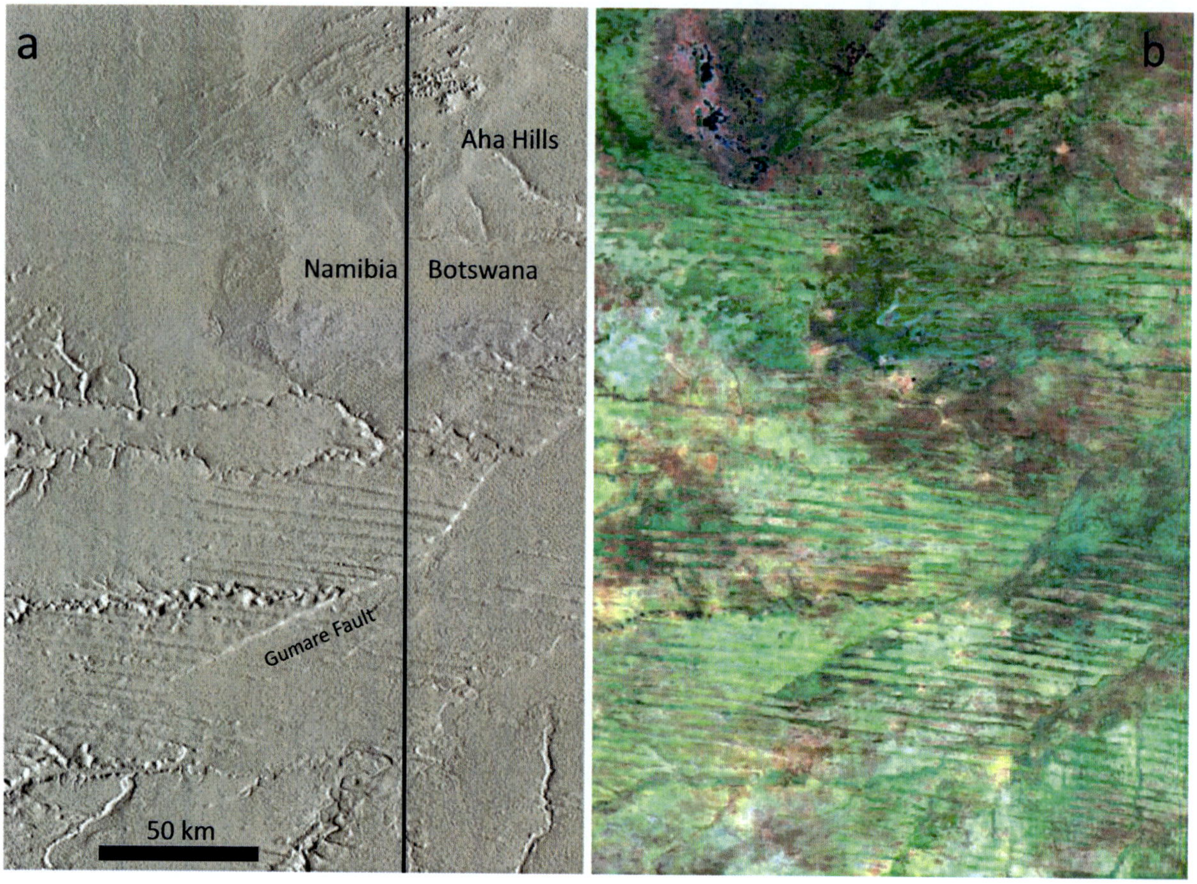

Fig. 9.6 Dunes and straats on the Namibia-Botswana border south of the Aha Hills. Dunes are dissected by fluvial processes, modified by tectonic movement and degraded with no residual topography being identified in the shaded Shuttle Radar Topography Mission elevation data (**a**) to the west. Stripes of vegetation remain visible in the Landsat 7 False Colour (742) Display (**b**). *Source* Frank D. Eckardt

from ridge crests towards the interdunes. McFarlane et al. (2005) analysed ridge planform patterns in northwest Ngamiland to suggest that this has resulted in the interdunes (or 'straats') becoming infilled with sand from the ridges, narrowing them and in places pinching them out entirely (Fig. 9.6).

The inter-digitation of fluvial and aeolian features in northern Botswana and western Zambia also contributes evidence of dune ridge degradation over time. Figure 9.4 shows how in western Zambia linear ridges to the west of the Zambezi appear to have been 'drowned' in fluvial sediments in the vicinity of the Mulonga Plain and further north where dambo (seasonal wetlands) appear to occupy former interdune areas (Williams 1986, see also Fig. 9.4). In Ngamiland, complex patterns of flooding and waterlogging of interdune areas (Grove 1969), as well as degradation of the closest ridges to the Okavango Panhandle and tributaries, have been mapped by McFarlane et al. (2005, Fig. 9.6). This includes locations where Nash et al. (2006) identified evidence of flooding and wetter conditions at c 9 ka and 2.3–1 ka from pollen and sediment analyses. At Tsodilo, low linear ridges have been truncated by the margins of a small lake that diatom, shell and shore margin deposits show existed to varying depths at c 40–32 ka and 27–12 ka, with dune sediments deposited 36–28 ka during a drier phase (Thomas et al. 2003).

Neotectonic movements along lineaments (McFarlane et al.2005) in Ngamiland and north of the Makgadikgadi basin (Eckardt et al. 2016) have also affected ridge forms and degradation. Ridges to the southeastern, subsided side of the Gomare Fault have been erased or submerged in fluvio-lacustrine sediments associated with a more extensive Okavango Delta (Jones 1962; Mallick et al. 1981). NE-SW trending lineaments associated with subtle horst and graben structures north of Ntwetwe Pan and Nata have been affected by vertical ridge displacements, with ridges accentuated on uplifted units (horsts) and degraded, to the extent of largely comprising vegetation stripes alone, on graben blocks (Eckardt et al. 2016, Fig. 9.7).

The sands of active dunes rarely contain more than 1% of fine silt and clay particles, as these are usually expelled into

Fig. 9.7 Dune ridges in the Makgadikgadi Rift Zone. Displacement, accentuation, and diminishment of linear ridges shown in a vertically exaggerated Shuttle Radar Topography Mission elevation data. Ridges are accentuated where displacement has elevated dune sections and diminished in blocks that have subsided. From Eckardt et al. (2016)

suspension during the saltation of moving sand grains. Sediments in the ridges of the northern Kalahari often contain in excess of 5% silt and clay, (Thomas 1984) which may derive from dust inputs, particle breakdown or organic processes since stabilisation. McFarlane et al. (2005) proposed, through geochemical analyses, that the fine component may have been derived, at least in Ngamiland, from the introduction of basement sediments through biological mixing. An alternative explanation is that fines may have been added to the dune sediments, since dune formation, through additions of atmospheric dust, since there are a number of significant mineral dust sources, including Makgadikgadi, in the region today (e.g. Ginoux et al. 2012; Vickery et al. 2013).

9.6 Conclusion

It is undoubtedly the case that the linear dune ridges of the northern Kalahari are landscape features that provide not only evidence of periods of aridity in the late Quaternary, and likely earlier for initial formation, but also of complex post-depositional histories. In some areas, the ridges are little more than forested vegetation lineations, while in others the wooded ridges attain at least 25 m altitude above grassy interdune straats. A series of processes involving fluvial and lacustrine erosion and sedimentation, tectonic displacement as well as sediment modifications have all contributed to the landscape that exists today, although the full developmental history of these features through the Quaternary has yet to be fully understood.

This chapter has focussed on the linear forms, which dominate the landscape, especially when observed from the air or in satellite imagery. The barchanoid forms that occur in Ntwetwe Pan, considered in Chap. 5, are significantly younger features, dating from the late Holocene (Burrough et al. 2015), but are also complex in their developmental history. Other aeolian forms in the region, including the 200 m wavelength linear sand streaks west of the Gidikwe Ridge of the Makgadikgadi basin that can be seen in satellite imagery do not have either vegetational or topographic representation on the ground (Massey 1974; Grey 1976), as well as the larger complex transverse ridges of the Central Kalahari Game Reserve, remain uninvestigated.

References

Bailey RM, Thomas DSG (2014) A quantitative approach to understanding dated dune stratigraphies. Earth Surf Proc Land 39:614–631

Burrough SL, Thomas DSG, Davies L, Bailey R (2012) From landform to process: morphology and formation of lake-bed barchan dunes, Makgadikgadi, Botswana. Geomorphology 161–162:1–14

Burrough SL, Thomas DSG, Orijemie EA, Willis K (2015) Landscape sensitivity and ecological change in western Zambia. The long-term perspective from a Zambian dambo. J Quat Sci 30:44–58

Chase B (2009) Evaluating the use of dune sediments as a proxy for palaeoaridity: a southern African case study. Earth Sci Rev 93:31–45

Chase BM, Brewer S (2009) Last Glacial Maximum dune activity in the Kalahari Desert of southern Africa: observations and simulations. Quatern Sci Rev 28:301–307

Cooke HJ (1984) The evidence from northern Botswana of climatic change. In: Vogel J (ed) Late Cenozoic palaeoclimates of the southern hemisphere. Balkema, Rotterdam, pp 265–278

Collins JA, Schefuss E,Govin A, Mulitza S, Tiedemann R (2014) Insolation and glacial-interglacial control onsouthwestern African hydroclimate over the past 140000 years. Earth Planet Sci Lett 398:1–10

Eckardt FD, Cotteril, FPD, Flugel TJ, Kahl B, McFarlane M, Rowe C, (2016) Mapping the surface geomorphologyof the Makgadikgadi Rift Zone (MRZ). Quat Int 404:115–120

Fitzsimmons KE (2007) Morphological variability in the linear dunefields of the Strekecki and Tirari Deserts, Australia. Geomorphology 91:146–190

Flint RF, Bond G (1968) Pleistocene sand ridges and pans in western Rhodesia. Bull Geol Soc Am 79:299–314

Ginoux P, Prospero JM, Gill TE, Hsu NC, Zhao M (2012) Global scale attribution of anthropoghenic and natural dust sources and their emission rates based on MODIS Deep Blue aerosol products. Rev Geophys 50:RG3005. https://doi.org/10.1029/2012RG000388

Grey DRC (1976) The prospecyting of the Mopipi area. Unpublished report of the Anglo-American Corporation

Grove AT (1969) Landforms and climatic change in the Kalahari and Ngamiland. Geogr J 135:191–212

Jones MT (1962) Report on a photogeological study of an area between the Okavango Swamps and the South west Africa border. Unpublished report, Botswana Geological Survey

Jones CR (1982) The Kalahari of southern Africa. In: Smiley TL (ed) The geological story of the world's deserts. Straie 17:20–34

King LC (1952) Southern African scenery, 2nd edn. Oliver and Boyd, London

Lancaster N (1981) Palaeoenvironmental implications of fixed dune systems in southern Africa. Palaeogeogr Palaeoclimatol Palaeoecol 33:327–346

Lancaster N (1982) Linear dunes. Progr Phys Geogr 6:475–504

Lewis AD (1936) Sand dunes of the Kalahari within the borders of the Union. South Afr Geogr J 19:22–32

Mallick DIJ, Habgood F, Skinner AC (1981) A geological interpretation of Landsat imagery and air photography of Botswana. Overseas Geol Miner Resour 56. Institute of geological Sciences, London 36 pp

Massey NWD (1974) Contribution to the Kalatraverse 1 report. Geological Survey of Botswana NWDM/13/74

McFarlane MJ, Eckardt FD (2007) Palaeodune morphology associated with the Gumare fault of the Okavango graben in the Botswana/Namibia borderland: a new model of tectonic influence. S Afr J Geol 110:535–542

McFarlane MJ, Eckardt FD, Ringrose S, Coetzee SH, Kuhn JR (2005) Degradation of linear dunes in Northwest Ngamiland, Botswana and the implications for luminescence dating of periods of aridity. Quatern Int 135:83–90

McFarlane MJ, Eckardt FD, Coetzee SH, Ringrose S (2010) An African surface weathering profile in the Kalahari of North West Ngamiland, Botswana: processes and products. Zeiutschrift Fur Geomorphologie 54:273–303

Nanson et al (1995) Aeolian and fluvial evidence of changing climate and wind patterns during the past 100 ka in the western Simpson Desert, Australia. Palaeogeogr Palaeoclimatol Palaeoecol 113:87–102

Nash DJ, Meadows ME, Gulliver VL (2006) Holocene environmental change in the Okavango Panhandle,northwest Botswana. Quat Sci Rev 25:1302–1322

O'Connor PW, Thomas DSG (1999) The timing and environmental significance of Late Quaternary linear dune development in western Zambia. Quatern Res 52:44–55

O'Connor PW (1997) Aeolian activity and environmental change in the central Mega Kalahari: implications for the timing, nature and Causes of Late Quaternary aridity. Unpublished PhD Thesis, University of Sheffield

Passarge S (1904) Die Kalahari. Dietrich Reamer, Berlin

Roskin J, Porat N, Tsoar H, Blumberg DG, Zander AM (2011) Age, origin and climatic controls on vegetated linear dunes in the northwestern Negev desert (Israel). Quatern Sci Rev 30:1649–1674

Stokes S, Thomas DSG, Washington R (1997) Multiple episodes of aridity in Africa south of the equator since the last interglacial cycle. Nature 388:154–158

Stokes S, Haynes G, Thomas DSG, Horrocks JL, Higginson M, Malifa M (1998) Punctuated aridity in southern Africa during the last glacial cycle: the chronology of linear dune construction in the northeastern Kalahari. Palaeogeogr Palaeoecol Palaeoclimatol 137:305–322

Telfer MW, Hesse PP, Perez-Fernadez M, Bailey RM, Bajkan S, Lancaster N (2017) Morphodynamics, boundary conditions and pattern evolution within a vegetated dunefield. Geomorphology 290:85–100

Thomas DSG (1984) Ancient ergs of the former arid zones of Zimbabwe, Zambia and Angola. Inst Brit Geogr Trans NS 9:75–88

Thomas DSG, Bailey RM (2017) Is there evidence for global-scale forcing of Southern Hemisphere Quaternary desert dune accumulation? A quantitative method for testing hypotheses of dune system development. Earth Surf Proc Land 42:2280–2294

Thomas DSG, Bailey RM (2019) Analysis of Late Quaternary dunefield development in Asia using the Accumulation Intensity model. Aeol Res 39:33–46

Thomas DSG, Burrough SL (2016) Luminescence-based dune chronologies in southern Africa: analysis and interpretation of dune database records across the subcontinent. Quatern Int 410:30–45

Thomas DSG, Wiggs GFS (2008) Aeolian system responses to global change: challenges of scale, process and temporal integration. Earth Surf Proc Land 33:1396–1418

Thomas DSG, O'Connor P, Bateman MD, Shaw PA, Stokes S, Nash DJ (2000) Dune activity as a record of late Quaternary aridity in the Northern Kalahari: new evidence from northern Namibia interpreted in the context of regional arid and humid chronologies. Palaeogeogr Palaeoecol Palaeoclimatol 156:243–259

Thomas DSG, Brook G, Shaw P, Bateman M, Haberyan K, Appleton C, Nash D, McLaren S, Davies F (2003) Late Pleistocene wetting and drying in the NW Kalahari: an integrated study from the Tsodilo Hills, Botswana. Quatern Int 104:53–67

Thomas DSG, Shaw PA (1991) The Kalahari environment. Cambridge University Press, 284 p

Tsoar H (1985) Profiles analysis of sand dunes and their steady state significance. Geografiska Annalar 67A:47–59

Tyson PD (1986) Climatic change and variabilty in souithern Africa. Oxford University Press, 220 p

Vickery KJ, Eckardt FD, Bryant RG (2013) A sub-basin scale dust plume source inventory for southern Africa, 2005–2008. Geophys Res Lett 40:5274–5279

Williams GJ (1986) A preliminary Landsat interpretation of the relict landforms of western Zambia. In: Williams GJ, Wood AP (eds) Geographical perspectives on development in southern Africa. James Cooke University, Queensland, Commonwealth Geographical Bureau, pp 23–33

David S. G. Thomas is Professor of Geography and former Head at the School of Geography and Environment, University of Oxford, and has worked in the Kalahari and surrounding areas since 1981 on geomorphology, long-term Quaternary environmental change and human-environment interactions. He has published over 200 research papers and 10 books, including The Kalahari Environment with Paul Shaw. He is Honorary Professor at the University of the Witwatersrand and has held a similar position at the University of Cape Town. He has supervised almost 50 doctoral students, many now in their own academic posts and even researching in Botswana, with four contributing to this volume.

Kalahari Pans: Quaternary Evolution and Processes of Ephemeral Lakes

10

Irka Schüller, Lukas Belz, Heinz Wilkes, and Achim Wehrmann

Abstract

Pans, playas and similar morphological structures occur across all continents. They are small, closed sediment sinks that are characteristic of arid and semi-arid regions with low relief. In the Kalahari Basin, small pans are widespread and they function as ephemeral lakes and sedimentary sinks. They are characterised by their flat, salty to clayey surface and slightly elevated surroundings. Some principle sedimentary processes are distinctive for pans. Showers during the short rainy season deliver siliciclastic material by runoff from the local catchment. During the dry season, evaporites precipitate by evaporation of the surface water and supersaturated pore water and groundwater. Additionally, fine-grained sediments are transported on the dry pan surface by strong wind regimes (long-term) or by vortices/eddies from thermal uplifts (short-term). The marginal lunette dunes, a characteristic feature of many small Kalahari pans, resulted from temporary phases of significant aeolian sediment transport. However, there is no hint of longer sedimentation gaps or significant periods of erosion. The prevailing depositional processes contradict the former assumption that deflation is the most prominent process in salt pan formation. The pans contain sediment sequences that are several decimetres to metres in thickness, which reach back until the Late Pleistocene. They contain different kinds of proxy information that can be used for environmental reconstruction and climate modelling throughout the pans' history.

I. Schüller (✉)
Marine Research Department, Senckenberg am Meer, Südstrand 40, 26382 Wilhelmshaven, Germany
e-mail: irka-schueller@t-online.de

I. Schüller · L. Belz · H. Wilkes · A. Wehrmann
Institute for Chemistry and Biology of the Marine Environment (ICBM), Carl Von Ossietzky University, Carl-von-Ossietzky-Straße 9-11, 26111 Oldenburg, Germany

Keywords

Playas • Semi-arid environment • Palaeoclimate • Botswana • Namibia • South Africa

10.1 Introduction

Pans are small, closed, intracontinental sediment sinks that are characteristic of arid and semi-arid regions with low relief. They occur on a global scale and a variety of terms exist to describe these structures, e.g. playa, continental sabkha, salina, dry lake, saline lake, ephemeral lake, salar, salt or clay pan, boinka, groundwater discharge zone or complex, playa lake, etc. (Rosen 1994). In total, they make up around 1% of the dryland area (Holmgren and Shaw 1996).

In the Kalahari Basin, there are more than 4,000 pans (see Fig. 10.1) with a density of one per 22 km^2 in the 'Bakalahari Schwelle' (Lancaster 1978). Most of the small Kalahari pans developed in areas with an annual precipitation of 100–550 mm (Goudie and Wells 1995) but there is always a negative hydrological balance between evaporation and rainfall, often exceeding a ratio of 10: 1 (Shaw and Bryant 2011). Normally, precipitation occurs in the form of heavy rain showers during the short period of the rainy season. Pans develop at the lowest points of morphological basins that receive hydrological input by a combination of direct precipitation in their catchment areas and (sub-surface) inflow by groundwater migration (e.g. Shaw and Bryant 2011). In the rainy season, a shallow, ephemeral water coverage is possible. The persistence of surface water tends to be a function of the geological and hydrological setting and links to groundwater depth, groundwater recharge/discharge, etc. (Rosen 1994; Shaw and Bryant 2011). In the Kalahari, ephemeral lakes only persist for days or weeks, depending on the amount of precipitation. The local runoff plays a significant role in the pan evolution. It

F. D. Eckardt (ed.), *Landscapes and Landforms of Botswana*, World Geomorphological Landscapes,
https://doi.org/10.1007/978-3-030-86102-5_10

enables the transport of sediments and dissolved matter into morphological depressions. Complex geochemical processes at groundwater level, as well as at the pan surface, generate evaporitic deposits like salts, carbonates and gypsum (Mees 1999; Shaw and Bryant 2011). On the other hand, aeolian processes lead to sediment loss by deflation during the dry periods (Torgersen et al. 1986). Some studies (Eckardt et al. 2001; Eckardt and Kuring 2005; Bryant et al. 2007; Vickery et al. 2013; Nield et al. 2015) have shown that playa surfaces and their salt crusts are an important regional source for mineral dust emission which has a strong impact on atmospheric processes. Continuous wind regimes (long-term) or vortices/eddies from thermal uplifts (short-term) are able to lift up sediment particles from pan surfaces into the atmosphere and transport them over long distances. A study by Vickery et al. (2013) revealed that more than half of all dust plumes detected in southern Africa originated from pans. Finally, lunette dunes occur on the leeward sides of numerous Kalahari pans and demonstrate the important role of deflation (Lancaster 1978; Heine 1981; Thomas et al. 1993; Goudie and Wells 1995).

After Van Straten (1955) and Boocock and Van Straten (1962), the small pans of the Kalahari Basin can be divided into three different classes:

(1) Grass pans with dark-coloured sandy clay, low saline concentrations and a dominantly grassy cover;
(2) Ungrassed pans where the dark-coloured sandy clay supports a sparsely developed, halophyllic, scrubby vegetation and has a high water-retaining capacity;
(3) Saline pans which are, usually, deeply etched into calcareous sandstone. The pan sediments are composed of saline and highly alkaline, sandy clay with nearly no plant growth on the pan surface (apart from algae).

These pan classes are based on different groundwater characteristics that affect the formation of the individual pan. Grassed pans are probably fed by surface water with no opportunity for evaporative concentration. Ungrassed clay pans are affected by recharge systems with little or no evaporative concentration of salts and carbonates. Saline pans are characterised by a shallow groundwater level within the critical zone and they may be linked to regional discharge zones.

In their function as ephemeral lakes and sedimentary sinks, the pans of the central and southwestern Kalahari represent quasi-continuous geoarchives. Heine (1981, 1982, 2005), Beaumont et al. (1984), Cockcroft et al. (1987), Thomas and Shaw (2002), Hürkamp et al. (2011) and

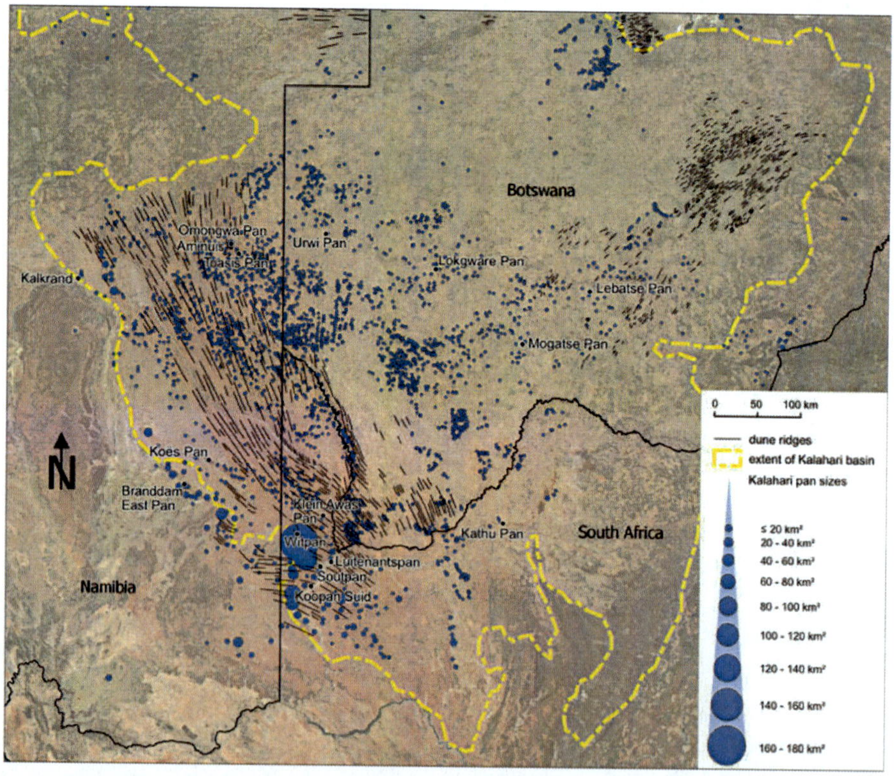

Fig. 10.1 Distribution and sizes of more than 4,000 small Kalahari pans and Kalahari dunes of the Kalahari Basin in Botswana, Namibia and South Africa. The boundary of the Kalahari is modified after the University of Cologne Acacia Project E1 (2004) and satellite images. Digitisation of dune ridges and pans after The Geological Map of Botswana 1:1 000 000 (Jones and Hepworth 1973) and high-resolution satellite images (image: Google, Landsat/ Copernicus; data: SIO, NOAA, U.S. Navy, NGA, GEBCO)

Burrough and Thomas (2013) specified a number of Holocene geoarchives that can potentially be used for reconstructing the palaeoenvironment in semi-arid to arid landscapes. These geoarchives include: pan sediments, fluvial deposits from ephemeral rivers, dunes and desert loess, cave deposits, soils, slope deposits, groundwater, trees (alive and dead), molluscs, ostracods, pollen, hyrax dung, marine sediments, and sabkha deposits. Even if these archives are diverse, many of them are rare in landscapes like the Kalahari and the topic still remains a research opportunity.

Some studies on the potential and limits of small Kalahari pans have already been carried out. Beaumont et al. (1984) studied environmental changes in Kathu Pan since 32,000 BP. Lancaster (1978, 1986) published general information on the origin and formation of pans, their sizes, distribution in Namibia and Botswana and the connection between pans and their surrounding dunes. Holmgren and Shaw (1996) reconstructed the paleoenvironment of Lebatse Pan from near-surface sediments. Mees (1999, 2002, 2003) studied the distribution patterns of salty and calcareous deposits in Namibian pans and along the pan margins. Telfer et al. (2009) dated the age of pan sediment by Optically Stimulated Luminescence (OSL) dating and reconstructed the paleoenvironment around Witpan. Genderjahn et al. (2017, 2018a, b) researched the fossil and recent microbial communities in salt pan sediments and their usage for paleoclimate and environmental reconstructions. Milewski et al. (2017) observed the recent sediment surface dynamics of the Omongwa Pan, based on Multi-temporal Landsat and Hyperspectral Hyperion Data. Schüller et al. (2018) developed a climate model of southern Africa with sedimentological and geochemical data from sediments of five Kalahari pans. Belz et al. (2020) analysed the first paleoclimate biomarker record from a Namibian salt pan and used this information for paleoenvironmental reconstructions.

10.2 Setting and Location

Most of the small pans are situated in the central, western and southwestern parts of the Kalahari Basin (Fig. 10.1). They occur in central and western Botswana, eastern Namibia and northern South Africa. The pans mostly developed in areas of low topographic gradients like water divide surfaces, elevated plains and areas without a surficial drainage system. There are more than 4,000 small pans scattered around the watershed between the Nossob-Molopo River and the Okwa River system. They vary in size between $\sim$10 m^2 and $\sim$200 km^2. The average size is around 1 km^2. They present a variety of shapes with nearly circular depressions in plains, elongated pans along dune ridges and river valleys and more irregular-shaped structures. Looking on a larger regional scale, an interrelation of location, size and shape of the pans is obvious. The largest pans occur in the belt alongside the southwestern rim of the Kalahari, roughly between Koës (Karas Region, Namibia) and Upington (Northern Cape Province, South Africa). An elongation of this belt and the Kalahari rim in a northwesterly direction (Namibia), shows that there are more, larger pans near Kalkrand. Many of them present strongly elongated shapes as they developed in the valleys of a landscape dominated by linear dunes, in karstic systems or along dry river beds. In areas of lower morphology, the wind direction also affects the shape and orientation of the pans. Another cluster of pans of remarkable size occurs around Aminuis (Omaheke Region, Namibia). Here, the pans are orientated along the palaeo river bed system of the former Nossob River (Lancaster 1986). In central Kalahari, mainly in Botswana, pan sizes are much smaller. They appear to be aligned between dune ridges as well as scattered in plains, where they often present nearly circular shapes (see Figs. 10.2c and 10.5).

Most pans are embedded in sands and/or calcretes of the Tertiary Kalahari Group, covering bedrocks of the Permo-Carboniferous to Jurassic Karoo Supergroup, mainly comprising shales and sandstones of the Stormberg Series and Dwyka Formation/Ecca Group (Lancaster 1976, 1986; Thomas and Shaw 1991). The Precambrian basement consists of a variety of sediments from the Ghanzi Formation, Waterberg System and Transvaal System as well as the felsites, serpentinites and gneisses of the Kanye Volcanic System Basement Complex (Lancaster 1976). In the southwestern Kalahari, dolerite and diabase dykes were probably intruded into the Ecca Group sediments (Kautz and Porada 1976) during the Jurassic (pre-Karoo and Karoo) and partly function as a basement for pans now. At the southwestern rim of the Kalahari, pans developed directly in glaciogenic shales and sandstones of the Permo-Carboniferous Dwyka Formation with no (or only a thin) Kalahari Group sediment coverage (Steyn and Nortman 1980; Verhagen 1990).

10.3 Formation of Pans

The formation of the Kalahari pan is caused by their specific preconditions and complex processes, i.e. their environmental conditions, groundwater situation, regional geology and climatic conditions (Fig. 10.2) (Boocock and Van Straten 1962; Heine 1981; Goudie and Thomas 1985; Goudie and Wells 1995). Pans exclusively represent the lowest points of morphological sinks. So, they can develop ubiquitously where environmental, geological, hydrological or climatic conditions form a basin that can be affected by groundwater or filled by surface water.

It is widely accepted that the development of some pans is linked to earlier drainage systems which repeatedly got

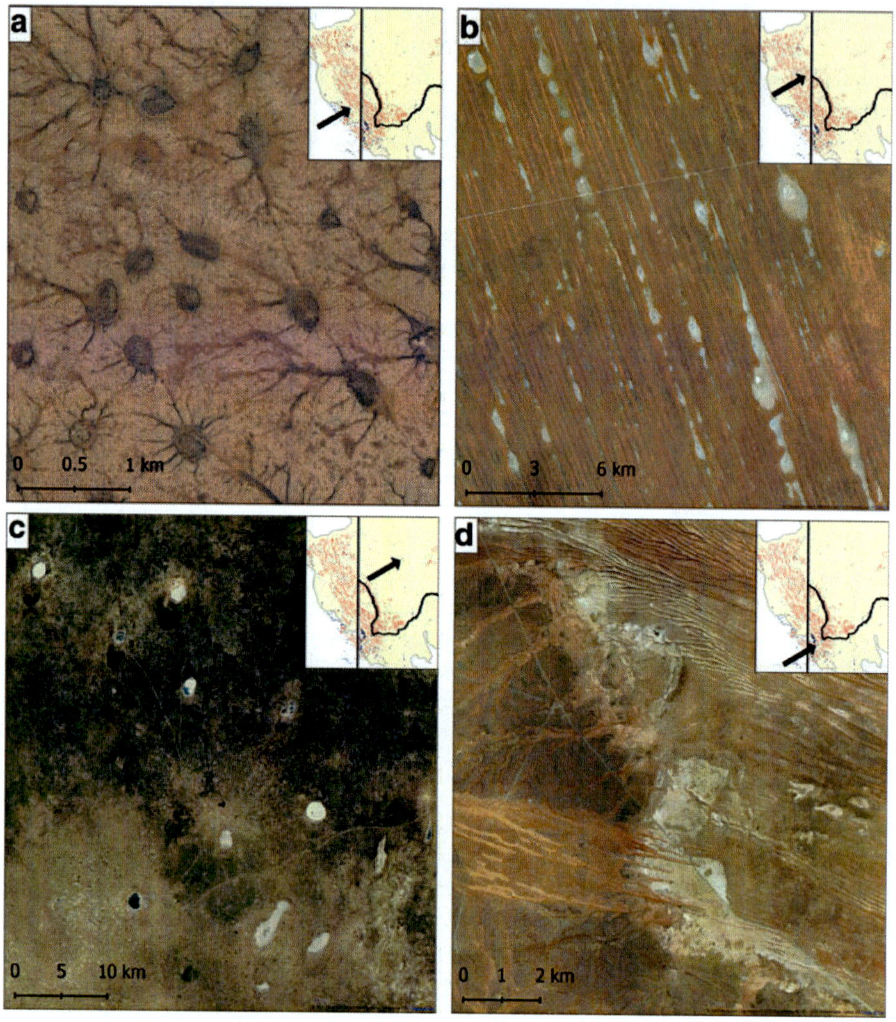

Fig. 10.2 Different shapes of pans related to their geological situation: **a** pans in nearly geometrically arranged karstic sinkholes (map data: Google, Maxar Technologies), **b** pans in inter-dune depressions of longitudinal Kalahari dunes (map data: Google, Maxar Technologies), **c** nearly round-shaped pans in a morphological flat landscape in the central Kalahari Basin (map data: Microsoft Bing Maps, Earthstar Geographics SIO, Tom Tom), **d** pans along the Molopo River system (map data: Microsoft Bing Maps, Maxar Technologies, Tom Tom)

barred and truncated by longitudinal Kalahari dunes during the Pleistocene and Holocene inter-pluvials (Boocock and Van Straten 1962; Goudie and Thomas 1985; Lancaster 1986; Verhagen 1990). In the topographically lower river beds, the typical pan formation started after they were partly or completely cut off by the migrating longitudinal Kalahari dunes. Normally, these drainage systems interact with local and regional groundwater that affects the persistence of the developing pans (Rosen 1994). Examples of these pans can be found along the Molopo River (South Africa, see Fig. 10.4g) and the former Kleiner Nossob River, near Aminuis (Namibia).

The southwestern part of the Kalahari is widely characterised by red, well-defined longitudinal dunes which were formed in the direction of the dominating wind systems. After the last glaciation, the dunes became inactive, vegetated and slightly cemented. In the inter-dune depressions, favourable conditions developed for the evolution of pans and led to the formation of thin and elongated pans or smaller pans, arranged in a row (Fig. 10.2b).

Calcretes occur all over the southwestern rim of the Kalahari. Especially around Koës (Namibia) they form widespread karstic sinkholes that are arranged in symmetrical patterns and regular distances (Fig. 10.2a). Every sinkhole contains pan sediments. They range in size, from m^2 up to tens of km^2. The largest basin is the Koës Pan, with distinct erosional drainage channels in the catchment area.

Another reason for pan evolution could be tectonic processes. In Death Valley, California (USA), tectonic processes induced a much higher thickness of sediments in the pan basin, enabling the deposition of some hundreds of metres, covering a 200 kaBP sedimentological record (Lowenstein et al. 1999). In Botswana, geochemical and geophysical studies have been carried out for Lokgware Pan and Mogatse Pan (Butterworth 1982; Farr et al. 1982). They showed that both pans developed above tectonic structures which caused more than 50 m of sedimentary infill. The tectonic basins are probably the reason for their origin and the specific regional groundwater situation inside these structures plays a part in the geochemistry of surface waters and the fact that a playa or pan sits in the basin today (McEwen 1982).

10.4 Processes in Pans

Recent principle processes in Kalahari pans are complex. The geochemistry of the pan influences many (sub)surface processes. So, pans with lunette dunes typically have a different geochemistry to those without. Nevertheless, many physical phenomena are comparable between pans of the same type.

Runoff

Periodic precipitation in the catchment area of a depression is an important factor for a continuous pan evolution. Many pans feature drainage channels around the depression. During precipitation, clastic sediment particles, organic debris and dissolved salts are transported from the catchment area into the pan by local runoff (Fig. 10.3).

Sedimentation

During transportation, coarser material like lithoclastic sediment particles and heavy minerals are deposited near the rim of the pan where the transport energy decreases significantly. The deepest areas of the depression, mainly the centre of the pan, are more frequently covered with water, which transfers its suspended and dissolved load into the pan (Milewski et al. 2017). In this way, every precipitation event that causes a significant runoff, results in net sedimentation. The dimensions of the pan sizes, in comparison to the catchment areas (ratio of catchment area/pan area), probably control the sedimentation rate in each pan (Schüller et al. 2018). Sediment cores, taken from different studies, revealed sharp boundaries between sediment layers with respect to soil colour and grain size. They mark varying environmental or climate conditions that are preserved in the pan deposits (Lowenstein and Hardie 1985; Genderjahn et al. 2017, 2018a, b; Schüller et al. 2018; Belz et al. 2020).

Evaporites

A high percentage of the salt-saturated water does not evaporate after a precipitation event but seeps into the pore space of the sediment. Depending on the groundwater status of the pan, an accumulation of local groundwater in the basin is possible. Authigenic minerals like salt, carbonates and gypsum crystals do not only occur at the pan surface but also in pan sediments below the surface (Mees 1999). In pans with a documented groundwater level, there have been distinct layers containing coarse gypsum crystals, up to several centimetres in diameter (Schüller et al. 2018). Rosen (1994) identified three chemical mechanisms for evaporite formation: (i) Surface crusts of evaporites form as the final stage of the evaporation of a standing water body or in the capillary zone or groundwater near the pan surface, (ii) subsurface growth of evaporite minerals is caused by precipitation from supersaturated saline groundwater and (iii) a standing brine body causes single crystals or aggregates to form on the water surface or at any depth in the water column. The total amount of carbonates and salts is the result of the local and regional groundwater hydrology (Rosen 1994) and differs distinctly in pans from different areas throughout the Kalahari.

Deflation

Deflation of pan sediment is also an important process. It may result in the export of sediment, reduced accumulation rates or discontinuities of the sediment record. Usually, a closed surficial salt or clay crust acts like a persistent shield that protects the pan effectively against wind erosion (Thomas and Burrough 2016). Disturbances of the pan surface (e.g. animal trails), evaporitic processes and desiccation cracks enable the export of fine-grained sediments from the pan (Baddock et al. 2011; Gill 1996). Clayey to sandy sediments are often redeposited directly behind the pan and lead to the development of lunette dunes on the leeward side of many Kalahari pans (Bowen et al. 2018). On the other hand, dusty material can be moved over long distances high up in the atmosphere. Therefore, playa (pan) surfaces are of importance as a significant regional source of mineral dust emissions (Nield et al. 2015). The dimensions of the deflation and erodibility of the sediments are a function of groundwater level and geochemistry. Deflation can only occur until the capillary fringe of the water table is reached in the pan sediments (Rosen 1994). Water coverage and wet phases during the year, with moist pan sediments, effectively protect against deflation. The age-depth models in the studies by Schüller et al. (2018) (on the basis of 57 AMS ^{14}C age determinations of sediments from different Kalahari pans) indicate that sediment input exceeds erosion in the sampled pans without noticeable gaps or rapid changes. However, there are other pans in the Kalahari that do not contain groundwater during relatively dry periods. Without the protecting capillary fringe of the water table, wind is able to deflate and erode sediments from the pan (Rosen 1994) and create sedimentation gaps, for example at Lebatse Pan, Botswana (Holmgren and Shaw 1996).

Lunette dunes are relicts of dry deflation periods and were formed along the downwind margin of some larger pans. In

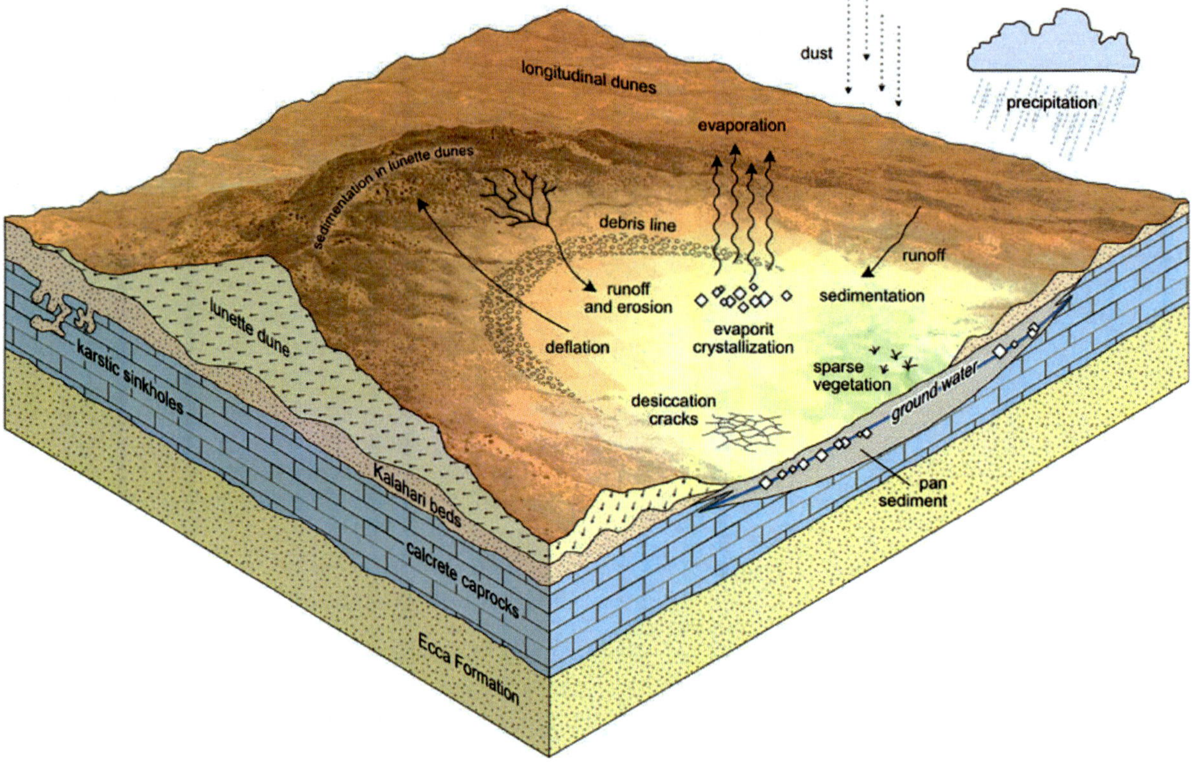

Fig. 10.3 Schematic southwestern Kalahari pan model with principle processes. Many pans developed in former river beds, in inter-dune depressions along longitudinal dunes, in karstic sinkholes of calcrete cap-rocks or in depressions caused by erosional differences between sedimentary rocks and dolerite dykes. An important factor of pan development is a high local or regional groundwater table in the structures. Precipitation by isolated showers in the local catchment area leads to sediment transport into the pan by runoff. The suspended loads, as well as evaporates, remain as a sediment layer after evaporation. Authigenic minerals, especially salts and gypsum, also recrystallise in the groundwater and lead to reworking in the critical zone. Significant deflation of fine material and the formation of lunette dunes on the leeward side of the pans are restricted to distinct episodes. This material can be transported back into the pan by fluvial erosion of the lunette dunes. Some pans also contain fossil root traces, from sparse, temporary vegetation in their deeper layers

the High Plains of Western Kansas, Bowen and Johnson (2012) and Bowen et al. (2018) found that lunettes in this area typically consist of reworked clay to fine sand-size particles and loess derived from playas and the Late Pleistocene High Plains loess mantle. However, not all lunette sediments necessarily originated from pans sediments. The geochemical analyses of Telfer and Thomas (2006) proved that a large percentage of the current lunette dunes of Witpan (South Africa) do not consist of primarily deflated sediments from the pan. Some parts of the sand of the lunettes were derived from the neighbouring linear dunes and from the recycled material of older lunette sediments.

So far, all studied Kalahari pans contain a sediment filling of several decimetres to metres in thickness (e.g. Beaumont et al. 1984; Lancaster 1978, 1986; Holmgren and Shaw 1996; Mees 1999; Telfer et al. 2009; Genderjahn et al. 2017, 2018a, b; Schüller et al. 2018).

Regional Geology

Regional geology affects the mineral and grain size composition of the pan sediments. In most cases, pan sediments primarily consist of varying amounts of quartz, feldspar and, to a minor degree, illite and smectite. So, there are differences in composition between different pan clusters throughout the Kalahari.

Climatic Variations

Climatic variations have a bearing on compositional changes between sediment layers of different periods. A change in transport energy (for example, stronger or weaker wind, increased or decreased inflow of surface water) causes grain size differences. A variation in the prevailing wind direction may cause a change in the chemical composition of the pan

Fig. 10.4 Characteristic surface features of small Kalahari pans: **a** dry surface of a clay pan with bird traces from the wet season and desiccation cracks, **b** white salt crust of a salt pan, **c** deeply eroded lunette dune, **d** drainage channel between grasses and shrubs in a lunette dune, **e** carbonate crusts at a pan rim with stromatolitic bedding, **f** close-up image of internal stromatolitic structures, **g** pans along the Molopo River as a result of a former barred river bed between longitudinal Kalahari dunes (red coloured in front) and the lunette dunes (bright coloured background), **h** debris line that developed from wind drift during a phase of water coverage along the leeward rim, mainly composed of plant fragments as well as terrestrial gastropod shells from the catchment area

sediments because of a different geological catchment area. Climatic variations can change the kind of erosion that occurs, as well. Drier conditions lead to an increase in physical erosion (rock weathering by temperature differences, fractionation and fragmentation of grains by sediment transport), whereas wetter periods are characterised by chemical weathering and a change in the chemical composition of the sediment. At the Witpan (South Africa), grain sizes vary within the sediment core, pointing to changing environmental conditions through time (Telfer et al. 2009; Genderjahn et al. 2017). Contrastingly, the sediment composition in the Branddam East Pan shows only minor changes, which indicate long-lasting and continuous sedimentation processes (Schüller et al. 2018).

Hydrology

The hydrology of the groundwater system beneath the pan plays an important role in its evolution. It defines the surface morphology as well as the sedimentation rate and the erodibility of the sediment. Whether evaporite crystals precipitate in the sediment or at the pan surface is a function of the height of the (ground)water table. Thick crusts of salts only develop in strongly water-influenced pans. Dry pans

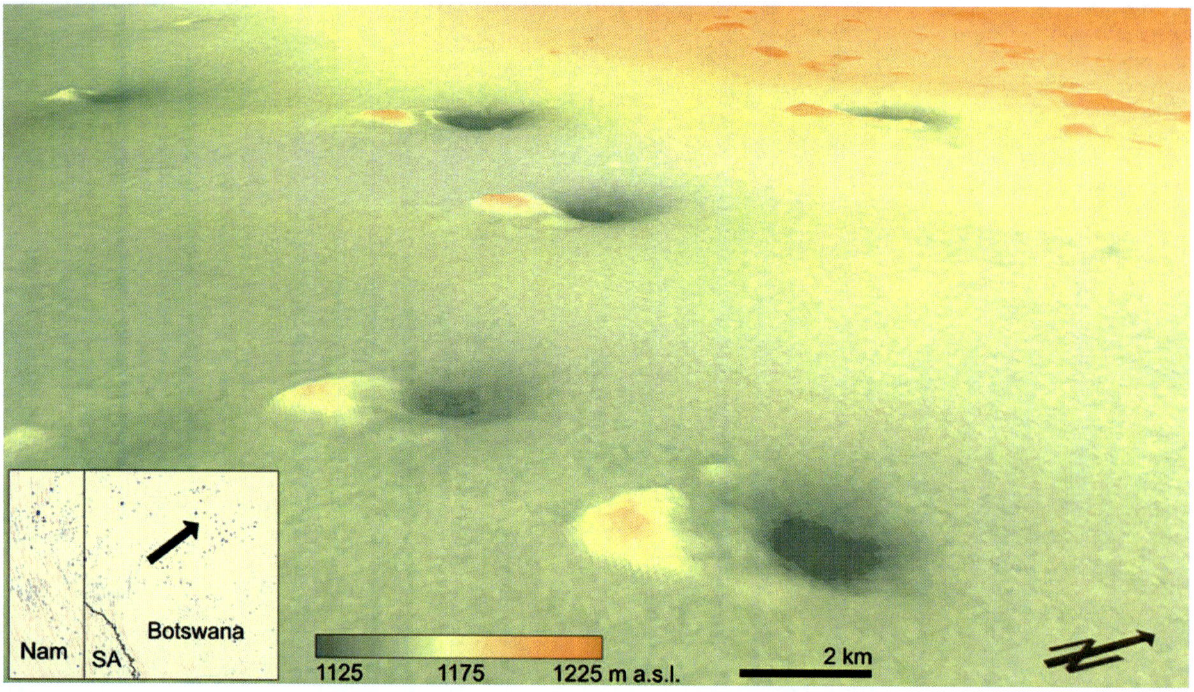

Fig. 10.5 Digital elevation model of round-shaped pans with their lunette dunes in the morphological flat landscape of the central Kalahari Basin (compare Fig. 10.2c). The lunette dunes reflect phases of significant aeolian transport (erosion). However, the Kalahari salt pans show a net accumulation over time. Super-elevation of the DEM: 15x; SRTM-data from NASA (Farr et al. 2007)

without their own water tables may act as a groundwater recharge zone to the deep aquifer after sufficient precipitation in the catchment area (Rosen 1994).

Charcoal

Charcoal fragments are a sedimentary component that can hint at the prevailing environmental situation during sedimentation. They indicate the occurrence of natural bush fires and can enable the estimation of their frequency (Patterson et al. 1987). However, fragments of micro charcoal are rare in pan sediments because their preservation potential is poor in the arid climate of the Kalahari. They mainly occur in very fine-grained pan sediments. This fact underpins the fact that charcoal is only preservable under oxygen-reduced conditions in clayey and fine silty sediments.

Diatoms and Pollen

For the same reason, it is often difficult to find diatoms and pollen in Kalahari pan sediments. The corrosive, salty water usually dissolves any pollen that was present before burial. Pollen was found in some pans that are located in areas with higher precipitation rates, for example in Kathu Pan (Northern Cape, South Africa) (Beaumont et al. 1984). Diatoms, with their filigree siliceous shells, are not normally preserved in pan sediments because of common conditions that support silcrete resolution (Holmgren and Shaw 1996).

Desiccation Cracks

Desiccation cracks are a very common evaporation feature of the majority of pan surfaces (Fig. 10.4a). The finer the surface sediments are, the better developed the cracks that can be found. In clay pans, they are deep and steady and frequently filled with younger sediment material. Salt pans have desiccation cracks as well because evaporites consolidate the surface sediments. Most pans have some sort of microbial component within the surface crust (Genderjahn et al. 2017, 2018a, b) that can bind sediment particles and affect the morphology. However, deflation mainly erodes the fine, unsolidified sediments from the crack fillings. Schüller et al. (2018) described the silty saltpan sediments of the Omongwa Pan and found fossil desiccation cracks in deeper layers, which were buried by younger undisturbed sediments.

Debris Lines

During the phases of water coverage of the pan, wind drift causes floating material to accumulate on the leeward side, in the form of a debris line (Fig. 10.4h). In the Omongwa Pan, the debris line was mainly composed of plant fragments and terrestrial snail shells, that live in the shrubby vegetation of the respective catchment area.

Vegetation

Plant growth on pan surfaces is rare and occurs as scattered small shrubs or as grass. In some deeper sediment layers of today's unvegetated pans, some fossil root horizons could be identified. They indicate periods of different climatic and environmental conditions that enabled more widespread vegetation to grow in the pans of the Kalahari. XRF analyses (Schüller et al. 2018) indicated sediment layers that are related to drier phases at the top levels of fossil root horizons. They interpreted that increased precipitation in the catchment areas leads to enhanced input of salt minerals that disturb plant growth. As given by the age-depth model of the study, these phases occurred at around 20.0 ka and 5.0 ka (Witpan), 25.0 ka (Omongwa Pan) and 31.0 ka (Toasis Pan).

Peat, Stromatolites

In the Kathu Pan (Northern Cape, South Africa), at the less arid southern margin of the Kalahari, Beaumont et al. (1984) described silty, calcareous sands and calcretes of a pan phase. On top of these sediments, they found a massive peat formation dating from the mid to late Holocene. Another hint for higher water availability in a pan's history is the occurrence of stromatolites along pan rims (Fig. 10.4e, f). These have been found in different pans, for example, at Urwi Pan (western Botswana, Lancaster 1977), Klein Awas Pan (South Africa, Hürkamp et al. 2011), Omongwa Pan and Toasis Pan (eastern Namibia, Schüller et al. 2018).

Secondary Processes

In the sedimentary record, primary structures have often been overprinted by secondary processes. These can be physical (e.g. temperature differences, water, wind), chemical (e.g. evaporite recrystallization or reworking within the critical zone) or biological processes. Even if pans are an extreme ecosystem with strong environmental gradients, some life forms are able to cope with them by their broad ecological tolerance. So, it is possible to find bioturbation, not only by plant roots, but also by burrows in the sediment, caused by salt beetles, etc. Microbial mats are also very common in pan sediments. Larger animals like antelopes, cows and horses cross pans on distinct animal trails that are clearly visible in the disturbed pan surface.

10.5 Lunette Dunes

Lunette dunes occur on the leeward side of many Kalahari pans in relation to the prevailing wind direction (Goudie and Thomas 1986), see Fig. 10.5. Mallick et al. (1981) counted more than 110 pans in the southern Kalahari with definable lunette dunes. Many pans developed two or more generations of dune ridges, differing in sediment characteristics.

The innermost dunes represent the youngest accumulations. Their composition is characterised by a higher amount of clay minerals and up to 15% $CaCO_3$ (Thomas and Shaw 1991). This causes a carbonate encrustation of the dune sediments (Heine 1981). This, together with a lack of sediment accumulation, has led to the current inactivity of this dune type. They are usually a bright greyish-brown or yellowish-brown colour and reach heights of only a few metres. They are situated directly at the leeward rim of the pan and, because of the recent inactivity, the inner dunes are predominantly disintegrated by erosional processes (Heine 1981), see Fig. 10.4c, d.

The outer dunes are up to 30 m high and are mainly composed of well-sorted sand (Heine 1981; Thomas and Shaw 1991). They are mostly light to dark red in colour, caused by iron oxide and iron hydroxide coatings on the quartz grains. The lunette dunes represent drier climatic phases with lower groundwater tables than today. Temporary erosion of the pan sediments leads to an accumulation of the material behind the pans, documenting palaeo wind directions. Stokes et al. (1997) dated the lunette dunes around Witpan (South Africa) to ages from 1.4 ± 0.3 to 1.1 ± 0.2 ka. Telfer and Thomas (2006) reported lunette dunes from the Witpan to be not older than 2.0 ka and found residuals from older dunes (32–23 ka and 6.5 ka) which are now buried by late Holocene sediments. Lawson and Thomas (2002) studied the lunette dunes of Luitenantspan (inner lunette: 3.23 ± 0.39 to 0.84 ± 0.08 kaBP; outer lunette: 14.29 ± 2.11 to 7.26 ± 1.64 kaBP), Soutpan (inner lunette: 8.02 ± 1.08 to 0.37 ± 0.04 kaBP; outer lunette: 16.1 ± 1.79 to 2.35 ± 0.22 kaBP) and Koopan Suid (inner lunette: 0.79 ± 0.1 to 0.09 ± 0.01 kaBP; outer lunette: 9.08 ± 1.02 to 2.17 ± 0.29 kaBP) in South Africa.

However, recently cemented lunette dunes become eroded by fluvial runoff into the pan, characterised by deep erosional channels (Fig. 10.4c, d). The lunette dune of

Witpan (Northern Cape, South Africa) is an example of strong lunette dune erosion.

10.6 Climatic and Environmental Records

Recent studies tested different proxies in Kalahari pan sediments to extract climatic and environmental information and prove the potential of salt and clay pans as Quaternary environmental archives (Holmgren and Shaw 1996; Genderjahn et al. 2017, 2018a, b; Schüller et al. 2018; Belz et al. 2020).

Holmgren and Shaw (1996) defined four different periods of deposition in the Lebatse Pan (Botswana), separated by depositional hiatuses. They detected phases of a permanent, shallow water body as well as ephemeral, evaporative conditions by comparing the shape and lustre of quartz grains, X-ray diffraction (XRD) and energy-dispersive X-ray (EDX) analysis as well as carbon measurements.

Changes in grain sizes inside a pan sediment profile reflect fluctuating transport energies and can be used as a proxy for the intensity of fluvial runoff and wind transport. During runoff, all grain sizes are delivered to the pan. Sandy material is deposited nearer to the rim, whereas clay has a more widespread distribution due to its suspension in the whole pan. Wind transport transfers sandy material by aeolian, saltatorial movement. Silty grain sizes can be transported by saltation (coarse silt) as well as dust emission (fine silt). Clay is typically transported over long distances as dust in the atmosphere. Heavy minerals, such as titanium and zircon, are a typical feature of wind transport. Secondary clay minerals that result from the chemical weathering of feldspars, point to periods of wetter conditions.

Long-chain *n*-alkanes, *n*-alkanols and *n*-fatty acids are constituents of cuticular plant waxes (Eglinton and Hamilton 1967) and can be used to reconstruct the affiliation of the occurring plants to different groups regarding their carbon assimilation pathways during photosynthesis, namely C_3 and C_4 plants (Belz et al. 2020). Variations in hydrological conditions, such as the amount of precipitation, the origin of water masses and evaporative conditions, can be assessed by the isotopic composition of hydrogen in plant wax constituents (Burdanowitz et al. 2018; Collins et al. 2013; Sachse et al. 2012). On this basis, plant wax-derived compounds (δD_{wax}) in pan sediments can be used to reconstruct the past vegetation assemblage.

Additionally, the content of total organic carbon (TOC), its isotopic composition ($\delta^{13}C$), the content of total inorganic carbon (TIC), and isotopic composition of carbon and oxygen in carbonates ($\delta^{13}C$, $\delta^{18}O$) can be used as environmental proxies to reconstruct past vegetation and ecology as well as hydrological changes (Belz et al. 2020).

Lipid biomarkers, such as bacterial membrane phospholipid life markers, saturated, branched and unsaturated phospholipid fatty acids are indicative of certain phyla of bacteria (such as gram-positive and gram-negative bacteria) and biomarkers for past microbial life (archaeol, branched and isoprenoid GDGTs), containing information about changes in biogeochemical processes and climate variation (Genderjahn et al. 2017). They reflect the influence of climate variation on the abundance and composition of microbial communities. Since water is an inevitable requirement for microbial life, in arid regions the abundance of past microbial biomarkers points to wetter periods in the past (Genderjahn et al. 2018a).

10.7 Conclusions

Pans are a characteristic landform of the southwestern Kalahari. They represent important sedimentary geoarchives in the semi-arid to arid landscapes of southwestern Africa, where other continuous archives are rare or absent. Due to their net accumulation, they can be classified as quasi-continuous geoarchives (Schüller et al. 2018). They are morphological sinks that function as ephemeral lakes. During the rainy season, numerous pans are temporarily filled with water by local runoff. The water coverage persists for some days or weeks. This leads to the gradual deposition of clayey to sandy siliciclastic sediments and evaporites. All of the pans studied contain a sediment sequence of a few decimetres to metres in thickness but they differ in their geological and hydrological background, the size of the catchment area and the local precipitation and evaporation rates. Consequently, the mineral composition and grain size distribution of the siliciclastic sediment components are mainly controlled by the regional geology, weathering intensities and climatic conditions. Root traces in distinct layers of pan sediments indicate temporal coverage of the pan surface by sparse vegetation. Bush fires cause the coalification of vegetation and the production of micro charcoal but the preservation potential in the arid pans is limited to scattered traces of charcoal in clayey and silty sediments.

Kalahari pans have the potential to function as geoarchives in the dry landscapes of the Kalahari Basin. Some work has already been done to explore the application possibilities of pan sediments for uses such as climatic and environmental reconstructions and modelling from a regional to a global scale. Nevertheless, there is a great potential for many interdisciplinary research approaches to work on Kalahari pans. The study of Botswanan pans is few compared to research that has been carried out on Namibian and Southern African pans. This would help to improve the understanding of southern hemispheric climate change and environmental evolution.

References

Digital Atlas of Namibia, Acacia Project E1, University of Cologne (http://www.unikoeln.de/sfb389/e/e1/download/atlas_namibia/index_e.htm), data base: Atlas of Namibia Project (2002) Directorate of Environmental Affairs, Ministry of Environment and Tourism (https://www.met.gov.na/)

Baddock MC, Zobeck TM, Van Pelt RS, Fredrickson EL (2011) Dust emissions from undisturbed and disturbed, crusted playa surfaces: cattle trampling effects. Aeolian Res 3(1):31–41

Beaumont PB, Van Zinderen EM, Vogel JC (1984) Environmental changes since 32,000 BP at Kathu Pan, northern Cape. In: Vogel JC (ed) Late Cainozoic Palaeoclimates of the Southern Hemisphere. Balkema, Rotterdam, pp 329–338

Belz L, Schüller I, Wehrmann A, Köster J, Wilkes H (2020) First leaf wax biomarker record of a Namibian salt pan reveals enhanced summer rainfall during the Last Glacial-Interglacial. Palaeogeogr Palaeoclimatol Palaeoecol 543. https://doi.org/10.1016/j.palaeo.2019.109561

Boocock C, Van Straten OJ (1962) Notes on the geology and hydrogeology of the central Kalahari region, Bechuanaland Protectorate. Trans Geol Soc South Afr 65:125–171

Bowen MW, Johnson WC (2012) Late Quarternary environmental reconstructions of playa-lunette system evolution on the central High Plains of Kansas, United States. Geol Soc Am Bull 124:146–161

Bowen MW, Johnson WC, King DA (2018) Spatial distribution and geomorphology of lunette dunes on the High Plains of Western Kansas: implications for geoarchaeological and paleoenvironmental research. Phys Geogr 39(1):21–37

Bryant RG, Bigg GR, Mahowald NM, Eckardt FD, Ross SG (2007) Dust emission response to climate in southern Africa. J Geophys Res Atmos 112(D9)

Burdanowitz N, Dupont L, Zabel M, Schefuß E (2018) Holocene hydrologic and vegetation developments in the Orange River catchment (South Africa) and their controls. The Holocene 28:1288–1300

Burrough SL, Thomas DSG (2013) Central southern Africa at the time of the African Humid Period: a new analysis of Holocene palaeoenvironmental and palaeoclimate data. Quatern Sci Rev 80:29–46. https://doi.org/10.1016/j.quascirev.2013.08.001

Butterworth JS (1982) The chemistry of Mogatse Pan-Kgalagadi District. Botswana Department of Geological Survey. Unpublished Report JSB/14/82. 15 p

Cockcroft MJ, Wilkinson MJ, Tyson PD (1987) The application of a present-day climatic model to the late Quaternary in southern Africa. Clim Change 10:161–181

Collins JA, Govin A, Mulitza S, Heslop D, Zabel M, Hartmann J, Röhl U, Wefer G (2013) Abrupt shifts of the Sahara-Sahel boundary during Heinrich stadials. Clim past 9:1181–1191. https://doi.org/10.5194/cp-9-1181-2013

Eckardt FD, Kuring N (2005) SeaWiFS identifies dust sources in the Namib Desert. Int J Remote Sens 26(19):4159–4167

Eckardt FD, Washington R, Wilkinson J (2001) The origin of dust on the West Coast of Southern Africa. Palaeoecol Afr 27:207–219

Eglinton G, Hamilton RJ (1967) Leaf Epicuticular waxes. Science 156:1322–1335

Farr JL, Peart RJ, Nellisse G, Butterworth JS (1982) Two Kalahari pans: a study of their morphology and evolution. Botswana Department of Geological Survey. Unpublished Report GS10/10. 21 p

Farr TG et al (2007) The shuttle radar topography mission. Rev Geophys 45:RG2004. https://doi.org/10.1029/2005RG000183

Genderjahn S, Alawi M, Kallmeyer J, Belz L, Wagner D, Mangelsdorf K (2017) Present and past microbial life in continental pan sediments and its response to climate variability in the southern Kalahari. Org Geochem 108:30–42. https://doi.org/10.1016/j.orggeochem.2017.04.001

Genderjahn S, Alawi M, Mangelsdorf K, Horn F, Wagner D (2018) Desiccation- and saline-tolerant bacteria and Archaea in Kalahari pan sediments. Front Microbiol 9:2082. https://doi.org/10.3389/fmicb.2018.02082

Genderjahn S, Alawi M, Wagner D, Schüller I, Wanke A, Mangelsdorf K (2018) Microbial community responses to modern environmental and past climatic conditions in Omongwa Pan, western Kalahari: a paired 16S rRNA gene profiling and lipid biomarker approach. J Geophys Res Biogeosci 123:1333–1351. https://doi.org/10.1002/2017JG004098

Gill TE (1996) Eolian sediments generated by anthropogenic disturbance of playas: human impacts on the geomorphic system and geomorphic impacts on the human system. Geomorphology 17(1–3):207–228

Goudie AS, Thomas DSG (1985) Pans in southern Africa with particular reference to South Africa and Zimbabwe. Zeitschrift für Geomorphologie, Neue Folge 29:1–19

Goudie AS, Thomas DSG (1986) Lunette dunes in southern Africa. J Arid Environ 10(1):1–12. https://doi.org/10.1016/S0140-1963(18)31260-6

Goudie AS, Wells GL (1995) The nature, distribution and formation of pans in arid zones. Earth Sci Rev 38:1–69. https://doi.org/10.1016/0012-8252(94)00066-6

Heine K (1981) Aride und pluvial Bedingungen während der letzten Kaltzeit in der Südwest-Kalahari (südliches Afrika) - Ein Beitrag zur klimagenetischen Geomorphologie der Dünen, Pfannen und Täler. Zeitschrift für Geomorphologie, Neue Folge, Suppl 38:1–37

Heine K (1982) The main stages of the Late Quaternary evolution of the Kalahari region, southern Africa. Palaeoecol Africa 15:53–76

Heine K (2005) Holocene climate of Namibia: a review based on geoarchives. African Study Monographs, Suppl 30:119–133

Holmgren K, Shaw P (1996) Palaeoenvironmental reconstruction from near-surface pan sediments: an example from Lebatse Pan, southeast Kalahari, Botswana. Geogr Ann 79 A(1–2):83–93

Hürkamp K, Völkel J, Heine K, Bens O, Leopold M, Winkelbauer J (2011) Late quaternary environmental changes from aeolian and fluvial geoarchives in the southwestern Kalahari, South Africa: Implications for past African climate dynamics. S Afr J Geol 114(3–4):459–474. https://doi.org/10.2113/gssajg.114.3-4.459

Jones CR, Hepworth JV (1973) Geological Map of Botswana. Director of Geological Survey and Mines with the authority of the Ministry of Commerce, Industry and Water Affairs, Gaborone

Kautz K, Porada H (1976) Sepiolite formation in a pan of the Kalahari, South West Africa. Neues Jahrbuch für Mineralogie-Monatshefte 12:545–559

Lancaster IN (1977) Pleistocene lacustrine stromatolites from Urwi Pan Botswana. Trans Geol Soc S Afr 80:283–285

Lancaster IN (1986) Pans in the southwestern Kalahari: a preliminary report. Palaeoecol Africa 17:59–67

Lancaster IN (1976) The Pans of the Southern Kalahari, Botswana. Dissertation, University of Cambridge

Lancaster IN (1978) The pans of the southern Kalahari, Botswana. Geogr J 81–98

Lawson MP, Thomas DSG (2002) Late Quaternary lunette dune sedimentation in the southwestern Kalahari desert, South Africa:

luminescence based chronologies of aeolian activity. Quatern Sci Rev 21:825–836

Lowenstein TK, Li J, Brown C, Roberts SM, Ku T-L, Luo S, Yang W (1999) 200 k.y. paleoclimate record from Death Valley salt core. Geology 27(1):3–6. https://doi.org/10.1130/0091-7613(1999)027<0003:KYPRFD>2.3.CO;2

Lowenstein TK, Hardie LA (1985) Criteria for the recognition of salt pan evaporites. Sedimentology 32:627–644. https://doi.org/10.1111/j.1365-3091.1985.tb00478.x

Mallick DIJ, Habgood F, Skinner AC (1981) Geological interpretation of Landsat imagery and air photography of Botswana. Stationery Office 56

McEwen G (1982) The geology of Mogatse Pan-Kgalagadi District. Botswana Department of Geological Survey. Unpublished Report GM/6/82

Mees F (1999) Distribution of gypsum and kalistrontite in a dry lake basin of the southwestern Kalahari (Omongwa Pan, Namibia). Earth Surf Proc Land 24:731–744

Mees F (2002) The Nature of calcareous deposits along pan margins in eastern central Namibia. Earth Surf Proc Land 27:719–735. https://doi.org/10.1002/esp.348

Mees F (2003) Salt mineral distribution patterns in soils of the Otjomongwa pan, Namibia. CATENA 54:425–437. https://doi.org/10.1016/S0341-8162(03)00135-8

Milewski R, Chabrillat S, Behling R (2017) Analyses of recent sediment surface dynamic of a Namibian Kalahari salt pan based on multitemporal landsat and hyperspectral hyperion data. Remote Sens 9:170. https://doi.org/10.3390/rs9020170

Nield JM, Bryant RG, Wiggs GFS, King J, Thomas DSG, Eckardt FG, Washington R (2015) The dynamisms of salt crust patterns on playas. Geology 43(1):31–34. https://doi.org/10.1130/G36175.1

Patterson WA, Edwards KJ, Maguire DJ (1987) Microscopic charcoal as a fossil indicator of fire. Quatern Sci Rev 6:3–23

Rosen MR (1994) The importance of groundwater in playas: A review of playa classifications and the sedimentology and hydrology of playas. Paleoclim Basin Evol Playa Syst Geol Soc Am Spec Pap 289:1–18. https://doi.org/10.1130/SPE289-p1

Sachse D, Billault I, Bowen GJ, Chikaraishi Y, Dawson TE, Feakins SJ, Freeman KH, Magill CR, McInerney FA, van der Meer MTJ, Polissar P, Robins RJ, Sachs JP, Schmidt H-L, Sessions AL, White JWC, West JB, Kahmen A (2012) Molecular paleohydrology: interpreting the hydrogen-isotopic composition of lipid biomarkers from photosynthesizing organisms. Annu Rev Earth Planet Sci 40:221–249

Schüller I, Belz L, Wilkes H, Wehrmann A (2018) Late quaternary shift in southern African rainfall zones: sedimentary and geochemical data from Kalahari pans. Z Geomorphol 61:339–362. https://doi.org/10.1127/zfg/2018/0556

Shaw PA, Bryant RG (2011) Pans, playas and salt lakes. In: Thomas DSG (ed) Arid zone geomorphology: process, form and change in drylands. Third Edition, John Wiley & Sons, Ltd, Chichester, UK. https://doi.org/10.1002/9780470710777.ch15

Steyn JH, Nortman LE, Drawing Office, Geological Survey, Pretoria (1980) Geological map 1 : 1,000,000, based on 1978 edition map published by the Surveyor-General, Windhoek

Stokes S, Thomas DSG, Washington R (1997) Multiple episodes of aridity in southern Africa since the last interglacial period. Nature 288:154–158

Telfer MW, Thomas DSG (2006) Complex Holocene lunette dune development, South Africa: Implications for paleoclimate and models of pan development in arid regions. Geology 34(10):853–856. https://doi.org/10.1130/G22791.1

Telfer MW, Thomas DSG, Parker AG, Walkington H, Finch AA (2009) Optically Stimulated Luminescence (OSL) dating and palaeoenvironmental studies of pan (playa) sediment from Witpan, South Africa. Palaeogeogr Palaeoclimatol Palaeoecol 273:50–60. https://doi.org/10.1016/j.palaeo.2008.11.012

Thomas DSG, Shaw PA (2002) Late Quaternary environmental change in central southern Africa: new data, synthesis, issues and prospects. Quatern Sci Rev 21:783–797. https://doi.org/10.1016/S0277-3791(01)00127-5

Thomas DSG, Nash DJ, Shaw PA, Van der Post C (1993) Present day lunette sediment cycling at witpan in the arid southwestern Kalahari desert. CATENA 20:515–527

Thomas DSG, Burrough SL (2016) Luminescence-based dune chronologies in southern Africa: Analysis and interpretation of dune database records across the subcontinent. Quatern Int 410(B):30–45. https://doi.org/10.1016/j.quaint.2013.09.008

Thomas DSG, Shaw PA (1991) The Kalahari Environment. Cambridge University Press, pp 157–162

Torgersen T, de Deckker P, Chivas AR, Bowler JM (1986) Salt lakes: a discussion of processes influencing palaeoenvironmental interpretation and recommendations for future study. Palaeogeogr Palaeoclimatol Palaeoecol 54:7–19. https://doi.org/10.1016/0031-0182(86)90115-X

Van Straten OJ (1955) The Geology and Groundwaters of the Ghanzi Cattle Route. Ann Rep Geol Surv Bech Prot 1955:28–39

Verhagen BT (1990) On the nature and genesis of pans A review and an ecological model. Palaeoecol Afr 21:179–194

Vickery K, Eckardt FD, Bryant RG (2013) A sub-basin scale dust plume source frequency inventory for southern Africa, 2005–2008. Geophys Res Lett 40:5274–5279. https://doi.org/10.1002/grl.50968

Irka Schüller is a Master in geoscience (Leipzig University). She works as an engineering consultant for hydrogeology. Alongside her job, she is a Ph.D. student at the Carl von Ossietzky University of Oldenburg. Her research interests are the geomorphological processes of geoarchives in the deserts and coastlines of southern Africa, and their applicability for climate research.

Lukas Belz is a graduate in marine geoscience. Currently, he is a Ph.D. student of marine environmental sciences at the Carl von Ossietzky University of Oldenburg. He also works in the field of occupational and maritime medicine.

Heinz Wilkes is a Professor for Organic Geochemistry at the Carl von Ossietzky University of Oldenburg in the Institute for Chemistry and Biology of the Marine Environment. His research focuses on the reconstruction of past environments and climates using organic-geochemical proxy parameters, on the origin, fate and transformation of hydrocarbons and petroleum in the environment and in the geosphere, and on the understanding of molecular mechanisms of biogeochemical key reactions and metabolic pathways in microorganisms.

Achim Wehrmann is a Geologist and Senior Scientist at the Research Institute Senckenberg am Meer in Wilhelmshaven. His main research topic are principle processes and controlling factors in biosedimentary systems. Beside interdisciplinary studies in modern environments he also works about palaeoenvironmental reconstruction of palaeozoic sedimentary sequences along the northern margin of the Gondwana supercontinent. He has a lectureship for Marine Geology and Biological Oceanography at the Carl von Ossietzky University of Oldenburg.

Check for updates

Dry Valleys (*Mekgacha*)

11

David J. Nash

Abstract

Conventional rivers are absent from much of Botswana, with only the Okavango, Chobe and Zambezi systems in the extreme north containing perennial flowing water. Ephemeral rivers occur in the eastern hardveld, but the most extensive components of the surface drainage are the networks of fossil or dry valleys (termed *mekgacha* in Setswana and *dum* in various San languages) that cross the sandveld. This chapter presents the first holistic review of current knowledge about these enigmatic landforms. It does so using a range of evidence types, from radar remote-sensing to the analysis of historical documents written by missionaries and explorers. The chapter considers dry valley distribution, morphology and contemporary and historical hydrology before discussing valley evolution over longer timescales. It concludes with a synthesis of the main arguments concerning how dry valley systems may have formed, including the balance between conventional fluvial incision and processes such as groundwater seepage erosion.

Keywords

Dry valley • Ephemeral river • Kalahari desert • Long-profile • Drainage evolution

11.1 Introduction

Perennial rivers are absent from much of Botswana, with the Okavango, Chobe and Zambezi systems in the extreme north of the country providing the only permanent drainage (see Chap. 7). Ephemeral rivers occur at the periphery of the Kalahari Basin in the east of the country and flow periodically during the summer rainy season (Shaw 1989). Most of these systems rise to the east of the Kalahari-Limpopo drainage divide and cross into South Africa, although some drain westwards towards the Makgadikgadi Basin and provide fluvial input to Sua Pan (see Fig. 11.1). The most extensive components of the drainage within Botswana, however, are the networks of dry valleys that form the focus of this chapter. These are termed *mokgacha* (singular; or *mekgacha*, plural) in Setswana, *dum* in various San languages, *omuramba* (singular; or *omiramba*, plural) in Otjiherero and *laagte* in Afrikaans.

The Kalahari Basin has experienced faulting and flexuring throughout the Cenozoic (e.g. du Toit 1933; Thomas and Shaw 1991; Haddon and McCarthy 2005; de Wit 2007; Moore et al. 2009, 2012), notably associated with the extension of the East African Rift System. There is also evidence for longer term climatic fluctuations about the present-day semi-arid mean, with parts of the Kalahari occupied by palaeolakes or covered by active linear dune systems at various periods during the Quaternary (Chaps. 5, 8and9). This combination of climate change and neotectonics has produced the complex drainage networks that exist today.

The aim of this chapter is to summarise the major characteristics of the Kalahari dry valley networks within Botswana, including their geomorphology, contemporary and past hydrology and the various theories surrounding their (still enigmatic) age and origins. Many dry valley systems are transboundary, rising either in South Africa or Namibia before crossing into Botswana (or, in the case of the Serorome, rising in eastern Botswana and crossing into South

D. J. Nash (✉)
School of Applied Sciences, University of Brighton, Brighton, UK
e-mail: D.J.Nash@brighton.ac.uk

D. J. Nash
School of Geography, Archaeology and Environmental Studies, University of the Witwatersrand, Johannesburg, South Africa

F. D. Eckardt (ed.), *Landscapes and Landforms of Botswana*, World Geomorphological Landscapes,
https://doi.org/10.1007/978-3-030-86102-5_11

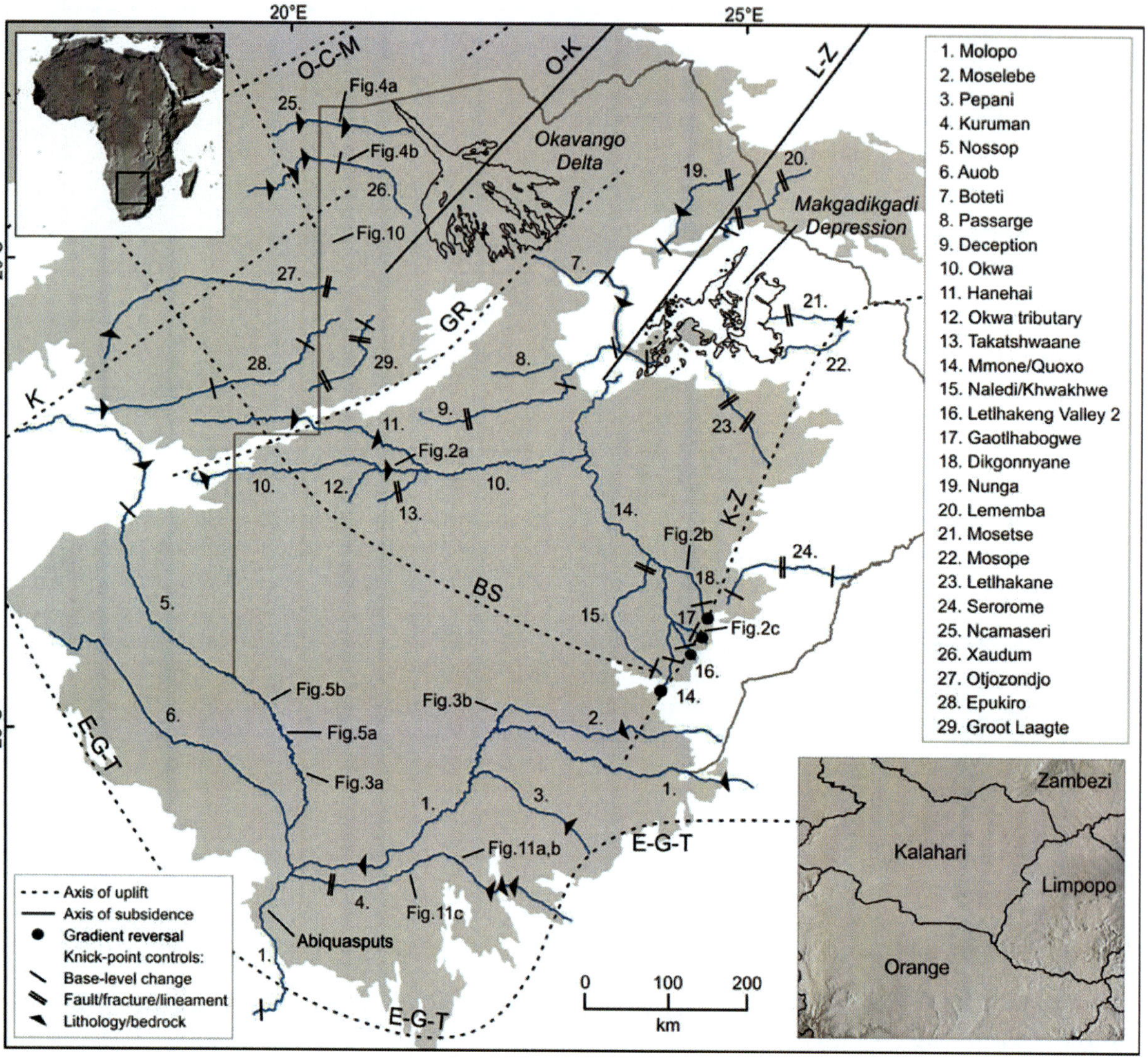

Fig. 11.1 Major dry valley and ephemeral river systems in Botswana and the wider Kalahari Basin (after Nash and Eckardt 2016). The Botswana border is delimited by a grey line and the area covered by Kalahari Group sediments shaded pale grey. Knickpoints and sections of valleys that exhibit gradient reversals are indicated. Knickpoints are distinguished according to their dominant control: base-level change; intersection with a fault, fracture or lineament; or differential rock resistance across a lithological boundary. The location of major flexural axes (after Haddon and McCarthy 2005) are also shown (BS, Bakalahari Schwelle; E–G–T, Etosha-Griqualand-Transvaal axis; GR, Ghanzi Ridge; K-Z, Kalahari-Zimbabwe axis; L-Z, Luangwa-Zambezi axis; O-K, Okavango-Kafue axis; O-C-M, Otavi-Caprivi-Mweru axis; K, Khomas axis). The numbers shown against individual valley segments are used in Figs. 11.6, 11.7 and 11.8. Inset maps (derived from SRTM30 digital terrain data at 1 km resolution) indicate the area covered by the main figure in relation to Africa (top left) and the boundaries between the Okavango-Kalahari, Orange, Limpopo and Zambezi drainage basins (bottom right; note that this map covers the same geographical area as the main figure)

Africa). Rather than focus only on the parts of these systems that fall within Botswana, characteristics of whole dry valley networks are considered where appropriate.

11.2 Major Dry Valley Systems

The main dry valley networks of the Kalahari are shown in Fig. 11.1 (after Nash and Eckardt 2016). Most systems rise in areas of exposed basement bedrock and traverse, and commonly terminate within, areas covered by Kalahari

Fig. 11.2 Views of internally-draining (endoreic) dry valleys in Botswana: **a** Okwa valley immediately west of Tswaane borehole; **b** Dikgonnyane valley; **c** Gaotlhobogwe valley immediately downstream of the major knickpoint ~60 km from the valley headwater area (see Figs. 11.1 and 11.7) with exposures of silcrete (photographs: David Nash)

Group sediments. Two groups of valleys can be identified, separated by the broad, featureless *Bakalahari Schwelle* interfluve in central Botswana at 1000–1100 m asl (Passarge 1904; Thomas and Shaw 1991; Nash et al. 1994b). The first includes internally draining (or endorheic) fossil systems such as the Okwa-Mmone/Quoxo, Deception, Letlhakane, Nunga and Lememba that formerly fed into the Makgadikgadi Depression, and the Ncamaseri, Xaudum, Epukiro, Groot Laagte and other smaller systems that drain eastwards from Namibia towards the Okavango Delta (Fig. 11.2). These drain the shield and platform desert areas of the eastern and northern Kalahari. The largest of these systems, the Okwa-Mmone, has a catchment area in excess of 90,000 km^2 (Thomas and Shaw 1991). Some of these systems, most notably those that cross northwest Botswana, rise at headwater springs. Many endoreic valleys contain ponded water during the rainy season; however, evidence for surface flow is scarce (see Sect. 11.4). One further valley, the Serorome in eastern Botswana, is considered here alongside the various endoreic systems. The Serorome is unusual in that it is externally draining and swings eastwards into the Notwane, a headwater of the Limpopo. However, it rises in an area of Kalahari Group sediments and is geomorphologically similar to other internally draining valley networks.

The second group of dry valleys includes the externally draining (or exoreic) Southern Kalahari networks—the Auob-Nossop, Molopo-Kuruman and their tributaries (Fig. 11.3). The Auob and Nossop rivers both rise in the Khomas Highlands east of Windhoek, Namibia, at ~1200 m and ~1800 m asl, respectively, and flow in a broadly northwest-southeast direction before joining in the Kgalagadi Transfrontier Park. The east–west flowing Molopo and Kuruman rivers both rise at ~1500 m asl in South Africa near to the towns of Mafikeng (Northwest Province) and Kuruman (Northern Cape), respectively. The Molopo demarcates the border between southern Botswana and South Africa, while the Nossop forms the border between the two countries in the Kgalagadi Transfrontier Park. Most of the Southern Kalahari systems originate in areas of basement bedrock beyond the margin of the Kalahari Group sediments (Boocock and van Straten 1962), although some tributary systems—including the Moselebe

Fig. 11.3 Views of externally-draining (exoreic) dry valleys: **a** the Nossop valley within the Kgalagadi Transfrontier Park (photograph courtesy of Philip Raggett); **b** the Moselebe valley north of its confluence with the Molopo (photograph: David Nash)

valley, a feeder to the Molopo in Botswana—originate in areas with a Kalahari sand cover (the uppermost unconsolidated component of the Kalahari Group). All four main systems converge in the southwest Kalahari and ultimately connect to the Orange River (and hence the Atlantic) downstream of Aughrabies National Park via a long-dry section of the Molopo. As discussed in Sect. 11.4, the Southern Kalahari valleys have flooded on a number of occasions within the past 150 years; as such, they may represent a transitional state between the endoreic Middle Kalahari systems and the ephemeral rivers of the semi-arid hardveld (Shaw 1989; Shaw et al. 1992).

11.3 Dry Valley Morphology

Until recently, the basic morphological characteristics of Kalahari dry valley systems, including fundamental properties such as their long-profile shape, were poorly understood. This was due primarily to a lack of topographic survey data for areas of Botswana away from the more densely populated eastern hardveld. This section considers the morphology of the dry valley systems, with processes of formation discussed in Sect. 11.6.

11.3.1 Older Accounts of Dry Valley Morphology

Many of the earliest accounts of the morphology of dry valleys were based on analyses of aerial photography—as such, they were relatively detailed in terms of the planform of many systems but provided limited information about topographic variability. The seminal study of Boocock and van Straten (1962) suggested that it was possible to identify three broad morphological stages within the internally draining *mekgacha* in Botswana. Headwater regions of systems (away from areas of outcropping basement bedrock) were noted as being typically flat, shallow, clay-floored and channel-less, sometimes with a dambo-like lobate morphology (Boocock and van Straten 1962; Cooke 1984). This generalisation is particularly true of the upper sections of the Mmone/Quoxo system south of Letlhakeng in southeast Botswana. This gentle form gives way, often abruptly, to an incised 'gorge-like' middle section with a rectilinear flat-bottomed form (Shaw and de Vries 1988; Shaw et al. 1992). Rarely, this middle section may contain evidence of abandoned meander channels (Fig. 11.4) but valley-floor sediments are hardly ever exposed. Note that the term 'gorge-like' is used here relative to the otherwise relentlessly flat Kalahari topography—endoreic dry valleys in Botswana are seldom incised more than 40 m into the surrounding landscape. The deeper sections of valleys often expose outcropping silcrete, calcrete and intergrade duricrusts such as cal-silcrete and sil-calcrete (Shaw and de Vries, 1988; Nash et al. 1994a; Kampunzu et al. 2007)—see Chap. 13. Many valleys eventually dwindle to a further broad and flat stage, often becoming completely sand-choked by wind-blown sediment in their lower reaches.

Investigations by Nash et al. (1994b), Bullard and Nash (1998) and Nash (2015) include further detail about the exoreic Auob, Nossop, Kuruman and Molopo valleys. The cross-sectional shape of these systems varies considerably, ranging from gently sloping convex/concave forms to steep-sided more incised sections (up to 30 m deep) where the valleys cut through the southwest Kalahari calcrete plateau (see also Chap. 13). Variability occurs not only between the different systems but also along the lengths of individual valleys. The Auob valley, for example, is narrow and barely distinguishable in its headwater sections but reaches a maximum width of 1.8 km south of Stampriet in Namibia before narrowing to around 0.5 km within the Kgalagadi Transfrontier Park (Bullard and Nash 1998). Wider sections of the Nossop also contain numerous abandoned channels and meander systems (Fig. 11.5), including where the valley forms the border between Botswana and South Africa (Nash et al. 1994b; Nash 2015).

11.3.2 Advances Since the Advent of Radar Remote Sensing

The release of Shuttle Radar Topographic Mission (SRTM) digital elevation data for Africa in 2003 made it possible to reconstruct the morphology of entire Kalahari drainage systems with an absolute vertical accuracy estimated at 3–5 m (Rodriguez et al. 2005). Using SRTM data, Nash and Eckardt (2016) produced the first spatially accurate location map of 29 exoreic and endorheic dry valley courses (see Fig. 11.1) and reconstructed their long profiles (Figs. 11.6, 11.7 and 11.8). Like river systems in other parts of the world, the majority of Kalahari dry valleys exhibit concave-up long profiles for at least part of their course, graded to local or regional base levels. Exceptions to this generalisation include the Auob, Passarge, Mosope and Letlhakane, which have broadly linear long profiles. Most dry valleys lack the pronounced concavity typical of temperate graded rivers (Hack 1973), with stream gradient indices lower than expected in headwaters and higher in downstream sections. Kalahari dry valley long profiles are similar to those of ephemeral rivers in Australia, Israel, Kenya and the southwest USA (e.g. Leopold and Miller 1956; Frostick and Reid 1989; Tooth 2000) but do not exhibit the marked convexities seen in some hyperarid systems, such as those in the Namib Desert (Goudie 2002).

Nash and Eckardt (2016) demonstrated that 25 of the 29 dry valley systems contain distinct knickpoints—locations where major changes in valley gradient occur—exceptions being the Auob, Passarge, Mosope and an unnamed tributary of the Okwa. A total of 55 knickpoints were identified, most of which occur as kilometre-scale transitional zones of gradient change in the valley long-profile. A detailed comparison of the knickpoint positions shown in Figs. 11.6, 11.7 and 11.8 with geological maps for Botswana and Namibia indicated that the majority of knickpoints occur at lithological boundaries. For example, the middle knickpoint on the Molopo (Fig. 11.6) is located where the valley passes from Kalahari Group sediments onto an inlier of more resistant Karoo Dwyka Group diamictites, siltstones and sandstones. The knickpoints on the middle Okwa and lower Hanehai (Fig. 11.7) coincide with locations where the valleys cross from inliers of more resistant Okwa Basement Complex and Ghanzi Group lithologies, respectively, onto Kalahari Group sediments. The middle knickpoint on the Nunga appears to be related to an outcrop of more resistant duricrusts within the Kalahari Group sediments. In general, Nash and Eckardt (2016) identified that knickpoints were steeper where valleys cross from less to more resistant lithologies, primarily due to the greater potential for headward erosion into softer materials. SRTM data reveal other knickpoints where valleys intersect exposed or shallow faults. For example, most

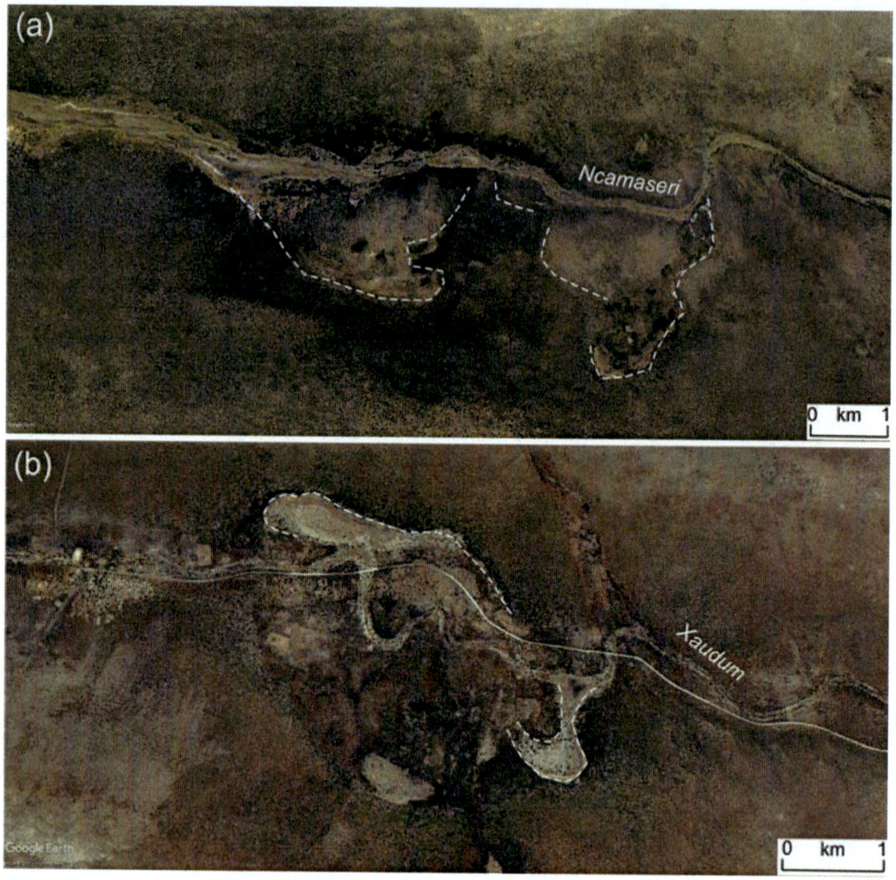

Fig. 11.4 Abandoned meanders (dashed line) along the course of the **a** Ncamaseri and **b** Xaudum dry valleys in northwest Botswana. See Fig. 11.1 for locations of images. Images courtesy of Google Earth, dated **(a)** 6 July 2008, **(b)** 19 October 2014

knickpoints on systems draining towards the eastern Makgadikgadi Depression (Fig. 11.8) occur where these valleys cross known or inferred faults or fractures, possibly linked to neotectonic movement within the basin.

The larger knickpoints within dry valley systems tend to be associated with areas where there is known evidence of base-level lowering and/or uplift along neotectonic flexural axes. Most pronounced is the distal knickpoint on the Molopo in South Africa (Fig. 11.6), where the valley floor drops from 750 to 460 m asl over a distance of 45 km. Nash and Eckardt (2016) argue that the knickpoint developed in response to headward migration of the Augrabies Falls on the Orange River past the lower Molopo-Orange confluence, presumably following subcontinental uplift between 75 and 65 Ma (McMillan 2003; Bluck et al. 2007). The lower knickpoint on the Serorome (Fig. 11.8) appears to have been generated in response to an incision along the Limpopo River. In contrast, the upper Serorome and middle Epukiro knickpoints and the lower knickpoints on the Lememba and Xaudum (Fig. 11.8) all fall at confluences with incised tributaries. The distal Otjozondjo, Epukiro and Groot Laagte knickpoints occur immediately up-valley of where these systems enter the half-graben occupied by the Okavango Delta and are most likely a product of tectonic base-level lowering. In the Otjozondjo, the distal knickpoint coincides with the Gumare fault, which has a 17 m southeast throw here (Kinabo et al. 2007; McFarlane and Eckardt 2008). The knickpoints on the Deception occur where the valley crosses palaeolake shorelines in the Makgadikgadi Depression and may be controlled by former lake levels.

The courses of four headwaters of the Mmone/Quoxo valley system straddle the Kalahari-Limpopo drainage divide (Fig. 11.7). These include the main Mmone/Quoxo, which rises at an altitude of c.1205 m asl and 'flows' uphill for 19 km before crossing the divide at 1286 m. Three tributaries to this system (the Dikgonnyane, Gaotlhabogwe and 'Letlhakeng Valley 2') also exhibit reversed gradients in their headwaters and can be traced for 2–5 km before crossing the drainage divide at 1173, 1233 and 1252 m, respectively (see Fig. 11.9). Analyses of aerial photographs of the headwater sections of each of these valleys by Nash and Eckardt (2016) indicate that the drainage lines were once continuous across the Kalahari-Limpopo divide but are now marked by a chain of small pans. Each of these valleys also

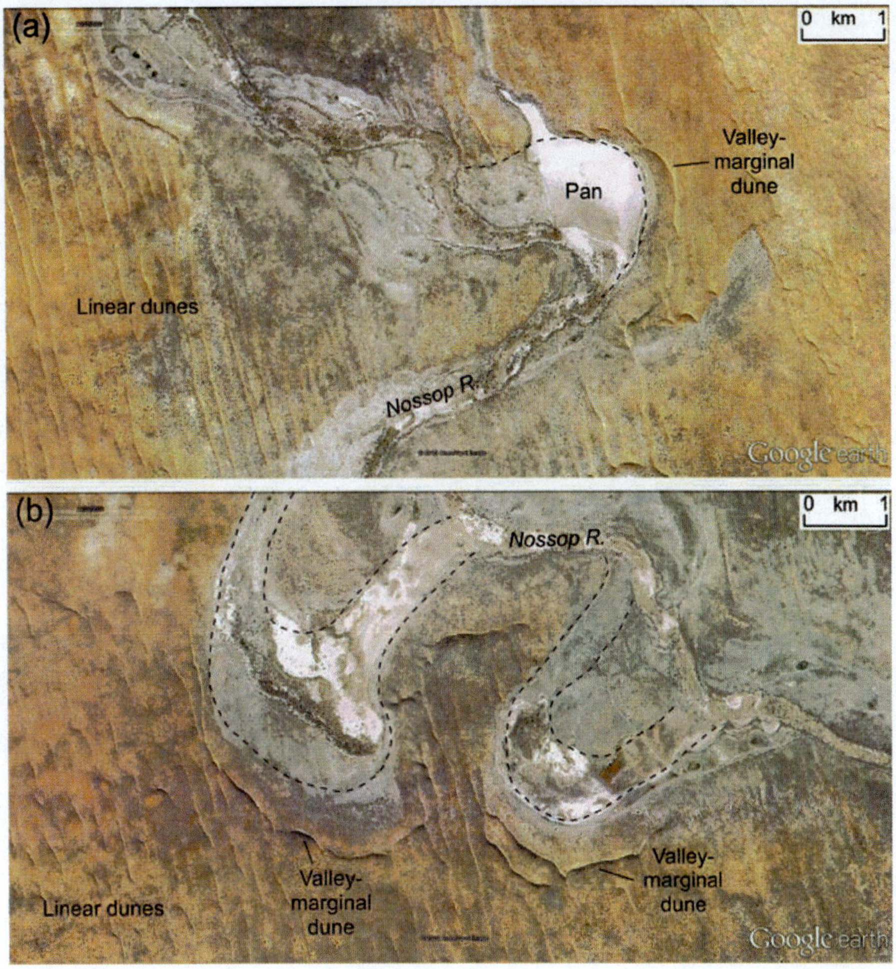

Fig. 11.5 Abandoned meanders (dashed line) and valley floor pans along the course of the Nossop River where it forms the boundary between Botswana and South Africa. Linear dunes and valley-marginal dunes (see Bullard and Nash 2000) are also indicated. See Fig. 11.1 for locations of images (a) and (b). Images courtesy of Google Earth (date of images, 22 November 2006)

contains major knickpoints (Fig. 11.7), the most pronounced occurring at 80 km from the source of the Mmone/Quoxo where the valley falls from 1150 to 1080 m over 11 km and extensive exposures of silcrete, calcrete and intergrade duricrusts occur (Shaw and de Vries 1988; Nash et al. 1994a; Kampunzu et al. 2007).

11.4 Hydrology and Palaeohydrology

11.4.1 Contemporary and Historical Hydrology —Endorheic Valleys

Surface flows have only been recorded within endorheic dry valley systems in Botswana on a handful of occasions, each time triggered by high-intensity summer rainfall events associated with the passage of convectional storms or tropical lows off the Indian Ocean (see Preston-Whyte and Tyson 1993). The Letlhakane has experienced the greatest number of documented flood events. Surface flow is known to have occurred over a section of the valley in 1969 in response to intense precipitation (Mazor et al. 1977). Short-lived flooding was also reported in January 2013 (Morrison and Morrison 2013) and February 2018 (Motlhabani 2018), with the former event triggered by ~100 mm precipitation falling near Letlhakane village in a single day (Morrison and Morrison 2013). Flooding has also been known to occur during the twentieth century in the Groot Laagte and Epukiro valleys (Nash 1996). The headwaters of valleys that rise at springs in the Aha Hills along the Namibia-Botswana border occasionally contain standing or gently flowing water (Fig. 11.10), but the flow is limited to bedrock sections of these valleys (Yellen and Lee 1976; Helgren and Brooks 1983).

Rather than transmitting surface flow, most endorheic valley systems normally remain dry throughout the year—the only exception being the occurrence of ephemeral pools in valley floor depressions in some systems during the summer rainy season. This is unsurprising, given the extent of unconsolidated sediments within valley floors and the lack of any significant valley gradient to create a slope that

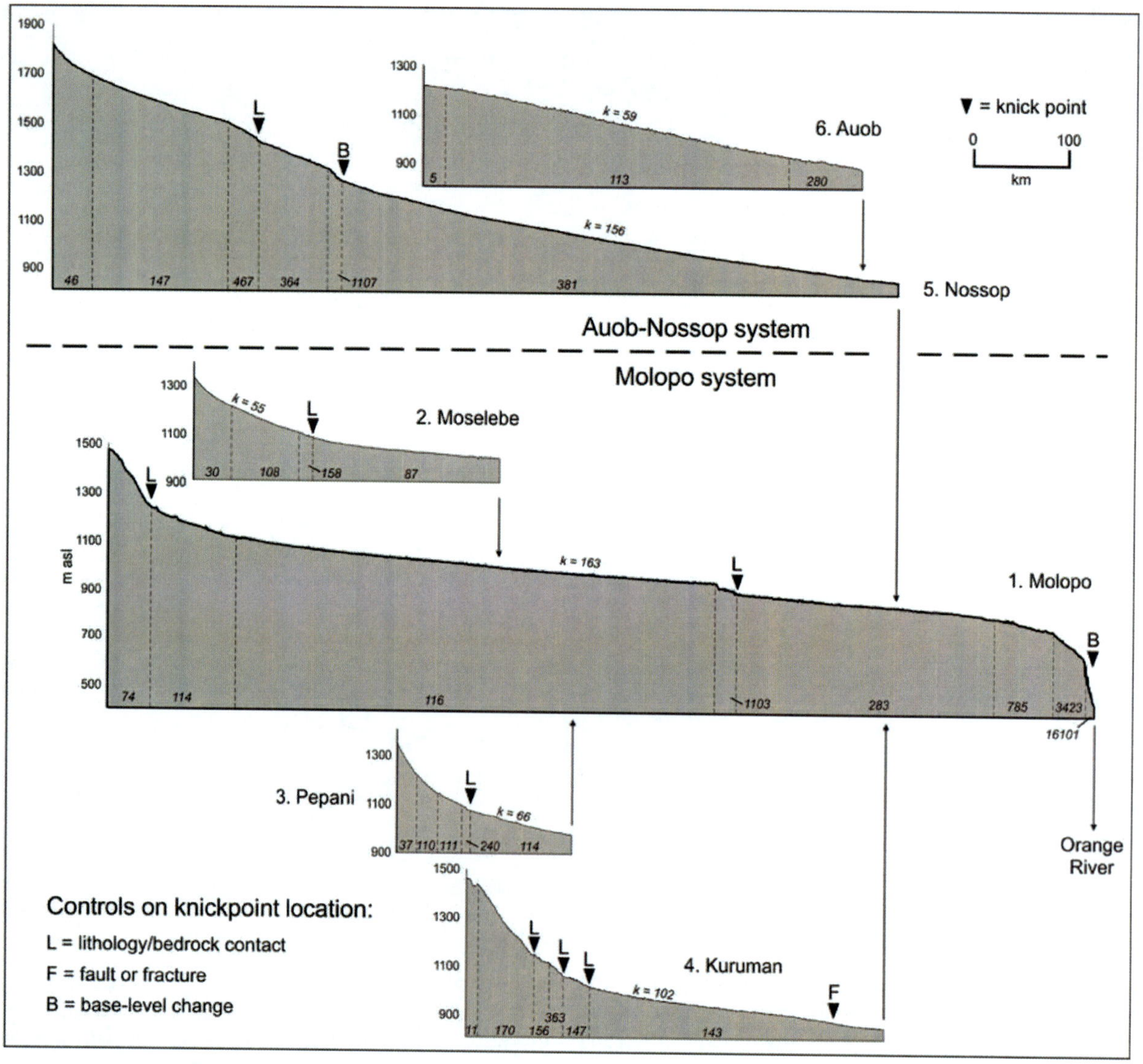

Fig. 11.6 SRTM-derived long-profiles for the externally-draining Southern Kalahari dry valley networks (after Nash and Eckardt 2016). Stream gradient index (k) values are shown for comparative purposes for entire long-profiles and individual valley segments. Stream gradient indices were calculated using the equation: $k = (H_i - H_j) / (\ln L_j - \ln L_i)$ where H is altitude and L the horizontal distance between two points i and j (after Hack, 1973). Vertical arrows indicate confluence points between valley segments and the connection with the Orange River. Long-profiles are plotted at the same horizontal scale in Figs. 11.5, 11.6 and 11.7, with 250-times vertical exaggeration. See Fig. 11.1 for locations of valley segments

would allow floodwater to flow faster and further. On the rare occasions that rainfall levels are sufficient to saturate valley-floor sediments, valleys may act as conduits for groundwater recharge (Farr et al. 1981; de Vries et al. 2000).

The general lack of surface hydrological activity within endorheic valley systems has persisted since at least the early nineteenth century when written descriptions of the Kalahari landscape first become available. There is, however, evidence that some valleys may have become drier over the past 150 years. Until the late nineteenth century, semi-permanent standing pools associated with near-surface water tables were common in many valley floors, only disappearing once human settlement and associated cattle spread throughout the region (Shaw et al. 1992). This mirrors the drying up of several perennial spring sites in Botswana during the nineteenth century evidenced by authors including David Livingstone (1857), Charles Andersson (1856) and Andrew Bain (Lister 1949). Bain indicated that Kang may once have had a spring (Campbell and Child 1971), whilst Andersson (1856) described a number of wells between Lake Ngami

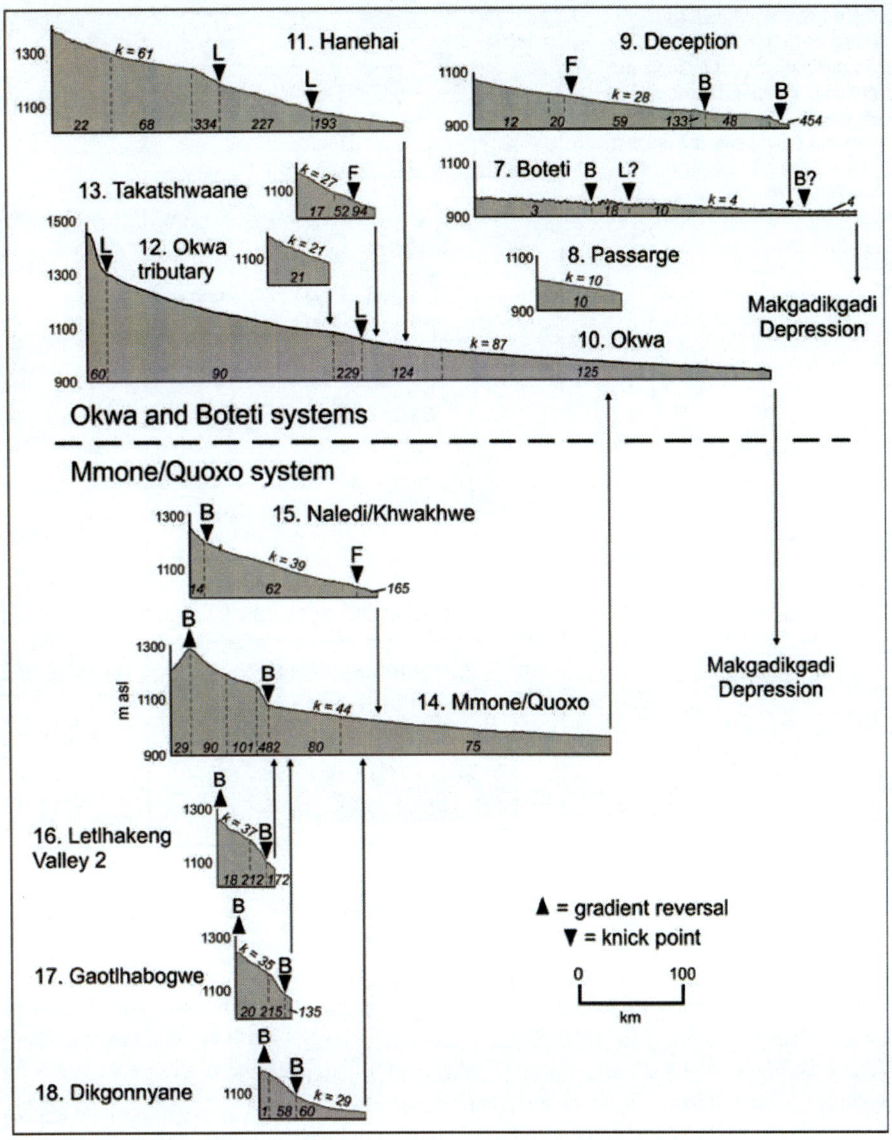

Fig. 11.7 SRTM-derived long-profiles for the internally-draining Okwa and Mmone/Quoxo dry valley networks (after Nash and Eckardt 2016). The profile for the ephemeral Boteti River is shown for comparison. See Fig. 11.6 for further details

and Ghanzi and strong spring-flow near to Mamuno. All these springs are now dry, although a map compiled by Stigand (1923) records spring sites near Ghanzi. It is also significant to note that many settlements in presently dry valleys (e.g. Letlhakeng and Letlhakane) have names meaning "place of reeds' in Setswana but lack swampy conditions today (Campbell and Child 1971). Whether this apparent regional lowering of water tables is a result of the over-consumption of groundwater (see Thomas and Shaw 1991) and/or reflects long-term drying trends over the southern African interior (see Neukom et al. 2014; Nash et al. 2016) is not known.

Only two accounts from the nineteenth century include direct descriptions of either standing or flowing water within an endorheic dry valley (see Nash and Endfield 2002). The most important of these is a flood reported in the Letlhakane to the south of the Makgadikgadi Depression in 1851. The flood is described within the unpublished and unedited version of David Livingstone's *Bechuanaland Journal* held in the Council for World Mission Archive (School of Oriental and African Studies, London). In this journal, Livingstone described his journey from the Chobe and Linyanti rivers to the Kuruman mission station in September 1851, immediately after crossing Ntwetwe Pan:

> While at Letlehan [Letlhakane] we had three days rain. This made the [valley] assume the appearance of a large river flowing

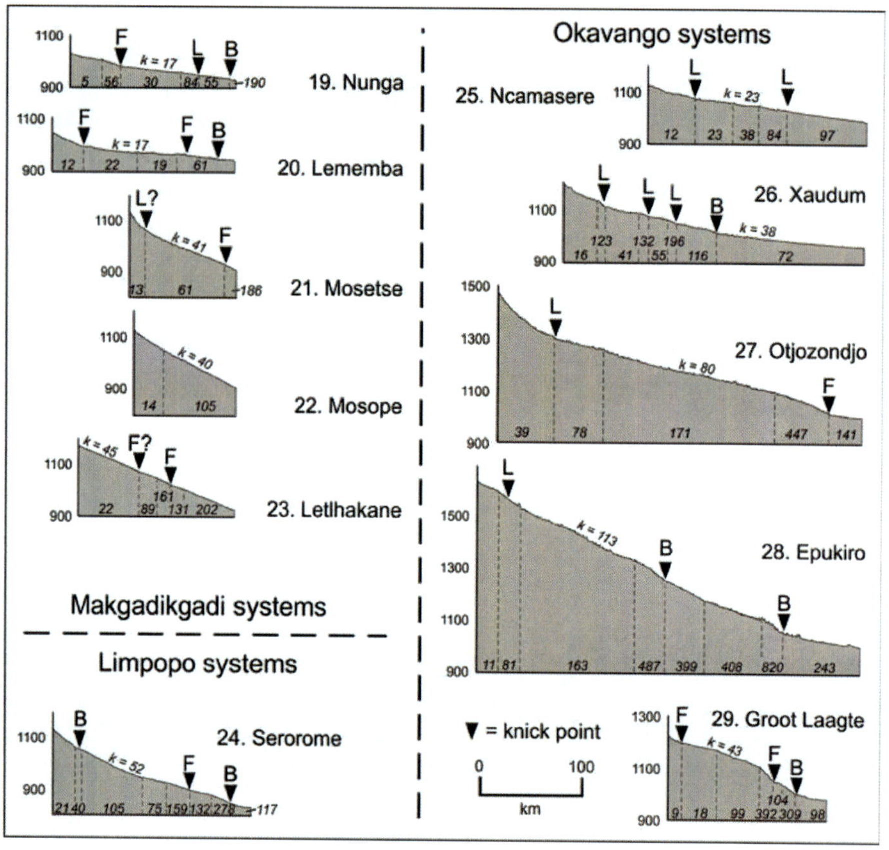

Fig. 11.8 SRTM-derived long-profiles for the externally-draining Limpopo and internally-draining Okavango and Makgadikgadi dry valley networks (after Nash and Eckardt, 2016). See Fig. 11.6 for further details

> northwards. The old people remember the time when it flowed as through the whole year…[1]

The other account is provided in an unpublished letter written from Shoshong to the London Missionary Society headquarters in London in 1880 by the missionary Rev. James D. Hepburn (Nash and Endfield 2002). This letter includes a second-hand report of an itinerating journey in 1879 by the Motswana missionaries Khukwe and Diphukwe from Lake Ngami to the north of Shakawe via the western side of the Okavango Delta. The journey followed the presently dry Thaoge river and then crossed a number of fossil valleys including the Xaudum, with Hepburn noting:

> In June they got to another river called Cadom. It has large pools of water all along its course in the dry season. They stayed here again for two weeks… Two treks from this river brought them to another lot of large pools in the riverbed and another trek brought them to a beautiful water called the waters of the Bakgalagadi. These are beautiful springs and a nice strip of ground lying below them. Again, they went on to another river called Kaudom. At Kaudom there is deep water in the riverbed all the way down, but it is under the surface and is very deceptive to anyone unacquainted with that fact. From what I can make out it must be almost like a quicksand for Khukwe says a man might easily drowned at this place. There is a place where it can be crossed by the wagon for all that…[2]

Khukwe and Diphukwe subsequently travelled to just north of Shakawe and then returned to Lake Ngami via the same route. Hepburn's letter noted that they travelled from:

> …Kaudom and from Kaudom to Chadom. They taught at these places Boers, Masarwa, Bechuana hunters and others gathered together at the waters[2]

These descriptions of water within presently dry valleys are potentially significant. It is, however, difficult to identify from Hepburn's account exactly which systems were crossed by Khukwe and Diphukwe's party, as several valleys in northwest Botswana are referred to by similar names in regional languages. It would appear likely from the accounts of distances travelled that the Cadom (or Chadom) is the present-day Xaudum whilst the Kaudum is the Ncamaseri (known as the Xeidum in westernmost Botswana and the Xaudum or Khaudum in Namibia). Regardless, that both

[1] Council for World Mission Archive, London Missionary Society, David Livingstone's Bechuana Journal p.151, September 1851.

[2] Council for World Mission Archive, London Missionary Society, South Africa Correspondence Box 40, Folder 3, Jacket C, Rev. J.D. Hepburn, Shoshong, June 1880.

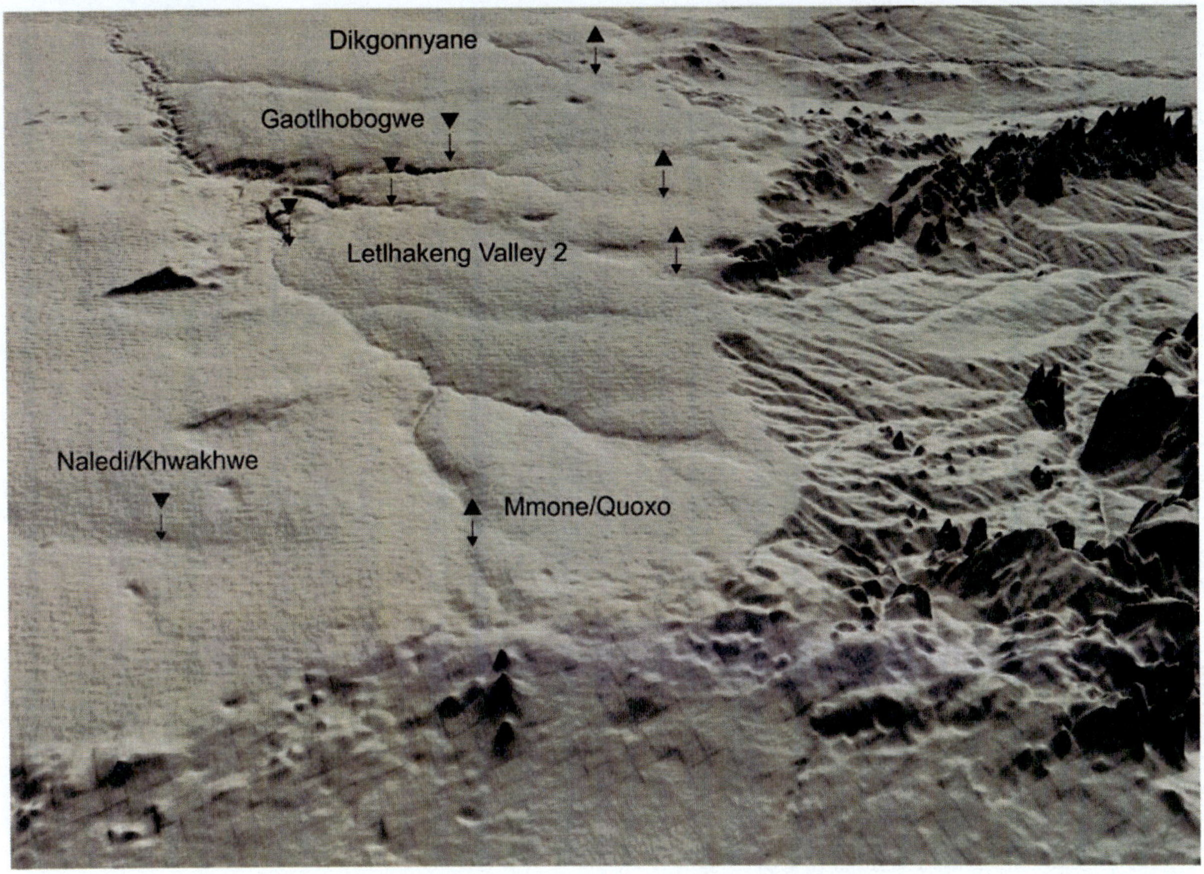

Fig. 11.9 Oblique SRTM-derived view of the headwaters of the Mmone/Quoxo system, showing areas of gradient reversal and major knickpoints (looking NNE; field of view approximately 70 km wide) (after Nash and Eckardt, 2016). Inverted black triangles indicate the position of the knickpoints in the Dikgonnyane, Gaotlhobogwe, Letlhakeng Valley 2, Mmone/Quoxo and Naledi/Khwakhwe systems shown in Figs. 11.1 and 11.7. Upright black triangles mark the approximate positions of the gradient reversals in the headwater sections of the Dikgonnyane, Gaotlhobogwe, Letlhakeng Valley 2, and Mmone/Quoxo systems. The smooth terrain on the left of the image is blanketed with Kalahari Group sediments, whilst the more dissected terrain to the right marks the western edge of the Limpopo catchment. The image demonstrates the indistinct nature of the drainage divides between Kalahari valley systems

valleys contained surface pools or shallow sub-surface water in 1879 is in marked contrast to the present day. Caution is needed, however, as it is possible that the sites described were adjacent to the Okavango Delta where relatively shallow groundwater occurs today beneath the floors of dry valleys where they join the Delta floodplain (Nash and Endfield 2002; Nash et al. 2006).

Wetter conditions in the Ncamaseri during the late nineteenth century may be corroborated in oral histories reported by Robbins et al. (1994). Samuchau, a Mumbukushu elder living at Tsodilo Hills, 7 km south of the Ncamaseri, described what he had been told by his grandparents sometime around 1930:

> Long ago during heavy rains, the Xeidum flowed from the west. When the big floods built up in the Okavango, water backed up in the Ncamaseri creating a continuous stretch from the Okavango to a place west of Tsodilo". Samuchau's ancestors could "...paddle or pole their mekoro [dugout canoes] from the Okavango up the Ncamaseri until it became the Xeidum where they hunted and fished (Robbins et al.1994, p.262).

Other nineteenth-century accounts provide more general descriptions of endorheic dry valleys, including both primary and second-hand references to the presence of water. The earliest of these is by the trader Joseph McCabe (as an appendix to Holden 1855), who crossed what is likely from his route description to be the Okwa and Hanehai valleys in July 1852 whilst travelling from Kang to LakeNgami. In his journal McCabe described crossing the Okwa:

> On the 8th July I reached a large and rather deep valley, called... Mugube Magoolo. This was the first appearance of a valley we had met since we left Sentuhe's [in Kolobeng]. This valley stretches from N.W. to N.E. and in many seasons holds water. On the south side it is lined with heavy sand hills (Holden 1855, p.419).

Fig. 11.10 Spring-fed waters ponded within an area of massively bedded calcrete at the head of the Gcwihabe valley, east of Xai-Xai, northwest Botswana (photograph: David Nash)

References to the Hanehai include that of Charles Andersson who travelled along the "sandy and dry water course' of the "Otjiombinde' on 15 May 1853 noting:

> The soil consisted of fine, white sand, reflecting a light dazzling and painful to the eyes... the grass was still green and plentiful, and the vegetation, in general, was rank. We passed several vleys containing small quantities of muddy water, alive with loathsome reptiles (Andersson 1856, p. 379).

Thomas Baines also followed the course of the Hanehai, noting that:

> ... it's breadth was from 100 to 150 yards, with low banks and ridges of sandstone here and there; and the grass in it was as dry, white and feathery as if water had never flowed there, and never could (Baines 1864, p.119).

Along with McCabe, Arnold Hodson was one of the few trans-Kalahari travellers who recorded detailed landscape information. Whilst crossing the Mmone/Quoxo valley near to the present-day village of Khudumalapye in January 1905 he noted:

> The last place of interest we passed was a dry river bed... whence we could obtain beautiful water at any point by digging. My own theory is that this place, like many similar dry river beds in the Kalahari has an underground river, otherwise it would be difficult to account for the presence of water so close to the surface (Hodson 1912, pp.79-80).

Other information about endorheic valleys is less detailed, with, for example, Hodson crossing the Okwa to the south of Kalkfontein in late May or early June 1905 describing it as "a river only in name as it was quite dry' (Hodson 1912, p.96). The Groot Laagte in northwest Botswana is mentioned by Passarge (1899, p. 312) who crossed the valley at a point where it was 3 km wide and contained a channel over 400 m wide, leading him to suggest that it must have once carried a body of water comparable to that of the Okavango River.

The only other endorheic dry valley mentioned explicitly in historical sources is the Serorome. The village of Boatlaname within the valley was a staging point on the Missionary Road from Kuruman (South Africa) northwards through eastern Botswana and is mentioned on a number of occasions by David Livingstone (1857) and Leyland (1866). Livingstone described Boatlaname as:

> ... a lovely spot in the otherwise dry region. The wells from which we had to lift out the water for our cattle are deep, but they were well filled *(*Livingstone, 1857, *p.54).*

However, he also noted (p. 54) that wells in the village of Lephephe provided:

> another proof of the desiccation of the country. The first time I passed it, Lopepe was a large pool with a stream flowing out of it to the south; now it was with difficulty we could get our cattle watered,

an observation that Leyland echoed when visiting Boatlaname on 16th April 1851, finding:

> several wells sixteen to seventeen feet deep, but not sufficient water for half the oxen *(*Leyland 1866, *p.140).*

William Finnaughty Tabler (1916), Chapman (1886) and Anderson (1887) also crossed the Serorome, with the latter noting water within calcrete pits at Boatlaname (Anderson 1887, p. 206).

11.4.2 Contemporary and Historical Hydrology – Exoreic Valleys

Surface flow is much more common in the externally draining Auob-Nossop and Molopo-Kuruman systems. The Auob and Nossop both rise in the highlands of central Namibia and have a flashy regime. Flow in the Nossop regularly reaches as far as Aranos, upstream of the Kgalagadi Transfrontier Park (Leistner 1967), and Gochas for the Auob (Range 1912), but may extend further following heavy rainfall. The main Kuruman and Molopo valleys both rise at spring sites, with surface flows and occasional floods triggered by the effects of either heavier precipitation over higher elevation areas or groundwater discharge (Nash et al. 1994b; Nash 1996). The Kuruman (Fig. 11.11) has the most regular flow of all Southern Kalahari networks and is effectively a permanent river over the first 10 km of its course due to the 750 m^3 of water per hour supplied by its main spring, the "Eye' at Kuruman; it is also more responsive to precipitation than other rivers in the network (Shaw et al. 1992). The Molopo only rarely contains water as far west as Bray on the Botswana border (Grove 1969) with the furthest recorded historical flow being at Watersend Farm (Shaw et al. 1992). Present-day flow conditions in all four systems have been greatly influenced by groundwater extraction. The waters of the Kuruman, for example, are used for irrigation and public supply, with the "Eye" and other springs being exploited since at least the 1820s (Livingstone, 1857 p. 8).

Floodwater has not reached the Orange-Molopo confluence during the historical period (Nash 1996; Nash and Endfield 2002). However, large-scale flood events are known to have occurred in all major Southern Kalahari systems. The Kuruman, for example, experienced extensive flooding in 1817–18, 1819–20, 1891–92, 1894, 1896, 1915, 1917–18, 1920, 1933–34, 1974–77, 1988–89 and 2021 (when floodwater reached to within a few kilometres of its confluence with the Molopo). Floods occurred in the Molopo in 1871, 1891–92, 1896, 1915, 1917, 1933–34, 1988, 1999–2000, 2014 and, most recently, 2017 in the aftermath of tropical cyclone Dineo (Clement 1967; Verhagen 1983; Thomas and Shaw 1991; Nash 1996; Nash and Endfield 2002). Records for the Auob and Nossop are more scarce, but the former is known to have flooded in 1933–34, 1973–74 and 2012 and the latter in 1806, 1933–34, 1963–64, 1987 and 2012 (Clement 1967; Verhagen 1983; Thomas and Shaw 1991; Nash 1996).

With a few notable exceptions, most historical floods appear to have had only limited long-term effects upon valley geomorphology. For example, the Molopo broke its banks in the section of its valley below the Kuruman-Molopo confluence in 1894 because of flooding in the Kuruman. Floodwaters were unable to follow the main course of the Molopo as it was partially blocked by sediment, and instead formed a large lake at Abiquasputs (Range 1912) (Fig. 11.1). Nash and Endfield (2002) detail a letter suggesting that flooding may have extended south of this lake, possibly as far as the confluence with the Orange River, although corroboration is required to confirm this. The lake was refilled in 1934 as a result of widespread flooding that affected all four Southern Kalahari drainage systems, and is reported to have had a surface area of over 12,000 ha (Barrow 1974). On this occasion, floodwater extended as far as Noenieput, approximately 27 km south of Abiquasputs, whilst in 1976 flooding reached Springbok Vlei, 15 km north of Abiquasputs.

Other evidence for the historical hydrology of the exoreic dry valleys comes from the writings of early missionaries and explorers. Although hunters and traders are known to have penetrated the southern Kalahari by the end of the eighteenth century (Campbell and Child 1971), descriptions of the Molopo and Kuruman only date from the early nineteenth century (Nash 1996; Nash and Endfield 2002). The first recorded accounts of the valleys are given by William Burchell who visited one of the minor sources of the Kuruman on 28 June 1812 noting that:

> the spring was in its lowest state, as its waters were too weak to run more than two hundred yards from the spot where they rose out of the ground (Burchell 1824, Vol. 2, p.209).

However, he also described crossing the main river on 29 June 1812 at an unknown point downstream where it was:

> a beautiful little river running in a plentiful stream of the clearest water. At this point of its course it was fifteen feet broad and abounded in tall reeds (urchell 1824, Vol. 2, p.214).

Burchell also provided accounts of other valleys within the Kuruman system. For example, he noted that the Matlhwaring was:

> merely a ditch about twenty feet broad, without a tree, or even reeds, to mark its course... There was an abundance of water in the deeper hollows of its bed; and at two or three hundred yards below our station, it ran in a plentiful stream (Burchell 1824, Vol. 2, p. 222)

and that the bed of the Moshaweng (at a point approximately 10 km down valley from Lobotsane) was:

> ...but a few yards wide, and of this the water occupied but a small part; yet as it flows constantly during the whole year, it is regarded... as a considerable stream (Burchell 1824, Vol. 2, p. 253).

Fig. 11.11 The Kuruman River in flood January/February 2021: **a**–**b** at Thota Lodge, 60 km northwest of Hotazel (photograph [a] courtesy of Marelise Theart and [b] Melissa Delport, Thota Lodge); **c** 30 km from Van Zylsrus (photograph courtesy of Nicholas Pattinson, University of Cape Town). See Fig. 11.1 for locations of images

The Rev. John Campbell described crossing the Molopo river in its headwater section on 2 May 1820, at the start of the dry season, noting that it:

> ... was about ten yards wide and in some parts two feet deep; the bottom was stony, but the water clear and well tasted. No trees grew on either side nearer than 500 yards but reeds were in great abundance (Campbell 1822, Vol. 1, p.208).

On the return leg of his journey, Campbell crossed the Kuruman on 21 June 1820, describing it as "a considerable stream'. Later the same day, having passed the confluence of the Kuruman and Matlhwaring at approximately 55 km down-valley from Kuruman, he noted:

> At three p.m. came to the bed of the Krooman River, which was dry, the stream having sunk into the sand nearly opposite Letakka (Campbell 1822, Vol. 2, p.87).

This suggests that water extended beyond the Kuruman-Matlhwaring confluence in this year, some 45 km further than usual.

The LMS missionary Robert Moffat crossed the Molopo at a point approximately three days journey west of Pitsane on 22 July 1824 and described it at as being "as dry as the neighbouring plains' (Schapera 1951, p.127). In a separate account of the same journey, Moffat noted finding water in the river bed approximately 20 km west of Pitsane (Moffat 1842, p. 387). Andrew Bain crossed the Molopo on 12

Fig. 11.12 Fault-controlled section of the Okwa Valley close to Tswaane borehole. Approximate positions of faults shown as red lines, dashed when inferred. Fault patterns redrawn after Aldiss and Carney (1992) and Ramokate et al. (2000)

August 1826 (Lister 1949) approximately 20 km east of Bray, describing it as having the appearance of a vlei and being dry except for a few muddy pools. He noted, however, that further upstream the Molopo was described by local inhabitants as being a ‵fine running stream' with herds of hippo. Bain also crossed three valleys of the Moselebe system, describing deep wells in the main Moselebe at Lorolwane village, two small lakes in the Sekhutlane and springs within the Selokolela.

A number of travellers provide information about the Molopo and Kuruman between 1835 and 1836. Amongst these are Andrew Smith, who crossed the Molopo river near Pitsane twice, once at the start of the dry season and again prior to the onset of seasonal rains. Following the first crossing on 18 May 1835, Smith noted that there were three or four permanent springs near the fording point and further that:

> for several years the strength of these springs have undergone little change though, previously, they were much stronger, when the stream was also larger and flowed to a greater distance westward than it does at present (Lye 1975, p.212).

When Smith crossed the Molopo again on 7 October 1835, he recorded ‵no difference in the quantity of water since we passed' (Kirby 1940, p. 257), which suggests constant spring outflow in the headwater regions. Smith also described travelling along the Kuruman in 1835 noting that ‵both water and grass were scanty… and, where water occurred, it was found in holes dug… in the bed of the river' (Lye 1975, p.178). He also noted reaching a ‵fine spring' in the form of a 6 m deep well in the river bed at the junction of the Kuruman and Matlhwaring. This spring was considered by Smith to be of importance to local inhabitants ‵since the drying up of the two streams' (Lye 1975, p. 178). Given Campbell's earlier description of water sinking into the river bed well beyond the Kuruman-Matlhwaring confluence, it can be assumed that the extent of water within the channel had diminished between 1820 and 1835.

Captain Sir William Cornwallis Harris crossed the Molopo near its headwaters sometime after 14 October 1836, noting that:

> ...this river... exhibits a broad shallow bed, covered with turf, traversed by a deep stream about ten yards wide, completely overgrown with high reeds (Cornwallis-Harris 1852, p.66).

Gordon-Cumming also crossed the Molopo some 7 years later in 1843 recording that the:

> darling little river is here completely concealed by lofty reeds and long grass which clothe its margins to a distance of at least a hundred yards (Gordon-Cumming 1909).

In comparison with the Molopo and Kuruman, the Auob rarely features in historical documents despite being one of the major travel routes between Upington and Windhoek in the late nineteenth and early twentieth centuries (du Toit 1926; Weinberg 1975). Sir James Alexander (1838, pp. 159–160) described stopping to shoot elephant in the Nossop River in 1837, noting that the river bed contained many holes dug by elephants to gain access to shallow groundwater. Charles Andersson (1856) documented permanent springs in the headwaters of the Nossop and suggested, based on common fish species, that it was once linked to the Orange. Thomas Baines (1864) and James Chapman (1886) both described crossing the Olifants and Nossop valleys, with Baines depicting the Olifants as being "three or four times as broad as a turnpike road' (Baines 1864, p. 64). William Leonard Hunt (known as Farini) travelled up the Nossop to at least as far as the confluence with the Auob in 1885, where he described a 30 m deep well in the valley floor (Farini 1886). Other references to the Nossop include that of Lieutenant Arnold Hodson in November 1904 who described the valley to the north of Union's End as:

> ...very broad and quite dry; in fact there were a number of trees growing in the middle of it. The Hottentots had dug a fairly deep pit here, but they had not succeeded in getting any water (Hodson 1912, pp. 58-59).

In contrast, Herbst (1908) noted from studies of wells in the floor of the Nossop and Molopo near their confluence that:

> their beds have... been raised considerably, for in those wells that have been sunk therein the white shingle, beautifully smooth and rounded by running water in time past, is found fifty feet below the present surface. Below this shingle... water is found (Herbst 1908, pp. 207-208).

11.4.3 Pleistocene and Holocene Fluvial Activity

Relatively little is known about the Quaternary history of the endorheic Kalahari dry valleys within Botswana. This is primarily because they contain limited sedimentary or geomorphological evidence for past fluvial activity—although conventional wisdom suggests they must have contained water at various times in the past (Shaw et al. 1992). A small number of studies provide evidence to suggest former standing or flowing water. Jack (1980), for example, reported undated shell beds and lignite deposits to a depth of 55 m against a faultline beneath the floor of the Xaudum, whilst Brook (1995) suggested that the upper Ncamaseri contained sufficient water during the late Holocene to allow the accumulation of peat deposits. Extensive delta features are also present at the distal ends of both the Okwa valley (Cooke and Verstappen 1984) and Groot Laagte (Thomas and Shaw 1991).

To date, only three studies have provided age estimates for late Quaternary fluvial activity in endorheic systems. The first two of these focussed on the evolution of landforms and sediments in the Gcwihabe Hills in northwest Botswana, through analyses of calcretes in the Gcwihabe valley and speleothems and flowstones within the adjacent Gcwihabe (Drotsky's) Cave. Cooke and Verhagen (1977) used radiocarbon dating of cave deposits to identify that vadose conditions occurred within Gcwihabe Cave from 45,000–37,000, 34,000–29,000 and 16,000–13,000 yr BP, with Holocene sinter formation at $\sim$2000 and $\sim$750 yr BP (all uncalibrated ages). Radiocarbon ages from calcretised sands and gravels within the Gcwihabe valley interpreted by Cooke (1984) as representing ephemeral river flow under sub-humid/semi-arid conditions—fall at 34,700 $\pm$ 2000 and 22,700 $\pm$ 500 yr BP, with four dates clustering between 11,000 $\pm$ 100 and 9800 $\pm$ 200 yr BP. Given that the major passages within Gcwihabe Cave have a mean elevation of 12–15 m above the valley floor, Cooke (1984) postulated that the water table would have been at a similar height during periods when vadose conditions existed in the cave. During two of these vadose phases ($\sim$30,000 and 16,000–13,000 yr BP), he suggested "the Kwihabe river must have flowed in a valley about 25 m above the present level" (Cooke 1984, p. 272).

The third study of Pleistocene fluvial activity in an endorheic dry valley (Shaw et al. 1992) dated shell material and calcified sediments from sites in the Xaudum, middle Okwa (close to where the Trans-Kalahari Highway crosses the valley) and distal Okwa (where the valley cuts through the Gidikwe palaeoshoreline before entering the southwest Makgadikgadi Depression). The sites in the Xaudum and middle Okwa were similar in context, comprising exposures through low (1.0–1.5 m) calcrete-cemented terraces occupying the valley floors. In both cases, the calcrete consisted of cemented silty alluvium, with the near-surface section of each terrace containing intact and comminuted shells of the freshwater snail *Lymnaea natalensis*. Shaw et al. (1992) inferred still, near-permanent water during the later stages of terrace sediment accumulation from the presence of this species. A sample of combined calcrete and shell material from the Xaudum terrace yielded an uncalibrated radiocarbon age of 14,570 $\pm$ 160 yr BP, with an equivalent sample from the middle Okwa terrace dated to 11,890 $\pm$ 60 yr BP. In the distal Okwa, two shell-bearing samples were taken for dating from a depth of 80 cm within the floor of the valley (radiocarbon age estimate 14,490 $\pm$ 150 yr BP) and at 70 cm depth on the eastern face of the nearby Gidikwe Ridge (14,070 $\pm$ 150 yr BP). Shells in these samples included the gastropod *Melanoides tuberculata* and the bivalve *Corbicula africana*, with isolated specimens of *Bulinus* and *Lymnaea* spp also identified. Calcified reed stems (casts of algal growth on former reeds) were also sampled within the floor of the Okwa where it cuts the Gidikwe Ridge—these were radiocarbon dated to 11,980 $\pm$ 130 yr BP. The combined dates from the Xaudum and Okwa all fall within the range 14,570–11,890 yr BP, and are interpreted by Shaw et al. (1992) as evidence for a prolonged episode of greater moisture availability in the period 16,000–13,000 ka followed by a lowering of water tables and calcrete formation at 12,000–11,000 ka.

Given the large number of flood events documented within exoreic Southern Kalahari systems during the historical period, equivalent floods or periods of more permanent flow are highly likely to have occurred in prehistory. Certainly, oral histories reported by Campbell (1822) and corroborated by Andrew Smith (see Lye 1975) suggest that the Kuruman was a stronger-flowing river during the mid- to late-eighteenth century. Wider sections of the Nossop within the Kgalagadi Transfrontier Park, for example, contain abandoned channels and meander bends (see Fig. 11.5) that appear to have been cut off by floodwaters to form now-dry 'oxbow lakes' (Nash et al. 1994b; Bullard and Nash 1998; Nash 2015).

A provisional framework for the late Quaternary fluvial history of the Molopo system has been put forward by Heine (1982) through analyses of sediments within valley floors in the southwest Kalahari. This has been criticised by Thomas and Shaw (1991) as it confuses terrace levels and, in part, interprets radiocarbon dates derived from the land snail *Xeroceratus* as humid indicators. However, those dates conducted on *Unio*, *Corbicula* and *Bulinus* shells—indicative of perennial or semi-perennial flow—suggest wetter conditions in the dated range 16,600–12,500 yr BP. Several episodes of Holocene flooding have been dated in the Kuruman. Shaw et al. (1992) identified two terrace levels along the Kuruman from the vicinity of Hotazel to its confluence with the Molopo. At Groot Drink, the uppermost terrace, at a height of ~ 8 m above the riverbed, is presumed to predate Middle Stone Age artefacts found upon it. *Ceratophallus*, *Lymnaea* and *Burnupia* shells found within the top 30 cm of the lower (~ 3 m) terrace were radiocarbon dated to 320 ± 150 yr BP. Two sites downstream of the confluence with the Moshaweng River provide good vertical exposures of these lower deposits, with at least six cycles of deposition represented. Charcoal from a layer at 1.4 m height within one exposure at Aansluit farm was radiocarbon dated to 1780 ± 60 yr BP, while ostrich eggshell fragments from heights of 3.4 m and 4.8 m within an equivalent sequence at nearby Bella Vista farm were dated to 2840 ± 110 yr BP and 540 ± 30 yr BP respectively.

11.5 Ages of Dry Valley Networks

Elements of many Kalahari dry valleys are of considerable antiquity (Thomas and Shaw 1991) and, in some cases, may pre-date the deposition of the Jurassic to Recent Kalahari Group sediments; sections of the Black Nossop may even pre-date the Permo-Carboniferous Dwyka glaciation (Nash et al. 1994b). However, constraining the precise ages of networks is not straightforward. Neither is it easy to contextualise valley formation in terms of past climate since, with the exception of records from the Zambezi Delta (e.g. Walford et al. 2005; Ponte et al. 2019) and the mouth of the Limpopo (e.g. Iliffe et al. 1991), regional palaeoenvironmental information extending back to the Neogene and Paleogene is limited.

It may be possible to estimate the relative ages of some valleys and their associated knickpoints (see Sect. 11.3) in relation to known episodes of tectonic flexuring within southern Africa—with the obvious caveat that the timing of key tectonic events is itself not precisely constrained. The headwaters of the Mmone/Quoxo, for example, must pre-date the uplift along the Kalahari-Zimbabwe axis (broadly coincident with the Kalahari-Limpopo drainage divide; Fig. 11.1) that led to the valley gradient reversals shown in Figs. 11.7 and 11.9. Geomorphological evidence suggests that there were two uplift episodes along the Kalahari-Zimbabwe axis; a minor Miocene phase and a more significant uplift in the Pliocene (du Toit 1933; Partridge and Maud 2000). These phases coincide with evidence for increased sedimentation in the Zambezi Delta from the Oligocene onwards (Walford et al. 2005) and off the mouth of the Limpopo from the Miocene (Iliffe et al. 1991), likely assisted by wetter conditions during the Late Oligocene and Late Miocene (Ponte et al. 2019). Pollen evidence from the Zambezi Delta, for example, suggests that rainfall over the Zambezi catchment may have reached 2000 mm per annum during the Late Oligocene (Ponte et al. 2019). Given the scale of the upper Mmone/Quoxo knickpoints, Nash and Eckardt (2016) suggested that valley incision was most likely to have been initiated during the Pliocene uplift phase (du Toit 1933; Haddon and McCarthy 2005), which coincides with evidence from the Zambezi Delta of wetter conditions during the Zanclean and Piacenzian (Ponte et al. 2019).

For the dry valley systems that terminate in the Makgadikgadi Depression, the history of knickpoint development (and hence valley evolution) can be interpreted in the context of known Quaternary hydrological changes that led to the rise and fall of various palaeolake phases within the basin (Nash and Eckardt 2016). The middle knickpoint on the Deception, for example, falls where the valley dog-legs through a former shoreline of Palaeolake Deception (McFarlane and Eckardt 2008), the highest elevation ($\sim$980–990 m asl) palaeolake identified in the Makgadikgadi region (Moore et al. 2012). The Passarge Valley terminates at, and therefore possibly predates, this shoreline. The initiation of the knickpoint on the lower Deception probably relates to a period of neotectonic tilting during the latest Pleistocene that also caused the southerly migration of the Boteti River to its present course (Cooke and Verstappen 1984). The absence of palaeolake-related knickpoints on valleys entering the eastern and southern Makgadikgadi may be because these systems cross more resistant basement lithologies and were therefore less sensitive to short term lake level changes.

The timing of knickpoint formation in dry valleys entering the Okavango half-graben in northwest Botswana is less clear. However, given the contemporary levels of tectonic activity associated with the half-graben (Gumbricht et al. 2001), and evidence that movement along the Gumare fault has truncated Pleistocene dune sequences to the west of the Okavango Delta (McFarlane and Eckardt 2007), tectonically driven base-level lowering and valley incision has probably been ongoing throughout the Quaternary (Nash and Eckardt 2016).

11.6 Theories About the Origins of Dry Valley Systems

Conventional wisdom suggests that the various dry valley networks were formed by fluvial erosion during periods of wetter climate and are therefore inherited landscape features. Certainly, the deltaic landforms, abandoned channels, fluvial terraces, and lignite, peat and freshwater shell deposits described in Sect. 11.3.2 indicate increased river flow in the past. The concave-upward long-profiles common to most dry valleys (Figs. 11.6, 11.7 and 11.8) and localised adjustments in valley gradient to lithological variations and base-level changes are also typical of drainage features formed by fluvial erosion elsewhere. Based on a simple comparison of long-profile shape, it is most probable that valley systems were the product of episodic fluvial erosion over protracted periods of time. However, even the most active valley systems today are not hydrologically connected along their full length, and it is likely that most have been in this disconnected state for hundreds if not thousands of years.

While fluvial erosion has clearly played a major role in shaping Kalahari dry valley systems, it cannot alone explain the morphology of some systems in their middle 'gorge-like' sections (Thomas and Shaw 1991). The low bifurcation ratios of tributaries, increases in valley width over relatively short distances, and lack of channels in many valley floors are not typical of conventional river systems. Instead, various authors have suggested that processes such as deep weathering along fracture zones and groundwater seepage erosion may have played a role in valley formation and evolution. Sections of many valleys systems—including the Okwa (Nash 1995), Deception (Coates et al. 1979), Xaudum (Wright 1978), and various tributaries to the Mmone/Quoxo (Mallick et al. 1981; Nash 1995)—are aligned with faults and other geological lineaments (Fig. 11.12). The relationship reaches an extreme in the Serorome, which is oriented west and north before being diverted eastwards by the Zoetfontein Fault to join the Limpopo system (Thomas and Shaw 1991). There has also been a long association between Kalahari dry valleys and groundwater availability (Chapman 1886) with, as noted in Sect. 11.4.1, *mekgacha* acting as important foci for groundwater recharge (Farr et al. 1981; de Vries et al. 2000). Nash (1995) has suggested that deep weathering (as groundwater moves vertically and laterally along subsurface faults and fracture zones) may have permitted rock weakening, preferential erosion and surface lowering to form linear depressions. The significance of this process for valley development is that surface lowering can occur in the absence of surface fluvial activity and does not necessarily require seasonal recharge. It is likely that surface lowering would proceed more rapidly if there was an alternation between periods of dominant deep weathering and dominant fluvial activity. However, as Nash (2011) argues, there still remains the unresolved 'chicken and egg' question of what came first, the valley or the deep weathering.

Arguments that groundwater sapping or seepage erosion played a significant role within Kalahari valley development put forward by Shaw and de Vries (1988), Nash et al. (1994b) and Nash (1995) seem less likely in light of the new evidence provided by SRTM data (see Sect. 11.3.2). The valley long-profiles shown in Figs. 11.6, 11.7 and 11.8 differ from the flat or stepped longitudinal forms typical of systems generated by groundwater sapping (see Higgins 1984; Howard et al. 1988; Nash 2011); indeed, only four valley networks (the Auob, Passarge, Mosope and Letlhakane) exhibit predominantly flat/linear longitudinal profiles. Instead, it now appears most likely that the pronounced steps in the long profiles of valley systems such as the Mmone/Quoxo and its headwater tributaries are relict knickpoints resulting from the response of the fluvial system to regional uplift.

In addition to their scientific interest, the origins of the dry valleys of the Kalahari also feature in traditional knowledge. The G/wi San of central Botswana, for example, have a belief that the Okwa Valley was formed by a supernatural shape-shifting being named G//awama (Silberbauer 1981; Main 1987). Oral traditions suggest that, while hunting one day to the west of the Kalahari near Gobabis (in eastern Namibia), G//awama was bitten on the leg by a python. The bite was a bad one, and G//awama soon became feverish and very thirsty, so he headed east towards the Boteti River in search of water. As he walked, he dragged his injured leg and his trailing foot gouged out the course of the Okwa Valley. Wild animals harassed him *en route*, taking advantage of his weakness; this caused his course to waver, creating the meanders present in central sections of the valley. The fever induced by the python bite made him nauseous and he vomited frequently, with the dried vomit forming the calcrete- and diatomite-floored tributary valleys to the Okwa. As G//awama neared the waters of the Boteti he found new strength and increased his walking speed, thus dragging his leg less heavily. He was eventually able to walk almost upright, dragging his leg only slightly. To the G/wi San, this explains why the course of the Okwa almost disappears as it nears the Makgadikgadi Depression and the Boteti River. Geomorphological explanations would instead attribute the disappearance of the Okwa to a lack of hydrological connectivity with its headwaters in Namibia since the late Pleistocene and a resultant progressive infilling by aeolian sediments over time in its distal sections.

References

Aldiss DT, Carney JN (1992) The geology and regional correlation of the Proterozoic Okwa Inlier, western Botswana. Precambr Res 56:255–274

Alexander JE (1838) An expedition of discovery into the interior of Africa. H. Colburn, London

Andersson CJ (1856) Lake Ngami; or explorations and discoveries during four years wanderings in the wilds of south western Africa. Hurst and Blackett Publishers, London

Anderson AA (1887) Twenty-five years in a waggon in the Gold Regions of Africa, 2 Volumes. Hurst and Blackett Publishers, London

Baines T (1864) Exploration in Southwest Africa. Longman, Greens and Co., London

Barrow B (1974) Song of a dry river. Purnell, Cape Town

Bluck BJ, Ward JD, Cartwright J, Swart R (2007) The Orange River, southern Africa: an extreme example of a wave-dominated sediment dispersal system in the South Atlantic Ocean. J Geol Soc 164:341–351

Boocock C, van Straten OJ (1962) Notes on the geology and hydrogeology of the Central Kalahari region, Bechuanaland Protectorate. Trans Geol Soc South Africa 65:125–171

Brook GA (1995) Quaternary environments in the Somali-Chalbi and Kalahari deserts of Africa. In: Alsharhan AS, Glennie KW, Whittle GL, Kendall CGSC (eds) Abstracts of the international conference on quaternary deserts and climatic change, 1995. United Arab Emirates University, Al Ain, UAE, Al Ain, p 6

Bullard JE, Nash DJ (1998) Linear dune pattern variability in the vicinity of dry valleys in the southwest Kalahari. Geomorphology 23:35–54

Bullard JE, Nash DJ (2000) Valley-marginal sand dunes in the southwest Kalahari: their nature, classification and possible origins. J Arid Environ 45:369–383

Burchell WJ (1824) Travels in the Interior of southern Africa, vol 2. Longman, Hurst, Rees, Orme, Brown and Green, London

Campbell AC, Child G (1971) The impact of man on the environment of Botswana. Botswana Notes Records 3:91–109

Campbell J (1822) Travels in South Africa, 2 Volumes. London Missionary Society, London

Chapman J (1886) Travels in the interior of South Africa, 2 Volumes. Bell and Daldy, London

Clement AJ (1967) The Kalahari and its lost city. Longman, Cape Town

Coates J, Davies J, Gould D, Hutchins D, Jones C, Key R, Massey N, Reeves C, Stansfield G, Walker I (1979) The Kalatraverse one report. Botswana Geol Surv Bull 21

Cooke HJ (1984) The evidence from northern Botswana of late Quaternary climatic change. In: Vogel JC (ed) Late Cainozoic palaeoenvironments of the southern hemisphere. Balkema, Rotterdam, pp 265–278

Cooke HJ, Verhagen BT (1977) The dating of cave development: an example from Botswana. In: Proceedings of the seventh international speleological congress, Sheffield.

Cooke HJ, Verstappen BT (1984) The landforms of the western Makgadikgadi Basin of northern Botswana. Zeitschrift Für Geomorphologie 28:1–19

Cornwallis-Harris W (1852) The wild sports of southern Africa; being the narrative of a hunting expedition from the Cape of Good Hope, through the territories of the Chief Moselekatse, to the Tropic of Capricorn. Henry G. Bohn, London

de Vries JJ, Selaolo ET, Beekman HE (2000) Groundwater recharge in the Kalahari, with reference to paleo-hydrologic conditions. J Hydrol 238:110–123

de Wit M (2007) The Kalahari Epeirogeny and climate change: differentiating cause and effect from core to space. S Afr J Geol 110:367–392

du Toit AL (1926) Report on the Kalahari reconnaissance of 1925. Department of Irrigation, Union of South Africa, Pretoria

du Toit AL (1933) Crustal movement as a factor in the geographical evolution of South Africa. S Afr Geogr J 16:3–20

Farini GA (1886) Through the Kalahari Desert. Sampson Low, Marston, Searle and Rivington, London

Farr JL, Cheney C, Baron J, Peart R (1981) GS10 Project: evaluation of underground water resources, final report. Botswana Geol Surv Lobatse

Frostick LE, Reid I (1989) Climatic versus tectonic controls of fan sequences: Lessons from the Dead Sea, Israel. J Geol Soc 146:527–538

Gordon-Cumming R (1909) The lion hunter. John Murray, London

Goudie AS (2002) Great warm deserts of the world. Oxford University Press, Oxford

Grove AT (1969) Landforms and climatic change in the Kalahari and Ngamiland. Geogr J 135:191–212

Gumbricht T, McCarthy TS, Merry CL (2001) The topography of the Okavango Delta, Botswana, and its tectonic and sedimentological implications. S Afr J Geol 104:243–264

Hack JT (1973) Stream-profile analysis and stream-gradient indices. USGS J Res 1:421–429

Haddon IG, McCarthy TS (2005) The Mesozoic-Cenozoic interior sag basins of Central Africa: the Late-Cretaceous-Cenozoic Kalahari and Okavango basins. J Afr Earth Sc 43:316–333

Heine K (1982) The main stages of the late Quaternary evolution of the Kalahari region, southern Africa. Palaeoecol Africa 15:53–76

Helgren DM, Brooks AS (1983) Geoarchaeology at Gi, a Middle Stone Age and Later Stone Age site in the northwest Kalahari. J Archaeol Sci 10:181–197

Herbst JF (1908) Report on the Rietfontein area, Colonial reports—miscell., Number 55, Cape Colony. HMSO, London

Higgins CG (1984) Piping and sapping: Development of landforms by groundwater outflow. In: La Fleur RG (ed) Groundwater as a geomorphic agent. Allan and Unwin, London, pp 18–58

Hodson AW (1912) Trekking the Great Thirst - sport and travel in the Kalahari Desert. T. Fisher Union, London

Holden WC (1855) History of the Colony of Natal. Alexander Heylin London

Howard AD, Kochel RC, Holt HE (1988) Sapping features of the Colorado Plateau: A comparative planetary geology fieldguide (Report SP-491). NASA, Washington, DC

Iliffe JE, Lerche I, Debuyl M (1991) Basin analysis and hydrocarbon generation of the South Mozambique Graben using extensional models of heat-flow. Mar Pet Geol 8:152–162

Jack DJ (1980) UCEX-CEGB Botswana uranium in calcrete project. Unpublished final report of the Union Carbide Exploration Corporation, Union Carbide Exploration Corporation

Kampunzu AB, Ringrose S, Huntsman-Mapila P, Harris C, Vink BW, Matheson W (2007) Origins and palaeo-environments of Kalahari duricrusts in the Moshaweng dry valleys (Botswana) as detected by major and trace element composition. J Afr Earth Sc 48:199–221

Kinabo BD, Atekwana EA, Hogan JP, Modisi MP, Wheaton DD, Kampunzu AB (2007) Early structural development of the Okavango rift zone, NW Botswana. J Afr Earth Sc 48:125–136

Kirby PRE (1940) The Diary of Dr. Andrew Smith, director of the Expedition for Exploring Central Africa, 1834–1836, Volume 2. Van Riebeeck Society, Cape Town

Leistner OA (1967) The plant ecology of the southern Kalahari. Botanical Research Institute Botanical Memoir 38. Republic of South Africa Department of Agricultural Technical Services, Pretoria

Leopold LB, Miller JP (1956) Ephemeral streams: Hydraulic factors and their relation to the drainage net (Professional Paper 282-A). USGS, Washington, DC

Leyland J (1866) Adventures in the far interior of South Africa. George Routledge, London

Lister MH (1949) The journals of Andrew Geddes Bain. Van Riebeck Society, Cape Town

Livingstone D (1857) Missionary travels and researches in South Africa. John Murray, London

Lye WFE (1975) Andrew Smith's journal of his expedition into the interior of South Africa, 1834–36. A.A. Balkema, Cape Town

Main M (1987) Kalahari: Life's variety in dune and delta. Southern Book Publishers, Johannesburg

Mallick DIJ, Habgood F, Skinner AC (1981) A Geological Interpretation of LANDSAT Imagery and Air Photography of Botswana. HMSO, London

Mazor E, Verhagen BT, Sellschop JPF, Jones MT, Robins NE, Hutton L, Jennings CMH (1977) Northern Kalahari groundwaters: hydrologic, isotopic and chemical studies at Orapa, Botswana. J Hydrol 34:203–234

McFarlane MJ, Eckardt FD (2007) Palaeodune morphology associated with the Gumare fault of the Okavango graben in the Botswana/Namibia borderland: A new model of tectonic influence. S Afr J Geol 110:535–542

McFarlane MJ, Eckardt FD (2008) Lake Deception: A new Makgadikgadi palaeolake. Botswana Notes Records 38:195–201

McMillan IK (2003) Foraminiferally defined bilostratigraphic episodes and sedimentation pattern of the Cretaceous drift succession (Early Barremian to Late Maastrichtian) in seven basins on the South African and southern Namibian continental margin. S Afr J Sci 99:537–576

Moffat R (1842) Missionary Labours and Scenes in South Africa. J. Snow, London

Moore A, Blenkinsop T, Cotterill F (2009) Southern African topography and erosion history: plumes or plate tectonics? Terra Nova 21:310–315

Moore AE, Cotterill FPD, Eckardt FD (2012) The evolution and ages of Makgadikgadi palaeo-lakes: Consilient evidence from Kalahari drainage evolution south-central Africa. S Afr J Geol 115:385–413

Morrison P, Morrison S (2013) Letlhakane flooded. Pete and Sue's Botswana. https://ourbots.wordpress.com/2013/01/04/letlhakane-flooded/. Accessed 16 April 2020

Motlhabani C (2018) Schools in Letlhakane area closed due to heavy floods. The Voice. https://thevoicebw.com/schools-letlhakane-area-closed-due-heavy-floods/. Accessed 22 April 2019

Nash DJ (1995) Structural control and deep-weathering in the evolution of the dry valley systems of the Kalahari, central southern Africa. Africa Geosci Rev 2:9–23

Nash DJ (1996) On the dry valleys of the Kalahari: Documentary evidence of environmental change in central southern Africa. Geogr J 162:154–168

Nash DJ (2011) Groundwater controls and processes. In: Thomas DSG (ed) Arid zone geomorphology: process, form and change in drylands, 3rd edn. John Wiley and Sons Ltd., Chichester, pp 403–424

Nash DJ (2015) Of dunes, depressions and dry valleys: the arid landscapes of the Kalahari Desert. In: Grab S, Knight J (eds) Landscapes and Landforms of South Africa. Springer, Berlin, pp 129–137

Nash DJ, Endfield GH (2002) Historical flows in the dry valleys of the Kalahari identified from missionary correspondence. S Afr J Sci 98:244–248

Nash DJ, Eckardt FD (2016) Drainage development, neotectonics and base-level change in the Kalahari Desert, southern Africa. S Afr Geogr J 98:308–320

Nash DJ, Shaw PA, Thomas DSG (1994a) Duricrust development and valley evolution: process-landform links in the Kalahari. Earth Surf Proc Land 19:299–317

Nash DJ, Thomas DSG, Shaw PA (1994b) Timescales, environmental change and dryland valley development. In: Millington AC, Pye K (eds) Environmental change in drylands: biogeographical and geomorphological perspectives, pp 25–41

Nash DJ, Meadows ME, Gulliver VL (2006) Holocene environmental change in the Okavango Panhandle, northwest Botswana. Quatern Sci Rev 25:1302–1322

Nash DJ, Pribyl K, Klein J, Neukom R, Endfield GH, Adamson GCD, Kniveton DR (2016) Seasonal rainfall variability in southeast Africa during the nineteenth century reconstructed from documentary sources. Clim Change 134:605–619

Neukom R, Nash DJ, Endfield GH, Grab SW, Grove CA, Kelso C, Vogel CH, Zinke J (2014) Multi-proxy summer and winter precipitation reconstruction for southern Africa over the last 200 years. Clim Dyn 42:2713–2716

Partridge TC, Maud RR (2000) Macro-scale geomorphic evolution of southern Africa. In: Partridge TC, Maud RR (eds) The Cenozoic of Southern Africa. Oxford University Press, Oxford, pp 3–18

Passarge S (1899) Dr. Passarge's journeys in South Africa. Geogr J 14:310–313

Passarge S (1904) Die Kalahari. Dietrich Riemer, Berlin

Ponte J-P, Robin C, Guillocheau F, Popescu S, Suc J-P, Dall'Asta M, Melinte-Dobrinescu MC, Bubik M, Dupont G, Gaillot J (2019) The Zambezi delta (Mozambique channel, East Africa): High resolution dating combining bio-orbital and seismic stratigraphies to determine climate (palaeoprecipitation) and tectonic controls on a passive margin. Mar Pet Geol 105:293–312

Preston-Whyte RA, Tyson PD (1993) The atmosphere and weather of Southern Africa. Oxford University Press, Cape Town

Ramokate LV, Mapeo RBM, Cuorfu F, Kampunzu AB (2000) Proterozoic geology and regional correlation of the Ghanzi-Makunda area, western Botswana. J Afr Earth Sc 30:453–466

Range P (1912) Topography and geology of the German South Kalahari. Trans Geol Soc South Africa 15:63–73

Robbins LH, Murphy ML, Stewart KM, Campbell AC, Brook GA (1994) Barbed bone points, paleoenvironment, and the antiquity of fish exploitation in the Kalahari Desert, Botswana. J Field Archaeol 21:257–264

Rodriguez E, Morris CS, Belz JE, Chapin EC, Martin JM, Daffer W, Hensley S (2005) An assessment of the SRTM topographic products (Technical Report D-31639). Jet Propulsion Laboratory, Pasadena, CA

Schapera IE (1951) Apprenticeship at Kuruman, being the journals and letters of Robert and Mary Moffat 1820–1828. Chatto and Windus, London

Shaw PA (1989) Fluvial systems in the Kalahari: A review. In: Yair A, Berkowicz S (eds) Arid and semi-arid environments - Geomorphological and pedological aspects. Catena Supplement 14, pp 119–126

Shaw PA, de Vries JJ (1988) Duricrust, groundwater and valley development in the Kalahari of southeast Botswana. J Arid Environ 14:245–254

Shaw PA, Thomas DSG, Nash DJ (1992) Late Quaternary fluvial activity in the dry valleys (mekgacha) of the middle and southern Kalahari, southern Africa. J Quat Sci 7:273–281

Silberbauer GB (1981) Hunter and habitat in the Central Kalahari Desert. Cambridge University Press, Cambridge

Stigand AG (1923) Ngamiland. Geogr J 62:401–419

Tabler EC (1916) The recollections of William Finnaughty, Elephant hunter 1864–1875. J.B. Lippincott Company, Philadelphia

Thomas DSG, Shaw PA (1991) The Kalahari environment. Cambridge University Press, Cambridge

Tooth S (2000) Downstream changes in dryland river channels: the Northern Plains of arid central Australia. Geomorphology 34:33–54

Verhagen BT (1983) Environmental isotope study of a groundwater supply project in the Kalahari of Gordonia. In: Proceedings of the International Atomic Energy Agency (IAEA) Symposium. Vienna, pp 1–15

Walford HL, White NJ, Sydow JC (2005) Solid sediment load history of the Zambezi Delta. Earth Planet Sci Lett 238:49–63

Weinberg C (1975) Fragments of a desert land: memoirs of a South West African doctor. Robert Hale & Co., London

Wright EP (1978) Geological studies in the northern Kalahari. Geogr J 144:235–250

Yellen JE, Lee RB (1976) The Dobe-/Du/da environment: Background to a hunting and gathering way of life. In: Lee RB, De Vore I (eds) Kalahari hunter-gatherers: studies of the !Kung San and their neighbors. Harvard University Press, Cambridge, MA, pp 27–46

David J. Nash is Professor of Physical Geography at the University of Brighton and Honorary Research Fellow at the University of the Witwatersrand (South Africa). He has worked on the geomorphology of Botswana since 1989, when he arrived in Gaborone for the first field season of his PhD research on the fossil valleys of the Kalahari. In the intervening years he has published over 100 articles on silcrete, calcrete and Holocene-to-recent environmental change, mostly with reference to southern Africa. His present research focusses on the use of geochemical data to determine the provenance of silcrete stone tools at archaeological sites in the Kalahari.

12 Landscape Evolution of the Stampriet Transboundary Basin and Relation to the Groundwater System: The Land of Duricrusts, Pans, Dry Valleys and Dunes, and the Relation to the Groundwater System

Abi Stone

Abstract

The Stampriet Transboundary Aquifer System (STAS) covering western Botswana, eastern Namibia and part of the Northern Cape of South Africa is an interesting location to explore landscape evolution and the interaction of a range of surface and subsurface geomorphological processes. This chapter aims to introduce the importance of the STAS. It outlines the geological and hydrogeological framework, the range of surface geomorphological features and considers how these relate to each other and the hydrogeology of this system. The long-term tectonic and sedimentary history of the STAS provided the geological sediments and structures upon which the interactions between surface and subsurface geomorphological processes have occurred. Water–rock interactions control the quantity and quality (natural baseline) of groundwater. Where groundwater reaches the surface zone it influences the persistence of water at the surface and pan sediment chemistry and also reduces the potential for deflation and abrasion of the surface by the wind. The main geomorphological components of the STAS region are the dunes and sand sheet cover of the Kalahari Erg, dry valleys and ephemeral rivers, pans, and calcrete plateaus and other duricrusts. This chapter also explores the possible links between duricrusts and incised valleys, between the incised valleys and dunes of the Kalahari Erg and the geomorphic influence of the valleys on the current wind regime and sediment supply. The role of groundwater in the development of the geomorphology of this system lies particularly within the formation of duricrusts, pans and dry valleys. The exploration of the relationship and interactions between duricrusts, valleys and the dunes of the Kalahari Erg shows stratigraphic evidence for an incision of the valleys that post-dates the formation of duricrusts. It is also possible that valley incision post-dates an early generation of aeolian linear dunes at the surface, although we are not able to offer a timeline for, or confirmation of, the later based on dating sediments because of the upper-age limits of our available chronological techniques.

Keywords

Stampriet transboundary aquifer system • Kalahari • Groundwater processes • Groundwater • Surface interactions • Duricrusts • Pans • Dry valleys • Dunes • Landscape evolution

12.1 Introduction

At the global scale, there are up to 445 recognised transboundary aquifers (Wada and Heinrich 2013), 72 within the African continent (Nijsetn et al. 2018) and 18 located in southern Africa (south of 15 °S). Sharing of groundwater across country boundaries involves challenges both for assessing and managing the aquifer. Maintaining groundwater quality and access is particularly important in southern Africa, with 60–90% of the rural population reliant on groundwater (Adams 2009) and a lack of perennial surface water in the southwestern portion, which makes agriculture and industry also reliant on groundwater.

The Stampriet Transboundary Aquifer System (STAS), often referred to as the Stampriet Basin (SAB) or the Kalahari/Karoo Aquifer (Puri 2001), is a multi-layered aquifer within Botswana, Namibia and South Africa (Figs. 12.1 and 12.2). Most is currently known about the Namibian portion of the basin, where it is thought that the majority of recharge occurs (Puri et al. 2001), although the GGRETA (Governance of Groundwater Resources in Transboundary Aquifers) project of the United Nations Educational, Scientific and Cultural Organisation

A. Stone (✉)
School of Environment, Manchester Environmental Research Institute, FSE Research Institutes, Manchester, UK
e-mail: abi.stone@manchester.ac.uk

F. D. Eckardt (ed.), *Landscapes and Landforms of Botswana*, World Geomorphological Landscapes,
https://doi.org/10.1007/978-3-030-86102-5_12

(UNESCO) International Hydrological Programme (IHP), in cooperation with IGRAC (International Groundwater Resources Assessment Centre) has been providing greater insights into the Botswanan and South African portions (UNESCO-IHP 2017). The STAS has a population of around 50,000 and the dominant use of abstracted groundwater is for irrigation and livestock, and a smaller proportion for domestic use and tourism (see Sect. 12.3.2, JICA 2002; CSO 2009; NamWater 2014; CGS 2015; NSA 2015; UNESCO-IHP 2017).

The geology, hydrogeology and surface geomorphology of this aquifer system are interconnected. The consideration of a basin such as the STAS offers an opportunity to explore overall landscape evolution and the connections between the surface geomorphological features as well the connections between geomorphological and hydrogeological features and processes. The aims of this chapter are to (i) consider the geological framework and how this translates to hydrogeological units, (ii) think about surface geomorphology and how it has evolved, (iii) consider how the geomorphology relates to hydrogeology (sites of inputs from recharge and outputs, any artesian springs) and (iv) outline some of the water use and water quality issues for this region. There are interesting points of intersection and cross-reference for this chapter (e.g. Chap. 8 on southwest Kalahari dunes (Thomas), Chap. 10 on Kalahari Pans (Schüller) Chap. 13 on Duricrusts (Nash) and Chap. 11 on Dry River Valleys (Nash)).

Fig. 12.1 Location and extent of the Stampriet Transboundary Aquifer System (STAS) in western Botswana, eastern Namibia and the north of South Africa (using the delineation by UNESCO-IHP (2017) that is based on the occurrence of geological formations belonging to the Ecca Group within the Auob and Nossob River basins)

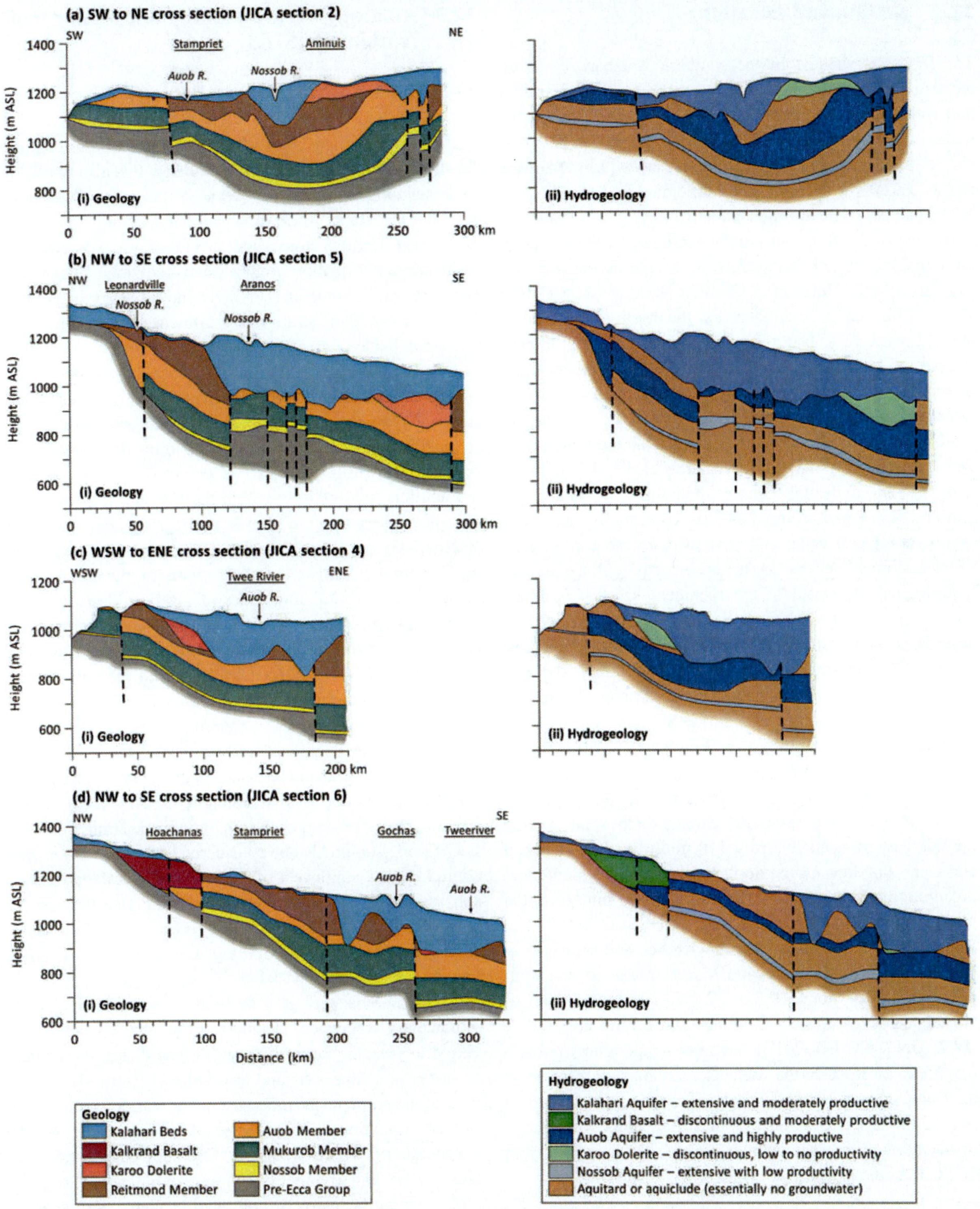

Fig. 12.2 Cross sections of simplified (i) geology and (ii) hydrogeology of the STAS for four cross sections **a** southwest to northeast, **b** northwest to southeast, **c** west-south-west to east-north-east and **d** northwest to southeast (redrawn from JICA 2002)

12.2 Setting and Location

The STAS is situated in the southern Kalahari, covering western-central Botswana, eastern- central-southern Namibia and some of the northern Cape of South Africa, with estimates of a total area of ~87,000 km^2 (UNESCO-IHP 2017) (Fig. 12.1), with ~65,000 km^2 of that within Namibia (JICA 2002). The overall surface topography of the STAS is fairly flat, with an average gradient of 1/500 m from the northwest (~1450 m a.s.l.) to the southeast (~900 m a.s.l). Around 75% of the basin surface is covered by the Kalahari Erg (Stone and Edmunds 2012; JICA 2002; UNESCO-IHP 2017). This erg is dominated by the northwest-southeast aligned southern Kalahari dunefield, which covers the central part of the STAS in Namibia and into the Northern Cape of South Africa. This gives way to sand sheet cover at the eastern margin into western Botswana and to a largely sand-free surface at the western margin. This western margin includes the Weissrand plateau (calcrete-rich Kalahari group deposits), and in the northwest of the basin basalt bedrock is exposed. The extent of the STAS is delimited by geological (hydrogeological) units, and is often the case the groundwater catchment does not coincide perfectly with the surface hydrological catchment of the ephemeral rivers (see Hiscock and Bense 2014). Following the UNESCO-IHP (2017) delineation, the STAS is taken to coincide with the extent of the Ecca Group and its formations (Table 12.1) The two major ephemeral river drainage basins, the Nossob and the Auob, have their sources outside the STAS in the Khomas region of Namibia, northeast of Windhoek (~2,000 m a.s.l.). The groudwater drainage is broadly from northwest to southeast following the decline in surface topography.

The STAS has a semi-arid climate with mean annual rainfall varying spatially from 140 mm/y in the southwest and increasing towards the north and west to 310 mm/y, and potential evapotranspiration reaching 3,500 mm/y in the south and 3,000 mm/y in the north (UNESCO-IHP 2017). Rainfall tends to occur between October and April, concentrated between January and March. Mean annual temperature varies from 19 to 22 °C, warmest west of Hoachanas and lowest in the north and central area (JICA 2002; UNESCO-IHP 2017). Summer temperatures reach a maximum of 36 °C in the southern margins, and these areas have the greatest annual contrasts, occasionally falling to ~2 °C in the winter, and by contrast, the temperature range in the northern reaches is from around 12 °C in winter to 26 °C in summer.

12.3 Geology, Hydrogeology and Water Use within the STAS

12.3.1 Geology and Hydrogeology

The geological framework of the STAS sets the foundation for the overlying surface geomorphology. It is also crucial to understanding groundwater because the geological units provide the static foundation (notwithstanding endogenic processes) for the stores and fluxes of groundwater and influence water quality through water–rock interactions (van der Gun 2012; Stone et al. 2019). The lithology of geological units and their stratigraphy influence the physical and chemical behaviour of the groundwater system, and they also mediate the operation of subsurface geomorphological processes that create subsurface landforms, some of which have an expression in the surface geomorphology. The geological units and their corresponding hydrogeological units are shown in Table 12.1. Three of the geological units are aquifers, two entirely within the Karoo Supergroup and the third spanning the Karoo Supergroup and overlying Kalahari Group (Fig. 12.2e–h). The deepest is the Nossob aquifer in the sandstones at the base of the Ecca group sediments, overlying aquicludes of glacial sediments in the Dwyka group, whilst above it is the Auob aquifer, in a second sandstone-rich unit, between two shale-rich aquiclude/tards (JICA 2002; Miller 2008). The uppermost aquifer is a discontinuous hydrogeological unit known as the Kalahari aquifer, which is unconfined. The upper parts of this aquifer are composed of unconsolidated gravels and sands, containing layers of calcretes and silcretes, all part of the Kalahari Group, whilst borehole data by JICA (2002) suggests the upper part of the Rietmond formation within the Ecca Group should also be considered as part of the Kalahari aquifer. In the southwest of Botswana, the Kalahari Group and associated hydrostratigraphic unit are also depicted as encompassing/overlapping with layers down to the base of the Ecca and into Dwyka Group (e.g. Leukla et al. 2018; Lekula and Lubczynski 2019).

The STAS is part of a spatially more extensive, thick sedimentary sequence spanning from the Carboniferous Period to present, containing sandstones, shales, mudstones, siltstones and limestone and also units of basalt. The thickness of these deposits increased in an easterly and south-easterly direction within the STAS (Fig. 12.2). In the northwest, there is a wedge of Kalkrand basalt with very little, or no underlying Ecca Group sediments. The basalt often contains interbeds of well-bedded pan deposits

Table 12.1 Simplified geological and hydrogeological stratigraphy of the STAS (modified after Kent 1980; Smith 1984; JICA 2002; Miller 2008 and UNESCO-IHP 2017)

Geological		Super-group	Group	Formation			Rock type/lithology (broad)	Hydrogeological unit
Era	**Period**			Botswana (B)	Namibia (N)	South Africa (SA)		
Cenozoic	Quaternary		Kalahari				Superficial sands	Unsaturated zone of Kalahari aquifer
	Neogene and Palaeogene						Sands with silcrete, calcrete, gravels, sandstones, marls, clayey-gravels	Kalahari aquifer(s)
Mesozoic	Cretaceous and Jurassic	Karoo	Country-specific	Stromberg-lava	Neu Loore (Kalkrand basalts and Etjo)	Drakensberg lavas	Basalt and dolerite	
	Triassic		Country-specific	Lebung (Ntane and Mosolotsane)		Clarens	Sandstone	
Paleozoic	Permian		Ecca	Kule	Whitehill	Whitehill	Shale and limestone	
				Rietmond Upper	Rietmond Upper	Prince Albert	Shale and sandstone	
				Rietmond lower	Rietmond lower			Aquitard/clude
				Otshe	Auob		Sandstone with interbedded shale and coal	Auob Aquifer
				Kobe	Mukorob		Shale, mudstone, siltstone	Aquitard/clude
				Ncojane	Nossob		Sandstone	Nossob Aquifer
			Dwyka				Glacial sediments	Aquitard/clude
	Carboniferous	Pre-Karoo	Nama					

containing sandstones and gritstones (derived from the basalt) and some freshwater limestones (Miller 2000; Tredoux et al. 2002) with a sporadic, thin cover of Kalahari Group sediments, including calcretes and sand dunes.

12.3.2 Groundwater Use and Quality

The STAS has a fairly small population of ~50,000 (about the same as the population of Maun), with 91%, 8% and 1% distributed in Namibia, Botswana and South Africa, respectively (estimates in the UNSECO-IHP 2017) report based on national census data for all three countries). The population is mostly low density, with 63% within rural farming areas including remaining San, or Basawra, (Bushmen) communities in Botswana (Chanda et al. 2003) and 37% within urban settlements (UNESCO-IHP 2017). A minimum estimate for groundwater abstraction within the STAS is 20 million cumecs per year, unevenly drawn from the three aquifer units (Fig. 12.2) with estimates from JICA (2002) for the Namibian portion suggesting 66% from the Kalahari aquifer(s), 33% from the Auob aquifer and 1% from the Nossob aquifer. The breakdown by use is 47% for irrigation, 37% for stock watering, 15% for domestic use and <1% for tourism (JICA 2002; MAWG 2014; NSA 2012; CSO 2009; CGS 2015). However, all of these estimates come with the recognition that abstraction data is incomplete and based largely on the assumption that abstraction is in compliance with legal permits.

The majority of the irrigated crop area is for animal fodder, such as lucerne (~50% of the irrigated area), whilst ~23% and ~4% are used for vegetables and fruits, respectively (Fig. 12.6b), which provide the highest potential income. Oats (7%) and maize (12%) are the other quantifiable types of irrigated crops in the STAS (UNESCO-IHP 2017). 90% of the ~1200,000 small livestock (mostly sheep, also goats and some pigs, horses, donkeys, ostriches and poultry) are farmed within Namibia, with a more even distribution of large stock (mostly cattle) across the STAS (UNESCO-IHP 2017). The irrigated agricultural areas make up a small proportion of the STAS, estimates for the Namibian portion are ~600 hectares (6 km^2), but where this occurs the surface is altered (Fig. 12.6c). It is expected that this use will alter the geochemical and hydrogeological properties of the sand-rich substrate, such as additional nutrient input from fertilisers and higher moisture contents from irrigation.

Water–rock interactions influence the natural groundwater quality, or baseline geochemistry (Edmunds and Shand 2009). Groundwater quality data for Botswana, Namibia and South Africa is collated for ~1000 boreholes in the UNESCO-IHP (2017) report. The overall spatial pattern in groundwater quality decreases towards the southeast of the STAS, following the direction of groundwater flow and reflecting a baseline geochemical pattern. This includes levels of TDS (total dissolved solids), sulphate and fluoride and the pattern is repeated in the Kalahari, Auob and Nossob aquifer units. The only variant is for nitrate, where concentrations drop with a depth of the aquifer units, owing to denitrification within the deeper, confined aquifers (Heaton et al. 1983). The region of lowest quality water in the southeast is known as the 'salt block' (Christelis and Struckmeier 2001), and here TDS, nitrate and sulphate not only exceed World Health Organisation (WHO 2014) guidelines for human consumption but are also unsuitable for stock watering. There are also locations within the basin where the groundwater is at risk from anthropogenic pollution from a range of pollutants including fertilisers and pesticides, livestock excreta, pit latrines, septic tanks and sewage works, burial sites and storage and disposal of oil and other fuels. The shallower Kalahari aquifer(s) are at the greatest risk from these pollutants (JICA 2002; UNSECO-IHP 2017).

Nitrate enrichment in groundwater is a particular focus of research in the STAS, and across dryland environments worldwide because it is one of the major pollutants for drinking water that poses serious health risks (e.g. Singelar and Milkowski 2012). There is also interest in bioavailable nitrogen in arid regions as part of the global nitrogen cycle (Walvoord et al. 2003). Widespread mapping of elevated nitrate across the STAS has been undertaken (e.g. Tredoux et al. 2009; UNESCO-IHP 2017). In addition, studies characterising the age and $\delta^{15}N$ ratios of the high nitrate concentration waters (Heaton et al. 1983; Heaton 1984) show that sources include the leaching of naturally produced nitrate from the soil nitrogen cycle, in addition to any anthropogenic pollution sources (fertilisers, livestock excreta, pit latrines and larger sewage infrastructure). A small number of site-specific studies also reveal details of the source and infiltration regimes of elevated nitrate in the unsaturated zone within the Kalahari aquifer (s) of the STAS, or nearby aquifers to the north and east in Botswana (e.g. Stadler et al. 2012; Stone and Edmunds 2014). These studies reveal not only variation in the source of nitrate (e.g. natural from soil nitrogen fixation or from septic tanks and manure), but that the infiltration regimes depend on aquifer properties (lithological properties and depth of the groundwater table). In the STAS Heaton et al. (1983) reported natural sources of elevated nitrate in the confined groundwater Auob and Nossob units (with calculations to account for denitrification) dated between 5.7 and >40.0 kyr B. P. Measurements from pore moisture in the USZ of the Kalahari aquifer by Stone and Edmunds (2014) reveal there are pulses of high concentration nitrate in the soil zone, and unpublished $\delta^{15}N$ and $\delta^{18}O$ analyses place the possible sources of the nitrate at the interface of the signatures from natural nitrogen-fixation and those from manure.

12.4 Evolution and Development of the STAS Landscape

12.4.1 Development of the Kalahari Basin

The STAS is on the southwestern edge of the Kalahari basin, which formed within the interior of southern Africa, following down-warping during the Late Cretaceous and early Palaeogene Periods. This down-warped tilted drainage towards the interior, and weathering and erosion of sandstones of the Ecca Group may have provided much material for the Kalahari Group sediments which have filled the basin (Haddon and McCarthy 2005). Periods of uplift in the Oligocene and Miocene Epochs led to erosion of Kalahari Group sediments from elevated areas and deposition in lower parts of the basin (Haddon and McCarthy 2005). Support for available surface water for erosion comes from floral and faunal evidence for relatively humid conditions from 19 to 17 Ma (Bamford 2000; Vrba 2000), perhaps driven by global oceanic warming (Tyson and Partridge 2000). In contrast, the calcretes and sil-calcretes of the Kalahari Group, which are exposed on the Weissrand Plateau, are thought to have formed during semi-arid climatic conditions within the Palaeogene and Neogene Periods via calcrete multiplication (Eitel 1993, 1995). From ~14 Ma, after the mid-Miocene expansion of the East Antarctic ice-sheet, the cold Benguela current established and this created dry conditions along the southwest coast of southern Africa (Tyson and Partridge 2000).

Early in the Pliocene (5 to 3 Ma) most uplift appears to have occurred on the eastern side of southern Africa, there may have been a rejuvenation of uplift axes in the interior and western edge (Etosha-Griqualand-Tansvaal axis), resulting in the Molopo being cut off from the Orange River (Moore 1999). Later uplift caused the capture of the upper Fish River by the lower Fish River north of Keetmashoop (Wellington 1955). The topographical high of the BaKalahari Schwelle (Fig. 12.1) may also be a line of crustal flexure (Moore 1999) that occurred after the deposition of Kalahari group sediments (Haddon and McCarthy 2005). The unconsolidated sands at the top of the Kalahari Group sediments may have been transported first by sheet wash and fluvial transport before being later shaped by the wind (Baillieul 1975; Moore and Dingle 1998; Lancaster 2000; Vainer et al. 2018a).

12.4.2 Main Geomorphological Components of the STAS

The modern ephemeral rivers of the STAS are misfit streams within wider (varying from 0.5 to 1.8 km), flat-bottomed channels (Shaw et al. 1992; Bullard and Nash 2000). The Auob River has a small western tributary, the Witvlei, and the main Olifants tributary joins at Tweerivier, Namibia (Fig. 12.1). The Nossob River tributaries (Swart Nossob and Wit Nossob), meet ~80 km southwest of Gobabis (Fig. 12.1). To the southeast, the Nossob demarcates the three-way border between Namibia, Bostwana and South Africa (at 20 °E) and south of here is used to mark both the Botswana-South African border and the eastern boundary of the STAS until the southeast corner of the STAS at Twee Rivieren Camp at the confluence of the Auob and Nossob rivers (Fig. 12.1). To the south of this is the confluence between the Nossob and Molopo rivers, which then drains into the Orange River, and hence the ephemeral STAS rivers are exorheic, as opposed to the endorheic valleys of the Middle Kalahari (see Shaw et al. 1992; and Chap. 11 of this volume). These wide, flat-bottomed channels have a typical incision depth of 25 m and the morphological characteristics of these valleys include sections that are steep-sided and gorge-like and sections that are more gently sloping (Bullard and Nash 1998). The slopes at the valley edges are often duricrust capped (Nash et al. 1994a) (Fig. 12.4a). There is a good deal of evidence to support a role for groundwater processes (sapping and deep weathering) in the formation of the valleys alongside fluvial processes (Shaw and deVries 1988; Nash 1992; Nash et al. 1994a, b; Nash/Chap. 11 of this volume), which demonstrates a long-term interaction between groundwater and surface water in the STAS. Much of the Auob is incised into a calcrete-covered plateau, whilst the Swart-Nossob is incised into quartzites, limestone and shales.

A calcrete plateau surface is exposed on the western edge of the STAS where it is called the Weissrand (Fig. 12.1), and geologically it is part of the Kalahari Group sediments overlying the Ecca Group (Table 12.1). It has a clear escarpment at its western edge (Goudie 2007). Geochemical analysis of material just south of Gochas shows that it is a silicified calcrete (sil-calcrete) (Nash and Shaw 1998; and see Chap. 13 by Nash in this volume for further details). The surface has been modified by karstic processes and aeolian erosion and contains pans up to 50 m deep, such as Koës Pan (Eitel and Blümel 1997) and a substantial region of aligned drainage (Goudie 2007) (Fig. 12.4b). The aligned drainage features reach a density of 2 per square kilometre and are arranged in lines running northwest to southeast (with the gentle decline of topography). Goudie (2007) suggests a four-phase evolutionary model: (i) depressions start to form in interdunes swales during a phase of dune coverage, (ii) subaerial exposure of the aligned drainage occurs after dune cover stripping caused by sediment starvation or changes to wind regime, (iii) karstic processes develop the depressions further, and they become the focal point for a dendritic pattern of shallow surface drainage and (iv) development of some depressions into deep pans through salt weathering and deflation. The deepest weathering reaches the uppermost shale and sandstone-rich units of the Ecca Group (Table 12.1), such as at Koës Pan (Eitel and Blümel 1997). The calcrete surface also caps some parts of the Kalkrand basalt in the northwest of the STAS, with similar karst-like sinkholes that have been shown to be an important conduit for surface water to reach the Auob aquifer within the STAS (Miller 2000; Tredoux et al. 2002).

Pans are widespread across the STAS, including between Koës, Namibia and Upington, South Africa, where the Kalahari Group cover is thin (Fig. 12.4b, c, d). The northeast part of the STAS lies within a region of ~1000 pans clustered along the watershed between the Nossob and Okwa River catchments (Goudie and Thomas 1985; Lancaster, 1986), with a high concentration of pans from Ncojane (in the far northeast of the STAS) to the edge of the Kalagadi Transfrontier Park (Lancaster 1978). Calcretes and calcareous sandstones of the lower portions of the Kalahari Group are exposed in some of these pans, such as Ukwi Pan, and 20 km northwest there is a rare example of a shoreline deposit including spheroidal stromatolites at Uwri Pan (Lancaster 1978). Further west within the Nossob catchment a high density of pans, some 2 to 5 km in diameter, are found near Aminuis (Lancaster 1986a). Pans are also found at a lower density within many of the interdunes corridors of the southern Kalahari dunefield. The presence of pans relates to an absence of an integrated fluvial drainage system, coupled with low surface gradient and suitable surface materials for pan formation (deflatable Kalahari Group sediments, soluble calcretes and sil-calcretes of the Kalahari Group and soluble salt-rich shales of the Ecca Group). This erosion gets accelerated by salt-weathering and perhaps also via excavation by biogeomorphological agents such as herbivores (Goudie and Thomas 1985; Lancaster 1986b; Goudie and Wells 1995). Further details about pans across Botswana can be found in Chap. 10 of this edition by Schüller. There is a strong interaction between the pans and groundwater within the uppermost Kalahari aquifer unit of the STAS. After rains, the pans may act as temporary recharge playas (sensu Rosen 1994) whilst when the groundwater tables are higher after longer rains, or in a localised manner associated with perched water tables, the moisture within the capillary fringe acts as a barrier to wind scour, known as a Stoke surface (Stokes 1968; Tyler et al. 2006).

The largest geomorphological component of the STAS is the sand-rich units of the Kalahari Group, which constitute the southwestern portion of the Kalahari Erg (eastern side of Fig. 12.4b). The geomorphology is dominated by linear dune forms, which give way to sand-sheet cover at the eastern margin of the STAS. This area of dunes has been referred to as the southwest Kalahari (cf Bullard et al. 1995; Stone and Thomas 2008) or as the western and south Kalahari dunefields as two separate regions (cf Thomas and Burrough 2016). In the hills of the northwest STAS (near Rehoboth) there are some topographically constrained (climbing and falling) dunes. The eastern sand sheet cover includes some intriguing patchy patterns km to 10's km wide, observable on GoogleEarth, yet to be described in the literature, which bears some resemblance to degraded megabarchans or barchanoid ridges (Fig. 12.5a). In contrast, there are detailed descriptions of the linear dunes (Fig. 12.5b) and their regional trends (Lewis 1936; Goudie 1969, 1970; Grove 1969; Breed et al. 1979; Thomas 1988; Lancaster 1986a, 1988; Bullard et al. 1995). The pioneering, large-scale mapping of Bullard et al. (1995) confirmed that dune patterning complexity increases from the northwest to southeast, with another axis of increased complexity towards the northeast, and a pattern of wider-spacing toward the transitional margin of the dunefield (Bullard and Livingstone 1995). The map of five simplified linear dune planimetric classes was refined by Bullard and Nash (1998). The dune heights and spacing of the southwester Kalahari have been studied using ASTER GDEM (global digital elevation model) data, and interestingly these data reveal that the southwest Kalahari does not follow the morphometric pattern of an increase in dune spacing with increase dune height (White et al. 2015), as is seen in many dunefields, including the largely vegetation-free dunes of the Namib Sand Sea (Lancaster 1989; Ewing et al. 2006; Bullard et al. 2011). It is possible that this relates to their present-day vegetated nature. Dunes above the STAS are mostly well-vegetated, with modern-day activity limited to dune crests and at a maximum when vegetated is reduced after clusters of drier years or disturbance by fire (Wiggs et al. 2005). In the southwest of the STAS, between the Auob and Nossob Rivers, where dune patterning is more complex and reticulate, there are also a small number of parabolic dune patches (Erikson et al. 1989). These are associated with possible disturbance to vegetation within the Kalagadi Transfrontier Park. Vegetation-free, mobile dune forms are much rarer. For example, there are some small areas of barchan dunes on river valley floors, although most of those reported by Lewis (1936) and Bullard and Nash (1998) lie beyond the southern and eastern margins of the STAS within the Molopo valley floors. River valley incisions also provide topography and occasional climbing and falling dunes are found (Bullard 1994). Source-proximal dunes types are also found. These occur as valley-marginal dunes, particularly where river valleys have a west–east orientation (Bullard and Nash 2000), and also as lunette dunes fringing the downwind side of pans (e.g. Lancaster 1978; Goudie and Thomas 1986; Eitel and Blümel 1997; Hürkamp et al. 2011). These often have only a low level of vegetation cover.

12.4.3 Evolution of Duricrusts, Valleys and Dunes above the STAS

12.4.3.1 Duricrust Formation

The presence of exposures of duricrusts (calcrete, sil-calcrete and silcrete) within incised valleys in the STAS region has facilitated studies that explore the link between the development of the duricrusts within the Kalahari Group sediments and the evolution of the incised valleys (Shaw and de Vries 1988; Nash et al. 1994a, b). For duricrusts to form there needs to be a source of calcium and/or silica, moisture to transport the solutes and mechanism, or process, to precipitate these solutes (Ollier 1991). There is a continuum of ways that the exposures of duricrust within the STAS valleys may have formed: (i) the duricrust precipitated via a pedogenic process in which solutes were vertically transferred before precipitation, and this occurred on a surface that was later incised by valley downcutting (Fig. 12.3a); (ii) the duricrust precipitated due to groundwater-related processes beneath, or at, the water table, and this occurred within sediment that was later incised by valley downcutting (Fig. 12.3b), (iii) the valley incision had begun and actually led to the precipitation of the duricrust due to the accumulation of minerals within the groundwater discharge zones at the valley flanks, after which further valley incision then separated the duricrust from the valley floor (Fig. 12.3c); (iv) the duricrust precipitated under pan conditions driven by changes in water pH as it entered the alkaline pans, and the pans were located in a proto-valley, which has further incised (Fig. 12.3d) (Nash et al. 1994b). Along large lengths of the Auob (from Stampriet down to the confluence with the Nossob) the incised valley walls have a well-defined and comparable stratigraphy, in which softer white calcrete is overlain by iron-stained calcareous duricrusts, and this is similar in the Olifants tributary (Nash et al. 1994a, b). This consistency suggests that the Auob and tributaries have incised a pre-existing duricrust surface, and the processes of precipitation of the duricrust may have included both pedogenic processes (position (i) on the continuum) and shallow water-table zone processes (position (ii) on the continuum) (Nash et al. 1994b).

12.4.3.2 Dunefield Formation and Valley Incision

The association between river valleys and dunes (Fig. 12.5c) also allows us to consider the antiquity of the aeolian-formed dunefield compared to the initiation of valley incision. Miller (2014) suggests that the relatively rare occurrence of dunes extending into the valleys indicates that valley incision post-dates the dunes. In terms of dune initiation, Miller (2014) infers that this occurred during the 3.2 Ma expansion of the Greenland ice-sheet (Moran et al. 2006), as this may have increased aridity over southern Africa. He supports this inference with evidence from northern Namibia where dunes are superimposed over megafan deposits, derived from Angolan highlands, thought to be ~4 Ma old (Miller 2008; Miller et al. 2010). Dates for incisions of the STAS river valleys are not well constrained. However, the incision must pre-date the aggradation of fluvial gravel terraces within the wider channels, and these gravel terraces are associated with Acheulean stone age tools (e.g. Korn and Martin 1955), which elsewhere on the African continent are constrained up to ~1.76 Ma in age (Lepre et al. 2011).

Evidence from luminescence dating of full-depths of dunes within the STAS demonstrates phases of dune accumulation during the Late Quaternary (over the last 200 ka), much later than a 4 Ma to 1.76 Ma age for fluvial incision. However, these are not at odds with Miller's (2014) ideas, as the luminescence dune accumulation chronology may represent a reworking of, or additions to, an earlier dunefield (Stokes et al. 1997; Blümel et al. 1998; Bateman et al. 2003; Telfer and Thomas 2007a, b; Stone and Thomas 2008). Above the STAS, there are basal dune ages of 49.0 ± 4.5 ka (just west of the Witvley River), ~122 ka (with the luminescence signal approaching saturation to the east of the Witvley River) and ~186 ka (with the luminescence signal approaching saturation, 20 km northeast of Stampriet) (Stone and Thomas 2008). Luminescence dating of dune sediments provides us with an estimate of the time elapsed since the sediment was last exposed to sunlight (and subsequently buried by more sediment). Therefore, luminescence dune accumulation chronologies provide an age estimate for the last phase of dune accumulation and stabilisation of that vertical depth of sediment above bedrock (this phase may be characterised by slow, relatively continuous accumulation or episodic accumulation). The resetting of luminescence signals with exposure to sunlight means that the timing of any earlier dune accumulation phases that have been entirely turned over (a surface sedimentary cycle) cannot be estimated by luminescence dating methods. In this way, the luminescence ages cannot be used to test, or help demonstrate, Miller's (2014) idea that there was a dunefield from ~4 to 1.7 Ma prior to fluvial incision.

However, cosmogenic nuclide (CN) dating approaches offer potential here because the signal is not reset on exposure to sunlight. Instead, the concentration of CN such as ^{26}Al and ^{10}Be reflects a series of exposure events (when the signal is produced and accumulates at the surface and with a decreasing rate with burial depth) and burial events (when the sediment is sheltered from cosmic-ray-induced CN production and the CN concentrations instead start to decay with known half-lives). CN are utilised by Vainer et al. (2018b) to establish estimates for the longer burial histories of sand in the southern Kalahari dunefield, using samples at Mamatwan Mine (a site outside the margin of the STAS)

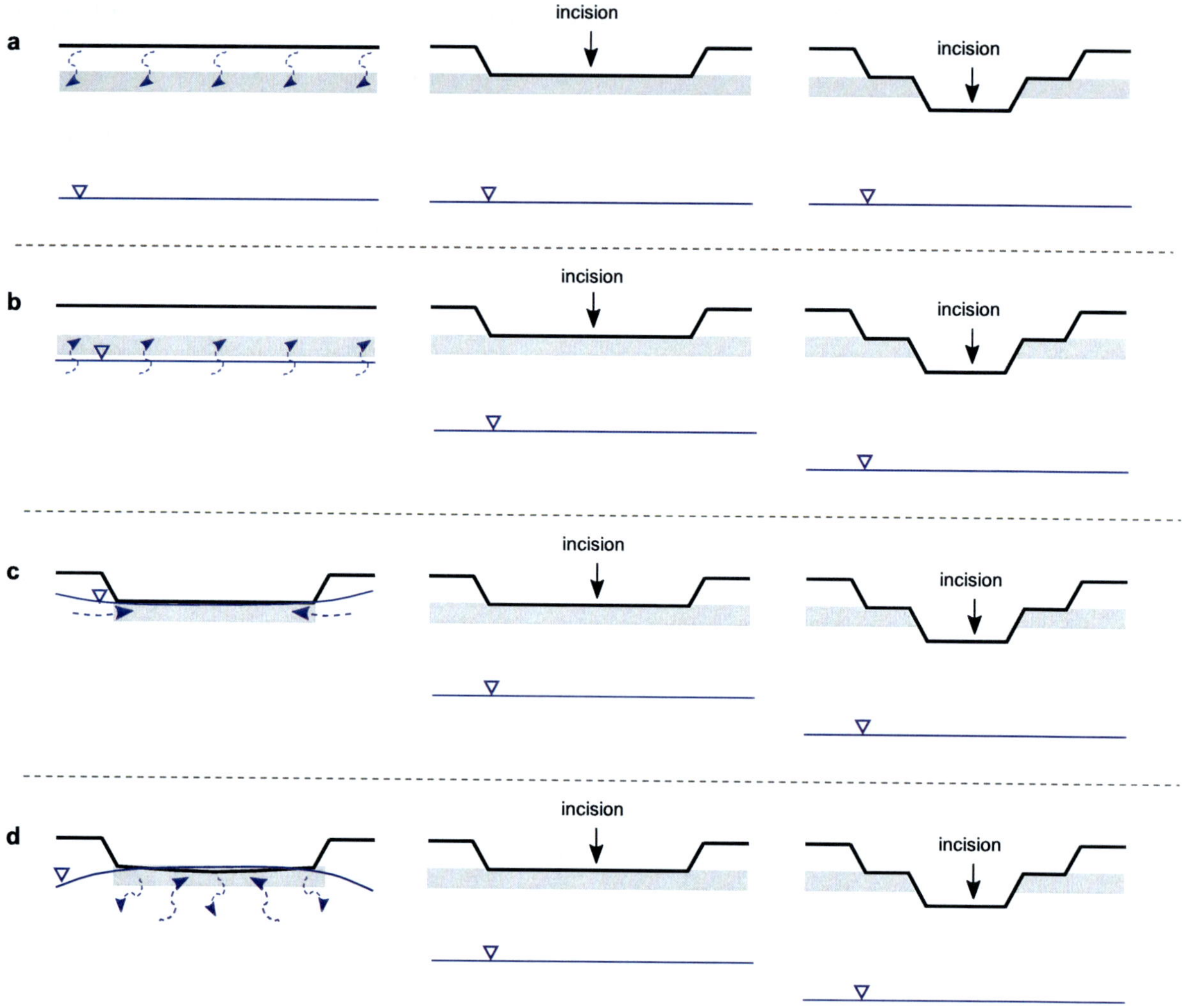

Fig. 12.3 Simple line diagram to illustrate the continuum of ways that the duricrusts may have been forming, relating to the incision of the rivers, so that there are now visible on the river terraces

(Fig. 12.1). However, this method is not straightforward because the production of the CN does not occur during one set interval, and this means that the measured nuclide concentration when sampled is not the result of decay of CNs from a single event. Rather, as the sediment is re-exposed, re-buried and re-exposed during surface sedimentary cycles the sample experiences a number of phases of CN production and decay, leading to a signal composed from superimposed multiple events. Vainer et al. (2018b) apply a novel and ingenious approach to estimating the varying CN production rates through time. They utilise the luminescence database for dunes in the Kalahari and globally (Lancaster et al. 2016) to estimate the 'pace' of vertical displacements of sediment during any sediment reworking cycles from the dune accumulation rate for each 20 cm of sediment. This 'pace' is used to model the range of depths the sediment spends time at during sediment reworking cycles and therefore the production and attenuation rates of the CN. Simulations of a range of possible scenarios of exposure and burial are run in the model (2,880,000 simulations derived from 3 samples, 96 scenarios and 10,000 model iterations) to find those that are capable of producing the measured CN concentrations. These simulations suggest that on average sediment in the southern Kalahari dunefield would have experienced 22 cycles of exposure and burial in order to produce the measured CN concentrations. This represents 22 overturnings (or full-depth luminescence-based dune accumulation profiles). This pioneering study of Vainer et al. (2018b) provides support for Miller's (2014) idea (indeed 22×186 ka ≈ 4 Ma).

12.4.3.3 Syntheses of the Record of Dunefield Accumulation

In the digital dune database (INQUA Project 0704: Digital Atlas of Quaternary Dune Fields and Sand Seas) (Lancaster et al. 2016), the dune field above the STAS is referred to as the western Kalahari, whilst the adjacent area to the east and south is known as the southern Kalahari (Thomas and Burrough 2016). It makes sense to consider the palaeoenvironmental record they contain together, and when this is done, it shows that the long-held view of a peak of aridity during the Last Glacial Maximum (LGM) has not stood up to the test of a larger geochronological dataset (Stone and Thomas 2008; Chase 2009; Thomas and Burrough 2012, 2016). If the past 50,000 years are considered as a series of 1,000-year time windows, there is dune accumulation recorded in the combined western and southern Kalahari during 32 of the 50 time windows, including the LGM (Thomas and Burrough 2012). If age-frequency plots are produced from the data, there are apparent 'peaks' in accumulation ages. However, this mode of data presentation is subject to bias, particularly given the larger number of samples collected from the first 4 m below the surface which drive a 'young' sample bias (Stone 2012), and the demonstration that changing initial sampling depth (below the surface) and vertical sampling interval produces a different temporal record (Stone and Thomas 2008). As demonstrated by a number of authors and reiterated in the INQUA Digital Atlas data overview (Thomas and Burrough 2016), as the number of dated dune samples has increased there has been a transition in the apparent record of system behaviour from one that has discrete phases of dune accumulation (cf. Stokes et al. 1997) to one where there is a near-continous record of dune accumulation through time (c.f. Bateman et al. 2003; Stone and Thomas 2008).

Records of aeolian accumulation are also preserved within lunette dunes. A fluvially incised, 10 m thick, lunette on the southeast of Koppieskraalpan accumulated between 36.0 ± 2.0 ka and 28.8 ± 1.9 ka (Hürkamp et al. 2011). Other dated lunette dunes represent some of the most recent periods of sediment accumulation in the dunefield, with up to 6 m of lunette deposition over the past ~600 years at Wiptan (Telfer and Thomas 2006). There are also young packages of sand (<400 years) at the top of a linear dune (down to a depth of 4 m) which is extending from northwest to southeast onto Bettenstadtpan in the Northern Cape, South Africa (Telfer 2011). Dunes extending from northwest to southeast across pans on the Weissrand Plateau are also reported by Eitel and Blümel (1997). These dunes are an indication that the groundwater table has not risen sufficiently to fill these pans with more permanent water, which could erode these aeolian sediments from the pan surface. Basal linear dune ages at Bettenstadtpan are 7.5 ± 0.6 ka furthest onto the pan and 17.6 ± 1.3 ka on a pan 600 m to the northwest (Telfer 2011).

12.4.3.4 Geomorphic Effects of the Valleys

The patterning of dunes around the river valleys (Fig. 12.5c) also provides an interesting opportunity to explore the geomorphic effects of the valleys upon wind regime and sediment supply in the more recent past (Bullard and Nash 1998). Aerial photographs described by Bullard and Nash (1998) and Miller (2014) show regions where linear dunes have been deflected from northwest-southeast to west–east along the north bank of the river, and also transverse dune ridges on the south side (Fig. 12.5c), some distance from the south bank. To the south of river valley-marginal dunes are found, which have a lighter, pale-grey colour than the red linear dunes (Thomas and Martin 1987), and whilst there is no chronological control for these features, they show that aeolian dune activity has occurred (continued) since incision of the river valleys. In some locations, where river-marginal dunes are parallel to the river channels and transverse to the dominant wind direction, there are linear dunes that extend at right angles from (and out of) these transverse dune forms (Bullard and Nash 1998).

12.4.3.5 Fluvial Sedimentary Records within the Valleys

There have also been studies of sedimentary stratigraphy and composition in the Molopo River (e.g. Ramisch et al. 2017). Whilst the sampling locations are beyond the southern border of the STAS, found in the Molopo Canyon close to the confluence with the Orange River, it is reasonable to assume that the valley fill records at that site are representative of other points of the Molopo catchment that falls within the STAS, given the location is <150 km from the boundary of the STAS, and falls within the same climatic setting. Luminescence dating of the preserved 1 to 2 m thick sedimentary profiles of valley fills (and also samples from alluvial fans and slope sediments) provides a chronology and these sequences record: (i) a stage of valley floor aggradation from 9 to 6 ka during flood events, followed by erosion and incision into this unit, (ii) a stage of fan aggradation from 6 to 1.5 ka generated by that erosion and deposited by further flood events of lower intensity (Ramisch et al. 2017). The composition of these sediments is dominated by mineral assemblages found in tributary rivers, reflecting Namaqua-Belt geology and the quartz-rich aeolian Kalahari sands are in low concentration, or absent, leading Ramisch et al. (2017) to conclude that the southern Kalahari drainage (including that in the STS) remained endorheic during the Holocene and disconnected from the Orange River.

12.5 Links between Hydrogeology and Surface Geomorphology

12.5.1 The Geomorphological Roles of Groundwater

Groundwater flowing in the Kalahari aquifer unit of the STAS may have been instrumental within the formation of a number of the surface geomorphological components observed, including the duricrusts, pans and dry river valleys. Higher groundwater tables (either regionally, or within local small-scale systems (e.g. Fig. 12.6a)) may have contributed to the precipitation of calcrete-rich duricrusts within the Kalahari Group sediments during the Palaeogene and Neogene Periods (Mann and Horwitz 1979; Eitel 1993, 1995; Nash et al. 1994a, b). This includes those exposed on the Weissrand Plateau and, as explored in Sect. 12.4.3.1, groundwater processes as well as pedogenic process are thought to have contributed to those duricrusts located at the upper banks of the Auob and Nossob Rivers prior to river incision (Nash et al. 1994a, b).

The pans within the STAS region (e.g. Fig. 12.4c, d) are an excellent example of a surface geomorphological feature where both groundwater (Rosen 1994; Yechieli and Wood 2002) and surface water play a role along with interactions with other surface processes, such as deflation and deposition (e.g. Fryberger et al. 1988; Reynolds et al. 2007), and aeolian influences on water and salt loss (e.g. Torgersen 1984). Studies of pans around the world demonstrate that groundwater is particularly important for the accumulation of evaporites (Rosen 1994) and influences the composition of brines (e.g. Hardie et al. 1978; Jankowski and Jacobson 1989; Herczeg and Lyons 1991; Rosen 1994). The relative contribution of groundwater and surface water can be considered along a continuum, with a well-known hydrological classification model proposed by Bowler (1986) for Australian lake basins, which depicts typical morphologies and sediment chemical compositions associated with five basin types from those with permanent surface water cover (playa lakes) through to entirely groundwater controlled. Rosen (1994) uses the depth of the groundwater table to classify the hydrology of pans and playas and also to highlight that the formation of evaporites requires the input of groundwater. Studies of mass budgets in playas have also revealed that the salt budget may have a lagged response to hydrological conditions of up to thousands of years (e.g. Yechieli and Wood 2002), showing that past groundwater circulation patterns and chemical composition can impact the chemical composition of pan surfaces in the present.

Lancaster (1986b) reported direct field observations of groundwater discharge in pans in the Aminius area (Fig. 12.1) in the 1980s. There are also a handful of studies with geochemical analysis of the pan floor sediment in this part of the Kalahari (e.g. Lancaster 1978; Mees 2002; Schüller et al. 2018). Within a study that focusses on the spatial distribution and morphology of pans Lancaster (1978) described the pan sediments composition as calcium-carbonate rich silty and sandy clay down to around 1 m depth, and this carbonate-rich composition is indicative of pan systems with a mix of groundwater and surface water control, but toward the surface water end of the continuum set out by Bowler (1986). In contrast, the presence of higher levels of gypsum, halite and bitterns would indicate a greater influence of groundwater in the formation of the pan floor sediment chemistry (Bowler 1986). At Omongwa Pan in the north of the STAS, there are gypsiferous minerals, and Mees (1999) related this to the presence of a shallow groundwater table and interpreted the gypsum distribution patterns through profiles as indicating a lowering of the groundwater level during a period of increasing groundwater salinity. Further pan floor sediment geochemical analysis from five pans across the STAS, or nearby (Omongwa, Toasis, Koës (Fig. 12.4c), Branddam East and Witpan), by Schüller et al. (2018) also indicate the influence of groundwater, with gypsum, halite, sepiolite and dolomite recorded at all five sites. In addition, at four of these sites, coarse gypsum crystals were observed in sampled groundwater at the water table (Schüller et al. 2018). The level of the groundwater table within pans also controls whether and when deflation has occurred from these surfaces through time, with the depth to which deflation occurs being controlled by the depth of the capillary fringe (e.g. Stokes 1968). In addition, a wet sediment surface is effective at trapping deposited aeolian material. Pan surfaces also act as a sink for aeolian input, and Schüller et al. (2018) use the Ti/Al ratio within sediment profiles (2 m and 3 m thick profiles), from the Omongwa and Branddam East Pans, respectively, as a proxy for trapped aeolian input over the past 40 ka (Fig. 12.5).

There is further evidence that the pan systems of the STAS region have been influenced by a range of geomorphic controls and sedimentary processes. For example, detailed petrographic analyses of pan margin deposits in the Gobabis region just outside the northern margins of the STAS by Mees (2002) revealed a possible lacustrine origin for calcareous material along these margins in addition to a pedogenic origin from groundwater and surface water. This may be evidence of shifts through time across the continuum of playa (dry for >75% of the years) and playa lake (neither dry >75% or wet >75% of the time) (Briere, 2000), driven by changes in the overall moisture balance of this region and the level of the groundwater table. Figure 12.6a shows the short-term (weeks to months) surface water ponding that occurs in this region over modern-day timeframes, in this case after the heavy rains of the 2010–2011 rainy season

Fig. 12.4 Some of the key geomorphological features of the STAS **a** Photo taken looking southeast at 24° 36′22.93″S, 18°37′9.13″ E showing the wide, flat valley bottoms and steep sides with a duricrust cap in the Auob River, **b** patterns of aligned drainage on the Weissrand Plateau, also showing the boundary with the cover of the western Kalahari linear dunefield and Köess Pan in the bottom right of the image, **c** the south east corner of Köess Pan, **d** a small (180 m diameter) interdune pan on farm Terra Rouge located at 25°25′58.18″S. 19°43′22.3″E

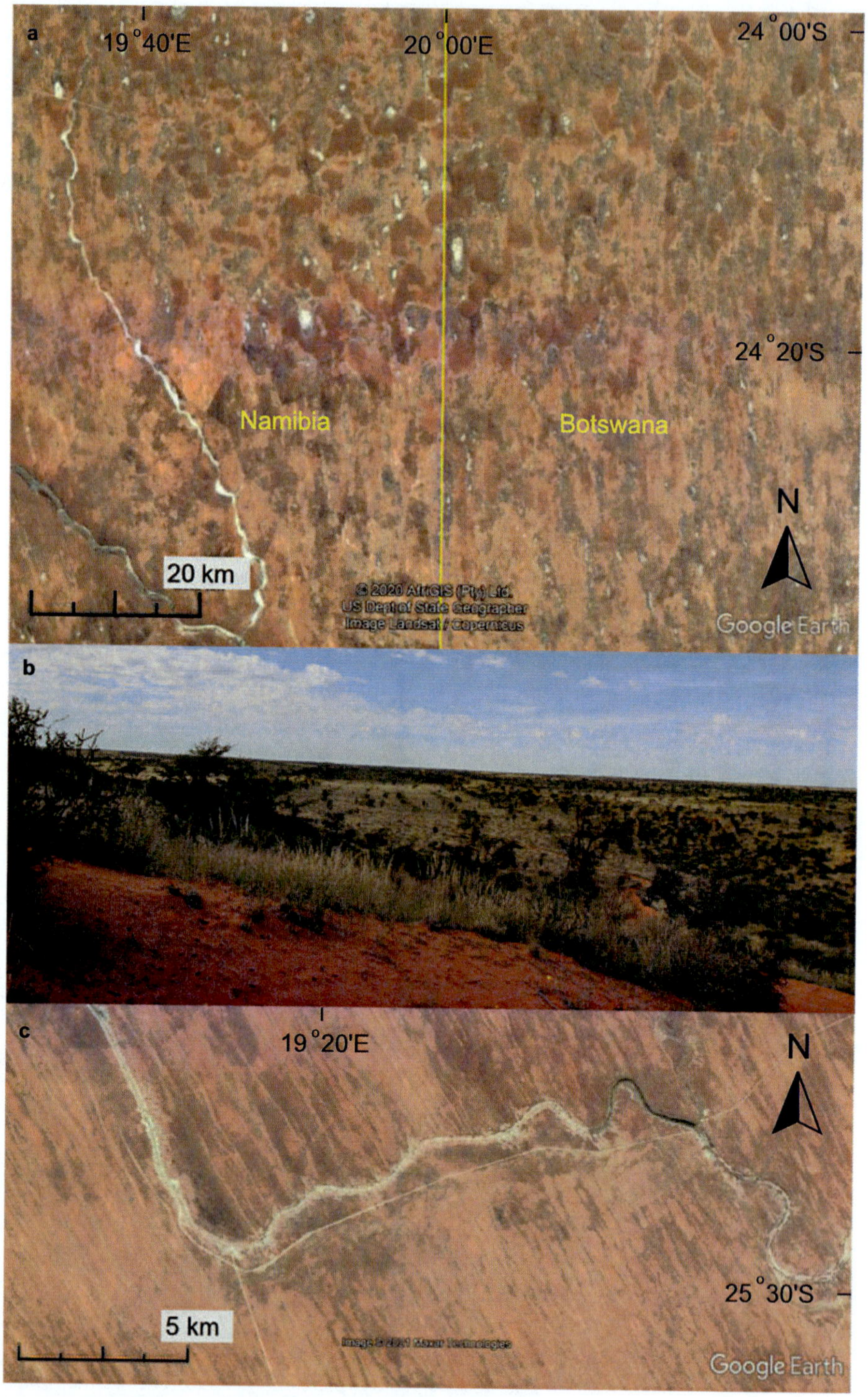

Fig. 12.5 More key geomorphological features of the STAS: **a** patchy patterns observed in sand sheet at the eastern margin of the STAS, **b** looking south down an interdunes between two vegetated linear dunes south of the Auob River at 23°31′59.88"S, 19°40′23.15", 25 km southeast of Tweerivier **c** the association between dunes and the dry valley of the Auob River, showing the disruption to dune patterning where the river course changes abruptly from a NW-to-SE to a SW-to-NE orientation

Fig. 12.6 Some key expressions of groundwater at the surface in the STAS **a** surface ponding of water within an interdune at 24°17′ 47.78‶S, 18°35′30.52′E in April 2011 (photo taken on the D1043 road, ~7 km southeast of the irrigated fields shown in C, **b** seepage of rainwater into sand dune surface, taken on 4 April 2016 after ~9 mm of rain. **c** small-scale irrigated agriculture for vegetables and fruits 16 km northeast of Stampriet on the C20 road

(455 mm during 2010 and 282 mm during 2011 averaged across 3 rain gauges on the farm Terra Rouge (Möller, pers comm).

Groundwater sapping and in situ deep-weathering may have played a role alongside surface water in the development of the flat-bottomed, wide valley systems of the Auob and Nossob and their tributaries (Nash et al. 1994a, b). Sapping by groundwater occurs when concentrated fluid flow in a seepage path enhances weathering and erosion and as a result weakens the basal support of a valley head or side wall (Laity and Malin 1985). The weathering may be both chemical, such as salt weathering, and mechanical, such as freeze–thaw. The process of sapping requires a setting where there is a near-vertical (or fairly steeply-sloped) surface from which groundwater can emerge, and the near-horizontal groundwater flow predicted by the Dupuit-Forchhemier assumption is further aided by the presence of lithological inhomogeneities with a stratigraphy, particularly if the lower one is an aquiclude or aquitard (Howard et al. 1988). In the case of the river valleys in the STAS, the contact between the calcrete and other lithological units in the Kalahari Group sediments provides such a point for groundwater emergence. Nash (1992) suggested that at the whole valley scale in the Kalahari basin there is a role for groundwater processes in valley formation, whilst the intra-valley landforms and sediments have a fluvial origin. The clearest evidence for the role of groundwater comes from an alignment of valley orientation along geological structures (e.g. the Okwa/Hanehai system aligns closely with the regional strike of the beds within the Ghanzi Formation) and borehole observations of weathering and replacement of bedrock at depth (e.g. parts of borehole profiles were granitoid-gneiss has been replaced by silcrete precipitation) (Nash 1992). In the Kalahari basin, this is most apparent for dry valley networks to the north and east of the STAS, including Deception, Okwa, Mmone/Quaxo and Serome valleys (Nash 1992). However, groundwater may have also played a role in the formation of rivers in the STAS Auob and Nossob catchments. For example, the silification of calcretes into sil-calcretes is a form of diagenetic alteration that implies exchanges of water between the surface and groundwater when water tables are less than 10 m below the surface (Arakel et al. 1989; Nash 1992).

The groundwater of the STAS (particularly that in the uppermost Kalahari aquifer unit) is important for biological life, in addition to that pumped and used for irrigation for agriculture (Fig. 12.6c). When the groundwater table is within ~10 m of the surface, this is particularly useful for roots of trees including Acacia species. Through supporting vegetation, groundwater facilitates the biogeomorphic role of trees, shrubs and grasses in this landscape. These roles include: (i) sediment trapping in roots and above-ground biomass; (ii) a bioprotective role, where vegetation cover reduces the erodibility of the surface and by increasing the aerodynamic roughness length of the surface also reduces the erosivity of the wind (e.g. Wiggs 2011) and (iii) modification to sediment moisture retention and the direction of moisture and vapour flux within the unsaturated zone of groundwater (above the water table) (e.g. Yang et al. 2018). Other biogeomorphic interactions in the Kalahari sand soils, such as those found above the STAS, include the associations between biological soil crusts, surface nutrients and vegetation (e.g. Dougill and Thomas 2004).

12.5.2 Influence of Surface Geomorphology on Groundwater Recharge

12.5.2.1 Focused Recharge (Sinkholes and River Valleys)

Geological cross sections show that the thick Kalahari Group sediments wedge out on the western edge of the STAS along with some of the aquitard/clude layers within the Ecca formation (Fig. 12.2). This means that calcrete karstic sinkholes on the Weissrand Plateau and above the Kalkrand basalts are a direct link for meteoric water to infiltrate and reach the confined Auob Aquifer through a more efficient hydrogeological connection (Christelis and Struckmeier 2001; Tredoux et al. 2002). The infamous Uhlenhorst cloudburst of 24th-25th February 1960, produced up to 489 mm of rain over 24 h, which as well as widespread surface flooding, led to flow in the nearest artesian borehole (Klein Swarmoddern farm) with a head of 1.4 m within three weeks (Schalk 1961). Tredoux et al. (2002) have investigated this further, finding variations in the piezometric levels of the Auob aquifer over longer cycles (groups of years) that follow periods of higher rainfall. For example, there is a lagged response of the piezometric level within the Auob aquifer at Olifantswater West (~15 km south of Blumfelde), in the Kalkrand Basalt area, to the exceptionally high rainfall during late 1999 to February 2000 and again in April 2001 (Tredoux et al. 2002). Tredoux et al. (2002) consider the major north–south fault in the northwest of the STAS and other faults in the eastern margin of the Kalkrand basalt to be pathways for direct recharge to the Auob aquifer (transmitted from the surface). Borehole data reveals there is no piezometric overlap between the confined Auob aquifer and the overlying unconfined Kalahari aquifer at the Olifantswater West borehole, therefore suggesting the direct hydraulic link to the surface must be some distance away from this site (Tredoux et al. 2002). This site is close to the ephemeral Olifants river, which provides a focussed recharge mechanism in addition to that which can occur through sinkholes.

12.5.2.2 Diffuse Recharge through Sand Dune and Sand Sheet Cover

The groundwater composition analysis of the Kalahari aquifer unit during the JICA (2002) study suggests that ~0.5% of precipitation during average levels of precipitation went on to recharge the Kalahari aquifer (this is something like 0.7 to 1.5 mm/y). The pathway, other than focussed recharge via sinkholes, pans and river valleys is diffuse recharge through the sand dune cover (Fig. 12.6b). de Vries et al. (2000) used chloride mass balance (CMB) tracer approaches and groundwater flow modelling near Letlhakeng in eastern Botswana, east of the STAS, to suggest that, at best, recharge of 5 mm/y was occurring in the fringes of the Kalahari where rainfall is $\geq$ 450 mm/y and there is a strong possibility for an upward flux of 1 mm/y for water table depths up to 20 m in the drier areas in central Botswana, to the east of the STAS. In northern regions of Namibia with 250–350 mm/y rainfall CMB tracer approaches suggest recharge rates of 1–15 mm/y for Omatjete/Omaruru, 1–18 mm/y for Omasati/Oshana and 2–16 mm/y for Okananguati/Kunene regions (Wanke et al. 2014).

Stone and Edmunds (2012, 2015) applied the CMB balance approach to interstitial water within the part of the unsaturated zone represented by dune cover, at a site east of Stampriet and further south ~40 km northwest of the Mata-Mata border with South Africa (at Terra Rouge Farm). For sediment profiles collected in 2011, using a concentration input of chloride in precipitation of 0.77 mg/L (an average for a year of precipitation chemistry measurements for rain gauges at Terra Rouge Farm) and mean precipitation of 181 mm/y, this yielded estimates of 4 and 5 mm/y in two dunes, one just north and one just south of the Auob River (2.2 to 2.8% of modern average precipitation). The profiles near Stampriet had higher values of 20 to 22 mm/y where rainfall is only slightly higher at 190 mm/y (so that recharge represents 10.5 to 14.2% of modern average precipitation). 11 further profiles (5 sampled in 2013 and 6 in 2016) from dunes near Terra Rouge Farm yielded a range of estimates from 3 to 16 mm/y (unpublished) giving an average this southern part of the STAS of 7 $\pm$ 4 mm/y (at 1 standard deviation), which is 4% of modern average precipitation. However, these figures should be treated with caution, given the lack of reproducible hydrostratigraphies (consistent variations in chloride concentrations with depth) in the 10–12 m profiles sampled between 2011 and 2016 (Stone and Smith 2018). The author is awaiting outputs from modelling of liquid and vapour flux in these sediments under current conditions, and these insights will help determine the likelihood of quantifiable recharge via diffuse infiltration. A further challenge is a lack of clear understanding of whether there is an extensive, less-permeable duricrust layer between the sand dunes and deeper parts of the Kalahari aquifer (and regional water table), that prevents an effective hydraulic connection. A numerical modelling approach (using MODFLOW-NWT) for the Central Kalahari Basin (just north and east of the STAS, between ~21 and 24 °S) characterised variations in water flux dynamics between 2002 and 2014 and showed that evaporation is effective at restricting gross recharge to just a few mm/y. For example, (+) 3 mm/y with 644 mm/y rainfall (0.5% of annual rainfall) and (-) 3.5 mm/y with 287 mm/y rainfall (upward flux of order of 1.2% compared to rainfall) (Lekula and Lubczynski 2019).

12.6 Conclusions

The STAS is an important transboundary water resource, which is the vital, primary source of water for ~ 50,000 residents and tourists who visit the region. The geological units present in the STAS influence the physical and chemical behaviour of the groundwater system. The varying porosity, permeability and hydraulic conductivities of different rock and soft sediment types controls the recharge rates and also groundwater storage, flow rates and directions within the aquifer units. Rock type and solubility, in addition to stratigraphy and geological structural, impacts groundwater flow rates and influences the natural baseline groundwater quality in the STAS, with a decline in quality towards the southeast of the basin. Where groundwater reaches the surface, it influences the geochemistry of many of the pan floor sediments and through increasing the moisture content of sediments reduces their susceptibility to aeolian erosion, with the water table controlling the depth of such erosion.

There are strong connections between the surface geomorphological features and the subsurface hydrogeology of the STAS, and these operate in both directions. Tectonic uplift and down warping in the Late Cretaceous, and to a lesser extent again in the Oligocene and Miocene, drove deposition of the Kalahari Group Sediments within the Kalahari basin. Groundwater processes, alongside pedogenic processes, played a role in the development of calcrete and silcrete duricrusts, and the formation of pans, both of which are widespread in the STAS. Groundwater may have also been important in the formation of the flat-bottomed and wide Auob and Nossob valley systems and their tributaries via sapping and deep weathering processes.

The relationship between major surface geomorphological features, including the duricrusts, dry river valleys and dunes suggests it is likely that the incision of the dry river valleys post-date the formation of the duricrusts and possibly also a generation of aeolian linear dunes on the calcrete surface. Luminescence dating chronologies of dune accumulation indicate significant dune activity and accumulation over the past ~180 ka, which may represent a later phase of reworking and accumulation of the unconsolidated sands at the surface of the Kalahari Group Sediments. The dune sediments are the dominant surface geomorphological cover of the STAS and small amounts of diffuse recharge to the uppermost Kalahari aquifer occur through these sands. In contrast, recharge to the Auob aquifer unit appears to be dominated by focused recharge through sinkholes in the calcrete-rich duricrusts, which are exposed on the northern and western margins of the STAS.

References

Adams S (2006) Basement aquifers of southern Africa: Overview and Research Needs. In Titus, R., BeekmanF H, Adams S, Stachan L (eds) The Basement Aquifers of Southern Africa Report to the Water Research Commission Water Research Commission, p 1–4. ISBN 978–1–77005–898–9

Arakel AN, Jacobson G, Salehi M, Hill CM (1989) Silicification of calcrete in basins of the Australia arid zone. Aust J Earth Sci 36:73–89. https://doi.org/10.1080/14400958908527952

Baillieul TA (1975) A reconnaissance survey of the cover sands in the republic of Botswana. J Sediment Petrol 45(2):494–503. https://doi.org/10.1306/212F6D9D-2B24-11D7-8648000102C1865D

Bamford MK (2000) Cenozoic Macro-Plants. In: Partridge TC, Maud RM (eds) Cenozoic of Southern Africa, Oxford Monographs on Geology and Geophysics, vol 40. Oxford University Press, New York, pp 351–356

Bateman MD, Thomas DSG, Singhvi AK (2003) Extending the aridity record of the southwest Kalahari: current problems and future perspectives. Quatern Int 111:37–49. https://doi.org/10.1016/S1040-6182(03)00013-2

Blümel WD, Eitel B, Lang A (1998) Dunes in southeastern Namibia: evidence for Holocene environmental changes in southwestern Kalahari based on thermoluminescence data. Palaeogeogr Palaeocl 138:139–149. https://doi.org/10.1016/S0031-0182(97)00114-4

Bowler JM (1986) Spatial variability and hydrological evolution of Australian lake basin: analogue for Pleistocene hydrological change and evaporite formation. Palaeogeogr Palaeocl 54:21–41. https://doi.org/10.1016/0031-0182(86)90116-1

Breed CS, Fryberger SG, Andrews S, McCauley C, Lennartz F, Gebel D, Hortsman K (1979) Regional studies of sand seas observed by remote sensing. In McKee ED (ed.) A Study of Global Sand Seas. USGS Professional Paper 1052. p 305–397.

Brier PR (2000) Playa, playa lake, sabkha: proposed definitions for old terms. J Arid Environ 45:1–7

Bullard JE (1994). An analysis of the morphological variation of linear sand dunes and of their relationship with environmental parameters in the south-west Kalahari. Unpublished Ph.D. thesis, University of Sheffield. 327 pp.

Bullard JE, Nash DJ (1998) Linear dune pattern variability in the vicinity of dry valleys in the south-west Kalahari. Geomorphology 23:35–54. https://doi.org/10.1016/S0169-555X(97)00090-1

Bullard JE, Nash DJ (2000) Valley-marginal sand dunes in the south-west Kalahari: their nature, classification and possible origins. J Arid Environ 45:369–383. https://doi.org/10.1006/jare.2000.0646

Bullard JE, Thomas DSG, Livingstone I, Wiggs GFS (1995) Analysis of linear sand dune morphological variability, southwest Kalahari Desert. Geomorphology 11:189–203. https://doi.org/10.1016/0169-555X(94)00061-U

Bullard JE, White K, Livingstone I (2011) Morphometric analysis of aeolian bedforms in the Namib Sand Sea using ASTER data. Earth Surf Proc Land 36:1534–1549. https://doi.org/10.1002/esp.2189

Chase B (2009) Evaluating the use of dune sediments as a proxy for palaeo-aridity: A southern African case study. Earth-Sci Rev 93:31–45. https://doi.org/10.1016/j.earscirev.2008.12.004

Chanda R, Totolo O, Moleele N, Setshogo M, Mosweu S (2003) Prospects for subsistence livelihood and environmental sustainability along the Kalahari Transect: The case of Matsheng in Botswana's Kalahari rangelands. J Arid Environ 54:425–445. https://doi.org/10.1006/jare.2002.1100

Christelis G, Stuckmeier W (2001) Groundwater in Namibia: an explanation to the hydrogeological map. ISBN No. 0–86976–571-X

Council for Geosciences - CGS (2015) Harmonized assessment of the South Africa segment of the Stampriet Transboundary Aquifer System. UNESCO-IHP Governance of Groundwater Resources in Transboundary Aquifers project. Internal report.

Central Statistics Office - CSO (2009) Botswana Water Statistics. Department of Printing & Publishing Services, Gaborone, Botswana

Edmunds WM, Shand P (2009) Natural Groundwater Quality. Blackwell, Oxford, p 489

Eitel B (1993) Kalkkrustengenerationen in Namibia: Carbonatherkunft und genetische Beziehungen. Erde 124:85–104

Eitel B (1995) Kalkkrusten in Namibia und ihre paläklimatische Interpretation. Geomethodica 20:101–124

Eitel B, Blümel WD (1997) Pans and dunes in the southwestern Kalahari (Namibia): geomorphology and evidence for Quaternary paleoclimates. Zeitschrift Für Geomorphologie NF, Supplementband 111:73–95

Eriksson PG, Nixon N, Snyman CP, Bothma J, duP, (1989) Ellipsoidal parabolic dune patches in the southern Kalahari desert. J Arid Environ 16:111–124. https://doi.org/10.1016/S0140-1963(18)31019-X

Fryberger SG, Schenk CJ, Krystinik LF (1988) Stoke surfaces and the effects of near-surface groundwater-table on aeolian deposition. Sedimentology 35:21–41. https://doi.org/10.1111/j.1365-3091.1988.tb00903.x

Goudie AS (1969) Statistical laws and dune ridges in southern Africa. Geographic Journal 135: 404–406. https://www.jstor.org/stable/1797329

Goudie AS (1970) Notes on some major dune types in southern Africa. S Afr Geogr J 52:93–101. https://doi.org/10.1080/03736245.1970.10559469

Goudie AS (2007) Desert landforms in Namibia – a Landsat interpretation. In Goudie AS, Kalvoda J (eds) Geomorphological Variations. Nakladateslstvi P3 K, Prague. p 19–36.

Goudie AS, Thomas DSG (1985) Pans in southern Africa with particular reference to South Africa and Zimbabwe. Zeitschrift für Geomorphologie N.F. 29(1): 1–19.

Goudie AS, Thomas DSG (1986) Lunette dunes in Southern Africa. J Arid Environ 10:1–12. https://doi.org/10.1016/S0140-1963(18)31260-6

Goudie AS, Wells GL (1995) The nature, distribution and formation of pans in arid zones. Earth-Sci Rev 38(1):1–69. https://doi.org/10.1016/0012-8252(94)00066-6

Grove AT (1969) Landforms and climatic change in the Kalahari and Ngamiland. Geogr J 135: 191–212. https://www.jstor.org/stable/1796824

Haddon IG, McCarthy TS (2005) The Mesozoic-Cenozoic interior sag basin of Central Africa: The Cretaceous-Cenozoic Kalahari and Okavango basin. J Afr Earth Sci 43:316–333. https://doi.org/10.1016/j.jafrearsci.2005.07.008

Hardie LA, Smoot JP, Eugster HP (1978) Saline lakes their deposits: a sedimentological approach. In Matter A, Tucker W (eds) Modern and Ancient Lake Sediments. International Association of Sedimentologists, Special Publication 2: 7–41.

Heaton THE (1984) Sources of the nitrate in phreatic groundwater in the western Kalahari. J Hydrol 67:249–259. https://doi.org/10.1016/0022-1694(84)90244-0

Heaton THE, Talma AS, Vogel JC (1983) Origin and history of nitrate in confined groundwater in the western Kalahari. J Hydrol 62:243–262. https://doi.org/10.1016/0022-1694(83)90105-1

Herczeg AL, Lyons WB (1991) A chemical model for the evolution of Australian sodium chloride lake brines. Palaeogeogr Palaeoclimatol 84:43–53. https://doi.org/10.1016/0031-0182(91)90034-O

Howard AD, Kochel, RC, Holt, HE (1988) Sapping features of the Colorado Plateau – a Comparative Planetary Geology Fieldguide. NASA Publication SP-491.

Jankowski J, Jacobson G (1989) Hydrochemical evolution of regional groundwater to playa brines in central Australia. J Hydrol 108:123–173. https://doi.org/10.1016/0022-1694(89)90281-3

JICA (Japan International Cooperation Agency) (2002). The Study on the Groundwater Potential Evaluation and Management Plan in the Southeast Kalahari (Stampriet) Artesian Basin in the Republic of Namibia – Final Report.

Kent LE (1980). Part 1: Lithostratigraphy of the Republic of South Africa, South West Africa/Namibia and the Republics of Bophuthatswana, Transkei and Venda. SACS, Council for Geosciences, Stratigraphy of South Africa. 1980. South African Committee for Stratigraphy. Handbook 8, Part 1, pp 690

Korn H, Martin H (1955) The Pleistocene in South West Africa. In Clarke JD, Cole S. (eds) Proceedings of the 3rd Pan-African Congress on Prehistory, Livingstone. London, Chatto and Windus. pp. 14–22.

Laity JE, Malin MC (1985) Sapping processes and the development of theatre-headed valley networks on the Colorado Plateau. Geol Soc Am Bull 96(2):203–217. https://doi.org/10.1130/0016-7606(1985)96%3C203:SPATDO%3E2.0.CO;2

Lancaster N (1978) The Pans of the southern Kalahari, Botswana. Geog J 144: 81–98. DOI: https://doi.org/10.2307/634651 https://www.jstor.org/stable/634651

Lancaster N (1986) Grain-size characteristics of linear sand dunes in the southwest Kalahari. J Sediment Petrol 56:395–400. https://doi.org/10.1306/212F8927-2B24-11D7-8648000102C1865D

Lancaster N (1986) Pans in the southwestern Kalahari: a preliminary report. Palaeoeco A 17:59–67

Lancaster N (1988) Development of linear dunes in the southwestern Kalahari. J Arid Environ 14:233–244

Lancaster N (2000) Eolian deposits. In: Partridge TC, Maud RM (eds) Cenozoic of Southern Africa, Oxford Monographs on Geology and Geophysics, vol 40. Oxford University Press, New York, pp 73–87

Lekula M, Lubczynski MW, Shemang EM (2018) Hydrogeological conceptual model of large and complex sedimentary aquifer systems: central Kalahari Basin (Botswana). Phys Chem Earth A/b/c. 106:47–62. https://doi.org/10.1016/j.pce.2018.05.006

Lekule M, Lubczynski MW (2019) Use of remote sensing and long-term in-situ time-series data in an integrated hydrological model of the Central Kalahari Basin, Southern Africa. Hydrogeol J 27:1541–1562. https://doi.org/10.1007/s10040-019-01954-9

Lepre CJ, Roche H, Kent DV, Harmand S, Quinn RL, Brugal J-P, Texier PJ, Lenoble A, Feibel CS (2011) An earlier origin for the Acheulian. Nature 477:82–85. https://doi.org/10.1038/nature10372

Lewis AD (1936) Sand dunes of the Kalahari within the borders of the Union. S Afr Geog J 19:22–32. https://doi.org/10.1080/03736245.1936.10559174

Mann AW, Horwitz R (1979) Groundwater calcrete deposits in Australia: some observations from Western Australia. J Geol Soc Aust 26:293–303. https://doi.org/10.1080/00167617908729092

Mees F (1999) Distribution patterns of gypsum and kalistrontinite in a dry lake basin of the southwestern Kalahari (Omongwa Pan, Namibia). Earth Surf Proc Land 24:731–744

Mees F (2002) The nature of calcareous deposits along pan margins in Eastern Central Namibia. Earth Surf Proc Land 27:719–735. https://doi.org/10.1002/esp.348

Miller R McG (2014) Evidence for the evolution of the Kalahari dunes from the Auob River, southeastern Namibia. T Roy Soc S AFr 69 (3):195–204. https://doi.org/10.1080/0035919X.2014.955555

Miller R McG (2008) The Geology of Namibia. Ministry of Mines and Energy, Geological Survey of Namibia, Windhoek Namibia.

Miller R McG, Pickford M, Senut B (2010) The geology, palaeontology and evolution of the Etosha Pan, Namibia: implications for terminal Kalahari deposition. S Afr J Geol 113:307–334. https://doi.org/10.2113/gssajg.113.3.307

Moore AE (1999) A reappraisal of epeirogenic flexure axes in southern Africa. S Afr J Geol 102:363–376

Moore AE, Dingle RV (1998) Evidence for fluvial sediment transport of Kalahari sands in central Botswana. S Afr J Geol 101(2):143–153

Moran K, Backman J, Brinkuis H, Clemens SC, Cronin T, Dickens GR, Eynaud F, Gattacceca J, Jakobsson M, Jordan RW, Kaminski M, King J, Koc N, Krylov A, Martinez N, Matthiessen J, McInroy D, Moore TC, Onodera J, O'Regan M, Pälike H, Rea B, Rio D, Sakamota T, Smith DC, Stein R, St John K, Suto I, Suzuki N, Takahashi K, Watanabe M, Yamamoto M, Farrell J, Frank M, Kubik P, Jokat W, Kristoffersn Y (2006) The Cenozoic palaeoenvironment of the Arctic Ocean. Nature 441:601–605. https://doi.org/10.1038/nature04800

Nash DJ (1992) The development and environmental significance of dry valley systems (*mekgacha*) in the Kalahari, central southern Africa. Department of Geography, University of Sheffield, PhD theses

Nash DJ, Shaw PA (1998) Silica and carbonate relationships in silcrete-calcrete intergrade duricrusts from the Kalahari of Botswana and Namibia. J Afr Earth Sc 27(1):11–25. https://doi.org/10.1016/S0899-5362(98)00043-8

Nash DJ, Thomas DSG, Shaw PA (1994) Timescales, environmental change and dryland valley development. In: Millington AC, Pye K (eds) Environmental Change in Drylands. John Wiley, Chichester, pp 25–41

Nash DJ, Shaw PA, Thomas DSG (1994) Duricrust development and valley evolution: process-landform links in the Kalahari. Earth Surf Proc Land 19:299–317. https://doi.org/10.1002/esp.3290190403

NamWater (2014) Aminuis, Aranos, Kriess, Gochas, Leonardville, Onderombapa and Stampriet Production Schemes. Production data for April 2013 to May 2014.

National Statistic Agency - NSA (2015). Namibia Agriculture Sector Census 2013/14 Report. Windhoek, Namibia.

Ramisch A, Bens O, Bulaert J-P, Eden M, Heine K, Hürkamp K, Schwindt D, Vökel J (2017) Fluvial landscape development in the southwestern Kalahari during the Holocene – Chronology and provenance of fluvial deposits in the Molopo Canyon. Geomorphology 281:94–107. https://doi.org/10.1016/j.geomorph.2016.12.021

Reynolds RL, Young JC, Reheis M, Goldstein H, JrP C, Fulton R, Whitney J, Fuller C, Forester RM (2007) Dust emission from wet and dry palays in the Mojave Desert, USA. Earth Surf Proc Land 32:1811–1827. https://doi.org/10.1002/esp.1515

Rosen MR (1994) The importance of groundwater in playas: a review of playa classifications and the sedimentology and hydrology of playas. In Rosen, M. R. (Ed) Palaeoclimate and Basin Evolution of Playa Systems, Geological Society of America Special Paper 289: 1–18. https://doi.org/10.1130/SPE289-p1

Schalk K (1961) The water balance of the Uhlenhorst cloudburst in South West Africa. Inter-African Conf. on Hydrology. CCTA Publication 66:443–449

Schüller I, Blez L, Wilkes H, Wehrmann A (2018) Late Quaternary shift in southern African rainfall zones: sedimentary and geochemical data from Kalahari pans. Zeitschrift Für Geomorphologie 61 (4):339–362. https://doi.org/10.1127/zfg/2018/0556

Shaw PA, de Vries JJ (1988) Duricrust, groundwater and valley development in the Kalahari of south-east Botswana. J Arid Environ 14:245–254

Sindelar JJ, Milowski AL (2012) Human safety controversies surrounding nitrate and nitrite in the diet. Nitric Oxide 26(4):259–266

Smith RA (1984) The lithostratigraphy of the Karoo Supergroup in Botswana. Bulletin 26 of the Geological Survey Department, Botswana, pp 239.

Stadler S, Osenbrück K, Knöller K, Suckow A, Sültenfuß J, Oster H, Himmelsbach T, Hötzl H (2008) Understanding the origin and fate of nitrate in groundwater of semi-arid environments. J Arid Environ 72:1830–1842. https://doi.org/10.1016/j.jaridenv.2008.06.003

Stadler S, Talma AS, Tredoux G, Wrabel J (2012) Identification of sources and infiltration regimes of nitrate in the semi-arid Kalahari: regional differences and implications for groundwater management. Water SA 38(2):213–224. https://doi.org/10.4314/wsa.v38i2.6

Stokes WL (1968) Multiple parallel truncation bedding planes – a feature of wind-deposited sandstone formations. J Sediment Petrol 38:510–515. https://doi.org/10.1306/74D719D3-2B21-11D7-8648000102C1865D

Stokes S, Thomas DSG, Washington R (1997) Multiple episodes of aridity in southern Africa since the last interglacial period. Nature 388:154–158. https://doi.org/10.1038/40596

Stone A (2012) The peaks and troughs of dune records: What to the frequency based distributions of luminescence ages from dunefields actually mean? Quatern Int 279–290:470

Stone A, Edmunds WM (2012) Sand, salt and water in the Stampriet Basin, Namibia: Calculating unsaturated zone (Kalahari dunefield) recharge using the chloride mass balance approach. Water SA 38 (2):367–378. https://doi.org/10.4314/wsa.v38i3.2

Stone A, Edmunds WM (2014) Naturally-high nitrate in unsaturated zone sand dunes above the Stampriet Basin, Namibia. J Arid Environ 105:41–51. https://doi.org/10.1016/j.jaridenv.2014.02.015

Stone A, Smith A (2018) Using tracers in the semi-arid vadose zone of the Stampriet Basin, southern Africa: insights into the origin and fate of nitrate, and challenges for hydrostratigraphies. 20th EGU General Assembly, EGU2018, Proceedings from the conference held 4–13 April, 2018 in Vienna, Austria, p.14523

Stone AEC, Thomas DSG (2008) Linear dune accumulation chronologies from the southwestern Kalahari, Namibia: challenges of reconstructing the late Quaternary palaeoenvironments from aeolian landforms. Quaternary Sci Rev 27:1667–1681. https://doi.org/10.1016/j.quascirev.2008.06.008

Stone A, Smedley P, Lanzoni M (2019) Groundwater resources: past, present and future. In Dadson S, Garrick D, Hughes J, Hall J, Hope R, Penning-Rowsell E (eds) Water Science Policy and Management: The Global Challenge. https://doi.org/10.1002/9781119520627.ch3

Telfer MW (2011) Growth by extension, and reworking, of a south-western Kalahari linear dune. Earth Surf Proc Land 36:1125–1135. https://doi.org/10.1002/esp.2140

Telfer MW, Thomas DSG (2006) Complex Holocene lunette dune development, South Africa: Implications for paleoclimate and models of pan development in arid regions. Geology 34(10):853–856. https://doi.org/10.1130/G22791.1

Telfer MW, Thomas DSG (2007) Late Quaternary linear dune accumulation and chronostratigraphy of the southwestern Kalahari: implications for aeolian palaeoclimatic reconstructions and predictions of future dynamics. Quaternary Sci Rev 26:2617–2630. https://doi.org/10.1016/j.quascirev.2007.07.006

Thomas DSG (1988) The nature and depositional setting of arid and semi-arid Kalahari sediments, southern Africa. J Arid Environ 14:17–26. https://doi.org/10.1016/S0140-1963(18)31092-9

Thomas DSG (1997) Sand seas and aeolian bedforms. In: Thomas DSG (ed) Arid zone geomorphology. Belhaven Press, London, pp 373–412

Thomas DSG, Burrough SL (2012) Interpreting geoproxies of late Quaternary climate change in African drylands: Implications for understanding environmental change and early human behaviour. Quatern Int 253:5–17. https://doi.org/10.1016/j.quaint.2010.11.001

Thomas DSG, Burrough SL (2016) Luminescence-based chronologies in southern Africa: Analysis and interpretation of dune database records across the subcontinent. Quatern Int 410(B): 30–45. https://doi.org/10.1016/j.quaint.2013.09.008

Thomas DSG, Martin HE (1987) Grain-size characteristics of linear dunes in the southwestern Kalahari: a discussion. J Sediment Petrol 57:231–242. https://doi.org/10.1306/212F8B9D-2B24-11D7-8648000102C1865D

Torgersen T (1984) Wind effects on water and salt loss in playa lakes. J Hydrol 74:137–149. https://doi.org/10.1016/0022-1694(84)90145-8

Tredoux G, Kirchner J, Miller R McG, Yamasaki Y, Christelis GM, Wierenga A (2002) Redefining the recharge behaviour of the Stampriet Artesian Basin, Namibia. IAH 'Balancing the Groundwater Budget' Conference. Darwin, Australia, May 2002

Tredoux G, Engelbrecht JP, Talma AS (2005) Nitrate in groundwater in arid and semi-arid parts of southern Africa. In: Vogel H, Chilume C (eds), Environmental Geology in Semi-arid Environments. Department of Geological Survey, Lobatse, pp 121–133

Tyler SW, Munoz J, Wood W (2006) The response of playa and sabkha hydraulics and mineralogy to climate forcing. Groundwater 44:329–338. https://doi.org/10.1111/j.1745-6584.2005.00096.x

UNSECO-IHP (2017) Stampriet Transboundary Aquifer System Assessment. Governance of Groundwater Resources in Transboundary Aquifers (GGRETA) - Phase 1, Technical Report. DOI https://doi.org/10.29104/WINS.D.0040.2018

van der Gun J (2012) Groundwater and Global Change: Trends, Opportunities and Challenges. United National World Water Assessment Programme Side Publication Series. Paris: UNESCO.

Vainer S, Erel Y, Matmon A (2018) Provenance and depositional environments of Quaternary sediments in the southern Kalahari Basin. Chem Geol 476:352–369. https://doi.org/10.1016/j.chemgeo.2017.11.031

Vainer S, Dor YB, Matmon A (2018) Coupling cosmogenic nuclides and luminescence dating into a unified accumulation model of aeolian landform age and dynamics: The case study of the Kalahari Erg. Quat Geochronol 48:133–144. https://doi.org/10.1016/j.quageo.2018.08.002

Vrba ES (2000) Major features of Neogene mammalian. In: Partridge TC, Maud RM (eds) Cenozoic of Southern Africa, Oxford Monographs on Geology and Geophysics, vol 40. Oxford University Press, New York, pp 277–304

Wada Y, Heinrich L (2013) Assessment of transboundary aquifers of the world – vulnerability arising from human water use. Environ Res Lett 8:024003. https://doi.org/10.1088/1748-9326/8/2/024003

Walvoord MA, Phillips FM, Stonestrom DA, Evans RD, Hartsough C, Newman BD, Striegl RG (2003) A reservoir of nitrate beneath desert soils. Science 302(7):1021–1024

Wanke H, Nakwafila A, Hamutoko JT, Lohe C, Neumbo F, Petrus I, David A, Beukes H, Masule N, Quinger M (2014) Hand dug wells in Namibia: An underestimated water source of a threat to human health? Phys Chem Earth A/b/c 76–78:104–113. https://doi.org/10.1016/j.pce.2015.01.004

Word Health Organisation – WHO (2014) Guidelines for drinking water quality.

Wiggs GFS (2011) Sediment mobilisation by the wind. In: Thomas DSG (ed) Arid Zone Geomorphology: Process, Form and Change in Drylands. Wiley-Blackwell, Chichester, pp 445–486

Yang T, Ala M, Zhang Y, Wu J, Wang A, Guan D (2018) Characteristics of soil moisture under different vegetation coverage in Horqin Sandy Land, northern China. PLoS ONE 13(6): e0198805. https://doi.org/10.1371/journal.pone.0198805

Yechieli Y, Wood WW (2002) Hydrogeological processes in saline systems: playas, sabkhas and saline lakes. Earth-Sci Rev 58:343–365. https://doi.org/10.1016/S0012-8252(02)00067-3

Abi Stone is a Geographer, geochronologist (luminescence dating and U-series) and senior lecturer at the University of Manchester, with a PhD in Quaternary Environments of southern Africa (University of Oxford) and a MSc in Quaternary Science (Royal Holloway, University of London). She is interested in the connections between changing hydroclimate, groundwater and landscapes in dryland environments, as well as the implications of these changes for the archaeological record. This involves work at the interface between Quaternary science, geomorphology, hydrogeology, geology and archaeology. Abi has been studying the drylands within southern Africa for 15 years, with a particular interest in the Stampriet Aquifer that stretches beneath Botswana, Namibia and South Africa in the Kalahari, as well as the Namib Sand Sea.

Calcretes, Silcretes and Intergrade Duricrusts 13

David J. Nash

Abstract

Calcretes and silcretes are the most widely encountered 'rocks' in the Kalahari sandveld that covers much of Botswana. This chapter presents the first holistic overview of current knowledge about these duricrusts at a national scale. It does so by considering the distribution, classification, macromorphology, geochemistry and mineralogy of each duricrust type in turn, alongside various models used to explain their formation. The chapter then reviews our understanding of a variant of duricrust encountered more in the Botswana Kalahari than anywhere else in the world—the silcrete–calcrete intergrade duricrust. The chapter concludes with a summary of knowledge about the age of duricrusts in Botswana before pointing to potential directions for future research.

Keywords

Calcrete • Silcrete • Geochemical sediments • Kalahari group • Botswana

13.1 Introduction

Travellers visiting Botswana for the first time could be forgiven for thinking that much of the surface geology of the country consisted of sand. While it is true that the Kalahari Sands—the stratigraphically youngest component of the Kalahari Group sediments (Thomas 1981; Malherbe 1984; Haddon and McCarthy 2005)—blanket much of the landscape away from the eastern hardveld, other rock types are also present. For example, Precambrian formations protrude through the sand cover in west and northwest Botswana, while inliers of the Mesozoic Karoo Supergroup occur in south-central regions. However, the most widely encountered 'rocks' in areas of present-day sandveld are various forms of geochemical sediment or *duricrust*. Of these, the most widespread (Fig. 13.1a) are calcium carbonate-cemented *calcretes* (Sect. 13.2) and silica-cemented *silcretes* (Sect. 13.3).

Research has been conducted on calcrete and silcrete in Botswana for over 100 years, commencing with Passarge's (1904) detailed descriptions of the regional geology in his monograph *Die Kalahari*. Examples from Botswana feature in Goudie's (1973) seminal book on *Duricrusts in Tropical and Subtropical Landscapes*, while Watts' (1980) work on pedogenic calcrete in the Kalahari remains one of the most cited on the subject. Other syntheses are included in the volume *The Kalahari Environment* by Thomas and Shaw (1991) and in the wide-ranging review of duricrusts in southern Africa by Nash (2012).

The aim of this chapter is to provide a state-of-the-art overview of research on duricrusts in Botswana. It does so by blending global research into the origins of calcrete and silcrete with regional studies from the Botswana Kalahari. It then considers a third, hybrid, variety of duricrust termed 'silcrete-calcrete intergrade duricrust' by Nash and Shaw (1998), which has been described extensively in Botswana. Iron oxide-cemented laterites and ferricretes have been documented in the eastern hardveld of Botswana (e.g. Goudie 1973; Nash et al. 1994a; Dorland 1999; Yang and Holland 2003; Yamaguchi et al. 2007). However, despite their use in the construction of low-volume roads (Association of Southern African National Roads Agencies 2014; Paige-Green et al. 2015), these duricrusts have not been studied in sufficient breadth to be considered further here.

D. J. Nash (✉)
University of Brighton, School of Applied Sciences, Brighton, United Kingdom
e-mail: D.J.Nash@brighton.ac.uk

D. J. Nash
University of the Witwatersrand, School of Geography, Archaeology and Environmental Studies, Johannesburg, South Africa

F. D. Eckardt (ed.), *Landscapes and Landforms of Botswana*, World Geomorphological Landscapes,
https://doi.org/10.1007/978-3-030-86102-5_13

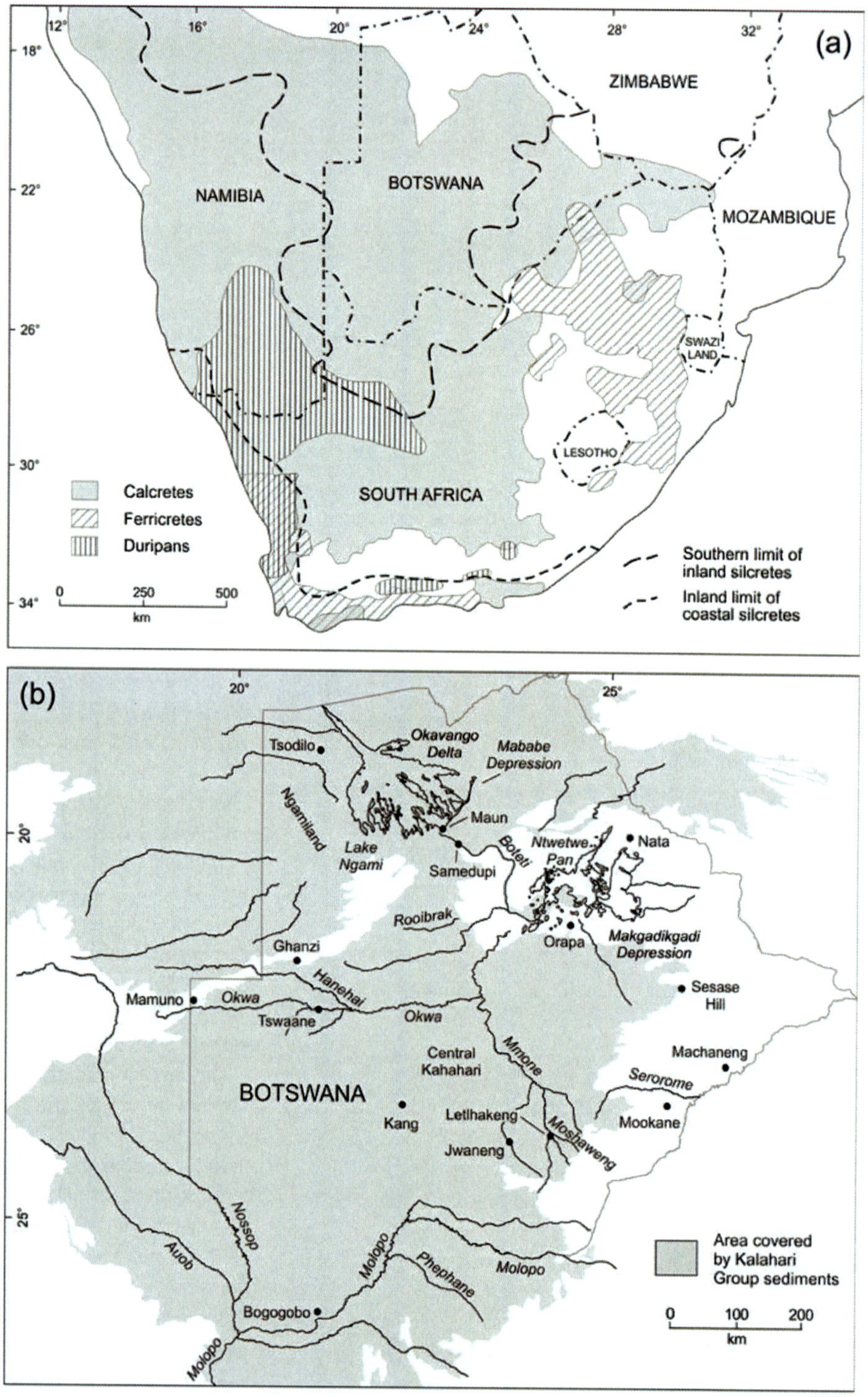

Fig. 13.1 a General distribution of pedogenic duricrusts in southern Africa (after du Toit 1954; Weinert 1980; Ellis and Schloms 1984; Schloms and Ellis 1984; Netterberg 1985; Partridge 1997; Botha 2000; Fey 2010; Nash 2012); **b** Main areas and locations mentioned in the text

Fig. 13.2 Examples of pedogenic and non-pedogenic calcrete profiles in Botswana: **a** Powdery to nodular pedogenic calcrete exposed near New Hanehai (image © David Nash); Calcrete profile exposed around the margin of a pan near Sese, southeast of Jwaneng (image © Frank D. Eckardt); **c** Valley calcrete exposed within a river terrace in the floor of the Okwa Valley (image © David Nash); **d** Thick pedogenic calcrete exposure at Damtshaa Mine (image © Frank D. Eckardt)

13.2 Calcrete

Botswana's generally flat topography, combined with the limited extent of landscape incision, restricts the exposure of thick sub-surface duricrusts. However, the available literature suggests that calcrete occurs beneath the Kalahari Sand cover across much of the country (Fig. 13.1a). The Botswana Kalahari contains some of the thickest calcrete sequences in the world (Watts 1980; Linol 2013). Calcrete is found both within sedimentary successions and at the surface (Goudie 1973; Nash et al. 1994a; Linol et al. 2015). Calcretes reach their greatest thickness, and frequently outcrop, in the vicinity of fossil valleys and pans—see Chaps. 10 and 12. Some of the most extensive outcrops are seen around the Makgadikgadi and Mababe depressions, Lake Ngami and various smaller pans in central Botswana (Grove 1969; Mallick et al. 1981; Lawrence and Toole 1984; Nash et al. 1994a; Ringrose et al. 2005, 2009; White and Eckardt 2006)—see Fig. 12.1b for locations. Considerable thicknesses are also exposed along the Okwa-Mmone drainage systems. These sequences may merge with similar deposits exposed along the Molopo-Auob-Nossop valleys in northern South Africa and southeastern Namibia (e.g. Boocock and van Straten 1962; Goudie 1973; Watts 1980; Mallick et al. 1981; Nash et al. 1994a; Ringrose et al. 2002; Nash and McLaren 2003; Haddon and McCarthy 2005) to form part of a wider calcrete-capped surface referred to as the 'Kalahari Limestone' by Range (1912) or the Mokalanen Formation by Thomas et al. (1988). Borehole data indicate that calcretisation (and silicification; Sects. 13.3 and 13.4) is widespread deep within the Kalahari Group sediments, with thick calcretes identified many tens of metres beneath the Kalahari Sand cover (e.g. Meixner and Peart 1984; Thomas and Shaw 1990; 1991; du Plessis 1993; Nash et al. 1994a; Haddon and McCarthy 2005; Linol et al. 2015).

For readers seeking more general information about calcretes, reviews of the characteristics, formation and significance of calcareous duricrusts within desert landscapes are provided by Alonso-Zarza (2003), Wright (2007), Dixon and McLaren (2009) and Nash (2011). Detailed overviews of southern African calcretes are provided by Netterberg (1969a), Watts (1980), Nash and McLaren (2003) and Nash (2012). The unpublished PhD thesis by Shaw (2009) also contains detailed descriptions of calcrete exposures from across the Kalahari.

13.2.1 Definitions

Before focussing on studies of calcrete in Botswana, it is useful to consider some definitions. The word 'calcrete' was first used by Lamplugh (1902) to describe calcium carbonate-cemented gravels in the vicinity of Dublin, Republic of Ireland. The term was applied subsequently to geochemical sediments exposed in the vicinity of the Zambezi Valley where it forms the border between Zambia and Zimbabwe (Lamplugh 1907). Numerous definitions now exist. Working mostly in South Africa, Netterberg (1969a p. 88) used the term calcrete rather broadly to describe almost any terrestrial material that has been cemented and/or replaced by calcium carbonate. Goudie (1983, pp. 94–95) tightened up this definition, noting that calcretes may contain cementing agents in addition to calcium carbonate and that carbonate accumulation occurs primarily in the vadose zone (i.e. in the region above the water table). Machette (1985) urged that the term calcrete should be reserved for indurated duricrusts and that 'calcic soil' be used to describe weakly carbonate-cemented soils—which begs the obvious question, 'when is a material sufficiently indurated to be termed a calcrete'? The most widely used definition today is that introduced by Wright, who, building on Goudie (1983), identifies calcrete as '…a near surface accumulation of predominantly calcium carbonate, which occurs in a variety of forms from powdery to nodular, laminar and massive. It results from the cementation and displacive and replacive introduction of calcium carbonate into soil profiles, sediments and bedrock, in areas where vadose and shallow phreatic groundwaters are saturated with respect to calcium carbonate' (Wright 2007, p. 10).

Calcrete is a non-genetic term—in other words, its application to a carbonate-cemented material does not imply that the material formed in a particular way. Rather, various sub-types of calcrete have been identified that suggest specific genetic origins. The most fundamental distinction made in the literature is between varieties that develop as a horizon (or multiple horizons) within the vadose zone of soil profiles (these are termed *pedogenic calcretes*) and those that form at or around the water table or in the capillary fringe (often grouped for simplicity as *non-pedogenic calcretes*). Beneath this top-level classification are a range of different types of calcrete, defined according to their geomorphic context and whether the duricrust is considered primary or secondary. One way in which the different sub-types can be grouped is shown in Table 13.1 (after Carlisle 1983). The bipartite division between pedogenic and non-pedogenic calcrete is not without its problems, since some calcretes may result from a combination of pedogenic and non-pedogenic processes. For example, a non-pedogenic valley calcrete that developed in association with a fluctuating water table may be exhumed and modified by pedogenesis (Machette 1985). However, the terms are widely used, so they are adopted here.

13.2.2 General Characteristics of Calcretes in the Botswana Kalahari

The majority of surface calcretes in Botswana are white, cream or grey in colour (Fig. 13.2), though pinkish mottling and banding is common where the calcrete is highly indurated and/or relatively mature. Calcretes exhibit a variety of forms, including weakly calcified, chalky, powdery, rhizocretionary, nodular, honeycomb, platy, laminar, stringer, pisolitic, brecciated, conglomeratic, massive and hardpan (cf. Wright 2007). Of these, massive and hardpan forms are most commonly encountered in Botswana, though this may be more a product of where calcretes are exposed than their true geographic extent; as discussed in Sect. 13.2.3, massive hardpan calcretes are more likely to be developed in association with valleys and pans and this is also where they are most visible. Most calcretes have developed within sandy components of the Kalahari Group sediments, although there

Table 13.1 A classification of calcrete types (Carlisle 1983)

Calcrete classification	Incorporated calcrete types
Pedogenic calcrete	Caliche; Kunkar; Nari
Non-pedogenic superficial calcrete	Laminar crusts; Case hardening
Non-pedogenic gravitational zone calcrete	Gravitational zone calcrete
Non-pedogenic groundwater calcrete	Valley calcrete; Channel calcrete; Deltaic calcrete; Alluvial fan calcrete
Non-pedogenic detrital and reconstituted calcrete	Recemented transported calcrete; Brecciated and recemented calcretes

are exposures at the periphery of the Kalahari Group where calcretised gravels are present (e.g. Thomas, 1981; Shaw and de Vries, 1988; Nash et al. 1994a; Shaw 2009).

Over much of Botswana, surface calcrete outcrops rarely exceed 2–3 m thickness. However, extremely thick stacked calcretes have been identified beneath the Molopo valley in southern Botswana and northern South Africa, where borehole records, confirmed by exposures, reveal sequences reaching up to 80 m in total (Goudie 1973; Thomas et al. 1988). In Ngamiland, a recent drilling programme has identified widespread calcrete layers at the base of the Kalahari Group sediments; these range between 10 and 60 m thick, and have been termed the Nxau-Nxau Calcrete Formation by Linol (2013). Stratigraphic correlation of the Nxau-Nxau Calcrete Formation with the sequence in another Ngamiland core (described by McFarlane et al. 2010) is problematic owing to evidence for extensive deep weathering in the latter.

The macromorphology of calcrete profiles varies between pedogenic and non-pedogenic types. Well-developed pedogenic calcretes are distinct in having a highly organised profile dominated by a hardpan horizon (Fig. 13.3). This hardpan is typically capped by a laminar calcrete and—if recently exposed and not subject to subaerial erosion—brecciated calcrete clasts and a layer of weakly calcified sediment. Below the hardpan, the profile becomes less well-cemented with depth, grading into a nodular horizon, then a chalky or powdery horizon before passing into non-calcified sediment. In contrast, non-pedogenic calcretes (e.g. Fig. 13.2b,c) commonly comprise a hardpan layer only, sometimes capped by a laminar calcrete, but with a sharp basal boundary with underlying uncemented sediment or bedrock.

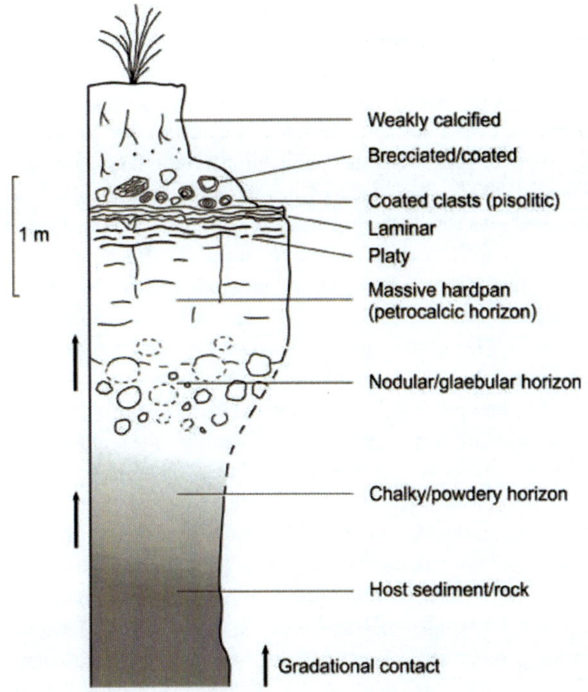

Fig. 13.3 Idealised pedogenic calcrete profile showing a range of macroforms (after Wright 2007; Nash 2012)

Regardless of whether a calcrete formed as a result of pedogenesis or other processes, its chemistry will be dominated by calcium carbonate. A sample of 300 bulk chemical analyses of calcretes published by Goudie (1972), including many from southern Africa, comprised on average 79.28% calcium carbonate (42.62% CaO), 12.30% silica, 3.05% MgO, 2.03% Fe_2O_3 and 2.12% Al_2O_3. Further data for the subcontinent are included within Netterberg (1969a). Few analyses have been published specifically for calcretes from Botswana. However, Gwosdz and Modisi (1983) report calcretes from around Ghanzi, Letlhakeng, Machaneng, Mamuno, Maun, Mookane and Nata, and from the Bonwapitse River and Serorome valley, with $CaCO_3$ contents of up to 96.3%.

In terms of mineralogy, the most detailed investigations of calcretes in Botswana are from exposures in 'borrow pits' and valley flanks by Watts (1980). His analyses of pedogenic calcretes around the Makgadikgadi Basin and in southeast Botswana show that the carbonate component is dominated by low-Mg calcite, which occurs as micrite, microspar and sparite. Minor quantities of high-Mg calcite, aragonite and, in well-developed calcrete profiles, dolomite may also be present. Profiles incorporate additional authigenic silica (Sect. 13.3) and various authigenic silicates including palygorskite, sepiolite, montmorillonite, illite, mixed-layer clays, glauconite and, more rarely, chlorite, kaolinite and the Na-Ca zeolite clinoptilolite. Palygorskite and sepiolite are more abundant in mature calcrete profiles, while montmorillonite with minor palygorskite and traces of other clays characterise less developed profiles. Clinoptilolite was only identified by Watts (1980) in calcretes associated with pans.

13.2.3 Processes of Calcrete Formation

The degree of development, or *maturity*, of an individual calcrete profile is related to a series of interdependent factors: time, climate, host sediment characteristics, carbonate source(s) and supply, geomorphological context, organic influences, sedimentation (or erosion rates) and a variety of localised conditions. As Watts (1980) notes, 'the interplay of such a number of parameters operating over an area as large as the Kalahari… results in a highly diverse suite of calcrete types' (p. 89).

Carbonate sources, transfer mechanisms and drivers of precipitation

The formation of pedogenic and non-pedogenic calcrete requires three things: (i) a source of calcium carbonate, (ii) a mechanism for transferring this carbonate to the site of calcrete accumulation, and (iii) a means of triggering carbonate precipitation (Goudie 1983). The carbonate sources and precipitation mechanisms described in this section are common to most calcrete types. However, the mechanisms by which carbonate is transferred to the site of accumulation differ between pedogenic and non-pedogenic sub-types.

Potential carbonate sources for any calcrete include solid and dissolved carbonate introduced into the soil or sediment from above (in Botswana this includes introduction via atmospheric dust, rainfall, surface runoff, plant materials and shells) or below (via groundwater) (Netterberg 1969a; Goudie 1983; Cailleau et al. 2004). The contribution of specific carbonate sources to calcrete formation will vary spatially and with calcrete type. By comparison with other desert and semi-desert regions, atmospheric dust and rainfall inputs are likely to be the most important for the formation of pedogenic calcretes in Botswana (e.g. Goudie 1973; Watts 1980; Capo and Chadwick 1999; Chiquet et al. 1999) and subsurface sources for non-pedogenic calcretes (e.g. Arakel and McConchie 1982; Arakel 1986, 1991; Nash and McLaren 2003; Nash and Smith 2003).

Carbonates are concentrated into specific horizons of a soil or sediment by different dominant mechanisms in pedogenic and non-pedogenic calcretes. Pedogenic calcrete formation is characterised by vertical transfers of dissolved carbonate, largely as a result of illuvial-eluvial pedogenetic processes. Carbonate redistribution is driven by the leaching of dissolved carbonate in downward percolating soil solutions, with the zone of maximum accumulation controlled by the depth of penetration of the wetting front following rainfall events (Goudie 1983; Klappa 1983). If the water table is relatively close to the surface, dissolved carbonate may also move upwards to the site of accumulation due to capillary action. In contrast, much of the carbonate for non-pedogenic calcrete formation is provided by lateral transfers of carbonate-rich surface or groundwater (e.g. within lacustrine or channel-margin settings), with vertical movements of dissolved carbonate only really coming into play once close to the site of precipitation (Goudie 1973, 1983; Bachman and Machette 1977; Arakel 1986).

Once at the horizon of carbonate accumulation, calcium carbonate dissolved in water may be precipitated by a variety of processes. These are well-described in general accounts of calcrete development (e.g. Wright 2007), and include evaporation, evapotranspiration, pH shifts to above pH 9, organic life processes (Goudie 1983), a decrease in the partial pressure of soil CO_2 (Schlesinger 1985), CO_2 loss by degassing (e.g. as temperature increases; Barnes 1965), and the common ion effect (Wigley 1973).

Pedogenic calcretes

A mature pedogenic calcrete profile, such as the one shown schematically in Fig. 13.3, is the end product of a long period of pedogenesis that may span hundreds of thousands of years. Calcic soil and pedogenic calcrete development follows a well-established morphological sequence, with many of the carbonate morphologies described in Sect. 13.2.2 falling into an evolutionary continuum that relates to the relative maturity, or *stage*, of the calcrete profile (see Gile et al. 1966; Reeves 1970; Bachman and Machette 1977; Netterberg 1980; Goudie 1983; Netterberg and Caiger 1983; Machette 1985).

There are six widely recognised stages of pedogenic calcrete development (Bachman and Machette 1977; Machette 1985); these are summarised in Table 13.2 and illustrated in Fig. 13.4. The upper routeway through Fig. 13.4—relating to host sediments with a low gravel content—is most relevant to the sandy sediments of the Kalahari. Under this idealised continuum, Stage I calcified soils and chalky or powder calcretes may develop into Stage II nodular calcretes over time as calcium carbonate concretions increase in size. The concretions may coalesce to form Stage III honeycomb calcrete, with a Stage IV-V hardpan calcrete developing as surface horizons become plugged with carbonate. Degradation of a hardpan as a result of solutional processes may lead to the development of a Stage VI brecciated (or boulder) calcrete. Laminar calcretes consisting of finely banded carbonate often cap hardpan and brecciated calcretes, and the whole sequence may be buried by upper soil horizons containing carbonate pisoliths (Wright 2007). The stage of development may be used in local regions for correlation and the determination of relative ages of soils and geomorphic surfaces (Bachman and Machette 1977). However, attempting to describe non-pedogenic calcretes using the scheme shown in Table 13.2 is problematic since non-pedogenic calcretes may not go through the same phases of development.

The seminal work by Watts (1980) includes logged pedogenic calcrete profiles from various sites along the Molopo valley and its tributaries (Fig. 13.5). The simplest depicted profiles comprise a single hardpan layer with underlying nodular calcrete and calcic soil; these represent a single cycle of calcrete development. In contrast, the thickest profile may—based on the presence of at least two hardpan calcrete horizons—represent two or more cycles of pedogenesis separated by periods of sediment accumulation. These sections are part of the 'Kalahari Limestone' first described by Passarge (1904) and Range (1912), which has undergone karstification beneath the Kalahari Sand cover in

Table 13.2 Stages in the morphogenetic sequence of carbonate deposition during calcrete formation (Bachman and Machette, 1977)

Stage	Diagnostic carbonate morphology
I	Filaments or faint carbonate coatings, including thin discontinuous coatings on the underside of pebbles
II	Firm carbonate nodules few to common but isolated from one another. The matrix between nodules may include friable interstitial carbonate accumulations. Continuous pebble coatings present
III	Coalesced nodules in disseminated carbonate matrix
IV	Platy, massive, indurated matrix, with relict nodules visible in places. The profile may be completely plugged with weak incipient laminar carbonate coatings on upper surfaces. Case hardening is common on vertical exposures
V	Platy to tabular, dense and firmly cemented. Well-developed laminar layer on upper surfaces. Scattered incipient pisoliths may be present in the laminar zone. Case hardening common
VI	Massive, multilaminar and brecciated profile, with pisoliths common. Case hardening common

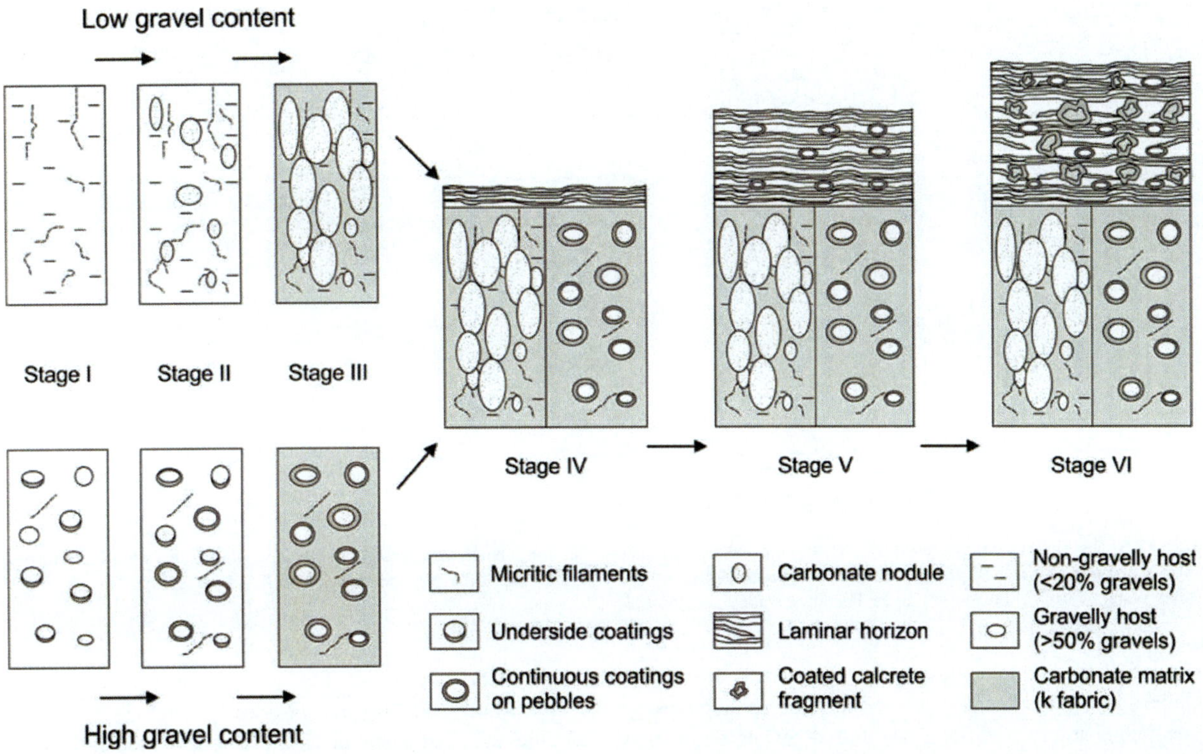

Fig. 13.4 Stages in the development of a typical pedogenic calcrete profile formed in gravel-poor and gravel-rich sediments (based on Machette 1985; Alonso-Zarza 2003). The low gravel content genetic sequence is most relevant in the context of the sand-dominated Kalahari Group sediments, but calcretised gravels have been described in the Letlhakeng area of Botswana (e.g. Nash et al. 1994a)

adjacent areas of Namibia (e.g. in the Weissrand area; Goudie and Viles 2015).

Watts (1980) includes a flow diagram illustrating the main processes operating during the formation of pedogenic calcretes in the Kalahari (Fig. 13.6). Calcite precipitation from percolating solutions is thought to involve either rapid or slow evaporation and/or CO_2 loss. Slow evaporation results predominantly in the precipitation of low-Mg calcite with a consequent gradual increase of Mg concentrations in the resulting solutions. Rapid evaporation may precipitate high-Mg calcites that are in thermodynamic disequilibrium with the low Mg/Ca ratio vadose percolating waters. Both passive (void-filling) and replacive (of silicate grains) calcite may be formed, with silica being released as a result of replacement and migrating down-profile to accumulate in lower calcrete horizons. Under more saline conditions, optically length-slow chalcedony (when viewed in thin-section under cross-polarised light) and potentially

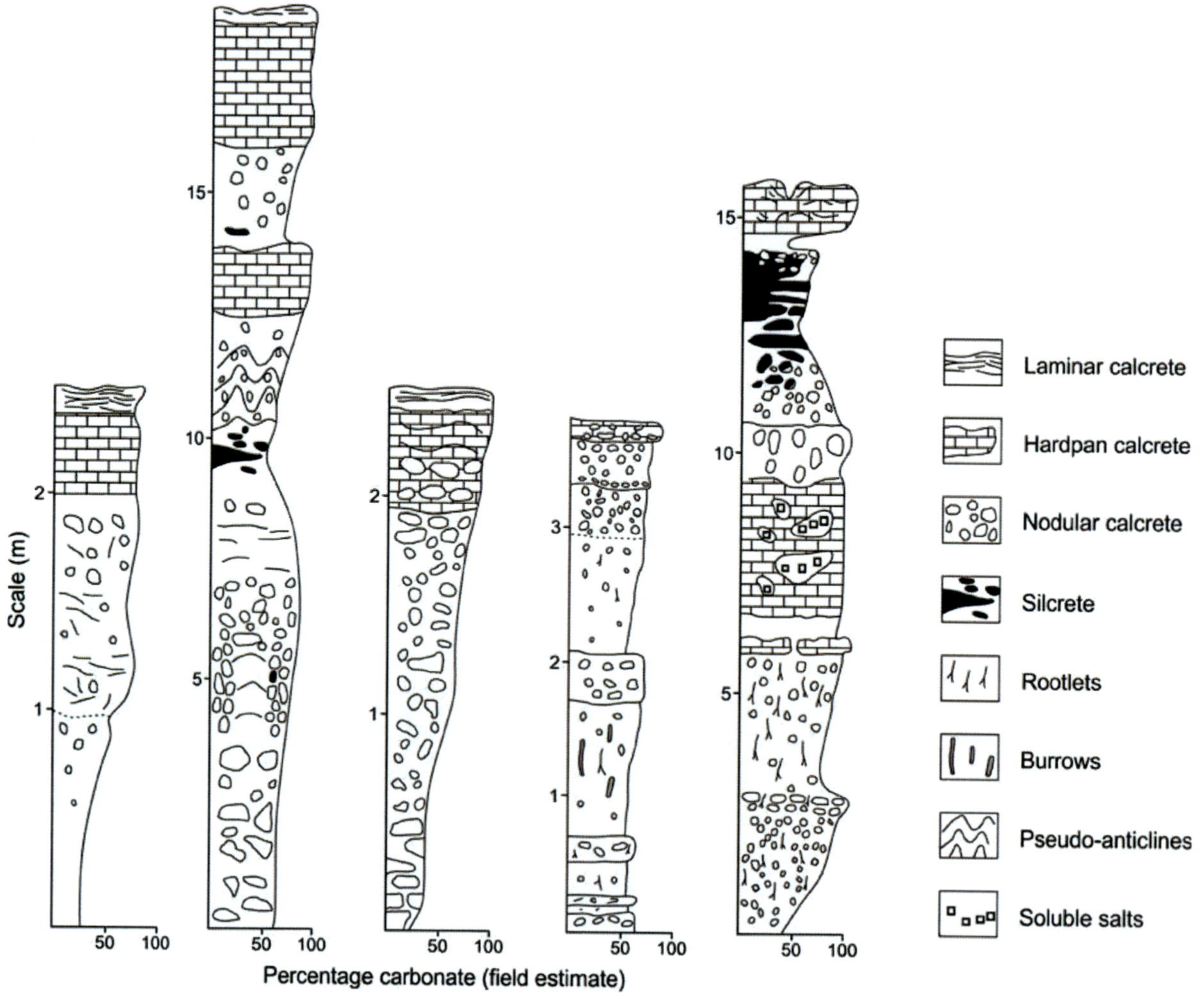

Fig. 13.5 Simple and composite logged pedogenic calcrete profiles from the southern Kalahari along the Botswana-South Africa border (after Watts 1980): (1) borrow pit in the Phephane valley just south of the confluence with the Molopo valley; (2–5) cliff exposures on the south side of the Molopo valley near Bogogobo

clinoptilolite will precipitate, whereas optically length-fast chalcedony and/or mega-quartz cements are precipitated in non-saline micro-environments. Rapid evaporation may produce solutions that are highly supersaturated with respect to calcite, resulting in the displacive growth of calcite crystals. Neomorphism of high-Mg to low-Mg calcite takes place rapidly and leads to the release of Mg. This, combined with the increased Mg/Ca ratio due to low-Mg calcite precipitation, increases the Mg concentration of pore fluids to the degree that authigenic Mg-rich silicates may precipitate. Clay authigenesis also occurs during calcrete development. Clay minerals such as palygorskite may form through the reaction of magnesium with montmorillonite or may be precipitated directly from solution in association with neoformed sepiolite and dolomite.

The flow diagram in Fig. 13.6 assumes a dominantly closed system. However, episodes of high rainfall may obscure this and flush the system. Local precipitation of high-Mg calcite is also thought to occur as a result of capillary rise from groundwater in and around saline depressions such as the Makgadikgadi Basin and larger pans (Watts 1980).

Non-pedogenic calcretes

Few studies have considered the development of non-pedogenic calcretes in Botswana. One exception is work by Nash and McLaren (2003), which discusses the formation of late Pleistocene to Holocene valley calcretes in the Kalahari. Valley calcretes occur exclusively within the base

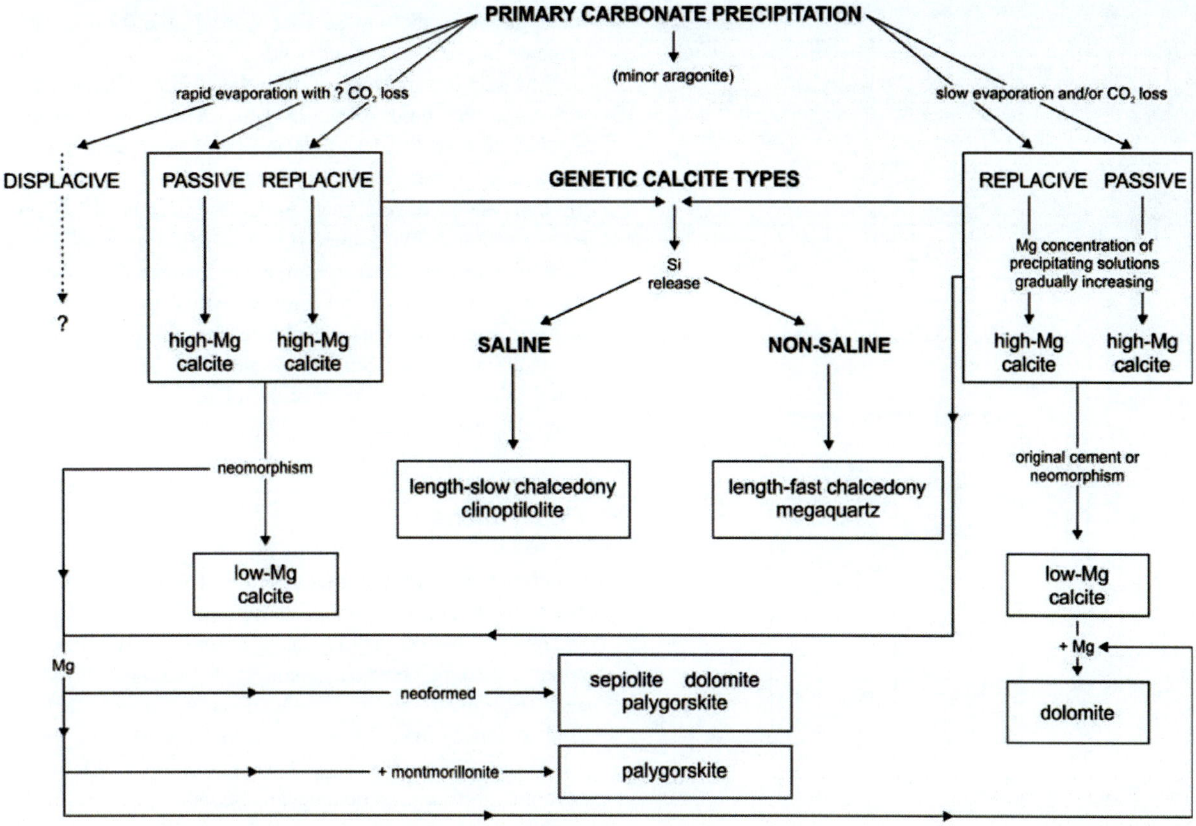

Fig. 13.6 Schematic flow diagram summarising the major pedogenic and diagenetic processes within pedogenic calcretes in the Kalahari (modified after Watts 1980)

of dry valleys (see Chaps. 11 and 12) and are commonly exposed either as a lag of calcrete gravel (where valley floors have not been incised) or within low terraces on valley sides. Calcrete-supported terraces are particularly prominent in the Hanehai and Okwa valleys where a single low terrace up to 4 m high can be traced along much of the valley courses west of the Trans-Kalahari Highway.

The majority of valley calcretes are massive in appearance with little evidence of profile organisation or vertical structures. Many thicker outcrops in the Hanehai and Okwa valleys are well-cemented in their uppermost sections and underlain by slightly more friable cemented sands. Almost all exposures exhibit a thin (<5 mm) laminar calcrete on their upper surfaces. Outcrops occasionally show abundant vertical to sub-horizontal tubular structures of up to 1.5 cm diameter, lined with laminar calcrete and partially infilled by sand, which are interpreted as either root or reed casts (Nash and McLaren 2003). It is not possible to tell from field observations whether valley calcretes cement the full width of alluvial valley fills or 'pinch out' towards valley flanks. However, analyses of borehole transect across dry valleys in the Central Kalahari (e.g. Fig. 13.7) would tend to favour the latter (Nash et al. 1994a).

The micromorphology of Kalahari valley calcretes provides important insights into their development (Fig. 13.8). The majority are highly indurated, with total cement contents ranging from 70–90% of the calcrete bulk volume. Massive, structureless, pore-filling micrite is the most important component of the cement (Fig. 13.8a, b), suggesting that carbonate precipitation was relatively rapid. Relatively few of the samples analysed by Nash and McLaren (2003) contained more coarsely crystalline forms of $CaCO_3$ cement, reinforcing the relative rapidity of precipitation. Void spaces provide further clues into formational processes. In addition to uncemented inter-granular spaces, many elongate, crack-like voids are present. These tend to occur towards the top of profiles and may be a product of the final stages of desiccation of the rivers or ponds that formerly occupied the valleys. Evidence of biotic activity is also abundant (Fig. 13.8c, d), including the presence of rhizoliths, root hairs, fungal hyphae and needle fibre cements, most commonly within cracks and other void

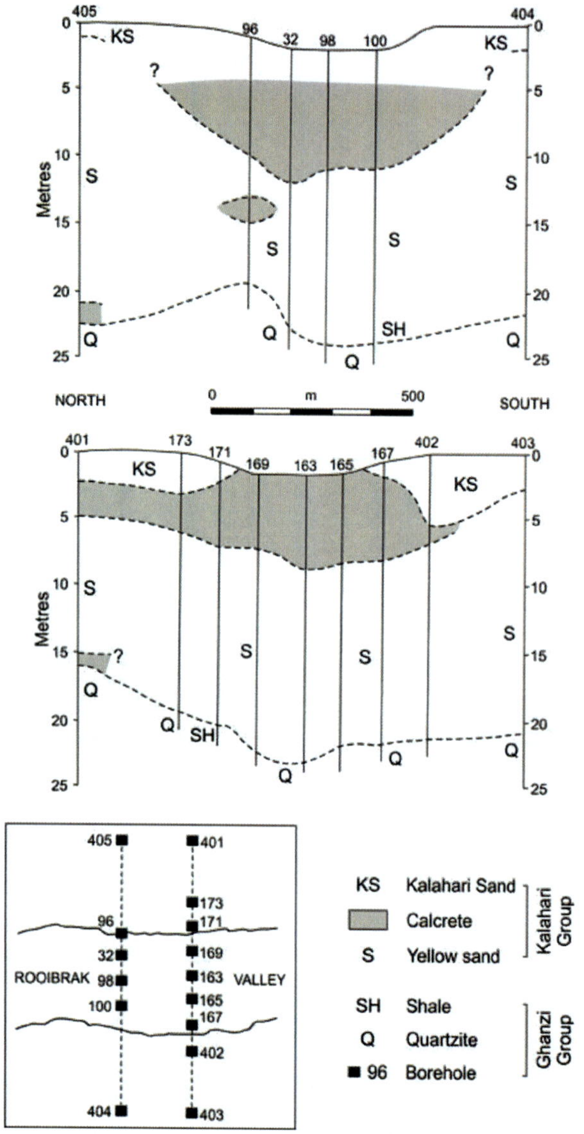

Fig. 13.7 Variability in thickness of valley calcrete development beneath the Rooibrak Valley identified from boreholes drilled during uranium prospecting by Union Carbide (after Nash et al. 1994a)

spaces. Authigenic silica cements may occur as void linings and rare void fills.

Kalahari valley calcretes appear to display at least two major phases of development (Nash and McLaren 2003). It is likely that the main cementation of alluvial sediments by micrite occurred either immediately following the cessation of surface flows or, more probably, whilst flow was in decline and valleys were shifting towards their present-day hydrological 'fossil' status during the late Pleistocene (see Chap. 11). Cracks developed within the calcrete and became the focus of biogenic activity. The final phase of precipitation in the vadose zone came in the form of silica cements in pore spaces and voids.

Thin-section analyses of the Nxau-Nxau Calcrete Formation at the base of the Kalahari Group sediments in Ngamiland reveal facies characteristic of both lacustrine and pedogenic environments (see Alonso-Zarza 2003), suggesting that calcrete development took place in or near a shallow lake or wetland environment (Linol 2013; Linol et al. 2015). Examples include extensive biogenic features (e.g. alveolar structures, rhyzoliths and burrows) that indicate intense biological activity during calcrete formation, as well as fenestral fabrics and circum-granular cracks within more muddy facies suggesting episodes of immersion and desiccation (Linol et al. 2015).

13.3 Silcrete

In comparison with other parts of the world, relatively little has been written on silcrete in Botswana. The situation is changing, however, mainly as a by-product of interdisciplinary research between geomorphologists and archaeologists seeking to identify the locations of silcrete 'quarries' used to supply materials for stone tool manufacture during the Middle Stone Age (e.g. Nash et al., 2013, 2016). For example, the large geochemical datasets arising from this work have been used by Webb and Nash (2020) to constrain the environmental controls upon silcrete formation in the Botswana Kalahari (see Sect. 13.3.3).

The distribution of silcrete in Botswana has not been mapped. However, like calcrete, published studies are restricted geographically to areas covered by Kalahari Group sediments (Fig. 13.9). The main areas of outcrop include the flanks (and occasionally beds) of drainage lines and the margins of ephemeral lakes and pans (e.g. Boocock and van Straten, 1962; Smale, 1973; Mallick et al., 1981; Summerfield, 1982; Shaw and de Vries, 1988; Nash et al., 1994a,b, 2004; Shaw and Nash, 1998; Ringrose et al., 2005, 2009; Kampunzu et al., 2007). Silcretes also occur as rare escarpment caprocks at the easternmost margin of the Kalahari Group sediments (Summerfield, 1982; Nash and Hopkinson, 2004), in some places forming a distinct unit overlying weathered basalt (Smale, 1973; Summerfield, 1982; see Fig. 13.9e). Recent drilling work in the vicinity of Tsodilo Hills has identified a 4 m thick silcrete at the interface between the basal Kalahari Group and underlying pre-Karoo bedrock (Linol et al., 2015), so it is possible that extensive silcretes occur at depth elsewhere.

For readers seeking general information about silcrete, useful reviews are provided by Milnes and Thiry (1992), Nash and Ullyott (2007), Nash (2011, 2012) and, from an archaeological perspective, Thiry and Milnes (2017).

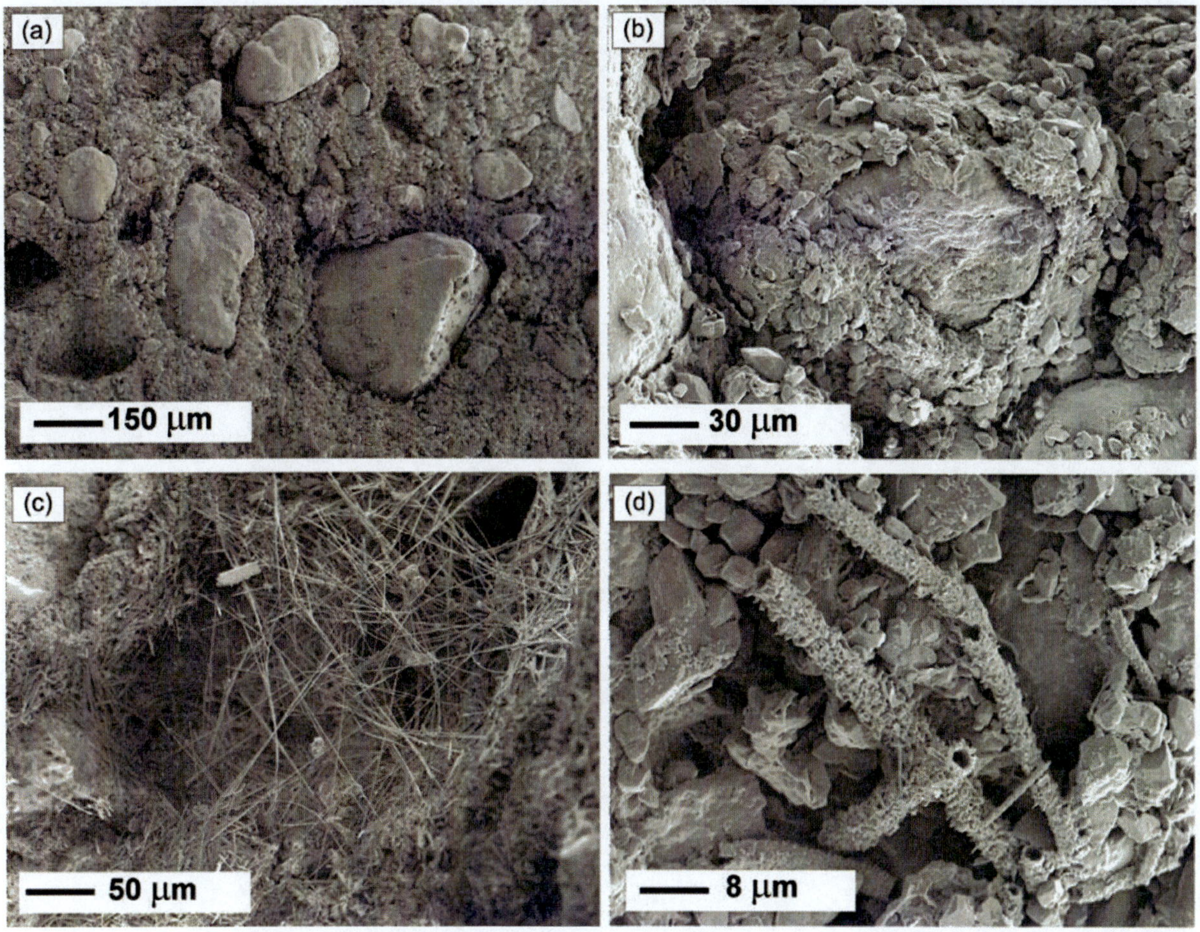

Fig. 13.8 Scanning electron microscope images of samples from non-pedogenic valley calcretes exposed in the Okwa and Hanehai valleys: **a** rounded quartz grains and minor shell fragments cemented by microcrystalline calcite (micrite); **b** quartz grain coated by micrite and microspar cement; **c** needle-fibre and calcified fungal filaments (plus possible fruiting spores) lining a void with needle-fibres crossing the void; **d** hollow fungal filaments encrusted with micrite and small bladed and acicular calcite crystals. Images **(a)** and **(b)** represent typical cements developed under evaporitic conditions whilst **(c)** and **(d)** suggest organic activity under vadose conditions (after Nash and McLaren 2003)

Overviews that focus on silcretes in Botswana are limited to early studies such as Goudie (1973) and Summerfield (1982; 1983b; 1983a) and the more recent syntheses by Nash (2012) and Webb and Nash (2020).

13.3.1 Definitions

The term silcrete was introduced by Lamplugh (1902) to describe the products of near-surface processes by which 'silica accumulates in and/or replaces a soil, sediment, rock or weathered material to form an indurated mass' (Nash and Ullyott 2007 p. 95). Silcretes sensu stricto are defined as containing >85 wt% SiO_2 (as determined by bulk chemical analysis), with many comprising >95 wt% SiO_2 (Summerfield, 1983b). Most silcretes require stable geomorphological conditions to develop, although the formation of some may be related to actively evolving landscapes (see, for example, Thiry 1999).

Like calcrete, silcrete is a non-genetic term, with various sub-types identified that suggest specific genetic origins. The most relevant scheme to classify these sub-types as they occur in Botswana is shown in Fig. 13.10 (after Nash and Ullyott 2007). This scheme builds upon the classification proposed by Milnes and Thiry (1992), but includes additional sub-types only documented in the Kalahari. Echoing Carlisle's (1983) classification for calcrete, silcretes are first subdivided into pedogenic and non-pedogenic varieties. Pedogenic silcretes are then divided into more opaline 'duripans' in which clay and iron are retained and those with abundant microquartz and titania that lack clay minerals and iron oxides (Thiry 1999). Non-pedogenic silcretes are

Fig. 13.9 Examples of silcrete outcrops in Botswana: (a) Massive drainage-line silcrete at Samedupi Drift at the margin of the Boteti River; (b) Massive to blocky pan/lacustrine silcrete from the southern margin of Ntwetwe Pan; (c) Nodular drainage-line silcrete in the bed of the Nata River downstream of Nata village; (d) Massive drainage-line silcrete outcropping in the flanks of the Moshaweng Valley, southeast of Letlhakeng; (e) Massive silcrete caprock overlying basalt at the eastern margin of the Kalahari Group sediments, Sesase Hill. All images © David Nash

grouped into *groundwater*, *drainage-line* and *pan/lacustrine* types according to their geomorphological context. To the knowledge of the author, pedogenic silcrete has yet to be identified in Botswana; as such the remainder of this chapter discusses non-pedogenic silcretes only.

13.3.2 Characteristics of Silcretes in the Botswana Kalahari

Surface silcrete occurrences in Botswana are usually localised and vary from small outcrops or gravel lags around the margins of pans (e.g. Summerfield 1982; Nash and Shaw 1998; see Fig. 13.9b) to layers up to several metres thick in the floors of rivers (Shaw and Nash 1998), the flanks of dry valleys (e.g. Summerfield 1982; Shaw and de Vries 1988; Nash et al., 1994b; Kampunzu et al. 2007; Fig. 13.9d) and within cemented palaeolake shorelines (Ringrose et al. 2005, 2009). The colours of individual outcrops vary considerably, from pale to dark green in the case of silcretes at the northern and southern margins of the Makgadikgadi Basin (Boocock and van Straten 1962; Summerfield 1982) and in the Okwa valley (Nash et al. 2004), to white, grey, cream, buff, brown, reddish-brown and black elsewhere (see Nash et al. 2013). These colours are rarely uniform, with many silcretes including darker or lighter flecks, veins and inclusions or—

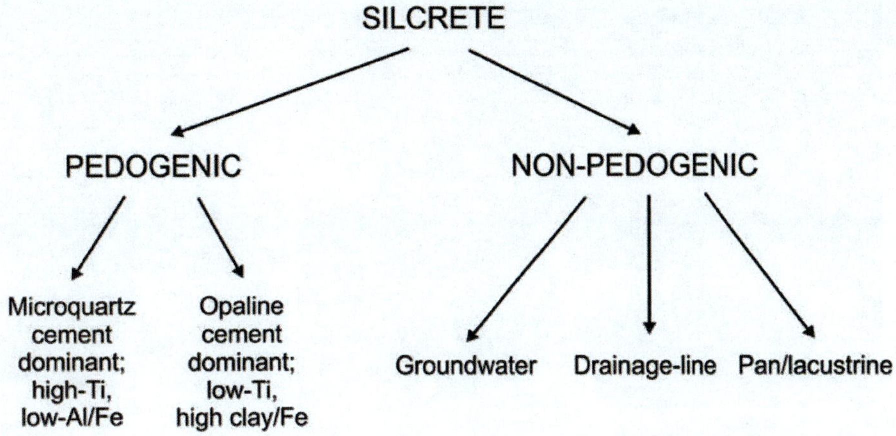

Fig. 13.10 Geomorphological classification of silcrete (after Nash and Ullyott 2007)

in the case of silcretes along the Boteti River—'whorl'-like patterns consisting of undulating, enclosed, parallel bands produced by variations in the degree of iron oxide staining within the cement (Summerfield 1982).

Silcrete outcrops in Botswana exhibit a wide variety of morphologies. In part, this depends upon whether the silcrete formed as a primary precipitate within aeolian, fluvial or lacustrine sediments or whether it developed via the replacement of a pre-existing calcrete (see Sect. 13.4). Primary silcretes in pan/lacustrine and drainage-line settings, such as those exposed along the Boteti (Shaw and Nash, 1998) and Nata rivers (Summerfield 1982) and within the Makgadikgadi depression (Ringrose et al. 2009), are typically nodular, sheet-like or massive in appearance (Fig. 13.9). Those that formed by replacement of calcrete have more irregular morphologies and, where exposed in palaeolake strandlines, may be lenticular in appearance. The range of silcrete morphologies encountered across Ntwetwe Pan alone—including unusual tube-like silcretes likely formed via the silicification of root structures—is illustrative of the wider variability (Fig. 13.11).

Most silcretes in Botswana consist of a framework of quartz grains, inherited from the parent Kalahari Sands, cemented by chalcedonic silica and/or microquartz (Summerfield 1982). Most exhibit grain-supported fabrics (using the terminology of Summerfield 1983b) but floating fabrics are common where silicification of a precursor calcrete has taken place. Silica void fills are widespread and comprise silica polymorphs including α-quartz, opal-CT or opal-T and moganite (e.g. Summerfield 1982; Nash et al. 1994a; Nash and Hopkinson 2004).

Several studies of Botswana silcretes incorporate geochemical data (Summerfield 1982; Nash et al., 1994b, 2004, 2013; Nash and Shaw 1998). Meta-analysis of major and trace element data for silcretes within the Okavango-Makgadikgadi system (Webb and Nash 2020) shows that the median composition comprises 94 wt% SiO_2 with small amounts (<1.1 wt%) of Al_2O_3, K_2O, Fe_2O_3, MgO, CaO and TiO_2 (in decreasing order of abundance). Barium is the most abundant trace element (~1000 ppm), with high levels of chromium (~200 ppm), strontium and zirconium (50 and 65 ppm respectively), low levels (3–20 ppm) of rubidium, cerium, lanthanum, neodymium and yttrium (in decreasing order of abundance), and very low levels (mostly <1 ppm) of other elements. The mineralogy of Botswana silcretes is dominated by quartz, with small amounts of calcite, Fe oxide/hydroxides and various heavy minerals also present. Glauconite has been identified in many silcretes (Summerfield 1982; du Plessis 1993; Webb and Nash 2020), and imparts the green colour apparent in some outcrops. The occurrence of glauconite may explain the small amounts of Al_2O_3, K_2O, Fe_2O_3 and MgO present in most silcretes (Webb and Nash 2020). Concentrations of rare earth elements vary spatially (Nash et al. 2013), and are very strongly correlated with each other (r^2 = 0.77–0.91) but not with any major elements or other trace elements; this suggests they are likely present as rare earth minerals such as bastnaesite and/or xenotime (Webb and Nash 2020).

13.3.3 Processes of Silcrete Formation

Silica sources, transfer mechanisms and drivers of precipitation

Research by Webb and Nash (2020) has shown that the formation of non-pedogenic silcretes in the Botswana Kalahari was approximately isovolumetric (i.e. there was little or no volume change during silicification). More importantly, it required a substantial addition of SiO_2. Some of this silica may have been derived locally from within the Kalahari Group sediments, but much was transported from more distant sites (Nash and Ullyott 2007). Few chemical elements were lost during silicification; Si and K were

Fig. 13.11 Variations in silcrete morphology across Ntwetwe Pan: **a** Silicified tube-like structures suggestive of former roots; **b** thin 'honeycomb' silcrete sheet formed by the merger of individual silcrete nodules; **c** decimetre-scale sheet-like silcrete with vertical hollows; **d** massive outcrops of black silcrete surrounded by a lag of black silcrete fragments. (**a-c**) North-central Ntwetwe Pan, (**d**) south-central Ntwetwe Pan. All images © David Nash

gained as microquartz precipitated in the sediment porosity and glauconite formed in suboxic groundwater conditions. This contrasts with other areas of southern Africa, such as the Cape coastal zone, where most silica for silcrete formation was derived from intense in situ weathering of bedrock (Webb and Nash 2020).

Weathering of silicate minerals provides the ultimate source for most silica within surface and subsurface waters, the atmosphere and plants and animals (Summerfield 1983b). Enhanced silica availability may occur for a range of reasons. One of the most important controls in the context of Botswana is environmental pH. Silica solubility is relatively stable under weakly acidic to neutral pH but increases rapidly above pH 9 (Dove and Rimstidt 1994). Such high pH levels are not uncommon in arid and semi-arid environments, where alkaline conditions may occur as a result of evaporation (Chadwick et al. 1989). Monitoring of hydrochemical conditions in Sua Pan over two wet seasons (1999–2000 and 2000–2001), for example, showed that the pH of lake waters ranged between pH 8.6 and 10, with maximum values recorded during the initial phases of seasonal flooding when pre-existing salts in riverbeds or pan margins were dissolved (Eckardt et al. 2008; McCulloch et al. 2008). This is consistent with other saline lakes in southern Africa rich in HCO_3^- + CO_3^{2-} ions.

In addition, there are a number of biological silica sources, with many plants, including grasses, reeds and palms, concentrating amorphous silica within their tissues (Goudie 1973); this may be released upon decay or temporarily stored as phytoliths. Diatoms, common in many fluvial and lacustrine environments in the Kalahari (Schmidt et al. 2017), extract silica from surface waters for their tests and may subsequently become a silica source upon death (McCarthy and Ellery 1995; Ringrose et al. 2014; Struyf et al. 2015).

Most of the silica required for silcrete formation within the Okavango-Makgadikgadi system was likely provided by the Okavango River. The Okavango has an average annual discharge of $\sim 1 \times 10^{10}$ m^3 where it enters Botswana, with an annual flood pulse generated by seasonal rains over its headwater areas in the Angolan highlands. In most years, floodwaters dissipate within the Okavango Delta. However, in years of exceptional flood, waters may extend along the Boteti River and into the Makgadikgadi depression. Substantially higher flows during the late Quaternary formed extensive palaeolakes linking the now discrete Ngami, Mababe and Makgadikgadi basins to the Okavango Delta (cf. Burrough et al. 2009). The Okavango River presently delivers an estimated 360,000 tonnes of solutes to the Okavango Delta per year, ~50% of which can be dissolved silica, with concentrations up to 20 mg/L (McCarthy and Ellery 1998; McCarthy 2006); there is negligible solute input from rainfall (Milzow et al. 2009). Silica transport occurs mainly in solution as undissociated monosilicic acid, either as the monomer H_4SiO_4 or the dimer $H_6Si_2O_7$ (Dove and Rimstidt 1994), although organic or inorganic complexes may also be formed. As the Okavango-Makgadikgadi system is endorheic, most of the silica supplied during low flows and large floods (other than that lost via deflation) is retained in the basin and is therefore available for silcrete formation.

Silica precipitation is controlled by the concentration of silica in solution, itself related to silica availability and a range of other factors (Nash and Ullyott 2007). The most soluble silica species will be precipitated first from a supersaturated solution; this is usually amorphous silica (Millot 1960). Silica monomers polymerise and aggregate to form colloids and these colloids may precipitate to form opal-A (Williams et al. 1985). If solutions are supersaturated with respect to quartz then precipitation can occur in the absence of other changes, providing solutions are slow moving (Summerfield 1982).

Silica precipitation is influenced by pH, Eh, evaporation, the presence of other dissolved constituents and life processes. A fall in the pH of surface and/or groundwater to below pH 9 may initiate silica precipitation due to its effect upon silica solubility. Eh has a complex effect. As discussed by Nash and Ullyott (2007), quartz placed in solutions containing ferrous iron has been shown to become more soluble following exposure to oxidising conditions. Silica solubility is reduced if adsorption onto iron or aluminium oxides occurs. In contrast, the presence of NaCl may enhance solubility due to the ionic strength effect (Dove and Rimstidt 1994). A range of solutes, including Fe^{3+}, UO^{2+}, Mg^{2+}, Ca^{2+}, Na^{2+} and F^{-}, react with dissolved silica to form complexes, thereby increasing silica solubility (Dove and Rimstidt 1994). Biological influences upon silica precipitation have been proposed and may be of local significance. Shaw et al. (1990), for example, have documented the role of cyanobacteria in silcrete development in Sua Pan, whilst McCarthy and Ellery (1995) identified transpiration from aquatic grasses as a factor in the precipitation of amorphous silica from groundwater beneath islands in the Okavango Delta.

Groundwater, drainage-line and pan/lacustrine silcretes

As noted in Sect. 13.3.1, non-pedogenic silcretes occur as three main sub-types: groundwater, drainage-line and pan/lacustrine. The only location where a groundwater silcrete has been described in the Botswana Kalahari is within a core drilled near Tsodilo Hills (Linol et al. 2015). This core intersects a 42 m thick Kalahari Group sequence comprising: (i) a 2 m calcrete that caps the present-day land surface, underlain by; (ii) 8 m of sandy carbonate; and (iii) a further 32 m of sandy fluvial sediments. The succession overlies pre-Kalahari sandstones and green mudstones that are increasingly silicified upward to form a 4 m thick silcrete at the interface with the Kalahari Group. Linol et al. (2015, p. 197) propose that the silcrete 'likely formed by deep weathering and groundwater activity at the discontinuity that defines the base of the Kalahari Group [and] possibly correlates with the Nxau-Nxau (and Etosha) Calcrete Formation'. No further details are provided of the silcrete chemistry or mineralogy. However, as identified for groundwater silcretes developed in arenaceous sediments elsewhere, silica precipitation was likely due to meteoric and groundwater mixing (e.g. Milnes et al. 1991; Milnes and Thiry 1992; Ullyott et al. 1998; Thiry 1999; Basile-Doelsch et al. 2005; Lee and Gilkes2005; Ullyott and Nash 2006).

Drainage-line silcretes are closely related to groundwater sub-types but develop *within* alluvial fills in current or former fluvial systems, as opposed to within bedrock *marginal* to valley systems (as is the case with many groundwater silcretes in France and Australia). Few models for the formation of drainage-line silcrete exist, in part due to a lack of such silcretes cropping out in the geomorphological context in which they formed (Nash and Ullyott 2007). Perhaps the best-documented example occurs in the Boteti River (Fig. 13.9a). Here, analyses of surface exposures and samples from two ~20 m long cores drilled into the riverbed and adjacent floodplain at Samedupi Drift reveal a complex sequence of duricrusts (Fig. 13.12). Massive silcrete layers reaching 1–2 m thick, exposed in the riverbed, are suggested by Shaw and Nash (1998) to have developed via the accumulation of clastic silica transported by annual floodwaters alongside precipitation of amorphous silica from shallow groundwater. Silica precipitation is likely to have been induced by near-surface evaporation and transpiration from aquatic grasses within seasonal pools in the channel (cf. McCarthy and Ellery 1998). In contrast, massive and pisolitic silcrete layers deep within the channel alluvium likely formed in response to salinity and pH shifts associated with vertical and lateral movements of the wetting front during flood events. Shaw and Nash (1998) propose that conditions beneath the channel floor shifted from unsaturated to saturated during groundwater recharge by floodwater, with

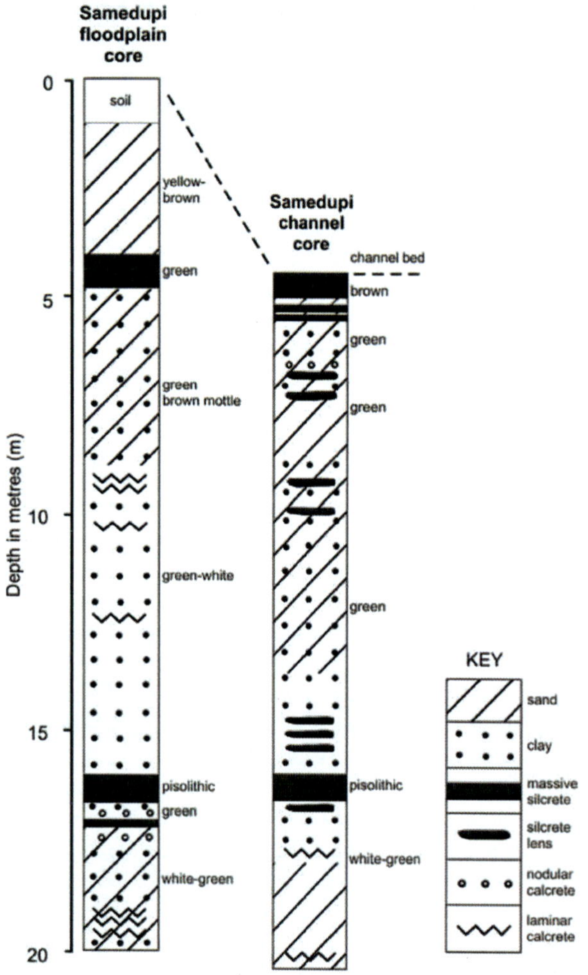

Fig. 13.12 Cores extracted from the bed of the Boteti River at Samedupi Drift, showing a range of geochemical sediments developed beneath the channel floor, including massive and pisolithic drainage-line silcretes (after Shaw and Nash 1998)

associated drops in salinity due to dilution effects; silica precipitation subsequently occurred as the water table fell.

As noted in Sect. 13.1, silcretes are commonly found in ephemeral pan and lake settings in Botswana. In most modern evaporitic lacustrine environments, silica precipitation is driven by changes in salinity and pH (Thiry 1999), both of which can vary spatially and temporally. An ingress of rain- or flood-water into an ephemeral lake, as noted above, can lead to fluctuations in salinity and pH and generate phases of silica mobility interspersed with periods of precipitation (Summerfield 1982). The zone of maximum mixing of 'fresh' and more saline lake waters is likely to occur around the borders of the evaporitic depression and immediately above the water table (Thiry 1999). This may explain why silicification often occurs marginal to ephemeral lakes and pans and may be linked to regression or to near-surface groundwater trapped by impermeable substrates (e.g. Summerfield 1982; Bustillo and Bustillo 1993, 2000; Armenteros et al. 1995; Alley 1998; Ringrose et al. 2005). This is well-illustrated in the Kalahari (Fig. 13.13), where silcrete lenses within lake-marginal duricrust sequences often formed in association with former water tables (Summerfield 1982). Biological fixing of silica may also occur in lake environments. Sua Pan, for example, contains sheet-like silcretes developed as a result of the desiccation of formerly floating colonies of the silica-fixing cyanobacteria *Chloriflexus* (Shaw et al. 1990).

13.4 Silcrete-Calcrete Intergrade Duricrusts

So far in this chapter, the main categories of duricrust in Botswana have been discussed as discrete forms. However, silcrete and calcrete are only the chemical end members of a complex spectrum of silica- and carbonate-cemented duricrust types. For example, calcretes often contain evidence for localised or more pervasive silicification while, less commonly, silcretes may be partly calcified. Further, some duricrusts may contain silica and carbonate cements in close to equal proportions and cannot be easily classified as either silcrete or calcrete per se under existing definitions. Nash and Shaw (1998) proposed the term *silcrete-calcrete intergrade duricrust* to describe such hybrid materials.

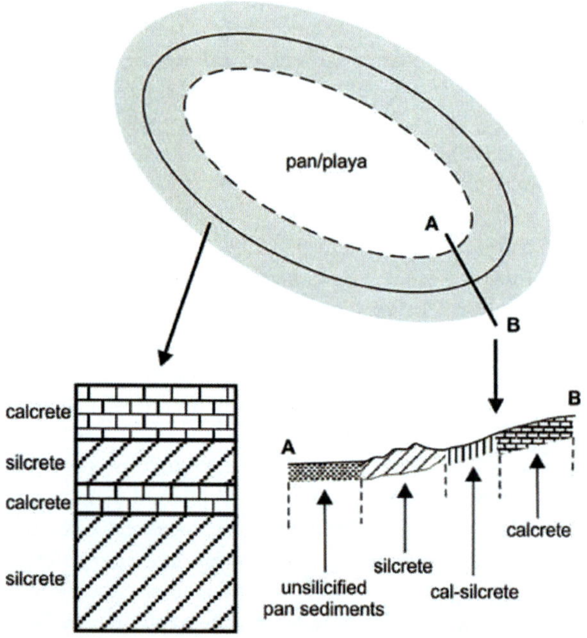

Fig. 13.13 Schematic representation of geochemical sedimentation patterns in the vicinity of a pan or ephemeral lake (after Summerfield, 1982)

Silcrete-calcrete intergrade duricrusts have been identified across the world, including North and South America, North Africa and Australia (see Nash et al. 2004, for a review). However, one of the most important areas of occurrence is southern Africa, where intergrade duricrust varieties form a conspicuous component of the Kalahari Group sediments (e.g. Passarge 1904; MacGregor 1931; Wright 1978; Watts 1980; Summerfield 1982; Shaw and de Vries 1988; Nash et al. 1994a, b; Nash and Shaw 1998; Ringrose et al. 2002, 2014; Kampunzu et al. 2007).

13.4.1 Definitions

Much of the primary research into silcrete-calcrete intergrade duricrusts has taken place in Botswana. Here, three main types of intergrade material have been defined on the basis of silica-carbonate associations within the duricrust cement (Nash and Shaw 1998). These are: (a) *sil-calcrete*, a term normally used where extensive secondary silicification has occurred within a pre-existing calcrete (but see below); (b) *cal-silcrete*, where secondary carbonate has been precipitated within a pre-existing silcrete; and (c) materials where layers of silica and carbonate cement appear to have been precipitated either contemporaneously or in close succession—these are defined according to the dominant chemical cement present. Further sub-types can be identified dependent upon whether secondary materials have either replaced pre-existing cements or have precipitated in pore spaces and voids, normally during the latter stages of development of a pre-existing duricrust.

Nash and Shaw (1998) offer chemical criteria for distinguishing the varieties of intergrade duricrust. As noted in Sect. 13.3, where the SiO_2 content of a duricrust is $\geq$ 85 wt % it is termed a silcrete. Providing a chemical definition for calcrete is more problematic since—as considered in Sect. 13.2—the $CaCO_3$ content may vary from a few percent when weakly developed up to ~ 100% for an indurated hardpan. Following Goudie (1973), Nash and Shaw (1998) propose a $CaCO_3$ content of 50 wt% as the lower boundary for a well-indurated calcrete. If bulk chemical analysis shows that a duricrust sample falls into an intermediate range it is classified according to the dominant cementing agent; cal-silcrete where silica dominates the cement but $\geq$ 15 wt % $CaCO_3$ is also present, or sil-calcrete in a calcrete where $\geq$ 50 wt% silica is present. For samples falling close to these boundaries, point-count analysis of the percentage of silica or carbonate cement present is recommended to overcome the effect of host material influences upon bulk chemical composition.

It should be noted that the terms cal-silcrete and sil-calcrete do not necessarily indicate that a cement is primary or secondary in a stratigraphic sense. For example, cal-silcrete may refer to a silcrete with secondary $CaCO_3$ cement but may also be applied to an extensively silicified calcrete where secondary silica has replaced carbonate to become the dominant cement type (Nash et al. 2004).

13.4.2 Processes of Silcrete-Calcrete Intergrade Duricrust Development

Limited research has been undertaken into the geomorphological factors controlling the distribution of silcrete-calcrete intergrade duricrusts. Experience from Botswana, however, suggests that cal-silcrete and sil-calcrete are more prevalent in association with drainage features (Mallick et al. 1981; Summerfield 1982; Shaw and de Vries 1988; Nash et al. 1994a, 2004; Ringrose et al. 2002, 2014; Nash and McLaren 2003; Kampunzu et al. 2007). This is likely because such landscape settings offer the greatest potential for pre-existing calcrete and silcrete to be subjected to chemical alteration as episodic solute-bearing surface waters infiltrate into the sub-surface and/or groundwater tables fluctuate under wetter/drier conditions.

Much of the work on silcrete-calcrete intergrade crusts in Botswana has involved the analysis of small outcrops or grab samples. Where extensive outcrops of silcrete or calcrete are present, notably in fossil valley systems and around lakes and pans, the exposures have been invariably subject to long periods of weathering, making the interpretation of cement interrelationships problematic. There are, however, locations where relatively 'fresh' exposures of intergrade duricrusts have been described and their origins interpreted.

Intergrade duricrusts in pan-lacustrine settings

One of the most detailed process-related analyses of silcrete-calcrete intergrade duricrusts has been undertaken at Kang Pan (Nash et al. 2004). Here, a ~50 by 30 m roadstone quarry cut into the pan floor exposes a maximum of 5.5 m of duricrust. The exposure exhibits a massive jointed blocky appearance near the base and thinner layers towards the surface. Thin-section evidence indicates that the duricrust was originally a groundwater calcrete but has been extensively altered by silicification. The exposure is now dominated by silica-rich duricrusts in lower sections that grade upwards into materials with more $CaCO_3$-rich cements (Fig. 13.14). This gradation is not, however, progressive. Silicification occurs in distinct zones in lower parts of the quarry, both at the edges and towards the centres of inter-joint blocks, with chalcedony and cryptocrystalline silica cements replacing the original sparry calcite matrix, and chalcedony and opaline silica precipitating in void spaces. These zones have sharp boundaries with adjacent areas of unaltered calcrete and are rimmed by thin (<10 mm) zones of powdery calcite, which appears to have been mobilised during silicification and then precipitated. Late-stage silicification has focused along cracks and fractures, mainly in upper parts of the profile, leading to the development of pale green silica-carbonate precipitates within and adjacent to some joints. Nash et al. (2004) suggest that the degree of silicification at the macroscale was determined by the position of the water table, with more extensive calcite replacement near the base of the profile associated with periods when groundwater levels were higher relative to the present day. At a local scale, the specific locations of silica replacement appear to be controlled by the degree of penetration of the wetting front into the pre-existing calcrete, both along joints and through the calcrete matrix.

Other studies of intergrade crusts in pan settings have combined sediment geochemistry with petrography to draw inferences about the environment of duricrust formation. For example, silcretes, calcretes and sil-calcretes formed within palaeoshoreline sediments have been examined at eleven sites in northern Sua Pan. These shorelines mark lake highstands within the Makgadikgadi sub-basin of the Makgadikgadi-Okavango-Zambezi rift depression and range in elevation from 945 m asl (~50 m above the present pan floor) to 904 m asl (close to the present pan margin) (Ringrose et al. 2005, 2009). Silcretes and sil-calcretes appear to have formed through the silicification of pre-existing calcretes, with the degree of silicification varying from 'silicified litho-clasts', lenses and nodules within calcrete profiles to the complete silicification of already part-silicified

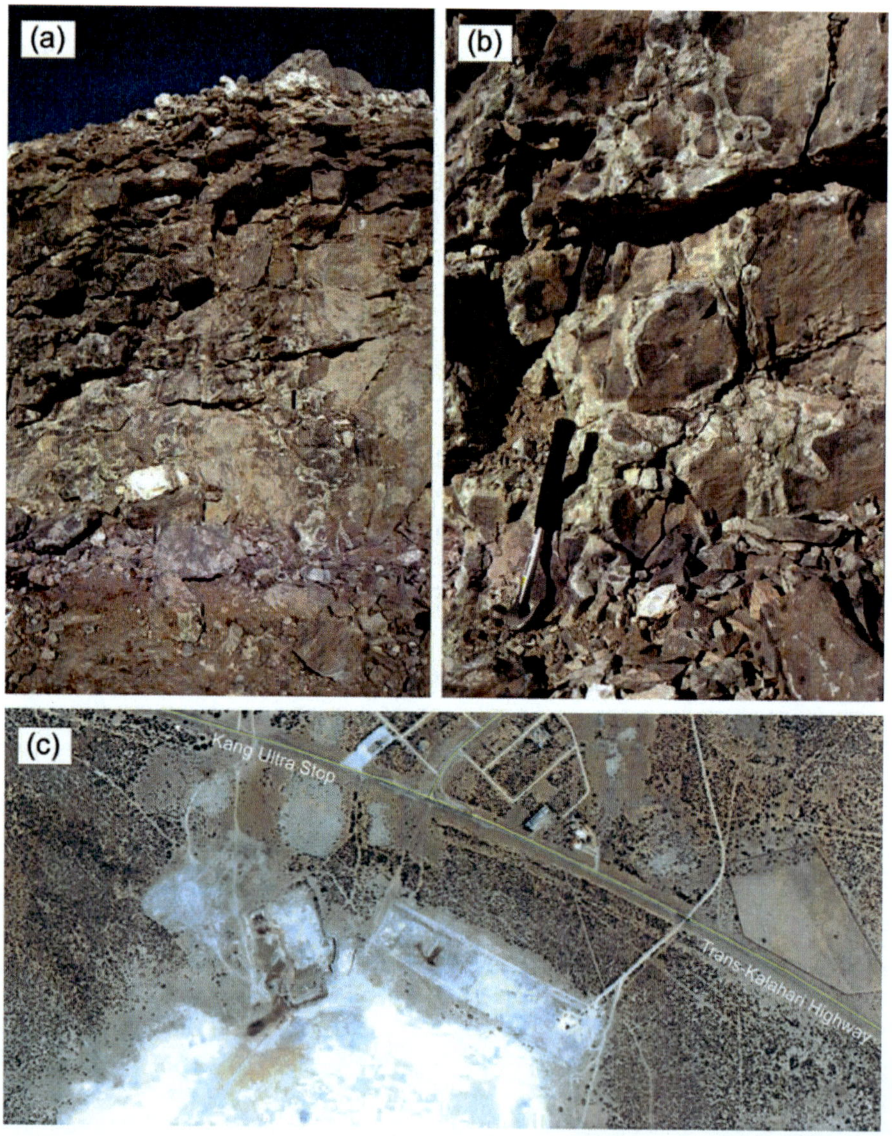

Fig. 13.14 Views of the aggregate quarry cut into the northern rim of Kang Pan: **a** Overview of the 5.5 m sequence of silcrete, calcrete and silcrete-calcrete intergrade duricrusts; **b** Close-up showing patchy silicification and alteration of pre-existing calcrete. **c** Aerial view of the quarry, showing its proximity to the Trans-Kalahari Highway. Images (**a**) and (**b**) © David Nash. Image (c) courtesy of Google Earth (date of image, 12 June 2016)

calcrete (Ringrose et al. 2005, p. 283). Chemical data show that the precursor calcrete formation took place under closed basin type evaporative conditions. As at Kang Pan, silicification occurred during periods of elevated water tables when the normally saline groundwater was diluted by relatively minor rainfall or inflow events (Ringrose et al. 2005). Repeated wet-dry cycles, with associated higher-lower water tables, are argued to have led to the progressive replacement of calcite cements within the precursor calcretes (Ringrose et al. 2009). The geochemistry of the groundwater during these events more closely resembled present-day Na-CO_3-SO_4-Cl-type brines (Ringrose et al. 2005).

Intergrade duricrusts in river and valley contexts

The other locations where intergrade duricrusts have been investigated are associated with dry valleys (see Chap. 11) and palaeo-estuaries. The aggregate quarry at Tswaane (Fig. 13.15), in the Okwa valley east of the Trans-Kalahari Highway, exposes a 3.5 to 4 m thick sequence of calcrete, silcrete and silcrete-calcrete intergrade duricrusts overlying >8 m of bedrock (see Nash et al. 2004). The duricrust exposure comprises (from the surface down) 0.5–1 m of calcrete overlying 3–3.5 m of pale green silcrete and cal-silcrete. Inherited calcrete textures within the silcrete, however, suggest that much of the silcrete and cal-silcrete formed via the replacement of a pre-existing micritic calcrete. The uppermost calcrete is suggested to have developed on top of this silcrete and post-dates the silicification process. This is evidenced by brecciation of the uppermost silcrete by calcite, fretting of the margins of brecciated silcrete fragments, and the presence of vertically-oriented strings of sparite crystals along cracks and late-stage calcite void fills within the silcrete. Nash et al. (2004) suggest that the initial replacement of calcrete was by groundwater silicification mechanisms, most likely beneath a former valley floor prior to incision of the Okwa. This was followed by a further phase (or phases) of non-pedogenic calcrete development, accompanied by the gravitational precipitation of calcite within cracks and voids to produce cal-silcretes.

Cal-silcretes, sil-calcretes and cal-ferricretes have been described in the flanks of the fossil valleys that converge around Letlhakeng (Boocock and van Straten 1962; Shaw and de Vries 1988; Nash et al. 1994a; Ringrose et al. 2002; Kampunzu et al. 2007). These valleys form the southeast part of the Okwa-Mmone drainage system, rising at the Kalahari-Limpopo watershed and flowing northwards towards the Makgadikgadi Basin (Nash and Eckardt 2016). Silcretes and sil-calcretes are exposed extensively along the Goatlhabogwe-Moshaweng valley north and east of Letlhakeng; Nash et al. (1994a) have described the nature of outcrops along the main valley flanks and Ringrose et al. (2002) sequences within lower (sub-rim) valley surfaces. The thickness of the duricrust sequence is ~ 18.5 m (Gwosdz and Modisi 1983), and, taking outcrops on both valley flanks into consideration, the total length of duricrust exposure exceeds 80 km (Nash et al., 1994a). Ringrose et al. (2002) report that duricrusts can be identified in three distinct levels: an upper level containing abundant sil-calcrete; an intermediate level in which sil-calcrete underlies nodular calcrete; and a lowest level where sil-calcrete and calcrete are interbedded. However, systematic profile mapping by Nash et al. (1994a) indicates that the situation may be more complex.

The environmental conditions required to generate the suite of duricrusts around Letlhakeng remain only partly understood. Ringrose et al. (2002) proposed that early calcrete formation in the area was initiated during wet periods when abundant Ca^{++}-rich groundwater flowed along the structurally aligned valley systems to converge at Letlhakeng. During subsequent drier periods, water table fluctuations led to the precipitation of nodular calcretes. For reasons unknown, the geochemistry of the valley hydraulic system then changed and became increasingly saline, leading to the preferential silicification of the nodular calcrete deposits to form sil-calcretes. Kampunzu et al. (2007) used trace element data to constrain the pH associated with silicification to between pH 6.5 and 8 under alternating wetter and drier environments. However, the patchy distribution of silcrete and sil-calcrete along the Goatlhabogwe-Moshaweng valley (Nash et al., 1994a) implies that pH must also have varied spatially. Little consideration has been given to changing environmental conditions as the valleys incised over time and regional groundwater systems were increasingly focussed towards the Letlhakeng area.

The most recent analyses of silcrete-calcrete intergrade crusts come from the palaeo-estuary of the Boteti River where it enters the Makgadikgadi Basin. Using examples from the proximal and distal palaeo-estuary, Ringrose et al. (2014) provide insights into the pre-cursor environmental conditions required for the formation of silcrete-calcrete intergrade crusts, with particular attention paid to those crusts where SiO_2 and $CaCO_3$ cements appear to have formed penecontemporaneously. The study again highlights the importance of cyclical freshwater inflows into saline depressions, at timescales from seasonal to centennial, to trigger changes in groundwater pH and hence silica precipitation in carbonate-rich settings. The most interesting insight from the analyses, however, is the importance of micro-fossils in providing a source of more soluble silica and carbonate. Intergrade crusts comprising irregular zones of siliceous sediment formed within otherwise calcareous deposits are suggested to relate to the irregular occurrence of biogenic silica in the source sediments, inferring a source for local mobilisation of silica.

Taken as a whole, these studies illustrate the crucial role of groundwater salinity/pH and the relative height of the

Fig. 13.15 Views of the **a** eastern and **b** northern face of the aggregate quarry at Tswaane on the Okwa Valley, 12 km east of the Trans-Kalahari Highway, where a complex sequence of calcrete, silcrete and silcrete-calcrete intergrade duricrusts are exposed above weathered Precambrian bedrock. Note figure for scale, bottom left of image (**b**). **c** Aerial view of the quarry, showing its proximity to the Okwa Valley. Images (**a**) and (**b**) © David Nash. Image (**c**) courtesy of Google Earth (date of image, 11 February 2006)

groundwater table in controlling if and where the replacement of silica and $CaCO_3$ cements takes place to generate intergrade duricrusts. Equivalent studies of the silicification of limestone (e.g. in the Paris Basin; Thiry 1999) indicate that there is a strong relationship between voids and silicification. This also appears to be the case for sil-calcrete development, where joints and other structural features within the pre-cursor calcrete act as foci for local or more pervasive silicification. What is now needed is a greater understanding of the causes of spatial variations in groundwater pH that lead to the patchy silicification of calcrete (or calcretisation of silcrete). Could it be, for example, that changes in pH associated with the wetting/drying of sediments beneath relatively short-lived pools are required, as proposed by Ringrose et al. (2002) for the formation of sil-calcretes at Sua Pan? Or maybe the periodic flushing of saline groundwater systems by infiltrating floodwaters, as discussed by Shaw and Nash (1998) in relation to silcretes in the Boteti River? Or the presence of amorphous biogenic silica in the host sediments, as discussed for the Boteti palaeo-estuary (Ringrose et al. 2014)? Botswana, with its substantial outcrops of intergrade duricrusts, offers a natural laboratory to address these questions in the future.

13.5 Determining the Age of Kalahari Duricrusts

Determining the timing of calcrete and silcrete formation in the Botswana Kalahari is not easy. No radiometric dating techniques are available to ascertain the age of silcrete. Mallick et al. (1981, pp. 25–26) suggested that duricrusts in

Botswana could be divided into two broad groups. The first, which they termed 'older' duricrusts, are mostly found in peri-Kalahari locations and include the calcretes that outcrop high up the flanks of drainage systems such as the Molopo, Auob, Nossop and Okwa-Mmone (Shaw and de Vries 1988; Nash et al. 1994a; Ringrose et al. 2002). Pedogenic calcretes in intervening areas may also fall into this 'older' grouping. 'Younger' calcrete duricrusts occur on the floors or as low terraces towards the centre of drainage lines below these older calcrete cliffs.

The ages of the 'older' duricrusts remain unknown. Silcretes and calcretes found at depth within the Kalahari Group are suggested on the basis of stratigraphic correlation to be of mid-Miocene age (Haddon and McCarthy 2005). Linol et al. (2015) use lithostratigraphic comparisons with the 'Polymorph Sandstones' at the base of the Kalahari Group in the Congo Basin to argue that the Nxau-Nxau Calcrete Formation may even be Eocene in age. However, attempts at correlation are complicated. Calcretes and silcretes necessarily develop within pre-existing bodies of sediment and therefore make poor stratigraphic markers. Further, while pedogenic calcretes form in association with palaeosurfaces, groundwater calcretes and silcretes can, in theory, develop at any depth, since cementation is linked to the position of the water table (Netterberg, 1969b; Nash and Smith 1998; Basile-Doelsch et al. 2005). Duricrusts have nonetheless been used as markers in some stratigraphic schemes for the Kalahari Group in Botswana (e.g. du Plessis 1993) and adjoining areas of South Africa (e.g. Thomas 1981). Bulk samples from 'younger' surface calcrete outcrops have been radiocarbon dated to the late Pleistocene (e.g. Cooke and Verhagen 1977; Cooke 1984; Shaw et al. 1992). Such dates can only be regarded as minimum age estimates, however, given that calcretes commonly contain multiple generations of carbonate cement that build up within sediment pore-spaces over time.

13.6 Conclusions

This chapter has provided an overview of the development of calcrete, silcrete and related intergrade duricrusts in Botswana. It is apparent from the text that the processes responsible for the formation of many silcrete and calcrete sub-types are well understood. There are, however, areas where future investigations in Botswana could help address globally significant research questions. For example, the formation of non-pedogenic calcretes and silcretes is less well understood compared with their pedogenic cousins. Detailed petrographic and geochemical analyses of silcretes in and around the Makgadikgadi Basin, for example, could not only shed new light on pan/lacustrine silcrete formation but also provide a window on processes of basin evolution over time. Studies of calcretes found in association with Kalahari pans could yield similar insights for smaller evaporitic basins. The widespread exposure of duricrusts in Botswana offers real potential for understanding the factors influencing calcrete and silcrete distribution at a landscape scale. However, to answer this question, much better mapping of duricrusts sub-types is needed at a national and, ideally, regional level. At present, our understanding of the distribution of silcrete and calcrete sub-types is limited to 'joining the dots' between outcrops in published studies.

References

Alley NF (1998) Cainozoic stratigraphy, palaeoenvironments and geological evolution of the Lake Eyre Basin. Palaeogeogr Palaeoclimatol Palaeoecol 144:239–263

Alonso-Zarza AM (2003) Palaeoenvironmental significance of palustrine carbonates and calcretes in the geological record. Earth Sci Rev 60:261–298

Arakel AV (1986) Evolution of calcrete in palaeodrainages of the Lake Napperby area, central Australia. Palaeogeogr Palaeoclimatol Palaeoecol 54:283–303

Arakel AV (1991) Evolution of Quaternary duricrusts in Karinga Creek drainage system, central Australian groundwater discharge zone. Aust J Earth Sci 38:333–347

Arakel AV, McConchie D (1982) Classification and genesis of calcrete and gypsum lithofacies in palaeodrainage systems of inland Australia and their relationship of carnotite mineralization. J Sediment Petrol 52:1149–1170

Armenteros I, Bustillo MA, Blanco JA (1995) Pedogenic and groundwater processes in a closed Miocene basin (Northern Spain). Sed Geol 99:17–36

Association of Southern African National Roads Agencies, 2014. Review of specifications for the use of laterite in road pavements. ASANRA, Lilongwe.

Bachman GO, Machette MN (1977) Calcic soils and calcretes in the southwestern United States. US Geological Survey Open File Report, pp 77–794

Barnes I (1965) Geochemistry of Birch Creek, Inyo County, California. Geochim Cosmochim Acta 29:85–112

Basile-Doelsch I, Meunier JD, Parron C (2005) Another continental pool in the terrestrial silicon cycle. Nature 433:399–402

Boocock C, van Straten OJ (1962) Notes on the geology and hydrogeology of the Central Kalahari region, Bechuanaland Protectorate. Transactions of the Geological Society of South Africa 65:125–171

Botha GA (2000) Paleosols and duricrusts. In: Partridge TC, Maud RR (eds) The Cenozoic of Southern Africa. Oxford University Press, Cape Town, pp 55–72

Burrough SL, Thomas DSG, Singarayer JS (2009) Late Quaternary hydrological dynamics in the Middle Kalahari: forcing and feedbacks. Earth Sci Rev 96:313–326

Bustillo MA, Bustillo M (1993) Rhythmic lacustrine sequences with silcretes from the Madrid Basin, Spain: Geochemical trends. Chem Geol 107:229–232

Bustillo MA, Bustillo M (2000) Miocene silcretes in argillaceous playa deposits, Madrid Basin, Spain: petrological and geochemical features. Sedimentology 47:1023–1037

Cailleau G, Braissant O, Verrecchia E (2004) Biomineralization in plants as a long term carbon sink. Naturwissenschaften 91:191–194

Capo RC, Chadwick OA (1999) Sources of strontium and calcium in desert soil and calcrete. Earth Planet Sci Lett 170:61–72

Carlisle D (1983) Concentration of uranium and vanadium in calcretes and gypcretes. In: Wilson, R.C.L. (Ed.), Residual Deposits: Surface Related Weathering Processes and Materials. Geological Society of London Special Publication 11, pp 185–195

Chadwick OA, Hendricks DM, Nettleton WD (1989) Silicification of Holocene soils in Northern Monitor Valley, Nevada. Soil Sci Soc Am J 53:158–164

Chiquet A, Michard A, Nahon D, Hamelin B (1999) Atmospheric input vs in situ weathering in the genesis of calcretes: An Sr isotope study at Galvez (Central Spain). Geochim Cosmochim Acta 63:311–323

Cooke HJ (1984) The evidence from northern Botswana of late Quaternary climatic change. In: Vogel JC (ed) Late Cainozoic palaeoenvironments of the southern hemisphere. Balkema, Rotterdam, pp 265–278

Cooke HJ, Verhagen BT (1977) The dating of cave development: an example from Botswana, Proceedings of the Seventh International Speleological Congress, Sheffield

Dixon JC, McLaren SJ (2009) Duricrusts. In: Parsons AJ Abrahams AD (eds) Geomorphology of Desert Environments, 2nd edn. Springer, Berlin, pp 123–151

Dorland HC (1999) Paleoproterozoic laterites, red beds and ironstones of the Pretoria Group with reference to the history of atmospheric oxygen. Unpublished Magister Scientae thesis. Rand Afrikaans University, Johannesburg

Dove PM, Rimstidt JD (1994) Silica-water interactions. In: Heaney, P. J., Prewitt, C.T., Gibbs, G.V. (Eds.), Silica: Physical Behaviour, Geochemistry and Materials Applications. Reviews in Mineralogy 29. Mineralogical Society of America, Washington DC, pp 259–308

du Plessis PI (1993) The sedimentology of the Kalahari Group in four study areas in northern Botswana. Unpublished MSc Thesis, University of Stellenbosch

du Toit AL (1954) The Geology of South Africa, 3rd edn. Oliver and Boyd, Edinburgh

Eckardt FD, Bryant RG, McCulloch GP, Spiro BF, Wood WW (2008) The hydrochemistry of a semi-arid pan basin case study: Sua Pan, Makgadikgadi, Botswana. Appl Geochem 23:1563–1580

Ellis F, Schloms BHA (1984) Distribution and properties of dorbanks (duripans) in south and south-western South Africa. In:Poster at the 12th Congress of the Soil Science Society of South Africa, Bloemfontein

Fey M (2010) Soils of South Africa. Cambridge University Press, Cape Town

Gile LH, Peterson FF, Grossman RB (1966) Morphological and genetic sequences of carbonate accumulation in desert soils. Soil Sci 101:347–360

Goudie AS (1972) The chemistry of world calcrete deposits. J Geol 80:449–463

Goudie AS (1973) Duricrusts in Tropical and Subtropical Landscapes. Clarendon Press, Oxford

Goudie AS (1983) Calcrete. In: Goudie AS, Pye K (eds) Chemical Sediments and Geomorphology. Academic Press, London, pp 93–131

Goudie AS, Viles HA (2015) Landscapes and Landforms of Namibia. Springer

Grove AT (1969) Landforms and Climatic Change in the Kalahari and Ngamiland. The Geographical Journal 135:191–212

Gwosdz W, Modisi MP (1983) The Carbonate Resources of Botswana. Mineral Resources Report 6. Botswana Geological Survey, Lobatse

Haddon IG, McCarthy TS (2005) The Mesozoic-Cenozoic interior sag basins of Central Africa: The Late-Cretaceous-Cenozoic Kalahari and Okavango basins. J Afr Earth Sc 43:316–333

Kampunzu AB, Ringrose S, Huntsman-Mapila P, Harris C, Vink BW, Matheson W (2007) Origins and palaeo-environments of Kalahari duricrusts in the Moshaweng dry valleys (Botswana) as detected by major and trace element composition. J Afr Earth Sc 48:199–221

Klappa CF (1983) A process-response model for the formation of pedogenic calcretes. In: Wilson, R.C.L. (Ed.), Residual deposits: Surface related weatherng processes and materials. Geological Society of London, Special Publication 11, pp 211–220

Lamplugh GW (1902) Calcrete. Geol Mag 9:575

Lamplugh GW (1907) The geology of the Zambezi Basin around the Batoka Gorge (Rhodesia). Quarterly Journal of the Geological Society of London 63:162–216

Lawrence CJ, Toole T (1984) The location, selection and use of calcrete for bituminous road construction in Botswana. TRRL Laboratory Report 1122. Transport and Road Research Laboratory, Crowthorne, Berkshire

Lee SY, Gilkes RJ (2005) Groundwater geochemistry and composition of hardpans in southwestern Australian regolith. Geoderma 126:59–84

Linol B (2013) Sedimentology and sequence stratigraphy of the Congo and Kalahari Basins of south-central Africa and their evolution during the formation and break-up of West Gondwana. Unpublished PhD thesis. Nelson Mandela Metropolitan University.

Linol B, de Wit MJ, Guillocheau F, de Wit MCJ, Anka Z, Colin J-P (2015) Formation and collapse of the Kalahari duricrust ('African Surface') across the Congo Basin, with implications for changes in rates of Cenozoic offshore sedimentation. In: de Wit MJ, Guillocheau F, de Wit MCJ (eds) Geology and Resource Potential of the Congo Basin. Springer-Verlag, Berlin, pp 193–210

MacGregor AM (1931) Geological notes on a circuit of the Great Makarikari Salt Pan, Bechuanaland Protectorate. Transactions of the Geological Society of South Africa 41:211–224

Machette MN (1985) Calcic soils of the south-western United States. Geol Soc Am Spec Pap 203:1–21

Malherbe SJ (1984) The geology of the Kalahari Gemsbok National Park. Koedoe 27(Supplement):33–44

Mallick DIJ, Habgood F, Skinner AC (1981) A Geological Interpretation of LANDSAT Imagery and Air Photography of Botswana. HMSO, London

McCarthy TS (2006) Groundwater in the wetlands of the Okavango Delta, Botswana, and its contribution to the structure and function of the ecosystem. J Hydrol 320:264–282

McCarthy TS, Ellery WN (1995) Sedimentation on the distal reaches of the Okavango Fan, Botswana, and its bearing on calcrete and silcrete (ganister) formation. J Sedimentary Res Sect A: Sedimentary Petrol Process 65:77–90

McCarthy TS, Ellery WN (1998) The Okavango Delta. Trans R Soc South Africa 53:157–182

McCulloch GP, Irvine K, Eckardt FD, Bryant RG (2008) Hydrochemical fluctuations and crustacean community composition in an ephemeral saline lake (Sua Pan, Makgadikgadi Botswana). Hydrobiologia 596:31–46

McFarlane MJ, Eckardt FD, Coetzee SH, Ringrose S (2010) An African surface weathering profile in the Kalahari of North West Ngamiland, Botswana: processes and products. Zeitschrift Für Geomorphologie 54:273–303

Meixner HM, Peart RJ (1984) The Kalahari Drilling Project. Bulletin 27. Botswana Geological Survey, Lobatse

Millot G (1960) Silice, silex, silicifi cations et croissance des cristaux. Bulletin De Service Carte Geologique, Alsace Lorraine 13:129–146

Milnes AR, Thiry M (1992) Silcretes. In: Martini IP, Chesworth W (eds) Weathering, soils and palaeosols. Developments in Earth Surface Processes 2. Elsevier, Amsterdam, pp 349–377

Milnes AR, Wright MJ, Thiry M (1991) Silica accumulations in saprolites and soils in South Australia. In: Nettleton WD (ed) Occurrence, Characteristics and Genesis of Carbonate, Gypsum, and Silica Accumulations in Soils. Soil Sci Soc Am Special Publ 26:121–149

Milzow C, Kgotlhang L, Bauer-Gottwein P, Meier P, Kinzelbach W (2009) Regional review: the hydrology of the Okavango Delta, Botswana - processes, data and modelling. Hydrogeol J 17:1297–1328

Nash DJ (2011) Desert crusts and varnishes. In: Thomas DSG (ed) Arid Zone Geomorphology, 3rd edn. Wiley-Blackwell, Oxford, pp 131–180

Nash DJ (2012) Duricrusts. In: Holmes PJ, Meadows ME (eds) Southern African Geomorphology: Recent Trends and New Directions. Sun Media, Bloemfontein, pp 191–229

Nash DJ, Eckardt FD (2016) Drainage development, neotectonics and base-level change in the Kalahari Desert, southern Africa. S Afr Geogr J 98:308–320

Nash DJ, Hopkinson L (2004) A reconnaissance laser raman and fourier transform infrared survey of silcretes from the Kalahari Desert, Botswana. Earth Surf Proc Land 29:1541–1558

Nash DJ, McLaren SJ (2003) Kalahari valley calcretes: their nature, origins, and environmental significance. Quatern Int 111:3–22

Nash DJ, Shaw PA (1998) Silica and carbonate relationships in silcrete-calcrete intergrade duricrusts from the Kalahari of Botswana and Namibia. J Afr Earth Sc 27:11–25

Nash DJ, Smith RF (1998) Multiple calcrete profiles in the Tabernas Basin, southeast Spain: their origins and geomorphic implications. Earth Surf Proc Land 23:1009–1029

Nash DJ, Smith RF (2003) Properties and development of channel calcretes in a mountain catchment, Tabemas Basin, southeast Spain. Geomorphology 50:227–250

Nash DJ, Ullyott JS (2007) Silcrete. In: Nash DJ, McLaren SJ (eds) Geochemical Sediments and Landscapes. Blackwell, Oxford, pp 95–143

Nash DJ, Shaw PA, Thomas DSG (1994) Duricrust development and valley evolution: process-landform links in the Kalahari. Earth Surf Proc Land 19:299–317

Nash DJ, Thomas DSG, Shaw PA (1994) Siliceous duricrusts as palaeoclimatic indicators: evidence from the Kalahari Desert of Botswana. Palaeogeogr Palaeoclimatol Palaeoecol 112:279–295

Nash DJ, McLaren SJ, Webb JA (2004) Petrology, geochemistry and environmental significance of silcrete-calcrete intergrade duricrusts at Kang Pan and Tswaane, Central Kalahari, Botswana. Earth Surf Proc Land 29:1559–1586

Nash DJ, Coulson S, Staurset S, Ullyott JS, Babutsi M, Hopkinson L, Smith MP (2013) Provenancing of silcrete raw materials indicates long-distance transport to Tsodilo Hills, Botswana, during the Middle Stone Age. J Hum Evol 64:280–288

Nash DJ, Coulson S, Staurset S, Ullyott JS, Babutsi M, Smith MP (2016) Going the distance: Mapping mobility in the Kalahari Desert during the Middle Stone Age through multi-site geochemical provenancing of silcrete artefacts. J Hum Evol 96:113–133

Netterberg F (1969a) The geology and engineering properties of South African calcretes. CSIR Monograph, Pretoria.

Netterberg F (1969b) Ages of calcretes in southern Africa. South African Archaeological Bulletin 24:88–92

Netterberg F (1980) Geology of southern African calcretes. I. Terminology, description, macrofeatures and classification. Transactions of the Geological Society of South Africa 83:255–283

Netterberg F (1985) Pedocretes. In: Brink ABA (ed) Engineering geology of southern Africa, Vol. 4. Building Publications (CSIR Reprint RR 430), Silverton, pp 286–307

Netterberg F Caiger JH (1983) A geotechnical classification of calcretes and other pedocretes. In: Wilson RCL (ed) Residual deposits: Surface related weathering processes and materials. Geological Society Special Publication 11, Blackwell, London, pp 235–243.

Paige-Green P, Pinard M, Netterberg F (2015) A review of specifications for lateritic materials for low volume roads. Transportation Geotechnics 5:86–98

Partridge TC (1997) Cainozoic environmental change in southern Africa, with special emphasis on the last 200 000 years. Prog Phys Geogr 21:3–22

Passarge S (1904) Die Kalahari. Dietrich Riemer, Berlin

Range P (1912) Topography and geology of the German South Kalahari. Transactions of the Geological Society of South Africa 15:63–73

Reeves CC (1970) Origin, classification, and geological history of caliche on the southern High Plains, Texas and eastern New Mexico. J Geol 78:352–362

Ringrose S, Kampunzu AB, Vink BW, Matheson W, Downey WS (2002) Origin and palaeo-environments of calcareous sediments in the Moshaweng dry valley, southeast Botswana. Earth Surf Proc Land 27:591–611

Ringrose S, Huntsman-Mapila P, Basira Kampunzu A, Downey W, Coetzee S, Vink B, Matheson W, Vanderpost C (2005) Sedimentological and geochemical evidence for palaeo-environmental change in the Makgadikgadi subbasin, in relation to the MOZ rift depression, Botswana. Palaeogeogr Palaeoclimatol Palaeoecol 217:265–287

Ringrose S, Harris C, Huntsman-Mapila P, Vink BW, Diskins S, Vanderpost C, Matheson W (2009) Origins of strandline duricrusts around the Makgadikgadi Pans (Botswana Kalahari) as deduced from their chemical and isotope composition. Sed Geol 219:262–279

Ringrose S, Cassidy L, Diskin S, Coetzee S, Matheson W, Mackay AW, Harris C (2014) Diagenetic transformations and silcrete-calcrete intergrade duricrust formation in palaeo-estuary sediments. Earth Surf Proc Land 39:1167–1187

Schlesinger WH (1985) The formation of caliche in soils of the Mojave Desert, California. Geochim Cosmochim Acta 49:57–66

Schloms, B.H.A., Ellis, F., 1984. Distribution of silcretes and properties of some soils associated with silcretes in Cape Province, South Africa. Proceedings of the 12th Congress of the Soil Science Society of South Africa, Bloemfontein.

Schmidt M, Fuchs M, Henderson ACG, Kossler A, Leng MJ, Mackay AW, Shemang E (2017) Paleolimnological features of a mega-lake phase in the Makgadikgadi Basin (Kalahari, Botswana) during Marine Isotope Stage 5 inferred from diatoms. J Paleolimnol 58:373–390

Shaw AL (2009) The characterisation of calcrete based on its environmental settings within selected regions of the Kalahari, southern Africa. University of Oxford, unpublished PhD thesis, Oxford

Shaw PA, de Vries JJ (1988) Duricrust, groundwater and valley development in the Kalahari of southeast Botswana. J Arid Environ 14:245–254

Shaw PA, Nash DJ (1998) Dual mechanisms for the formation of fluvial silcretes in the distal reaches of the Okavango Delta Fan, Botswana. Earth Surf Proc Land 23:705–714

Shaw PA, Cooke HJ, Perry CC (1990) Microbialitic silcretes in highly alkaline environments: some observations from Sua Pan, Botswana. S Afr J Geol 93:803–808

Shaw PA, Thomas DSG, Nash DJ (1992) Late Quaternary fluvial activity in the dry valleys (mekgacha) of the middle and southern Kalahari, southern Africa. J Quat Sci 7:273–281

Smale D (1973) Silcretes and associated silica diagenesis in southern Africa and Australia. J Sediment Petrol 43:1077–1089

Struyf E, Mosimane K, Van Pelt D, Murray-Hudson M, Meire P, Frings P, Wolski P, Schaller J, Gondwe MJ, Schoelynck J, Conley DJ (2015) The Role of Vegetation in the Okavango Delta Silica Sink. Wetlands 35:171–181

Summerfield MA (1982) Distribution, nature and genesis of silcrete in arid and semi-arid southern Africa. Catena Supplement 1:37–65

Summerfield MA (1983) Petrography and diagenesis of silcrete from the Kalahari Basin and Cape coastal zone, southern Africa. J Sediment Petrol 53:895–909

Summerfield MA (1983) Silcrete. In: Goudie AS, Pye K (eds) Chemical Sediments and Geomorphology. Academic Press, London, pp 59–91

Thiry M (1999) Diversity of continental silicification features: examples from the Cenozoic deposits in the Paris Basin and neighbouring basement. In: Thiry M, Simon-Coinçon R (Eds), Palaeoweathering, Palaeosurfaces and Related Continental Deposits. International Association of Sedimentologists, Special Publication No 27. Blackwell Science, Oxford, pp 87–127

Thiry M, Milnes A (2017) Silcretes: Insights into the occurrences and formation of materials sourced for stone tool making. Journal of Archaeological Science-Reports 15:500–513

Thomas DSG, Shaw PA (1990) The deposition and development of the Kalahari Group sediments, central southern Africa. J Afr Earth Sc 10:187–197

Thomas DSG, Shaw PA (1991) The Kalahari Environment. Cambridge University Press, Cambridge

Thomas MA (1981) The geology of the Kalahari in the Northern Cape Province (areas 2620 and 2720). Unpublished Msc thesis, University of the Orange Free State, Bloemfontein

Thomas RJ, Thomas MA, Malherbe SJ (1988) The geology of the Nossob and Twee Rivieren areas. Republic of South Africa, Pretoria, Department of Mineral and Energy Affairs

Ullyott JS, Nash DJ (2006) Micromorphology and geochemistry of groundwater silcretes in the eastern South Downs, UK. Sedimentology 53:387–412

Ullyott JS, Nash DJ, Shaw PA (1998) Recent advances in silcrete research and their implications for the origin and palaeoenvironmental significance of sarsens. Proc Geol Assoc 109:255–270

Watts NL (1980) Quaternary pedogenic calcretes from the Kalahari (southern Africa): mineralogy, genesis and diagenesis. Sedimentology 27:661–686

Webb JA, Nash DJ (2020) Reassessing southern African silcrete geochemistry: implications for silcrete origin and sourcing of silcrete artefacts. Earth Surf Proc Land 45:3396–3413

Weinert HH (1980) Natural road construction materials of Southern Africa. CSIR, Pretoria

White K, Eckardt F (2006) Geochemical mapping of carbonate sediments in the Makgadikgadi basin, Botswana using moderate resolution remote sensing data. Earth Surf Proc Land 31:665–681

Wigley TML (1973) Chemical evolution of the system calcite-gypsum-water. Can J Earth Sci 10:306–315

Williams LA, Parks GA, Crerar DA (1985) Silica diagenesis, I. Solubility controls. J Sediment Petrol 55:301–311

Wright EP (1978) Geological studies in the northern Kalahari. Geogr J 144:235–250

Wright VP (2007) Calcrete. In: Nash DJ, McLaren SJ (eds) Geochemical Sediments and Landscapes. Blackwell, Oxford, pp 10–45

Yamaguchi KE, Johnson CM, Beard BL, Beukes NJ, Gutzmer J, Ohmoto H (2007) Isotopic evidence for iron mobilization during Paleoproterozoic lateritization of the Hekpoort paleosol profile from Gaborone, Botswana. Earth Planet Sci Lett 256:577–587

Yang W, Holland HD (2003) The Hekpoort paleosol profile in Strata 1 at Gaborone, Botswana: Soil formation during the Great Oxidation Event. Am J Sci 303:187–220

David J. Nash is Professor of Physical Geography at the University of Brighton and Honorary Research Fellow at the University of the Witwatersrand (South Africa). He has worked on the geomorphology of Botswana since 1989, when he arrived in Gaborone for the first field season of his PhD research on the fossil valleys of the Kalahari. In the intervening years he has published over 100 articles on silcrete, calcrete and Holocene-to-recent environmental change, mostly with reference to southern Africa. His present research focusses on the use of geochemical data to determine the provenance of silcrete stone tools at archaeological sites in the Kalahari.

Geodiversity of Caves and Rockshelters in Botswana

14

Mark Stephens, Mike de Wit, and Senwelo M. Isaacs

Abstract

Caves and rockshelters occur in dolomite, siliciclastic and granite rocks of varying ages in Botswana and this leads to an interesting variety of entrance, passage and cavern types controlled by fracturing, bedding, groundwater levels, springs and slope processes. The dolomitic caves include the Gcwihaba Caves in northwestern Botswana and Lobatse Caves in southeastern Botswana. The siliciclastic caves/rockshelters occur at: Tsodilo Hills in northwestern Botswana; Tswapong Hills and Shoshong in eastern Botswana; Molepolole, Lekgolobotlo (Mmalogage), Manyana, Otse and Lobatse in southeastern Botswana. The granitic caves occur within the Lepokole Hills in eastern Botswana. The formation of the caves/rockshelters and their deposits are overviewed along with description of previous studies and the potential for further research is outlined in terms of landscape evolution and geoarchaeology.

Keywords

Geodiversity • Cave • Rockshelter • Botswana • Dolomitic • Siliciclastic • Granitic

M. Stephens (✉)
School of Chemistry, Environmental and Life Sciences, Faculty of Pure and Applied Sciences, University of The Bahamas, PO Box N-4912 Nassau, Bahamas
e-mail: mark.stephens@ub.edu.bs

M. Stephens · S. M. Isaacs
Department of Environmental Science, Faculty of Science, University of Botswana, P/Bag UB00704, Gaborone, Botswana

M. de Wit
Earth Sciences Department, University of Stellenbosch, Private Bag X1, Matieland 7602, South Africa

S. M. Isaacs
Department of Archaeology, Faculty of Humanities, University of Botswana, P/Bag UB00703, Gaborone, Botswana

14.1 Introduction

Geodiversity describes the variety of geological, geomorphological and soil features within a given spatial context (Gray 2008, 2013) and is important as it helps to understand the range of abiotic features of the natural environment. It is also important as it forms the context for geoheritage—those physical aspects of an area's geodiversity that are assessed as worthy of conservation—and also geoconservation—selecting sites that are representative of an area's geodiversity that may also provide geotourism opportunities (Stephens et al. 2013; Knight et al. 2015; Ngwira 2015; Gray 2018; Stephens 2020a; Tlhapiso and Stephens 2020; Isaacs and Stephens, this book).

A cave is a natural rock cavity that often acts as a conduit for water flow between an input point, such as a streamsink, and an output point, such as a spring or seep (White 1984). Once a conduit has a diameter larger than 5–15 mm, the basic form and hydraulics do not change much to allow turbulent flow, optimizing the solution of rock, and effective sediment transport (Gillieson 1996). Small conduits < 5 mm diameter but connected to an input or output or both, are nascent cave systems carrying seepage water in the vadose zone (above the water table) or groundwater in the phreatic zone (below the water table), and at the output may allow the formation of hollows. These may coalesce to form initial rockshelters that may also or alternatively be excavated by differential erosion of softer strata or stream scour at the base of cliffs (Barton and Clark 1993). A rockshelter is thus an overhanging cliff face typically with a shallow opening. A non-scientific definition of a cave can be a natural rock cavity which is enterable by people (Gillieson 1996). This implies a minimum size of ~0.5 m diameter and are the caves that can be mapped and studied and is the minimum size used for the practical purposes of this chapter.

Caves are part of larger scale karst systems, formed principally by the solution of rock, most commonly limestone and other carbonate rocks such as dolomite. Carbonate rocks underlie some 2% of southern Africa (Marker 1988).

F. D. Eckardt (ed.), *Landscapes and Landforms of Botswana*, World Geomorphological Landscapes,
https://doi.org/10.1007/978-3-030-86102-5_14

Solution of rocks occurs in other lithologies such as evaporites (e.g. gypsum and halite), siliciclastics (e.g. sandstone and quartzite) and in some granites and basalts where conditions are favourable (e.g. Hedges 1969; Gilbert 1984; Gillieson 1996; Wray 1997; Gaál and Bella 2008). Evacuation of molten rock/meltwater, granular disintegration, hydraulic plucking and tectonic movements may also produce cave/rockshelter landforms termed 'pseudokarst' (Gillieson 1996). The geodiversity of pseudokarst caves as a reason for their scientific importance and protection has been described by Urban and Otęska-Budzyn (1998).

Caves have long been important to humankind for a wide variety of uses, including: shelter, burial, collecting bird's nests, cave art, defence, water supply, mining, hydroelectricity, sanatoria, cheese making, classical music concerts, tourism and spirituality (e.g. Gillieson 1996 and references therein; Barker et al. 2002). At present, certain caves in Botswana are used for tourism as the country aims to diversify its tourism product, such as the large caves at Gcwihaba (Mbaiwa and Sakuze 2009) and rock art at the rockshelters and caves at Manyana, Lepokole and the Tsodilo Hills (Fig. 14.1). Caves are also used for spiritual purposes as observed at Athlone Cave, Kobokwe Cave and where springs occur in the Otse Caves (Brook 2017). Brook (2015) also describes how the Maredi Rockshelter at Lepokole is used for spiritual purposes (Fig. 14.1). The Lobatse Caves were used in recent times by South African freedom fighters hiding there during the apartheid era (Campbell and Main 2003; Brook 2017). Caves and rockshelters are also very important for geoarchaeological and palaeoclimatic research and thus also have a strong educational value; they are one of the few situations where geological structures may be viewed in three dimensions.

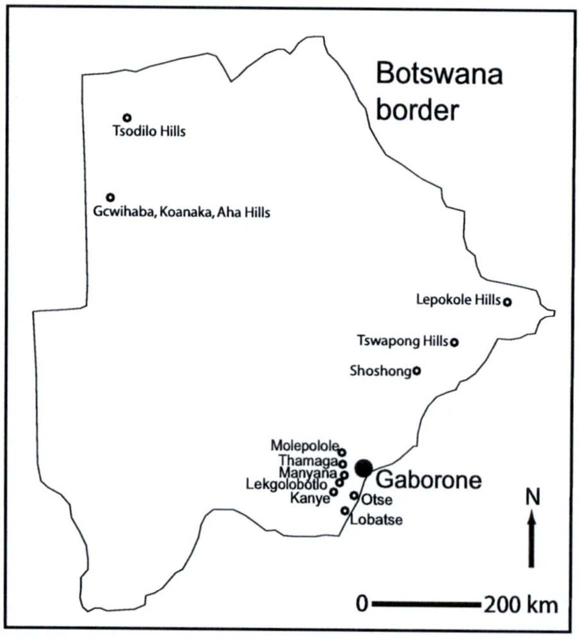

Fig. 14.1 Map of Botswana showing place names mentioned in the text where caves and rockshelters occur

This chapter provides the first compiled description of caves in Botswana from the authors' own descriptions, published and unpublished literature and includes some caves and rockshelters documented and located with GPS coordinates for the first time. The focus herein is on naturally-formed caves and not those excavated by humans, for example for human burial in river sediments/calcrete in the bank of the Thamalakane River and those that occur in manganese mines at Otse (Brook 2017) and specularite mines at the Tsodilo Hills (Murphy et al. 2010; de Wit and Main 2016).

14.2 Geological Context

Most of the metamorphic and igneous basement rocks of the Kaapvaal and Zimbabwe cratons, separated by the Limpopo Mobile Belt, are at least 2,500 Ma in age and formed during the Archean with surface outcrops of schists, gneisses and granites occurring in eastern Botswana. A younger cover of Karoo rocks (ranging from approximately 320 to 180 Ma in age) and Cretaceous and Cenozoic Kalahari Group sediments conceal the western margins of these older rocks and largely conceal Proterozoic orogenic belts. This younger stratum was laid down in the Kalahari Basin which underlies the centre and western part of the country (Thomas and Shaw 1991).

Sedimentary rocks were deposited on the basement rocks; the earliest sediments would be the Black Reef and Vryburg Formations ($\sim$2,650 Ma) belonging to the Transvaal Supergroup and sediments of the Waterberg Group ($\sim$2,025 Ma). Transvaal-age dolomites are found in the Lobatse and Ramotswa areas. Most Waterberg-age outcrops occur in the east and southeast, e.g. Tswapong Hills and the area between Gaborone and Lobatse; including sandstones, shales, conglomerates and quartzites. In the northwest of Botswana, more recent sediments overlie rocks of the Neoproterozoic and possibly early Palaeozoic rocks of the Damara orogen ($\sim$850–600 Ma).

Caves in Botswana have formed in three geological contexts; within siliciclastic (sandstones, and quartzites), dolomitic and granitic rocks as outlined in Table 14.1, along with their precise locations. The formation processes of the caves/rockshelters and their deposits are overviewed along with descriptions of previous studies. In addition, the potential for further research is outlined where appropriate in terms of landscape evolution and geoarchaeology.

Table 14.1 List of caves and rockshelter locations in Botswana, including GPS locations (in degrees and decimal minutes, where available), lithology, entrance geomorphology and stratigraphical information. '–' denotes information unavailable

Name	S	E	Lithology	Entrance geomorphology	Stratigraphic information
Lobatse:					
Cave I	25° 09.771'	25° 42.680'	Dolomitic	Roof collapse	Transvaal Supergroup
Cave II	25° 07.751'	25° 43.537'	Dolomitic	Initial quasi-horizontal passage from weathering/erosion of vertical fracture	Transvaal Supergroup
Cave III	25° 09.956'	25° 42.228'	Dolomitic	Weathering/erosion of a fracture producing a crevice	Transvaal Supergroup
Cave IV (Mohu)	25° 10.065'	25° 42.122'	Dolomitic	Roof collapse	Transvaal Supergroup
Athlone Cave	25° 12.400'	25° 40.213'	Siliciclastic	Weathering/erosion at intersection of fractures	Segwagwa Group
Lepokole Caves	–	–	Granitic	Weathering/erosion of horizontal fractures	Central Zone of Limpopo Mobile Belt
Manyana Rockshelter and Cave	24° 45.831'	25° 35.372'	Siliciclastic	Undercutting of horizontally-bedded rocks and rock fall (rockshelter); weathering/erosion at intersection of bedding/vertical fracture (cave)	Waterberg Group
Thamaga rockshelters	–	–	Siliciclastic	Weathering/erosion at base of kopje	Waterberg Group
Mmalogage (Lekgolobotlo)	25° 54.944'	25° 32.936'	Siliciclastic	–	Waterberg Group
Kobokwe Cave (Molepolole)	24° 25.889'	25° 32.132'	Siliciclastic	Weathering/erosion of vertical fracture	Waterberg Group
Otse:					
Malladiepe Cave	25° 01.962'	25° 45.354'	Siliciclastic	Weathering/erosion at intersection of bedding/vertical fracture	Waterberg Group
Malladiepe rockshelter & spring	25° 02.012'	25° 45.306'	Siliciclastic	Weathering/erosion of a bedding plane	Waterberg Group
Mannyelanong Cave	25° 2.669'	25° 46.131'	Siliciclastic	Weathering/erosion at intersection of horizontal bedding/vertical fracture	Waterberg Group
Mannyelanong (vulture cliff) Cave	25° 3.610'	25° 45.504'	Siliciclastic	Weathering/erosion at intersection of horizontal bedding/vertical fracture	Waterberg Group
Baratani –Segorong Cave	25° 00.467'	25° 43.449'	Siliciclastic	Weathering/erosion at intersection of horizontal bedding/vertical fracture	Waterberg Group
Shoshong Cave	–	–	Siliciclastic	–	Waterberg Group
Tswapong Hills:					
Dimomo Cave	–	–	Siliciclastic	–	Waterberg Group
Cave in Phato Ye Sebilo Gorge	22° 36.676'	27° 26.255'	Siliciclastic	–	Waterberg Group
Gcwihaba Caves	20° 1.487'	21° 21.276'	Dolomitic	Roof collapse	Xaudum Group, Damara Belt
Aha Hills	20° 1.233'	21° 21.573'	Dolomitic	Roof collapse	Damara Belt
Koanaka Hills:	19° 46.685'	21° 3.421'	Dolomitic	–	Koanaka Group, Damara Belt
Tsodilo Hills:	18° 43.848'	21° 43.864'	Siliciclastic	Majority from weathering/erosion of bedding planes and at intersection of horizontal bedding/vertical fracture; also talus-type caves	Tsodilo Hills Group Damara Belt

14.3 Caves in Dolomite

14.3.1 Caves in Ngamiland

In the Ngamiland District of northwestern Botswana, caves are formed in dolomite rocks that stratigraphically form part of the Otavi Group of the Damara orogenic belt (Selfe 2005). There are three separate cave systems: Gcwihaba Caves, Koanaka Hill Caves (Blue Cave and Bone Cave) and sink-hole type caves in the Aha Hills (Waxhu North and Waxhu South) (Garner and Ritter 1994) (Fig. 14.1).

The Gcwihaba Caves ('Gcwihaba' means 'Hyena's lair' in the local San language), also known as Drotsky's Caves (named after one of the early European settlers who visited the caves) are located ~150 km southwest of the Okavango Delta ('G1', Fig. 14.2). The caves and tube-like passages were developed horizontally at two levels (as also found in Lobatse) from water-table levels in past humid phases such as during the Pleistocene. They are the largest caves discovered in Botswana so far, with a maximum vertical height of ~25 m and are formed in one of six hills (dolomitic outcrops) in the vicinity of a dry river valley (Cooke 1975b). Both entrances to Gcwihaba's main cave were formed from rockfall by roof collapse and evidenced by rocky debris there; evidence for movements also comes from horizontal cracks in speleothem pillars (Wayland 1944; Cooke 1975b). The cave entrances were settled by people around the Late and terminal Pleistocene as evidenced by dated deposits (e.g. Yellen et al. 1987) containing Late Stone Age (LSA) lithics and faunal remains including an aquatic *Xenopus* frog and side neck turtle confirming that conditions were mostly moister than at present between approximately 30,000 and 11,000 B.P. (Robbins et al. 1996). Rich concentrations of micromammals and skeletons of larger mammals such as bovid have also been found from deposits in the Gcwihaba Caves (Pickford 1990). Due to the diverse cave formations and deposits with a high density of fossils, the Gcwihaba Caves are currently on a tentative list to become a UNESCO World Heritage listed site.

The caves contain a variety of speleothems (e.g. Wayland 1944; Mann and Ritter 1995) including a new form where thin films of $CaCO_3$ are deposited around fine lateral roots that grow upwards from the cave floor (Du Preez et al. 2015). Speleothems from the Gcwihaba Caves have been used to derive important palaeoclimatic and palaeoenvironmental information. For example, Cooke (1975b), Cooke and Verhagen (1977) and Cooke (1984) performed radiocarbon dating, and Shaw and Cooke (1986) used radiocarbon dating and

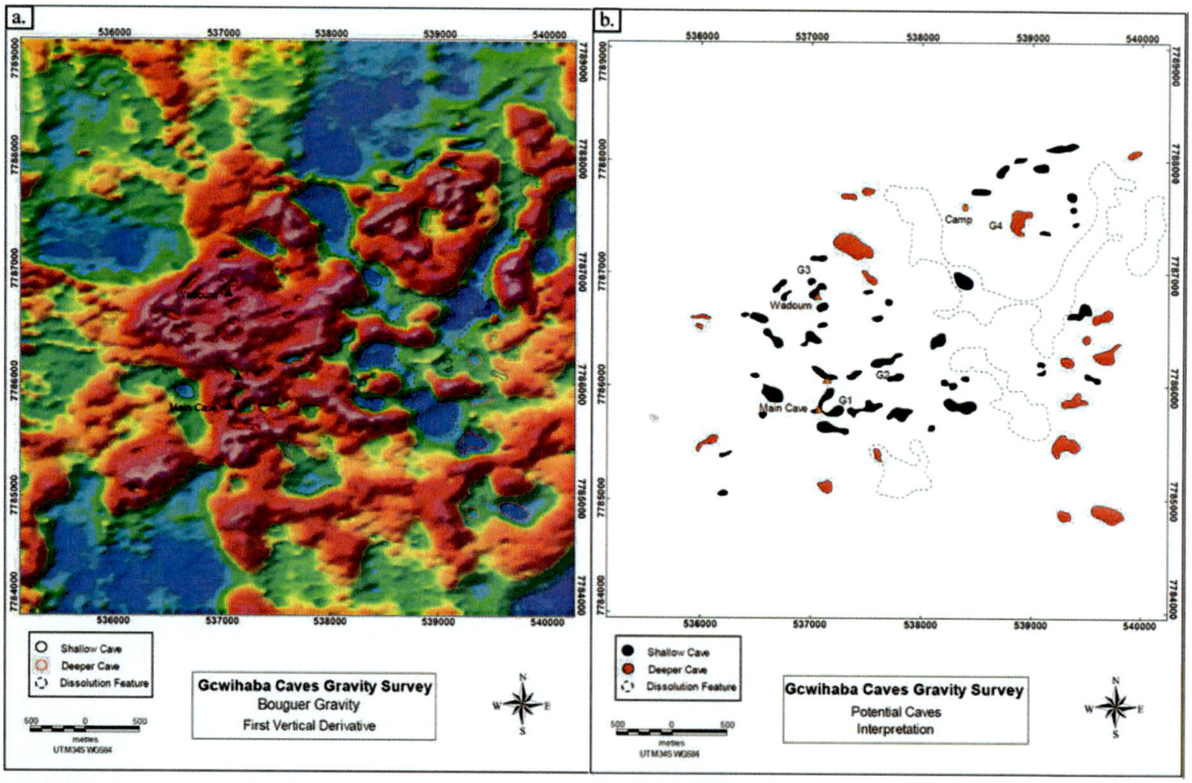

Fig. 14.2 a Gravity survey of the Gcwihaba Caves area; b Interpretation of the gravity survey for clarity. (Maps taken from Selfe 2005)

U/Th determination on stalagmites and found growth associated with humid phases at 45,000–37,000 B.P., 34,000–29,000 B.P., 16,000–14,000 B.P., 6000–5000 B.P., 4000 B.P, 2000 B.P. and 750 B.P. (Thomas and Shaw 1991). Petrological studies of stalagmites have found that thickness of calcite layers correlates with rainfall (Railsback et al. 1994) and increased detrital sand and silt is inversely correlative with periods of decreased rainfall (Railsback et al. 1999).

A new cave, named '!WaDoum Cave' (a Basarwa phrase translating as "Stone Throat", and likely due to zones of high CO_2 there) was discovered in 1992 in an adjacent hill to the north ('G3', Fig. 14.2) and first described by Ritter and Garner (1994) and resurveyed by Mann and Ritter (1995). A 5 km by 5 km gravity survey was undertaken in the Gcwihaba Caves area by Selfe (2005) with detection of known caves (triangle symbols shown in Fig. 14.2a, b) and delineation of many new potential caves (black and red lines/polygons in Fig. 14.2a, b). Black polygons mark the outline of shallow features in the estimated 0–40 m range. Red features, some of which are quite large, mark the outlines of deeper features, in the estimated 40–80 m range, whereas the dashed lines outline linear gravity lows which are interpreted to represent dissolution features along joints in the dolomite, but are not necessarily caverns such as Gcwihaba's main cave (Selfe 2005).

Selfe (2005) summarized that nearly all of the new caves that were found are buried or show no sign of a surface entry point and recommended that any attempt at creating an entrance requires an approach of minimal disturbance due to the sensitive nature of the cave ecosystems (e.g. Du Preez et al. 2015). A Digital Elevation Model is presented in Fig. 14.3 and shows black patches that represent depressions, located at ~1029 m asl, and likely relate to the shallow caves in the vicinity of 'G2', as indicated in Fig. 14.2b (Selfe 2005).

Bone Cave and Blue Cave in the Koanaka Hills contain fossiliferous deposits of breccia (Pickford and Mein 1988; Pickford 1990; Pickford et al. 1994; Gabadirwa 1995; Ritter and Mann 1995) and speleothems are also present in the caves. The Bone Cave complex is ~300 m long and ~100 m wide with a large central chamber ~75 m in diameter and up to 7 m high; the Blue Cave complex is ~150 m long and ~75 m wide (Brook 2017). The Aha hills are situated ~50 km northwest of Gcwihaba, cover an area of ~245 km^2 and contain sinkholes at Waxhu north ~70 m deep, and Waxhu South ~50 m deep ('Waxhu' means 'house of god' in the local language) (Cooke and Baillieul 1974; Brook 2017). Elsewhere to the north of Botswana caves have also been reported in Maun (Sealetsa 2009).

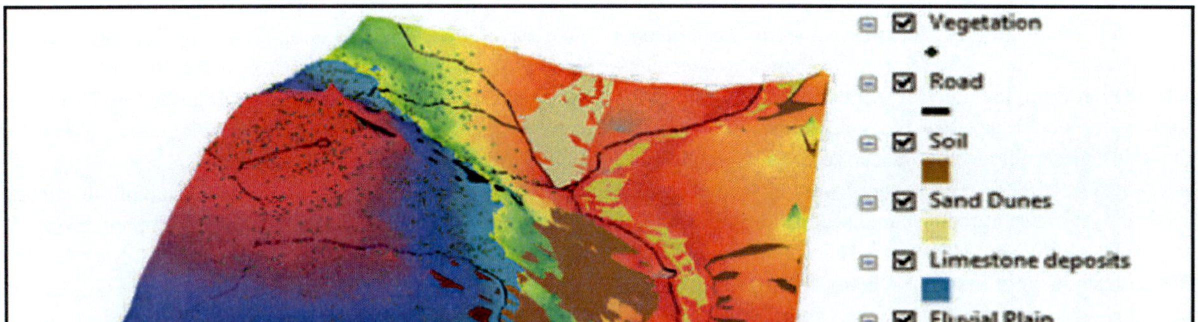

Fig. 14.3 Digital Elevation Model of the Gcwihaba site 'G2', located at 20 1.228' S 21 21.574' E. The model uses a surface interpolation algorithm to manipulate the surface and subsurface features of the landscape (DEM by S. M. Isaacs)

14.3.2 Lobatse Caves

The Lobatse Caves are located ~10 km northeast of Lobatse Town in southern Botswana on the Lobatse Estates private land (Fig. 14.1). They are formed within dolomite, a Ca-Mg carbonate rock not dissimilar to limestone, that is interbedded with chert and stratigraphically are part of the Transvaal Supergoup (Key 1983). King (1951) related the development of the caves to the African erosion cycle and assigned them to a Middle Pliocene age. There are several karst features in the area including caves containing speleothems, a dry tributary of the Peleng River, and dolines. Previous studies by Cooke (1975a), Holmgren et al. (1995) and Holmgren and Shaw (1999) have largely focused on two cave systems named Cave I and Cave II, although at least two further caves exist; herein named Cave III and Cave IV (also termed Mohu Cave; 'Mohu' means 'wasp' in Setswana).

Cave I contains a vertical entrance drop of ~15 m with a talus cone of debris at its base from roof fall, guano from resident bats and animal bones (Cooke 1975a; MLGNSS 2006). The main passage is a straight fissure of width varying from 1–14 m (Cooke 1975a). Speleothems are secondary mineral deposits, such as stalactites and stalagmites, found in caves from the dripping of mineral-rich water. Rainwater in the soil zone reacts with soil CO_2 to create weakly acidic water; as this water travels through the carbonate rock it dissolves the rock and when the solution reaches a cave, degassing due to lower cave CO_2 pressures drives the precipitation of carbonate minerals. Speleothems are present in Cave I in the form of flowstones (produced by precipitation from flowing water) and stalagmites. Along the northwest wall is a level of old sandy flowstone ~3–4 m above the present floor and indicative of past groundwater levels (Cooke 1975a). At its northwest end, the cave is blocked by a brown to red coloured sedimentary fill which is unsorted and thought to be largely uncontaminated with outside material prior to the cave being opened to the outside (Cooke 1975a). At the southeast end, the passage becomes a narrow tube and Cooke (1975a) notes hot and stagnant air in this location in the cave, likely from high levels of CO_2 as found by reconnaissance investigations by MLGNSS (2006) who also recorded three species of bat living there, thought to include the Horseshoe Bat (*Rhinolophus*) and Egyptian Slit-Faced Bat (*Nycteris thebaica*) that may number several thousand in the cave. Brook et al. (1998) produced three U-series ages on speleothems from Cave I with ages of approximately 103 ka, 83 ka and 69 ka, indicating more humid conditions at those times.

The orientation of joints exhibited at the entrance to Cave I is NW–SE (Fig. 14.4a). The entrance to Cave I is 18.5 m above the adjacent valley floor (Cooke 1975a). In comparison to Cave I, the entrance to Cave II is oriented NNE-SSW and is significantly higher at ~50 m above the valley floor indicating a lack of contemporaneity between the two cave systems (Cooke 1975a). It has a quasi-horizontal initial entrance passage that at present is too tight to enter and has presumably filled in with soil and wind-blown dust since the investigations of Holmgren and Shaw (1999) (Fig. 14.4b). At the entrance to Cave II is a vertical shaft of ~17.85 m beneath of which there is a talus cone of debris from roof collapse (Cooke 1975a). In general, the sedimentary fill of the cave is composed of weathered bedrock, guano and skeletal remains of bats and insects indicating a largely autochthonous origin (Cooke 1975a; Holmgren and Shaw 1999). King (1951) suggested that the caves at Lobatse had been opened to the atmosphere for a very much shorter time than caves in rocks of the same geological formation in nearby South Africa, albeit based on the absence of 'older red sand'. One shaft in Cave II was described by Holmgren and Shaw (1999: 73) as having toxic CO_2 levels, likely due to decomposing organic materials there.

This cave is more complex than Cave I and has developed at two levels indicating formation through epiphreatic (water table level) processes (Cooke 1975a). Cave II also exhibits a large high-roofed cavern containing stalagmites, stalactites

Fig. 14.4 **a** Vertical entrance to Lobatse Cave I; **b** Initial horizontal passage to Lobatse Cave II. (Photographs by M. Stephens)

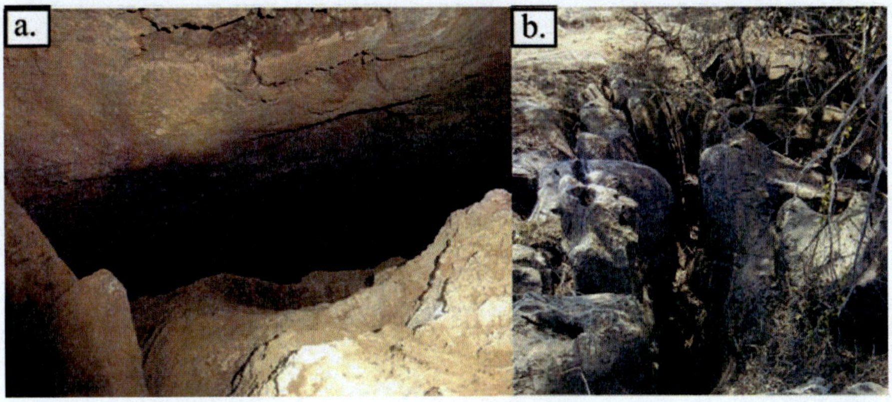

Fig. 14.5 **a** Looking downwards into the Lobatse Cave III crevice; **b** Solutional doline features in the vicinity of Lobatse Cave III. (Photographs by M. Stephens)

and flowstones. Speleothems from this cave have been subject to stable carbon and oxygen isotope analysis and uranium-series dating (Holmgren et al. 1994, 1995; Holmgren and Shaw 1999). The record indicates a gradual decrease in temperature from 50 ka towards the Last Glacial Maximum at 21 ka. Episodes of increased precipitation occurred at 49, 47, 27, 25 and 23.5 ka. The stable carbon isotope record suggests that a significant proportion of the increase was due to greater winter rainfall (Holmgren and Shaw 1999).

Cave III is a crevice-type cave (e.g. Vitek 1983), located beneath a large tree, ~10 m deep and ~30 m long; the fracture that the crevice is formed in being oriented NNE-SSW (Fig. 14.5a). The cave is formed in an area of several dolines, whose more linear features are oriented NW–SE and NNE-SSW (Fig. 14.5b), the latter being consistent with the orientation of the entrance to Cave II and the crevice of Cave III. The relationship between fracture patterns and karst in SE Botswana has previously been alluded to by Bons and Van Loon (1985) and Van Haren and Spaans (1988).

Cave IV has horizontal passage development as also found in Caves I and II although a larger part of the ceiling has collapsed, creating an opening ~10 m long and ~4 m wide (Fig. 14.6). As a result of this collapse, a large talus pile of rockfall and sediment occurs within the cave, including several animal bones (Fig. 14.7a). Flowstones occur in narrow passages to the west and north of the cave, including a small rimstone dam (Figs. 14.6 and 14.7b). This cave is named 'Mohu Cave' by MLGNSS (2006) after the mud dauber and paper wasps and their nests that occur on the ceiling of the southern part of the cave (Fig. 14.6).

Sediments within the talus mounds of Caves I, II and IV have high archaeological/palaeontological potential and as pointed out by King (1951), Cooke (1975a) and MLGNSS (2016), are in the same geological formation as the more famous caves of Sterkfontein and Kromdraai in South Africa (e.g. Stratford 2015). Cave IV in particular, has the best access, and excavation could take place in winter months when wasps are less active, although it would likely take excavation of several metres of deposits before Pleistocene layers are found. Other possible studies include the novel analysis of metagenomics within dated speleothem from the caves to provide information on ancient African populations (e.g. Gallego Llorente et al. 2015). It should be noted that no archaeological lithic material has been observed by the authors on the surface around the Lobatse Caves, although it does not remove the possibility of material occurring in buried layers in the cave deposits.

Cosmogenic dating of the Lobatse cave sediments would provide useful information on the timing of cave passage development (e.g. Wagner et al. 2011; Calvet et al. 2015) and correlation to surficial geomorphology (Cooke 1975a). MLGNSS (2006) note an additional small rockshelter and small cave occurring in the area near to Cave II; suggesting the potential for further discoveries there. In Kanye ~40 km to northwest of Lobatse the former presence of inter-stratal dolomitic caves has been suggested by Modisi (1995) to explain layering and apparent slump structures in chert breccia in Kanye. In addition, underground caverns have been encountered in boreholes to the south of Kanye Village (BRGM 1986, 1988).

14.4 Caves in Siliciclastic Rocks

14.4.1 Caves and Rockshelters of the Tsodilo Hills

All siliciclastic caves observed, except for a talus-type cave described below, were formed from weathering and erosion of pre-existing weaknesses in the rock such as along bedding planes and intersections of these with vertical fractures (Table 14.1). Caves and rockshelters occur within the siliciclastic Tsodilo Hills of the Neoproterozoic–Lower Palaeozoic that are an eastward extension of the Damara

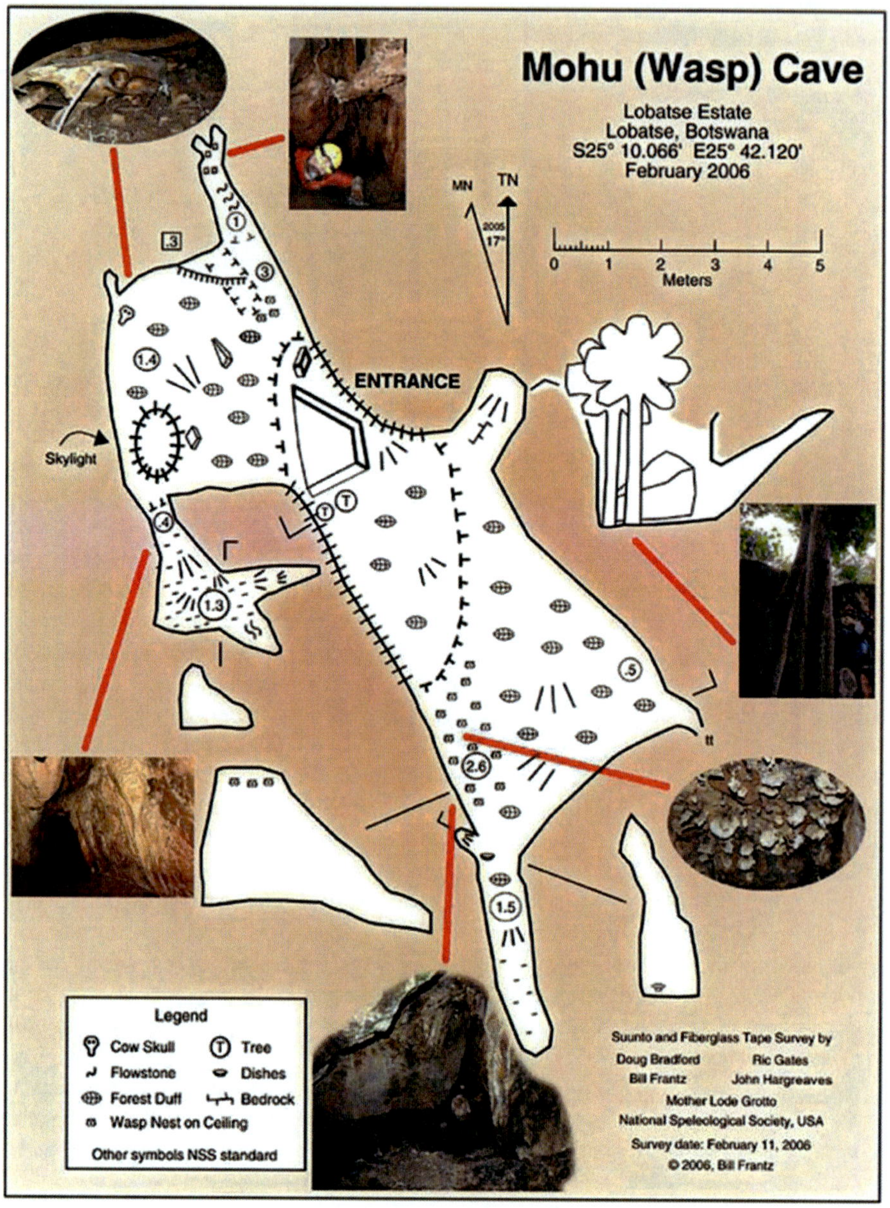

Fig. 14.6 Map showing the explored extent of Lobatse Cave IV. (Map by B. Frantz)

Supergroup (~500–600 Ma) (Wendorff 2005) in northwestern Botswana (Fig. 14.1). This is a UNESCO World Heritage listed site and protected as such. It consists of four inselbergs, known locally as the male, female, child and grandchild. According to Campbell et al. (1980), the name 'Tsodilo' was taken from the Hambukushu name 'Sorile', which means 'shear'. The male hill is ~400 m, the female hill ~300 m, the child hill ~40 m and the grandchild hill even lower above the surrounding plain (Roodt 2010). The Tsodilo Hills are of great significance to the San peoples of the Kalahari, and they believe that the caves of the female hill are the resting place for their gods (e.g. Fig. 14.8a) (Brook 2017). There are over 4500 rock art paintings (e.g. Fig. 14.8b, c) in an area of 10 km^2, including around 20 sites exhibiting rock carvings (Roodt 2010). Most of the images likely date back to between 850–1100 AD (Brook 2017). Also observed is a talus-type cave (e.g. Vitek 1983), a type suggested to be also classified as 'quasikarst' by Grimes (1997), formed in the large voids between a deposit of huge boulders at the base of a steep hillslope, on the 'Rhino Trail' (Fig. 14.8d).

In the Depression Rock Shelter on the eastern side of Female Hill, Robbins and Campbell (1988) derived a radiocarbon dated sequence indicating a sophisticated

Fig. 14.7 **a** Talus pile including animal bones in Lobatse Cave IV; **b** Speleothem in Lobatse Cave IV. (Photographs by M. Stephens)

Fig. 14.8 **a** Cave on the 'Rhino Trail' (Female Hill) of the Tsodilo Hills, formed from the intersection of a bedding plane and vertical fracture in siliciclastic rocks; **b** Rock art on the 'Rhino Trail' indicating contact of San peoples with the Atlantic Ocean, ~1000 km to the west; **c** Rock art on the 'Cliff Trail' from Rhino Cave showing a white rhino that has been overpainted by a red giraffe; **d** Talus-type cave on the 'Rhino Trail' (Female Hill) of the Tsodilo Hills (18°45.671' S 21°44.794' E). (Photographs by M. Stephens)

microlithic technology significantly earlier than 19,000 B. P. At the Rhino Cave, perched high on the northernmost ridge of Female Hill, excavations by Coulson et al. (2011) found a Middle Stone Age (MSA) assemblage characterized by a large number of colourful points, mostly produced in non-locally acquired raw materials. The assemblage was recovered directly beneath a massive, virtually free-standing rock face that has been carved with hundreds of cupules of varying sizes and shapes. Coulson et al. (2016) suggested ritualized behavior from two unusual sets of artefacts, the base of a pointed-bottom pot and a cluster of impact-fractured armatures from the upper deposits of Rhino Cave and Corner Cave (south side of Male Hill) during the Later Stone Age/Early Iron Age.

In the White Paintings Rock Shelter, on the west side of the Male Hill, Robbins et al. (2000) uncovered 7 m of sediments that are at least 100,000 B.P. in age at the base, including a lengthy record of LSA and MSA artefacts. Also excavated were numerous fish bones, wetland mammals and barbed bone points indicating past periods of moister conditions (Robbins et al. 2000). Staurset and Coulson (2014) conducted refitting analysis of a MSA lithic assemblage from White Paintings Rock Shelter and reported an admixture caused by sloping deposits and post-depositional processes with implications for dating of the site.

14.4.2 Siliciclastic Caves and Rockshelters of Southern Botswana

Athlone Cave (herein named after the nearby Athlone Hospital) is situated in hills that overlook the town of Lobatse (Fig. 14.1) and has formed in dipping quartzites, including quartz veins of the Pretoria Group of the Transvaal Supergroup (Beger 2001). The cave occurs within the north-facing side of an ENE-oriented gorge (Stephens, this book) and has an opening oriented ~20° NNE (Fig. 14.9a) that can be entered in a crouched position. Outside the cave entrance is a sloping platform with rock debris. After an initial short entrance passage, the cave is oriented ~335° NNW and its width/height increases to ~10 m/2 m and has a maximum walkable length of ~20 m at an incline of ~15° before it becomes too narrow to pass. Exposed rock surfaces within the cave are smooth and likely from water flow and human treadage. The cave contains a sedimentary fill of brown, largely sandy-silty matrix with gravels and cobbles (Fig. 14.9b). Waste droppings and urine-stained rocks also occur from Rock Hyrax (*Procavia capensis*) middens. Human activity inside the cave is evidenced from candle wax on rocks associated with spiritual practices there.

At the periphery of the Kolobeng Hills and just at the northern edge of Manyana Village (Campbell and Main 2003) (Fig. 14.1) is an 8 m high rockshelter formed in sandstone that is stratigraphically part of the Mesoproterozoic aged Waterberg Group (Carney et al. 1994). The rockshelter was possibly formed from undercutting by the nearby Kolobeng River and subsequent rockfall at some point in antiquity (Fig. 14.10a). The site is best known for its prehistoric rock art and is the only known site in Southern Botswana that contains rock paintings of the same era as

Fig. 14.9 **a** Athlone Cave (just right of centre) formed in steeply dipping quartzite rocks; **b** Inside Athlone Cave looking towards its entrance passage that has formed at the intersection of fractures. (Photographs by M. Stephens)

Fig. 14.10 **a** Manyana Rockshelter; **b** Rock art panel at Manyana; **c** Manyana Cave. (Photographs by M. Stephens)

those of the Tsodilo Hills in the northwest of Botswana (Brook 2017). The rock paintings at Manyana are slowly fading, however, due to weathering (Fig. 14.10b).

The Manyana site is also known for a cave with a very small entrance (Fig. 14.10c) where it is said that the pregnant Mma Sechele, Queen mother and third wife of the Chief of the Bakwena Tribe, hid for a month during the battle of Dimawe in 1852 (Brook 2017). Modern Rock Hyrax (*Procavia capensis*) activity is evident from white staining associated with its waste nearby the entrance to the small cave (Fig. 14.10c). Small test excavations were undertaken below the painted areas by Robbins (1985), whose findings included LSA artefacts and Iron Age (IA) pottery.

Approximately 10 km NNW from Manyana, at nearby Thamaga (Fig. 14.1), Robbins (1986) made test excavations beneath one of three rockshelters that occur there. As found at Manyana, findings included LSA artefacts associated with IA pottery; with a radiocarbon age on charcoal of 1190 ± 90 B.P. (Robbins 1986). The site is located just outside of Thamaga Village on the southestern hardveld; the rockshelters occur from the weathering and erosion of a small kopje known locally as 'Radiepolong' (Robbins 1986).

Kobokwe Cave is situated ~4 km southeast of Molepolole (Fig. 14.1) and lies just outside the southeastern edge of the semi-arid sandy savannah of the Kalahari Desert (Campbell and Main 2003; Tlhapiso and Stephens 2020). The cave has formed in horizontally-bedded sandstones of the Waterberg Group (Carney et al. 1994) (Fig. 14.11a). The cave entrance is formed through vadose processes, occurring tens of metres up from the ground surface on the southern facing wall of Kobokwe Gorge (Fig. 14.11a, b) (Stephens, this book) and can be entered horizontally for ~10 m. The cave contains a sedimentary fill of a dark brown, largely sand-silt matrix, with occasional gravels, cobbles, boulders and plastic refuse and candle wax (Tlhapiso and Stephens 2020).

Kobokwe Cave (also known as 'Sechele's Cave' or 'Livingstone's Cave'; Campbell and Main 2003) was named after a man called Kobokwe after he was thrown into the cave for having struck Kgosi (Chief) Kgari II when Kgosi Sebele II and Kgosi Kgari II were contesting chieftainship of the Bakwena Tribe in the 1930s. However, it is also well known as the site responsible for the Christian 'conversion' of Kgosi Kgari Sechele I, the first and only such occurrence in Botswana by the British missionary David Livingstone in 1847 (Campbell and Main 2003). The name Kobokwe Cave is preferred in this chapter to denote that it is the heritage of the local people that is being conserved and to foster better community engagement in its conservation and use (Keitumetse 2016; Tlhapiso and Stephens 2020).

Mmalogage, meaning 'mother of caves' in the Setswana language, is a collection of four caves that occur within sandstone rocks that stratigraphically form part of the Waterberg Group. The caves reportedly occur below the summit of a south-facing steep hillslope near to the village of Lekgobotlo (Fig. 14.1). Three caves occur to the west of a minorly incised gorge in the hillslope produced by a perennial stream and one at the gorge head, just to the east of a spring that feeds the stream (Stephens, this book).

Caves occur in the vicinity of Otse Village (Fig. 14.1) within the largely sandstone rocks of the Waterberg Group at Mannyelanong Hill, Malladiepe Hill, Baratani Hill and Segorong Gorge (Segorong means 'curve' in Setswana) (Stephens, this book) on the hillslope of Otse Hill (the highest point in Botswana at 1491 m asl). Mannyelanong, which means in Setswana, 'where vultures make their droppings' occurs to the southeast of Otse, and where the cliff faces ~NW/NNW small caves occur (Table 14.1 and Fig. 14.12a). The SSW facing cliff is stained white in many places due to the waste from roosting vultures who inhabit crevices and small caves in the rocks (Fig. 14.12b).

Malladiepe Hill occurs to the east of Otse and contains tilted sedimentary layers dominated by sand, gravels, cobbles and even boulder-size materials. Two small rockshelters and one cave occurs on this hill. Near the foot of the

Fig. 14.11 **a** Kobokwe Cave entrance formed at the intersection of bedded sandstone and a vertical fracture; **b** The entrance to Kobokwe Cave, formed from the widening of a vertical fracture. (Photographs by M. Stephens)

Fig. 14.12 **a** Mannyelanong cave, Otse, formed at the intersection of a horizontally bedded conglomerate with a vertical fracture; **b** Mannyelanong vulture cave. (Photographs by M. Stephens)

hill, ~30–40 m above the floodplain, is a rockshelter ~4 m wide and ~2 m high, facing SSE and contains a spring that is used by local villagers for spiritual purposes. The spring here flows only during the wet season and continues for around one month afterwards, although in the past it continued for longer (B. Fologang and O. Sanka, pers. comm., 2018). The spring has clearly altered the surrounding rocks giving a whitish-green staining and its occurrence has likely weakened the rocks and assisted in the natural excavation of the rockshelter along a bedding plane (Fig. 14.13a).

Malladiepe Cave occurs further up the hillslope facing SSE to Mannyelanong Hill, that occurs across the floodplain. The entrance to the cave is 5.7 m wide by 2.7 m high and is aligned with the dip of the sedimentary layers. The cave is elongated upslope and downslope (Fig. 14.13b) and narrows to unknown depths. A rockshelter with dimensions of 7.8 m wide and 1.5 m high occurs a few metres below the cave and its location is likely associated with a horizontal fracture at the back of the shelter (Fig. 14.13b).

Segorong Gorge, the highest elevation gorge in Botswana (Stephens this book), contains Segorong Cave on the southern side of the gorge and other smaller caves (Fig. 14.14a) and rockshelters with openings mainly on the northern side of the gorge (Brook 2017). The entrance to Segorong Cave (Fig. 14.14b) is located ~20 m above the floor of the gorge and within the cave is candlewax attesting to spiritual practices there. Caves can also be observed from a distance near the peak of Baratani Hill. King (1951) described a cave occurring in Otse, as 'a series of curious zigzag dip and joint openings…(and) found in the farthest recesses a number of large round native pots said to have been left by refugees from Moselikatse'.

14.4.3 Siliciclastic Caves of Eastern Botswana

A cave is described by Knight (2014) as occurring in the vicinity of Shoshong Village (Fig. 14.1) that the Bakaa Tribe hid in during raids by the Matebele Tribe in 1842/3. The cave presumably occurs stratigraphically in the Shoshong Formation of the Palapye Group (Carney et al. 1994), which is a correlate of the Soutpansberg Group (Ermanovics et al. 1978). Caves are also noted to occur in the Tswapong Hills (Fig. 14.1) and stratigraphically in the Tswapong Formation of the Palapye Group (Carney et al. 1994). One cave occurs in the Phato Ye Sebilo Gorge (Main 2008) and another named Dimomo Cave is located on the summit of the Tswapong Hills about 3 km south of Malaka Village and was used as a place of refuge during the 'Matebele wars'

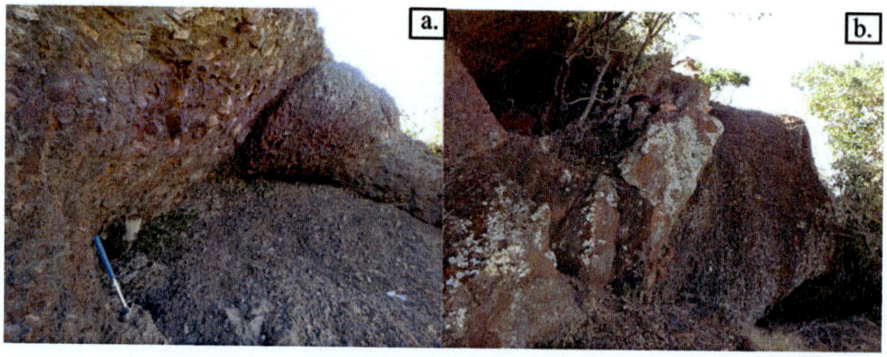

Fig. 14.13 **a** Malladiepe rockshelter, Otse, formed by the action of spring water seepage; **b** Malladiepe Cave with a rockshelter below. (Photographs by M. Stephens)

Fig. 14.14 a Cave formed in the north-facing sandstone wall of the Segorong Gorge, Otse; **b** Segorong Cave. (Photographs by M. Stephens)

(Difaqane) of the mid-nineteenth century (Segadika 2010). Approximately 250 km north of Shoshong in the Makgadikgadi Heritage Area is Black Semaroba Hill that is listed as a national monument and described to contain a 'Late Stone Age Cave' with a 'natural sandstone arc' (Pheto 2006).

14.5 Granitic Caves of Lepokole

Lepokole Caves are located ~25 km northeast of Bobonong and form part of the southernmost extension of the granitic Matopos Hills in Zimbabwe (Fig. 14.1). Stratigraphically the Lepokole Granites are part of the Archean Phikwe Complex which occurs within the Limpopo Mobile Belt (Carney et al. 1994; Holzer et al. 1999). The most well-known feature is the 'Maredi Rockshelter' (Fig. 14.15 a) containing prehistoric rock art; the small cave has two areas where water drips from two crevices to create a white impression on which the paintings of animals are found (Fig. 14.15b) (Brook 2015). Figure 14.15b also exhibits hollows associated with seepages in the granite, likely associated with a horizontal joint and possible concentration of moisture in the scarp-foot piedmont zone (e.g. Twidale and Vidal Romaní 1995; Migoń and Goudie 2000; Twidale and Bourne 2008). Caves with rock paintings are noted by Westphal (1975) to occur close to Babirwa settlements around Bobonong and are inferred to be those that occur at Lepokole.

14.6 Conclusions

Geodiversity is important as it helps to understand the range of abiotic features of the natural environment. It is also important as it forms the context for geoheritage and geoconservation. Caves have throughout history and prehistory provided a wide range of uses for humans, and at present for

Fig. 14.15 a 'Maredi Rockshelter' at the base of a cliff in the granitic Lepokole Hills; **b** Hollows forming along seepages in the granite of the 'Maredi Rockshelter'. (Photographs by M. Stephens)

geoarchaeological and palaeoclimatic research, sustainable geotourism and educational purposes.

Caves occur across Botswana in a variety of rock types, broadly categorized here as dolomitic, siliciclastic and granitic, and within rocks of varying ages and stratigraphic contexts. The dolomitic caves include the Gcwihaba Caves in northwestern Botswana and Lobatse Caves in southeastern Botswana. The siliciclastic caves/rockshelters occur at: Tsodilo Hills in northwestern Botswana; Tswapong Hills and Shoshong in eastern Botswana; Molepolole, Lekgolobotlo (Mmalogage), Manyana, Otse and Lobatse in southeastern Botswana. The granitic caves occur within the Lepokole Hills in eastern Botswana. In terms of new cave research in Botswana, the dolomitic caves in Lobatse have particular potential for information on landscape evolution and geoarchaeology as they contain passages at different levels and large talus mounds containing bones. This paper provides the first compiled description of caves in Botswana and some are documented here for the first time.

Acknowledgements Otse guides Benjamin Fologang and Oabona Sanka are gratefully thanked for a tour of the caves there. Kgosi Boikanyo and Thabang Setlalekgomo (both UB) are also thanked for their assistance with cave visits in Lobatse. Selfe (2005) and De Beers are thanked for the report on the Gcwihaba Caves and for enabling the reproduction of maps therein. Bill Frantz and John Hargreaves are thanked for enabling reproduction of B. Frantz's Cave IV (Mohu Cave) map. Access to the caves on the Lobatse Estates was generously provided by Philip Fischer and Annie Janda and was skillfully guided by Mogomotsi Pheto. At the Lepokole Hills generous guidance was provided by Mr. Johannes. Professor Alexander Proyer (formerly of UB) suggested the potential for analysing landscape evolution through cosmogenic dating of sediments in the cave passages at Lobatse. The University of Botswana is thanked for providing a vehicle for fieldwork. Gratitude is shown to the reviewer who provided helpful comments on an earlier draft. This chapter is dedicated to the memory of Leslie Stephens whose encouragement brought MS to live and work in Botswana.

References

Barker G, Barton H, Beavitt P, Bird M, Daly P, Doherty C, Gilbertson D, Hunt C, Krigbaum J, Lewis H, Manser J, McLaren S, Paz V, Piper P, Pyatt B, Rabett R, Reynolds T, Rose J, Rushworth G, Stephens M (2002) Prehistoric foragers and farmers in South-east Asia: renewed investigations at Niah Cave, Sarawak. Proc Prehist Soc 68:147–164

Barton, CM, Clarke GA (1993) Cultural and natural formation processes in Late Quaternary cave and rockshelter sites of Western Europe and the Near East. In: Goldberg P, Nash DT, Petraglia MD (eds) Formation Processes in Archaeological Context. Monographs in World Archaeology, no. 17, Prehistory Press, Madison, WI, 33–60.

Beger K (2001) Environmental hydrogeology of Lobatse, Southeast District, Republic of Botswana. Unpublished MSc thesis, Bundesanstalt fur Geowissenschaften und Rohstoffe (BGR), Hannover, Germany

Bons CA, Van Loon JAWM (1985) Water resources evaluation of a Dolomite ground water basin in Southeast Botswana. Unpublished report, Institute of Earth Science, Free University, Amsterdam

Brook GA, Cowart JB, Brandt SA (1998) Comparison of quaternary environmental change in eastern and southern Africa using cave speleothem, tufa and rock shelter sediment data. In: Alsharhan G, Whittle K (eds) Quaternary Deserts and Climatic Change. Balkema, Rotterdam, pp 239–249

Brook M (2015) Wild about Botswana. Michael C. Brook, Gaborone

Brook M (2017) Botswana Monuments, Heritage Sites & Museums. Michael C. Brook, Gaborone

Bureau de Recherches Géologiques et Minieres (BRGM) (1986) Hydrological survey of transvaal supergroup Dolomite in the Kanye area. Final report volume 1, 86 BWA 020 EAU. Unpublished report, Department of Water Affairs, Botswana, Ministry of Mineral Resources and Water Affairs

Bureau de Recherches Géologiques et Minieres (BRGM) (1988) Kanye dolomite groundwater basin, hydrogeological investigation, resources assessment and development project. Phase II, Main report, volume 1, TB 10/1/85–86. Unpublished report, Geological Survey of Botswana, Ministry of Mineral Resources and Water Affairs

Campbell AC, Hitchcock R, Bryan M (1980) Rock art at Tsodilo, Botswana. S Afr J Sci 76(10):476–478

Campbell A, Main M (2003) Guide to Greater Gaborone: a historical guide to the region around Gaborone including Kanye, Lobatse. Main in association with the Botswana Society, Mochudi and Molepolole. A. Campbell and M. Main

Carney J, Aldiss D, Lock NP (1994) The geology of Botswana. Geological Survey Department, Lobatse, p 113

Calvet M, Gunnell Y, Braucher R, Hez G, Bourlès D, Guillou V, Delmas M (2015) Cave levels as proxies for measuring post-orogenic uplift: Evidence from cosmogenic dating of alluvium-filled caves in the French Pyrenees. Geomorphology 246:617–633

Cooke HJ (1975a) The lobatse caves. Botswana Notes and Records 7:29–34

Cooke HJ (1975b) The palaeoclimatic significance of caves and adjacent landforms in western Ngamiland. Botswana. Geographical J 141(3):430–444

Cooke HJ, Baillieul T (1974) The caves of Ngamiland: an interim report on explorations and fieldwork 1972–74. Botswana Notes & Records 6(1):147–156

Cooke HJ, Verhagen BTh (1977) The dating of cave development-an example from Botswana. Proc, 7 Intern. Spel. Congr. edn., (Sheffield)

Cooke HJ (1984) The evidence from northern Botswana of climatic change. In: Vogel JC (ed) Balkema, Late Cainozoic Palaeoclimates of the Southern Hemisphere. Rotterdam, pp 265–278

Coulson S, Segadika P, Walker N (2016) Ritual in the Hunter-gatherer/early pastoralist period: Evidence from Tsodilo Hills. Botswana. Afr Archaeol Rev 33(2):205–222

Coulson S, Staurset S, Walker N (2011) Ritualized behavior in the middle stone age: evidence from Rhino Cave, Tsodilo Hills, Botswana. PaleoAnthropology 18–61

de Wit M, Main M (2016) The Tsodilo Hills of Botswana. In: Viljoen R, Viljoen M, Anhaeusser C (eds) Africa's Top Geological Sites. Chapter 26. Struik Nature

Du Preez GC, Forti P, Jacobs G, Jordaan A, Tiedt LR (2015) Hairy Stalagmites, a new biogenic root speleothem from Botswana. Int J Speleol 44(1):37–47

Ermanovics I, Key RM, Jones MT (1978) The Palapye group, central-eastern Botswana. Trans Geol Soc S Afr 81:61–73

Gaál L, Bella P (2008) Granites and granite caves in the Western Carpathians. Cadernos Lab. Xeolóxico De Laxe Coruña 33:11–18

Gabadirwa M (1995) Fossil deposits of Koanaka Cave. Zebra's Voice 22(4):8–9

Gallego Llorente M, Jones ER, Eriksson A, Siska V, Arthur KW, Arthur JW, Curtis MC, Stock JT, Coltorti M, Pieruccini P, Stretton S, Brock F, Higham T, Park Y, Hofreiter M, Bradley DG, Bhak J, Pinhasi R, Manica A (2015) Ancient Ethiopian genome reveals extensive Eurasian admixture in Eastern Africa. Science 350(6262):820–822

Garner RA, Ritter RC (1994) Re-Survey of Gcwihaba cave and exploration of the Aha Sinkhole caves. Botswana Notes Rec 26:183–188

Gilbert T (1984) Limestone and volcanic caves of the Fiji Islands. Cave Sci 11(2):105–118

Gillieson D (1996) Caves: processes, development, and management. Blackwell, Oxford

Gray, M. (2008) Geodiversity: the origin and evolution of a paradigm. In: Burek CV, Prosser CD (eds) The history of geoconservation. Geological Society, London, Special Publications, 300, pp 31–36

Gray M (2013) Geodiversity: valuing and conserving abiotic nature, 2nd edn. UK, Wiley Blackwell, Chichester

Gray M (2018) Geodiversity: the backbone of geoheritage and geoconservation. In: Reynard E, Brilha J (eds) Geoheritage: assessment, protection and management. Elsevier, Amsterdam, pp 13–25

Grimes KG (1997) Redefining the boundary between karst and pseudokarst: a discussion. Cave and Karst Sci 24(2):87–90

Hedges J (1969) Karst caves in silicate rocks. D.C. Speleo, 3–4

Holmgren K, Karlén W, Shaw P (1995) Paleoclimatic significance ot variations m stable isotopie composition and petrology of a late Pleistocene stalagmite from Botswana. Quatern Res 43:320–328

Holmgren K, Lauritzen SE, Possnert G (1994) $^{230}Th/^{234}U$ and ^{14}C dating of a late Pleistocene stalagmite in Lobatse II Cave, Botswana. Quaternary Geochronology (QSR) 13:111–119

Holmgren K, Shaw P (1999) A late Pleistocene palaeoenvironmental record from Lobatse II Cave. Botswana Notes Rec 31:73–81

Holzer L, Barton JM, Paya BK, Kramers JD (1999) Tectonothermal history of the western part of the Limpopo Belt: tectonic models and new perspectives. J Afr Earth Sc 28(2):383–402

Keitumetse SO (2016) Interpretation: dealing with multiple Identities. In: African Cultural Heritage Conservation and Management. Springer, pp 113–133

Key RM (1983) The geology of the area around Gaborone and Lobatse. District Mem Geol Surv Botswana 5:230

King LC (1951) The geology of Makapan and other caves. Trans Roy Soc S Afr 33:121–150

Knight J (2014) Shoshong - a short history. Kwangu, Gaborone

Knight J, Grab S, Esterhuysen AB (2015) Geoheritage and Geotourism in South Africa. In: Grab S, Knight J (eds) Landscapes and landforms of South Africa. World geomorphological landscapes, pp 165–173

Main M (2008) Report on Tswapong Hills archaeological surveys 1996–2002. Botswana Notes Rec 40:55–59

Mann PM, Ritter R (1995) Further exploration and resurvey of WaDoum Cave. Botswana Notes Rec 27:13–18

Marker ME (1988) Karst. In: Moon BP, Dardis GF (eds) The Geomorphology of Southern Africa. Southern, Johannesburg, pp 175–197

Mbaiwa JE, Sakuze LK (2009) Cultural tourism and livelihood diversification: The case of Gcwihaba Caves and XaiXai village in the Okavango Delta, Botswana. J Tour Cult Chang 7(1):61–75

Migoń P, Goudie A (2000) Granite landforms of the Central Namib. Acta Universitatis Carolinae, Geographica, XXXV, Supplementum, pp 17–38

Modisi MP (1995) The origin of chert breccia around Kanye. Botswana Notes Rec 27:299–308

Mother Lode Grotto (MLG) National Speleological Society (NSS) (USA) (2006). The drought in Botswana is over: Lobatse cave reconnaissance, Feb 8 to 11, 2006. Unpublished report, retrieved from: https://www.chilembwe.org/botswana/reports.html

Murphy M, Robbins L, Campbell A (2010) The prehistoric mining of specularite. In: Campbell A, Robbins L, Taylor M (eds) Tsodilo Hills. Michigan State University Press and The Botswana Society, East Lansing and Gaborone, Copper Bracelet of the Kalahari, pp 82–94

Ngwira PM (2015) Geotourism and geoparks: Africa's current prospects for sustainable rural development and poverty alleviation. In: Errami E, Brocx M, Semeniuk V (eds) From Geoheritage to Geoparks—Case Studies from Africa and Beyond. Geoheritage, Geoparks and Geotourism. Springer International Publishing, Switzerland, pp 25–33

Pheto M (2006) Declaration of national monuments order, 2006. Statutory instrument No. 10 of 2006 Monuments and Relics Act (Cap. 59:03). Ministry of Labour and Home Affairs, Government of Botswana, L2/7/246II (37)

Pickford M (1990) Some fossiliferous plio-pleistocene cave systems of Ngamiland. Botswana. Botswana Notes Rec 22(1):1–15

Pickford M, Mein P (1988) The discovery of fossiliferous Plio-Pleistocene cave fillings in Ngamiland, Botswana. C R Acad Sci Paris Ser II 307:1681–1686

Pickford M, Mein P, Senue B (1994) Fossiliferous Neogene karst fillings in Angola, Botswana and Namibia. S Afr J Sci 90:227–230

Railsback LB, Brook GA, Chen J, Kalin R, Fleisher CJ (1994) Environmental controls on the petrology of a late Holocene speleothem from Botswana with annual layers of aragonite and calcite. J Sediment Res 64(1):147–155

Railsback BL, Brook GA, Webster JW (1999) Petrology and paleoenvironmental significance of detrital sand and silt in a stalagmite from Drotsky's Cave. Botswana. Phys Geogr 20(4):331–347

Ritter RC, Garner RA (1994) Discovery and preliminary exploration of a new cave in the Gcwihaba Valley. Botswana Notes Rec 26:55–65

Ritter R, Mann PM (1995) Discovery and exploration of two new caves in the northwest district. Botswana Notes Rec 27(1):1–12

Robbins LH (1985) The Manyana rock painting site. Botswana Notes Rec 17:1–14

Robbins LH (1986) Recent archaeological research in Southeastern Botswana: the Thamaga site. Botswana Notes Rec 18(1):1–13

Robbins LH, Campbell AC (1988) The depression rock shelter site, Tsodilo Hills. Botswana Notes Rec 20:1–3

Robbins LH, Murphy ML, Stevens NJ, Brook GA, Ivester AH, Haberyan KA, Klein RG, Milo R, Stewart KM, Matthiesen DG, Winkler AJ (1996) Paleoenvironment and Archaeology of Drotsky's Cave: Western Kalahari Desert, Botswana. J Archaeol Sci 23:7–22

Robbins LH, Murphy ML, Brook GA, Ivester AH, Campbell AC, Klein RG, Milo RG, Stewart KM, Downey WS, Stevens NJ (2000) Archaeology, Palaeoenvironment, and chronology of the Tsodilo hills white paintings rock shelter, Northwest Kalahari Desert. Botswana. J Archaeol Sci 27(11):1085–1113

Roodt V (2010) The shell tourist travel and field guide of Botswana. 6th edn. Vivo Energy Botswana (Shell Licensee) (Pty) Ltd, PO Box 334, Gaborone

Sealetsa N (2009) Maun's Mysterious Caves. *Kutlwano* 47(3):16–19

Segadika P (2010) The domestication of landscape through naming and symbolic protection among the Batswapong peoples of eastern Botswana: fullness and emptiness of landscapes in the eyes of the beholder. In: Smith A, Gazin-Schwartz A (eds) Landscapes of Clearance: Archaeological and Anthropological Perspectives. One World Archaeology (Book 57), Routledge, pp 139–153

Selfe GR (2005) Mapping of potentially new cave formations in the Drotsky's (Gcwihaba) Cave Area, NW Botswana. In: Final Report.

Selfe GR (ed) GRS Consulting, on behalf of De Beers Prospecting Botswana Ltd

Shaw PA, Cooke HJ (1986) Geomorphic evidence for the late Quarternary palaeoclimates of the middle Kalahari of northern Botswana. Catena 13:349–359

Staurset S, Coulson S (2014) Sub-surface movement of stone artefacts at White Paintings Shelter, Tsodilo Hills, Botswana: Implications for the Middle Stone Age chronology of central southern Africa. J Hum Evol 75:153–165

Stephens M (2020a) Geoconservation for Sustainable Development. In: Filho WL (ed) Encyclopedia of the UN Sustainable Development Goals. Springer

Stephens M, Hodge S, Paquette J (2013) Geoconservation of Volivoli Cave, Fiji: a prehistoric heritage site of national significance. Geoheritage 5(2):123–136

Stratford D (2015) The Sterkfontein Caves: Geomorphology and Hominin-Bearing Deposits. In: Grab, S., Knight, J. (eds.) Landscapes and Landforms of South Africa. Chapter 17. World Geomorphological Landscapes. Springer International Publishing, Switzerland

Thomas DSG, Shaw PA (1991) The Kalahari environment. Cambridge University Press, Cambridge

Tlhapiso M, Stephens M (2020) Application of the Karst Disturbance Index (KDI) to Kobokwe Cave and Gorge, SE Botswana: implications for management of a nationally important geoheritage site. Geoheritage, 12 (39). https://doi.org/10.1007/s12371-020-00461-8

Twidale CR, Bourne JA (2008) Caves in granitic rocks: types, terminology and origins. Cadernos Lab. Xeolóxico De Laxe Coruña 33:35–57

Twidale CR, Vidal Romaní JR (1995) Landforms and geology of granite terrains. Balkema, London

Urban J, Otęska-Budzyn J (1998) Geodiversity of pseudokarst caves as the reason for their scientific importance and motive of protection. Geologica Balcanica 28(3–4):163–166

Van Haren I, Spaans S (1988) Aspects of recharge in a small basin in SE Botswana. Unpublished report, Institute of Earth Sciences, Free University of Amsterdam

Vitek J (1983) Classification of pseudokarst forms in Czechoslovakia. Int J Speleol 13:1–18

Wagner T, Fritz H, Stüwe K, Nestroy O, Rodnight H, Hellstrom J, Benischke R (2011) Correlations of cave levels, stream terraces and planation surfaces along the River Mur—Timing of landscape evolution along the eastern margin of the Alps. Geomorphology 134 (1–2):62–78

Wayland EJ (1944) Drotsky's Cave. Geogr J 103(5):230–233

Wendorff M (2005) Outline of lithostratigraphy, sedimentation and tectonics of the Tsodilo Hills Group, a Neoproterozoic-Lower Palaeozoic siliciclastic succession in NW Botswana. Ann Soc Geol Pol 75:17–25

Westphal EOJ (1975) Notes on the Babirwa. Botswana Notes and Records 7:191–194

White WB (1984) Rate processes: chemical kinetics and karst landform development. In: La Fleur RG (ed) Groundwater as a Geomorphic Agent. Allen & Unwin, Boston, pp 227–248

Wray RAL (1997) Quartzite dissolution: Karst or pseudokarst? Cave Karst Sci 24(2):81–86

Yellen JE, Brooks AS, Stuckenrath R, Welbourne R (1987) A terminal Pleistocene assemblage from Drotsky's Cave, western Ngamiland, Botswana. Botswana Notes Rec 19:1–6

Mark Stephens is Associate Professor of Geography at the University of The Bahamas and was previously Senior Lecturer in Geomorphology at the University of Botswana. He holds a M.Sc. in Quaternary Science, and Ph.D. in Geography (both University of London). He mainly investigated landforms of the hardveld in Botswana, including the geodiversity and geoconservation of caves and gorges. His research interests include tropical environmental change, natural hazards, karst geomorphology, geoarchaeology, and geoheritage.

Mike de Wit is a research associate at the University of Stellenbosch and was previously an exploration geologist for several exploration companies working in many places in Africa. He holds MSc's in Geophysics (University of Pretoria, RSA) and in Sedimentology (Reading University, UK), and a PhD in Geology (University of Cape Town, RSA). He spent many years researching the distribution of heavy minerals in several parts of Africa while exploring for alkaline rocks. His main research interests include geomorphology and erosion history of southern Africa, palaeo-drainage studies and the distribution of alluvial diamonds.

Senwelo M. Isaacs is Director Geospatial in the Directorate of Intelligence and Security, Office of the President and was Lecturer in Geographic Information Systems in Archaeology at the University of Botswana. She was principal Land Use Planner in the Ministry of Lands and Housing, Department of Town and Regional Planning. She was also an Environmental Impact Assessment Practitioner and Private Archaeologist. She was a Senior Teacher in the Ministry of Education and served in various schools mainly in the Greater Gaborone area. She holds an MSc and MPhil in Environmental Science and is a PhD candidate Environmental Science (All from University of Botswana). She mainly investigated landforms of the hardveld in Botswana, including the geodiversity and geomorphology of caves and gorges and geoarchaeology. Her research interests include geoarchaeology, karst geomorphology, geoheritage, remote sensing and geographic information systems, land use planning and sustainable socio-economic development of local communities.

15 Kimberlites, Kimberlite Exploration, and the Geomorphic Evolution of Botswana

Andy Moore and Mike Roberts

Abstract

An understanding of the regional geomorphic framework of Botswana played a critical role in the discovery by De Beers of economic kimberlite pipes in the Orapa area in central Botswana. Early prospecting in the eastern Botswana hardveld in the late 1950s to early 1960s resulted in the recovery of a train of diamonds in the east-flowing Motloutse River, extending to the headwaters of this drainage line. De Beers concluded that the Motloutse had previously extended further west, based on an appreciation that the eastern Botswana watershed represented an axis of epeirogenic flexure, which had beheaded the former headwaters. The implication was that the source of the diamonds recovered in this river could be located in the Kalahari sandveld to the west of the drainage divide. Subsequent extension of sampling to the west into the Kalahari was rewarded by the discovery of the Orapa kimberlites in 1967, a year after Botswana achieved independence. We discuss how these sampling programmes in turn provided a feedback loop, making it possible to refine understanding of the regional geomorphic evolution. Because kimberlite pipes typically taper in size with depth, they also provide a gauge for estimating the depth of post-emplacement erosion, providing an important constraint on landscape evolution. Soil sampling in the Kalahari environment demonstrated that bioturbation has translocated kimberlite indicator minerals (KIMs) vertically from sub-Kalahari kimberlite pipes, resulting in their concentration in the Kalahari surface sand directly above the buried kimberlite. A consequence of this vertical mixing of the Kalahari sedimentary section is that luminescence dating of Kalahari quartz grains may produce spurious "mixed" ages. In the south of Botswana, mid-Cretaceous sub-Kalahari kimberlites are overlain by fluvial conglomerates, while in the northeast of the country, pipes of comparable age are overlain by a calcretized silcrete, which has been linked to the development of the African erosion surface. The basal conglomerates and African Surface duricrusts are unconformably overlain by semi-consolidated Kalahari sand. However, both conglomerate clasts and fragments of the duricrust have been identified within craters sediments forming the upper section of the respective kimberlite pipes. This implies that deposition of the conglomerates, and development of the African surface overlapped the episode of kimberlite eruption. This, in turn, provides an important mid-Cretaceous temporal constraint on initiation of Kalahari deposition, and of the African erosion cycle.

Keywords

Kimberlite • Erosion • Crustal flexures • Kalahari deposition and age • Calonda Formation conglomerate • Kimberlite indicator minerals (KIMS) • Bioturbation • African surface

15.1 Introduction

In 2018, Botswana was ranked as the world's second highest producer of rough diamonds, after Russia, both in terms of the volume of carats mined (24.37 MCts.) and the value ($3.53Bn) thereof (https://kimberleyprocessstatistics.org/public_statistics). All of the Botswana diamond production is extracted from a group of volcanic rocks known as kimberlites. From an exploration perspective, a total of 405 kimberlites have been discovered in Botswana to date and the country remains an attractive investment destination for diamond exploration companies.

A. Moore (✉)
Department of Geology, Rhodes University, Grahamstown, South Africa
e-mail: andy.moore.bots@googlemail.com

M. Roberts
De Beers Exploration, Gaborone, Botswana
e-mail: mike.roberts@debeersgroup.com

F. D. Eckardt (ed.), *Landscapes and Landforms of Botswana*, World Geomorphological Landscapes,
https://doi.org/10.1007/978-3-030-86102-5_15

Fig. 15.1 Artist's impression of a kimberlite intrusion (Maggie Newman, courtesy of De Beers)

Kimberlites commonly occur as volcanic pipes, either flared or conical in shape, increasing in area towards the surface (Fig. 15.1). In terms of surface expression, following emplacement, a temporary crater lake may develop over the pipe, resulting in the deposition of a sequence of crater infill sediments. In addition, the crater vent may be encircled by a tephra cone or ring comprised of excavated volcaniclastic material. At depth, kimberlites pipes are typically associated with sets of dykes and sills in the upper crust, either as precursor systems or post-eruption intrusions. Kimberlites have been divided on the basis of their isotopic characteristics into two sub-populations, referred to respectively as Group I and Group II kimberlites (Smith 1983).

As a broad generalization, the magma (or molten lava) which gave rise to these different kimberlite bodies originated from depths >150 km within the earth's mantle, where diamond is the stable polymorph of carbon. During ascent to the earth's surface, the intruding kimberlite magma incorporated fragments (xenoliths) derived from the mantle, which included diamonds, as well as crustal rocks. This entrained cargo of mantle- and crust-derived samples provides a direct "window" to deeper portions of the planet, and has been extensively studied, leading to major advances in our understanding of the composition of the mantle and its processes. However, kimberlites and kimberlite exploration also have important implications for interpreting surface processes and landscape evolution. This study reviews some of the important links between kimberlites and kimberlite exploration and our understanding of the geomorphic evolution of Botswana.

15.2 Kimberlite Distribution in Botswana and Surrounding Areas

Kimberlites tend to be gregarious, occurring as fields or clusters of pipes and associated intrusions (sills, dykes). Figure 15.2 illustrates the distribution and ages of the major kimberlite clusters in Botswana and immediate surroundings. Many of the Botswana kimberlite clusters appear to lie within a roughly southwest–northeast orientated belt, extending approximately through the centre of the country. This forms part of a more extensive lineament of kimberlites and related alkaline volcanic pipes that extend to the northeast into Zimbabwe and southwest to South Africa (Moore et al. 2008). The majority of the pipes associated with this lineament within Botswana have Upper Cretaceous ages ($\sim$70–100 Ma), and over 80 individual kimberlites have been identified in some of these clusters. Several of the Cretaceous kimberlites in Botswana have unusually large surface areas. Thus, for example, the M1 kimberlite, in the Tsabong cluster, has an estimated surface area of $\sim$140 ha, making it one of the largest in the world. To the northeast of Tsabong, several of the kimberlites in the Kokong cluster have diameters in excess of 500 m (areas $\sim$20 ha), while the AK1 pipe in the Orapa cluster has an estimated surface area of 110 ha. The large size of these Cretaceous kimberlites is in part linked to the preservation of original crater sediments, indicating relatively limited erosion, although some are composite bodies. For example, the AK1 (Orapa) kimberlite comprises two pipes which coalesce near the surface.

Straddling the Botswana–South Africa border, are a number of older kimberlites, with two broad groups of ages reported—one approximately 500 Ma (early Palaeozoic), and the second around 1300 Ma (Mesoproterozoic) (Jelsma et al. 2004) (Fig. 15.2). These have been interpreted to reflect an earlier roughly west-southwest–east-northeast oriented kimberlite trend (Jelsma et al. 2004). However, this remains the subject of considerable debate, and their apparent alignment may simply be a result of their discovery in an area of largely exposed basement terrain (Fig. 15.2). These older clusters are characterized by fewer (sometimes <10) kimberlites than are typical of the Cretaceous kimberlite clusters. Generally, only the deeper root zones are

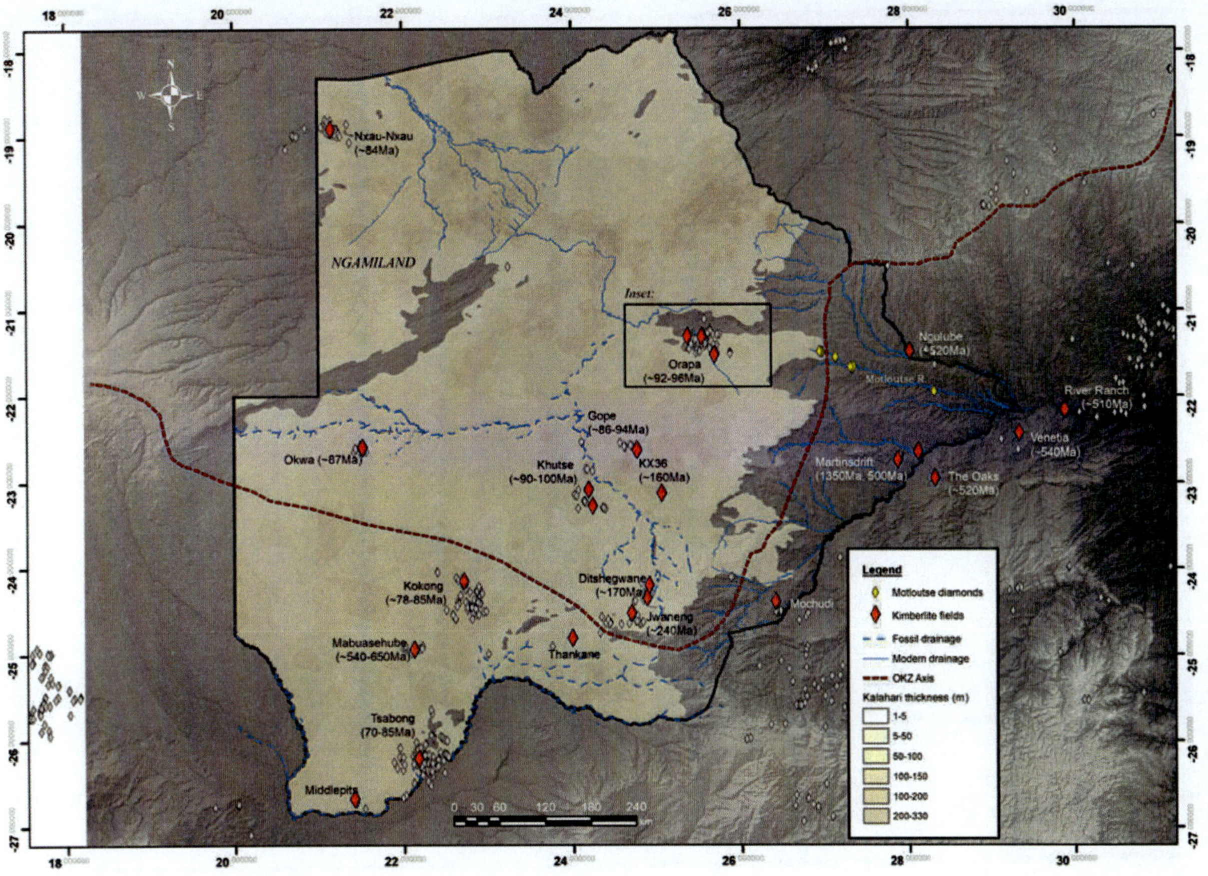

Fig. 15.2 Distribution of kimberlite clusters in Botswana and immediately surrounding areas. Black lines show country boundaries. Dashed red line denotes the Owambo-Kalahari-Zimbabwe (OKZ) Axis of Moore (1999). In eastern Botswana, this forms the watershed between west bank tributaries of the Limpopo and fossil drainage lines which formerly emptied into the Makgadikgadi basin of central Botswana. Rectangle: location of Fig. 15.5

preserved in these older kimberlites, often with complex morphology, or as dykes and sills. Consequently, these earlier kimberlites are typically much smaller than the large Cretaceous kimberlites. Thus, for example, the largest of the Martin's Drift kimberlites has a surface area of ~2.3 ha, while The Oaks kimberlite, just to the east in South Africa, has a surface area of ~1.3 ha.

However, it should be noted that not all kimberlites in Botswana are associated with these two "lineaments". For example, the Jwaneng (~240 Ma) (Allsopp et al., 1989), Ditshegwane (~170 Ma., De Beers, unpublished data), the Nxau-Nxau (84 ± 4 Ma, Farr et al. 2018), Okwa (~87 Ma; De Beers, unpublished data) kimberlites, and the ~140–160 Ma KX36 kimberlite in the Central Kalahari Game Reserve (Vines et al. 2017) do not appear to be associated with any obvious major lineaments (Fig. 15.2).

15.3 Broad Geomorphic Framework of Botswana

In a prescient paper, which is arguably one of the most important contributions to understanding the geomorphology of southern Africa, du Toit (1933) proposed that the major drainage divides in southern Africa reflect axes of Plio-Pleistocene epeirogenic flexure (Fig. 15.3). Du Toit envisaged that the watershed which extends through central Zimbabwe and the east of Botswana, which he designated the Kalahari-Rhodesia (now Kalahari-Zimbabwe) Axis, marks a line of crustal uplift that was coupled with subsidence to the west, to initiate the Kalahari basin. He concluded that this epeirogenic crustal tectonism beheaded former links between the Limpopo and the original

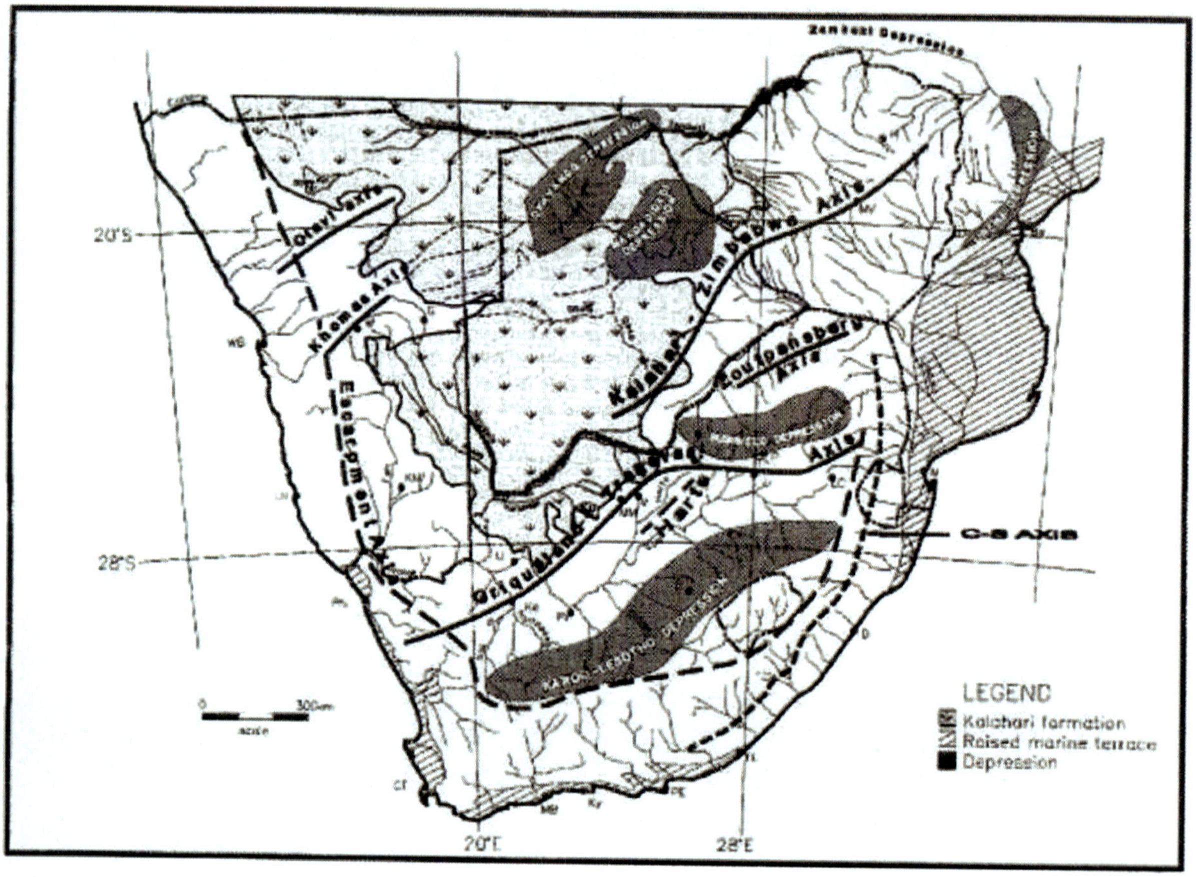

Fig. 15.3 Axes of epeirogenic flexure of inferred Plio-Pleistocene age proposed by du Toit (1933) (heavy black lines), together with the Escarpment flexure Axis (King 1963) and the Ciskei-Swaziland Axis (C-S; Partridge and Maud 1987) (heavy dashed lines). Modified from The South African Geographic Journal vol. XVI: 3–20 (1933)

headwaters of tributaries—such as the Motloutse—as well as the former link between the Okavango and the Limpopo. Du Toit (1927) noted that "It can be conjectured that had it not been for an insidious enemy (the epeirogenic flexure), the Okavango River might now be flowing through the thirsty Kalahari…".

At the time du Toit developed his model of epeirogenic buckling of southern Africa, little was known about the fossil drainage system of the central Kalahari, which formerly emptied into the Makgadikgadi Pans of northern Botswana. By integrating the topology of this fossil drainage system with the regional network, Moore (1999) proposed that the major watersheds in southern Africa formed three major concentric divides, roughly parallel to the coastline (Fig. 15.4). These were designated, from the coast inlands; the Escarpment, the Etosha-Griqualand-Transvaal (EGT) and the Owambo-Kalahari-Zimbabwe (OKZ) Axes. The Escarpment Axis, originally recognized by King (1963), was ascribed to uplift associated with the disruption of Gondwana, while the EGT and OKZ Axes were interpreted to respectively reflect mid-Cretaceous and late Palaeogene flexures (Moore et al. 2009). The latter authors proposed that the tectonism associated with the development of the EGT and OKZ Axes reflected far-field intra-plate stresses linked to changes in the spreading regimes of oceanic ridges surrounding southern Africa. The interplay between these flexures and drainage evolution in southern Africa is discussed in detail by Moore and Larkin (2001).

The EGT axis marks the eastern, southern, and western extremes of the Kalahari basin, while further to the north, the east and west of the basin are delineated by the OKZ Axis (Fig. 15.4). Basement outcrops occur along the southern portion of the latter axis. Moore (1999) followed du Toit (1933) in interpreting these divides to represent axes of epeirogenic flexure, responsible for initiating the Kalahari basin, as discussed in greater detail ahead.

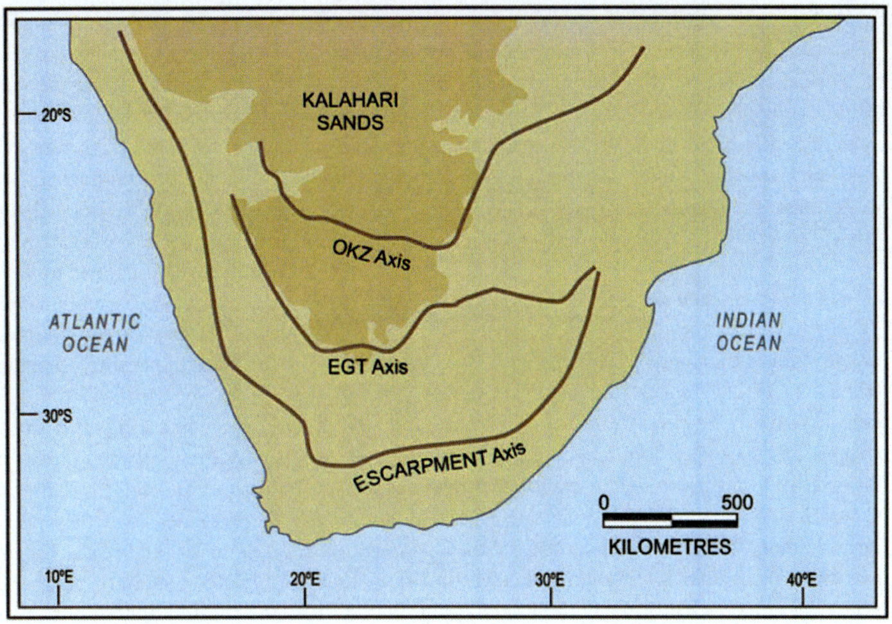

Fig. 15.4 The Escarpment, the Etosha-Griqualand-Transvaal (EGT), and the Owambo-Kalahari-Zimbabwe (OKZ) Axes (Moore 1999)

15.4 Kimberlites, Kimberlite Exploration, and the Geomorphic Evolution of Botswana

15.4.1 Evidence for Regional Botswana Erosion

Kimberlite exploration, and in particular the extensive diamond exploration drilling and sediment sampling work undertaken since the mid-1950s, has contributed significantly to advancing our understanding of the geomorphic evolution of Botswana and the depositional environment of the Kalahari Group sediments.

While kimberlites, by way of their internal geology and morphology, have the potential to provide important information of the erosional history of the surrounding area, this is by no means a perfect science. Nevertheless, where crater sediments are preserved, this would imply relatively limited erosion subsequent to emplacement. Conversely, areas which have experienced extensive erosion subsequent to pipe emplacement would be characterized by kimberlites with only the narrower, deeper portions of the pipe preserved. Kimberlite dykes and sills tend to be more common in areas of deeper post-emplacement erosion, although dykes also occur associated with the upper sections of large, relatively uneroded kimberlites.

For example, in Botswana (Fig. 15.2), the Tsabong (70–85 Ma), and Kokong (~80 Ma) kimberlite fields exhibit fully preserved craters and have had very little, if any, erosion. The kimberlites of the Orapa field (~90 Ma) are considered to have <200 m erosion (Field et al. 1997; Gernon et al. 2009a, b). Thus, the plateau of central Botswana has experienced limited erosion since the mid-Cretaceous. In contrast, in the Kimberley area of the South African highveld plateau, it is postulated that the Group 1 kimberlite pipes (~90–100 Ma) have experienced ~850 m of erosion, while the original surface associated with the older Group II (~120 Ma) pipes has been lowered by ~1250 m of erosion (Hanson et al. 2009). In all cases, these southern African pipes are associated with dykes and sills, including late-stage dikes present within some of the pipes and surrounds.

The small sizes of the early Palaeozoic and Mesoproterozoic kimberlites in eastern Botswana, and presence of associated sills and dykes, points to relatively deep erosion of this area subsequent to the emplacement of these kimberlites. Across the border in South Africa, the Venetia kimberlite field (Fig. 15.2), dated at 519.2 ± 5.8 Ma (Phillips et al. 1999) Ma is estimated to have been eroded to depths of ~500 m (Kurszlaukis and Barnett 2003). The Martin's Drift kimberlites in eastern Botswana (Fig. 15.2) reflect two distinct emplacement ages (De Beers, unpublished data). Those in the east of the cluster have ages of 500 Ma, closely comparable to the age of 506 ± 8 Ma for the Marnitz kimberlite cluster immediately to the east in South Africa, which includes "The Oakes" kimberlite pipes. The kimberlites in the west of the Martin's Drift cluster are older (~1300 Ma). The Martin's Drift kimberlites in eastern Botswana, and especially the 1300 Ma pipes in the western part of this cluster, probably experienced comparable or greater depths of post-emplacement erosion than has been estimated for the Venetia. Two small (<1 ha) parakimberlites near Mochudi (Fig. 15.2), dated at ~1395 Ma (Mothibedi and Duncan 2000), probably also reflect significant post-emplacement erosion. Just to the south of Mochudi, in the vicinity of the village of Sikwane, small pipes and sills dated at ~501 Ma (Mothibedi and Duncan 2000) probably also reflect significant, but poorly quantified post-emplacement erosion.

15.4.2 Discovery of the Orapa Kimberlite Field

In addition to diamonds, kimberlites carry a very characteristic suite of dense indicator minerals, including picro-ilmenite (Mg-rich ilmenite), Cr-spinel and purple pyrope, and orange eclogitic garnets. These minerals are dispersed into the surrounding drainage systems as a result of weathering of kimberlites subsequent to emplacement. Early diamond prospectors in southern Africa soon realized that these distinctive kimberlitic indicator minerals (today often referred to as KIMs) could be used as "pathfinders" to locate the source rocks. Indeed, they appreciated at an early stage that purple-mauve pyrope garnets, which they termed "geldrubyne" (money rubies) were good indicators of a diamond-bearing source. In contrast, red and yellow metamorphic garnets, which provided no encouragement for a kimberlite source, were dismissed as "bankrotrubyne" (bankrupt rubies). Sampling of drainage systems became a highly successful prospecting technique, entailing relatively wide-spaced initial sampling of major drainages, followed by more detailed follow-up of secondary and tertiary tributaries, aimed at identifying the drainage basin or sub-basin where the source kimberlites were located.

During the late 1950s, stream sediment sampling carried out by Consolidated African Selection Trust (Pty) Ltd. (CAST) in the hardveld of eastern Botswana resulted in the recovery of a number of diamonds in the Motloutse River—a Limpopo tributary (Willis 1960) (Fig. 15.2). However, despite intensive follow-up sampling, extending to the headwaters of the Motloutse, no kimberlite source which might account for the alluvial stones was discovered in this drainage basin. In the Final Report on the CAST prospecting activities, Willis (1960) highlighted the evidence presented by du Toit (1933) for beheading of the former headwaters of the Motloutse River by uplift along the Kalahari-Zimbabwe axis of flexure. He suggested that the diamonds recovered in this drainage might therefore reflect a source to the west of the watershed. However, despite this inspired geomorphic interpretation, no further work was carried out by CAST, as it was thought that prospecting in the remote Kalahari would be prohibitively costly.

De Beers subsequently re-investigated eastern Botswana in the mid-1960s, and their sampling programme resulted in the recovery of further diamonds in the Motloutse River. However, Dr. Gavin Lamont, who was in charge of the company's diamond exploration at the time, was also aware of the significance of the watershed flexure, and successfully motivated for the sampling programme to be extended to the west into the Kalahari (De Wit 2019). This De Beers regional reconnaissance work included sampling of the surface soils (termed loam sampling). This programme resulted in the identification of a major spread of kimberlitic ilmenites and garnets to the south of the Ntwetwe and Sowa pans (Fig. 15.5), collectively known as the Makgadikgadi Pans. Follow-up sampling in this area led to the discovery of the first kimberlite, designated BK1, in March 1967, followed by identification of the economic 110 ha AK1 or Orapa Mine kimberlite in April 1967 (Fig. 15.6).

With hindsight, it was appreciated that success in the intrepid De Beers venture into the geological unknown was in part aided by a very favourable local geomorphic setting. The cover sand of the Kalahari Group varies considerably in thickness, attaining depths in excess of 300 m in the south of the basin, although thicknesses of 80–100 m are more common (Haddon 2000, 2005). However, in the area where the first of the Orapa field kimberlites were discovered, the Kalahari cover proved to be negligible (Fig. 15.5), and the BK1 kimberlite formed a low positive topographic feature, while several of the other kimberlites, including the AK1 pipe, stood out as clear photo-features in aerial photographs taken prior to their discovery (Fig. 15.6). Ironically, prior to its discovery, the large AK1 kimberlite had long been recognized as a prominent large vegetation anomaly, used by pilots as a navigation beacon.

Detailed follow-up sampling around the Orapa kimberlite field identified a complex distribution of KIMs, which points to the interplay of a number of different geomorphic processes. Relatively elevated numbers of ilmenites and garnets were recovered in close proximity to the Orapa kimberlites, but both the numbers of grains and the ilmenite/garnet ratio showed a broad decrease eastwards (Fig. 15.5). Detailed sampling around individual kimberlites (Ustinov et al. 2019) has shown that the prevailing northeasterly wind in the area has resulted in limited, localized westward dispersion of KIMs (on the order of a few 100 s of metres to 1–2 km). However, aeolian processes do not provide a satisfactory explanation for the extensive halo to the east of the Orapa kimberlite field.

Pyrope garnet has a lower specific gravity (3.5) than ilmenite (4.5), and is thus more readily transported in both fluvial and aeolian environments. Consequently, as a broad generalization, increasing dispersion of KIMs from source kimberlites would result in decreasing overall KIM concentrations, coupled with a progressive increase in the garnet/ilmenite ratio (Grey 1976). It is necessary to caution however, that while these relationships appear to hold in Botswana, the reverse has been documented in the tropical

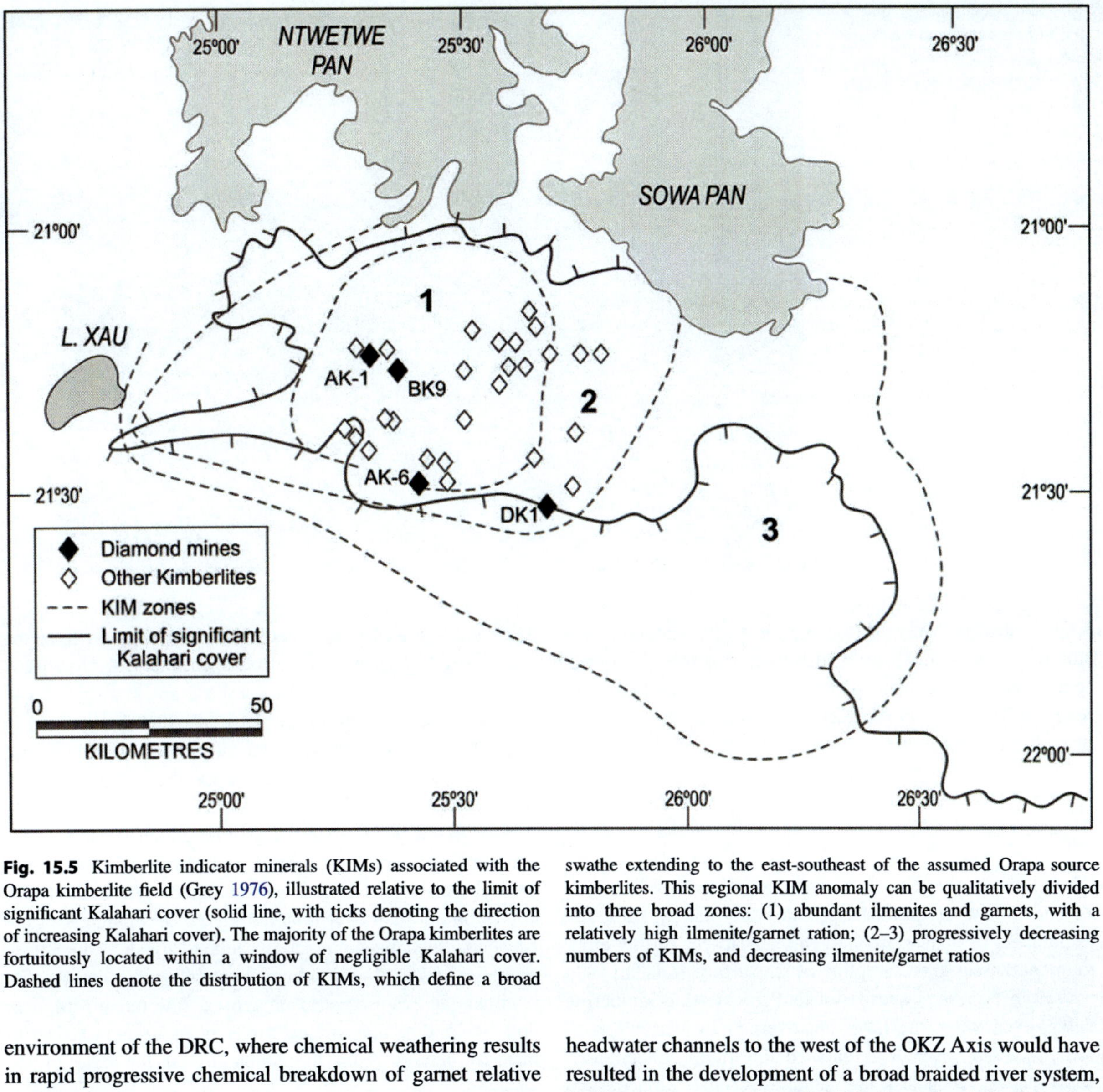

Fig. 15.5 Kimberlite indicator minerals (KIMs) associated with the Orapa kimberlite field (Grey 1976), illustrated relative to the limit of significant Kalahari cover (solid line, with ticks denoting the direction of increasing Kalahari cover). The majority of the Orapa kimberlites are fortuitously located within a window of negligible Kalahari cover. Dashed lines denote the distribution of KIMs, which define a broad swathe extending to the east-southeast of the assumed Orapa source kimberlites. This regional KIM anomaly can be qualitatively divided into three broad zones: (1) abundant ilmenites and garnets, with a relatively high ilmenite/garnet ration; (2–3) progressively decreasing numbers of KIMs, and decreasing ilmenite/garnet ratios

environment of the DRC, where chemical weathering results in rapid progressive chemical breakdown of garnet relative to ilmenite.

The anomalous easterly dispersion of indicators from the Orapa kimberlites is down the original palaeo-slope prior to uplift along the OKZ flexure, and thus seems most readily explained in terms of a dominant fluvial dispersion process, modified by very secondary aeolian reworking. However, this raises the question of why fluvial dispersion should have produced the broad KIM halo identified to the east of the Orapa kimberlites. Uplift of the OKZ axis across the headwaters of former Limpopo tributaries such as the Motloutse would have resulted in a progressive decrease in river gradient. Prior to final dismemberment of the original drainage line, it is most likely that decreasing gradient of former headwater channels to the west of the OKZ Axis would have resulted in the development of a broad braided river system, leading in turn to a broad KIM dispersion halo. We thus interpret the broad dispersion pattern of KIMs to the east of the Orapa kimberlite field to be consistent with the flexure model proposed by du Toit (1933).

The De Beers follow-up programme, which included surface loam sampling and pitting, also identified a linear concentration of KIMs to the west of the Orapa kimberlite, with a westwards decrease in the ilmenite/garnet ratio (Grey 1976). This anomalous concentration of KIMs was best developed at ~920 m, which corresponds to one of the former shorelines of the Makgadikgadi paleo-lake. It was concluded that it reflected westwards longshore drift along this former shoreline of KIMS dominantly derived from the

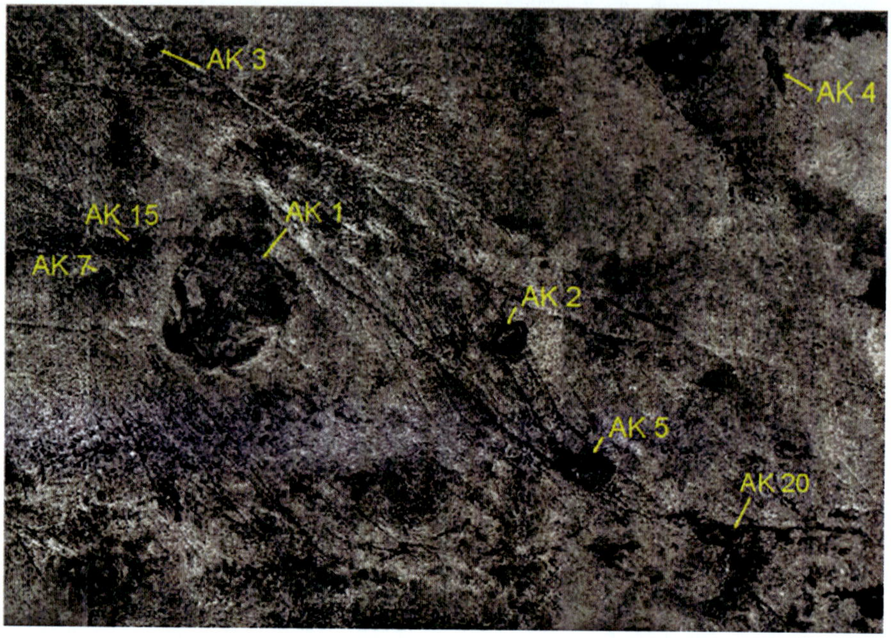

Fig. 15.6 Aerial photograph (ca 1957) of kimberlites in the Orapa field, taken prior to their discovery (courtesy of De Beers)

AK1 kimberlite (Grey 1976). The De Beers kimberlite exploration programme has thus made an important feedback contribution to the understanding of the geomorphic processes which formerly operated in the area.

15.4.3 Discovery of the Jwaneng Kimberlite Field

When kimberlite exploration expanded into areas covered by Kalahari sand, it became necessary to modify sampling techniques, as active river systems are absent in this environment. Direct loam sampling of the surface Kalahari sand proved to be a very successful tool for locating kimberlite fields, even when the latter are overlain by 40–100 m of Kalahari sand cover. This entailed the initial collection of wide-spaced reconnaissance samples of the surface sand (termed "loam samples"), with progressively more close-spaced follow-up sampling in the vicinity of any initial samples containing KIMs. Such detailed follow-up sampling demonstrated that there is often a relative concentration of dense kimberlitic minerals, and particularly kimberlite picro-ilmenites (Mg- and Cr-rich ilmenite), in the surface sands immediately overlying kimberlites, with an intervening very diffuse scatter of KIMs. This is illustrated for kimberlites designated DK1-9 in the Jwaneng kimberlite cluster of southern Botswana (Fig. 15.7, Data from Lock 1985; see Fig. 15.2 for locality of the Jwaneng kimberlite cluster). Note that some of the surface KIM halos, such as those over DK4 and DK5 are elongate to the southwest, and parallel to the prevailing north-easterly wind. This suggests that there has been limited aeolian dispersion away from the underlying kimberlite source rocks, in line with the observation by Ustinov et al. (2019) for the Orapa kimberlite field.

The Jwaneng sampling results imply that the kimberlitic minerals have been moved vertically from the underlying source kimberlite through the entire Kalahari sand column. It is generally assumed that transport from the base to the surface of the Kalahari reflects bioturbation, with termites considered to be a significant contributing factor in the vertical transport of the kimberlitic indicator minerals. Other key factors at play over time during the development of the Kalahari sequences include aeolian-fluvial sedimentary processes, mineral grain chemical stability and density differentials and hypsometric influences. The overall implication being that the surface sands have been mixed with material derived from the lithologies at the base of the Kalahari.

15.4.4 Discovery of the Gope Kimberlites

In 1980, Falconbridge Explorations Botswana (FEB) carried out helicopter-supported reconnaissance loam sampling over the central Kalahari (Fig. 15.8) on a very wide-spaced ($\sim$13 km $\times$ 13 km) grid. This programme, initiated by geologists Chris Jennings, John Lee, John Blaine and Roger Billington, resulted in the recovery of 4 ilmenites within an area 4000 km^2 in the east of the reconnaissance sampling block. An aeromagnetic survey, was flown over this ground with a wide (500 m) line-spacing, and a number of priority "bulls-eye" magnetic features were selected for ground

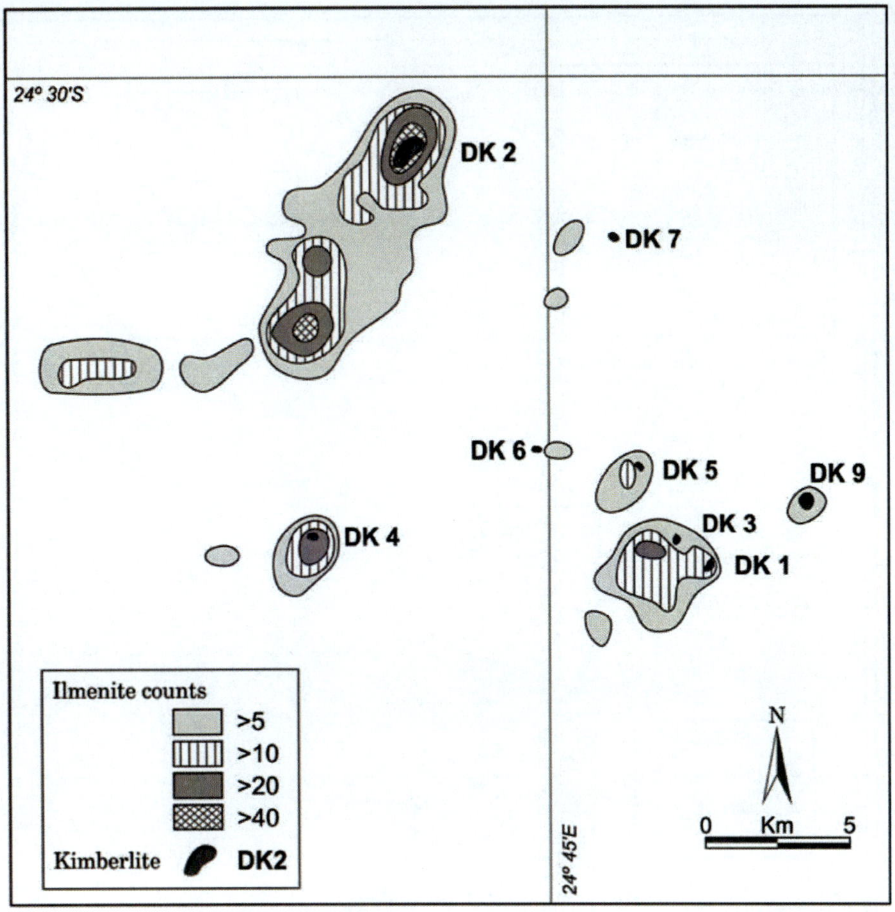

Fig. 15.7 Soil sampling results from the Jwaneng kimberlite field area. Adapted from Lock (1985)

follow-up. Drilling of the first of these magnetic target intersected a kimberlite beneath approximately 80 m of Kalahari sand (Lee 2009). This kimberlite was designated Gope—the Setswana word for "nowhere".

The Falconbridge Central Kalahari reconnaissance sampling programme underlined the efficiency of bioturbation in translocating kimberlite minerals from a source at the base of the Kalahari to the surface, and thus extensive mixing of the sand forming the Kalahari section. An unexpected further finding was that the texture of Kalahari surface sand within the area sampled is not uniform, but shows a very systematic pattern of variation. This is illustrated in Fig. 15.8, which shows variations in the proportion of the coarse (>425 μm) sand fraction.

The Gope kimberlites proved to be located in an area characterized by a very low proportion ($\sim 1\%$) of the coarse sand fraction (Fig. 15.8), which in part explains the very subtle associated KIM soil anomaly. The remarkably systematic pattern of textural variation of the Kalahari sand has important implications for understanding the depositional processes, as discussed in greater detail in the following section.

This evidence for extensive vertical mixing of Kalahari sediments has critical implications for the use of luminescence dating techniques to determine the age of Kalahari landforms such as fossil dunes and former shorelines of the Makgadikgadi palaeo-lake, and their use as palaeo-climate proxies (e.g. Thomas and Burrough 2012). The fundamental assumption of luminescence dating is that quartz, once buried, remains in-situ, and that the age reflects the time of development of the landform. However, the evidence for pervasive bioturbation of the Kalahari sequence implies that this assumption is invalid in this environment. This in turn raises the question of the significance, if any, of quartz luminescence ages of landforms in the Kalahari sedimentary environment.

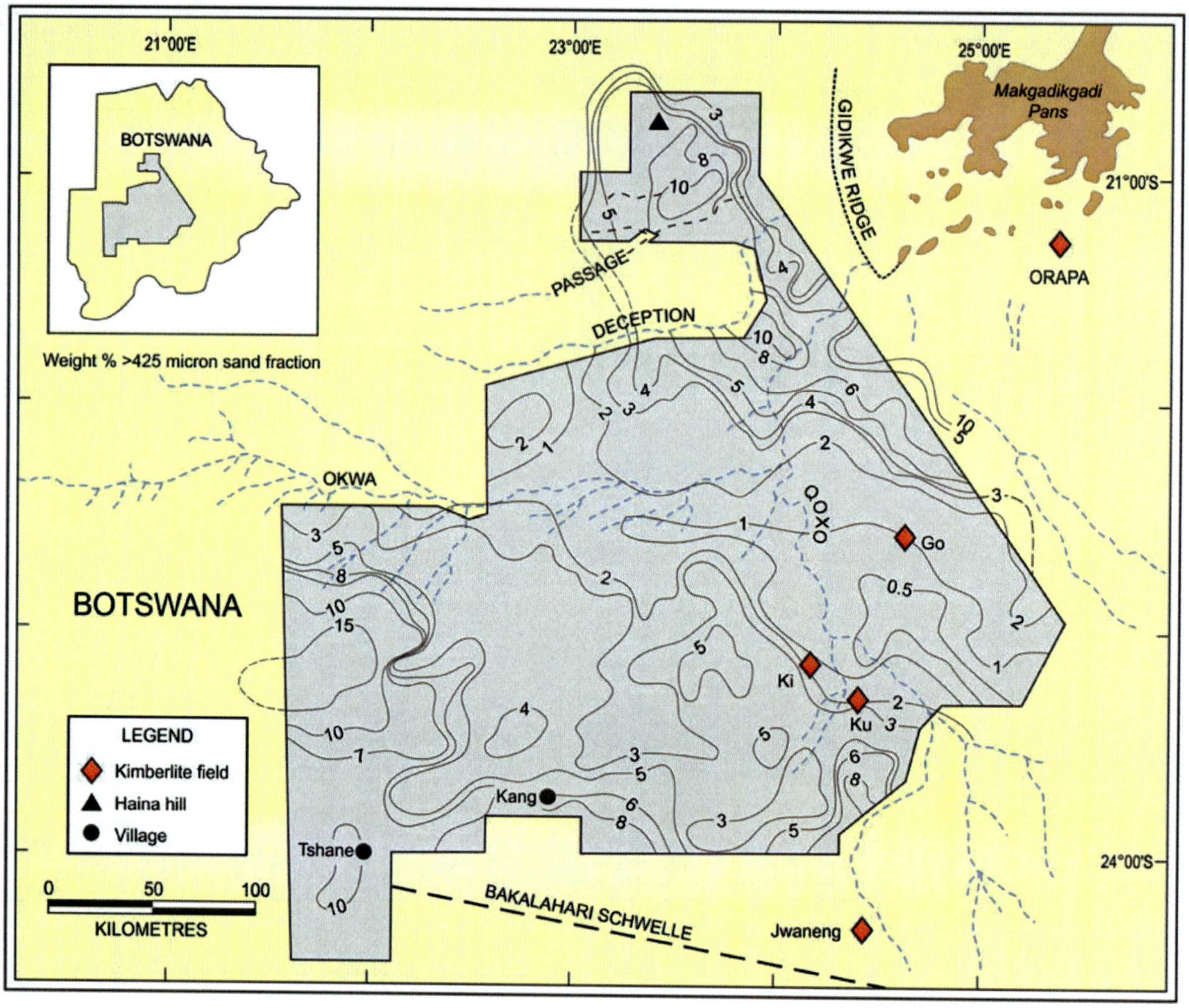

Fig. 15.8 Variation in the proportion (weight %) of the coarse (>425 μm) sand fraction within the Central Kalahari area sampled by Falconbridge Explorations Botswana. Kimberlite clusters: Go—Gope; Ki—Kikao; Ku—Khutse. Dashed blue lines denote fossil drainages which formerly emptied into the Makgadikgadi Pans

15.5 Kimberlites Exploration and the Kalahari Formation

15.5.1 Evolution of the Kalahari Basin

South-central Africa is blanketed by post-Gondwana continental sediments which extend from the Orange River in South Africa in the south, to the DRC in the north, and are termed the Kalahari Group by Haddon (2000, 2005) (Fig. 15.9).

Isopachs of these post-Gondwana continental sediments (Fig. 15.9) indicate that they define a number of sub-basins, with a major depo-centre in northern Angola, a second straddling the Angola-Namibia border, designated the Etosha basin, and a third, designated the Okavango basin, which extends northwards from the Okavango area of northern Botswana into southwest Zambia. The Bakalahari (or Bakgalagadi) Schwelle, recognized by Passarge (1904), which forms part of the OKZ Axis, reflects shallow Kalahari cover, associated with basement outcrops. This flexure separates a Central Kalahari sub-basin from the Morokweng depo-centre in the extreme south of Botswana. A further subsidiary depo-centre, in east-central Namibia, was designated the Aranos sub-basin (Fig. 15.9).

The regional stratigraphy and ages of these post-Gondwana continental deposits are not tightly constrained because of poor exposures and a general lack of age diagnostic fossils, but selected sequences are illustrated in Fig. 15.10. The uppermost sections are dominated by friable

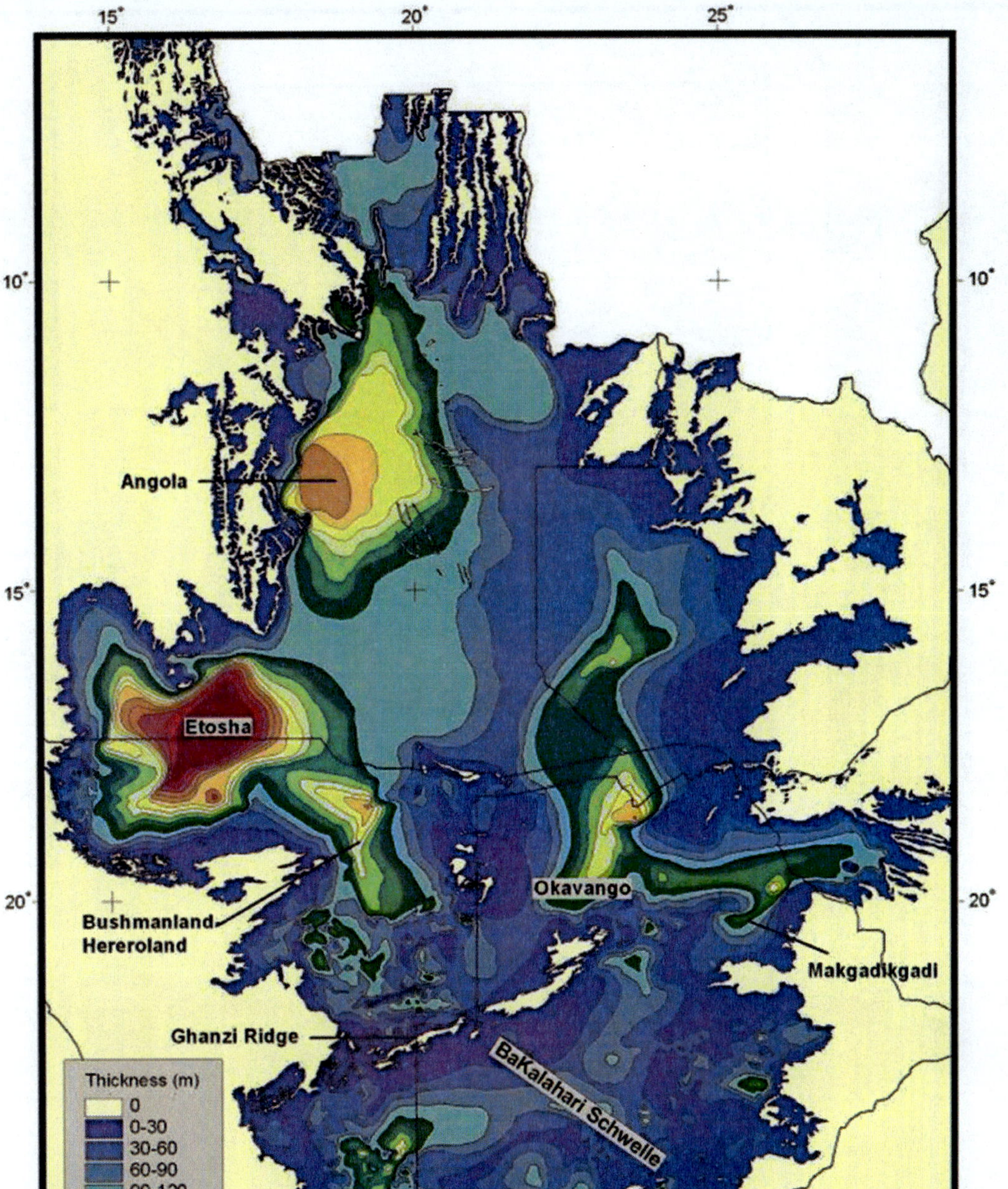

Fig. 15.9 Isopachs of post-Gondwana terrestrial sediments in southern and central Africa (from Haddon 2005, and included with kind permission of the author)

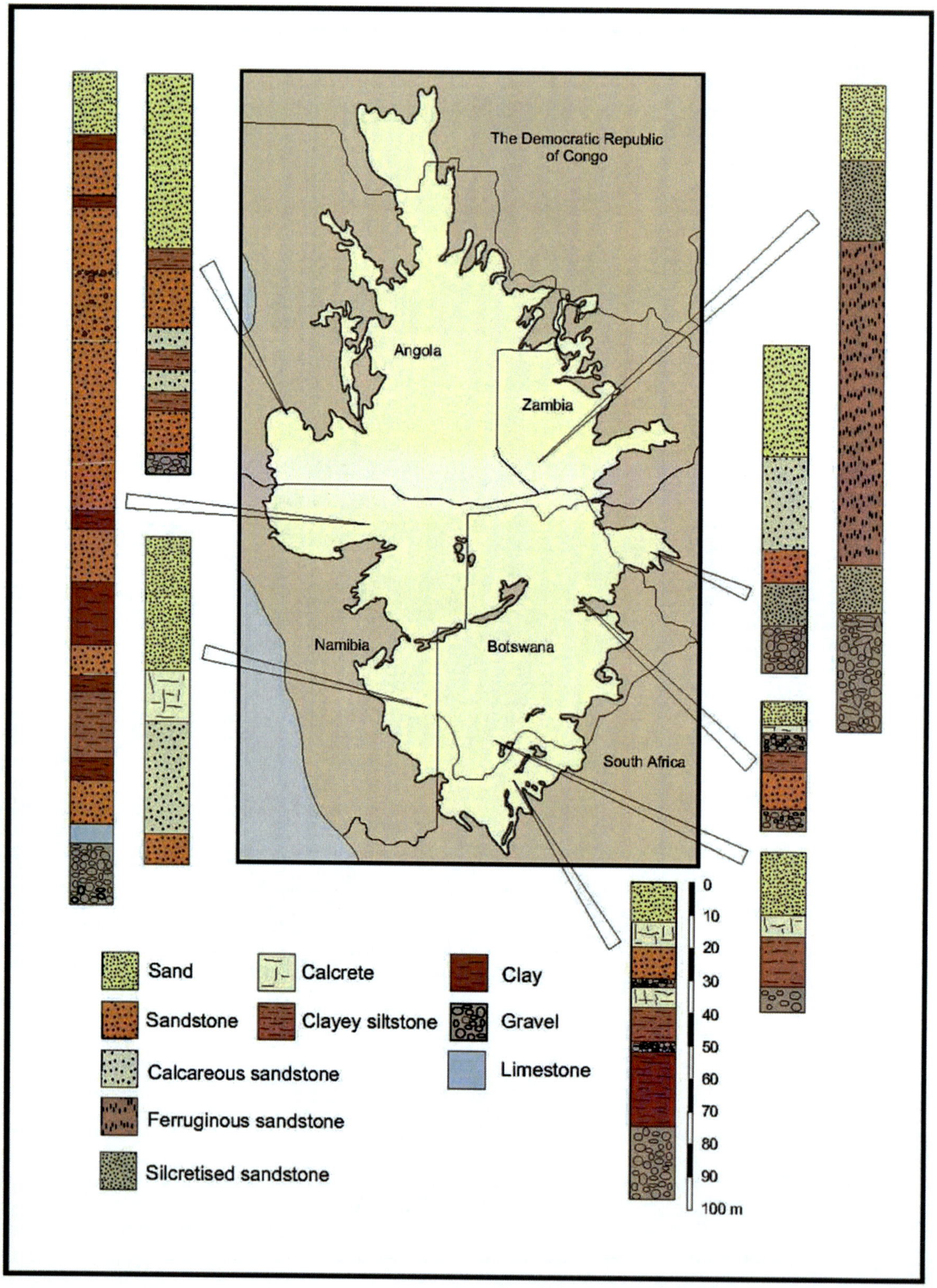

Fig. 15.10 Selected stratigraphic sections of the post-Gondwana terrestrial sedimentary basin of southern and central Africa (from Haddon 2005, included with kind permission of the author)

to poorly consolidated arenaceous sediments with interbedded calcrete and silcrete and duricrust horizons, which are generally considered to be of Cenozoic age. In Northern Angola, these arenaceous sediments overlie mid-Cretaceous gravels of the Calonda Formation of the Kwango Group (Pereira et al. 2003), the two separated by a late-Cretaceous unconformity. The Kwango Group is in turn inferred to be separated from underlying Jurassic sediments by an early Cretaceous unconformity. However, Roberts et al. (2015) interpret the mid-Cretaceous Kwango and Cenozoic Kalahari sedimentary units of the Kasai Region of the DRC to the north of Angola to be a conformable sequence, which they classified as Formations rather than Groups.

In Botswana, where the Kalahari sediments blanket the western two-thirds of the country, drilling carried out in the course of water and mineral (particularly diamond) exploration programmes offer an extensive database to characterize regional lithological variations of this cover sequence. Although a detailed synthesis of this invaluable database is not yet available, a number of broad observations are highlighted.

In general, the upper sections of the Kalahari sequence in Botswana is characterized by unconsolidated sand overlying friable sandstones which have been interpreted to reflect a fluvio-lacustrine depositional sequence (du Toit 1927; Du Plessis and Le Roux 1995). Degraded linear dunes in the north of Botswana and Namibia, which extend into northwest Zimbabwe and southern Zambia, are interpreted to reflect aeolian reworking of the surface sediments during Plio-Pleistocene periods of hyper-aridity (Miller et al., 2010). The textural characteristics of the surface sand of the Central Kalahari (Fig. 15.8) are consistent with these conclusions. Coarser sand is associated with the margins of the Central Kalahari basin, and grain size fines into the interior of the basin. This pattern is not readily explained in terms of aeolian processes, dominated by the prevailing easterly to northeasterly wind, but rather suggests fluvial redistribution from the margins to the distal central portions of the basin (Moore and Dingle 1998).

The arenaceous upper Kalahari sediments are in places underlain by conglomerates with a poorly defined, patchy distribution. This basal unit appears to be more widespread in the southern third of Botswana. Figure 15.11a illustrates the massive, polymict basal Kalahari conglomerate which is exposed in the Kgome quarry, located a few kilometres to the west of Jwaneng. This basal conglomerate is dominated by matrix-supported angular to sub-rounded cobbles which in places show a poorly developed imbricate structure (Fig. 15.11b). Black manganese-iron oxide occurs as a discontinuous cement in the upper portions of the conglomerate. The cobbles are dominated by quartzites and subordinate banded ironstones, which are probably ultimately locally derived from Transvaal and Waterberg Supergroup sediments. However, they may represent the winnowed relics of glacial basal Karoo sediments, which outcrop to the north of the quarry.

Detailed prospecting shows that the basal gravels have a very patchy, discontinuous distribution in the Jwaneng area (Debridge 1986). This points to an episode of erosion prior to the deposition of the overlying arenaceous upper Kalahari sediments. Basal gravels appear to be rare or absent in the northern two-thirds of Botswana. However, it is not clear whether they were formerly present over a wider area, and were removed by erosion prior to deposition of the arenaceous upper Kalahari sediments.

Ilmenite and other KIMs dispersed from the primary kimberlite source would have been redistributed and concentrated in the basal Kalahari gravels. As a result, the presence of these gravels at the base of the Kalahari is often reflected by secondary (redistributed) KIM anomalies in the overlying surface sands. These are often very difficult to distinguish from the primary anomalies which overlie sub-Kalahari kimberlites. The ilmenites recovered from surface sands to the southwest of the DK-2 (Jwaneng) kimberlite (Fig. 15.7) appear to be an example of such secondary (spurious) anomalies, as they are not associated with known kimberlites.

The Kalahari sequence of the Ngamiland region of northwest Botswana is notably different from that which has been described in the southern Kalahari. In the Ngamiland area, basal conglomerates are absent, and the lowermost Kalahari unit is a fine-grained sandstone, up to 20 m thick, which has experienced post- depositional silicification, and subsequent calcrete replacement. This basal Kalahari duricrust, ascribed to the African Surface (McFarlane et al. 2007), is overlain by about 40 m of unconsolidated and friable, semi-consolidated Kalahari sand, which must therefore have been deposited unconformably following development of the African Surface duricrust. The silcretized basal Kalahari sandstone is exposed in the Eiseb valley of northeast Namibia (Fig. 15.12a, b). From a distance, the unit appears to be well-bedded, but it is highly brecciated as a result of the silicification and subsequent calcrete replacement. As a result, bedding is difficult to discern from close range. The study by McFarlane et al. (2007) demonstrated that this basal Kalahari duricrust has a wide but discontinuous distribution in northwest Botswana, pointing to erosion of the unit prior to deposition of the less indurated overlying Kalahari sand. This is analogous to our observations on the basal gravels in southern Botswana, indicating that there was a widespread erosional break in the Kalahari Group sediments in southern Africa.

The age of the Kalahari Group in southern Africa is not tightly constrained because of the general lack of age-diagnostic fossils. An early proposal (Mabbutt 1955) was that the basal units in the south of the basin were of late

Fig. 15.11 a Massive polymict basal Kalahari conglomerate exposed in the Kgome quarry to the east of Jwaneng, b Detail of imbricate structure in the Kgome quarry basal Kalahari conglomerate

Cretaceous age. The mid/late Cretaceous sub-Kalahari kimberlites are generally considered to provide a lower age limit for the basal Kalahari. However, a number of lines of evidence suggest that the basal gravels may significantly pre-date the arenaceous upper Kalahari sediments.

Recent exploration drilling carried out by De Beers in the Tsabong kimberlite field in the southwest of Botswana (Fig. 15.2) has shown that in this area, the kimberlites are typically capped by a silcretized basal Kalahari conglomerate (Fig. 15.13a). However, a closely similar conglomerate

a

b

Fig. 15.12 a Basal Kalahari sandstone, Eiseb Valley, Namibia. From afar, the unit appears to be well-bedded. **b** Detail of the basal Kalahari sandstone shown in Fig. 15.12a (Eiseb Valley, northeast Namibia), considered by McFarlane et al. (2007) to be the duricrust above the African Surface. Bedding is difficult to discern at close range because of intense brecciation of the unit related to silcrete formation and subsequent calcrete replacement

was encountered in drill core at a depth of 246 m, within the crater sequence of one of the Tsabong kimberlites (Fig. 15.13b). This was presumably a country rock fragment incorporated into the crater sequence, indicating that this unit was also present at the time of eruption of some of the pipes. This in turn suggests that in southern Botswana, silcretization of a basal Kalahari conglomerate occurred over a period broadly coeval with and following the eruption of the Tsabong kimberlites, which have yielded zircon ages of 70–85 Ma (De Beers, unpublished data). This points to a mid- to

Fig. 15.13 **a** Silcretized polymict conglomerate in drill core (~77 m depth) at the base of the Kalahari, overlying kimberlite crater sediments (courtesy De Beers). **b** Similar conglomerate in drill core (~246 m depth) from within kimberlite crater-filled sediments, Tsabong area, southern Botswana (Courtesy De Beers)

late-Cretaceous age for the basal Kalahari conglomerate, suggesting that it is broadly coeval with the basal Calonda conglomerates of northern Angola and Kasai (Pereira et al. 2003).

In contrast to the evidence for a mid- to upper-Cretaceous age for the basal Kalahari in southern Botswana, a series of boreholes drilled to the west of the Makgadikgadi Pans, in central Botswana (Fig. 15.8), produced fossil pollens of Miocene and probably late Miocene age from pyritic mudstones and sandstones, overlying Karoo basalt (Du Plessis 1998 cited by Moore et al. 2012). Du Plessis (1998) interpreted this basal Kalahari unit to reflect deposition in a deep perennial lake This evidence points to a much younger Miocene and probably Late Miocene vintage for the basal Kalahari in this area, which is probably part of the Central Kalahari sub-basin, bounded in the south by the flexure of the OKZ Axis (Fig. 15.4), and separated from the Etosha basin to the northwest by the shallow Kalahari associated with the NE-SW-trending Ghanzi Ridge (Fig. 15.9).

The drilling work carried out by De Beers in the Tsabong kimberlite field therefore suggests that the basal Kalahari in the south of the basin may be of middle Cretaceous age, in line with early suggestion by Mabbutt (1955), but that the basal Kalahari sediments are much younger in the Central Kalahari Basin. This interpretation is consistent with the mid-Cretaceous age of the EGT Axis (Moore et al. 2009). Uplift along this axis severed the former link between the Molopo River and the Orange, initiating Kalahari sedimentation in the south of the basin (Moore 1999). The earliest sediments filled major south-flowing former Molopo tributaries, which in turn exploited earlier Permian glacial valleys. Continued sedimentation would be expected to result in an onlap, and younging of the basal Kalahari to the north.

The reason for the absence of basal Kalahari conglomerates in central Botswana is speculative. One possibility is that they were removed by erosion associated with the intra-Kalahari unconformity identified in southern Botswana. However, the processes controlling Kalahari sedimentation may have been more complex, and involved the interplay of later epeirogenic flexuring. The northern portion of the Kalahari basin is bounded in the east and west by the late Palaeogene OKZ axis (Fig. 15.4), while outcrops of basement rocks are associated with the southern portion of this axis. This flexure would have severed the former links to the ocean of major drainage lines such as the Zambezi, Okavango and Cuando to the Indian Ocean. In turn, this would have initiated a major endoreic drainage system, which emptied into the northern section of the Central Kalahari sub-basin, bounded in the south by the OKZ Axis. As was the case with uplift along the EGT Axis, this would have resulted in onlap of the basal Kalahari sediments to the north. This provides a possible alternative explanation for the young age of the basal Kalahari in the north of the basin.

In the Ngamiland area of northwest Botswana, the thick basal silcretized sandstone, which shows the imprint of subsequent calcrete replacement, caps the 84 ± 4 MaNxau-Nxau kimberlites (Farr et al. 2018). However, de Wit (2012) reports the recovery of inferred basal Kalahari silcrete xenoliths from both these kimberlites, and the Sikereti kimberlites, which represent the extension of the Nxau-Nxau cluster to the west in Namibia. This is analogous to the relationship between kimberlite emplacement and deposition of the basal Kalahari in southern Botswana. De Wit (2012) therefore suggests that the basal Kalahari in the Ngamiland area is Cretaceous in age.

Ngamiland is located near the extreme eastern part of the Etosha basin, in an area with very young tectonism, which is reflected by the faulted truncation of deflated late Kalahari dunes (McFarlane and Eckardt 2007) The Etosha sub-basin is separated from the Central Kalahari sub-basin by a NE-SW zone of shallow Kalahari sand associated with the Ghanzi Ridge (Fig. 15.9). The sediments in these two sub-basins may therefore record contrasting geological records. The sub-Kalahari surface deepens from northwest Botswana westwards towards the Etosha sub-basin. A speculative possibility is that the basal silcretized and calcretized sandstone was deposited in the eastern extremity of the Etosha basin, but was exposed by tectonism in the area, followed by duricrust formation linked to the African Surface. The latter has been linked to an erosion cycle which extended into the Miocene (Partridge and Maud 1987). This, in turn, would imply a late- or post Miocene age for the later unconformable deposition of the arrenaceous upper Kalahari in northwest Botswana. However, a further complexity in deciphering the geological history of the area is that Pleistocene dunes in northwest Botswana have been displaced by recent faulting (McFarlane and Eckardt 2007).

The evidence that the arenaceous upper Kalahari sediments are relatively young, dating from the Miocene ($\sim$23 Ma) or even late Miocene ($\sim$5 Ma) flags a possible intriguing relationship. An expansion of savannah grasslands at the expense of woody cover commenced in the late Miocene to early Pliocene (8–4 Ma) (Osborne 2007; Cerling et al. 2011). The Cenozoic was characterized by major radiations of termites (family Termitidae), with increasing diversification following $\sim$40 Ma (mid-Eocene). This is reflected in a subsequent significant increase in the number of termite individuals in the fossil record, together with termite proportions increasing from <1% of all insects to $\sim$10% over the period of 40 Ma to the present (Engel et al. 2009). These increases in numbers of termites may, at least in part, have been enhanced by the late Miocene expansion of grasslands. Irrespective of such a linkage, increasing numbers of termites in the late Cenozoic would enhance bioturbation of the Kalahari sequence, and in turn, transport of kimberlitic minerals from the base of the Kalahari to the surface.

15.6 Secondary or False Kimberlitic Indicator Mineral Anomalies

While surface soil sampling programmes have been remarkably successful in locating virgin kimberlite fields, there are pitfalls to the method, which are often directly linked to the local geomorphic setting.

Sampling carried out by De Beers during the early 1970s identified a broad swath of kimberlitic garnets (shown in green in Fig. 15.14) to the east of the boundary of the Central Kalahari Game Reserve, and to the southwest of the Makgadikgadi Pans complex.

Subsequent sampling by Kalahari Explorations (Kalex) showed that the kimberlitic garnet anomaly identified by De Beers extended to the west (Fig. 15.16), but despite the detailed follow-up sampling, coupled with high resolution (200 m line spacing) aeromagnetic surveys, no local kimberlites have been discovered. A review of the geomorphic setting of this regional garnet anomaly provides a framework to account for this broad regional garnet anomaly.

Grove (1969) recognized a major curvilinear sand ridge, with a crest elevation close to 945 m, to the west of the Makgadikgadi Pans (Fig. 15.15a). He interpreted this sand ridge to represent a fossil offshore bar, associated with a former shoreline of the Makgadikgadi palaeo-lake. McFarlane and Eckardt (2006) subsequently recognized a comparable, but more degraded sand ridge further to the west, which they designated the Deception Ridge. They concluded that it was related to an earlier, and previously unrecognized higher fossil shoreline at an elevation of $\sim$995 m. Figure 15.15b is a cross-section across the Gidikwe and Deception Ridges. Note that both were separated from their respective former shorelines by a shallow lagoon.

The major kimberlitic garnet anomaly (Fig. 15.14) is also broadly symmetrical about the lower reaches of the Okwa-Quoxo fossil drainage line, which formerly emptied into the Makgadikgadi palaeo-lake. The Kalex sampling identified a linear zone of relatively elevated garnet proportions, which proved to be symmetrical about the Deception Ridge, which reflects a former relatively high energy environment associated with the 995 m shoreline. In this geomorphic

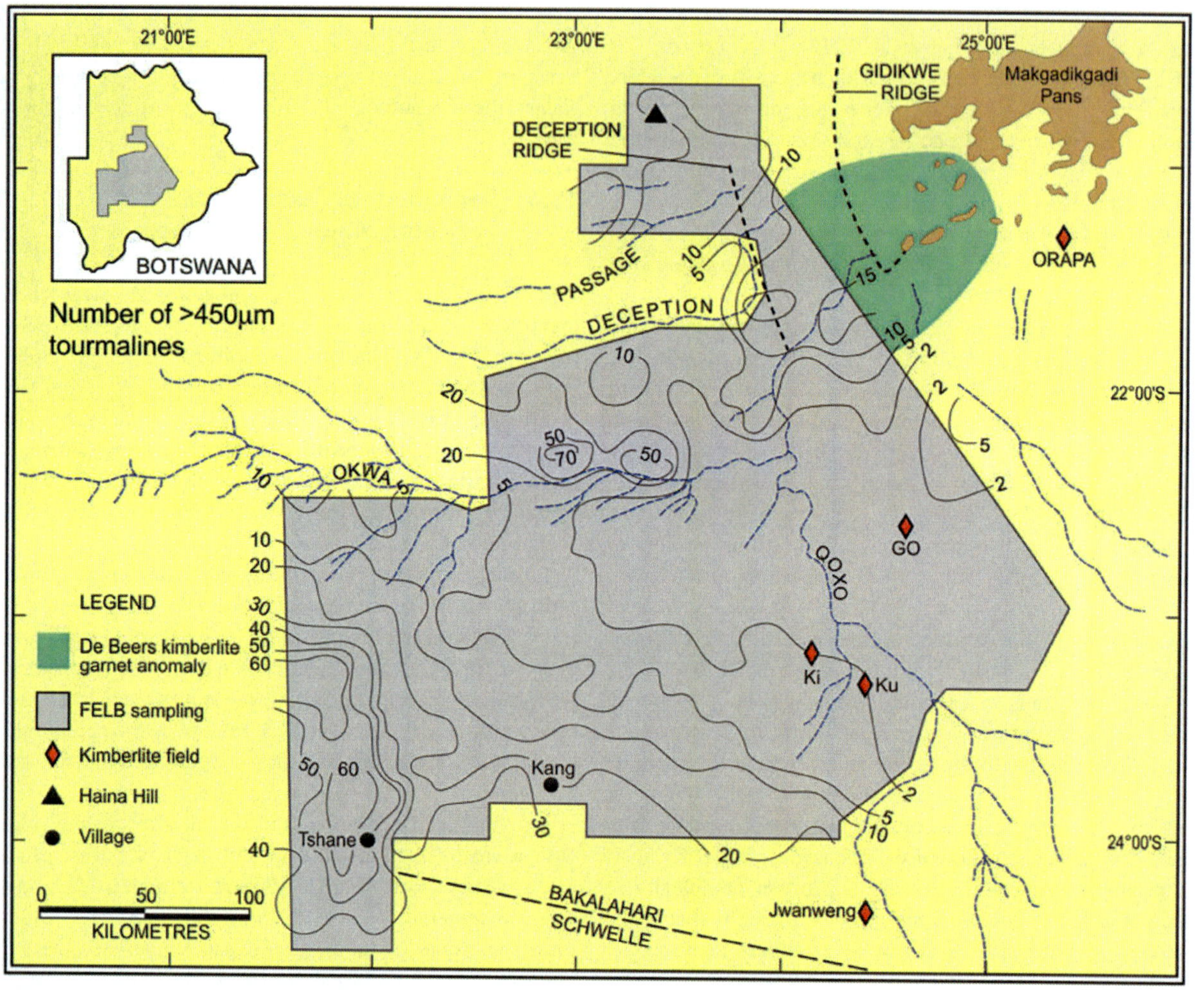

Fig. 15.14 Relative numbers of tourmalines recovered from the coarse (>425 μm) sand fraction during a sampling programme carried out by Falconbridge Explorations Botswana within Reconnaissance Permit RP80/1 covering the central Kalahari (grey colour) (data from Moore and Dingle 1998). The green shading shows the distribution of kimberlitic garnets recovered by De Beers. Dashed lines denote the locations of the Gidikwe and Deception sand Ridges, which are shown in detail in Fig. 15.15

environment, winnowing of less dense minerals would have resulted in a relative concentration of higher density phases.

Figure 15.14 shows the number of tourmalines (SG ~3.1) recovered by Falconbridge Explorations Botswana (Pty.) Ltd., during a reconnaissance soil sampling programme over the Central Kalahari. This was carried out at a much broader sample spacing (13 km × 13 km) than the semi-detailed Kalex sampling. While the Falconbridge sampling did not extend to the Gidikwe Ridge, elevated numbers of tourmalines were recovered immediately to the west or the ridge. This tourmaline concentration straddles the Deception Ridge and the lower reaches of the Okwa-Quoxo, in the area where this drainage formerly emptied into the Palaeo-Makgadikgadi lake system during the tenure of the 995 m and 945 m shorelines (designated Palaeolake Deception and Palaeolake Makgadikgadi respectively by McFarlane and Eckardt 2006. This, in turn, shows that the elevated numbers of kimberlitic garnets and tourmalines are both located within a relatively high energy geomorphic environment favourable for the concentration of dense minerals.

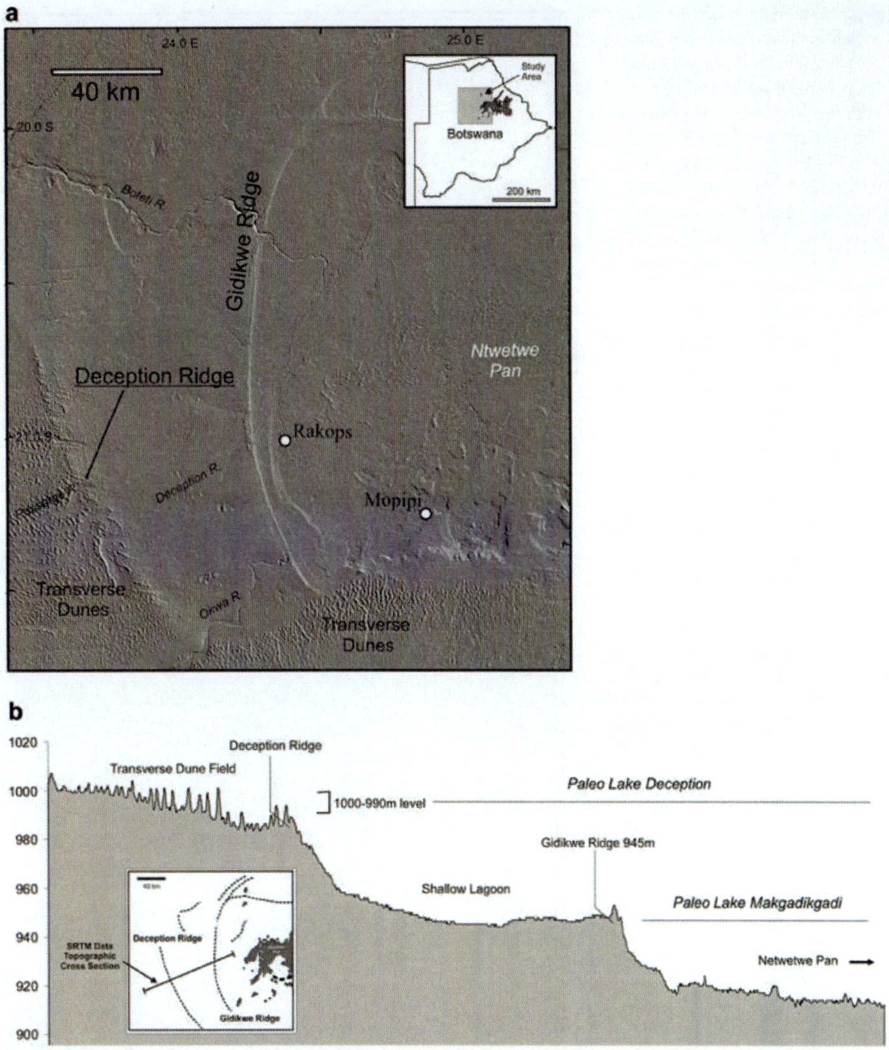

Fig. 15.15 **a** From McFarlane and Eckardt (2006), published with kind permission of Prof. Frank D. Eckardt). Digital terrain model of the southwest corner of the Makgadikgadi Pan complex. The Deception and Gidikwe Ridges are interpreted to represent offshore bars linked to the 995 m and 945 m palaeo-lake shorelines, respectively. **b** Cross-section across the Deception and Gidikwe Ridges (see inset for location). Published with kind permission of Frank D. Eckardt

Moore (2013) concluded that the regional spread of garnets (and associated tourmalines) was a secondary anomaly, ultimately derived from distal kimberlites, probably including Jwaneng, located in the lower reaches of the Okwa-Quoxo drainage system (Fig. 15.14). Secondary heavy mineral concentration would have occurred where this fossil drainage line once debouched into the Makgadikgadi palaeo-lake. The formation and recognition of secondary KIM anomalies is discussed in greater detail by Moore (2013).

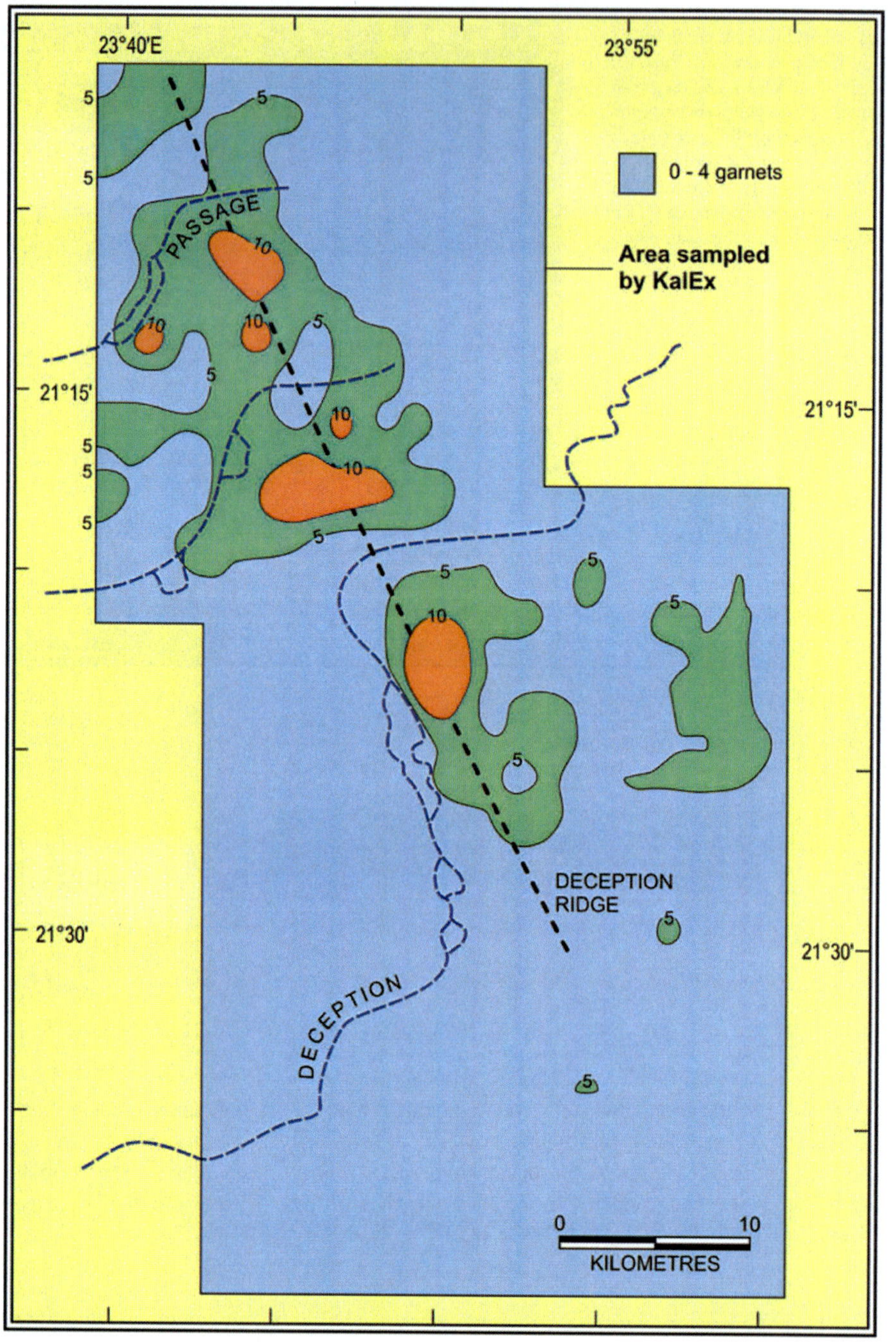

Fig. 15.16 Illustrates the results of semi-detailed sampling (500 m × 500 m) carried out by KalEx (summarized by Moore 2013), with the position of the Deception Ridge (dash line) superimposed. The highest garnet counts show a remarkable correspondence with the ridge, indicating that they were concentrated in the high energy zone associated with a former shoreline of the Makgadikgadi at ~995 m

a

b

Fig. 15.17 **a** View of the Orapa mine pit, looking down the long axis (approximately 1.5 km in length). The breadth of the pit is roughly 1 km. Country rocks are approximately 180 Ma Karoo basalts. **b** View of the sampling cut excavated as part of evaluating the BK-9 kimberlite, in the Orapa Cluster, which forms part of the Damtshaa mine. Note the thick calcrete overlying the weathered kimberlite

15.7 Kimberlites

This narrative would be incomplete without illustrations of the kimberlites which provided the stimulus for our geomorphic study. The massive Orapa pit is shown in Fig. 15.17a, while Fig. 15.17b illustrates the bulk sampling cut which was excavated on the smaller, associated BK-9 kimberlite, which now forms part of the Damtshaa mine.

15.8 Conclusions

Our study has highlighted how an appreciation of the geomorphic evolution of southern Africa provided the framework for planning of the successful De Beers exploration programme, which ultimately resulted in the discovery of the Orapa kimberlite field—one of the world's major diamond resources. At a more detailed level, geomorphology further provides a powerful tool for understanding the origin of concentrations of kimberlitic minerals which do not appear to have a local kimberlite source. However, there is a feedback loop to kimberlite prospecting, because the results of detailed sampling programmes can lead to a more refined appreciation of the geomorphic processes responsible for the dispersion of kimberlitic minerals.

Acknowledgements Susan Abraham is thanked for contributing a number of the Figures. De Beers Exploration is thanked for permission to publish Fig. 15.1 and for making unpublished kimberlite geochronological data available. Prof. Frank D. Eckardt is thanked for permission to reproduce Fig. 15.15a, b and for providing the images of the Orapa (AK-1) and DK-9 kimberlites. Constructive, critical comments by the reviewer (Mike de Wit) considerably improved the manuscript.

References

Allsopp HL Bristow JW Smith CB Brown R Gleadow AJW Kramers JD Garvie OG (1989) A summary of radiometric dating methods applicable to kimberlites and related rocks. In: Ross J (ed.) Proceedings of the Fourth International Kimberlite Conference vol 2, 343–357

Cerling TE et al. (2011) Woody cover and hominin environments in the past 6 million years. Nature 476

Debridge Exploration Company (1986) Final Report PL14/81 Final Report plus 3 Appendices. Botswana Geological Survey, Open File

De Wit MCJ (2012) The Xaudum kimberlite province straddling the southern margin of the Angolan craton. Geol Soc Africa Colloquium of African Geology no, 24, Ethiopa, Extd Abstracts P142

De Wit MCJ (2019) Prospecting history leading to the discovery of Botswana's diamond mines: from artefacts to Lesedi la Rona. Mineral Petrol 112(S1):7–22

Du Plessis PI (1998) Palaeo-lake Makgadikgadi: an African rift lake in Botswana. De Beers Prospecting Botswana Ltd Internal Report, unpublished, 6pp

Du Plessis PI, Le Roux JP (1995) Late Cretaceous alkaline saline lake complexes of the Kalahari Group in northern Botswana. Jour Afr Earth Sci 20:7–15

du Toit AL (1927) The Kalahari and some of its problems. S Afr J Sci 24:88–101

du Toit AL (1933) Crustal movements as a factor in the geographical evolution of South Africa. S Afr Geogr Jour 1:4–20

Engel MS Grimaldi DA Krishna K (2009) Termites (Isoptera): their phylogeny, classification and rise to ecological dominance. Novitates No 3650, 27pp

Farr H, Phillips D, Maas R, de Wit MCJ (2018) Petrography, Sr-isotope geochemistry and geochronology of the Nxau-Nxau kimberlites, north-west Botswana. Mineral Petrol S112:S625–S638

Field M, Gibson JG, Wilkes TA, Gababotse J, Khujwe P (1997) The geology of the Orapa A/K1 kimberlite Botswana: further insight into the emplacement of kimberlite pipes. Russian Geol Geophysics 38:24–39

Gernon TM, Fontana G, Field M, Sparks RSJ, Brown RJ, Mac Niocaill C (2009a) Pyroclasticflow deposits from a kimberlite eruption: the Orapa south crater, Botswana. Lithos 112:566–578

Gernon TM, Fontana G, Field M, Sparks RSJ (2009b) Depositional processes in a kimberlite crater: the Upper Cretaceous Orapa South Pipe (Botswana). Sedimentology 56:623–643

Grey DRC (1976) Prospecting in the Mopipi Area, north-central Botswana. Internal Anglo American Report, Open File, Botswana Geological Survey, 97pp

Grove AT (1969) Landforms and climate change in the Kalahari and Ngamiland. Geogr Jour 135:191–212

Haddon IG (2000) The Kalahari Group sediments. In: Partridge TC Maud RR (eds) The Cenozoic of southern Africa, Oxford monographs on geology and geophysics no. 40, Oxford University Press, pp173–181

Haddon IG (2005) The sub-Kalahari geology and tectonic evolution of the Kalahari basin, southern Africa. Unpubl PhD Thesis, University of the Witwatersrand, South Africa 360pp

Hanson EK, Moore JM, Bordy EM, Marsh JS, Howarth G, JvA R (2009) Cretaceous erosion in central South Africa: evidence from upper crustal xenoliths in kimberlite diatremes. S Afr J Geol 112:125–140

Jelsma HA, de Wit MJ, Thiart C, Viola DPHGM, Basson G, Anckar IJ, E, (2004) Preferential distribution along transcontinental corridors of kimberlites and related rocks of southern Africa. S Afr J Geol 107:301–324

King LC (1963) South African scenery. Oliver Boyd, 308pp

Kurszlaukis S, Barnett WP (2003) Volcanological and structural aspects of the Venetia kimberlite cluster – a case study of South African kimberlite maar-diatreme volcanos. S Afr J Geol 106:165–192

Lee JE, Jennings CMH, Blaine JW (2009) The Gope 25 kimberlite discovery, Botswana, predicated on four Mg-ilmenite grains from reconnaissance sampling: a case history. Explore 143:1–7

Lock NP (1985) Kimberlite exploration in the Kalahari region of southern Botswana with emphasis on the Jwaneng kimberlite province. The Institution of Mining and Metallurgy, Prospecting in areas of desert terrain, 183–190

Mabbutt JA (1955) Erosion surfaces in Namaqualand and the ages of surface deposits in the south-western Kalahari. Trans Geol Soc S Afr 58:13–29

McFarlane MJ, Coetzee SH, Kuhn JR, Vanderpost CMH, Eckardt FD (2007) In situ rounding of quartz grains within an African Surface weathering profile in northwest Ngamiland, Botswana. Z Geomorph 51:269–286

McFarlane MJ, Eckardt FD (2006) Lake Deception: a new Makgadikgadi palaeo-lake. Botswana Notes Records 38:195–201

McFarlane MJ, Eckardt FD (2007) Palaeodune morphology associated with the Gumare faultof the Okavango graben in the Botswana/Namibia borderland: a new model of tectonic influence. S Afr J Geol 110:535–542

Mothibedi M Duncan DJ (2000). De Beers Prospecting Botswana, Final Report for Prospecting Licences 157–161/94 (Mochudi-Sikwane)

Miller RMcG Pickford M Senut B, (2010) The geology, palaeontology and evolution of the Etosha Pan, Namibia: implications for terminal Kalahari deposition. S Afr J Geol 113:307–334

Moore AE (1999) A reappraisal of epeirogenic flexure axes in southern Africa. S Afr J Geol 102:363–376

Moore AE (2013) Anatomy of a distal kimberlite indicator mineral (KIM) anomaly in the central Kalahari, Botswana. S Afr J Geol 116:67–78

Moore AE, Blenkinsop TG, Cotterill FPD (2008) Controls of post-Gondwana alkaline volcanism in southern Africa. Earth Planet Sci Lett 268:151–164

Moore AE, Blenkinsop TG, Cotterill FPD (2009) South African topography and erosion history:plumes or plate tectonics. Terra Nova 21:310–315

Moore AE, Dingle RV (1998) Evidence for fluvial sediment transport of Kalahari sand in central Botswana. S Afr J Geol 101:145–153

Moore AE, Eckhardt CFPD, FD, (2012) The evolution and ages of Makgadikgadi palaeo-lakes: consilient evidence from Kalahari drainage evolution. S Afr J Geol 115:385–413

Moore AE, Larkin PA (2001) Drainage evolution in south-central Africa since the break-up of Gondwana. S Af J Geol 104:47–68

Osborne CP (2008) Atmosphere, ecology and evolution: what drove the Miocene expansion of C_4 grasslands. J Ecol 96:35–45

Partridge TC, Maud RR (1987) Geomorphic evolution of southern Africa since the Mesozoic. S Afr J Geol 90:179–208

Passarge S (1904) Die Kalahari. Dietrich Reimer, Berlin

Perreira E, Rodrigues J, Reis B (2003) Synopsis of Lunda geology NE Angola: implications for diamond exploration. Communic Do Instit Geol e Mineiro 90:189–202

Phillips D Kieviets GB Barton ES Smith CB Viljoen KS Fourie LF (1999) $^{39}Ar/^{40}Ar$ dating of kimberlites and related rocks: problems and solutions. In: Gurney JJ et al. (eds) Proceedings of the VIIth International Kimberlite Cnference vol 2: 677–688.

Roberts EM Jelsma HA Hegna T (2015) Mesozoic sedimentary cover sequences of the Congo basin in the Kasai region, Democratic Republic of Congo. In: de Wit MG Guillocheau F de Wit MCJ (eds) Geology and Resource Potential of the Congo basin Springer, Dordrecht pp163–191

Smith CB (1983) Pb, Sr and Nd isotope evidence for sources of southern African Cretaceous kimberlites. Nature 304:51–54

Thomas DG, Burrough SL (2012) Interpreting geoproxies of late Quaternary climate change in African drylands: implications for understanding environmental change and early human behaviour. Quat Internat 253:5–17

Ustinov VN, Mosigi B, Kukui IM, Nokolaeva EV, Campbell JAH, Stegnitskiy YB, Congo basinAntashcuk MG, (2019) Aeolian indicator mineral transport haloes from the Orapa kimberlite cluster. Botswana. Mineral Petrol 112(S2):775–783

Vinès M Tappe S Stracke A Wilson A Rogers A (2017). Discovery of an orangeite magmatic event in the central Kalahari: Implications for the origin of southern African kimberlites. 11th International Kimberlite Conference Extended Abstract No. 11IKC-4535, 2017

Willis JHA (1960) Final report on the Bechuanaland investigation 1959/1960. Consolidated African Selection Trust. Botswana Geological Survey open file reports, 7pp

Andy Moore Has a Ph.D. in geochemistry from the University of Cape Town, South Africa. He has been based in Botswana since 1980, mainly involved in kimberlite exploration in the Kalahari environment. His interests include the factors controlling the dispersal of kimberlitic indicator minerals in the Kalahari environment, and also the interface between geomorphology and biology. He is an honorary research associate in the Dept. of Geology, Rhodes University, Grahamstown (South Africa).

Mike Roberts Senior Exploration Manager for De Beers, responsible for the company's exploration ventures in Africa. He has been with the company for 21 years; all of which have been spent on exploration programmes across India and Africa including South Africa, Namibia, Gabon, Zimbabwe, Lesotho, Botswana (where he resided for 11 years), Angola and the DRC.

16 Geomorphology and Landscapes of the Limpopo River System

Jasper Knight

Abstract

The Limpopo River system is a major drainage pathway in southern Africa but very little is known about its catchment-scale geomorphology and dynamics. This study maps its geomorphology using the River Styles Framework as an interpretive tool. This analysis highlights that river geomorphology varies spatially, in particular between bedrock-controlled and overbank floodout reaches which are found throughout the river system. In addition, much of the present reach-scale geomorphology does not correspond to present climate forcing—even though flood events are important drivers of geomorphic change, flood effects are spatially variable. This is due in part to different substrate types and valley width, but also to flow constrictions that compartmentalize the integrity of the river sediment system into different morphodynamic zones. The ephemeral tributaries drawing water from Botswanan territory into the Limpopo are strongly affected by human activity, including river damming and groundwater abstraction. This significantly impacts on river discharge and thus sediment dynamics. The net outcome is that there is limited streamwise sediment connectivity through these tributaries and thus through the Limpopo River system as a whole.

Keywords

Climate change • Flood events • Limpopo • River geomorphology • River Styles • Semiarid

J. Knight (✉)
School of Geography, Archeology and Environmental Studies, University of the Witwatersrand, Johannesburg, 2050, South Africa
e-mail: jasper.knight@wits.ac.za

16.1 Introduction

The Limpopo River is a major drainage system of southeast Africa (Fig. 16.1). It has a total drainage area of 412,938 km^2 and draws water from Botswana, Zimbabwe, South Africa and Mozambique. The Limpopo drains 14% of the land surface area of Botswana. This drainage system in this part of southern Africa has ancient origins: the Limpopo River system became isolated from adjacent systems at the end of the Cretaceous as a result of epeirogenic uplift (Moore and Larkin 2001; Moore et al. 2012). This uplift in headwater areas reduced catchment area and possibly flow and sediment contribution to the lower parts of the Limpopo system, rejuvenated bedrock incision and, in combination with climate changes, gave rise to variations in sediment provenance from the Limpopo and adjacent river systems (Setti et al. 2014; Schüürman et al. 2019). Thus, this river system has both evolved over long time periods and has experienced significant climatic and environmental changes over the Cenozoic (last 66 million years) (Knight and Grab 2016). From this, it can be inferred that river geomorphology may have also evolved in different ways over this time period, and that evidence for past river dynamics (terraces, flood boulder deposits, abandoned meanders, underfit channels, bedrock incision features, waterfalls) may exist along the river system. Previous studies from rivers elsewhere in the region have discussed the role of seasonal flooding in rapidly changing river sediment distribution and morphodynamics (e.g. Rowntree et al. 2000; Heritage et al. 2015), and this is also seen from the Limpopo River (Spaliviero et al. 2014; Mvandaba et al. 2018). However, the Limpopo River is understudied in comparison with other semiarid rivers in southern Africa, such as the Sabie (Heritage and Moon 2000; Heritage et al. 2001a), Olifants (Rountree et al. 2001) and Letaba (Heritage et al. 2001b; Moon and Heritage 2001). Thus, the interpretation of channel geomorphology and dynamics of the Limpopo River undertaken in this study is set into this wider regional

F. D. Eckardt (ed.), *Landscapes and Landforms of Botswana*, World Geomorphological Landscapes,
https://doi.org/10.1007/978-3-030-86102-5_16

context. A good overview of the properties and dynamics of the Limpopo River and its catchment is given by FAO (2004). Here, this overview is supplemented with more detailed mapping and analysis of different aspects of catchment properties and processes.

In detail this study (1) describes the major properties of the Limpopo River catchment as a whole; (2) presents new evidence for river geomorphology based on reach-scale analysis of river channel patterns and sediment distributions throughout the river system, using the River Styles Framework; and (3) discusses this evidence in the context of river morphodynamics and the critical role of floods in geomorphic change, with a geographical focus on that part of the Limpopo River system that lies within Botswana. A critical outcome of this study is that, for the first time, the Limpopo drainage has been mapped and classified, providing the context for future more detailed studies.

16.2 Physical Background

The Limpopo Basin as a whole includes a range of rock types of different ages and origins. Geology in the wider region is summarized by Schlüter (2006) and encompasses part of the northern margin of the Witwatersrand basin. Rocks present in this region include conglomerates, sandstones and shales interbedded with basaltic lavas, as part of the Soutpansberg and Waterberg groups, and the Transvaal Supergroup. This

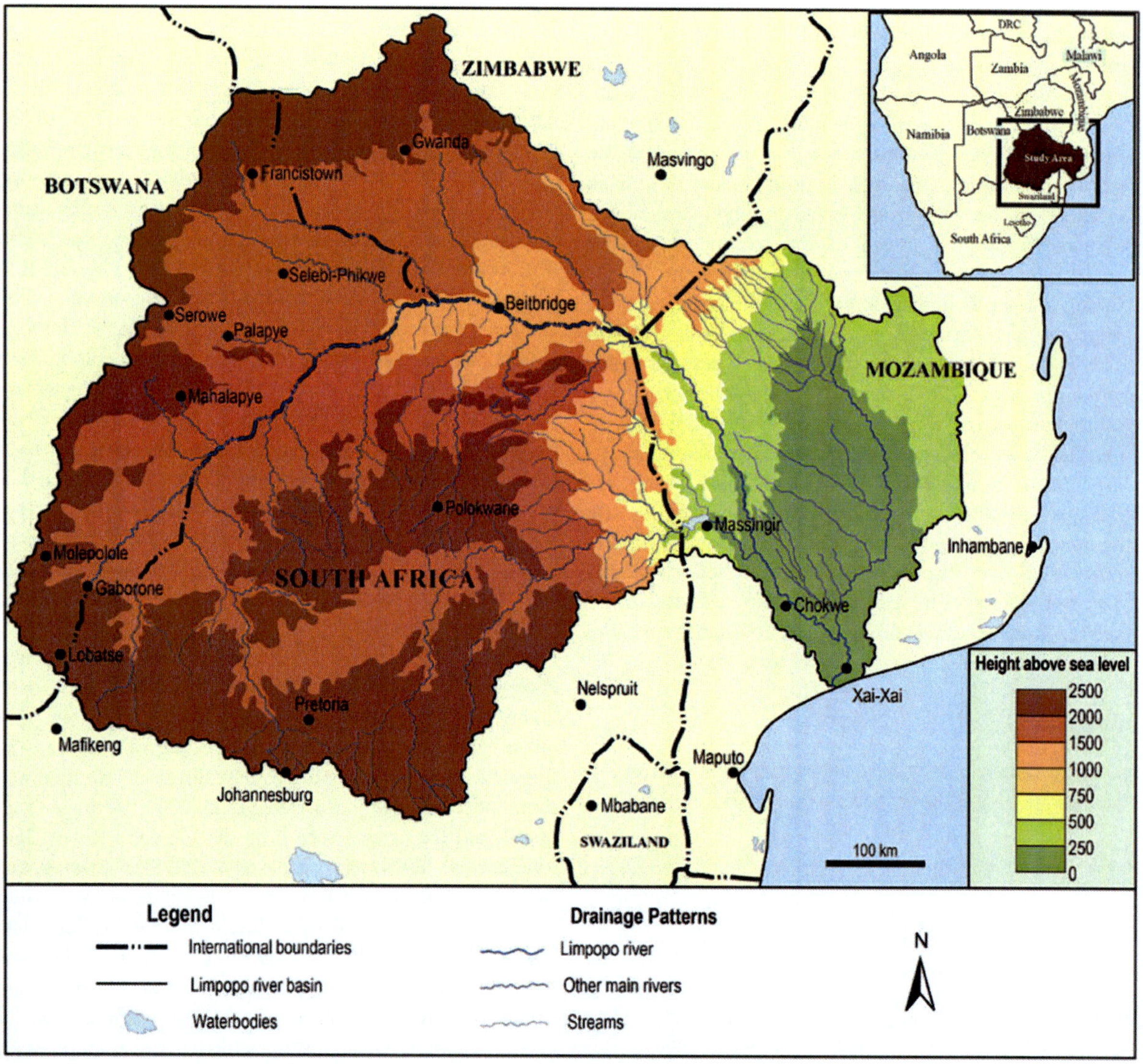

Fig. 16.1 Map of the Limpopo River catchment, showing catchment topography and major tributaries (named in Fig. 16.3)

also includes sandstones of the Tuli Basin, discussed below. Igneous and metamorphic rocks (mainly gneisses) are part of the Archaean Limpopo Belt and dominate in the Botswana part of the Limpopo catchment. The lowermost part of the catchment in Mozambique is dominated by mudstones and siltstones. These rock types have influenced regional soils and topography (FAO 2004). The topography of the region comprises higher relief headwater areas of the Drakensberg, Waterberg and Strydpoort mountains, with some areas above 2000 m asl. Large parts of the lower river course in Mozambique lie below 7 m asl. The river long profile (Fig. 16.2) broadly reflects this geological control, with steep, bedrock-dominated reaches corresponding to geological boundaries. It may also reflect episodic uplift of the southern Africa interior (Moore et al. 2012), consistent with rivers elsewhere in Botswana and in the Kalahari Desert (Nash and Eckardt 2016). FAO (2004) identifies different landscape physiographic types within the Limpopo catchment of which relatively flat plains are by far the most dominant throughout all areas of the catchment and along the main river channel. This reflects its long-term development by subaerial weathering and erosion (Knight and Grab 2016). Soil types are responsive to geology and landscape physiography. In the southeast Botswana sector of the catchment, soils around Francistown and Selebi Phikwe are clay-rich luvisols. Between Mahalapye and Molepolole sandy arenosols dominate, typical of savannas. Around Gaborone and in patches elsewhere are lixisols which are clay-dominated residual soils typical of weathered relict land surfaces. These soils are found in particular on higher elevation highveld plateau surfaces across this region (Runge 2016). Sandier soils are located dominantly at lower elevations on the lowveld and associated with floodplain elements.

The region of the Limpopo basin as a whole lies mainly within Köppen climate zone BSh (semiarid, dry, hot). Annual rainfall is in the region of 450 mm, but varies according to seasonal ingress of Indian Ocean cyclones during the summer, which commonly results in seasonal flood regimes on rivers in this region (Jury and Lucio 2004; Malherbe et al. 2012; Jury 2016), including on the Limpopo River itself (Spaliviero et al. 2014; Sitoe et al. 2015). Climate and soils also impact upon ecosystem types and land cover. Most of the catchment as a whole (53%) comprises cultivated agricultural land/grassland followed by savanna grassland (bushveld) (25%) and open/xeric savanna grassland (20%) (FAO 2004). Grassland is found throughout the Botswana part of the catchment, with a small region of xeric shrubland found between Gaborone and Lobatse. Montane grasslands are found in higher elevation areas of the catchment, outside of Botswana, with tropical/subtropical forests near the Limpopo River mouth in Mozambique.

Discharge patterns of the Limpopo River are discussed in several studies (Trambauer et al. 2014; Jury 2016; Mosase and Ahiablame 2018), but there is a general lack of information on river hydrodynamics. Rainfall patterns (1979–2013) vary across the basin from 160 mm/yr in the west (including Botswana) to 1152 mm/yr in the east and southeast at the coast (Mosase and Ahiablame 2018). Seasonality of rainfall increases to the east but variance increases to the west, indicating higher rainfall variability in the Botswana part of the catchment. In turn, this can result in more variable river responses. To counter the general absence of detailed river discharge evidence, the Pitman rainfall–runoff relationship has been used to model runoff for the Mokolo, Mzingwani (Mvandaba et al. 2018), Luvuvhu (Oosthuizen et al. 2018a), Mogalakwena and Shashe rivers, which are tributaries of the Limpopo (Oosthuizen et al. 2018b). The results show that there is high uncertainty in this rainfall–runoff relationship, due mainly to the role of dams and irrigation in diverting water from river systems.

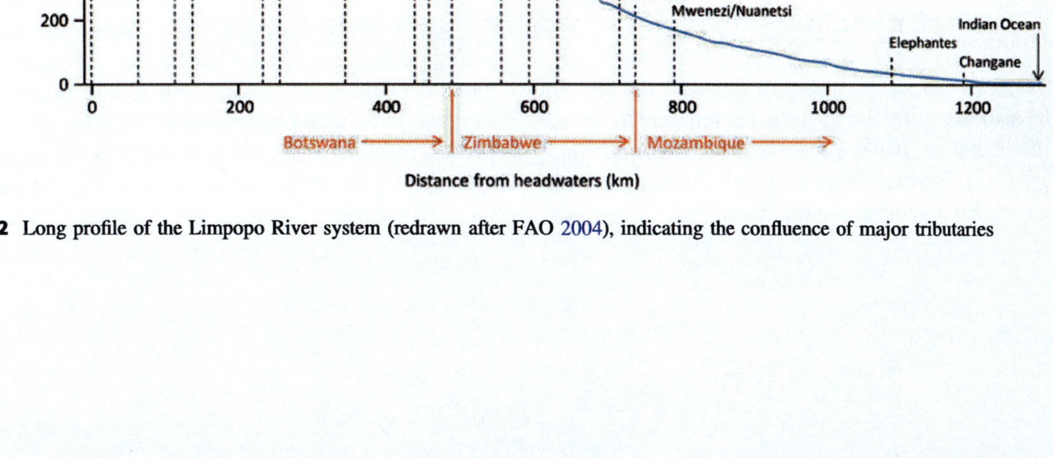

Fig. 16.2 Long profile of the Limpopo River system (redrawn after FAO 2004), indicating the confluence of major tributaries

16.3 Methods

This study undertakes a first-pass baseline survey of river geomorphology along the Limpopo River and its major perennial tributaries based on a simple geomorphic classification using Google Earth imagery (various dates in May 2019, austral winter, during low flow conditions when river geomorphology along perennial reaches is exposed). This methodology is based on the River Styles Framework (RSF) developed for semiarid Australian rivers by Brierley and Fryirs (2005). Implementation of the RSF has four successive stages: (1) baseline survey of river character and behaviour; (2) assessment of river evolution and geomorphic condition; (3) interpretation of the potential of river geomorphology to recover from flood events; and (4) application of this knowledge to river management. In this study, only stages 1 and 2 are undertaken for mapping reach-scale geomorphology through the Limpopo River system.

A river reach is defined as a certain length of river channel in which channel geomorphology (fluvial style, sediment distribution, landforms) is relatively uniform (Rowntree et al. 2000). Given the size of the Limpopo system, a standard reach length of 5 km is used in this study, providing a large-scale overview of the system. Six river styles are identified (Table 16.1). These have been shown to be useful in categorizing reach-scale geomorphology along the semiarid Sabie River, South Africa (Eze and Knight 2018). In this present study, 19 perennial tributaries of the Limpopo River were chosen (Table 16.2) because these are where year-round water flow is able to shape river geomorphology. Ephemeral rivers were excluded from this analysis. It is notable, however, that much of the drainage within the Botswana part of the catchment is by ephemeral rivers, including the Mahalatswe and Bonwapitse, and upper parts of the Lotsane and Ngotwane. The large Changane tributary in Mozambique was also excluded from this study because drainage is largely captured into a natural endorheic basin, which has some episodic and limited overflow into the Limpopo River. Various spellings of some of the main tributaries exist. The most commonly used spellings are the ones used here.

16.4 Limpopo River Analysis

16.4.1 Distribution of Channel Types

In total 931 5 km-length reaches were mapped (4655 km in total). The six river styles are identified in different locations through the Limpopo River system and in its major tributaries (Fig. 16.3). For illustrative purposes, different channel types are grouped together: channel types 1 and 2 (bedrock dominated reaches), types 3 and 4 (sand beds), and types 5 and 6 (palaeochannels, swamps and floodouts). Across the river system as a whole, planform-controlled sand beds are the most common river style (37% of total) with styles 3 and 4 (floodplain meanders), making up 63% of all reaches. However, styles 3 and 4 also include channels of different sinuosities, including straighter reaches found in a floodplain setting and where lateral bars rather than point bars and terraces are common. The river system is therefore sand-dominated and controlled by variations in water discharge that lead to periods of overbank deposition during floods and possibly meander migration. This conforms to other river systems in the region (e.g. Larkin et al. 2017). It is notable that the different tributaries have different geomorphic properties. Gorges are common in the Lephalala and Luvuvhu tributaries (Table 16.2). These rivers draw from mountain catchments. Bedrock dominated reaches (gorges, bedrock-forced meanders) are located in mid-river reaches within tributaries, reaching 21% in the Lephalala and 14% in the Ngotwane rivers. At the other end of the scale, floodouts are found in the upper part of the Limpopo catchment in the Marico and Matlabas tributaries, and in lower parts of the catchment in lowland regions of Mozambique, and have low long profile gradients. Floodouts are associated with highly sinuous and abandoned meanders, ox-bow lakes and overbank wetlands (swamps). The spatial distribution of reach styles highlights that different tributaries have different properties. For example, bedrock-dominated reaches are associated with more flashy and thus amplified flood responses downstream. Sand-dominated reaches, especially with wide floodplains, have higher potential for geomorphic change during floods where erosion and deposition buffer river discharge.

16.4.2 Examples from the Botswana Sector of the Limpopo River

There are few specific studies on Botswana rivers outside of the Okavango Delta region. Byakatonda et al. (2018b) evaluated rainfall–runoff regimes for some Limpopo River tributaries in Botswana, showing that there is a time lag between peak rainfall and peak river discharge (Fig. 16.4). This is attributed in part to water storage into, and flow regulation by, dams, and also to long-term changes in rainfall and temperatures (in particular during the summer), resulting in changes in water balance. For example, climate records at Francistown (since 1960) and Mahalapye (since 1971) show a linear temperature increase of ~ 0.15 °C/decade and 0.27 °C/decade, respectively; and rainfall at Mahalapye (since 1960) shows a linear decrease by 6%/decade (Byakatonda et al. 2018a). River channels are

Table 16.1 River Styles identified in this study (Brierley and Fryirs 2005) along the Limpopo River system

Channel type	River Style	Typical geomorphic features present
1	Gorge	Cascades, rapids, boulders, bedrock
2	Bedrock-forced meander	Bedrock, gravel, pools, runs
3	Low/moderate sinuosity, planform-controlled sand bed	Incised banks, pools, riffles
4	Meandering sand bed	Sand sheets, lateral bars, flood channels, palaeochannels
5	Low sinuosity, fine grained	Sand sheets, palaeochannels, point bars, point benches
6	Floodouts	Ponds, floodouts, swamps, crevasse splays

dominantly ephemeral; some specific examples from the Botswana sector of the Limpopo River system are shown in Fig. 16.5. Bedrock-controlled meanders of the perennial Ngotwane River (Fig. 16.5a) are uniformly incised into a relatively flat residual surface and framing the lateral extent over which the present underfit channel can meander (River Styles 1, 2). The presence of abandoned and cut-off meanders across this surface attests to past humid climate phases. These abandoned meanders are partly filled, similar to lowland Limpopo reaches (Sitoe et al. 2015) and other mixed sand–bedrock rivers (McCarthy 2004). Based on cosmogenic ^{3}He dating, Keen-Zebert et al. (2016) calculated bedrock erosion rates of 14–255 m/Ma and lateral erosion rates of 11–50 m/Ma on the mixed sand–bedrock Klip, Mooi and Schoonspruit rivers, South Africa. This is likely similar to the situation on the Ngotwane. The ephemeral Bonwapitse River is sand-dominated, and it is notable on Fig. 16.5b that water present within the channel drains through floodplain sediment before it reaches the Limpopo River confluence. Short ephemeral tributaries are also seen along the Bonwapitse. The confluence region is typical of a flood-dominated distributary plain, with meandering or braided channels during high flow and an exposed deflation plain during low flow conditions (River Styles 5, 6). This confluence area, of around 4 km^2, can be considered as a palaeo-delta similar to much larger versions identified on the Mozambique lowlands of the Limpopo catchment, formed by avulsion from the main channel and its meanders during past floods (Spaliviero et al. 2014). This behaviour is also well-marked at the Shashe–Limpopo confluence where there is a strong seasonal contrast in discharge and therefore river morphodynamics between the main channel (Limpopo) and incoming tributary (Shashe) (Fig. 16.6). The wide alluvial valley present near the ephemeral Mahalatswe River confluence with the Limpopo (Fig. 16.5c) shows a single low-sinuosity channel occupying different valley positions, outside of which are vegetated bars and floodplain surfaces (equivalent to River Styles 3, 4). This is the most common channel morphology within the Limpopo catchment (Table 16.2), in which a wide floodplain or alluvial-filled valley developed under more humid climates, and is now being reoccupied or reworked at different times in response to variations in discharge. This describes the most typical behaviour of such semiarid rivers in the region (e.g. Tooth et al. 2004, 2013; Larkin et al. 2017).

16.5 The Tuli Basin

Another important geomorphic element of the Limpopo River system within Botswana is in the area of the Tuli Basin, a Kalahari–Karoo Basin correlative of Triassic–Jurassic age found in southeast Botswana, broadly between Francistown and Gaborone. The Tuli Basin contains mainly sandstones with minor presence of siltstones and mudstones that accumulated in a fault-bounded basin. This region is of interest because its topography is significantly lower than elsewhere in Botswana (at ~ 500 m compared to ~ 1000 m asl), and this subdued surface may in turn have influenced river patterns within the Limpopo watershed. A well-known feature of the region today is Solomon's Wall, a prominent (30 m high) cliff-face and popular tourist site, developed along the lower Motloutse River by incision into a basalt dike (Fig. 16.7). It was likely that the more resistant dike once impounded water behind it, forming a lake and waterfall. This now-dry ancient system may correspond to a late Pleistocene pluvial phase. Although the sandstone geomorphology in the Tuli Basin region has not been fully described, this area is significant for its Late Stone Age archaeology (Forssman 2013) and rock art and engravings (van der Ryst et al. 2004). This evidence indicates a long cultural association of human activity with the Limpopo catchment landscape, extending farther east into the orbit of the Great Zimbabwe and Mapungubwe cultures (Forssman 2013).

Table 16.2 Proportion of different River Styles identified in different Limpopo River sectors. N = number of 5 km-long reaches examined. ***Italics bold*** indicate the most dominant class

	River style (%)						
	Gorge	Bedrock-forced meander	Low/moderate sinuosity, planform-controlled sand bed	Meandering sand bed	Low sinuosity, fine grained	Floodouts	N
Limpopo River (main channel, to Marico–Crocodile confluence)	0	12	25	***31***	19	13	244
Tributaries							
Marico	0	13	22	9	19	***37***	32
Crocodile	0	12	***66***	19	3	0	41
Ngotwane	14	***53***	25	0	3	5	36
Matlabas	0	18	***46***	0	18	18	17
Mokolo	9	34	***44***	9	4	0	32
Lephalala	21	25	***50***	4	0	0	28
Lotsane	0	18	38	0	***41***	3	34
Mogalakwena	0	15	***72***	13	0	0	47
Motloutse	0	6	24	32	***38***	0	47
Shashe	0	3	***57***	39	1	0	57
Mzingwani	0	8	***42***	27	19	4	52
Sand	8	16	27	***49***	0	0	37
Nzhelele	0	***84***	16	0	0	0	6
Bubi	0	12	***49***	35	4	0	51
Luvuvhu	31	19	***38***	8	4	0	26
Mwenezi	2	20	***40***	32	6	0	64
Elephantes (to Letaba–Olifants confluence)	10	0	16	***69***	5	0	19
Letaba	6	26	***45***	23	0	0	31
Olifants	0	17	26	***47***	10	0	30
Average across the entire catchment	***3***	***17***	***37***	***26***	***12***	***5***	

16.6 Discussion

16.6.1 Landscape Development in the Limpopo Catchment

The present geomorphology and dynamics of the Limpopo River and its major tributaries are driven by climate variability but strongly affected by geologic controls. Thus, this river system has evolved over long time scales in response to tectonic uplift, land surface denudation, and river and slope movement of weathered products through the river system and into the Indian Ocean (Setti et al. 2014; Schüürman et al. 2019). Variations in river long profile (Fig. 16.2) and the position of knickpoints and waterfalls may reflect the combination of geologic and tectonic forcing (Nash and Eckardt 2016). The varied distributions of different River Styles (Fig. 16.3) show that there is not a consistent relationship between reach position in the catchment context and its River Styles classification. Bedrock-controlled reaches are found along both the main and tributary channels but in topographically higher positions such as in the Waterberg Mountains around Polokwane (Fig. 16.1). A notable result is that lowland River Styles (low sinuosity channels and

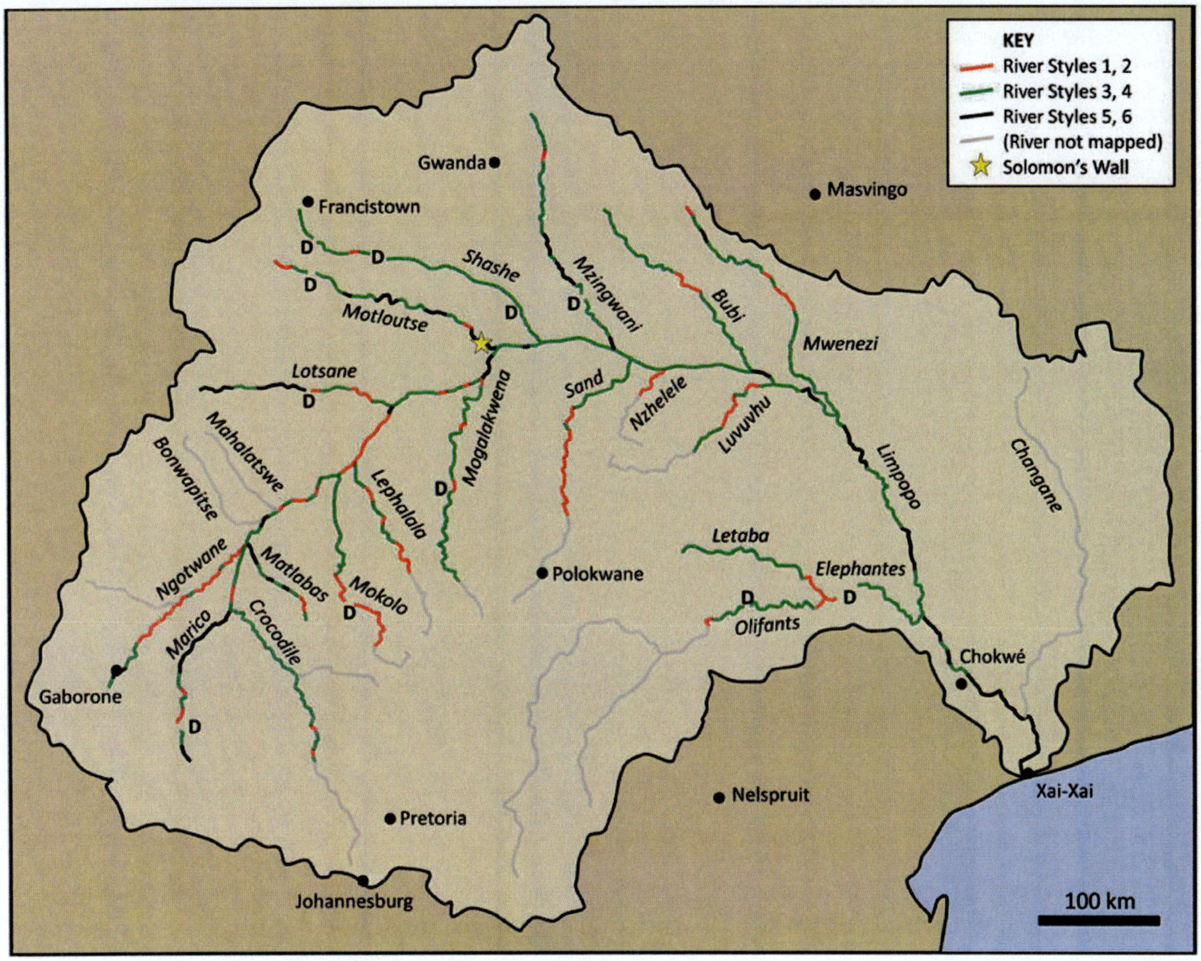

Fig. 16.3 Distribution of different River Styles through the Limpopo's perennial rivers (named in italics). River Styles are numbered according to Table 18.1. Gaps in the river network are where dams (**D**) are present

floodouts) are found even within some headwater tributaries (Fig. 16.3). Most present active channels are underfit relative to the size of their hosting valleys, and that episodic floods cause abandoned channels to be reoccupied along certain river reaches, with overbank and crevasse splay deposition. The large floodplain/valley width means that there is a lot of low-energy standing water after flood events, and generally quite restricted or compartmentalized patterns of downstream sediment transport. This is what is seen in the flood dynamics of similar semiarid rivers elsewhere in the region (Knight and Evans 2017, 2018). A key inference from this observation is that the Limpopo River is not a straightforward water and sediment transport system, but that both these elements vary significantly spatially and temporally with constrictions in transport taking place where extensive sandy floodplain sediments facilitate water infiltration and buffer flood response. The RSF approach taken in this study provides a useful context for, first, categorizing the macroscale geomorphology of the river system and, second, applying this knowledge to better understand its morphodynamics and evolution.

Some previous studies have examined geomorphic evidence for past flooding along the Limpopo. Dating of organic infills within abandoned ox-bow lakes in the lower part of the catchment shows episodes of flooding around mid-1200s AD, late 1300s, mid-1500s, and within the last century (Sitoe et al. 2015). Although the analysis in this study cannot inform on the timing of river floods or their specific impacts, the fact that partly-infilled meanders are present in the landscape 500–800 years after abandonment attests to the relict nature of at least some river landforms, and likely in headwater and middle reaches (Fig. 16.5) as well as in lowland areas.

Fig. 16.4 Averaged monthly rainfall and river discharge data for the Limpopo River system, 1975–2014 (redrawn from Byakatonda et al. 2018b). **a** Rainfall (mm/month) averaged from three stations within the Botswana part of the Limpopo catchment (at SSKA, Mahalapye and Selebi Phikwe), and **b** river discharge at Buffel's Drift

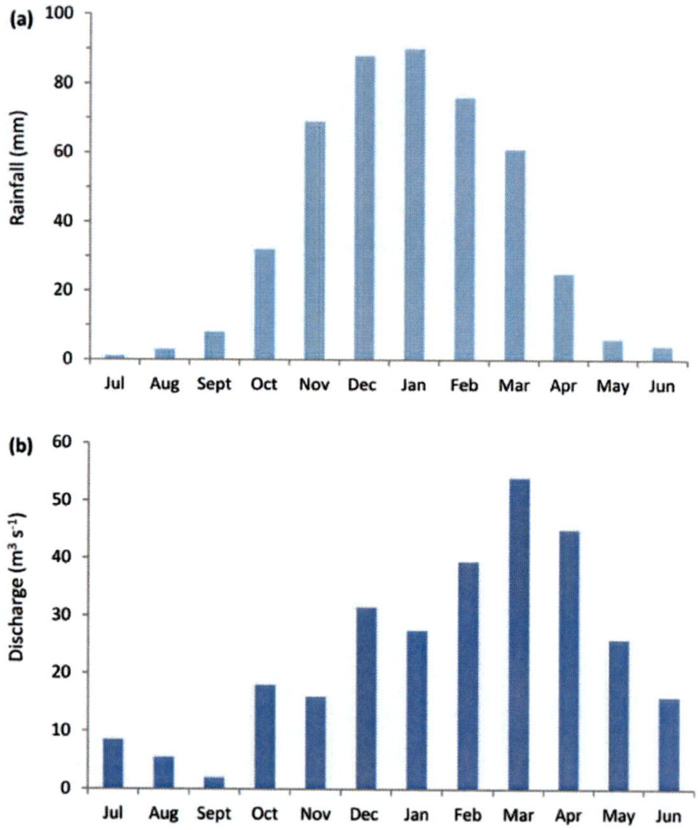

16.6.2 Water and Flood Management in the Limpopo Catchment

Many studies have been concerned with water management along the Limpopo and its tributaries (Boroto 2001; Zhu and Ringler 2012; Meissner and Ramasar 2015; Oosthuizen et al. 2018b), and river basins themselves are viewed as the most suitable scale of analysis for river system and water management (Warner et al. 2008). Naturalized runoff (river discharge unimpeded by dams, irrigation or extraction) is significantly greater than the total recorded river discharge along Limpopo tributaries, because of the high water demand in the region. For example, the Ngotwane catchment, including urban centres such as Gaborone, loses 56% of its water by abstraction (FAO 2004). There is also high groundwater extraction from sandstone aquifers, resulting in reduced river baseflow. However, the amount of naturalized water varies considerably between catchments, with some increasing and others decreasing. For example, within the Limpopo system, the Lephalala experienced a decrease of 2% in the volume of naturalized water between 1920–1989 and 1920–2009 averaged values, whereas the Nzhelele catchment experienced an increase of 6% between these time periods (Bailey and Pitman 2016). The main reason is recent climate and human activity, not different datasets. Monitoring naturalized water flow is an important baseline against which to evaluate changes in water resource availability. This is also affected by cross-boundary transfer: for example, water export from South Africa to Botswana through the Limpopo River system is 102% of naturalized streamflow (FAO 2004).

Another important catchment factor is soil erosivity, mapped throughout the catchment by FAO (2004). In the Botswana sector, erosivity was mapped as moderate to high, likely reflecting the dry land surface, sparse vegetation, and high seasonal rainfall intensity. High soil erosivity values have implications for surface sediment yield into and through river channel systems, corresponding to the key role of flood events in driving river geomorphic change (Rountree et al. 2000; Jury and Lucio 2004).

16.7 Conclusions

The Limpopo River system is a significant drainage system in southern Africa but its geomorphology varies significantly between different tributaries and along its 1200 km-long main channel. Categorization of reach-scale geomorphology

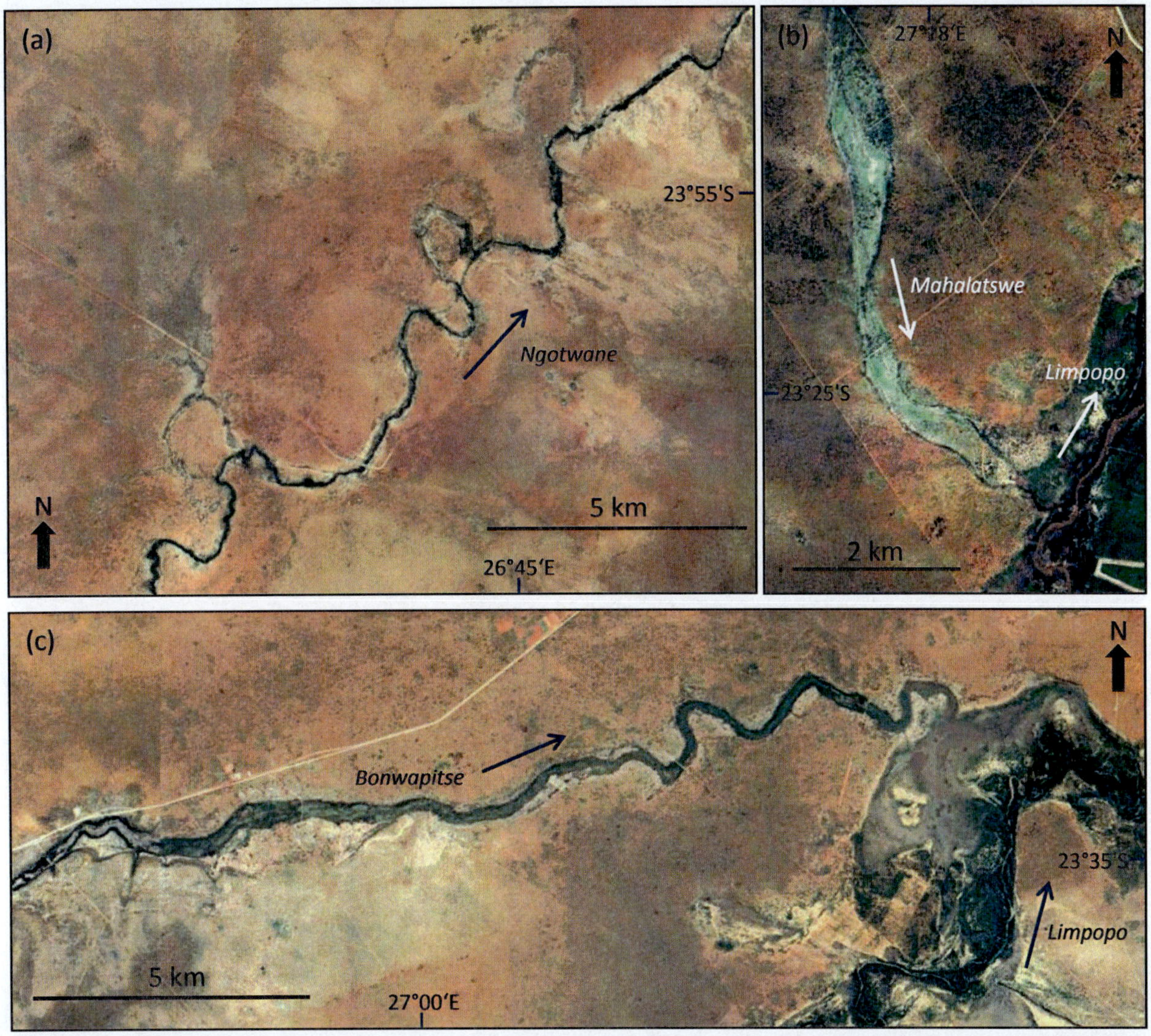

Fig. 16.5 Examples of river geomorphology within the Botswana sector of the Limpopo River catchment. Arrow indicates water flow direction. **a** Abandoned meanders along the Ngotwane River; **b** Mahalatswe River confluence with the Limpopo River; **c** Bonwapitse River confluence with the Limpopo River. *Image source* Google Earth, image date: 31 December 2016

using the River Styles Framework provides a baseline for evaluating both its morphodynamics and evolutionary history, and also for considering the potential impacts of flood events which may have significant negative impacts on both the river system and human activity. The approach adopted in this study can be used at a more detailed, local level with shorter reach lengths, and enables different rivers or reaches to be compared with one another. Limpopo River geomorphology in the Botswana sector of the catchment is also quite different to the Zambezi and Okavango drainages elsewhere within Botswana. This may have implications for how such diverse hydrological systems are compared, monitored and managed.

Acknowledgements I thank Bennie van der Waal for a critical review of this chapter, and Frank D. Eckardt for helpful suggestions.

Fig. 16.6 Panoramic view looking to the northeast at the Shashe–Limpopo confluence during **a** the dry summer season (date: 22 July 2014) and **b** the wet winter season (date: 13 February 2017). River flow is from left to right. Note the largely ephemeral Shashe (S) and perennial Limpopo (L). *Photos* Stefania Merlo

Fig. 16.7 View of Solomon's Wall in the lower part of the Motloutse River, Botswana. *Photo* Frank D. Eckardt

References

Bailey AK, Pitman WV (2016) Water Resources of South Africa, 2012 Study (WR2012), WR2012 Study Executive Summary. Water Research Commission, WRC Report TT 683/16, Gezina, 58pp

Boroto RAJ (2001) Limpopo River: steps towards sustainable and integrated water resources management. IAHS Publ 268:33–39

Brierley GJ, Fryirs KA (2005) Geomorphology and River Management, Applications of the River Styles Framework. Blackwell, Chichester, 398pp

Byakatonda J, Parida BP, Moalafhi DB, Kenabatho PK (2018a) Analysis of long term drought severity characteristics and trends across semiarid Botswana using two drought indices. Atmos Res 213:492–508

Byakatonda J, Parida BP, Kenabatho PK (2018b) Relating the dynamics of climatological and hydrological droughts in semiarid Botswana. Phys Chem Earth 105:12–24

Eze PN, Knight J (2018) A geomorphological characterisation of river systems in South Africa: a case study of the Sabie River. Phys Chem Earth 105:196–205

FAO (2004) Drought impact mitigation and prevention in the Limpopo River Basin. Land and Water Discussion Paper 4, FAO, Rome, 160pp.

Forssman T (2013) A preliminary report on fieldwork in the northern Tuli Game Reserve, northeastern Botswana. S Afr Arch Bull 68:63–71

Heritage GL, Moon BP (2000) The contemporary geomorphology of the Sabie River in the Kruger National Park. Koedoe 43:39–55

Heritage GL, Broadhurst LJ, Birkhead AL (2001a) The influence of contemporary flow regime on the geomorphology of the Sabie River, South Africa. Geomorphology 38:197–211

Heritage GL, Moon BP, Large ARG (2001b) The February 2000 floods on the Letaba River, South Africa: an examination of magnitude and frequency. Koedoe 44:1–6

Heritage G, Tooth S, Entwistle N, Milan D (2015) Long-term flood controls on semi-arid river form: evidence from the Sabie and Olifants rivers, eastern South Africa. IAHS Publ 367:141–146

Jury MR (2016) Climate influences on upper Limpopo River flow. Water SA 42:63–71

Jury MR, Lucio FDF (2004) The Mozambique floods of February 2000 in context. S Afr Geogr J 86:141–146

Keen-Zebert A, Tooth S, Stuart FM (2016) Cosmogenic ^{3}He measurements provide insight into lithologic controls on bedrock channel incision: examples from the South African interior. J Geol 124:423–434

Knight J, Evans M (2017) The sediment stratigraphy of a flood event: an example from the Sabie River, South Africa. Catena 151:87–97

Knight J, Evans M (2018) Luminescence dating, sediment analyses, and flood dynamics on the Sabie River, South Africa. Geomorphology 319:1–14

Knight J, Grab SW (2016) A continental-scale perspective on landscape evolution in southern Africa during the Cenozoic. In: Knight J, Grab SW (eds) Quaternary environmental change in southern Africa: physical and human dimensions. Cambridge University Press, pp 30–46

Larkin ZT, Tooth S, Ralph TJ, Duller GAT, McCarthy T, Keen-Zebert A, Humphries MS (2017) Timescales, mechanisms, and controls on incisional avulsions in floodplain wetlands: Insights from the Tshwane River, semiarid South Africa. Geomorphology 283:158–172

Malherbe J, Engelbrecht FA, Landman WA, Engelbrecht CJ (2012) Tropical systems from the southwest Indian Ocean making landfall over the Limpopo River Basin, southern Africa: a historical perspective. Int J Climatol 32:1018–1032

McCarthy TS (2004) Incised meanders along the mixed bedrock-alluvial Orange River, Northern Cape Province, South Africa. Z Geomorphol 48:273–292

Moon BP, Heritage GL (2001) The contemporary geomorphology of the Letaba River in the Kruger National Park. Koedoe 44:45–55

Moore AE, Larkin PA (2001) Drainage evolution in south-central Africa since the breakup of Gondwana. S Afr J Geol 104:47–68

Moore AE, Blenkinsop T, Cotterill FPD (2012) Dynamic evolution of the Zambezi-Limpopo watershed, central Zimbabwe. S Afr J Geol 115:551–560

Mosase E, Ahiablame L (2018) Rainfall and Temperature in the Limpopo River Basin Southern Africa: Means, Variations, and Trends from 1979 to 2013. Water 10:364. https://doi.org/10.3390/w10040364

Mvandaba V, Hughes D, Kapangaziwiri E, Mwenge Kahinda J-M, Oosthuizen N (2018) Modelling of channel transmission loss processes in semi-arid catchments of southern Africa using the Pitman Model. IAHS Proceedings 378:17–22

Nash DJ, Eckardt FD (2016) Drainage development, neotectonics and base-level change in the Kalahari Desert, southern Africa. S Afr Geogr J 98:308–320

Oosthuizen N, Hughes D, Kapangaziwiri E, Kahinda JM, Mvandaba V (2018a) Quantification of water resources uncertainties in the Luvuvhu sub-basin of the Limpopo river basin. Phys Chem Earth 105:52–58

Oosthuizen N, Hughes DA, Kapangaziwiri E, Kahinda J-MM, Mvandaba V (2018b) Parameter and input data uncertainty estimation for the assessment of water resources in two sub-basins of the Limpopo River Basin. IAHS Publ 378:11–16

Rountree MW, Rogers KH, Heritage GL (2000) Landscape state change in the semi-arid Sabie River, Kruger National Park, in response to flood and drought. S Afr Geogr J 82:173–181

Rountree MW, Heritage GL, Rogers KH (2001) In-channel metamorphosis in a semiarid, mixed bedrock/alluvial river system: implications for Instream Flow Requirements. IAHS Publ 266:113–123

Rowntree KM, Wadeson RA, O'Keefe J (2000) The development of a geomorphological classification system for the longitudinal zonation of South African rivers. S Afr Geogr J 82:163–172

Runge J (2016) Soils and duricrusts. In: Knight J, Grab SW (eds) Quaternary environmental change in southern Africa: physical and human dimensions. Cambridge University Press, pp 234–249

Schlüter T (2006) Geological Atlas of Africa. Springer, Berlin, p 307

Schüürman J, Hahn A, Zabel M (2019) In search of sediment deposits from the Limpopo (Delagoa Bight, southern Africa): Deciphering the catchment provenance of coastal sediments. Sed Geol 380:94–104

Setti M, López-Galindo A, Padoan M, Garzanti E (2014) Clay mineralogy in southern Africa river muds. Clay Miner 49:717–733

Sitoe SR, Risberg J, Norström E, Snowball I, Holmgren K, Achimo M, Mugabe J (2015) Paleo-environment and flooding of the Limpopo River-plain, Mozambique, between c. AD 1200–2000. Catena 126:105–116

Spaliviero M, De Dapper M, Maló S (2014) Flood analysis of the Limpopo River basin through past evolution reconstruction and a geomorphological approach. Nat Hazard 14:2027–2039

Tooth S, Brandt D, Hancox PJ, McCarthy TS (2004) Geological controls on alluvial river behaviour: a comparative study of three rivers on the South African Highveld. J Afr Earth Sc 38:79–97

Tooth S, Hancox PJ, Brandt D, McCarthy TS, Jacobs Z, Woodborne S (2013) Controls on the genesis, sedimentary architecture, and preservation potential of dryland alluvial successions in stable continental interiors: insights from the incising Modder River, South Africa. J Sediment Res 83:54–561

Trambauer P, Maskey S, Werner M, Pappenberger F, van Beek LPH, Uhlenbrook S (2014) Identification and simulation of space–time

variability of past hydrological drought events in the Limpopo River basin, southern Africa. Hydrol Earth Syst Sci 18:2925–2942
van der Ryst M, Lombard M, Biemond W (2004) Rocks of potency: engravings and cupules from the Dovedale Ward, southern Tuli Block, Botswana. S Afr Arch Bull 59:1–11
Warner J, Wester P, Bolding A (2008) Going with the flow: river basins as the natural units for water management? Water Policy 10:121–138
Zhu T, Ringler C (2012) Climate Change Impacts on Water Availability and Use in the Limpopo River Basin. Water 4:63–84

Jasper Knight Professor of Physical Geography in the School of Geography, Archaeology & Environmental Studies, University of the Witwatersrand, South Africa. His research interests focus on the dynamics of glacial, mountain, river and coastal environments in response to climate change. He has published over 120 peer-reviewed scientific articles based on research in diverse locations globally, and many books.

Dams in Botswana: Drying Times Ahead

17

Jeremy S. Perkins and Bhagabat P. Parida

Abstract

Botswana's dams and the sand rivers that run through them appear set for unprecedented levels of threat and uncertainty due to climate change this Century. Spatially specific impacts such as severe soil erosion, sand mining, industrial and agricultural pollution, invasive and alien species, dumping of rubbish and the disposal of sewage water, will be accentuated further by the predicted hotter and drier conditions and more extreme weather events. Harmful algal blooms will become more frequent and further threaten an already challenging water supply and management situation. High distribution losses, limiting reservoir safe yields, due in large part to their shallow depths and fluctuating water levels, seem unlikely to be offset by increased conjunctive use of ground and surface water within the existing constraints of supply networks. Transformative change within the water sector is required, that places greater emphasis on water use efficiency and that of green water, along with inter-regional water transfer programs. The arid conditions of the past seem set to return to Botswana, requiring Policy makers to look both inwards at how water is used and valued, and outwards, far beyond their borders, if sustainable water supply solutions to be found.

Keywords

Dams • Sand rivers • Environmental flows • Climate change

J. S. Perkins (✉)
Department of Environmental Science, University of Botswana, Private Bag 0022 Gaborone, UB, Botswana
e-mail: perkinsjs@ac.ub.bw

B. P. Parida
Department of Civil & Environmental Engineering, Botswana International University of Science & Technology, Private Bag 16, Palapye, Botswana
e-mail: paridab@biust.ac.bw

17.1 Introduction

The future for river ecosystems is in general uncertain (Tonkin et al. 2019) and in the case of southern Africa, particularly disturbing (De Wit and Stankiewicz, 2006). Unprecedented levels of flow regulation due to in-channel dam building are being exacerbated by the impacts of climate change, with increased temperatures, reduced rainfall totals, 'mega droughts' and increased flooding (Engelbrecht et al. 2015; Pohl et al. 2017), threatening an already declining water supply and challenging management situation. Streams and rivers are a vital link between the hydrological cycle and the natural water conveyance systems, which transports water to support a diverse array of ecological and human requirements (Larkin et al. 2020). The outlook for southern Africa appears particularly bleak, with a 10% drop in rainfall by 2050, resulting in regions receiving 500–600 mm year^{-1} losing 30–50% of their surface drainage, leading to major water shortages (De Wit and Stankiewicz 2006). It is clear that water management solutions will not reside in more dam building in Botswana, but will lie further afield in inter-basin water transfer schemes that connect with the catchments of central and eastern Africa and regions largely outside of severely desiccated southern Africa.

The semi-arid climate of Botswana, its deep sand cover, low relief (Fig. 17.1) and spatially and temporally highly variable rainfall have always meant that surface water is scarce. The perennial rivers of the Okavango and Zambezi basins are fed by higher rainfall outside of Botswana, while the ephemeral rivers of the Limpopo River system in the eastern part of the country, flow only after rainfall events, often for several days. It follows that flow regulation has been a priority of the Botswana Government, with the construction of dams in the Limpopo basin, linked increasingly to the North–South carrier pipeline that conveys water to the capital's growing population in Gaborone in the south-east of the country. The prime sites for the

F. D. Eckardt (ed.), *Landscapes and Landforms of Botswana*, World Geomorphological Landscapes,
https://doi.org/10.1007/978-3-030-86102-5_17

construction of large dams have all been used up, with even those for small and medium dams limited due to the country's low relief (Geoflux and Hendry 2001).

Dams have also undergone a major re-evaluation in the recent past with many raising major concerns over the fact that their downstream impacts run contrary to sustaining healthy ecosystems (Belletti et al. 2020). Dams have a number of detrimental impacts including disruption of the longitudinal flow of sediment and water, and changes to water temperature and streamflow (Poff and Schmidt, 2016) and loss of biodiversity (from diatoms to mammals). The latter authors estimate that globally there are 58,000 dams with a dam wall more than 15 m high, two-thirds of which have been geo-referenced and captured digitally (Mulligan et al. 2020). Dam removal is now significant in many developed countries (Habel et al. 2020), but in some regions such as Latin America and Africa dams are still seen as the panacea to water and energy supply problems (O'Connor et al. 2015).

Dam building in Botswana began just before Independence when Gaborone Dam was constructed, with the biggest dam, Dikgatlhong near Francistown, completed in 2012, taking the complement of large dams in Botswana to eight. Intriguingly on completing many of the dams or raising their walls, the respective engineers have repeatedly stated that they will take years to fill, or will never fill, only to find that they spill in the following rainfall season! Recreational tourism, commercial fishing (Nermark and Mmopelwa 1994; Cowx 2012) and irrigated agriculture projects linked to many of the country's larger dams have quite simply failed to materialise, with the water supply function dominating but compromised by high rates of open water evaporation, sedimentation and high transmission losses through pipeline infrastructure.

17.2 Basins

The large water supply reservoirs in Botswana are located exclusively in the Limpopo River Basin. The Okavango and Zambezi Basins are affected by dams outside of Botswana, and appear likely to undergo renewed phases of dam building over the next decade. Both the Okavango and the Zambezi appear to be under the influence of a quasi-80 year

Fig. 17.1 The limited relief in the Kalahari provides little opportunity for dams in Botswana

climatic oscillation (Wolski et al. 2006), raising its own challenges for conservation (Sandi et al. 2020), not least as they are likely to come under increasing utilization pressure (King and Chonguiça 2016). Plans to dredge the Boro River in 1988 in order to increase water supply through the Okavango delta to the diamond mine at Orapa were subject to international review (IUCN 1989) and outcry, and were subsequently shelved by the Government of Botswana. The latter has secured a 500 Mm^3 annual offtake (495 × 106 m^3 y^{-1}) (Hughes and Farinosi 2020) from the Zambezi Basin Commission, ostensibly for its Agro-Zambezi Commercial Agriculture Project at Pandamatenga and the augmentation of water supplies to eastern Botswana (DWA 2013).

Due to the extensive covering of Kalahari sand there is only highly localised surface runoff from the Orange-Senqu Basin into the Molopo River, which has remained dry for over a century in Botswana. There is evidence of perennial or semi-perennial flow within the Molopo valley system between 16 and 12,000 years Before Present with additional fluvial events during the mid-Holocene. Cornwallis-Harris crossed the river in 1836, describing it as having: *'a broad shallow bed, covered with turf, traversed by a deep stream about ten yards wide, completely overgrown with high reeds....[with] an abundance of hippo'* (Cornwallis-Harris 1852, p. 66; from Nash 1992, p. 143). Desiccation and river capture have been used to explain why the Molopo ceased flowing. Dam construction by South Africa, such as Disaneng Dam for irrigation in the early 1980s, certainly helped remove significant flows down the Molopo River, with inter-basin transfers now required to sustain water demand in the settlements along it. Cessation of flows, die-offs of wildlife, invasion by exotic species—such as *Prosopis*, and conjunctive use via inter-basin transfers, may well serve as a precursor as to what can be expected in Botswana's river systems to the north this Century (Perkins 2020).

17.2.1 Limpopo Basin

The main river catchments of the Limpopo River System in Botswana are characterised by considerable heterogeneity in terms of both their key biophysical and socio-economic characteristics. Granite rock outcrops dominate the landscape in and around the majority of dams around Francistown, particularly at Thune, Letsibogo and Dikgatlhong Dams. The greater part of the basin receives rainfall of less than 500 mm yr^{-1} between October and April (Bhalotra 1987). The catchments of all the dams are characterised by a mix of land tenure with private farms dominating the main channels of the Letsibogo (Motloutse), Shashe (Shashe) and Ntimbale (Tati) Rivers. While they were dominated by wildlife a century and a half ago, today domestic stock characterise the majority of the catchments, except for isolated game ranches and Game Reserves (as at Mashatu) and the Mapungwabwe Trans Frontier Conservation Area. Gaborone Dam (Fig. 17.2) has seen the City steadily encroach into the surrounding undeveloped areas, causing pollution concerns. Garzanti et al. (2014) refer to the southern Limpopo catchment as having a water discharge (~5 km^3/year) and a sediment load (~35 × 10^6 ton/year).

The Limpopo River Basin is shared by four SADC Member States, i.e. Botswana, South Africa, Zimbabwe and Mozambique (Refer to Chap. 16 and Fig. 16.3), with the commitment of the riparian states to shared management dating back to 1986 and since 2003 fostered through the Limpopo Water Course Commission (LIMCOM).

17.3 Dams

The large dams in Botswana (Dam wall > 15 m high) typically have an earth-filled core with an ogee spillway at the crest, that delivers water to a raw pump station, located immediately downstream of the dam. Gaborone dam is the oldest and among the largest dams in the country.

17.3.1 Large Dams

There are six large storage dams within the Limpopo basin of Botswana with an estimated total storage capacity of close to 750 Mm^3, with the inclusion of Gaborone Dam and Bokaa Dam the total storage capacity is over 910 Mm^3 (Table 17.1; Fig. 17.3). Safe yields are less than 40% of the total due to erratic river flows, high rates of evaporation (>2000 mm/year) and water depths that are dominantly shallow (DWA 2013), but range from several metres to >30 m.

Many water supply challenges lie ahead for Botswana this Century. Dams typically supply just over half of Botswana's annual water demand, with their potential to supply more limited by the issue of safe yields (DWA 2013). More than fifty per cent of this water demand lies in the south-east due to the capital's growing population, such that water from the main dams in the north must be pumped some 400 kms to the south along the north–south carrier pipeline. The latter has faced many challenges including a failure to operate at the desired pressurization with unacceptably high losses (12–46%) also plaguing the water distribution system throughout key urban centres and settlements (DWA, 2013). Currently, water is pumped from Letsibogo Dam down the North–South carrier (Figs. 17.4 and 17.5) to the water treatment plant at Mmamashia, via Bokaa Dam. Dikgathlong Dam also feeds into the North–South carrier under phase 2 of its development that is augmented by a number of wellfields along the way, such as Palla Road, Chepete, Masama, Makhujwane, Malotwane, and Palapye Wellfields (Lindhe

Fig. 17.2 Gaborone Dam the oldest in the country as been expanded by raising the wall (top) and extending it with a second wall at Notwane (bottom). Note the changing alignment and abandonment of roads and tracks which are a response to dam construction and expansion. (Photo: Frank D. Eckardt)

et al. 2020). The latter point to the optimization of safe yields by the linked up use of surface and groundwater sources (GoB 2016) along with managed aquifer recharge to help adapt to climate change (Lindhe et al. 2020).

Table 17.1 Storage capacity of the eight major dams in Botswana

Dam	Storage Mm^3	River	Date
Gaborone	144	Notwane	1966
Shashe	88	Shashe	1972
Bokaa	18.5	Metsimotlhabe	1990
Letsibogo	104	Motloutse	1997
Ntimbale	26	Tati	2005
Dikgatlhong	400	Lower Shashe	2012
Thune	90	Thune	2012
Lotsane	40	Lotsane	2012

Mogomotsi et al. (2018) emphasise that the 'lack of efficient and effective water supply and management institutions and water infrastructure' in Botswana has complemented its physical scarcity. The adaptability of many of Gaborone's residents was evident when Gaborone Dam failed in 2015 (Kadibadiba et al. 2018), with supplies from Molatedi Dam in Madikwe Game Reserve in South Africa also less than the annual water quota of 7.3 Mm^3 (DWA 2013), due to drought. The capital was spared a severe water shortage by the return of good rains the following season, although the crisis should serve as a forerunner as to what is coming.

17.3.2 Medium and Small Dams

Geoflux (Pty) Ltd., in association with Knight Hall and Associates in 2000 drew up a shortlist of 12 of the most promising small to medium dam sites drawn up from an initial 34 (Geoflux and Hendry 2001). Water supply is multi-faceted and for dominantly political reasons a dam was built on the Mosetse River despite it being part of the endoreic Makgadikgadi Pans drainage system. Such dams are important to livestock with little or no associated irrigated agriculture.

FAO (2004) identified close to 400 small, typically earth bund, dams mostly located in the Lotsane and Motloutse catchments, with this number likely to be well over a thousand today for the Botswana part of Limpopo River catchment. In South Africa, O'Connor (2001) showed that the riparian and associated vegetation within a sub-catchment of the Limpopo River had become desiccated as a result of the cumulative effect of many small farm dams within the lowest stream orders of the catchment. Indeed, the reduction of flow during low rainfall years was the most likely explanation of the mortality and die-back of species that required either substantial amounts of water (e.g. *Ficus albida*), or of individuals (e.g. *Combretum imberbe*) growing at their limits of water availability great distances from and high elevations above the river (O'Connor 2001).

Small dams have significant cumulative effects on the runoff into and the yield from large reservoirs, particularly in dry years and at the start of the hydrological year (Meigh 1995). Chanda et al. (2018) point out that the Notwane catchment has some 237 small agricultural dams which are essential to rural water supply and livelihoods, even though they impact significantly upon urban water supply as provided by Gaborone Dam for domestic/institutional and industrial water needs.

17.3.3 Wet and Dry Cycles

Byakatonda et al. (2018) identified the timescales at which various components of the hydrological cycle, especially stream flow, responds to climate variability. They defined climatological drought as below normal precipitation coupled with increase in potential evapotranspiration covering large spatial extents for prolonged time periods, and hydrological drought as being associated with below average surface water flow for prolonged periods. Over the period 1975–2014 they found that droughts, particularly moderate ones, showed an increasing trend in both the Okavango and Limpopo river systems towards drying conditions. Climatic droughts were found to take 6 months to propagate to hydrological droughts in the Okavango river system and one month longer in the Limpopo (Byakatonda et al. 2018).

Correlations between climatic and hydrological droughts were found to be mainly significant in March and April reaching the peak at 15 months timescale in the Okavango river system. Dry conditions generally followed for the next 40 months, with recovery to normal conditions expected in the last 20 months of the forecast period. The results were seemingly influenced by El Niño-Southern Oscillation (ENSO) episodes in the Equatorial Pacific, with dry periods recorded from 1981/82 to 1986/ 87 repeated in 1994/95–1998/99 (Byakatonda et al. 2018). Similarly, a wet spell recorded over the study area of 1975/76 to 1979/80 also was preceded by La Niña year of 1973/74 (Byakatonda et al. 2018).

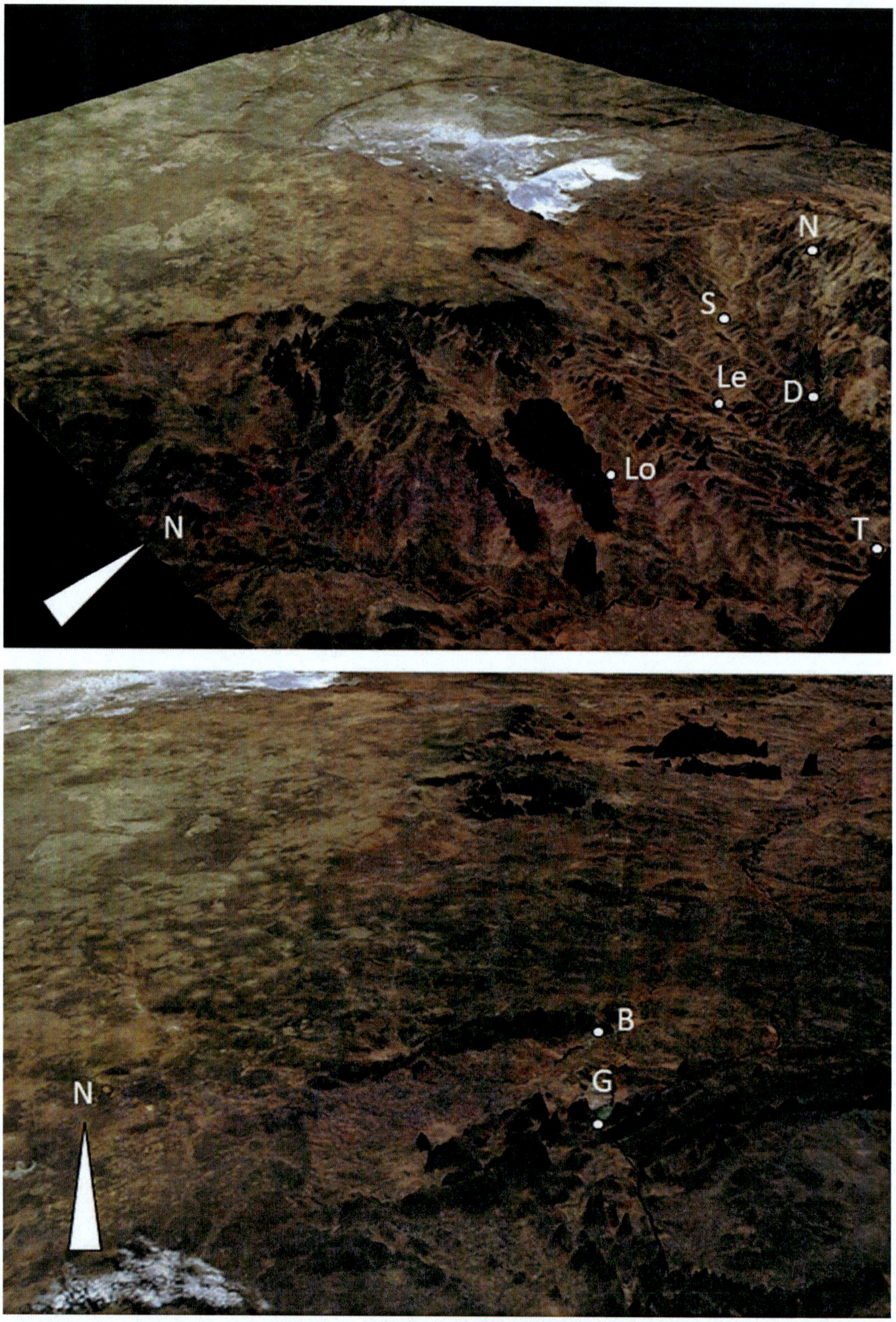

Fig. 17.3 Location of dams in Botswana, making use of the limited relief in eastern part of the country. From north to south Top: Ntimbale (N), Shashe (S), Letsibogo (Le), Dikgatlhong (D), Thune (T), Lotsane (Lo), Bottom: Boka (B) and Gaborone (G) dams, which form the backbone of the North–South Carrier (NSC) covering a distance of ca 400 km. (Compare to Catchment Fig. 16.3)

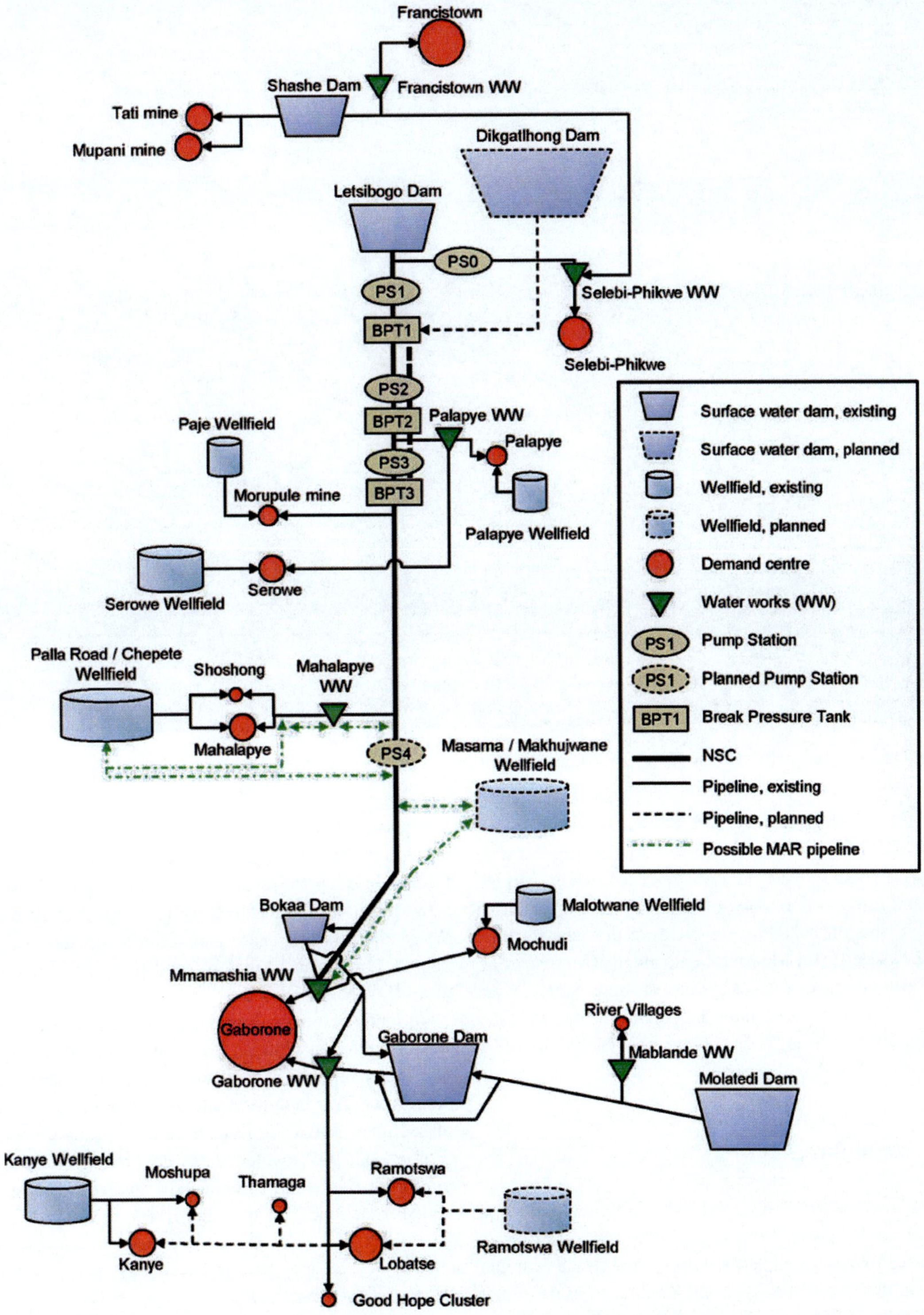

Fig. 17.4 Schematic illustration of the water supply system linked to the North–South Carrier (NSC) (Lindhe et al. 2020)

A number of studies have emphasised the water-stressed nature of much of the Limpopo River Basin and how this is set become even more accentuated under climate change scenarios (Midgley et al. 2013). Referring to streamflow as

Fig. 17.5 Letsibogo Dam feeds into the North–South Carrier (NSC) and can supply water to Gaborone some 300 km away

'blue water' Mosase et al. (2019a) showed that between 1984 and 2013 it varied from 0.2 to 570 mm year^{-1}, with alternating cycles of one to two years of water surplus periods and three to five years of dry periods. For example 1984–1986 and 2001–2005 were cycles of dry year for blue water for Notwane and Motloutse catchments (Mosase et al. 2019a). By contrast, surplus years of blue water were associated with extreme rainfall and flood events, typically influenced by cyclones, such as cyclone Elaine in 2000 (Mosase et al. 2019b).

17.3.4 Sand Rivers

Botswana is characterised by sand rivers in which the bed is lined with sand typically to a depth of 3–9 m. Sand rivers occur throughout East and North Africa, the Middle East and southern Africa, with Botswana and Zimbabwe in particular utilising them as a potential source of drinking water, for both people and livestock (Mpala et al. 2016). The latter cite Herbert (1998) who estimated that there were 900 km of sand rivers in Botswana, which in turn constitute a valuable source of water in often remote rural areas. Even during drought periods sip wells in the bed of sand rivers throughout the Limpopo system are dug by subsistence farmers in order to water livestock and exploit the water that flows through the sand body (Fig. 17.6). It is estimated that the sand rivers of eastern Botswana could sustainably yield 2.5–55 Mm^3 year^{-1} km^{-1} of sand river, primarily for local users (DWA 2013).

In many respects, the life blood of the rivers is the sand within them that sustains the water that runs through them, with water supply issues focussing upon the latter rather than the former. This is unfortunate as the longitudinal movement of sediment down the river is broken by dam construction (Poff et al. 1997), which when coupled with the excessive mining of river sand can severely impact the complex ecological functioning of the sand rivers.

17.4 Key Concerns

Human impact on rivers are a key part of the Anthropocene (Syvitski et al. 2020) and affect all life forms. Land use/land cover change has disturbed the landscape and in turn altered river discharge and load (water, particulate and dissolved

Fig. 17.6 Sip well in a riverine sand body which provide a valuable source of water in rural areas

constituents), with dams shifting climate and biogeochemical cycles (C, N, P and Si, in particular) (Maavaraa et al. 2020) and biodiversity. The predicted decreased rainfall, increased temperature and evapotranspiration and greater climate variability will impact significantly on dryland rivers, particularly in Australia, where downstream declines in bankfull discharge have been widely predicted (Larkin et al. 2020). Increased vegetation water use following future decreases in precipitation are a key impact in water-stressed environments such as Australia (Ukkola et al. 2016), with similar effects likely in Botswana.

The overall question of water security in Botswana this Century and beyond is a major concern as following an increase of 1.7 °C in mean temperature, the region's precipitation is expected to decrease by 5–20% in all major river basins, which in turn is expected to increase salt loads (Borrok and Engle 2014). Annual evaporation from open water in Botswana varies from about 1900 to 2200 mm $year^{-1}$, but may be as high as 2600 mm $year^{-1}$ in the Limpopo River Basin (LRB) (FAO 2004). Annual average recharge throughout the LRB is of the order of 44 mm yr^{-1} and is highly seasonal and localised (Mosase et al. 2019a). Mwenge Kahinda et al. (2016) emphasise that if the ecological flow requirement for water was met by all tributaries, the Limpopo River basin would be 'closed', with projected climate changes (rainfall reductions and temperature increases) set to further accentuate the seriousness of the situation.

17.4.1 Sedimentation

The nature and extent of sedimentation of Botswana's dams is an issue of major concern. Ephemeral rivers in arid and semi-arid regions are known to have considerable potential to transport sediment, perhaps 400 times that of perennial rivers in humid zones (Laronne and Reid 1993), particularly during flash floods (Hooke 2019). Alemaw et al. (2013) using the Revised Universal Soil Loss Equation (RUSLE) and GIS found the average annual sedimentation rate ranges of 12 dams in the Lotsane catchment to be between 0.54 and 85.29 ton ha^{-1} $year^{-1}$. The lower value concurs with that of Cave Klapwijk and Associates (1994) of 0.5 tons $ha^{-1}$of catchment per year for the Motloutse River, with Alemaw et al (2013) predicting an annual average soil loss rate of 26 tons ha^{-1} $year^{-1}$ (2,559 tons km^{-2} $year^{-1}$) in the Lotsane watershed. Franchi et al (2020) found that sediment in cores 15–35 cm deep around Notwane Dam correlated with flood events in 2006–2011.

The main income-earning activities of most rural inhabitants of the catchment are livestock rearing and dryland farming. The entire catchment is stocked at rates considerably above those recommended, based on range productivity (Government of Botswana 1991). At an assumed stocking rate of <4 Ha/LSU, the sedimentary and nutrient inputs to tributaries of the river systems from erosion, livestock manure and fertilisers can be anticipated to be high. The potential for siltation and eutrophication of the dams will therefore also be high (Fig. 17.7).

Soil erosion has been greatly accentuated by widespread vegetation clearance and cultivation for arable agriculture under successive agricultural subsidy programmes—ALDEP, ARAP and ISPAAD. While an unintended consequence many beneficiaries have actively harvested the subsidies rather than the crops, significantly damaging biodiversity and ecosystem resilience. Against such widespread clearance of the savanna climate change with mega-droughts and flooding will clearly accentuate soil erosion and dam sedimentation.

17.5 Sand Mining

Kalahari sand which covers 75% of Botswana, like desert sand in general, is too smooth to be used in construction, which in turn has put considerable pressure on more angular river sand. It is a global problem that shows no sign of abating (Torres et al. 2017; Bendixen et al. 2019; Hackney

Fig. 17.7 Bank erosion on the Metsemotlhabe River just outside Gaborone

et al. 2020). Botswana's sand rivers are no exception with Shaw et al (1994) pointing out that 100,000 m^3 $year^{-1}$ of sand had been removed since 1974 from the Motloutse River near Selebi-Phikwe for smelting flux at the BCL Mine (MacDonald and Partners 1989; from Shaw et al. 1994). WCS (2010a) emphasised that the removal ('mining') of large volumes of sand from the Motloutse sand river body, especially immediately downstream of the dam, was having serious effects on the ability of the sand river to contribute towards the environmental water demand—particularly with respect to the human and livestock demands which are totally dependent on it. Sand mining by removing the very 'life-blood' of the river, the sand itself, can be particularly damaging, especially when it occurs downstream of dams where there is no chance of replenishment of the sand river body (e.g. Petts and Gurnell 2005).

During the construction of the Lotsane Dam at Maunatlala a shortage of construction sand, some 60,000 m^3, led the contractor to seek exploitation of river sand just below the dam wall. However, under the guidance of the Environmental Management Plan (WCS 2010b) the sand was sourced from the nearest downstream tributary, the Tshokana River. Significantly, the sand that was removed from an approximately 10 kms section in the winter was largely replenished at the end of the following wet season.

17.5.1 Fly Dumping

Illegal or fly dumping of rubbish has long been a problem in Botswana and is most noticeable, but far from confined to, the major urban centres of Gaborone and Francistown. Dumping miscellaneous waste in the rivers, and even around the dams themselves, is all too common, with much of the waste undoubtedly being transported to South Africa, Mozambique and the Indian Ocean. It is a growing problem as urban centres expand and many appear unwilling to dispose of waste at official landfill sites. Close to two-thirds of the 250,000 tons of household waste produced in Botswana annually does not make it to official landfill sites (Urio and Brent 2006; from Kgosiesele and Zhaohui 2010). Indeed, one may argue that rivers are regarded by many as 'mobile landfills' as the problem is simply exported somewhere else.

17.5.2 Elephants

Botswana's burgeoning elephant population is largely confined to the north, although since the onset of flows down the Boteti River in 2008/09, some have dispersed southwards with over one thousand now resident in the Central Kalahari Game Reserve and utilising the artificial water points found there. Lone bulls and small bull herds have long impacted upon local communities around Masunga dam, while the Central Limpopo Valley elephant population is known to occur around Letsibogo Dam and the Motloutse, Shashe and Thune Rivers (Selier 2010). With the predicted mega-droughts this Century it is conceivable that elephants could increasingly impact upon the dam infrastructure as they take up residence around them, in turn causing considerable human-wildlife conflict.

17.5.3 Pollution

Pollution takes many forms which are inevitably case specific. Cave Klapwijk (1994) emphasised that the major 'chemical' polluter in the Central area of eastern Botswana was undoubtedly BCL mine with a smelting plant which emits SO_2 into the atmosphere, a coal burning power station smelting CO_2 and SO_2 to the atmosphere and with ash dumps, slag heaps and tailings dams (Fig. 17.8) with a high pyrite content which weathers to release acid mine drainage water (AMD). While production at BCL has ceased, AMD continues to leach into the Letlhakane River (Fig. 17.9) and then the Motloutse. The sand–river water is in its natural state is calcium, magnesium and sodium bicarbonate dominant which shifts drastically to a non-carbonate (sulphate) hardness from the introduction of mine effluent (MacDonald 1990).

All discharges to the environment in Botswana, including rivers, must comply with BOS 93: 2012. It is clear however that key sections of Botswana's ephemeral rivers are facing unprecedented pollution threats from return flows from mining (Fig. 17.8), farming (chicken farms, cattle grazing and horticulture) and sewage treatment plants, all of which contribute to emerging contaminants (Gomes et al. 2020), with cascading effects for waterborne diseases (Semenza 2020). Sewage releases into rivers is a global challenge and one that appears set to worsen as urban populations grow (Strokal et al. 2021), the climate warms, river flows become more sporadic and wastewater fluxes increase (Woodward et al. 2021). Sewage treatment plants are typically located adjacent to the key rivers with discharges of largely untreated effluent not uncommon. For this reason, reed beds have become permanently established in the beds of the rivers downstream of sewage treatment plants and centres such as Gaborone, Palapye, Francistown and Selebi-Phikwe are all too frequently permeated with the stench of sewage disposal into the environment. Numerous reports have emphasized the need to implement the country's 'Polluter Pays' Principle more effectively and while innovative concepts such as 'green infrastructure' offer a clear route to livable and sustainable cities (Lund et al. 2019), they are regrettably the exception rather than the rule.

Fig. 17.8 Runoff from tailing at the Selebi-Phikwe mine

17.5.3.1 Phytoplankton Blooms

Dams disrupt the biogeochemical cycles of river networks by impeding the flow of essential nutrients (C, P, N and Si) along them (Maavara et al. 2020). As the latter authors emphasise damming by altering nutrient ratios can lead to changes in phytoplankton composition, with toxic algae or cyanobacteria now dominating, while invertebrate richness (including that of diatom communities), abundance and density and organic matter decomposition are reduced (Sabater et al. 2018). Algal blooms often occur during late winter and early spring when seasonal thermoclines break down allowing the mixing of nutrient-poor surface waters with the nutrient-rich waters below (Paerl and Huisman 2008). The latter authors emphasise that global warming seems likely to accentuate this effect, with high temperatures coupled with more intense precipitation flushing nutrients into dams and droughts increasing their residence time and overall load, all serving to increase the likelihood of harmful algal blooms. Attempts to control such outbreaks by opening dam sluices may even serve to accentuate the problem by increasing the residence time of the outbreaks (Paerl and Huisman 2008). Indeed, it is clear that the predicted mega-droughts, heatwaves and intense precipitation and flooding events offer much more than increasingly erratic flows, but come with potentially dire impacts for the ecology of the system and human health (Semenza 2020).

17.5.3.2 Aquatic Weeds

Most attention on aquatic weeds in Botswana has focussed upon *Salvinia molesta* and attempts to control it in the Okavango and Chobe-Linyanti-Kwando river systems through the use of weevils (Kurugundla et al. 2016). Water Hyacinth (*Eichhornia crassipes*) is found in Hartbeesport Dam on the Crocodile River in South Africa (Auchterlonie et al. 2021) and threatens the dams and rivers on the Limpopo River system in Botswana (Kurugundla et al. 2016). *Ludwigia stolonifera* was found in the Shashe and smaller dams of the region but was not thought to pose a danger of spreading uncontrollably (Cave Klapwyck and Associates 1994). *Phragmites mauritianus* may also be considered a

Fig. 17.9 Degraded State of the Letlhakane River

potential invader which has been shown to cause shoreline encroachment in some dams (Cave Klapwyck and Associates 1994). The aquatic weed problem on Botswana's dams is likely to become more challenging if recreational boating is permitted upon them, as currently the use of motorboats is banned.

17.5.4 Impact on the Morphometric Characteristics of a Catchment

In addition to the above, one of the other key concerns is the impact of the reservoir on the geomorphology of the catchment upstream, which is responsible for the water collection and supply into the reservoir. In one such study on the Shashe reservoir located in north-eastern Botswana and a major source of water supply to major settlements of the country, it was found that the morphometric characteristics such as the Horton Order Ratios, Drainage Density and Stream Frequency had changed considerably after the dam construction (Parida et al. 2010).

The study was untaken in four major sub-catchments namely, Shekwe-Maeroro, Lonye, Mooke and Kgonyane for the years 1970 and 1990, which are the years before and after the dam construction respectively. The details of analyses are given in Table 17.2.

The overall results show that the Horton Oder stream characteristics such as: R_A, R_B, and R_L, have changed ever since the dam was constructed.

(i) R_A overall has decreased by over 40% indicate that areas drained by lower order streams (headstreams) have increased may be due to upstream degradation and the watershed has now more uniform soil pattern.
(ii) R_B has decreased by over 20% suggesting reduction in drainage efficiency.
(iii) R_L has increased by about 8% suggesting that type of soil covering the drainage basin have changed due to

Table 17.2 Average values of stream characteristics for the years 1970 and 1990

Horton order parameters	1970	1990	Change (%)
R_A	7.5	4.4	−41
R_B	5.8	4.6	−21
R_L	2.5	2.8	+7.5
Stream frequency	0.4	0.5	+10.7
drainage density	0.7	0.9	+26.5

soil erosion which is supported by about 10% increase in stream frequency.

(iv) The value of drainage density has increased by more than 26% suggesting that more area of the watershed are now required to maintain 1 km of the channel.

17.5.5 Environmental Flows

'The effects of flow regulation can be ecologically catastrophic for dryland river ecosystems, for they are adapted to unregulated and highly variable flow, sediment and chemical regimes. The spatial and temporal variability of dryland river ecosystems suggests that the time required to respond to, and to recover from, environmental change may be more prolonged than for rivers in more stable environments' (Davies et al. 1994; p. 494).

Botswana has no statutory requirement to make water releases from dams for environmental flow purposes, although ratification of a number of international conventions and mandatory requirements of international dam construction financiers now make this a pressing issue. The National Water Master Plan (DWA 2006) includes recommendations for estimating the Environmental Flow Release (EFR) and that EFRs should be 'done in such a way as to mimic the natural hydrograph'. However, despite high evaporative losses, there is an understandable reluctance on the part of many key stakeholders to deliberately release water from dams that were built to store it.

In Zimbabwe Tennant's (1976) estimate that 30% of the mean annual runoff (MAR) was sufficient to maintain good habitats matches that of Smakhtin et al. (2004) and Mazvimavi et al (2007) (Table 17.3).

Environmental flow releases from dams have for decades sought, as far as is possible, to mimic the natural flow regime the dam disrupted, although the concept of designer flows has recently emerged as a 'holistic' approach to reconciling ecological and societal needs (Chen and Olden 2017; Palmer and Ruhi 2019).

In order to illustrate the effects of regulation on the nature of flows in the Motloutse River at Letsibogo, WCS (2010a) made the assumption that the dam was built in 1922 and the nature of spills derived from the Pitman model.

17.5.6 Impact of Letsibogo Dam on Motloutse Flow Regime

The results of the effects of regulation on flows show that the pre-dam wet season months of flow is almost equivalent to the post dam wet season months of no flow (WCS 2010a) (Table 17.4). This reversal in fortunes undoubtedly has profound implications for the functioning of the river ecosystem itself. There is also a 95% reduction in the number of months with low flows and an 85% reduction in the number of months with medium flows. Very high flows are also drastically reduced, while exceptional flows, such as occurred in Sept 2009, also register as an exceptional spill. Before regulation, the maximum number of consecutive wet season months of no flow was only 5 but this alters dramatically to 81 consecutive months of no wet season flow in the post-regulation regime.

Letsiboga dam as expected has resulted in much fewer spills than flows. Significantly the early wet season (Oct-Dec) spills are made up of very few events, Oct = 0 spills, Nov = 3 spills in 87 months, and Dec = 7 spills in 87 months, as compared to Oct = 58 flows, Nov = 77 flows and Dec = 83 flows (Table 17.4) (WCS 2010a). The average monthly spills and flows can therefore be misleading as very few spills (Fig. 17.10) contribute to the average, whereas pre-regulation many flows occurred. The functioning of the river system has therefore changed markedly.

The second half of the wet season Jan-April shows a similar pattern and while large spills still contribute disproportionately to the average, a greater number of flows occur (Table 17.5) (WCS 2010a). The flows in the Motloutse River, and the ecological triggers associated with them, therefore occur later in the wet season, than was the case before the dam was built. This has been a reality for almost the last two decades.

Botswana's rivers are therefore experiencing an unprecedented level of regulation and it is clear that future management of the dams will have to be 'adaptive' and

Table 17.3 Environmental flows in Botswana

Dam	EFR characteristics
Thune MNRC (2000), Geoflux (2005)	Total of 0.98 Mm^3/year, which is equivalent to 6% of the Mean Annual Runoff When the storage is <50%, only one EFR release of 0.18 Mm^3 should occur in March EFR releases are related to the actual inflow during any specific month. Set as:- November 0.432 MCM, December 0.972 MCM and January/February 1.296 MCM
Mosetse (WRC 2010)	Four releases of 1 million m3 each at a rate of 20 m^3/s for 14 h, made during months of December, January, February and March when natural flow occurs. When there is no natural inflow to the dam, no release should be made, and if there is more than one natural run-off event in any month, only one release should be made
Letsibogo WCS (2010a)	EFR = 2.94 Mm^3/year, released in Nov and Feb (0.49 Mm^3 and 2.49 Mm^3 respectively) Releases are controlled by a series of 'triggers' relating to dam stage levels (rising or falling) and whether any spills occur between Nov and Feb
Ntimbale (WCS 2011)	FSL 1 103.5–1 101.4 m in November, outlet is opened fully for 3 days *FSL 1 103.5–1 101.4 m in February the outlet is opened fully for 9 days FSL 1 101.4–1 098.8 m in November the outlet is opened fully for 2 days and FSL 1 101.4–1 098.8 m in February the outlet is opened fully for 6 days When the dam level is between 1 098.8 m and the MOL (1 086.25 m) in November the outlet is opened fully for 3 days and when the dam level is between 1 098.8 m and the MOL (1 086.25 m) in February the outlet is opened fully for a duration of 7 days When the dam level is below MOL (1 086.25 m) no releases are made

From WCS (2011)
*On the 16/02/2015 such an EFR was effected from Ntimbale Dam (https://www.facebook.com/BotswanaGovernment/posts/notification-of-environmental-flow-release-from-ntimbale-damfor-immediate-releas/1580133718735947/)

Table 17.4 Summary of observations relating to pre and post-regulation flows

Parameter	Pre-dam	Post-dam
Wet season months of flow	543	67
Wet season months of no flow	66	547
Low flows 0.5–5	339	16
Medium flows 5–20	130	20
High 20–100	64	27
Very high flows 100–200	7	3
Unusually high flows 200–205	2	0
Exceptionally high flows (Sept 2009) > 650	1	1
Wet season months of consecutive no flow	5	81

guided by both relevant monitoring data and sound scientific principles and will undoubtedly include deliberate environmental flow releases from the dam (King et al. 2003; Petts 2008; Catelotti et al. 2015; Ramirez-Hernandez et al. 2015; Aguilar and Polo 2016; Poff and Schmidt 2016) and perhaps even designer flow releases (Chen and Olden 2017; Poff and Olden 2017).

17.6 Way Forward

Botswana has long recognised that the conjunctive use of surface water and groundwater is essential for sustaining water quantity and quality requirements of users (DWA 2006, 2013). Flow regime alteration and water abstraction

Fig. 17.10 Ntimbale dam spilling in 2011

Table 17.5 Average monthly flow and spill statistics (1922–2008, n = 87)

	Oct	Nov	Dec	Jan	Feb	Mar	Apr
Flows	58	77	83	83	84	78	72
Spills	0	3	7	15	20	15	4

remain the key stressors to southern African River systems (Fouchy et al. 2019). Rising atmospheric CO_2 and climate change have on the one hand been projected to cause large scale afforestation via woody cover increases, reducing both biodiversity and freshwater over large areas, while on the other there is also the likelihood of greater fire frequencies and severity due to the hotter and drier conditions (Midgley and Bond 2015). Increased fire frequencies and severity (Ruffault et al. 2020) seems likely to increase sedimentation rates as it leads to increased surface runoff and sedimentation rates.

17.6.1 Inter-Regional Water Management

Climate change scenarios effectively mean that Botswana's water supply challenges will not be resolved within its own borders. Inter-regional water transfer schemes will undoubtedly be the key to adapting to climate change, with collaboration across other key sectors such as energy, agriculture and conservation also likely to be critical. Best (2018) points out that in 1977 Luna Leopold appealed that we adopt a 'reverence for rivers' to hold them in the esteem they truly deserve (Leopold 1977, from Best 2018) and it is with considerable regret that globally this has not happened. The need to increase water harvesting in Africa's drylands is abundantly clear (Rockström and Falkenmark 2015) with Di Baldassarre et al. (2018) hypothesizing that reservoir building and extended periods of abundant water supply actually serve to generate an increasing dependence on water infrastructure, which in turn increases vulnerability and economic damage when water shortages eventually occur.

17.7 Conclusion

Considerable uncertainty surrounds the future of Botswana's dams and the rivers that supply them. The economic value of the river sand is widely realised and despite increasing regulation remains extensively exploited. Dryland rivers globally have become convenient conduits for waste disposal (Vidal-Abarca et al. 2020), as flash floods and rapid flows in the Limpopo Basin, removes sewage and all manner of rubbish through South Africa and Mozambique into the Indian Ocean. As in the Northern hemisphere rivers in Botswana seem likely to become increasingly polluted despite the wealth of scientific evidence as to how damaging this is to ecosystem health, the hydrological cycle and attainment of the UN's Sustainable Development Goals. Tabari (2020) emphasises the fact that 'it never rains, but it pours' will affect dry regions such as southern Africa which will experience extreme precipitation increases and yet total precipitation declines.

Botswana's dams will be increasingly unable to meet the growing requirements of its population and economy under such a scenario and will undoubtedly have to look beyond its borders, to other regions and inter-basin transfers, if water supply solutions are to be found. As 'blue water' becomes increasingly scarce 'green water' will have to be utilised by collecting run-off; improving the infiltration of rain in soils; and managing land, water and crops across watersheds to increase water storage in soils, wetlands and the water table (Rockström and Falkenmark 2015; Tortajada and van Rensburg 2020). It is clear that tough choices between competing sectors for water use, particularly agriculture, mining, health and sanitation, and river ecosystem health, will need to be made.

The increasing influence of the El Nino Southern Oscillation (ENSO) (Engelbrecht et al. 2015; Byakatonda et al. 2020) on Southern Africa's river regime (Zambezi, Limpopo, Orange) is evident in the early Holocene (11.5–9.0 ky B.P.) and 4.2 and notably 2.0–1.5 ky B.P. (Macklin and Lewin 2018). The aridity of the past should serve as a dire warning of the urgent need to adapt to heat and dessication of the future and to recognise the increasingly limited role dams can play in solving Botswana's increasing water scarcity. Ironically, floods will also increase this Century and St George et al. (2020) quote the American writer Toni Morrison who once wrote that 'rivers do not flood; rather, they on occasion remember where they used to be' (Morrison 1998 from St George 2020).

Acknowledgements This chapter benefitted from a review by Dr A. E. Palmer.

References

Aguilara C, Polo MJ (2016) Assessing minimum environmental flows in non-permanent rivers: the choice of thresholds. Environ Model Softw 79:120–134

Alemaw BF, Majauale M, Simaleng T (2013) (2013) Assessment of sedimentation impacts on small dams—A case of small reservoirs in the lotsane catchment. J Water Resour Prot 5:1127–1131. https://doi.org/10.4236/jwarp.2013.512118

Auchterlonie J, Eden C-L, Sheridan C (2021) The phytoremediation potential of water hyacinth: a case study from hartbeespoort dam, South Africa. S Afr J Chem Eng 37:31–36. https://doi.org/10.1016/j.sajce.2021.03.002

Belletti B, Garcia de Leaniz C, Jones J, Bizzi S, Börger L, Segura G, Castelletti A, van de Bund W, Aarestrup K, Barry J, Belka K, Berkhuysen A, Birnie-Gauvin K, Bussettini M, Carolli M, Consuegra S, Dopico E, Feierfeil T, Fernández S, Garrido P, Garcia-Vazquez E, Garrido S, Giannico G, Gough P, Jepsen N, Jones PE, Kemp P, Kerr J, King J, Łapińska M, Lázaro G, Lucas MC, Marcello L, Martin P, McGinnity P, O'Hanley J, del Amo RO, Parasiewicz P, Pusch M, Rincon G, Rodriguez C, Royte J, Schneider CT, Tummers JS, Vallesi S, Vowles A, Verspoor E, Wanningen H, Wantzen KM, Wildman L, Zalewski M (2020) More than one million barriers fragment Europe's rivers. Nature 588:436–441. https://doi.org/10.1038/s41586-020-3005-2

Bendixen M, Best J, Hackney C, Iversen LL (2019) Time is running out for sand. Nature 571(7763):29–31. https://doi.org/10.1038/d41586-019-02042-4

Best J (2018) Anthropogenic stresses on the world's big rivers. Nat Geosci. https://doi.org/10.1038/s41561-018-0262-x

Bhalotra YPR (1987) Climate of Botswana—Part II: elements of climate. Department of Meteorological Services. Ministry of Works and Communications. Gaborone

Borrok D, Engle M (2014) The role of climate in increasing salt loads in dryland rivers. J Arid Environ 111:7–13. https://doi.org/10.1016/j.jaridenv.2014.07.001

Byakatonda J, Parida BP, Kenabatho PK (2018) Relating the dynamics of climatological and hydrological droughts in semiarid Botswana. Phys Chem Earth 105:12–24. https://doi.org/10.1016/j.pce.2018.02.004

Byakatonda J, Parida BP, Moalafhi DB, Kenabatho PK, Lesolle D (2020) Investigating relationship between drought severity in Botswana and ENSO. Nat Hazards 100:255–278. https://doi.org/10.1007/s11069-019-03810-113

Catelotti K, Kingsford RT, Gilad B, Bacon BP (2015) Inundation requirements for persistence and recovery of river red gums (*Eucalyptus camaldulensis*) in semi-arid Australia. Biol Conserv 184. https://doi.org/10.1016/j.biocon.2015.02.014

Cave Klapwyck and Associates (1994) An Environmental impact, compensation and relocation assessment and other associated works for the Motloutse River at Letsibogo. Government of Botswana, Ministry of Mineral Resources and Water Affairs, Department of Water Affairs

Chan EK, Timmermann A, Baldi BF, Moore AE, Lyons RJ, Lee SS, Kalsbeek AM, Petersen DC, Rautenbach H, Förtsch HE, Bornman MR (2019) Human origins in a southern African palaeowetland and first migrations. Nature 575(7781):185–189

Chanda R, Mosetlhi BT, Sakuringwa S, Makwati M (2018) Water for Urban development or rural livelihoods: is that the question for Botswana's Notwane river catchment? Botsw Notes Rec 50:123–137

Chen W, Olden JD (2017) Designing flows to resolve human and environmental water needs in a dam-regulated river. Nat Commun 8:2158. https://doi.org/10.1038/s41467-017-02226-4

Chen S-A, Michaelides K, Grieve SWD, Singer MB (2019) Aridity is expressed in river topography globally. Nature 573:573–577. https://doi.org/10.1038/s41586-019-1558-8

Cornwallis HW (1852) The wild sports of Southern Africa published by Bohn HG, London

Cowx I (2012) Fish Stock Assessment in Major Dams in Botswana. Project Funded by the European Union. Project ref. N° SA-3.2-B15. Landell Mills Consultants

Davies BR, Thomas MC, Walker KF, O'Keeffe JH, Gore JA (1994) Dryland rivers: their ecology, conservation and management. In: Calow P, Petts GE (eds) The rivers handbook: hydrological and ecological principles, vol 2. Blackwell Scientific Publications. Oxford, UK, pp 484–511

De Wit M, Stankiewicz J (2006) Changes in surface water supply across africa with predicted climate change. Science 311 (5769):1917–1921. https://doi.org/10.1126/science.1119929

Di Baldassarre G, Wanders N, Kouchak A, Kuil L, Rangecroft S, Veldkamp TIE, Margaret Garcia M, van Oel PR, Breinl K, Van Loon AF (2018) Water shortages worsened by reservoir effects. Nat Sustain 1(11):617–622. https://doi.org/10.1038/s41893-018-0159-0

DWA (2006) Botswana National Water Master Plan (BNWMP) Review. Department of Water Affairs, Gaborone, Botswana

DWA (2013) Botswana integrated water resources management & water efficiency plan. In: Dikobe L (ed) Department of water affairs, ministry of minerals, energy and water resources. Gaborone, Botswana: Government of Botswana

Engelbrecht F, Adegoke J, Bopape MJ, Naidoo M, Garland R, Thatcher M, McGregor J, Katzfey J, Werner M, Ichoku C, Gatebe C (2015) Projections of rapidly rising surface temperatures over Africa under low mitigation. Environ Res Lett 10(8):085004

FAO (2004) Drought impact mitigation and prevention in the Limpopo River Basin: a situation analysis, land and water discussion Paper 4, Prepared by the FAO Subregional Office for Southern and East Africa, Harare. http://www.fao.org/3/y5744e/y5744e00.htm#Contents

Fouchy K, McClain ME, Conallin J, O'Brien G (2019). Multiple stressors in African freshwater systems. In: Multiple stressors in river ecosystems, pp 179–191. https://doi.org/10.1016/b978-0-12-811713-2.00010-8

Franchi F, Ahad JME, Geris J, Jhowa G, Petros AK, Comte J-C (2020) Modern sediment records of hydroclimatic extremes and associated potential contaminant mobilization in semi-arid environments: lessons learnt from recent flood-drought cycles in southern Botswana. J Soils Sediments 20:1632–1650. https://doi.org/10.1007/s11368-019-02454-9

Garzanti E, Padoan M, Setti M, López-Galindo A, Villa IM (2014) Provenance versus weathering control on the composition of tropical river mud (southern Africa). Chem Geol 366:61–74. https://doi.org/10.1016/j.chemgeo.2013.12.016

Geoflux and Hendry KH (2001) Engineering consultancy for the feasibility study of small to medium dams in Eastern Botswana, Draft Feasibility Report, Geoflux (Pty) Ltd. in association with Knight Hall Hendry, Botswana

Geoflux (2005) Environmental supervision of the construction of the Thune dam & associated works on the thune river. Water Affairs Gaborone

Gomes IB, Maillard JY, Simões LC, Simões M (2020) Emerging contaminants affect the microbiome of water systems—Strategies for their mitigation. NPJ Clean Water 3:39. https://doi.org/10.1038/s41545-020-00086-y

Government of Botswana (1991) National Policy on Agricultural Development. Government Paper No. 1. of 1991. Government Printer. Gaborone

Government of Botswana (2016) Botswana Water Accounting Report 2016. WAVES, CAR and MMEWR. Gaborone

Habel M, Mechkin K, Podgorska K, Saunes M, Babiński Z, Chalov S, Absalon D, Podgórski Z, Obolewski K (2020) Dam and reservoir removal projects: a mix of social-ecological trends and cost-cutting attitudes. Sci Rep 10:19210. https://doi.org/10.1038/s41598-020-76158-3

Hackney CR, Darby SE, Parsons DR et al (2020) River bank instability from unsustainable sand mining in the lower Mekong River. Nat Sustain 3:217–225. https://doi.org/10.1038/s41893-019-0455-3

Herbert R (1998) Water from sand rivers in Botswana. Q J Eng Geol Hydrogeol 31:81–83

Higginbottom TP, Adhikari R, Dimova R, Redicker S, Foster T (2021) Performance of large-scale irrigation projects in sub-Saharan Africa. Nat Sustain. https://doi.org/10.1038/s41893-020-00670-7

Hooke JM (2019) Extreme sediment fluxes in a dryland flash flood. Sci Rep 9:1686. https://doi.org/10.1038/s41598-019-38537-3

Hughes DA, Farinosi F (2020) Assessing development and climate variability impacts on water resources in the Zambezi River basin. Simulating future scenarios of climate and development. J Hydrol: Region Stud 32:100763. https://doi.org/10.1016/j.ejrh.2020.100763

Kadibadiba T, Roberts L, Ronlyn D (2018) Living in a city without water: a social practice theory analysis ofresource disruption in Gaborone, Botswana. Glob Environ Change 53:273–285. https://doi.org/10.1016/j.gloenvcha.2018.10.005

Kahinda JM, Meissner R, Engelbrecht FA (2016) Implementing integrated catchment management in the upper Limpopo river basin: a situational assessment. Phys Chem Earth, Parts A/B/C 93:104–118

Kgosiesele E, Zhaohui L (2010) An evaluation of waste management in Botswana: achievements and challenges. J Am Sci 6(9)

King J, Brown C, Sabet H (2003) A scenario-based holistic approach to environmental flow assessments for rivers. River Res Appl 19:619–639

King J, Chonguiça E (2016) Integrated management of the Cubango-Okavango River Basin. Ecohydrol Hydrobiol 16(4):263–271. https://doi.org/10.1016/j.ecohyd.2016.09.005

Kurugundla NC, Mathangwane B, Sakuringwa GK (2016) Alien invasive aquatic plant species in botswana: historical perspective and management. Open Plant Sci J 9(1):1–40. https://doi.org/10.2174/1874294701609010001

Larkin ZT, Ralph TJ, Tooth S, Fryirs, K.A. and A.J.R. Carthey (2020) Identifying threshold responses of Australian dryland rivers to future hydroclimatic change. Sci Rep 10:6653. https://doi.org/10.1038/s41598-020-6362

Laronne J, Reid l (1993) Very high rates of bedload sediment transport by ephemeral desert rivers. Nature 366:148–150. https://doi.org/10.1038/366148a0

Leopold LB (1977) A reverence for rivers. Geology 5:429–430

Lindhe A, Rosén L, Johansson P-O, Norberg T (2020) Dynamic water balance modelling for risk assessment and decision support on MAR potential in Botswana. Water 12:721. https://doi.org/10.3390/w12030721

Lund NSV, Borup M, Madsen H, Mark O, Arnbjerg-Nielsen K, Mikkelsen PS (2019) Integrated stormwater inflow control for sewers and green structures in urban landscapes. Nat Sustain. https://doi.org/10.1038/s41893-019-0392-1

Maavara T, Chen Q, Van Meter K, Brown LE, Zhang J, Ni J, Zarfl C (2020) River dam impacts on biogeochemical cycling. Nat Rev Earth Environ 1:103–116. https://doi.org/10.1038/s43017-019-0019-0

MacDonald SM, Partners (1989) The motloutse dam feasibility and preliminary design study (Dept of Water Affairs, Government of Botswana/Sir MMacDonald and Partners, Gaborone, 4 Vols)

MacDonald M (1990) Motloutse dam feasibility/preliminary design study. 6: Annex K: Environmental impactstudies. Department of water affairs. Botswana. Unpublished report

Macklin MG, Lewin J (2018) River stresses in anthropogenic times. Progress Phys Geogr: Earth Environ. 030913331880301. https://doi.org/10.1177/0309133318803013

Matthews MW, Bernard S (2015) Eutrophication and cyanobacteria in South Africa's standing water bodies: a view from space. S Afr J Sci 111(5/6). 10.17159/ sajs.2015/20140193

Mazvimavi D, Madamombe E, Makurira H (2007) Assessment of environmental flow requirements for river basin planning in Zimbabwe. Phys Chem Earth 32:995–1006

Meigh J (1995) The impact of small farm reservoirs on urban water supplies in Botswana. Natl Resour Forum 19(1):71–83

Midgley SJE, Petrie B, Martin L, Chapman RA, Whande W, Parker R (2013) The limpopo river basin system: climate impacts and the political economy. In: Chemonics international, USAID, Technical Report, Pretoria, South Africa

Midgley G, Bond W (2015) Future of African terrestrial biodiversity and ecosystems under anthropogenic climate change. Nature Clim Change 5:823–829. https://doi.org/10.1038/nclimate2753

Mantswe Natural Resources Consultants (MNRC) (2000) Environmental impact assessment for feasibility study of a dam on the thune river, vol 1, Ministry of Mineral Resources and Water Affairs, Department of Water Affairs

Mogomotsi PK, Mogomotsi GEJ, Matlhola DM (2018) A review of formal institutions affecting water supply and access in Botswana. Phys Chem Earth (105):283–289. https://doi.org/10.1016/j.pce.2018.03.010

Morrison T (1998) Inventing the truth: the art and craft of memoir. Houghton Mifin Harcourt, USA

Mosase E, Ahiablame L, Park S, Bailey R (2019a) Modeling potential groundwater recharge in the Limpopo River Basin with SWAT-MODFLOW. Groundw Sustain Dev 100260. https://doi.org/10.1016/j.gsd.2019.100260

Mosase E, Ahiablame L, Srinivasan R (2019b) Spatial and temporal distribution of blue water in the Limpopo River Basin, Southern Africa: a case study. Ecohydrol Hydrobiol. https://doi.org/10.1016/j.ecohyd.2018.12.002

Mpala SC, Gagnon AS, Mansell MG, Hussey SW (2016) The hydrology of sand rivers in Zimbabwe and the use of remote sensing to assess their level of saturation. Phys Chem Earth Parts a/b/c 93:24–36. https://doi.org/10.1016/j.pce.2016.03.004

Mulligan M, van Soesbergen A, Sáenz L (2020) GOODD, a global dataset of more than 38,000 georeferenced dams. Sci Data 7:3. https://doi.org/10.1038/s41597-020-0362-5

Murray-Hudson M, Wolski P, Ringrose S (2006) Scenarios of the impact of local and upstream changes in climate and water use on hydro-ecology in the Okavango Delta, Botswana. J Hydrol 331:73–84. https://doi.org/10.1016/j.jhydrol.2006.04.041

Naidu CK, Mathangwane B, Sakuringwa S, Katorah G (2016) Alien invasive aquatic plant species in Botswana: historical perspective and management. Open Plant Sci J 9(1):1–40. https://doi.org/10.2174/1874294701609010001

Nash D (1992) The development and environmental significance of the dry river valley system (Mekgacha) in the Kalahari, central Southern Africa. PhD thesis. University of Sheffield. UK

Nermark UP, Mmopelwa TG (1994) Utilisation of small water bodies, Botswana: report of activities towards fisheries exploitation, 1992–93. ALCOM. Field dcument. No 29, 29 pp

O'Connor TG (2001) Effect of small catchment dams on downstream vegetation of a seasonal river in semi-arid African savanna. J Appl Ecol 38:1314–1325

O'Connor JE, Duda JJ, Grant GE (2015) 1000 dams down and counting. Science 348(6234):496–497. https://doi.org/10.1126/science.aaa9204

Paerl HW, Huisman J (2008) CLIMATE: blooms like it hot. Science 320(5872):57–58. https://doi.org/10.1126/science.1155398

Palmer M, Ruhi A (2019) Linkages between flow regime, biota, and ecosystem processes: implications for river restoration. Science 365 (6459). eaaw2087. https://doi.org/10.1126/science.aaw2087

Parida BP, Moalafhi DB, Sebokolodi K (2010) Impact of reservoirs on morphometric characteristics of streams: a case of shashe Damsin north-eastern Botswana. In: Proceedings of the 3rd IASTED African conference water resources management (Africa WRM 2010), Gaborone, Botswana

Perkins JS (2020) Take me to the river along the African drought corridor: adapting to climate change. Botswana J Agric Appl Sci 14 (1):60–71. https://doi.org/10.37106/bojaas.2020.77

Petts GE (2008) Instream flow science for sustainable river management. FLOW 2008: State of the Art—Science

Petts GE, Gurnell AM (2005) Dams and geomorphology: research progress and future directions. Geomorphology 71:27–47

Poff NL, Allan JD, Bain MB, Karr JR, Prestegaard KL, Richter BD, Stromberg JC (1997) The natural flow regime: a paradigm for river conservation and restoration. Bioscience 47:769–784

Poff NL, Olden JD (2017) Can dams be designed for sustainability? Science 358(6368):1252–1253. https://doi.org/10.1126/science.aaq1422

Poff NL, Schmidt JC (2016) How dams can go with the flow. Science 353(6304):1099–1100. https://doi.org/10.1126/science.aah4926

Pohl B, Macron C, Monerie P-A (2017) Fewer rainy days and more extreme rainfall by the end of the century in Southern Africa. Sci Rep 7:46466. https://doi.org/10.1038/srep46466

Ramírez-Hernández J, Rodríguez-Burgueno JE, Zamora-Arroyo F, Carreón-Diazconti C, Pérez-González D (2015) Mimic pulse-base flows and groundwater in a regulated river in semiarid land: Riparian restoration issues. Ecol Eng 83:239–248

Reynolds S, Marston C, Hassani H, King GCP, Bennettet MR (2016) Environmental hydro-refugia demonstrated by vegetation vigour in the Okavango Delta, Botswana. Sci Rep 6:35951. https://doi.org/10.1038/srep35951

Rockström J, Falkenmark M (2015) Increase water harvesting in Africa. Nature 519:283–285. https://doi.org/10.1038/519283a

Ruffault J, Curt T, Moron V, Trigo RM, Mouillot F, Koutsias N, Pimont F, Martin-St Paul N, Barbero R, Dupuy J-L, Russo A, Belhadj-Khedher C (2020) Increased likelihood of heat-induced large wildfires in the Mediterranean Basin. Sci Rep 10:13790

Sabater S, Bregoli F, Acuña V, Barceló D, Elosegi A, Ginebreda A, Marcé R, Muñoz I, Sabater-Liesa L, Ferreira V (2018) Effects of human-driven water stress on river ecosystems: a meta-analysis. Sci Rep 8:11462. https://doi.org/10.1038/s41598-018-29807-7

Sandi SG, Rodriguez JF, Saintilan N, Wen L, Kuczera G, Riccardi G, Saco PM (2020) Resilience to drought of dryland wetlands threatened by climate change. Sci Rep 10:13232 (2020). https://doi.org/10.1038/s41598-020-70087-x

Selier J (2010) Report on the Aerial Census of the Central Limpopo River Valley, Southern Africa. https://doi.org/10.13140/RG.2.2.32575.71844

Semenza JC (2020) Cascading risks of waterborne diseases from climate change. Nat Immunol 21:484–487. https://doi.org/10.1038/s41590-020-0631-7
Shaw P, Shick A, Hassan M (1994) Bedload sediment transport in the sand rivers of Botswana. Botswana Notes and Records, vol 26, pp 115–127. http://www.jstor.org/stable/40959187
Smakhtin V, Revenga C, Döll P (2004) A pilot global assessment of environmental water requirements and scarcity. Water Int 29:307–317
St. George S, Hefner AM, Avila J (2020) Paleofloods stage a comeback. Nat Geosci. https://doi.org/10.1038/s41561-020-00664-2
Stankiewicz J, de Wit MJ (2006) A proposed drainage evolution model for Central Africa—Did the Congo flow east? J African Earth Sci 44 (1):75–84
Strokal M, Bai Z, Franssen W, Hofstra N, Koelmans AA, Ludwig F, Ma L, van Puijenbroek P, Spanier JE, Vermeulen LC, van Vilet MTH, van Wijnen J, Kroeze C (2021) Urbanization: an increasing source of multiple pollutants to rivers in the 21st century. npj Urban Sustain 1:24. https://doi.org/10.1038/s42949-021-00026-w
Symphorian GR, Madamombe E, van der Zaag P (2003) Dam operation for environmental water releases; the case of Osborne dam, save catchment, Zimbabwe. Phys Chem Earth 28:985–993
Syvitski J, Waters CN, Day J, Milliman JD, Summerhayes C, Steffen W, Zalasiewicz J, Cearreta A, Galuszka A, Hajdas I, Head MJ, Leinfelder R, McNeill JR, Poirier C, Rose NL, Shotyk W, Wagreich M, Williams M (2020) Extraordinary human energy consumption and resultant geological impacts beginning around 1950 CE initiated the proposed Anthropocene Epoch. Commun Earth Environ 1:32. https://doi.org/10.1038/s43247-020-00029-y
Tabari H (2020) Climate change impact on flood and extreme precipitation increases with water availability. Sci Rep 10:13768. https://doi.org/10.1038/s41598-020-70816-2
Tennant DL (1976) Instream flow regimens for fish, wildlife, recreation and related environmental resources. Fisheries 1:6–10
Tonkin JD, Poff NL, Bond NR, Horne A, Merritt DM, Reynolds LV, Ruhi A, Lytle DA (2019) Prepare river ecosystems for an uncertain future. Nature 570(7761):301–303. https://doi.org/10.1038/d41586-019-01877-1
Torres A, Brandt J, Lear K, Liu J (2017) A looming tragedy of the sand commons. Science 357(6355):970–971. https://doi.org/10.1126/science.aao0503
Tortajada C, van Rensburg P (2020) Drink more recycled wastewater. Nature 577:26–28. https://doi.org/10.1038/d41586-019-03913-6
Ukkola A, Prentice I, Keenan T, van Dijk AIJM, Viney NR, Myneni RB, Bi J (2016) Reduced streamflow in water-stressed climates consistent with CO_2 effects on vegetation. Nat Clim Change 6:75–78
Urio AF, Brent AC (2006) Solid waste management strategy in Botswana: the reduction of construction waste. J Civ Eng 48(2):18–22
Vidal-Abarca MR, Gómez R, Sánchez-Montoya MM, Arce MI, Nicolás N, Suárez ML (2020) Defining dry rivers as the most extreme type of non-perennial fluvial ecosystems. Sustainability 12:7202. https://doi.org/10.3390/su12177202
WCS (2010a) Study on the determination of environmental flow requirements downstream of Letsibogo Dam. Wellfield Consulting Services Pty Ltd., Gaborone
WCS (2010b) Environmental report on the proposed extraction plan for the Tshokana River Sand Deposit. Wellfield Consulting Services Pty Ltd, Gaborone
WCS (2011) Study on the determination of environmental flow requirements downstream of Ntimbale Dam. Wellfield Consulting Services Pty Ltd, Gaborone
WRC (2010) Environmental impact assessment study and preparation of an environmental management plan for the proposed Mosetse Dam. Water Resources Consultants, Gaborone, Botswana
Wolski P, Savenije HHG, Murray-Hudson M, Gumbricht T (2006) Modelling of the flooding in the Okavango Delta, Botswana, using a hybrid reservoir-GIS model. J Hydrol 331:58–72
Woodward J, Li J, Rothwell J, Hurley R (2021) Acute riverine microplastic contamination due to avoidable releases of untreated wastewater. Nat Sustain. https://doi.org/10.1038/s41893-021-00718-2

Jeremy S. Perkins is Associate Professor in Range Ecology at the Department of Environmental Science at the University of Botswana. He undertook his Ph.D. research on Kalahari cattle posts in 1988 and returned to Botswana in 1992 and has lived and worked there ever since.

Bhagabat P. Parida holds a master's degree in Hydrology from the National University of Ireland (Galway) and a Ph.D. in Water Resources Engineering from the Indian Institute of Technology, Delhi (India). He is Professor of Water Resources and Environmental Engineering, currently with the Botswana International University of Science and Technology after serving the Departments of Environmental Science and Civil & Environmental Engineering at the University of Botswana from 1999. His main academic and research activities include modelling of hydrological extremes for prediction in ungauged catchments (PUB), water resources and environmental modelling to determine the impact of land use land cover (LULC) and climate changes on water resources.

Gorges of Eastern Botswana

18

Mark Stephens

Abstract

Gorges in the hardveld of southeastern and eastern Botswana occur within fractured siliciclastic rocks likely influenced by uplift associated with the Ovambo–Kalahari–Zimbabwe (OKZ) Axis in the end-Cretaceous to early Tertiary. Nine (Peleng East Gorge, Lobatse Rock Paintings Gorge, Athlone Gorge, Segorong Gorge, Mmalogage Gorge, Mmamotshwane Gorge, Phataletshaba Gorge, Shoshong Gorge, and Goo-Moremi Gorge) of the eleven gorges discussed in this paper originate in a bedrock high with a mean length of 1839 m. Two (Pharing Gorge and Kobokwe Gorge, mean length 1454 m) of the eleven gorges appear to originate in small upstream catchments, and all gorges presented in this paper are wider than the streams they currently hold, implying inherited forms from past humid climatic phases (pluvials), in addition to the influences of modern aquifer flow and wet season runoff.

Keywords

Gorge • Southeastern Botswana • Eastern Botswana • Bedrock high • Uplift • Pluvial

18.1 Background to Gorge Formation in Botswana

A gorge is a vertical, steep-sided narrow and deep valley cut almost invariably in hard rock. The most common cause in the formation of a gorge is back-cutting of waterfall walls by rivers, e.g. the Victoria Falls gorges on the Zambezi River, which occur in a series for ~100 km downstream of the falls (Faniran and Jeje 1983; Moore et al. 2007). In highland areas, rivers are typically characterized by vertical erosion and deposition of large boulders, whereas their valleys are narrow and deep, with steep sides, potholes, rapids, waterfalls, and gorges. In lower gradient areas rivers typically display more lateral erosion and deposition of finer material, whereas their valleys are wide with e.g. meanders (Nash et al. 1994: Fig. 2.3), and swamps and deltas in coastal areas. The morphology of fluvial systems may also vary across different climatic zones, for example Grenfell et al. (2014) studied two catchments of similar geology but with contrasting climates in South Africa and found vastly differing channel patterns in each.

Gorges are only found in the few hilly areas in Botswana such as the Otse Hills, Shoshong Hills and Tswapong Hills and that predominantly occur in the hardveld, i.e. those rocky land surfaces to the east and southeast of the country not covered by sands of the Kalahari (Fig. 18.1) (Silitshena and McLeod 1998). Gorges of the hardveld of Botswana are the source of many of the country's rivers that play a useful role in groundwater recharge in semi-arid Botswana (e.g. De Vries et al. 2000). The hilly areas of eastern and southeastern Botswana receive more rainfall than surrounding flatter areas due to the orographic effect, with rainwater percolating slowly through rock strata.

Gorges in Botswana form in hills of the hardveld that are composed of siliciclastic rock types; these are the sandstones, conglomerates, shales, and quartzites of the Transvaal (earliest age ~2,650 Ma) and Waterberg (earliest age ~2,025 Ma) Supergroups. All gorges in this chapter occur in the Waterberg Supergroup except for those that occur in Lobatse that are in the Upper Transvaal Supergroup (Beger 2001) (for more information on the geological context of Botswana see Stephens et al. this book). Hilly areas in Botswana are particularly affected by the arcuate line of the Ovambo–Kalahari–Zimbabwe (OKZ) flexure axis; a

M. Stephens (✉)
School of Chemistry, Environmental and Life Sciences, Faculty of Pure and Applied Sciences, University of the Bahamas, PO Box N-4912 Nassau, Bahamas
e-mail: mark.stephens@ub.edu.bs

M. Stephens
Department of Environmental Science, Faculty of Science, University of Botswana, P/Bag UB00704 Gaborone, Botswana

F. D. Eckardt (ed.), *Landscapes and Landforms of Botswana*, World Geomorphological Landscapes,
https://doi.org/10.1007/978-3-030-86102-5_18

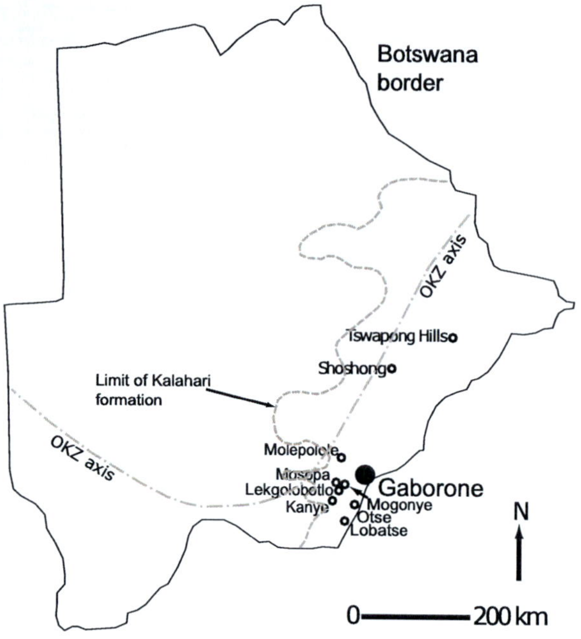

Fig. 18.1 Locations of gorges in the eastern and southeastern hardveld of Botswana. The approximate limit of Kalahari formation sand cover (sandveld) and Ovambo-Kalahari-Zimbabwe (OKZ) flexure axis are also shown

tectonically uplifted area produced in the end-Cretaceous to early Tertiary in the eastern and southeastern parts of Botswana (Moore 1999) (Fig. 18.1). Gorges in the hardveld of Botswana have thus probably been incising mainly since this time, with some form of channel initially present to focus flow to form the gorges.

The fractured nature of the rocks in the hardveld of Botswana, affected by the Limpopo Mobile Belt (e.g. Thomas and Shaw 1991), created pre-existing weaknesses in the landscape that running water collects in and flows through and thus, a potential for gorge development. Another factor important in gorge development is past humid climatic phases (pluvials) where increased fluvial activity and erosion took place (e.g. Thomas and Shaw 1991; Shaw et al. 1992). All gorges of the hardveld described in this chapter at the present day are wider than the streams they currently hold, except during times of extreme flooding that also contributes to their ongoing formation. In the Kalahari sandveld of Botswana are incised gorge-like valleys, for example around Letlhakeng (e.g. Nash et al. 1994), or the Boteti where it cuts the Gidikwe Ridge associated with Late Pleistocene flow into the Makgadikgadi basin (e.g. Ringrose et al. 1999). This is in notable contrast to the greater antiquity of incision inferred here for gorges found in the hardveld.

Other gorges in the Southern African region may be many orders of magnitude larger than those in Botswana, and with different geological histories such as the Fish River Canyon (Namibia) that flows into the Orange River (Simpson and Davies 1957; Goudie and Viles 2005), and the Orange River Gorge that occurs below the Augrabies Falls in Northern Cape Province, South Africa (Tooth 2015). Uplift of the Southern African sub-continent during the Miocene and Pliocene resulted in the incision of the lower courses of many rivers there and gorges are a common feature of South African river systems (Rowntree 2012: Fig. 5.12). To the northeast of Botswana in Zimbabwe and Zambia occurs the Victoria Falls and associated gorges that evolved during the Pleistocene (e.g. Moore and Cotterill 2010).

In addition to the hydrogeological importance of gorges in Botswana, there are associated fauna and flora (e.g. Kipping 2006; Vega-Hernandez and Caudales-Cepero 2008), archaeological resources (e.g. Huffman et al. 2016) and historical uses (e.g. Knight 2014). The gorges of Botswana are thus an important part of the rich geoheritage and biodiversity of the country and have clear implications for geotourism (e.g. Atlhopheng and Mulale 2009; Ngwira 2015; Tlhapiso and Stephens 2020), in addition to the spiritual visitations by traditional doctors and churches (e.g. Dichaba and Thebe 2016), and permanent protection by the state (Campbell 2004).

This chapter provides an overview of the known gorges in eastern and southeastern Botswana although judging by incised fluvial features in Google Earth imagery others exist in the region that have perhaps not been formally named. This paper provides the first compiled study of gorges in Botswana from the author's own descriptions and published literature and includes some gorges documented and located with GPS coordinates for the first time.

18.2 Geomorphology of Gorges in the Hardveld of Botswana

Gorges occur in the fractured hills surrounding Lobatse Town, ~65 km SSW of the capital city Gaborone, in southeastern Botswana (Fig. 18.1). Just to the south of Lobatse is Peleng East Gorge, that is ~595 m long with a ~91 m difference in elevation (~1319–1228 m), ~150 m wide, ~9 m deep, and is oriented ~SSW-NNE (Table 18.1). To the southwest of Lobatse is a gorge that is noted to exhibit rock paintings in yellow ochre of five animals (probably wildebeest) on a rock face about 20 m above the floor of the gorge (Campbell and Main 2003: Map 16). The Lobatse Rock Paintings Gorge is ~439 m long with a ~37 m difference in elevation (~1242–1205 m), ~240 m wide, ~26 m deep, and is oriented ~WNW-ESE (Table 18.1). To the west of Lobatse is Athlone Gorge, that is ~1000 m long with a ~143 m difference in elevation (~1384–1241 m), ~380 m wide, ~14 m deep, and is oriented ~WSW-ENE (Table 18.1). Athlone Gorge contains a perennial spring that feeds the gorge stream (S

Table 18.1 Location, dimensions, elevation, orientation, and origin/control of gorges in southeastern and eastern Botswana (data derived from measurements and observations in Google Earth). For gorges formed in a bedrock high, length is measured from the highest elevation point upstream within an oriented gorge section to the gorge mouth. Width and depth were both measured ~mid-section of the gorges using the elevation profile tool in Google Earth

Name	Gorge upstream	Gorge downstream	Approx. length (m)	Approx. width × depth (m)	Approx. elevation diff. (high-low) (m)	Approx. predominant orientation	Gorge origin/control
Peleng East Gorge (Lobatse)	S 25° 14.334′ E 25° 40.796′	S 25° 14.046′ E 25° 40.874′	595	150 × 9	91 (1319–1228)	SSW-NNE	Bedrock high/fracture
Rock Paintings Gorge (Lobatse)	S 25° 13.737′ E 25° 39.673′	S 25° 13.830′ E 25° 39.902′	439	240 × 26	37 (1242–1205)	WNW-ESE	Bedrock high/fracture
Athlone Gorge (Lobatse)	S 25° 12.580′ E 25° 39.838′	S 25° 12.330′ E 25° 40.349′	1000	380 × 14	143 (1384–1241)	WSW-ENE	Bedrock high/fracture
Segorong Gorge (Otse)	S 25° 1.029′ E 25° 43.009′	S 24° 59.702′ E 25° 43.817′	3097	600 × 63	233 (1365–1132)	SW-NE	Bedrock high/fracture
Pharing Gorge (Kanye)	S 24° 58.889′ E 25° 21.924′	S 24° 58.964′ E 25° 22.370′	848	266 × 16	27 (1310–1283)	WNW-ESE	Small upstream catchment/fracture
Mmalogage Gorge (Lekgolobotlo)	S 24° 54.876′ E 25° 32.865′	S 24° 54.950′ E 25° 32.942′	247	192 × 8	78 (1380–1302)	NW–SE	Bedrock high/fracture
Mmamotshwane Gorge (Mogonye)	S 24° 51.807′ E 25° 41.141′	S 24° 51.628′ E 25° 41.151′	376	205 × 14	75 (1280–1205)	SSW-NNE	Bedrock high/fracture
Phataletshaba Gorge (near Mosopa)	S 24° 46.065′ E 25° 32.258′	S 24° 45.540′ E 25° 31.580′	1520	395 × 24	93 (1243–1150)	SE-NW	Bedrock high/fracture
Kobokwe Gorge (Molepolole)	S 24° 25.374′ E 25° 30.287′	S 24° 25.943′ E 25° 31.188′	2060	713 × 39	7 (1122–1115)	NW–SE	Small upstream catchment/fracture
Shoshong Gorge	S 22° 57.734′ E 26° 32.684′	S 23° 1.548′ E 26° 31.090′	8060	970 × 90	160 (1252–1092)	NNE-SSW	Bedrock high/fracture
Goo-Moremi Gorge (Tswapong Hills)	S 22° 36.910′ E 27° 25.923′	S 22° 36.491′ E 27° 26.370′	1220	463 × 80	134 (1090–956)	SW-NE	Bedrock high/fracture
			Mean (all): 1769 **Mean (bedrock high):** 1839 **Mean (small upstream catchment):** 1454	**Mean:** 416 × 35	**Mean:** 98		

25°12′26.5″ E 25°40′05.8″) and a circular stonewall ruin (S 25° 12′23.9″ E 25°40′06.8″) occurs towards the head of the gorge. All three of the gorges in Lobatse originate within a bedrock high and are fracture controlled (Table 18.1).

Otse Village is located ~45 km SSW of Gaborone and contains Segorong Gorge that begins on the slopes of Otse Hill, the highest point in Botswana at 1491 m asl (Campbell and Main 2003: Map 14) (Fig. 18.1). Segorong Gorge has a permanent spring that emanates from the side of the gorge, just below a cave that is used for spiritual gatherings (Brook 2015, 2017). Segorong Gorge is ~3097 m long with a ~233 m difference in elevation (~1365–1132 m), ~600 m wide, ~63 m deep, and is oriented ~SW-NE (Table 18.1). The Gorge originates within a bedrock high with bedrock exposed in many places (Fig. 18.2a) is fracture controlled, and exhibits step-pools and cobble-boulder fills and a small waterfall that occurs in the rainy season (Brook 2015).

Pharing Gorge at Kanye occurs ~30 km west of Otse (Campbell and Main 2003: Map 9, and for further historical information) (Fig. 18.1) and is ~848 m long with a ~27 m difference in elevation (~1310–1283 m), ~266 m wide, ~16 m deep, and is oriented ~WNW-ESE (Fig. 18.2b). The Gorge has a small upstream catchment,

Fig. 18.2 a. Segorong Gorge; b. Pharing Gorge. (Photographs by M. Stephens)

Fig. 18.3 Satellite image (Google Earth) showing Pharing Gorge and its smaller perpendicular gorges. Another gorge is also observed just to the north and contains a road

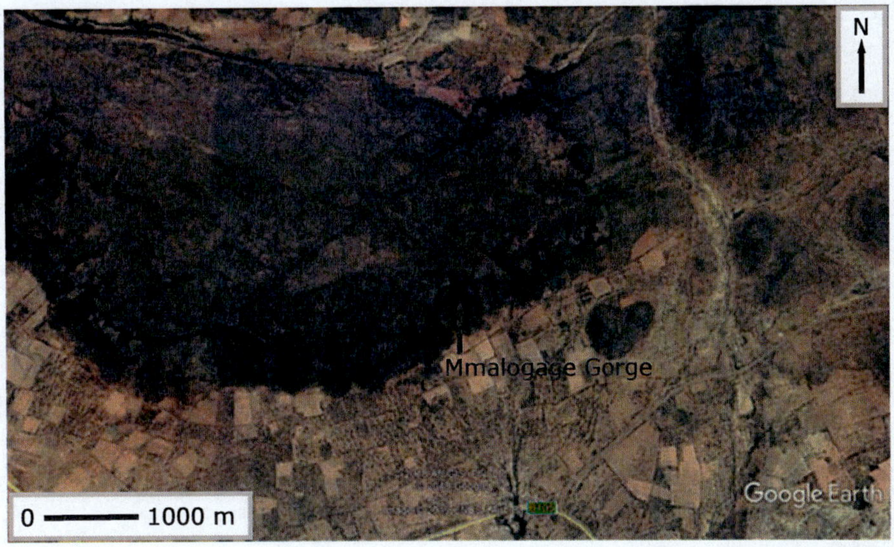

Fig. 18.4 Satellite image (Google Earth) showing the hill at Lekgolobotlo, with incipient gorges (including Mmalogage Gorge) at its periphery

is fracture controlled, and exhibits exposed, jointed bedrock and step-pools. There are also meanders associated with smaller perpendicular gorges from the steep valley sides (Fig. 18.3). Campbell and Main (2003) note that it almost always has some water running through it.

In nearby Ranaka, ~15 km northeast of Kanye, is Mmalogage Gorge which is the southernmost of several gorges found around the village of Lekgolobotlo, and contains one of the few known perennial streams that flow in a southerly direction in Botswana (Fig. 18.4). Mmalogage Gorge is ~247 m long with a ~78 m difference in elevation (~1380–1302 m), ~192 m wide, ~8 m deep, oriented ~NW–SE, and originates in a fracture-controlled bedrock high (Table 18.1). Gorges here are incipient (Fig. 18.5a) compared to those others described in the chapter and perhaps due to them forming post-uplift of the OKZ axis and/or initiated by relatively recent spring occurrence.

Mogonye Village is located ~31 km SW of Gaborone and contains seven gorges from which three rivers originate (Mmamotshwane, Metsimaswaane and Molapowamodimo) (Brook 2017) (Fig. 18.1). The main gorge that is open to the public is the Mmamotshwane Gorge, that is ~376 m long with a ~75 m difference in elevation (1280–1205 m), ~205 m wide, ~14 m deep, oriented ~SSW-NNE, and originates in a fracture controlled bedrock high (Table 18.1) (Fig. 18.5b). A stream that runs through the gorge is fed by springs and currently flows

Fig. 18.5 **a**. Incipient gorge at Mmalogage; **b**. Mogonye Gorge. (Photographs by M. Stephens)

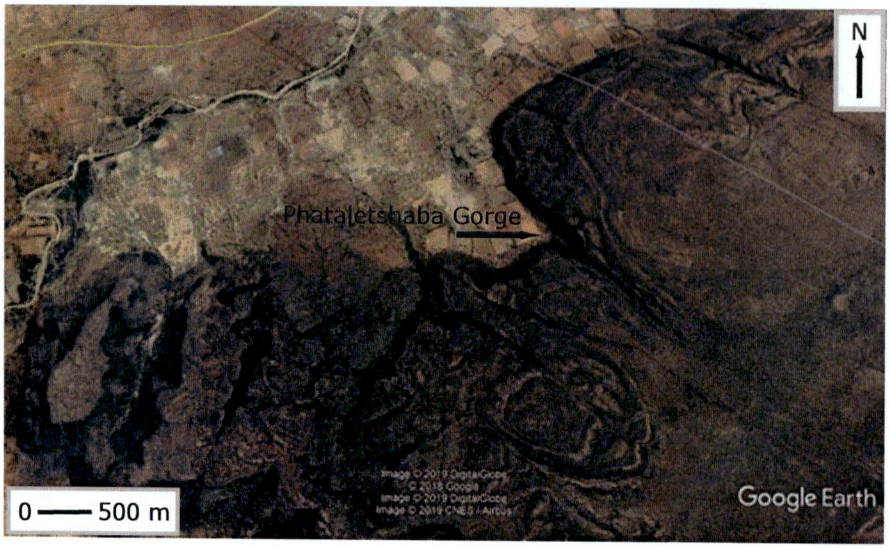

Fig. 18.6 Satellite image (Google Earth) showing Phataletshaba Gorge and smaller gorges nearby

only significantly during and for a few months after the wet season; in the winter flows can be at a minimum (Brook 2017). Small potholes are minimally present and weakly developed in the floor of the Mmamotshwane Gorge and rarely viewed elsewhere in the other gorges described in this chapter and perhaps due to the lack of any strong perennial flow. The other six gorges at Mogonye are found at Birds Spring, Marete Spring, Leopard Valley, Aardwolves Valley, Hammerskop's Valley and Tamboti Spring (Brook 2017). The local community treat the gorges as sacred places and further information on the history and archaeology of the Mogonye Gorges can be found in Brook (2015).

Approximately 6 km NNW of Mogonye, Campbell and Main (2003: Map 7) describe Fikeng ('Place of the Rock') as the name of both an area and a small river (Fikeng River) that flows north and has apparently cut a gorge through Phata ya Lefika ('Valley of the Rock') (Fig. 18.1). At Phataletshaba ('Fearful Ravine'), ~10 km ENE from nearby Mosopa are two gorges, that have cut into and descend from a plateau of the Sesitajwane Hills to a valley below (Campbell and Main 2003: Map 8) (Fig. 18.6). The source of water that has cut the gorges is a spring that can be seen in a cliff and often has water for many months of the year (Campbell and Main 2003). The largest of the two gorges is ~1520 m long with a ~93 m difference in elevation (~1243–1150 m), ~395 m wide, ~24 m deep, oriented ~SE-NW and originates in a fracture controlled bedrock high (Table 18.1).

Kobokwe Gorge (Fig. 18.7a) and an associated cave (Tlhapiso and Stephens 2020) are situated ~4 km southeast from Molepolole, itself ~50 km northwest from Gaborone (Campbell and Main 2003: Map 26) (Fig. 18.1). Kobokwe Gorge has been carved by the southeasterly flow of the Molepolole River (also locally known as Tshwaanyane stream whence it flows from the Gorge), and there is also a perennial spring located within the gorge valley floor (Tlhapiso and Stephens 2020). Kobokwe Gorge is ~2060 m long with a ~7 m difference in elevation (~1122–1115 m), ~713 m wide, ~39 m deep, oriented ~NW–SE with fracture control (Table 18.1), and along with Pharing Gorge is only the second gorge identified in this study that appears to originate from a small upstream catchment, and likely both gorges formed in combination with incision into the bedrock highs. The range and mean length (848–2060 m, 1454 m) of Pharing Gorge and Kobokwe Gorge is within the range and of a similar mean length to gorges formed in a bedrock high in this study (247–8060 m, 1839 m) (Table 18.1).

One night in 1962, a boulder fell down from the top of a cliff not far from Kobokwe Cave and as it hit the ground a huge noise associated by what felt like shaking from an earthquake was reported by villagers as far as ~10 km radius away (Tlhapiso and Stephens 2020). After falling the boulder rolled and rested near what is now the Molepolole-Thamaga road. The following day it was announced to the Bakwena tribe that Kgosi (Chief) Kgari II was late and had passed on that same night. The boulder is believed to have alerted the tribe on the passing of their paramount chief. The boulder was later carelessly destroyed during the construction of the Molepolole-Thamaga road, sometime between 1996–2001 (Tlhapiso and Stephens 2020).

Around 180 km NNE of the capital city Gaborone in eastern Botswana is Shoshong Gorge, that is ~8060 m long with a ~160 m difference in elevation (~1252–1092 m), ~970 m wide, ~90 m deep, oriented ~NNE-SSW, and originates in a fracture controlled bedrock high (Table 18.1). For historical information and images of Shoshong Gorge see (Knight 2014). The

Fig. 18.7 a. Kobokwe Gorge; b. Moremi Gorge. (Photographs by M. Stephens)

Tswapong Hills occupying an area of ~750 km^2, also in eastern Botswana, contain gorges that are noted by Brook (2015) to occur at an altitude between 850–1000 m asl. Goo-Moremi Gorge, located ~67 km east of Palapye in the Tswapong Hills is ~1220 m long with a ~134 m difference in elevation (~1090–956 m), ~463 wide, ~80 m deep, oriented ~SW-NE, and also originates in a fracture controlled bedrock high (Table 18.1). It is fed by springs that produce three permanent waterfalls (Fig. 18.7b), with the uppermost being 10 m high (Brook 2015, 2017).

As at Kobokwe Gorge, the occurrence of a boulder fall in Goo-Moremi Gorge has been associated with the passing of an important person. At 2 am on 13th July 1980 a loud rumbling was reported by villagers of Goo-Moremi Village as a giant boulder fell into the lower part of Goo-Moremi Gorge and on that same morning Sir Seretse Khama, the first President of Botswana, had passed on (Brook 2015, 2017).

Nearby to Goo-Moremi Gorge is the smaller Malaka Gorge which also contains a waterfall and perennial waters fed by a spring (Brook 2017). Motetane Gorge, also occurring in Malaka, includes a stream and waterfall; the stream apparently flows along a weakened fault structure there (Brook 2017). Nearby to Motetane Gorge is Borotelatshwene Gorge (translated from Setswana means 'where baboons urinate') (Brook 2015, 2017). Near Malaka at Old Palapye also occurs the Phothophotho Gorge and Bakwena Gorge that also have perennial waters (Brook 2015). See Brook (2015) for further information and images on the Tswapong Hills Gorges and Huffman et al. (2016) for archaeological resources there.

18.3 Summary

Gorges occur in many of the hilly parts of the hardveld of southeastern and eastern Botswana, with dimensions ranging from ~247 m long, ~192 m wide and ~8 m deep for the incipient Mmalogage Gorge to ~8060 m long, ~970 m wide and ~90 m deep for Shoshong Gorge. They occur within fractured/faulted siliciclastic rocks and their development is speculated here to be due to a combination of uplift associated with the Ovambo–Kalahari–Zimbabwe Axis in the end-Cretaceous to early Tertiary, past humid climatic phases, modern aquifer flow and wet season runoff. Two gorges appear to originate in small upstream catchments, and all gorges presented in this paper are wider than the streams they currently hold, implying inherited forms from past wetter conditions.

Acknowledgements S. M. Isaacs (University of Botswana, UB) is thanked along with B. Fologang and O. Sanka for guidance in the Segorong Gorge. M. Pheto and T. Setlalekgomo (UB) are thanked for guidance in Mmalogage Gorge, Pharing Gorge and the gorges in Lobatse. Mr. Kgosidintsi, M. Tlhapiso and P. Gaboutloeloe (all UB) are thanked for their help at Kobokwe Gorge. Tour guides at Moremi Gorge and Mmamotshwane Gorge are also thanked for their invaluable assistance and knowledge. The University of Botswana is thanked for providing a vehicle for fieldwork. Gratitude is also shown to the reviewer and to Prof. Piotr Migoń and Prof. Eckardt for useful comments that helped improve the manuscript.

References

Atlhopheng J, Mulale K (2009) Natural resource-based tourism and wildlife policies in Botswana. In: Saarinen J, Becker FO, Manwa H, Wilson D (eds) Sustainable tourism in Southern Africa: local communities and natural resources in transition. Aspects of tourism book, vol 39. Channel View, Bristol, 15, pp 134–149

Beger K (2001) Environmental hydrogeology of Lobatse, Southeast District, Republic of Botswana. Unpublished MSc thesis, Bundesanstalt fur Geowissenschaften und Rohstoffe (BGR), Hannover, Germany

Brook M (2015) Wild about Botswana. Michael C. Brook, Gaborone

Brook M (2017) Botswana monuments, heritage sites & museums. Michael C. Brook, Gaborone

Campbell A (2004) Establishment of Botswana's national park and game reserve system. Botsw Notes Rec 36:55–66

Campbell A, Main M (2003) Guide to Greater Gaborone: a historical guide to the region around Gaborone including Kanye, Lobatse, Mochudi and Molepolole. A. Campbell and M. Main in association with The Botswana Society

De Vries JJ, Selaolo ET, Beekman HE (2000) Groundwater recharge in the Kalahari, with reference to paleo-hydrologic conditions. J Hydrol 238(1–2):110–123

Dichaba T, Thebe PC (2016) In search of identity and ownership of the MmaMotshwane Gorge: heritage tourism and management of Mogonye cultural landscape. Botsw Notes Rec 47:56–68

Faniran A, Jeje LK (1983) Humid tropical geomorphology: a study of the processes and landforms in warm humid climates. Longman, London

Goudie A, Viles H (2015) The rivers. In: Goudie A, Viles H (eds) Landscapes and landforms of Namibia. Springer, Berlin

Grenfell SE, Grenfell MC, Rowntree KM, Ellery WN (2014) Fluvial connectivity and climate: a comparison of channel pattern and process in two climatically contrasting fluvial sedimentary systems in South Africa. Geomorphology 205:142–154

Huffman TN, Thebe PC, Watkeys MK, Tarduno JA (2016) Ancient metallurgy in the Tswapong Hills, Botswana: a preliminary report on archaeological context. South Afr Humanit 28:119–133

Kipping J (2006) The Odonata of Botswana – an annotated checklist. Cimbebasia Mem 5

Knight J (2014) Shoshong - a short history. Kwangu, Gaborone

Moore AE (1999) A reappraisal of epeirogenic flexure axes in Southern Africa. S Afr J Geol 102:363–376

Moore A, Cotterill F (2010) Victoria Falls: Mosi-oa-Tunya – the smoke that thunders. In: Migoń P (ed) Geomorphological landscapes of the world, Chapter 15. Springer, Berlin, pp 143–153

Moore AE, Cotterill FPD, Main MPL, Williams HB (2007) The Zambezi River. In: Gupta A (ed) Large rivers: geomorphology and management, Chapter 15. Wiley, Ltd, Hoboken

Nash D, Thomas D, Shaw PA (1994) Timescales, environmental change and dryland valley development. In: Millington AC, Pye K (eds) Environmental change in drylands: biogeographical and geomorphological perspectives. British geomorphological research group symposia series. Wiley-Blackwell, Hoboken

Ngwira PM (2015) Geotourism and geoparks: Africa's current prospects for sustainable rural development and poverty alleviation. In: Errami E, Brocx M, Semeniuk V (eds) From geoheritage to geoparks: case studies from Africa and beyond. Springer, Berlin, pp 25–33

Ringrose S, Downey B, Genecke D, Sefe F, Vink B (1999) Nature of sedimentary deposits in the western Makgadikgadi basin, Botswana. J Arid Environ 43:375–397

Rowntree KM (2012) Fluvial geomorphology. In: Holmes P, Meadows M (eds) Southern African geomorphology: recent trends and new directions. SUN MeDIA Bloemfontein, Bloemfontein

Shaw PA, Thomas DSG, Nash DJ (1992) Late Quaternary fluvial activity in the dry valleys (mekgacha) of the middle and Southern Kalahari, Southern Africa. J Quat Sci 7(4):273–281

Silitshena RMK, McLeod G (1998) Botswana: a physical, social and economic geography. Longman Botswana, Gaborone

Simpson ESW, Davies DH (1957) Observations on the Fish River Canyon in South West Africa. Trans R Soc S Afr 35(2):97–107

Thomas DSG, Shaw PA (1991) The Kalahari environment. Cambridge University Press, Cambridge

Tlhapiso M, Stephens M (2020) Application of the karst disturbance index (KDI) to Kobokwe Cave and Gorge, SE Botswana: implications for management of a nationally important geoheritage site. Geoheritage 12(39). https://doi.org/10.1007/s12371-020-00461-8

Tooth S (2015) The Augrabies Falls region: a fluvial landscape divided in flow but magnificent in spectacle. In: Grab S, Knight J (eds) Landscapes and landforms of South Africa. Springer, Berlin, pp 65–73

Vega-Hernandez E, Caudales-Cepero R (2008) An approach to the study of the flora and floristic of the Kanye Plateau, Southern District, Botswana. Botsw Notes Rec 40:100–112

Mark Stephens is Associate Professor of Geography at the University of The Bahamas and was previously Senior Lecturer in Geomorphology at the University of Botswana. He holds a M.Sc. in Quaternary Science, and Ph.D. in Geography (both University of London). He mainly investigated landforms of the hardveld in Botswana, including the geodiversity and geoconservation of caves and gorges. His research interests include tropical environmental change, natural hazards, karst geomorphology, geoarchaeology, and geoheritage.

19 Soil Development in the Eastern Hardveld

Peter N. Eze

Abstract

Variations in the nature and properties of soils (pedodiversity) on the Late Neogene hilly dryland and Quaternary erosional and depositional surfaces of the Hardveld, spanning the south-eastern to south-central parts of Botswana are well established. Following the updated IUSS Working Group World Reference Base (WRB) soil classification system, this chapter discusses the major soils found on the eastern Hardveld of Botswana. Ten Reference Soil Groups develop on the Hardveld: Regosols, Arenosols, Luvisols, Lixisols, Cambisols, Calcisols, Vertisols, Leptosols, Planosols, and Acrisols. Dominant pedogenic processes in the area include salinization, calcification and decalcification, illuviation, eluviation and erosion. Interbedded fossil soils (palaeosols) within alluvial deposits found in the Hardveld are indicators of environmental and climate fluctuations in the region. The complex interplay of parent material and climate as key active factors, and organisms, topography and time as passive factors, through material fluxes such as addition, removal, transfer and transformation are fundamentally responsible for the formation of different soils found on the Hardveld.

Keywords

Botswana • WRB soil classification • Soilscape • Pedogenesis • Palaeosols • Semi-arid climate

19.1 Introduction

The term *veld*, also written as *veldt*, which literally means "field" or "landscape" in southern Africa, has Afrikaans language origins. True to its name, the Hardveld (Bekker and De Wit 1991) refers to the south-eastern to east-central areas of Botswana (Fig. 19.1) characterized by rocky soils, savannah vegetation, and undulating plains with scattered rocky hill ranges (Fig. 19.2). Other geo-ecological landscapes found in Botswana include the aeolian Sandveld, which covers most of the southwestern part of the country and the Waterveld which is the smallest by size and found in Chobe and Okavango in the northeastern region (Fig. 19.1). This chapter, however, will focus on soil-landscape relationships in the eastern Hardveld. Soil as a non-renewable natural resource is increasingly being accorded its rightful place in national and international political schemata. Accurate mapping and inventorying of soil resources is fundamental to its sustainable use and management. The most comprehensive soil survey and mapping of Botswana remain the low resolution 1:100,000 map (FAO 1992). However, recent scientific advances have led to multiple revisions of the classification frameworks, descriptive terms and nomenclature used in classifying and mapping the soils (USDA 2010).

The eastern Hardveld is the hub of economic activities of Botswana, including mining and agriculture, and is home to about 80% of the nation's human population. The aim of this chapter is to provide a summary of the major soil resources of the eastern Hardveld in accordance with the IUSS Working Group on World Reference Base for Soil Resources and to show their relationships with landforms and landscapes.

P. N. Eze (✉)
Department of Earth and Environmental Science, Botswana International University of Science & Technology, Private Bag 16, Palapye, Botswana
e-mail: ezep@biust.ac.bw

F. D. Eckardt (ed.), *Landscapes and Landforms of Botswana*, World Geomorphological Landscapes,
https://doi.org/10.1007/978-3-030-86102-5_19

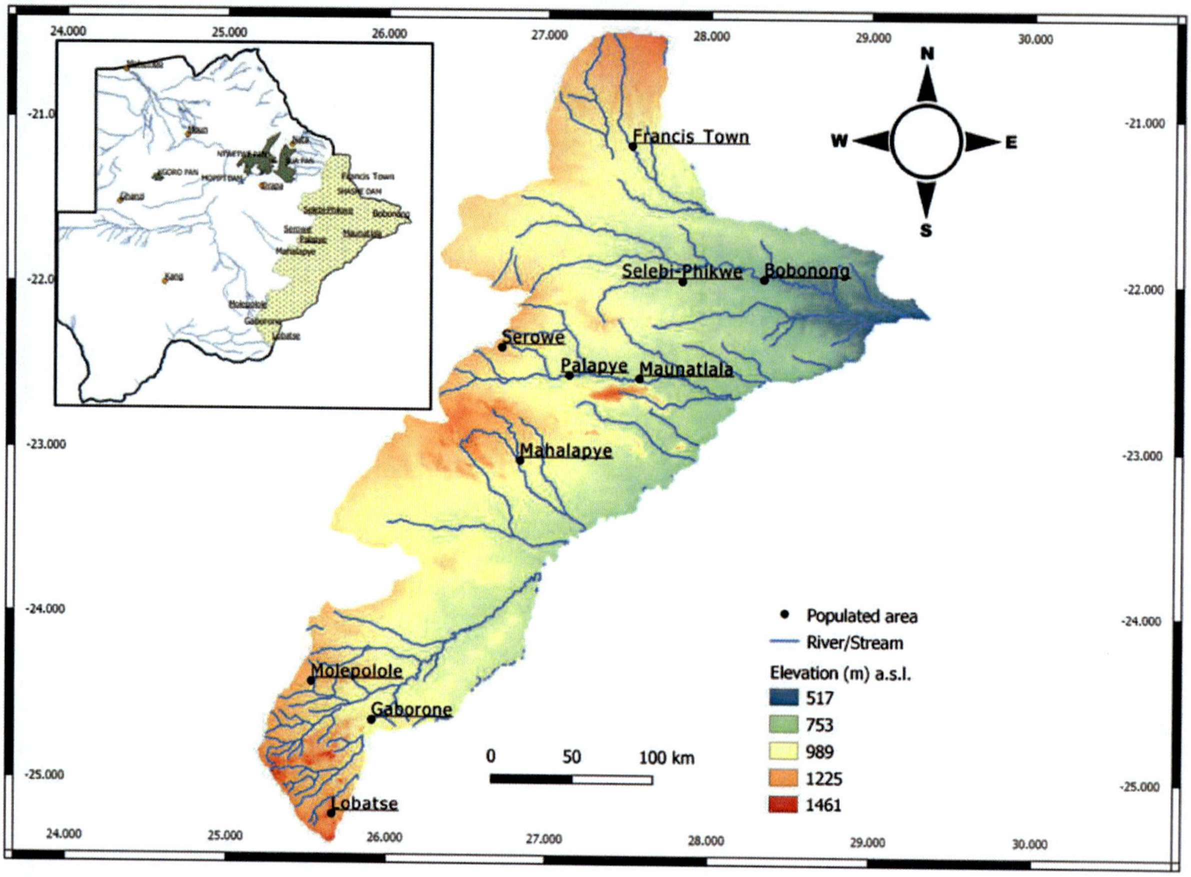

Fig. 19.1 Geographical map of the eastern Hardveld showing elevation and internal drainage pathways

19.2 Geographical/Environmental Settings

Located between latitudes 20° and 26° S and longitudes 24° and 30° E (Fig. 19.1), the eastern Hardveld presents a beautiful landscape characterized by a wide range of topography. Lobatse Hill in the southeast, with an elevation of about 1400 m.a.s.l., is the highest point, while the confluence of Shashe and Limpopo rivers in the eastern end of Botswana with an elevation of about 517 m.a.s.l. is the lowest point on the Hardveld (Fig. 19.1). The present-day configuration of the Hardveld landscape reflects a balance between erosion spanning Paleogene-Quaternary and depositional processes, which resulted in dots of hills (Fig. 19.2). A few shallow valleys indicating extinct drainage channels and a more humid palaeo-fluviatile environment are present in the Hardveld (De Wit and Bekker 1990; Bekker and De Wit 1991). With the exception of the Marico, Limpopo and Notwani rivers, none of the rivers in the Hardveld are perennial for more than a few days at a time. Being part of the Sashi-Limpopo catchment, most of the water from the Hardveld flows out of Botswana (See Chap. 16). The vegetation cover of the Hardveld can generally be described as woodland or tree savannah (Bekker and De Wit 1991) and is often strongly influenced by the nature of soil parent material, soil depth and physiography (Sebego et al. 2008).

19.3 Development and Distribution of Major Reference Soils Groups

19.3.1 Classification Principles

The mapping and classification of soils in Botswana follows the FAO-UNESCO system (FAO 1974) because of its simplicity, international recognition and the ability to complement the much more comprehensive USDA *Soil Taxonomy* (Soil Survey Staff 1975). The FAO-UNESCO soil classification system has evolved into the current, globally adopted, World Reference Base for Soil Resources (IUSS Working Group WRB 2015), made possible by the Working Group of the International Union of Soil Science (IUSS). The soils are classified using soil properties defined in terms of diagnostic horizons, properties and materials that are

Fig. 19.2 The Tswapong Hill ridge, Waterberg Supergroup, composed of sandstone, ironstone and quartzite, which imparts the characteristic rich hues to the colluvial soils (Maunatlala village)

measurable in the field (IUSS Working Group WRB 2015). In the WRB system, climate parameters are not applied to soil classification. However, for soil interpretation purposes, climate data and soil properties should be used, but should not be part of soil definition. In consequence, global or local climate change cannot render the name of any soil obsolete. Structurally, the WRB consists of two levels of categorical detail including the First Level having 32 Reference Soil Groups (RSGs); and the Second Level, consisting of the name of the RSG combined with a set of principal and supplementary qualifiers. With the exception of cases where special parent materials are of superseding importance, characteristic soil properties produced by pedogenic processes are used to separate classes at the first level (RSGs). Finer classification at the second level (RSGs with qualifiers) is done by grouping soils according to properties arising from any secondary pedogenetic process that has considerably affected the primary features. Also, soil properties that impart significantly on land use are often taken into account.

The dominant RSG on the Hardveld and transition zones are Regosols, Arenosols, Luvisols, Lixisols, Cambisols, Calsisols, Vertisols (IUSS Working Group WRB 2015). Patches of Leptosols, Planosols and Acrisols also develop on the Hardveld (Fig. 19.3), but are not covered in this chapter. Fey's 'Soils of South Africa' as a generally applicable reference material that can provide more detail at the 'popular science' level is recommended.

19.3.2 Regosols

The "Regosol" Soil Reference Group is a generic name for all the soils having little to no profile differentiation. It originates from the Greek word *rhegos*, meaning blanket. Technically, Regosols is a taxonomic class for weakly developed soils on unconsolidated material with only ochric horizon (having a mineral surface horizon, ≥ 10 cm thick, with ≥ 0.2 and <0.6% soil organic carbon (weighted average) (IUSS Working Group WRB 2015). Regosols do not have any diagnostic horizons other than ochric horizon over almost unaltered parent material (Fig. 19.4).

Regosols cover about 16,000 km^2 of the eastern Hardveld and are found in places like Bobonong, Francistown, Lobatse and Ramotswa (Fig. 19.3). They have low water-holding capacity and their high permeability to infiltrating water makes them sensitive to drought (Mogotsi et al. 2013). The formation of Regosols in the eastern Hardveld is facilitated by the semi-arid climate which slows the rate of pedogenesis. These soils could also form over short periods through the continuous addition of materials of aeolian

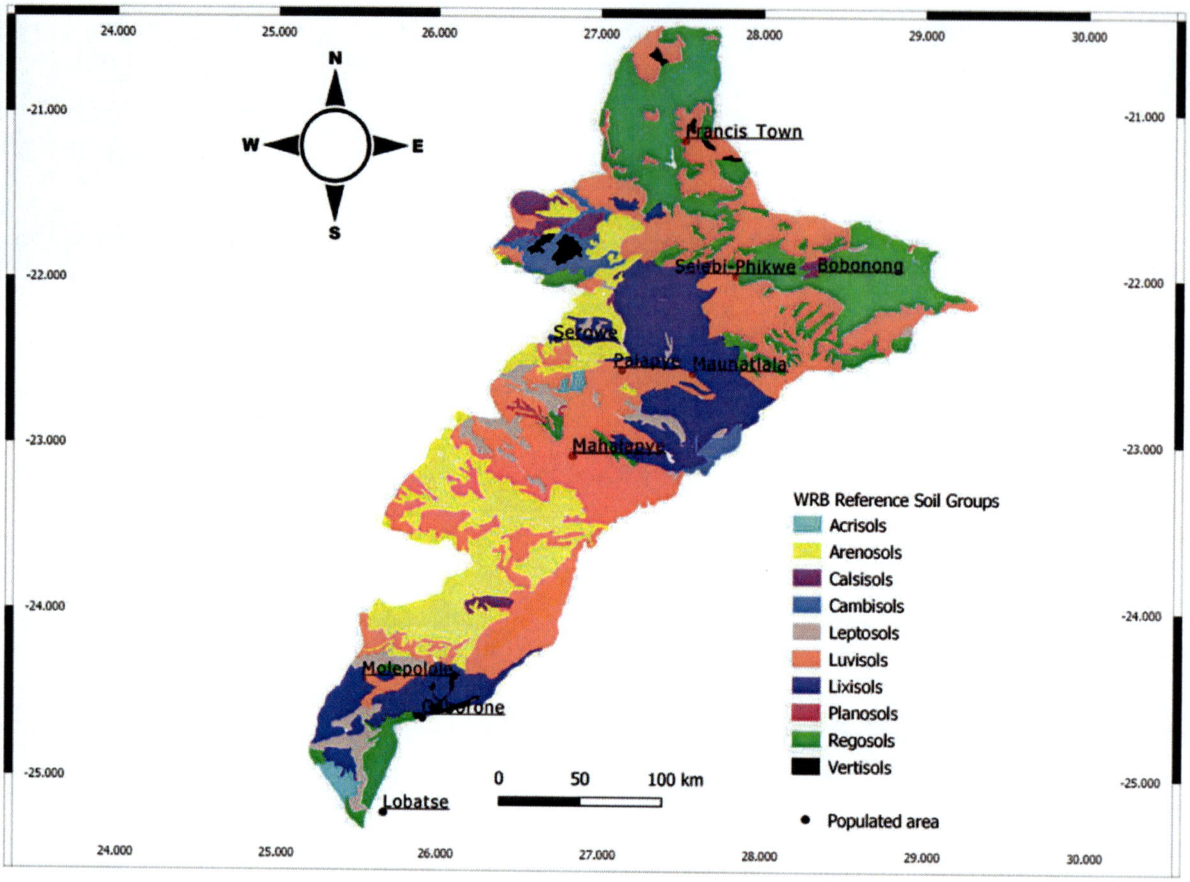

Fig. 19.3 A generalized soil map of the eastern Hardveld (Redrawn from FAO soil map of the Republic of Botswana 1990)

origin. Since evaporation exceeds precipitation in the Hardveld, it is possible to find some calcium carbonate and/or gypsum accumulates at shallow depths in Regosols, but not to the extent of forming calcic or gypsic horizons. Soils in the recent mine waste deposits (Fig. 19.5), landfills, urban waste and dredging that are still too young for pedogenesis to occur, are all included in this Reference Group (IUSS Working Group WRB 2015). Eutric Regosols have effective base saturation, exchangeable bases and exchangeable aluminium of $\geq 50\%$ are the most common Regosols on the Hardveld. In Kanye, the soils derived from the Kanye felsic volcanics have loamy sand and sandy loam texture and are generally shallow with much exposed rock, being good examples of Eutric Regosols (Abel and Stocking 1987; Fig. 19.6).

19.3.3 Arenosols

Derived from the Latin word *arena*, Arenosols means "sand". Approximately, two-thirds of the soils in Botswana belong to the Reference Soil Group Arenosols, with the vast majority occurring in the Kalahari Sandveld in the southeast of Botswana. In most scenarios, Arenosols in dryland environments have minimal soil profile development as opposed to those from humid tropical climates, because the prolonged period of seasonal drought slows pedogenesis and/or because the parent material is young due to high rates of aeolian deposition on the land surface. Coarse- to fine-grained, sometimes calcareous, sedimentary parent rocks are the leading sediment provenance for the formation of Arenosols in the eastern Hardveld and the transitional zones between the Hardveld and Sandveld ecological units. Characterized by predominantly sandy soil texture (more than 70% sand and less than 15% clay), high permeability, low water retention capacity, slow chemical weathering rates, low specific heat and often minimal nutrient contents, Arenosols are unfavourable for sustainable agriculture (van Wambeke 1992).

It is estimated that Arenosols make up about 23.1% of the total arable land in Botswana (Hartemink and Huting 2008), supporting open to sparse grassland and the scattered presence of shrubs (Fig. 19.7). Consistent with the IUSS Working Group WRB (2015) classification, different

Fig. 19.4 Debris flow within a Leptic Regosol near Gaborone with ochric epipedon (surface horizon). (Photo: F. Eckardt)

Fig. 19.5 Protic Regosol (Geoabruptic) developed on Monarch gold mine tailing dam in Francistown. Depth ~100 cm

qualifiers and specifiers can be used for the Second Level classification of various Arenosols of the eastern Hardveld. Rhodic Arenosols, having a subsurface horizon ≥ 30 cm thick within ≤ 150 cm of the soil surface, with a Munsell colour hue redder than 5 YR moist, a value of ≤ 3 moist and a value dry, no more than one unit higher than the moist value (Fig. 19.8), are one of the most prominent.

Arenosols are the third largest reference soil group on the Hardveld, covering about 14,000 km^2 northwards of Molepolole (Fig. 19.3) and parts of Palapye (Kebonye et al. 2017). They have regional/sub-continental significance as they form part of the "Kalahari Sands" that extend from the central African plateau of Congo to South Africa. Parts of Arenosols developed in Botswana are synonymous with the Kalahari basin which is known to hold the most extensive body of sands on earth (Ashkenazy et al. 2012).

19.3.4 Luvisols

Soils of the Luvisolic (from the Latin word *Luere,* meaning to wash) Reference Group generally have a prominent illuvial accumulation of clays in the sub-horizons. The dominant pedogenic process is the translocation of clay materials from the A horizon (top soil) to the Bt "argic" horizon (zone of accumulation) by water. This process occurs in three important stages: mobilization of clay-sized particles from the surface soil (A horizon), transportation of clay to the zone of accumulation (Bt horizon) and immobilization of the transported clays. Iron oxides produced during advanced chemical weathering generally impart brownish to reddish colour to soils. However, in the case of Luvisols, the clayey subsoils do not show the iron oxide colouration which indicates that the clay accumulation is not a product of chemical weathering but of mechanical migration from the top to the subsoils. Under field examination of Luvisols peds, the shiny films of oriented clays are observable on the sub-angular to the angular blocky structure of the underlying "Bt" horizons (Fig. 19.9). These clay skin's pore coating or cutans are a manifestation of the dominant soil-forming process which is the downward translocation of mobile clay-sized mineral particles. This process is responsible for the marked gradual differences in texture as seen in Luvisol profiles.

Luvisols are the most abundant soils on the eastern Hardveld (FAO 1987; Nkambwe and Totolo 2005), covering

Fig. 19.6 Rocky slope with Eutric Regosol developed on Kanye felsic lavas

about 29,000 km^2 (Fig. 19.3). Due to the presence of high cation exchange capacity clays, most sub-units of Luvisols are fertile and suitable for crop production. The major limitation of using Luvisols of the eastern Hardveld is their poor physical properties which make water retention very difficult, thereby denying plants of soil moisture storage needed to buffer them against seasonal drought (Ringrose and Matheson 1991; Akinyemi et al. 2019). In the eastern Hardveld, Luvisols are found in different physiographic positions in the undulating landscape, from bottomland flats to hill crests in Palapye, Selebe-Pikwe and Mahalapye (Fig. 19.10). Haplic Luvisols spread across the Hardveld and are described as Luvisols with a typical expression of certain features (typical in the sense that there is no further or meaningful characterisation) and only used if none of the preceding qualifiers in soil classification applies (Fig. 19.9).

19.3.5 Lixisols

The Fourth Draft of the Soil Map of the World legend in 1987, introduced the term Lixisols as a Soil Reference Group. The term originates from the Latin word *lixivia*, which means washed-out substances. Soils in this Reference Group are strongly weathered and have an eluvial horizon from which clay has been washed out and re-deposited in an illuvial argic horizon with a dominance of low activity clays and moderate to high percentage base (Ca^{2+}, Mg^{2+}, Na^+ and K^+) saturation. Lixisols are known to have an argic horizon starting ≤ 100 cm from the soil surface and a cation exchange capacity by 1 M NH_4OAc, pH 7 of less than 24 Cmol kg^{-1} clay, in some parts of the argic horizon within ≤ 50 cm below its upper limit (IUSS Working Group WRB 2015). Lixisols basically differ from Luvisols in two properties: CEC and base saturation. The evolution of many Lixisols is believed to have begun at times when the climate was wetter than the present. Strong chemical weathering during wetter periods was preceded by chemical enrichment brought about by evaporation being in excess of precipitation.

In the eastern Hardveld context, the presence of iron-rich soil horizons (plinthite, petroplinthite and pisoplinthite) in Lixisols (Fig. 19.11) points to wetter climatic conditions during formation than those that occur in the region today. The deposition of base-rich aeolian materials and bioturbation activities of soil organisms are the most likely sources of Lixisols enrichment.

Fig. 19.7 Haplic Arenosol soilscape in Ntlhantle, ~60 km southwest of Gaborone

Lixisols on the eastern Hardveld are associated with Pleistocene and older land surfaces and occur sporadically across old landscapes (Yamaguchi et al. 2007). It is common to find Lixisols occurring in association with Luvisols and Acrisols. It is estimated that Lixisols cover about 13,000 km^2 of the Hardveld, thereby making it the fourth most abundant RSG. The highly weathered soils in Maunatlala and the environs (Fig. 19.2) are home to the largest expanse of Lixisols on the Hardveld. Differentiating Lixisols from argic (clay-rich) soils (Luvisols) in the field could sometimes be challenging since the only basis for their differentiation is analytical properties and not physical appearance.

19.3.6 Cambisols

This originated from the Latin word *Cambiare*, which means to change, thus the nomenclature for soils with early signs of horizon differentiation through changes in colour, structure and/or texture. The Soil Reference Group Cambisols show visible evidence of initial horizon differentiation, but are generally weakly developed. The diagnostic subsurface horizon is Cambic, showing evidence of alteration relative to the underlying parent material with respect to leaching of carbonates, aggregate structure, formation of silicate clays and sesqui(hydr)oxides resulting from chemical weathering of primary minerals and other properties that reflect evidence of soil evolution processes. They have characteristic subsurface horizons with sandy loam or finer texture with at least 8% clay content by mass and a thickness of 25 cm or more (IUSS Working Group WRB 2015). Cambisols are sought after for dryland and irrigated agriculture because of their good aggregate structure and high content of weatherable primary minerals.

The general sequence of formation of Cambisols following the main soil-forming processes are: (a) dissolution and removal of carbonates, (b) alteration of especially easy-weatherable labile minerals such as mica and feldspar, (c) formation of silicate clay and precipitation of iron(hydr) oxides, (d) aggregation to form soil structure (Spaargaren

Fig. 19.8 Author examining a Rhodic Arenosol profile in Palapye

Fig. 19.9 Haplic Luvisol (Fluvic), more than 200 cm deep, developed on the bank of Lotsane River, Palapye

2006). In the eastern Hardveld, Cambisols develop across varying physiographic positions in the landscape, on medium to fine-textured parent materials in Serowe (Mweso 2003). Patches of Cambisols are also present in Palapye and Martin Drift (about 50 km west of Maunatlala) (Fig. 19.3). Although they may look similar, Cambisols differ from Leptosols and Regosols by possessing a greater depth and a finer texture. On landscapes, Cambisols develop in close association with Luvisols. An example of Cambisol in the eastern Hardveld includes Thionic (Typical Cambisols starting ≤ 100 cm from the soil surface, Fig. 19.12).

19.3.7 Calcisols (Aridisols containing soluble secondary carbonates (Calcids), *Soil Taxonomy*)

The term Calcisol originates from the Latin word *calcarius*, meaning calcareous. By definition, Calcisols are a group of soils having a petrocalcic horizon starting at less than or 100 cm from the soil surface; or soils with a calcic horizon starting ≤ 100 cm from the soil surface; and no argic horizon above the calcic horizon, unless the argic horizon is permeated throughout with secondary carbonate (IUSS Working Group WRB 2015). Generally, soils with a substantial accumulation of secondary calcium carbonates make up the Calcisols Reference Group (Fig. 19.13). The evolution of Calcisols in the eastern Hardveld could be attributed to the influence of calcareous parent material and semi-arid climate. The natural vegetation of Calcisols is generally sparse with an abundance of xerophytes and grasses at Tshimoyapula (5 km east of Serowe) and Mosomane (about 80 km northeast of Gaborone) (Fig. 19.3). In Calcisols with shallow to exposed petrocalcic sub-surface horizons (Fig. 19.14), agricultural land use is highly limited as root penetration and ease of water percolation is severely impaired under such conditions.

Many Calcisols are described as polygenetic, which means that they are old soils (by age, not advanced weathering) formed in different geologic eras with varying climatic conditions (Wright 1990). The main pedogenic process responsible for Calcisols formation is the precipitation of secondary carbonates from soil solution when it evaporates. Two factors, including the partial CO_2 pressure of the soil air

Fig. 19.10 Haplic Luvisol soilscape under tillage in Palapye

(this is maximum in the A horizon due to plant root activities and soil microbial respiration) and the concentration of dissolved ions in the soil moisture, are the principal factors responsible for the dissolution of carbonates and subsequent accumulation of calcic or petrocalcic horizon in the soil environment. A commonly occurring Calcisols in the eastern Hardveld include Luvic Calcisols (Fig. 19.13), having an argic horizon starting ≤ 100 cm from the soil surface. Soils found in close association with Calcisols include Leptosols (on high slope positions) and Vertisols (in bottomland positions).

19.3.8 Vertisols

Originating from the Latin word *Vertere*, meaning to turn, Vertisols are described as churning, heavy clay soils with a high amount of expansive (shrink-swell) phyllosilicate clay minerals. Vertisols are considered the most interesting soils of the world because of their unique morphology (Pal 2017). Among the spectacular properties of Vertisols are: 30% or more clay content in the upper 100 cm of the soil profile; wedge-shaped parallelepipeds or parallel-piped structural aggregates with the longitudinal axis tilted between 10° and 60° from the horizontal; intersecting slickensides; and cracks, which open and close periodically (Driessen et al. 2001). Although Vertisols occupy less than 1% of soils on the eastern Hardveld, they are regarded as an important soil Reference Group due to the inherent agricultural management problems caused by the abundance of 2:1 clay minerals in the fine fraction, poor subsoil porosity and aeration. The cracking of houses and the Stadium in Serowe, east-central Hardveld are examples of negative, socio-economic implications. Vertisols present significant geotechnical and

Fig. 19.11 Endocalcaric Lixisol (Aric), ~150 cm deep, developed on undulating terrain in Serowe

Fig. 19.12 Thionic Cambisol (Arenic, Ferric) formed on sandstone on the shoulder of Lecheng-Boseja catena. Depth is >150 cm

structural engineering challenges all over the world, with associated amelioration costs estimated to run into several billions annually (Jones and Jefferson 2012).

Environmental factors which favour the formation of the smectite-rich parent material of Vertisols are also responsible for the evolution of vertic soil properties. These factors are the presence of sufficient rainfall to trigger chemical weathering, but not so high as to cause leaching of basic cations and silicon, and a seasonal dry period long enough to enable crystallization of the first secondary clay minerals, smectite, that form upon sediment or rock weathering above neutral soil pH. Very high evapotranspiration relative to precipitation helps to keep the soil pH high and high temperatures on sedimentary lowlands, denudation plains and erosive uplands with limestone, claystone, marls or shale bedrock encourage Vertisols formation (FAO 2001; Fig. 19.15). The formation of Vertisols in the eastern Hardveld could be largely explained using self-mulching (see Buol et al. 2003) and shear failure models of Wilding and Tessier 1988 (see Coulombe et al. 1996). A vast proportion of Vertisols in Botswana occur on the Pandamanteng Plains of the Sandveld with localized occurrences in the

Fig. 19.13 Luvic Calcisol, ~1 m deep, developed on a flat terrain near Letlhakane

southern Hardveld (FAO 1987; Nkambwe and Totolo 2005) and the east-central Hardveld (Aynekulu 2003). The Gaborone dam arena, Tlokweng (about 30 km east of Gaborone), and Serowe are some of the few places where Vertisols developed on the Hardveld (Fig. 19.3). Pellic Vertisols (Fig. 19.16), having in the upper 30 cm of the soil a Munsell colour value of ≤ 3 and a Chroma of ≤ 2, both in a moist state, are dominant on the Hardveld. Vertisols are the most homogenous of all soil reference groups and as such do not often have many sub-units.

19.4 Palaeosols on the Eastern Hardveld

Palaeosols (fossil soils often found in pedo-sedimentary sequences) are found in alluvial floodplains and eroded landscapes in the eastern Hardveld. Unlike the present-day soils, the study of palaeosols (palaeopedology) requires unravelling the effects of concurrent pedogenic and sedimentary processes under changing environmental and climate conditions. As a consequence, the discontinuity of soil evolution at geological time scales and the palimpsest-like memory of palaeosols are the two basic operational principles of palaeopedology (Fedoroff et al. 2018). Over the past 50 years, the use of palaeosols for palaeoecological reconstruction has gained popularity amongst Earth scientists (Eze 2013; Beverly et al. 2018). Carbonate palaeosols are the most common type of palaeosols in the eastern Hardveld. Calcisols (Figs. 19.13 and 19.14), being polygenetic in nature, are good examples of palaeo-pedogenic products. They are designated as *Bk* if the accumulation of visible pedogenic calcium carbonate is less than 50% (Fig. 19.13) and *Bkk* if the pedogenic calcium carbonate is more than 50% by volume, occurring as filaments, soft masses, coating particles and filling pores, of subsurface mineral horizons. While *Bk* horizons imply seasonal wetness in a semi-arid to arid climate and could also occur in poorly drained alkaline soil environments (Eze et al. 2021), *Bkk* horizons indicate prolonged surface stability. However, care should be taken in the interpretation as this could be mistaken for calcium carbonate layers that precipitated in or near the phreatic groundwater zone (Beverly et al. 2018). Palaeosols occur in different forms in the Hardveld, as buried (Fig. 19.17 and 19.18), exhumed by natural or anthropogenic activities (Fig. 19.13), or relict profiles (Fig. 19.14). Proxies including clay mineralogy, total elemental composition and stable isotope geochemistry, structure and morphological

Fig. 19.14 Petric Calcisol (Arenic) on a palaeo-channel at Limpopo-Lipadi Nature Reserve

properties of palaeosols hold potential for unravelling the palaeoenvironments and palaeoclimates of the eastern Hardveld (Yamaguchi and Kosei 2010; Thomas 2013).

19.5 Soil–Landscape Relationships

Formation and development of soils on the eastern Hardveld result from a complex interplay between the dynamic exogenous and passive vectors of Jenny's soil-forming functional–factorial model, namely parent material, climate, topography, biotic organisms and time on the outermost veneer of the Earth's surface (Jenny 1994). The eastern Hardveld is characterized by a wide variety of bedrocks and sediments. The largest portion of the Hardveld is underlain by granitic-gneiss (associated with granulites, migmatites, paragneisses and older undifferentiated granites) and other rock types from the Pre-Cambrian Basement complex (De Wit and Bekker 1990). Flat to undulating plains are developed on these rocks and the soils have ochric epipedon. Meta-basalt lavas (amphibolites) with ironstone layers and minor ultramafic schists, serpentinites and meta-sediments occur locally on the northern edge of the Hardveld, supporting gently to undulating plains with occasional hills and bedrock outcrops (De Wit and Bekker 1990). Soils formed on the ancient crystalline basement parent material are generally deep to moderately deep (50–150 cm) and sandy. The metamorphosed basic and ultrabasic Karoo basaltic lavas of the Karoo Supergroup underlie the Hardveld from the eastern margin (Serowe) to the west, where these rocks

Fig. 19.15 Vertisols on gently rolling landscape 200 m northeast of Serowe Stadium

become buried by aeolian deposits. In depressions and level to undulating areas, where these mafic rocks occur, Vertisols are usually developed. Arenosols on the Hardveld typically formed in situ on the sedimentary Paleoproterozoic Palapye Supergroup rocks consisting of limestone, quartzite and greywacke.

Climate, particularly seasonality, temperature range and precipitation, exerts a strong influence on the soil evolution processes which occur within the Hardveld. Kööppen's classification of the Hardveld climate is *BSh* (semi-arid dry climate with mean temperature over 18 °C (64 °F). This climatic regime is characterized by strong seasonal contrasts: low precipitation, high temperature, a rainfall deficit due to high evaporation rates. These dryland conditions often result in high soil salt concentration that affects vegetation growth, and limits microbial activity.

The land surface morphology greatly influences soil evolution and the development of a toposequence. For

Fig. 19.16 Pellic Vertisol (Calcaric) developed on basalt in a sedimentary lowland, Serowe. Depth is >120 cm

instance, in nearly flat to undulating areas, excluding valley bottomland, the water table is deeper than 1 m, which is a good environment for the formation of highly saline soils. In depressions and valleys, water tables can be found near the surface (mostly within the upper 1 m), which provides an ideal environment for the formation of water-saturated soils. On ridges and hills, a thin lithosol layer (skeletal soils) form on the slopes. Soils that developed on higher elevations and sloping areas are generally efficiently drained.

Soil evolution is actively affected by flora and fauna in different ways. Their major influence is the effect on the chemical and physical environment of the soils (Mhete et al. 2020). In Hardveld, the role of vegetation in soil formation is severely limited due to the high rate of organic

Fig. 19.17 Quaternary palaeosol section (~10 m) witihn a river floodplain incision in Serowe

matter mineralization within the topsoil environment. Termites play an active role in soil formation in the Hardveld through bioturbation, with termite mounds of varying sizes dotting the soilscape (Fig. 19.19). The extensive use of the land for domestic livestock grazing encourages erosion, desertification and leaching in the Hardveld as the grass stratum which protects soils and improves soil structure is selectively removed.

The evolution of soils is a continuing process and generally takes several thousand years for significant changes to occur. Due to low, seasonal precipitation, the development of a soil profile in the semi-arid Hardveld is generally very slow.

19.6 Summary

Using the soil classification system revised by the International Union of Soil Science Working Group (IUSS Working Group WRB 2015), this chapter presents the various soils developed on the eastern Hardveld of Botswana according to the updated nomenclature. The dominant Reference Soil Groups include Regosols, Arenosols, Luvisols, Lixisols, Cambisols, Calcisols and Vertisols. Other, less pronounced RSG in the Hardveld geo-ecological zones are Leptosols, Planosols, and Acrisols. Soil development on the eastern Hardveld is mainly influenced by the nature of parent

Fig. 19.18 Author (right) and his student examining illite-rich palaeosol section (~4 m) in a Hardveld transition zone, Letlhakeng

material and climate, and to a lesser extent by topography, organisms and time. Consequently, the area has more variety of reference soil groups than the other land units present in Botswana such as the sandveld, lacustrine and alluvial landscape. Chemical weathering, an important process that modifies the earth's surface and influences the mobilization and redistribution of geochemical elements within the regolith profile, is most often impeded due to limited precipitation characteristics of the region. Illuviation/eluviation, calcification/decalcification and erosion are the key operational pedogenic processes responsible for the nature and properties of soils found on the hill-dotted dry landscape. Palaeosols occur in various forms on the Hardveld and due to their polygenetic nature, they can be systematically used as proxies for palaeoenvironmental and palaeoecological reconstructions.

Fig. 19.19 Termite mound (~1.8 m) in Palapye – a product of bioturbation. *Source* Eze et al. (2020)

Acknowledgements I would like to thank Gregory A. Botha for an in-depth review of this chapter, and Frank D. Eckardt for giving constructive suggestions.

References

Abdelfattah MA, Shahid SA (2007) A comparative characterization and classification of soils in Abu Dhabi coastal area in relation to arid and semi-arid conditions using USDA and FAO soil classification systems. Arid Land Res Manag 21:245–271

Abel N, Stocking M (1987) A rapid method for assessing rates of soil erosion from rangeland: an example from Botswana. J Range Manage 40:460–466

Akinyemi FO, Tlhalerwa LT, Eze PN (2019) Land degradation assessment in an African dryland context based on the composite land degradation Index and mapping method. Geocarto Int. https://doi.org/10.1080/10106049.2019.1678673

Ashkenazy Y, Yizhaq H, Tsoar H (2012) Sand dune mobility under climate change in the Kalahari and Australian deserts. Clim Change 112:901–923

Aynekulu E (2003) Analysis of Soil-vegetation interaction in relation to soil carbon sequestration: a case study in Serowe, Botswana. Unpublished MSc Thesis, International institute for geo-information science and earth observation (ITC), Enschede, The Netherlands, 112 pp

Bekker RP, De Wit PV (1991) Vegetation map of republic of Botswana. In: Soil mapping and advisory services project. AG: DP/BOT/85/011. Food and Agricultural Organisation and Government of Botswana

Beverly EJ, Lukens WE, Stinchcomb GE (2018) Paleopedology as a tool for reconstructing paleoenvironments and paleoecology. In: Croft DA, Su D, Simpson SW (eds) Methods in paleoecology: reconstructing cenozoic terrestrial environments and ecological communities, vertebrate paleobiology and paleoanthropology methods in paleoecology. Springer, Cham, pp 151–183

Buol SW (2003) Philosophies of soil classification: from is to does. In: Eswaran H, Rice T, Ahrens R, Stewart BA (eds) Soil classification: a global desk reference. CRC Press Boca Raton Florida, pp 3–10

Carney JN, Aldiss DT, Lock NP (1994) The geology of Botswana, vol 37. Geological Survey Department

Coulombe CE, Dixon JB, Wilding LP (1996) Mineralogy and chemistry of Vertisols. In: Developments in soil science, vol 24. Elsevier, pp 115–200

De Wit PV, Bekker RP (1990) Explanatory note on the land systems map of Botswana. FAO/UNDP/Government of Botswana. In: Soil mapping and advisory services project AG: BOT/85/011. Field Document 31. 43 pp, 1 map

Driessen P, Deckers J, Spaargaren O, Nachtergaele F (2001) Lecture notes on the major soils of the world. In: World soil resources reports No. 94. FAO, Rome 334 pp

Eze PN (2013) Reconstruction of environmental and climate dynamics using multi-proxy evidence from palaeosols of the Western Cape, South Africa. Unpublished Ph.D thesis, University of Cape Town. 203 pp

Eze PN, Kokwe A, Eze, JU (2020) Advances in nanoscale study of organomineral complexes of termite mounds and associated soils: a systematic review. App Environ Soil Sci 2020:8087273. https://doi.org/10.1155/2020/8087273

Eze PN, Molwalefhe LN, Kebonye NM (2021) Geochemistry of soils of a deep pedon in the Okavango Delta, NW Botswana: implications for pedogenesis in semi-arid regions. Geoderma Regional 24: e00352

Food and Agriculture Organization of the United Nations (1987) East and Southern African Sub-committee for soil correlation and land evaluation. In: Meeting. Seventh Meeting of the East and Southern African sub-committee for soil correlation and land evaluation. Gaborone, Botswana

Food and Agriculture Organization (FAO) (2001) Major soils of the world: lecture note on world reference base for soil resources: atlas. In: World soil resource report 94. FAO of the United Nations, Rome, Italy

FAO/UNESCO (1974–78). Soil maps of the world. UNESCO, Paris

Fedoroff N, Courty MA, Guo Z (2018) Palaeosoils and Relict Soils: A Conceptual Approach. In: Interpretation of micromorphological features of soils and regoliths. Elsevier, Amsterdam, pp 821–862

Fey M (2010) Soils of South Africa. Cambridge University Press. 287 pp

Hartemink AE, Huting J (2008) Land cover, extent, and properties of Arenosols in Southern Africa. Arid Land Res Manag 11:134–147

IUSS Working Group WRB (2015) World reference base for soil resources 2014, update 2015 international soil classification system for naming soils and creating legends for soil maps. World Soil Resources Reports No. 106. FAO, Rome

Jenny H (1994) Factors of soil formation: a system of quantitative pedology. Courier Corporation

Jones LD, Jefferson I (2012) Expansive soils. In: Burland J (ed) ICE manual of geotechnical engineering, geotechnical engineering principles, problematic soils and site investigation, vol 1. London ICE Publishing, pp 413–441

Kebonye NM, Eze PN, Akinyemi FO (2017) Long term treated wastewater impacts and source identification of heavy metals in semi-arid soils of Central Botswana. Geoderma Reg 10:200–214

Mhete M, Eze PN, Rahube TO, Akinyemi FO (2020) Soil properties influence bacterial abundance and diversity under different land-use regimes in semi-arid environments. Scientific African 7:e00246

Mogotsi K, Nyangito MM, Nyariki DM (2013) The role of drought among agro-pastoral communities in a semi-arid environment: the case of Botswana. J Arid Environ 91:38–44

Monger HC, Bestelmeyer BT (2006) The soil-geomorphic template and biotic change in arid and semi-arid ecosystems. J Arid Environ 65:207–218

Mweso E (2003). Evaluating the importance of soil moisture availability, as a land quality, on selected rainfed crops in Serowe area, Botswana. Unpublished M.Sc thesis. In: International institute for geo-information science and earth observation (ITC), Enschede, The Netherlands 136 pp

Nicholson SE, Farrar TJ (1994). The influence of soil type on the relationships between NDVI, rainfall, and soil moisture in semiarid Botswana. I. NDVI response to rainfall. Remote Sens Environ 50 (2):107–120

Nkambwe M, Totolo O (2005) Customary land tenure saves the best arable agricultural land in the peri-urban zones of an African city: Gaborone, Botswana. Appl Geogr 25:29–46

Pal DK (2017) Cracking clay soils (vertisols): pedology, mineralogy and taxonomy. In: A Treatise of Indian and Tropical soils. Springer, Cham, pp 9–42

Pal DK, Srivastava P, Bhattacharyya T (2003) Clay illuviation in calcareous soils of the semiarid part of the Indo-Gangetic Plains. India Geoderma 115(3–4):177–192

Ringrose S, Matheson W (1991) Characterization of woody vegetation cover in the south-east Botswana Kalahari. Global Ecol Biogeogr 1:176–181

Staff Survey Staff USA (2010) Keys to soil taxonomy. United States Department of Agriculture, Soil Conservation Service, Washington, DC

Soil Survey Staff USA (1975) Soil taxonomy: A basic system of soil classification for making and interpreting soil surveys. US Government Printing Office

Spaargaren O (2006) Mineral soils conditioned by limited age: Cambisols. In: 3rd European summer school on soil survey office for official publications of the European communities, p 21

Sebego RJ, Arnberg W, Lunden B, Ringrose S (2008) Mapping of *Colophospermum mopane* using Landsat TM in eastern Botswana. S Afr Geogr J 90:41–53

Thomas DS (2013) Reconstructing paleoenvironments and palaeoclimates in drylands: what can landform analysis contribute? Earth Surf Proc Land 38(1):3–16

USDA–Natural Resources Conservation Service (USDA-NRCS) (2010) Keys to soil taxonomy. 11th edn. Washington, DC

van Wambeke A (1992) Soils of the tropics-properties and appraisal. McGraw Hill Inc., New York

Wright VP (1990) Estimating rates of calcrete formation and sediment accretion in ancient alluvial deposits. Geol Mag 3:273–276

Yamaguchi KE, Kosei E (2010) Iron isotope compositions of Fe-oxide as a measure of water-rock interaction: an example from Precambrian tropical laterite in Botswana. Front Res Earth Evol (IFREE Report for 2003–2004)

Yamaguchi KE, Johnson CM, Beard BL, Beukes NJ, Gutzmer J, Ohmoto H (2007) Isotopic evidence for iron mobilization during Paleoproterozoic lateritization of the Hekpoort paleosol profile from Gaborone Botswana. Earth Planet Sci Lett 256(3–4):577–587

Peter N. Eze is a Soil Scientist (Pedologist) and Senior Lecturer at the Department of Earth and Environmental Science of Botswana International University of Science and Technology (BIUST). He is a Ph.D. in Environmental and Geographical Science and a Master in Soil Science. His academic and research activities focus on Soil and Landscape Dynamics, Digital Soil Mapping, Soil Geomorphology and Palaeopedology.

20 The Tsodilo Hills: A Multifaceted World Heritage Site

Marek Wendorff

Abstract

The Tsodilo Hills are a cluster of three inselbergs rising up over 400 m above the Kalahari Desert in the Ngamiland, NW Botswana, to the west of the Okavango swamps. The exposed succession consists of metamorphosed siliciclastic sedimentary rocks deposited on a marine shelf at a margin of the Congo Craton between ca. 1.90 and 1.1 Ga. The sediments formed a Gilbert-type delta grading towards the open shelf covered with large underwater dunes and influenced by tidal currents. The rocks contain specularite, which was mined and traded throughout Southern Africa from the Late Stone Age up to the nineteenth century. The creativity and culture of the local communities is reflected by over 4000 rock paintings and engravings. A permanent lake, up to 7 m deep, existed between 27,000 and 12,000 years ago adjacent to the hills. The geomorphic features of the Tsodilo Hills document processes of both the physical and chemical modifications of the rocks. Two geological-timescale erosional cycles sculpted the area: (i) continental Dwyka glaciation (the Carboniferous-Permian) when Tsodilo Hills formed nunataks, and (ii) post-Karoo formation of African Surface (the late Cretaceous). Steep slopes of Tsodilo Hills show stepped morphology and some cliffs have flared sides. Silica solution resulted in localised arenisation and karst-like features including phreatic zone-related subartesian well, and vadose zone vertical shaft, horizontal tube and karren.

Keywords

Botswana • Damara Belt • Palaeoproterozoic • Siliciclastic shelf • Geomorphic features • Ancient mining

20.1 Introduction

The Tsodilo Hills form a cluster of inselbergs (Fig. 20.1) located in the north-western corner of the Ngamiland District of Botswana, about 40 km to the west of the Okavango River and rising dramatically by ca. 400 m above a relatively monotonous expanse of the Kalahari Desert (Fig. 20.2). The Tsodilo rock succession is of the Palaeoproterozoic age, whereas the surrounding Kalahari sedimentary cover has been formed and evolving from the Cretaceous to Recent. Exposures of rocks older than the young Kalahari sediments are extremely rare in the region, and therefore the Tsodilo Hills form a unique "window" providing direct access to the ancient strata that belong to the basement complex underlying north-west Botswana.

The exceptional character of the Hills is also reflected by the long human history. This is documented by a staggering world-class collection of over 4,000 rock paintings, prehistoric specularite mining and archaeological discoveries, which testify to the human occupation, evolution of the way of life, work, crafts and arts that started with the first settlers who arrived there in the Middle Stone Age about 100 thousand years ago (Campbell and Robbins 2010). Importantly, the Hills continue to play an important role in the spiritual traditions of the San people. Considering all these unique aspects, the Tsodilo Hills have been inscribed as UNESCO World Heritage Site (International Council on Monuments and Sites (ICOMOS) 2001; Botswana National Museum 2000) and the National Museum of Botswana Complex at Tsodilo was opened by the then President of the Republic, His Excellency Festus Mogae in 2001. Readers

M. Wendorff (✉)
Faculty of Geology, Geophysics and Environmental Protection, AGH University of Science and Technology, al. A. Mickiewicza 30, 30-059 Kraków, Poland
e-mail: wendorff@agh.edu.pl

F. D. Eckardt (ed.), *Landscapes and Landforms of Botswana*, World Geomorphological Landscapes,
https://doi.org/10.1007/978-3-030-86102-5_20

Fig. 20.1 Low-level aerial photograph showing the Tsodilo inselbergs rising above the expanse of the Kalahari Desert seen from the ESE (Photograph courtesy Els Slots, with permission; all subsequent photographs in this chapter are by Marek Wendorff)

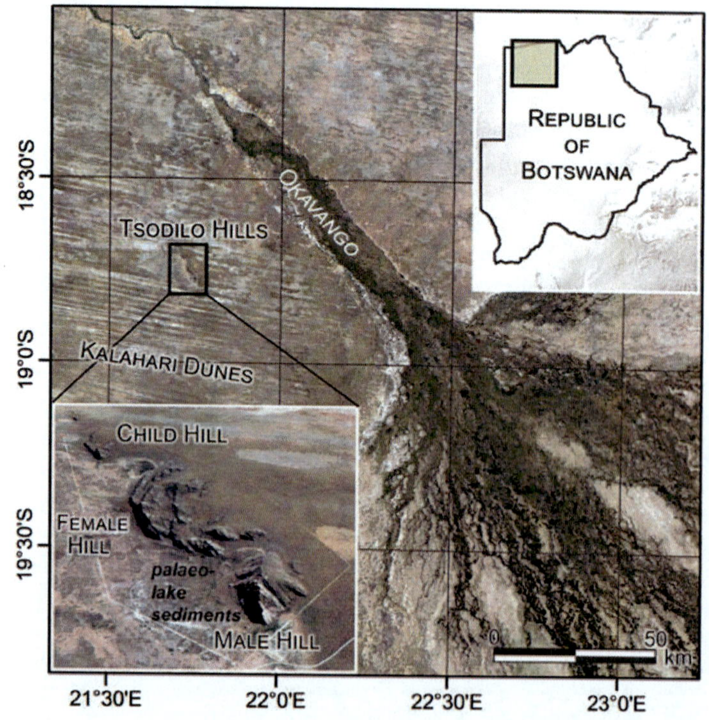

Fig. 20.2 Position of the Tsodilo Hills relative to the Okavango River in NW Botswana (Landsat 8 natural colours composition). Relict seif dunes of the Kalahari Beds are visible on the satellite image as parallel stripes oriented WNW-ESE. Wind shadow devoid of dunes extends to W of the Hills. Inset shows oblique view of the three Hills seen from SW (Sentinel 2 natural colours composition)

interested in details of the rock art, archaeology and historical aspects of human activity are directed to the excellent richly illustrated compendium edited and co-authored by Campbell et al. (2010). Highlights of the Tsodilo Hills appear in a volume published on the occasion of the 35th International Geological Congress, which presents fascinating geological sites of Africa (de Wit and Main 2016).

The aim of this chapter is to summarise the recent state of knowledge of the geological makeup of the Tsodilo Hills, with emphasis on features of their metamorphosed sedimentary rocks, which preserve a multitude of primary sedimentary structures and textures. These characteristics document the processes of deposition, their evolution in time and space, and therefore provide insights into the palaeolandscape and sedimentary environments in which Tsodilo Hills strata were formed. A brief discussion of geomorphic features and relations is followed by a summary of the ancient mining and rock art enclosed to emphasise the extent to which the geological resources combined with a period of hospitable climate influenced the ancient industry, economic activity, creativity and culture.

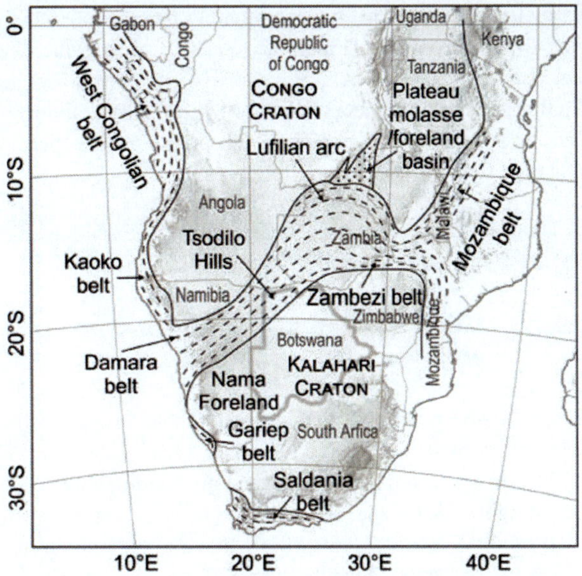

Fig. 20.3 Regional position of the Tsodilo Hills in the framework of the Neoproterozoic-Lower Palaeozoic Pan-African belts and cratons of Central and Southern Africa

20.2 Environmental Setting

The Tsodilo Hills represent a morphological feature unique to the landscape of NW Botswana. Steep slopes and precipitous cliffs of the inselbergs expose a succession of resistant to weathering metamorphosed sedimentary rocks of the Proterozoic age. By contrast, the surrounding relatively monotonous landscape of the Kalahari Desert is underlain by Cretaceous-to-Recent in age Kalahari sediments.

Three main hills of the Tsodilo cluster are called Nxum Ngxo (Male), Nxum Di (Female) and Picannin (Child) in the local San People language and occupy an area of about 3 × 10 kms. Male Hill is the highest peak rising 1,395 m a.s.l. and about 420 m above the surrounding Kalahari Desert, which extends at an elevation of ca 1000 m a.s.l. (Fig. 20.1). Female Hill occupies an area almost three times broader and the highest of its several peaks rises about 300 m above the surrounding plains. Child Hill, 1.7 km north of Male, is only 40 m high. About 2.1 km to the north-west lies a very small hill, called Grandchild.

Geologically, the Tsodilo Hills lie in the Damara Belt—one of the arms of the Pan-African system of several Neoproterozoic orogenic belts that traverse the African continent. The Damara Belt extends from Namibia in the southwest, crosses northwest Botswana and southern Angola, and continues to Zambia and the DRC to the northeast, where it is known as the Lufilian Arc, or the Katanga Belt (Fig. 20.3). These two orogenic segments are the erosional remnants of a large mountain belt that resulted from a continental collision between the Kalahari Craton in the south and the Congo Craton in the north at the time of assembly of the Gondwana Supercontinent, at the turn of the Precambrian and Palaeozoic eras (Rogers et al. 1995; Unrug 1996).

The surrounding Kalahari Group sediments rest directly upon the Proterozoic basement rocks and attain a maximum thickness in excess of 300 m in the Okavango graben to the east of the Tsodilo Hills (Fig. 20.2). Regionally, the Kalahari strata are formed by laterally and vertically variable facies of continental deposits: gravels and conglomerates, sands and sandstones, marls, evaporites, as well as duricrusts - calcrete, silcrete and ferricrete (Thomas and Shaw 1991). These are covered by now inactive, partly degraded and vegetated linear dunes, which rise up to 25 m in height, are usually spaced 1–2.5 km apart and constitute the predominant landscape feature surrounding Tsodilo Hills (Carney et al. 1994; Thomas and Shaw 1991). The linear dune sands of the north-west Ngamiland rest directly upon the Precambrian basement units (McFarlane et al. 2005, 2010; McFarlane and Eckardt 2006), which contrasts with central Botswana where they are underlain by the Karoo Supergroup strata spanning Carboniferous/Permian to Late Triassic Periods (Carney et al. 1994; Key and Ayers 2000).

The Tsodilo Hills presently have a semi-arid climate. However, the sedimentary record of a palaeolake adjacent to the Hills and archaeological research document much wetter periods, which are discussed later. The annual average rainfall can reach about 560 mm; November to March are the wettest months, with January's maximum of 134.9 mm. The dry period extends from April to October, with the driest

June–July (0 mm). The average monthly temperature varies very widely from a maximum of over 30 °C during the day between September and April (34.3 °C in November) to 26.0 °C in July Botswana National Museum 2000), and may fall down to below freezing point in the winter period (Brook 2010).

20.3 Geology of the Tsodilo Hills

20.3.1 Lithology, Stratigraphic Position and Age

In the earliest, unpublished accounts, the Tsodilo Hills metasedimentary rocks were described as an association of "micaceous schist, quartz-rich layers, grits and pebbly beds" by Wright (1956) and subdivided into four units called "stages" by Vermaak (1962). Both authors noted the occurrences of muscovite, kyanite, hematite, tourmaline and dumortierite as an assemblage of accessory minerals. The succession was named 'the Tsodilo Hillls Group' by Key and Ayers (2000), who suggested its Mesoproterozoic age. Subsequently, Wendorff (2005) presented preliminary observations on sedimentary features, facies trends, palaeocurrent patterns, mineral composition and metamorphism and suggested a Neoproterozoic age on the basis of regional position and lithological similarities with some Damara units in Namibia. Most recently published accounts presented a map and lithostratigraphic classification of the mappable rock sub-units (Wendorff and Świąder 2019) and age determinations by zircon geochronology (Mapeo et al. 2019).

Stratigraphically, the Tsodilo Hills Group metasedimentary strata rest non-conformably on the Quangwadum Basement Complex (Fig. 20.4) composed of granitoids and granite-gneisses, which originated between 2031 ± 1 Ma (Singletary et al. 2003) and 2024 ± 3.3 Ma (Mapeo et al. 2019). According to the most recent latter study, the spectrum of ages obtained on detrital zircons from the Tsodilo Hills Group rocks, and their comparison with ages of other rock units in the region, implies that deposition of the Tsodilo sedimentary succession occurred between approximately 1.90 and 1.1 Ga, i.e. during the youngest divisions of the Palaeoproterozoic-Mesoproterozoic. Hence, in spite of their occurrence within the Neoproterozoic Damara orogenic belt, the Tsodilo Hills Group strata are older and represent an exotic element of the unclear tectonic position. These rocks were involved in the Pan-African orogenic movements, which is testified by the 40Ar/39Ar age of 490 ± 2.3 Ma on metamorphic muscovite from Tsodilo Hills interpreted as the cooling age after regional metamorphism related to the Damara Belt orogeny (Singletary et al. 2003). Therefore, the non-conformable stratigraphic relation with the Quangwadum Complex and spectrum of numerical ages imply that the Tsodilo succession formed as a part of sedimentary cover of the Congo Craton, was first deformed during the Eburnean orogenesis (Mapeo et al. 2019) and then involved with the Neoproterozoic Pan-African orogenic movements. The Late Palaeoproterozoic-Mesoproterozoic depositional age of these rocks also shows that they are age-equivalent to a.o. the Muva Group in the basement of the Neoproterozoic Lufilian Belt succession of Zambia (Mapeo et al. 2019).

The Tsodilo Hills Group rocks are terrigenous detrital sedimentary rocks affected by low-grade metamorphism. The succession is dominated by micaceous and non-micaceous quartzites and contains quartz-mica schists, subordinate layers of meta-sandstones and meta-conglomerates, interbeds of meta-pelites and a red-bed-type association of siltstone, sandstone, mudstone and sedimentary breccia (Wendorff

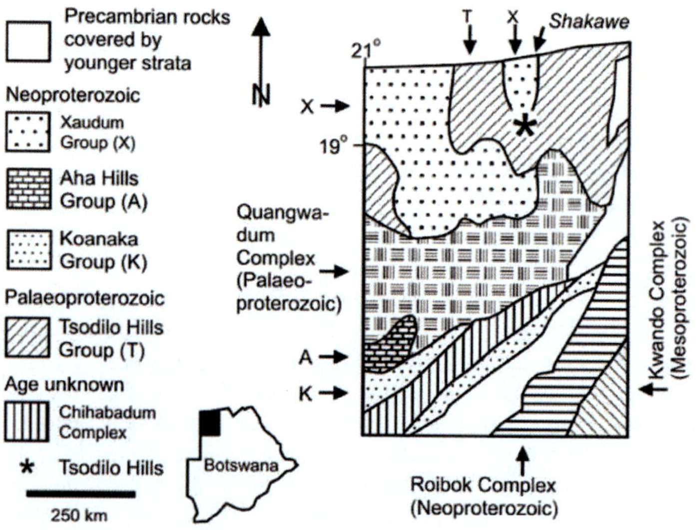

Fig. 20.4 Precambrian units in NW Botswana (modified from Carney et al. 1994; Key and Mothibi 1999; Key and Ayers 2000; Singletary et al. 2003; Wendorff 2005). Tsodilo strata age updated after Mapeo et al. (2019)

Fig. 20.5 Examples of mineral occurrences in Tsodilo Hills. **a** Quartz vein, ca. 30 cm thick, within sheared micaceous schist; **b** Entrance to Greenstone Mine, S cliff of Female Hill; **c** Two rosettes of dumortierite, tip of matchstick for scale; **d** Kyanite vein in the middle of quartz vein

2005). The succession contains a variety of accessory minerals: disseminated chlorite, ilmenite, titaniferous magnetite, titanite, hematite/specular hematite, tourmaline, biotite and dumortierite. The rock foliation is defined by the metamorphic minerals—muscovite and kyanite and some veins are filled with dumortierite or kyanite (Mapeo et al. 2019; Vermaak 1961, 1962; Wendorff 2005; Wright 1956) (Fig. 20.5).

Tsodilo quartzites/meta-arenites are light to medium grey and composed of usually well-sorted quartz grains cemented with silica. Recrystallisation of the cement ranges from moderate to very intense when the rock attains a glassy appearance and quartz overgrowths observed under polarising microscope show polygonal boundaries. The absence of both clay matrix and clay particles on arenite grains suggests deposition of quartzose sand in highly agitated waters responsible for removal of fine detritus. The mineral composition and very good sorting classify these rocks as super mature meta-arenites. This, coupled with sedimentary facies and bimodal-bipolar (bidirectional) palaeocurrents system recorded in the Tsodilo Hills Group suggest deposition and intense reworking and abrasion of arenaceous sediments on an open continental shelf under the combined influence of tidal currents and longshore littoral drift (Wendorff 2005).

The Tsodilo rock succession is deformed by several thrust faults and many shear zones, locally several metres thick, commonly indicating the direction of thrusting towards the southwest. Importantly, the deformations that would totally obliterate the primary features of meta-sandstones are not ubiquitous, and therefore intact primary sedimentary structures occur in abundance recording the sedimentary processes that shaped the Tsodilo Group strata.

20.3.2 Rock Complexes of the Tsodilo Hills

Lithostratigraphy, geographical distribution and correlation of the rock successions outcropping in the Male and Female

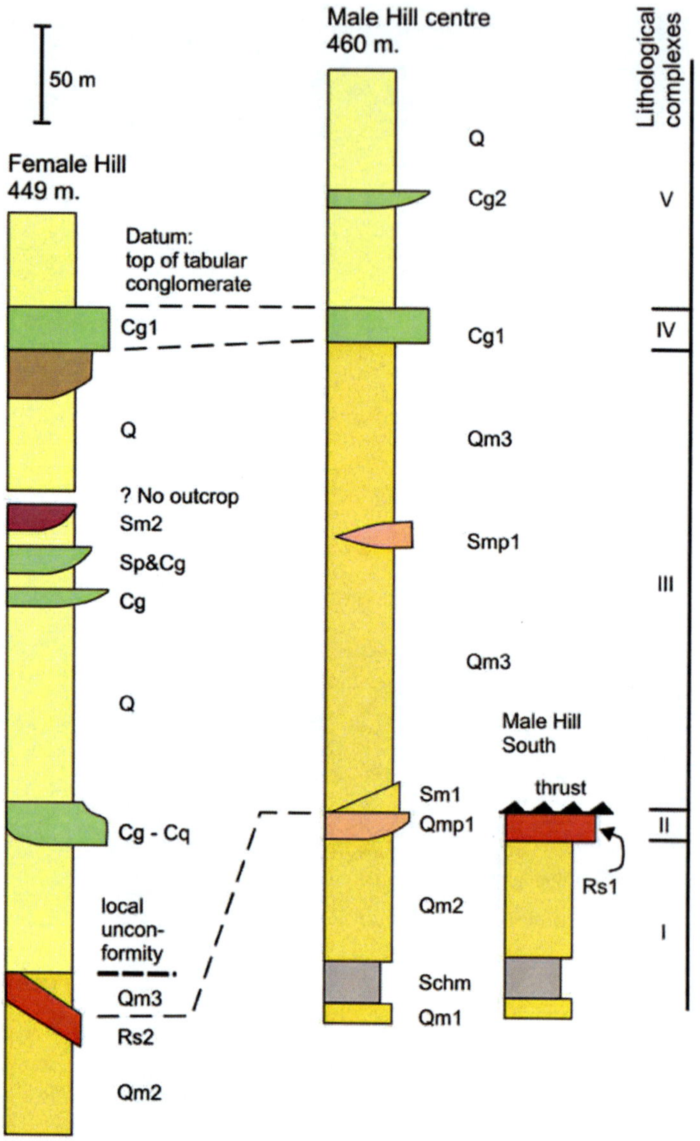

Fig. 20.6 Lithostratigraphic columns of the Tsodilo Hills Group strata exposed in Tsodilo Hills constructed along cross-section lines A-B for Male Hill and C-D for Female Hill shown in the geological map (Fig. 20.7)

Hills (Wendorff 2005; Wendorff and Świąder 2019) has been re-examined most recently and is presented here in a generalised form. The sequence is subdivided into five major, lithologically distinct complexes labelled I-V (Fig. 20.6). Their extent can be traced, or outcrops correlated, through the whole area of the Male and Female Hills (Fig. 20.7). Each complex is characterised by the predominant lithology or lithological association, as well as sedimentary structures and stratigraphic position in the succession. The features of some complexes show lateral variations and the prominent units contain mappable lenticular or wedge-shaped interlayers of different rock types. Some non-mappable occurrences are mentioned in the descriptions that follow, which summarise the features observed on both hills and outline their lateral variations with reference to Figs. 20.5 and 20.6.

Lithological complex I

Emerging from underneath the Kalahari Beds is ca. 90-m-thick oldest part (I) of the Tsodilo Hills Group succession. It is exposed in the lowermost slopes of the Male and Female Hills and consists of micaceous quartzite units Qm1 and Qm2 separated by an interlayer of micaceous schists Schm and terminating with pebbly micaceous quartzite.

The quartzites are coarse- to fine-grained and enriched in dark heavy minerals; high content of muscovite in the Male

Fig. 20.7 Geological map of the Tsodilo Hills (from Wendorff and Świąder 2019)

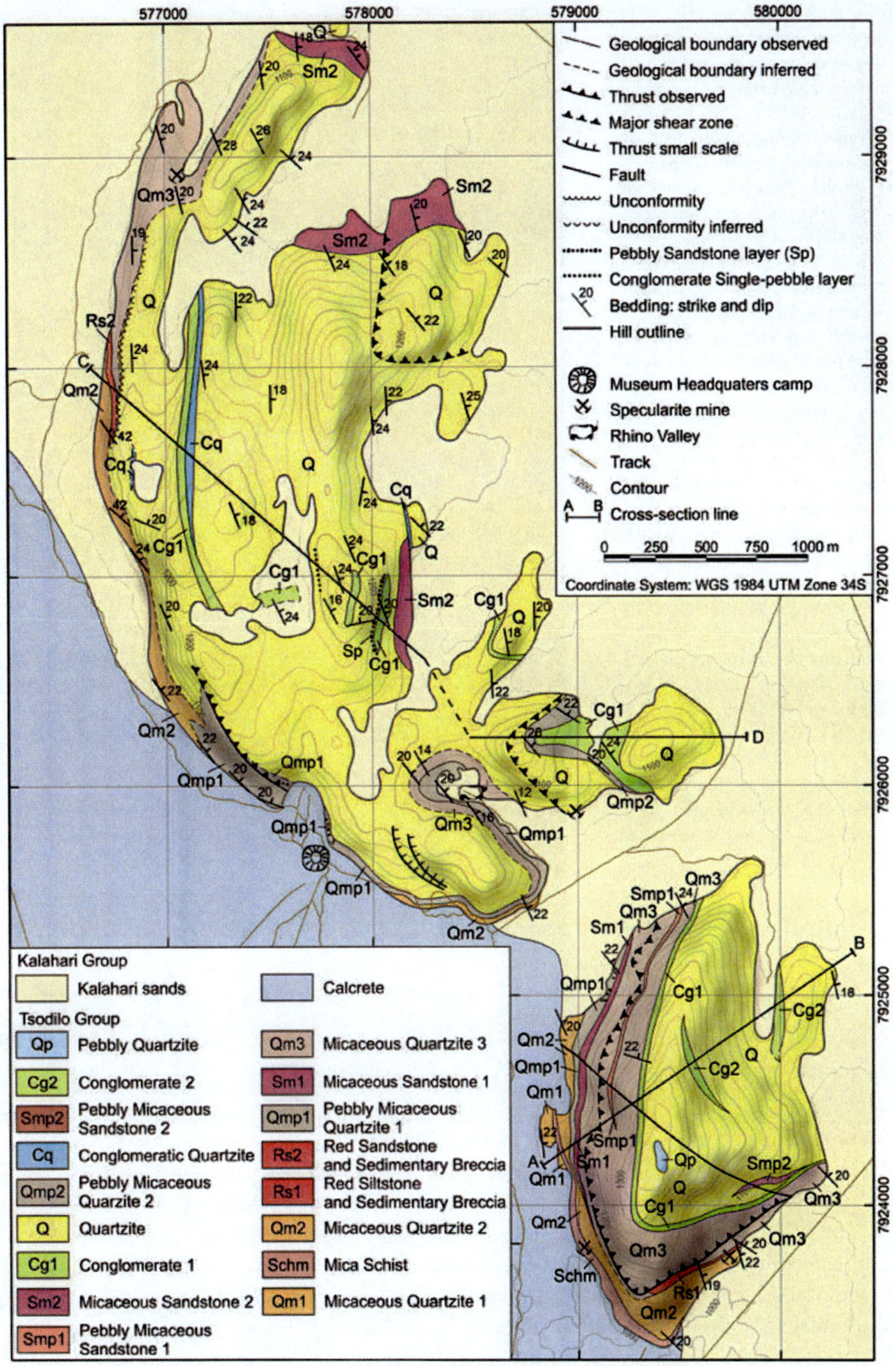

Hill decreases northwards through the Female Hill. Sedimentary structures are dominated by various types of cross-bedding occasionally associated with subordinate parallel-bedded layers (Fig. 20.8a). The cross-bedding is represented by low angle, swash-type and small to medium-scale tabular and trough cross-bedded sets, which often show alternating palaeocurrent directions towards the NW and the SE, contain reactivation surfaces and mud drapes (Fig. 20.8b). Large-scale cross-bedded units exceed 2 m in thickness, often contain normally graded foreset layers terminating with mud-drapes, and are occasionally deformed by slumping (Fig. 20.8c). The micaceous schist Schm forms a flat elongated lens maximum 20 m thick, present only in the Male Hill slopes between Qm1 and Qm2.

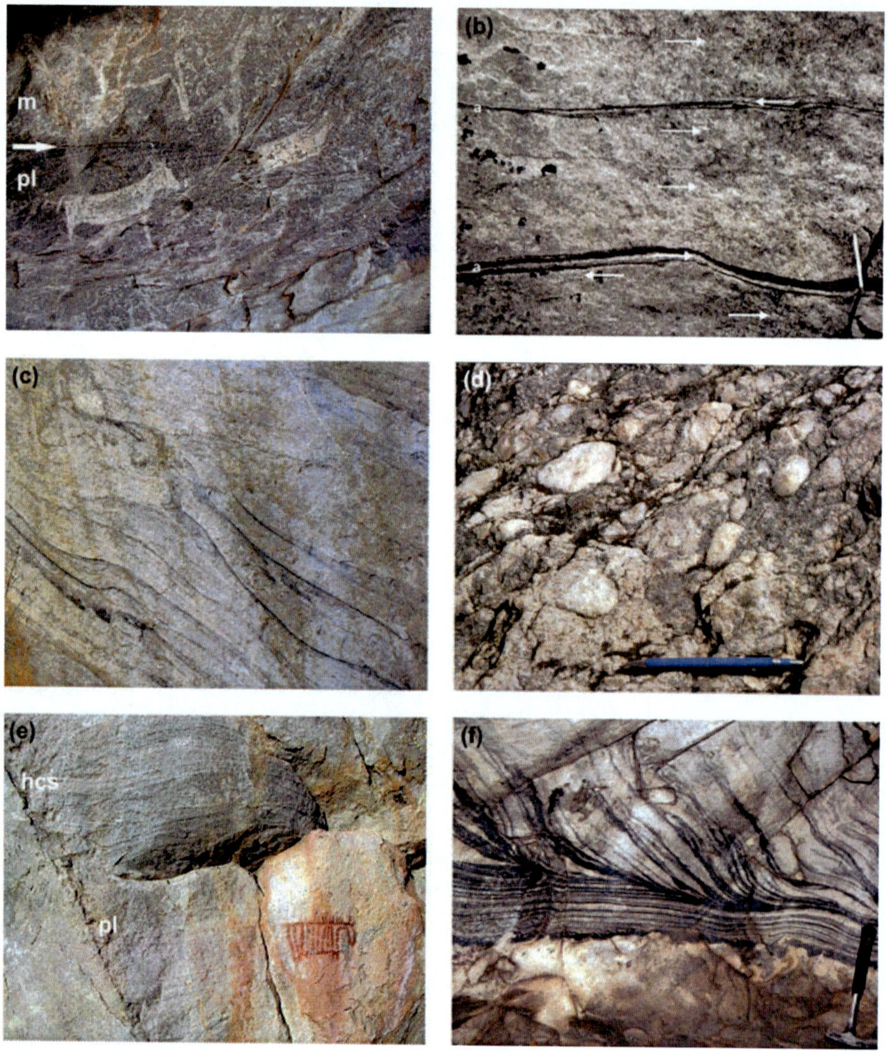

Fig. 20.8 Examples of sedimentary features of quartzites, meta-sandstones/arenites and meta-conglomerates exposed in Tsodilo Hills. **a** Parallel-lamination truncated by an erosional incision (arrowed) filled with massive quartzite (m); white paintings adorn the rock face (Snake Mine, Female Hill); **b** Small current ripplemarks showing two opposite current directions (arrowed) and interlayered with two couplets of mud drapes (black) typical for tidally-influenced environment (matchstick for scale); **c** Circa 2 m thick set of very large-scale cross-bedding composed of meta-arenite beds about 10 cm thick separated by black mudstone drapes and deformed by small slumps; **d** Massive conglomerate with cobbles and pebbles embedded in quartz arenite/quartzite matrix. **e** Hummocky cross stratification (hcs) overlying parallel and low angle cross-bedded set (pl) with red rock painting of zebra ca. 20 cm high; **f** Parallel lamination followed by large-scale cross bedding of quartz arenite/ quartzite containing dark laminae of specularite

The rock is grey in colour, contains a high proportion of muscovite and shows well-developed foliation. This layer is absent in the Female Hill outcrop probably due to facies pinch-out.

The pebbly micaceous quartzite Qmp1 has a limited lateral continuity. It wedges out southward along the western slope of the Male Hill, whereas in the south-western part of the Female Hill passes towards the north into quartzite Qm2. The unit displays significant lateral variations in the textural composition. In the Male Hill outcrops, the rocks range from micaceous quartzite containing several thin, discontinuous pebble layers to quartzite with scattered pebbles and cobbles. The coarse clasts size increases northwards from 0.5–3 cm in the south to 5–7 cm across in the Female Hill outcrops, and this trend is associated with an increase in the proportion of pebbles and cobbles, which are rounded- to well-rounded, of very low sphericity and represent vein quartz, red jasper and mudstone (Fig. 20.8d).

The most common sedimentary structures in Qm2 are: tabular planar and tangential cross-bedding, current ripples, mud drapes, tidal bundles and swash cross-stratification. The cross-bedded sets reach a thickness of 40 cm and are bounded by 1–2 cm-thick mudstone beds. Dip directions of cross-bedded layers indicate NW- and SE-directed alternating palaeocurrents.

Lithological complex II

Complex II is represented by red-bed facies (Rs). It is maximum 20 m thick, therefore thin by comparison with other rock complexes, but composed of distinctly different lithological associations. On the other hand, the internal organisation and specific characteristics of this unit vary between the Male and Female Hill outcrops.

The south-eastern part of the Male Hill, complex II (Rs 1) consists of red siltstone containing intercalation of sedimentary breccia, which extends over 3 m along outcrop face, and about 20-cm-thick interlayer of dark red to brownish silty mudstone hosting scattered phosphorite nodules. Sedimentary structures in the siltstone vary from parallel lamination to lenticular bedding expressed by isolated, small current ripplemarks composed of very fine-grained sandstone. The sedimentary breccia consists of randomly oriented angular to subrounded clasts of red siltstone embedded in reddish-brown, occasionally yellowish, silty matrix. The breccia is massive, but in places shows high porosity caused by leaching of clasts not resistant to weathering.

By contrast, the red bed facies exposed in the western slopes of the Female Hill (Rs2) are much coarser-grained. Here, the unit consists of ferruginous sandstone, which includes two broad lenses of sedimentary breccia in the lower part and a wedge of quartzite in the middle. The predominant ferruginous sandstone is coarse- to fine-grained, tabular cross-bedded and contains small subangular to rounded intraclasts of red siltstone. Each of the two sedimentary breccia lenses is maximum of 1–1.5 m thick and ca. 60 m long at the outcrop face. The breccia clasts represent three rock types: well-rounded fragments of ferruginous silty mudstone, subangular tabular fragments of reddish-brown mudstone, and whitish argillaceous flakes, 1–2 mm thick and up to 2 cm across, rich in phosphorous minerals (Wendorff 2005). The breccia matrix varies from light-grey sandstone to yellow siltstone. The quartzite wedge embedded in the ferruginous sandstone above the breccia lenses attains a maximum thickness of 6 m and consists of stacked tabular cross-bedded sets and mega ripplemarks up to 35 cm high. The upper part of the Rs facies unit contains a few shallow (≤ 2 m) erosional incisions filled with thin ripple cross-laminated sandstone and siltstone beds with forests indicating bipolar palaeocurrents oriented about the SW-NE axis (Wendorff 2005). The upper boundary of the Rs unit is conformable in the Male Hill outcrop but in the Female Hill occurs as an angular unconformity of 5°-10° below an erosional base of the overlying, large-scale cross-bedded quartzite Q of the succeeding lithological complex III.

Lithological complex III

Lithological complex III is dominated by meta-arenites, the composition of which grades laterally between the two hills. About 220-m-thick succession of mainly micaceous facies outcropping in the Male Hill (Qm3) evolves laterally to ca 305 m thick mainly non-micaceous strata in the Female Hill (Q) section. Subordinate are lenses of sandstone and pebble-bearing strata. The proportion of muscovite decreases laterally in the northern direction generally in the succession and within individual subunits. Relatively frequent are beds, or thick laminae, enriched in dark grains of heavy minerals.

Micaceous quartzite Qm3 exposed in the Male Hill slopes is medium- and fine-grained. The lower part contains stacked cosets of trough cross-bedding succeeded by a discontinuous layer of imbricated pebbles followed by a succession with current ripplemarks that show bidirectional/bipolar SE-NW palaeocurrent directions. The upper part is dominated by low-angle tabular planar cross-bedding alternating with swash-type cross-bedding. The Qm3 unit continues along the north-western slopes of the Female Hill where it displays hummocky cross-stratification (Fig. 20.8e), mud drapes, large-scale trough cross-beds and tabular planar cross-beds with reactivation surfaces. Parallel-laminated and large scale cross-bedded sets are sometimes enriched in specularite (Fig. 20.8f).

Up the Female Hill succession, Qm3 evolves into occasionally muscovite-bearing, coarse- to fine-grained quartzite Q containing some interbeds of weakly cemented sandstone; the boundary between these lithological varieties changes laterally from conformable to locally erosional associated with an angular unconformity (Fig. 20.7). Sedimentary structures are well-preserved, and the most common are: medium-scale current ripples, climbing ripples, a variety of cross-bedding (trough, tabular planar, swash, bi-directional/herringbone, hummocky), reactivation surfaces, mud drapes, double mudstone layers called tidal bundles and rare symmetrical wave ripplemarks. Completely preserved depositional bedforms range in size from large current ripplemarks, through subaqueous dunes, to sand waves from 2 to 10 m thick with composite internal structure and giant-scale cross-bedding.

Two laterally discontinuous rock units crop out at the western slopes of Male Hill. Micaceous sandstone Sm1 forms a wedge overlying the Qmp1 conglomerate, attaining a maximum thickness of 15 m, and pinching out below Qm3 in the western slopes of the Male Hill. The rock is massive ('structureless'), coarse- to very coarse-grained arenite, which weathers relatively easily because of a much lower proportion of silica cement than the quartzite facies present in other units. A lens of pebbly micaceous sandstone (Smp1), about 1200 m in length along strike and up to 15 m thick, crops out on the western slope of the Male Hill in the middle of the Qm3 unit. It is medium- to coarse-grained, and contains a subordinate proportion of well-rounded vein quartz pebbles 0.5–4 cm in diameter.

In the Female Hill slopes, a few single-pebble conglomerate layers and several thin and laterally discontinuous, not mappable conglomerate beds, are intercalated within the non-micaceous Q unit. They contain pebbles up to 6 cm across composed of vein quartz, accessory red jasper and rare mudstone intraclasts set in a matrix of arenite enriched in heavy minerals. Sedimentary structures range from massive to cross-bedded with pebbles dipping to the N, to quartzite normally graded from pebbly to medium- to fine-grained.

In the western part of the Female Hill, a broad lens ca. 25 m thick and extending over ca. 1600 m along strike is formed by very thick beds of conglomerate (Cg1) overlain by pebbly quartzite (Cq) (Fig. 20.8d). The lenticular form and sharp erosional lower boundary of this association suggest that it fills a shallow, broad erosional feature. Sedimentary structures preserved in the pebbly quartzite range from tabular to trough cross-bedding to ripplemarks.

In the eastern part of the Female Hill, there is another lens, ca. 400 m wide, composed of a 18-m-thick succession of pebbly sandstone (Sp) coarsening upwards to conglomerate (Cg1). The sandstone is poorly sorted, contains scattered pebbles up to 6 cm across. Tabular planar cross-bedded sets ca. 40-cm-thick overlain by current ripplemarks and 1-cm-thick mud drapes are the characteristic sedimentary structures.

At three localities, micaceous sandstone Sm2 overlies non-micaceous quartzite Q in the north and east of the Female Hill (Fig. 20.7). The sandstone is fine- to medium-grained with an occasional admixture of fine pebbles of vein quartz, jasper and mudstone. Medium-scale tabular planar cross-bedding sets, about 50 cm thick, and current ripplemarks are the main sedimentary structures.

Lithological complex IV

Complex IV consists of a tabular conglomerate body (Cg1) ranging from 11 to 20 m thick forming a marker bed between the Male and Female Hills, and underlain in the southeastern part of the Female Hill by a lens of pebbly micaceous quartzite (Qmp2). The quartzite is a medium- and coarse-grained meta-arenite containing randomly scattered, well-rounded pebbles of vein quartz and brown mudstone. The overlying tabular conglomerate unit consists of conglomerate beds associated with subordinate layers of quartzite and sandstone. The conglomerate beds range in thickness from 10 to over 200 cm, contain well-rounded pebbles of vein quartz, jasper and mudstone, and vary in sorting from clast- to matrix-supported. Structurally, the Cg1 unit is dominated by tabular cross-stratified sets ranging in thickness from 20 to 80 cm, whereas subordinate massive beds are 20–50 cm thick.

Lithological complex V

Complex V is maximum 120 m thick, dominated by medium- to coarse-grained, non-micaceous quartzite (Q), mostly grey but pinkish in places. Well-preserved sedimentary structures comprise current ripplemarks, hummocky cross-stratification (Fig. 20.8e) and a variety of cross-bedding: swash, tabular and trough, ranging from small to large scale.

A lens (Cg2) of pebbly conglomerates interbedded with quartzite layers crops out within the micaceous quartzite of complex V at the eastern slope of the Male Hill. Due to the slope undulations, it is exposed in both the upper and lower reaches of the slope. The geometry of the outcrops implies that sediments forming the lens fill an erosional channel ca. 500 m wide, maximum 12 m deep and minimum 700 m long. The infilling conglomerate beds are usually clast-supported, whereas some of the quartzite interbeds contain a varying admixture of pebbles. Tabular cross-bedded sets and massive bedding occur in both conglomerate and quartzite facies.

Tectonic structures

The succession is deformed by several thrust faults and many shear zones of various scale, commonly indicating the direction of dislocation towards the southwest. An impressive exposure of a prominent thrust occurs in the southern slopes of the Male Hill where it affects the upper part of the red bed complex (Rs) resulting in brecciation and south-west verging drag folds. Thrusting direction towards the south-west is also indicated by slickensides, striations, small drag folds and reverse faults that affect the underlying strata. Laterally, the thrust-related deformation decreases along strike across the western slope of the Male Hill, evolving towards the north into a thin shear zone extending into the south-eastern part of the Female Hill (Fig. 20.7).

At several other points, locally developed thrust zones of limited lateral extent, and small-scale reverse faults pass laterally into shear zones, which often contain lenticular quartz veins (Fig. 20.5a). The southwest-oriented direction of tectonic deformation is also indicated by occasionally observed slightly stretched pebbles, sigmoidal structures and deformed, or sheared, primary sedimentary structures. The micaceous units show a well-developed schistose fabric. Several joint systems cross the Tsodilo Group succession. Vertical joints with orientation changing between W-E and WNW-ESE and dipping up to 50° to NE occur in the Male Hill. The Female Hill contains a prominent NNW-SSE system and subordinate N-S sets of joints, both vertical to subvertical.

20.4 Geomorphic Relations in the Tsodilo Hills

20.4.1 Wind-Blown Sands

In the direct vicinity of the Hills, wind-blown sand is piled up at the foot of the topographic barrier of the Male and Female Hill, forming ramps adjacent to their E slopes. This is especially prominent at the foot of the eastern slope of Female Hill where such a ramp (Fig. 20.9) has been interpreted as an access route used by the ancient inhabitants of a settlement located at the Nqoma plateau (Denbow 2011). On the other hand, a zone devoid of dune sands (Fig. 20.2) that extends ca. 50 km to the W of the Hills is interpreted as 'wind shadow' (Campbell et al. 2010). These two features combined and associated with the E-W extension of the linear dune ridges in the region, suggest that prevailing easterly winds transported Kalahari sands over a geomorphologically significant period.

20.4.2 Upper Depressions

Two shallow sand-filled internal depressions in the upper part of the Female Hill provided important archaeological documentation of the times when the area was inhabited in the environmentally hospitable periods, before ca. AD 800 (Robbins et al. 2010; Denbow 2011). Smaller of the two, the Nqoma plateau is located in the S part of the hill, ca. 90 m above the Kalahari plain. Adjacent to the north, and 70 m higher, there extends a much larger plateau, the Divuyu that occupies the south-central part of the Female Hill (Fig. 20.9).

20.4.3 Palaeolake

Sedimentary and fossil records enabled the interpretation of a palaeolake that existed to the SW of the mouth of the valley, which separates Male and Female Hills (Figs. 20.2 and 20.9). The sediments and fossils of diatoms, molluscs and freshwater algae preserved in calcrete show that the lake was permanent, with a maximum radius of ca 4 km, up to 7 m in depth and laterally oscillating shoreline indicating variations in water levels. Radiocarbon dating on mollusc shells suggests a predominance of the lacustrine conditions during two stages, namely 27,000–22,000 and 19,000–12,000 years ago (Brook et al. 1992; Thomas et al. 2003). The lake was adjacent to the Female and Male Hills cliffs in the NE and its outer margin is marked by an abrupt termination of the linear dunes to the south and west against the lacustrine sediments; four palaeoshorelines identified by Geppert et al. (2021) document climatic variations. As there are no traces of any palaeostreams entering the area, seepage of groundwater hosted in the Kalahari beds is inferred as the main source of supply in addition to the runoff down the slopes of the adjacent hills during rainy periods. Even today, the groundwater table is very shallow remaining at ca 2 m depth, which results in springs, seeps and even long-lived ponds within the Hills. Of these, the most important source

Fig. 20.9 View to N of Female Hill from the top of Male Hill showing position of two ancient settlements accessible via sandy access ramp and palaeolake sediments adjacent to cliffs of both hills

of permanent spring water is known as the Water Hole Mine, at the foot of the western cliff of the Female Hill (Brook 2010; Campbell et al. 2010). Interestingly, the palaeolake area shows high uranium concentrations ascribed to the past surface runoff down the adjacent hill slopes (Bezuidenhout 2019).

20.4.4 Weathering and erosion of the Tsodilo Hills rocks

Geomorphic features of the Tsodilo Hills document processes of physical and chemical modifications of the rock succession. The character of angular, loose rock debris is influenced by lithology of the parent rock—boulders are derived from massive quartzite complexes and finer fragments result from disintegration of micaceous schist, meta-siltstone and mudstone. Some generally very steep slopes have stepped appearance; namely, cliffs formed on quartzite complexes are separated by benches of lower inclination underlain by micaceous schists, meta-siltstone or mudstone and vegetated by grasses and small bushes (e.g. Fig. 20.10a and b). Some of the cliffs rising above the surrounding Kalahari sediments surface have flared sides. The scale of this feature ranges from small, with overhangs occurring ca. 1.5–3 m above the adjacent ground level, to large, up to ca. 15–20 m in height (Fig. 20.10b). Some of the former occurrences are called 'shelters' (Campbell et al. 2010), in which the overhang protects paintings on the recessed lower face of the outcrop from the destructive elements.

Karst-like features suggestive of palaeokarstification and signs of recent dissolution of silica are infrequent and locally subtle. The 'Water Hole Mine' (also called "Python Spring"), holding water throughout the year and overflowing during rainy periods, occurs in the lower part of the Female Hill westernmost cliff. This feature is considered in the literature as one of the ancient specularite mines (Campbell et al. 2010). However, it appears to be a smooth-surfaced shaft inclined over 45 degrees into the rock face, elliptical in cross section and measuring 2.5 × 2 m across. These features suggest against its formation exclusively by ancient mining, which used to involve spalling and rock breakup that would have resulted in sharp-edged surfaces of the mine workings (Murphy et al. 1994). Instead, it is proposed here to be a karst feature—the epiphreatic oblique tube functioning as a subartesian well. Vadose zone-related features occur high up the western ridge of Female Hill, above the red bed (Fig. 20.7). The walls of ca. 12 m deep vertical shaft with some collapsed blocks of breakdown breccia at the Kalahari sand-covered bottom shows subhorizontal layer-parallel solution features at bed boundaries, and an array of small karren in the upper part (Fig. 20.11). In the eastern slope of this ridge, entrance to a horizontal tube occurs high up in the cliff face. The tube, or passage, is about 50 m long, roundish in cross-section, ca. 1–1.5 m in diameter, and with smooth walls; mirroring similar karst features known in the carbonate rock complexes. The floor of the passage is covered with the Kalahari sand.

Other effects of chemical weathering are reflected by the variations in quartzite hardness and rock-surface colouration. Laterally restricted zones of weakly cemented, crumbling parts of otherwise resistant to weathering, extremely hard, quartzite beds occur sometimes in recessed parts of cliffs and steep slopes. Such cases are interpreted here as a reflection of arenisation (Wray and Sauro 2017), i.e. preferential dissolution along crystal boundaries of quartz overgrowths by

Fig. 20.10 Examples of stepped appearance of the Tsodilo Hills steep slopes. **a** NW slope of Male Hill (height 410 m); **b** western slope of Female Hill with flared cliff side on the left (ca. 20 m high) and contrasting rock debris types: angular boulders beneath cliff-forming massive quartzite and finer fragments overlying benches on thinner beds of meta-siltstone, mudstone and schist

Fig. 20.11 Vertical shaft with two karst-like solution features parallel to bedding in the lower part and an array of karren in the upper part, on the right

seeping groundwater. On the other hand, deposition of iron oxides resulting from specularite weathering results in frequently occurring multi-coloured staining of rock faces.

20.5 Ancient Human Activity: Mining and Rock Art

Human activity and cultural evolution through millennia are documented by a world-class treasure of over 4,000 rock paintings and other items of rock art and mine workings bearing witness to prehistoric mining of specularite. Several archaeological sites provided evidence for the arrival of the earliest settlers about 100,000 years ago, in the Middle Stone Age, the use of pigments minimum 40,000 years ago, use of pottery, metal implements, fishing tools and other artefacts (Robbins et al. 2000). The paint was made from locally available minerals—red ochre in case of red paintings and chalk-like calcrete for white paintings (Kiehn et al. 2007).

Specularite was mined in the Tsodilo Hills and ground to produce hematite flakes of deep red shine traded for copper and iron implements as well as jewellery. Specular hematite, briefly called specularite, is a rock composed of light-reflecting tiny hematite crystals (Fe_2O_3)with metallic lustre, and reddish streak, which break up to form shining flakes when scratched or ground. The term is derived from specular reflection, i.e. mirror-like, as opposed to diffuse reflection in all directions from, e.g. a matte surface (Fox 2001). The mineral occurs in the Tsodilo schists, quartz veins and quartzites as disseminated crystals and specularite-enriched laminae, lenses and thin beds black in colour and often emphasising the primary structures, e.g. cross-bedding in meta-arenites.

Nineteen ancient mining sites of specularite are known, five of which in the Male Hill, thirteen in Female and one in Child (Murphy et al. 1994; Fig. 6.1 in Murphy et al. 2010). The earliest, occasional mining has been dated at 5300 ± 160 yr. B.P., and the radiocarbon dating of charcoal excavated from the floors of the mining sites provided dates in the range AD 850–1025 (Kiehn et al. 2007). However, a suggestion that the efficient mining activity might have had started even a few thousand years earlier is supported by the size of some of the mine excavations compared to the laborious mining method utilising fire spalling and stone tools to crush the ore-hosting rock (Murphy et al. 1994, 2010; Robbins et al. 1998). Nevertheless, the available data and historical accounts indicate that the specularite mining in Tsodilo started sometime before A.D. 850 and the last traces of this industry date back to the nineteenth century. It was traded throughout Southern Africa through the Late Stone Age and Iron Age up to the nineteenth century as a cosmetic since a mixture of ground specularite and grease applied to the hair and body gave the wearer a glittering appearance (Kiehn et al. 2007).

20.6 Summary

The age dating obtained most recently (Mapeo et al. 2019) sheds new light on the geological evolution of the Damara Belt in NW Botswana and highlights a special role of the Tsodilo Hills Group succession as a unit that may help to refine palaeogeographical reconstructions of the Late Paleoproterozoic-Mesoproterozoic continental masses, the remnants of which are now scattered and incorporated as parts of African and South American cratons. Marine siliciclastic strata deposited at continental margins hold information important for broad tectonostratigraphic reconstructions because they reflect the evolution of shallow marine depositional basins and elevated source areas. Therefore, detailed knowledge of the Tsodilo Hills rocks may support future mapping of age-equivalent, lithologically similar regions and their interregional correlation.

The Neoproterozoic-Lower Palaeozoic age of the last tectonic deformation suggests that during the Pan-African orogenesis a part of the older Quangwadum granitoid basement (Palaeoproterozoic) could have been sheared off the Congo craton together with the overlying sedimentary rocks of the Tsodilo Hills Group and emplaced as a tectonic unit lodged within the Damara Belt fold-and-thrust structure. Such tectonic position would be similar to the huge fragments of basement rocks known in the Domes Region of the Lufilian Belt in Zambia (Cosi et al. 1992).

The sedimentary features and regional position of the Tsodilo Hills succession suggest deposition in the open shelf environment at the Congo Craton margin. Male Hill was located closer to the source area as suggested by the deposits interpreted as Gilbert-type delta (Fig. 20.12a), and Female Hill facies suggest a more distal location. The shelf was influenced by tidal currents, littoral longshore currents, occasional storms and variations in sea level resulting in alternating regressive and transgressive trends within the Tsodilo suite. Of two regressions, the first one resulted in a tidal flat (Rs1 in Fig. 20.12a) incised by tidal creeks (Rs2 in Fig. 20.12b) and a break in the deposition of the Gilbert-type delta sequence, resulting in two stages of its evolution (Fig. 20.12a). The second regression is reflected by the tabular conglomeratic body truncating the Gilbert-type delta top in the uppermost reaches of the Male Hill (Fig. 20.12a), which records a substantial increase in the sediment supply, probably caused by uplift of the source area to the south. This conglomerate layer is succeeded by a fining-upwards succession of another transgressive event. Farther to the north, during submergence periods related to transgressive stages, giant cross-bedded sand waves were formed (Fig. 20.12b), migrated northward, away from a coastal zone, and accumulated to form the aggrading succession of the Female Hill.

In the context of regional stratigraphy, the relations observed in the Tsodilo Hills area suggest large-scale landscape-sculpting processes that resulted in two erosional cycles since the turn of Neoproterozoic and Palaeozoic when the Damara orogenic belt was elevated.

Fig. 20.12 Interpretation of depsositional complexes. **a** Male Hill, W face: Gd1—Gilbert-type delta stage 1, Rs1—red siltstone unit, Gd2—Gilbert-type delta stage 2, Cg1—tabular conglomerate; **b** Female Hill, W cliff: bsc—beach facies swash cross-bedded interval, Rs2—red siltstone, quartzite and mudstone of tidal flat (tf) with incised tidal channel (tfc), swg—shelf sand waves of giant-scale cross-bedded quartzite (cliff height ca. 60 m)

Weathering and erosion of the Damara mountains continued to the Carboniferous-Permian glaciation of Gondwanaland. It is suggested here that the Tsodilo Hills Group rocks formed nunataks within the Dwyka continental glacial cover. The glacially scoured landscape was covered by a succession of syn- and post-glacial deposits of the Karoo Supergroup. These were subsequently eroded in hot and humid late Cretaceous climate to form a regional peneplanation surface called the "African Surface" (Key and Ayers 2000; Miller 2008), which is an unconformity at the base of the succeeding Kalahari sediments.

Flared sides of some cliffs suggest prolonged evolution of the boundary between rock and regolith associated with topographic surface stability in the geological past, followed by regolith denudation. Karst-like features result from silica dissolution at a large scale, whereas this process progressing at a microscopic scale is reflected by the localised arenisation of some quartzite beds. Ongoing, recent fragmentation of the Tsodilo rocks produces debris ranging from fine clasts of micaceous schists, siltstone and mudstone to angular boulders of quartzite.

Frequent, strong easterly winds result not only in the wind shadow (mentioned earlier) but also carry abrasive quartz sand grains. No direct effects of the resulting erosion were studied in the Tsodilo Hills, but it is suggested here that especially the eastern rock faces, exposed to sand blasting, should be expected to act as a 'factory' enriching the Kalahari environment in detritus of quartz, muscovite and other minerals.

A limited selection of geomorphological features outlined here indicates that systematic research into the geomorphic characteristics and history of Tsodilo Hills is very much needed. The few karst-like features mentioned above suggest that several caves archaeologically investigated in the past (Campbell et al. 2010 and references therein) may represent karst features formed in quartzite, similar to the phenomena documented in Brazil, for example (Wray and Sauro 2017 and references therein). These should be studied in Tsodilo as well.

Considering the interesting geological and geomorphological features intertwined at Tsodilo with the record of ancient human activities and with the recently continuing religious and symbolic significance for the local communities, we may summarise by quoting a fragment of geological documentation by the present author and included in the Tsodilo Dossier prepared for UNESCO (Botswana National Museum 2000): 'The Tsodilo Hills have multifaceted values that combine uniquely, and in a fascinating way, several aspects of Earth's history with the cultural history of humankind. The area shows a most interesting interplay between a broad variety of geological processes spanning over a billion years of the Earth's history. At the same time, the Hills demonstrate how the minerals, products of slow natural processes, reflected on the cultural activity of our ancestors who fashioned the foundations of our history. Thus, the history of nature acting over many millions of years is intertwined with the history of people active in the Hills about a thousand years ago, and both are now being unravelled by modern science. These three elements can be perceived as a symbolic bridge between the past and present and make the Tsodilo Hills worthy of consideration for World Heritage Status'. This status was granted in 2001 by the World Heritage Committee.

References

Bezuidenhout J (2019) The relationship among naturally occurring radionuclides, geology, and geography: Tsodilo Hills, Botswana. J Radiation Res ApplSci 12(1):93–100

Botswana National Museum (2000) Tsodilo - Mountain of the Gods: World Heritage Nomination Dossier. The Government of the Republic of Botswana, Department of National Museums, Monuments and Art Gallery, Gaborone

Brook GA (2010) The Paleoenvironment of Tsodilo. In: Campbell AC, Robbins L, Taylor M (eds) Tsodilo Hills: Copper Bracelet of the Kalahari. Michigan State Univerity Press, East Lansing & The Botswana Society, Gaborone, Michigan & Gaborone, pp 30–49

Brook GA, Haberyan KA, De Filippis S (1992) Evidence of a shallow lake at Tsodilo Hills, Botswana, 17'500 to 15'000 yr BP:further confirmation of a widespread Late Pleistocene humid period in the Kalahari Desert. In: Heine K (ed) Palaeoecology ofAfrica and the Surrounding Islands, vol 23. CRC Press, pp 165–175

Campbell AC, Robbins L (2010) Rock Art at Tsodilo. In: Campbell AC, Robbins L, Taylor M (eds) Tsodilo Hills: Copper Bracelet of the Kalahari. Michigan State Univerity Press, East Lansing & The Botswana Society, Gaborone, Michigan & Gaborone, pp 94–115

Campbell AC, Robbins L, Taylor M (eds) (2010) Tsodilo Hills: Copper Bracelet of the Kalahari. Michigan State Univerity Press, East Lansing & The Botswana Society, Gaborone, Michigan & Gaborone

Carney J, Aldiss D, Lock N (1994) The Geology of Botswana, vol 37. Bulletin. Geological Survey of Botswana, Lobatse

Cosi M, De Bonis A, Gosso G, Hunziker J, Martinotti G, Moratto S, Robert JP, Ruhlmann F (1992) Structure, metamorphism and uranium mineralizations of the Domes area, Lufilian Arc (NW Zambia). Precambr Res 58:215–240

de Wit M, Main M (2016) The Tsodilo Hills of Botswana. In: Anhaeusser CR, Viljoen MJ, Viljoen RP (eds) Africa's top geological sites. Struik Nature, Cape Town, pp 176–180

Denbow J (2011) Excavations at Divuyu, Tsodilo Hills. Botswana Notes and Records 43:76–94

Fox M (2001) Optical properties of solids. Oxford University Press

Geppert M, Riedel F, Gummersbach VS, Gutjahr S, Hoelzmann P, Reyes Garzón MD, Shemang EM, Hartmann K (2021) Late Pleistocene hydrological settings at world heritage Tsodilo Hills (NW Kalahari, Botswana), a site of ancient human occupation. Quatern Sc Adv 3:100022. https://doi.org/10.1016/j.qsa.2021.100022

Key RM, Ayers N (2000) The 1998 edition of the National Geological Map of Botswana. J Afr Earth Sci 30 (3):427–451. https://doi.org/10.1016/S0899-5362(00)00030-0

Key RM, Mothibi D (1999) The national geological map of the Republic of Botswana 1:500,000. Botswana Geological Survey, Lobatse

Kiehn AV, Brook GA, Glascock MD, Dake JZ, Robbins LH, Campbell AC, Murphy ML (2007) Fingerprinting specular hematite from mines in Botswana, Southern Africa. In: Glascock M, Speakman RJ, Popelka-Filcoff RS (eds) Archaeological chemistry: analytical techniques and archaeological interpretation. ACS Symposium Series. American Chemical Society, vol. 968. Washington DC, pp 460–479

Mapeo RBM, Kampunzu AB, Armstrong RA (2000) Ages of detrital zircon grains from Neoproterozoic siliciclastic rocks in the Shakawe area: implications for the evolution of Proterozoic crust in northern Botswana. S Afr J Geol 103:156–161

Mapeo RBM, Wendorff M, Ramokate LV, Armstrong RA, Mphinyane T, Koobokile M (2019) Zircon geochronology of basement granitoid gneisses and sedimentary rocks of the Tsodilo Hills Group in the Pan-African Damara Belt, western Botswana: age constraints, provenance, and tectonic significance. J Afr Earth Sc 159:103576. https://doi.org/10.1016/j.jafrearsci.2019.103576

McFarlane MJ, Eckardt FD (2006) The "transparent" Linear Dunes of Northwest Ngamiland, Botswana. Botswana Notes &Records 36:136–139

McFarlane MJ, Eckardt FD, Coetzee SH, Berthold C, Ringrose S (2010) African surface weathering in the Kalahari of North West Ngamiland, Botswana: processes and products. Z Geomorphol 54 (3):273–303

McFarlane MJ, Eckardt FD, Ringrose S, Coetzee SH, Kuhn JR (2005) Degradation of linear dunes in Northwest Ngamiland, Botswana and the implications for luminescence dating of periods of aridity. Quatern Int 135:83–90

Miller RM (2008) The geology of Namibia, vol 3. Ministry of Mines and Energy, Geological Survey of Namibia, Windhoek

Murphy M, Murphy L, Campbell A, Robbins L (1994) Prehistoric mining of mica schist at the Tsodilo Hills, Botswana. J South Afr Inst Min Metall 94(5):87–92

Murphy M, Robbins L, Campbell AC (2010) The Prehistoric Mining of Specularite. In: Campbell AC, Robbins L, Taylor M (eds) Tsodilo Hills: Copper Bracelet of the Kalahari. Michigan State Univerity Press, East Lansing & The Botswana Society, Gaborone, Michigan & Gaborone, pp 82–93

Robbins L, Murphy M, Campbell AC (2010) Windows into the Past: Excavating Stone Age Shelters. In: Campbell AC, Robbins L, Taylor M (eds) Tsodilo Hills: Copper Bracelet of the Kalahari. Michigan State Univerity Press, East Lansing & The Botswana Society, Gaborone, Michigan & Gaborone, pp 50–63

Robbins LH, Murphy L, Campbell AC (1998) Intensive mining of specular hematite in the Kalahari ca. AD 800–1000. Curr Anthropol 39:144–150

Robbins LH, Murphy M, Brook GA, Ivester AH, Campbell AC, Klein RG, Milo RG, Stewart KM, Downey WS, Stevens NJ (2000) Archaeology, palaeoenvironment, and chronology of the Tsodilo Hills White Paintings rock shelter, northwest Kalahari Desert Botswana. J Archaeol Sci 27(11):1085–1113

Rogers JJ, Unrug R, Sultan M (1995) Tectonic assembly of Gondwana. J Geodyn 19(1):1–34

Singletary SJ, Hanson RE, Martin MW, Crowley JL, Bowring SA, Key RM, Ramokate LV, Direng BB, Krol MA (2003) Geochronology of basement rocks in the Kalahari Desert, Botswana, and implications for regional Proterozoic tectonics. Precambr Res 121:47–71

Thomas DGS, Shaw PL (1991) The Kalahari environment. Cambridge Uiversity Press, Cambridge

Thomas DSG, Brook G, Shaw P, Bateman M, Haberyan K, Appleton C, Nash D, McLaren S, Davies F (2003) Late Pleistocene wetting and drying in the NW Kalahari: an integrated study from the Tsodilo Hills Botswana. Quatern Int 104(1):53–67

Unrug R (1996) The assembly of Gondwanaland. Episodes 19:11–20

Vermaak CF (1961) Geological Map 1:10,000 of the Tsodilo Hills, Western Ngamiland (unpubl.). Johannesburg Consolidated Investment Company Ltd., Johannesburg

Vermaak CF (1962) Batawana Reserve - Bechuanaland Protectorate Geological Report. Johannesburg Consolidated Investment Company Ltd., Johannesburg

Wendorff M (2005) Outline of lithostratigraphy, sedimentation and tectonics of the Tsodilo Hills Group, a Neoproterozoic-Lower Palaeozoic siliciclastic succession in NW Botswana. Ann Soc Geol Pol 75(1):17–25

Wendorff M, Świąder A (2019) Lithostratigraphic classification of the Tsodilo Hills Group: a Palaeo-to Mesoproterozoic metasedimentary succession in NW Botswana. Geol Geophys Environ 45:305–318. https://doi.org/10.7494/geol.2019.45.4.305

World Heritage Committee (2001) World Heritage Committee Inscribes 31 New Sites on the World Heritage List (http://whc.unesco.org/en/news/143/, and, http://whc.unesco.org/en/list/1021). Accessed Dec 15 2020

Wright EP (1956) The Tsodilo Hills (Ngamiland). Geological Survey of Botswana, Lobatse

Wray RAL, Sauro F (2017) An updated global review of solutional weathering processes and forms in quartz sandstones and quartzites. Earth Sci Rev 171:520–557. https://doi.org/10.1016/j.earscirev.2017.06.008

Marek Wendorff is Full Professor with a PhD and DSc in Sedimentary Geology from the Jagiellonian University, Kraków, Poland. Worked at the Jagiellonian, Universities of Zambia and Botswana; now, is associated with AGH University of Science and Technology, Kraków. His research focuses on applied clastic sedimentology and interplays between tectonics and sedimentation. Major projects included the Carpathian flysch, continental and marine Carboniferous-to-Jurassic suites of Svalbard (Norwegian Arctic), Proterozoic and Phanerozoic successions of Central and SW Africa. He is an author of the new tectonostratigraphy of the Central African Copperbelt and co-discoverer of the PGE mineralisation in the Kalahari Copperbelt. Initiated and was a Leader of IGCP Projects No 302 and 419 (UNESCO/IUGS) and collaborated with several Mining Houses and Exploration Companies.

21 Geoconservation in Botswana

Senwelo M. Isaacs and Mark Stephens

Abstract

Geoconservation is action taken with the intent of conserving and enhancing geological, geomorphological and soil features, processes, sites and specimens, including associated promotional/awareness-raising activities and the recording and rescue of data or specimens from features and sites threatened with loss or damage. In Botswana, the geoconservation strategies are gatehouses, fencing, guides, information boards and construction of footpaths above nearby streams. Local communities are typically involved with geoheritage sites that occur close to their settlements and Community Trusts have been set up in several instances, that contribute to sustainable rural livelihoods. Geoconservation methods that have been implemented at geoheritage sites in Botswana are reviewed here, and all of the sites are protected by the Monuments and Relics Act, 2001 (Cap 56:02) that describes and contextualizes the preservation and conservation of an ancient monument and relic and ancient working which is known or believed to have constructed, erected or used in Botswana before 1st June 1902. Suggestions for the future improvements of geoconservation measures in the sites are made and include adequate planning for the peri-urban areas that incorporate geoheritage sites, preservation methods of rock art and more geoscientific information to be included in information boards and tours.

S. M. Isaacs (✉)
Department of Archaeology, Faculty of Humanities, University of Botswana, P/Bag UB00703, Gaborone, Botswana

S. M. Isaacs · M. Stephens
Department of Environmental Science, Faculty of Science, University of Botswana, P/Bag UB00704, Gaborone, Botswana

M. Stephens
School of Chemistry, Environmental and Life Sciences, Faculty of Pure and Applied Sciences, University of The Bahamas, PO Box N-4912 Nassau, Bahamas

Keywords

Geoconservation • Geoheritage • Botswana • Sustainable Development • Community Trust

21.1 Background to Geoconservation

Geoconservation has been defined as 'the conservation of geodiversity for its intrinsic, ecological and (geo)heritage values' (Sharples 1995, p. 38). Burek and Prosser (2008, p. 2) define it as 'action taken with the intent of conserving and enhancing geological and geomorphological features, processes, sites and specimens', and provide a useful summary of what is and what is not considered to be geoconservation. A more recent and systematic definition refers to 'action taken with the intent of conserving and enhancing geological, geomorphological and soil features, processes, sites and specimens, including associated promotional and awareness raising activities and the recording and rescue of data or specimens from features and sites threatened with loss or damage' (Prosser 2013, p. 568). Evolution of the definition reflects how geoconservation is now considered a specialized and comprehensive strategy, and an emergent geoscience discipline (Henriques et al. 2011).

Geodiversity is viewed as the foundation for geoconservation and geoheritage and is 'the natural range (diversity) of geological (rocks, minerals, fossils), geomorphological (landforms, topography, physical processes), soil and hydrological features', including 'their assemblages, structures, systems and contributions to landscapes' (Gray 2013) (see Stephens et al., this book). Geoheritage 'are those parts of the "Identified Geodiversity" of the Earth that are deemed to be worthy of (geo)conservation because of their importance/value' (Gray 2018). Geotourism is 'tourism which focuses on an area's geology and landscape as the basis of fostering sustainable tourism development' (Dowling 2015), that requires appropriate geoconservation

F. D. Eckardt (ed.), *Landscapes and Landforms of Botswana*, World Geomorphological Landscapes,
https://doi.org/10.1007/978-3-030-86102-5_21

approaches with present and future generations in mind (Stephens et al. 2013; Stephens 2020, this book).

Geoconservation is a relatively recent concept that has developed over the past 70 years and has grown rapidly in the past 40 years and is anchored in sustainable development (Brilha 2002; Burek and Prosser 2008; Stephens 2020, this book). There are various scales of geoconservation, e.g. from rock crystal to terrain-scale (Brocx and Semeniuk 2015) and the spatial scale of geoconservation projects have notably changed from small-scale to large-scale activities such as geoparks (Farsani et al. 2010; Brilha 2018b). Also, geoconservation is a temporally dynamic initiative as it involves working with 'geosites' (geological or geomorphological sites) of varying age from ancient to recent/modern, including natural processes of change such as weathering and erosion. For example, a geoconservation project may be concerned to retain a feature of interest, such as maintaining a clear exposure of a stratigraphic sequence in an eroding cliff (Burek and Prosser 2008).

Geoconservation also helps to combat environmental problems of social relevance, like those resulting from the over-exploitation of geological resources or improper land-use planning, which may endanger the physical integrity of geoheritage and associated cultural heritage (e.g. Larwood and Prosser 2008; Olson and Dowling 2018). In addition to guaranteeing the protection of nature, geoconservation is capable of promoting economic and social development at every scale from local to global (e.g. through geotourism) and relevant for mitigation against the socio-economic impacts of climate change (Dowling and Newsome 2006; Raharimahefa 2012). The benefits for education and science are also clear through study of well-preserved natural instruction sites, including geological and geomorphological type sections (stratotypes) (Brocx et al. 2019). Geosites often also contain rich natural archives of palaeoenvironmental, palaeontological and archaeological information (Crofts 2019).

Although an equal part of nature conservation, geoconservation has not been given the same attention as biological conservation (Brilha 2002; Crofts 2019) but there have been recent developments at the international level. The International Union for Conservation of Nature (IUCN) now recognises the importance of geological features as integral parts of nature at the same level as biological elements and in 2014 it established a Geoheritage Specialist Group within the IUCN World Commission on Protected Areas (WCPA) (Reynard and Brilha 2018). In addition, there are two main international programmes of relevance to geoconservation, both promoted by the United Nations Educational, Scientific and Cultural Organization (UNESCO): World Heritage Sites and Global Geoparks (Reynard and Brilha 2018; Gray 2019).

The creation of UNESCO World Heritage Convention in 1972 was the first international effort to select sites of paramount world importance due to their natural characteristics (Brilha 2018b; Migoń 2018). In 2004, 17 European and 8 Chinese geoparks came together to form the Global Geoparks Network (GGN) where national geological heritage initiatives contribute to and benefit from their membership of a global network of exchange and cooperation. The UNESCO Global Geoparks (UGGps) label was ratified in 2015 and the UGGps have three main aims: conservation of geoheritage, geological education for the public, and sustainable economic development mainly through geotourism (Farsani et al. 2010; Brilha 2018b; Gray 2019). A fundamental requirement is that they are rooted in local communities and Henriques and Brilha (2017) saw them as contributing to global understanding and sustainability.

Community involvement in geoconservation strategies remains pivotal to the sustainable management of the resource especially community involvement in all geoconservation actions (see below), e.g. (Tavares et al. 2015). It is also imperative that the public/local community perceptions and beliefs about the geoheritage resources are captured and integrated (Buckley 2003; Avelar et al. 2015; Brocx and Semeniuk 2015).

Geoconservation can be conducted through a Generic Geosite Conservation Framework (Prosser et al. 2018). This has three main parts: (1) geosite audit and selection, where valuation of each geosite (however this is defined) is assigned, and a geosite inventory is established; (2) conservation needs of the geosites are analysed and identified and; (3) conservation planning and action take place (Prosser et al. 2018). Of relevance to part 1 above, Brilha (2016) outlined procedures to achieve a numerical evaluation of the value of geosites, including proposal of new criteria. In addition, Brocx and Semeniuk (2015) developed the Geoheritage Resources Toolkit (GRT) to aid the inventorying of geosites. Of importance to part 3 above are improvements in the ability to measure and monitor geoconservation using digital tools (Cayla and Martin 2018). It is also important in part 3 above to assess optimal land utilization is taking place (e.g. Isaacs and Manatsha 2016).

Methods of geoconservation to physically protect, manage and provide education on geosites for visitors (part 3 above) may include: building of non-disruptive footpaths, interpretative trails, optional guided walks, information boards, informative websites, installation of drainage systems and amenities, netting/rock bolts to prevent mass movements, fenced reserves, gatehouses, state/community trust stewardship and a geopark (e.g. Gillieson 1996; Gray 2008, 2013; Stephens et al. 2013; Brocx and Semeniuk 2015). The choice and application of such geoconservation methods can be complex based on the particular circumstances of each geosite and Gray (2008, 2013) produced a useful provisional classification of methods, that include: secrecy, signage, physical restraint, reburial, excavation/curation, permitting/licensing, supervision, benevolent ownership, legislation, policy, site management and education (see Gray 2008, 2013, for more detailed discussion of each and for which elements of geodiversity they are recommended for use with).

21.2 Geoconservation in Africa

Many countries in Africa contain important sites displaying high geoheritage value (Schlüter 2006; Henriques et al. 2013; Billi 2015; Goudie and Viles 2015a: Grab and Knight 2015; Errami et al. 2015a, b; Viljoen et al. 2016; Embabi 2018), although the development of geoconservation varies throughout the continent. Geoconservation in southern Africa appears to be at least as well developed and researched as in northern Africa, judging by the research output from there, and in particular from Ethiopia and Morocco (Asrat et al. 2012; Asrat 2015, 2018; Errami et al. 2015a; Bouzekraoui et al. 2018; Mauerhofer et al. 2018; Thomas and Asrat 2018; Abioui et al. 2019; Hagos et al. 2019; Williams 2020), and relatively more than in central, west and east Africa (e.g. Boamah and Koeberl 2007; Żaba and Gaidzik 2011; Errami et al. 2015b; Henriques and Neto 2015; Schumann et al. 2015; Thomas and Asrat 2018). Thomas and Asrat (2018) noted that across most of the African continent there is an emphasis on wildlife and cultural tourism although geoconservation also inadvertently occurs in many countries within protected national parks and nature/game reserves. Reimold (1999) recommended that established National Geological Surveys could play a key role in the forefront of pro-active geoconservation and site protection in Africa.

Geoconservation is a growing movement in Africa (Errami et al. 2015a) and Ngwira (2015) indicated that it has the potential to improve socio-economic livelihoods in rural communities there. Ngwira (2019) outlined the key challenges for geotourism and geoparks in Africa and also proposed a way forward that included the need for further quantifiable research on their role in development, for African policy makers to develop and implement strategic policy guidelines and legal frameworks aimed at promoting geotourism and geoparks, and for geotourism practitioners and geopark managers to adopt positive policies towards stimulating locals' participation for local economic prosperity, through the creation of new products, employment and recreational/educational activities.

Two UGGps occur in Africa, in Morocco (M'Goun UGGp) and Tanzania (Ngorongoro Lengai UGGp), although there are others in different stages of formulation and the potential for geoparks and important heritage issues could be considered in the context of initiatives such as 'Africa Alive Corridors' (Toteu et al. 2010; Tavares et al. 2015; Thomas and Asrat 2018). Errami et al. (2015b) stated that to help to improve the situation regarding geoheritage in Africa, the African Association of Women in Geosciences (AAWG) created the African Geoparks Network (AGN) in 2009. The aims of the AGN include but are not limited to: identifying and inventorying geosites of outstanding value; creating and maintaining dynamic Geographic Information System (GIS) data bases; promoting and increasing the awareness amongst policy makers and the general public in Africa, in particular local communities; building the capacity of the local population in the field of geoheritage and geoconservation through a strong network including the organization of conferences seminars, symposia, training courses and workshops focusing on the non-professional community (Errami et al. 2015b).

Knight et al. (2015) overviewed geoheritage and geotourism in South Africa with examples and note the rich geodiversity and internationally important ecological and cultural/archaeological associations there. However, it is also noted that the potential of geoheritage tourism has not yet been fully exploited in rural South Africa (Reimold et al. 2006). A study by Taru and Chingombe (2016) suggested that the community at the Golden Gate Highlands National Park, Free State, would benefit from geotourism provided that geoconservation measures were in place. Another study from the Free State examined the Witsie Cave project and interviews and analysis of documents by Mukwada and Sekhele (2017) indicated the need to address intertwined endogenous and exogenous factors to enhance the viability of community-based geotourism projects there.

Errami et al. (2015b) noted that Namibia has a good inventory of its geological heritage (Schneider 2003) and information has been maintained for many years at the Geological Survey of Namibia. Namibia is also in the process of establishing its first geopark, the Gondwana Geopark in the central-west part of the country (Schneider and Schneider 2005). Goudie and Viles (2015a, b) state the importance of Namibia's geomorphological landscapes for being a foundation for biodiversity and geotourism there. However, they also note human activities that threaten the integrity of Namibia's geomorphological landscapes, and how conflicts between economic development and geoconservation are not always easy to resolve. The Geological Survey of Namibia and other bodies there are working to improve public understanding of the special and unique geomorphological landscape of the country (Goudie and Viles 2015b).

Tavares et al. (2015) used two case studies from southern Angola and argued that effective protection of geoheritage requires the local community's involvement in all geoconservation actions, i.e. inventory, evaluation, conservation, valuation and monitoring procedures, and not only at the final part of the process, when it is expected from local communities that the physical integrity of geoheritage is guaranteed.

21.3 Geoconservation Potential in Botswana

In Botswana there exist UNESCO World Heritage Sites, the Tsodilo Hills (inscribed under criteria i, iii and vi) and Okavango Delta (inscribed under criteria vii, ix and x), categorized as cultural and natural sites, respectively, and are thus legally protected by international treaties. The Gcwihaba Caves are currently on a tentative list to become a UNESCO World Heritage Site. Mbaiwa and Sakuze (2009) used Gcwihaba Caves and Xaixai Village in the Okavango Delta to demonstrate how cultural tourism can be a tool for rural livelihoods diversification in such areas.

With these exceptions, however, most geoheritage sites in Botswana have only partially been inventoried and assessed in the literature, and not always efficiently used for sustainable economic development (e.g. Campbell 2004; Brook 2015, 2017; Stephens 2020, this book; Stephens et al. this book; Tlhapiso and Stephens 2020). In general, local communities hardly use geomorphic resources except for growing/harvesting wild fruits and as pastureland for cattle and small stock. The traditional Batswana (people from Botswana) belief is that the geologic/geomorphic resources are ancestral grounds which should not be disturbed and are hence economically undervalued and under-used.

In attempting to attain sustainable peri-urban and rural development, Community Based Natural Resources Management (CBNRM) has proven to have potential to generate income for the rural communities and reduce poverty yet its impacts have not been measured and CBNRM has probably been underestimated as a result (Arntzen 2006). In Botswana there are Community-Based Development policies and action plans for Global Conventions, e.g. Community Based Rural Development Strategy (CBRDS) (1997), Revised National Policy for Rural Development (2002) and Botswana's National UNCCD (1995, ratified 1996). Although these policy instruments exist, the inclusion of geoconservation within them has been lacking.

The Monuments and Relics Act (MRA) (2001) was promulgated in 1968 and incepted in 1970 following the repeal of the Bushman's Relics and Ancient Ruins Protectorate Proclamation 68 of 1911 and subsequently, 1934. The MRA (2001) remains in force to date and helps with geoconservation in Botswana because it recognizes the attempts of the Bushman Relics Act (1934) of publishing the list of natural and historical monuments, relics and antiquities, e.g. stonewall ruins, prehistoric settlements, hills, rock paintings and graves. The MRA (2001) now recognizes monuments as buildings and mine workings created before 1902. The MRA (2001) also controls the use of monuments and relics that are now included under government budget. It has also contributed to the generation of considerable knowledge on archaeological and geological sites. The MRA (2001) also helps in the documentation of the location of archaeological, palaeontological and geologic sites in the Botswana National Museum and Monuments (BNMM) Site Register.

A unique range of landscape types occur in Botswana creating an interesting suite of geodiversity and geoheritage sites (Fig. 21.1) including one of the world's largest inland deltas (Fig. 21.2a), hills with rock art (e.g. Fig. 21.2b) and archaeological stone walls, vast salt pans (e.g. Fig. 21.2c), numerous fossil dry river valleys ('Mekgacha'), caves and rockshelters with archaeological and speleothem deposits, gorges (e.g. Fig. 21.2d), fossil stromatolites (e.g. Fig. 21.2e), and ancient and historical mines (e.g. Fig. 21.2f) (for further geochronological/archaeological details on the various sites see, e.g. Robbins et al. 1996; McFarlane et al. 2002; Campbell and Main 2003; Atlhopheng 2004; Mbaiwa and Sakuze 2009; Thomas and Shaw 2010; Grant 2012; Knight 2014; Brook 2015, 2017; De Wit and Main 2016; Dichaba and Thebe 2016; Stephens and Stephens et al. this book; Tlhapiso and Stephens 2020). Schlüter (2006) noted that the middle Cretaceous maar lake deposit of the Orapa kimberlite pipe, with its unique floral and faunal components should be considered an (important) geosite.

21.4 Geoconservation Practice and Approaches in Botswana

The main geoconservation approaches/strategies, in addition to the MRA (2001), observed to presently occur at geosites throughout Botswana are: gatehouses, fencing, an available guide, a community trust, footpaths and information panels/boards (Table 21.1). In relation to the classification of geoconservation methods by Gray (2008, 2013): gatehouse and fencing correlate to physical restraint; guide correlates to supervision; Community Trust correlates to benevolent ownership; footpaths to site management, and; information panel/board to education. Footpaths identified at geosites in Table 21.1 are those where treadage has been managed, e.g. raised wooden walkway, concrete/gravel pathway and/or with railings to hold on to. Well-worn trails occur at many of the geosites and although serve the purpose of encouraging people not to trample elsewhere, are often not a conscious management decision and so are not listed in Table 21.1.

Tlhapiso and Stephens (2020) applied the Karst Disturbance Index (KDI) to Kobokwe Cave and Gorge, near Molepolole, and found indication of relatively high cultural disturbance at the site. They recommended several of the methods mentioned above to conserve this nationally important site, including involvement of the local community in site management, raising awareness about and promotion of geoheritage sites, provision of guides, installation

Fig. 21.1 Map of Botswana showing the location of geoheritage sites mentioned in the text. **a** Provides a national overview and **b** Focus on the southeast and the area around Gaborone. For more information refer to Table 21.1

of litter bins, removal of graffiti, a gatehouse, constructed footpath to the cave and road to the gorge, development of relevant and effective conservation legislation, programs and strategies that involve all relevant stakeholders, continued monitoring and research. Also, of particular concern is the preservation of rock paintings at Tsodilo Hills, Manyana (e.g. Fig. 21.3), Lepokole Hills and Lobatse Gorge, and of the rock engravings at Matsieng Footprints and at Kopong. Digitization (e.g. Gallinaro et al. 2018) and ecology-based preservation methods (e.g. Bakkevig 2004) could be considered in this regard.

Guides are the most prevalent geoconservation method in use in Botswana and are available at the following 20 of the 45 listed sites (Table 21.1): Manyana rock shelters, Mogonye Gorge, Kobokwe Cave, Baratani Hill and Segorong Gorge, Mannyelanong Hill, Otse manganese mine, Matsieng rock engravings (e.g. Fig. 21.4), Peleng Hills, Baratani archaeological excavation site, Ditsotswane hills, Tswapong Hills, Lepokole Hills, Domboshaba Hills, Shoshong Hills, Gcwihaba Hills and Caves, Tsodilo Hills, Koanaka hills, Aha hills, Magapatona Hills and Okavango River and Delta system (Table 21.1). Often guides are from the local

Fig. 21.2 **a** A side channel and forested termite mound island of the 'panhandle' area of the Okavango River; **b** Tsodilo Hills; **c** Lekhubu Island with the Sua salt pan in the distance; **d** Goo-Moremi Gorge, Tswapong Hills; **e** Ramotswa Fossil Stromatolites; **f** former manganese mine at Otse. (Photos by M. Stephens)

Table 21.1 Geoconservation measures in place for the protection and education of geoheritage sites, including their coordinates and grouped by landform type, in Botswana. Where a site contains two landform types it is grouped under the type it is best known for in the country

Geoheritage site in Botswana	Coordinates (x, y)	Gate-house	Fenced	Guide	Comm. Trust	Foot-path	Info. panel
Hills and Gorges							
Peleng Hills and Gorges (Lobatse)	25°14′40.84″S 25°40′41.76″E			x			
Segorong Gorge and small caves (Otse)	25°00′31.30″S 25°43′20.35″E			x	x		x
Mannyelanong Hill, stone wall ruins, small caves (Otse)	25°03′33.71″S 25°45′29.51″E		x	x	x		x
Otse Hill, small gorges and small caves	25°00′48.13″S 25°44′15.95″E						
Ditsotswane Hills, small caves and small gorges (Otse)	24°55′31.97″S 25°49′23.94″E			x			
Mogonye Gorge(s)	24°51′44.64″S 25°41′06.61″E	x	x	x	x	x	x
Mmalogage Gorge and caves (Lekgolobotlo, Ranaka)	25°54′56.64″S 25°32′56.16″E						
Pharing Gorge (Kanye)	24°58′58.14″S 25°22′13.38″E						
Sekalaba Hills and Phata-ya-Mookana gorges (Kanye)	25°03′05.31″S 25°21′04.54″E						
Dithejwane Hills and stone wall ruins (Molepolole)	24°22′41.88″S 25°45′36.14″E						
Phataletshaba Gorges (near Mosopa)	24°45′42.42″S 25°30′48.90″E						
Kgale Hill and small gorges	24°41′41.75″S 25°52′0.90″E					x	
Modipe Hill and archaeological sites	24°38′24.78″S 26°09′33.18″E						
Shoshong Hills, Gorge and small cave	22°59′60.00″S 26°30′00.00″E			x			x
Majande Hills and small caves (Palapye)	22°44′37.13″S 27°00′40.07″E						x
Tswapong Hills, Moremi Gorge and small caves	22°37′15.65″S	x	x	x	x	x	x

(continued)

Table 21.1 (continued)

Geoheritage site in Botswana	Coordinates (x, y)	Gate-house	Fenced	Guide	Comm. Trust	Foot-path	Info. panel
	27°30′39.44″E						
Grootlagte/Qabo hills and small caves (Ghanzi)	21°01′13.13″S 21°42′36.09″E				x		
Caves and Rockshelters							
Lobatse Caves and Gorge (incl. rock paintings)	25°13′13.55″S 25°40′05.24″E		x				
Athlone Cave and Gorge (Lobatse)	25°12′38.15″S 25°39′55.55″E						
Kobokwe Cave and Gorge (Molepolole)	24°25′49.72″S 25°32′12.22″E			x			x
Manyana rock shelter (incl. rock art) and cave	24°46′01.48″S 25°34′46.75″E		x	x	x		x
Thamaga rock shelters	24°39′53.91″S 25°32′26.14″E						
Lepokole Hills, Caves (incl. rock art) and stonewall ruins	21°49′46.30″S 28°23′56.79″E	x	x	x	x		x
Gcwihaba Hills and Caves	20°01′23.76″S 21°21′16.92″E			x	x	x	x
Tsodilo Hills, mines and caves (incl. rock art)	18°44′37.92″S 21°43′59.33″E	x	x	x	x	x	x
Koanaka Hills (Nqcumtsa Hills) and caves	20°08′35.69″S 21°12′33 53″E			x	x		x
Aha Hills sinkholes	19°46′47″S 21°03′09″E	x		x	x		
Dunes/Pans/Fluvial							
Makgadikgadi, Nxai, Ntwetwe, Sua Pans	20°41′35.79″S 25°32′35.66″E				x		x
Kalahari sand dunes	26°28′34.41″S 21°43′33.07″E				x		
Kalahari 'Mekgacha'	24°58′22.18″S 21°59′48.27″E						
Okavango River and Delta system	18°56′46.04″S 22°27′24.85″E			x	x		x
Geoarchaeology History Other							
Seoke Hills and stone wall ruins (Lobatse)	25°10′52.13″S 25°44′42.41″E		x				x
Otse Manganese Mine	25°02′06.80″S 25°43′44.52″E			x	x		x
Baratani Hill excavated archaeological site (Otse)	25°00′12.60″S 25°43′50.54″E			x	x		
Motibatibo geoheritage site (Otse)	25°03′53.00″S 25°45′31.39″E						
Ramotswa fossil stromatolites	24°52′38.97″S 25°52′27.29″E						
Matsieng Footprints (engravings) and Loe Hole	24°35′13.52″S 26°09′32.10″E		x	x	x		x
Kopong rock engravings	24°11′32″S 25°32′26″E						

(continued)

Table 21.1 (continued)

Geoheritage site in Botswana	Coordinates (x, y)	Gate-house	Fenced	Guide	Comm. Trust	Foot-path	Info. panel
Lentsweletau specularite mine	24°17′45″S 25°53′36″E						
Sebilong specularite mine (Thamaga area)	24°44′27″S 25°31′59″E						
Magapatona Hill and stone wall ruins	20°27′57.77″S 27°08′33.14″E		x	x			x
Marapoatshwene Hills and stone wall ruins	25°05′41.01″S 25°46′38.70″E						
Mosenekatse Hill and stone wall ruins (Kanye area)	25°09′42.72″S 25°34′17.49″E						
Domboshaba Hills, stone wall ruins and small caves	20°42′50.14″S 27°24′21.53″E		x	x	x		x
	Total	5	11	20	18	5	19

Fig. 21.3 Faded rock painting of a gemsbok at Manyana (see Stephens et al., this bok, for further information on the site) (Photo by M. Stephens)

community associated with a Community Trust and provide excellent stewardship although can also be professional Batswana guides employed from outside by the BNMM, as is the case at Mogonye Gorge. Guides at Baratani Hill and Segorong Gorge are from the local community though they have been employed by the BNMM and have been trained in, e.g. archaeological excavation and locations of the geosites.

Information panels/boards have been erected at the following 19 of the 45 listed sites (Table 21.1): Manyana rock paintings, Mogonye Gorge, Kobokwe Cave, Segorong Gorge (e.g. Fig. 21.5), Mannyelanong Hill, Otse manganese mine, Seoke Hills, Tswapong Hills and Moremi gorge, Lepokole Hills, Makgadikgadi and Nxai Pans, Domboshaba Hills, Majande Hills, Shoshong Hills, Gcwihaba Hills and Caves, Tsodilo Hills, Koanaka Hills, Aha Hills sinkholes, Magapatona Hills and Okavango River and Delta system (Table 21.1). Typically, they are written in both English and Setswana (on reverse side) and provide very useful information on the history of the sites, including the flora and fauna (Fig. 21.5). It has been observed that accurate geoscientific information such as the age, rock types and geomorphological processes is typically lacking in the information panels at many of the sites. An atypical information board occurs at the Manyana rock art site, it has useful information on visiting hours although has no information about the ancient paintings or natural history there. At Tsodilo Hills, the information boards about the

Fig. 21.4 A guide at Matsieng instructing about the formation of the engraved footprints there and recording GPS coordinates of the rock engravings that have become less clear due to natural weathering and human treadage (Photo by S. M. Isaacs)

geo-cultural heritage of the area are found at the Tsodilo Site Museum and the rest of the site has directional boards to direct tourists, e.g. rock art sites, archaeological excavation sites and network of caves and rock shelters. Site data is presented to tourists by the site tour guides and is often focused on archaeological and historical perspectives probably because the custodian is the BNMM. Inside the gatehouse to Mogonye Gorge exist information panels on the archaeology, flora and fauna there although detailed information on the geomorphological history of the gorges is absent.

Community Trusts occur at the following 18 of the 45 listed sites (Table 21.1). The CgaeCgae Tlhabololo Trust and the Tsodilo Community Trust manage the Tsodilo geoheritage sites, including geoarchaeo-tourism and intangible cultural heritage (Segadika 2006) in collaboration with the BNMM. The XaiXai Community Trust manages the Gcwihaba and Aha Geoheritage sites and involves geoarchaeo-tourism. In the Kalahari, the Khawa Community Trust are stewards of the aeolian resources there, i.e. sand dunes and promote geotourism to improve the socio-economic livelihoods of the local San/Khoe community in Huiku and Kuru Community Trusts. In addition, they are engaged in managing the natural resources of the Ghanzi region as a whole and promote cultural festivals. According to Roodt (2010), the Gaing-o Community Trust from Mmatshumo village manage Lekhubu Island (granite boulders with archaeology) in the Sua (Sowa) Pan area. The Okavango Community Trust and the Okavango-Jakotsha Community Trust are involved in the management of the Okavango River-Delta system.

The Mmamotshwane Conservation Trust in Mogonye is involved in the stewardship of geoheritage there and generate revenue through geotourism at household and community level (Dichaba and Thebe 2016). The Matsieng Community Trust also engages in geotourism of the rock engraved footprints and preservation of intangible cultural heritage relating to creation myths at the site. Other community trusts involved with geotourism are: Moremi-Manonnye Conservation Trust of Moremi Gorge

Fig. 21.5 An information panel at the mouth of Segorong Gorge. The reverse of the panel is written in Setswana (Photo by S. M. Isaacs)

and other gorges in the Tswapong area; Lepokole Community Trust of the Lepokole Caves and rock art sites there, and; Manyana Community Trust which provides stewardship and guided tours to the rock paintings, rock shelter and cave, including intangible cultural heritage.

The Ba Ga Malete Development Trust consider Baratani-Otse Hills as sacred ancestral locations (intangible cultural heritage) and are engaged in annual litter picking on the talus of the Otse Hill, fundraising for site protection through proposals for fencing and a gatehouse. The Otse Vultures Community Trust (OVCT) is involved in the conservation of the Cape Vultures Colony whose niche (including small caves) is on the southern cliff of Mannyelanong Hill.

Fencing to protect the sites from human and animal disturbance has been installed at the following 11 of the 39 listed sites (Table 21.1): Manyana rock paintings (e.g. Fig. 21.3), Mogonye Gorge (Stephens 2020, this book: Fig. 21.1), Lobatse Caves, Mannyelanong Hills, Seoke Hills, Matsieng Footprints, Lepokole Hills, Domboshaba Hills, Tsodilo Hills and Magapatona Hills (Table 21.1). A large areal extent of fencing surrounds the three main hills at Tsodilo (Fig. 21.6). At Lobatse Caves fencing is erected around the Lobatse Estates, including agricultural fields, and not intentionally there to conserve the geosites.

Gatehouses occur at the following 5 of the 45 listed sites (Table 21.1): Mogonye Gorge (Stephens 2020, this book: Fig. 21.1), Moremi Gorge (Tswapong Hills), Lepokole Hills, Domboshaba Hills, Tsodilo Hills, Aha Hills and Magapatona Hills (Table 21.1). Due to the construction costs involved, they have been built at the sites that have most touristic value (e.g. Fig. 21.7). They typically contain information on the site and have ablution facilities. Those

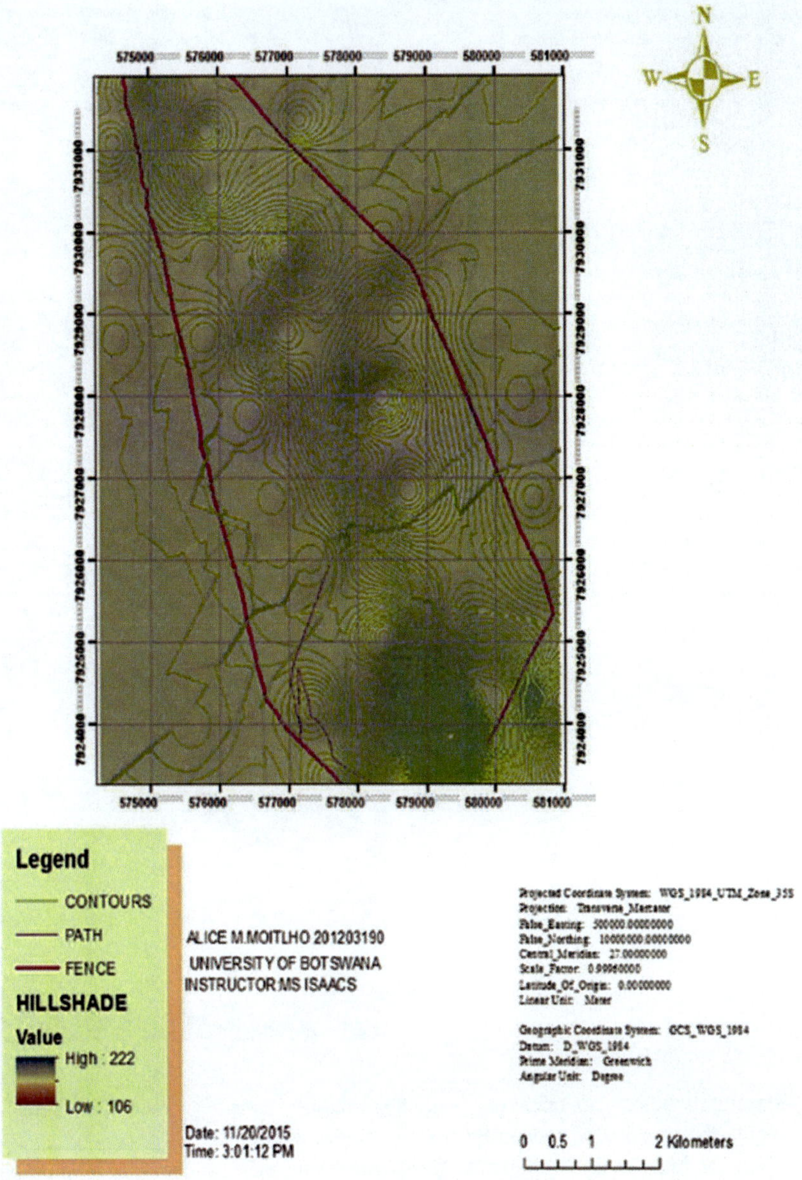

Fig. 21.6 Contour map showing the large area of fencing around the UNESCO World Heritage Site Tsodilo Hills

sites with gatehouses also have camping facilities. A gatehouse is also being planned for Baratani Hill and Segorong Gorge, Otse and an Environmental Impact Assessment (EIA) for the gatehouse was commissioned in 2019 by BNMM.

Constructed footpaths occur at the following 5 of the 45 listed sites (Table 21.1): Mogonye Gorge, Kgale Hill, Mogonye Gorge (Tswapong Hills), Tsodilo Hills and Gcwihaba Hills. Elevated wooden pathways have been built on the trails at Mogonye Gorge, Moremi Gorge (Tswapong Hills) and Tsodilo Hills and helps to improve safety there from slippery rocks. At Gcwihaba Hills also, trail routes have been created through vertical shafts that are used to enter the caves and carry tourists to view e.g. speleothems inside. A section of cemented footpath occurs on a steep portion of Kgale Hill and possibly associated with the nearby dolerite quarry there.

21.5 Summary and Recommendations

A summary of geoconservation initiatives generally being used in Botswana include gatehouses with camp sites, fencing, footpaths, guides, information panels/boards that require more geoscientific information, and involvement of local communities through non-governmental

Fig. 21.7 Large gatehouse structure at Lepokole Hills (Photo by M. Stephens)

community-based organization, e.g. Community Trusts. Whilst the practice of geoconservation in Botswana is at a relatively high level compared to most other countries in Africa, this chapter recommends that appropriate measures be put in place to minimize human impact and enhance visitor experience on those sites that have little or no geoconservation (Table 21.1): Thamaga rock shelters, Mmalogage Gorge and caves, Pharing (Khanye) Gorge, Kobokwe Cave and Gorge (also see Tlhapiso and Stephens 2020), Phataletshaba Gorges, Lentsweletau specularite mine, Sebilong specularite mine, Ramotswa fossil stromatolites, Kgale Hill, Modipe Hill (a very early gabbro outcrop with archaeological sites and geomythology, Campbell and Main 2003), Moritsane Hill, Mosenekatse Hill, Dithejwane Hill, Sekalaba Hill, Lobatse Caves, Athlone Cave and Gorge, Marapoatshwene Hill, Seoke Hill, Majande Hills, Motibatibo geoheritage site, Kalahari sand dunes and 'Mekgacha' and Grootlaagte/Qabo small caves. It is recommended that all geoheritage sites be promoted for geotourism. Geoconservation initiatives that are urgent to Botswana include inventorying the geoheritage (as initiated in this chapter and see: Brook 2015, 2017; Stephens et al. and Stephens, this book), appropriate planning for peri-urban areas with geoheritage resources (e.g. Isaacs and Manatsha 2016), establishing accurate scientific data and developing geoparks for sustainable socio-economic rural livelihoods (e.g. Henriques and Brilha 2017; Brilha 2018b). A candidate for at least national geopark status in Botswana suggested here is at Otse due to its rich concentration of stone-walled archaeological sites, cultural landscapes, highest peak in Botswana (Otse Hill at 1491 m asl) and associated geomyth (Lentswe la Baratani), ongoing geoarchaeological research and opportunity for sustainable community development.

Acknowledgements We wish to recognize the immense contribution and support provided by the Botswana National Museum and Monuments particularly Mr Vasco O. Baitsiseng who is the Principal Curator - Archaeology & Monuments. SMI wishes to thank her undergraduate research assistants (University of Botswana) especially in map productions, e.g. Katlego Motsumi, Etner Mathiba and Alice Moitlho. We also appreciate the efforts by local communities in validating the site locations and names as well as the Indigenous Knowledge System (IKS) associated with the geoheritage sites. We show gratitude to Mike de Wit (University of Stellenbosch) for communication on the geology of several of the sites during re-drafting. We thank private property owners for allowing us access into the farms to reach the geoheritage sites, e.g. Lobatse Estate. The University of Botswana has also provided support in terms of the resources needed to reach the various localities, e.g. funding and provision of site vehicles and accommodation.

References

Abioui M, M'Barki L, Benssaou M, Ezaidi A, El Kamali N (2019) Rock art conservation and geotourism: a practical example from Foum Chenna engravings site, Morocco. Geoconserv Res 2(1):1–11

Arntzen J (2006) Case study of the Community Based Natural Resource Management programme in Botswana. USAID Framework Project, Center for Applied Research IUCN, South Africa

Asrat A (2015) Geology, geomorphology, geodiversity and geoconservation of the Sof Omar Cave System, Southeastern Ethiopia. J Afr Earth Sc 108:47–63

Asrat A (2018) Potential geoheritage sites in Ethiopia: challenges of their promotion and conservation. Chapter 19. In: Reynard E, Brilha J (eds) Geoheritage: assessment, protection, and management. Elsevier, Amsterdam, pp 339–353

Asrat A, DeMissie M, Mogessie A (2012) Geoheritage conservation in Ethiopia: the case of the Simien Mountains. Quaestiones Geographicae 31(1):7–23

Atlhopheng JR (2004) Terrain evaluation and the botswana environment. Botswana Notes Rec 36:37–47

Avelar S, Mansur KL, Anjos SC, Vasconcelos GF (2015) Community perceptions for geoconservation of a coastal area in Rio de Janeiro, Brazil. Geoheritage 7:275–283

Bakkevig S (2004) Rock art preservation: improved and ecology-based methods can give weathered sites prolonged life. Nor Archaeol Rev 37(2):65–81

Billi P (ed) (2015) Landscapes and landforms of Ethiopia. World Geomorphological Landscapes Springer

Boamah D, Koeberl C (2007) The Lake Bosumtwi impact structure in Ghana: a brief environmental assessment and discussion of ecotourism potential. Meteorit Planet Sci 42(Nr 4/5):561–567

Bouzekraoui H, Barakat A, El Youssi M, Touhami F, Mouaddine A, Hafid A, Zwoliński Z (2018) Mapping geosites as gateways to the geotourism management in central high-atlas (Morocco). Quaestiones Geographicae 37(1). https://doi.org/10.2478/quageo-2018-0007

Brilha J (2002) Geoconservation and protected areas. Environ Conserve 29(3), 273–276

Brilha J (2016) Inventory and quantitative assessment of geosites and geodiversity sites: a review. Geoheritage 8:119–134

Brilha J (2018a) Geoheritage: inventories and evaluation. Chapter 4. In: Reynard E, Brilha J (eds.) Geoheritage: assessment, protection, and management. Elsevier, Amsterdam, pp 69–85

Brilha J (2018b) Geoheritage and geoparks. Chapter 18. In: Reynard E, Brilha J (eds) Geoheritage: assessment, protection, and management. Elsevier, Amsterdam, pp 323–335

Brocx M, Brown C, Semeniuk V (2019) Geoheritage importance of stratigraphic type sections, type localities and reference sites—review, discussion and protocols for geoconservation. Aust J Earth Sci 66(6):823–836

Brocx M, Semeniuk V (2015) Geoheritage and geoconservation – history, definition, scope and scale. J R Soc West Aust 90:53–87

Brook M (2015) Wild about Botswana. Michael C. Brook, Gaborone

Brook M (2017) Botswana monuments, heritage sites & museums. Michael C. Brook, Gaborone

Buckley R (2003) Environmental inputs and outputs in ecotourism, geotourism with a positive triple line? J Ecotour 2(1):76–82

Burek CV, Prosser C (2008) The history of geoconservation: an introduction. In: Burek CV, Prosser CD (eds) (2008) The history of geoconservation. The Geological Society, London, Special Publication 300, pp 1–5

Campbell A (2004) Establishment of Botswana's National Park and Game Reserve System. Botswana Notes Rec 36:55–66

Campbell A, Main M (2003) Guide to Greater Gaborone: a historical guide to the region around Gaborone including Kanye, Lobatse, Mochudi and Molepolole. A. Campbell and M. Main in association with The Botswana Society

Cayla N, Martin S (2018) Digital geovisualisation technologies applied to geoheritage management. Chapter 16. In: Reynard E, Brilha J (eds) Geoheritage: assessment, protection and management. Elsevier, Amsterdam, pp 289–304

Crofts R (2019) Linking geoconservation with biodiversity conservation in protected areas. Int J Geoheritage Parks. https://doi.org/10.1016/j.ijgeop.2019.12.002

De Wit M, Main M (2016) The Tsodilo Hills of Botswana. In: Viljoen R, Viljoen M, Anhaeusser C (eds) Africa's top geological sites. Chapter 26. Struik Nature

Dichaba T, Thebe PC (2016) In search of identity and ownership of the MmaMotshwane Gorge: heritage tourism and management of Mogonye cultural landscape. Botswana Notes Rec 47:56–68

Dowling R (2015) Geotourism. In: Jafari J, Xiao H (eds) Encyclopedia of tourism. https://doi.org/10.1007/978-3-319-01669-6_93-1

Dowling RK, Newsome D (2006) Geotourism: sustainability, impacts and management. Elsevier, Butterworth – Heineman, Oxford

Embabi NS (2018) Landscapes and landforms of Egypt. Springer, World Geomorphological Landscapes

Errami E, Brocx M, Semeniuk V (eds) (2015a) From geoheritage to geoparks: case studies from Africa and beyond. Springer

Errami E et al (2015b) Geoheritage and geoparks in Africa and the middle-east: challenges and perspectives. In: Errami E, Brocx M, Semeniuk V (eds) From geoheritage to geoparks: case studies from Africa and beyond. Springer, pp 3–23

Farsani NT, Coelho C, Costa C (2010) Geotourism and geoparks as novel strategies for socio-economic development in rural areas. Int J Tourism Res. https://doi.org/10.1002/jtr.800

Gallinaro M, Zerboni A, Solomon T, Spinapolice EE (2018) Rock art between preservation, research and sustainable development—a perspective from Southern Ethiopia. Afr Archaeol Rev 35:211–223

Gillieson D (1996) Caves: processes, development, and management. Blackwell, Oxford

Goudie A, Viles H (2015a) Landscapes and landforms of Namibia. Springer, World Geomorphological Landscapes

Goudie A, Viles H (2015b) Valuing, conserving and raising awareness of Namibia's landscapes and landforms. Landscapes and landforms of Namibia. World Geomorphological Landscapes. Springer, Dordrecht, pp 165–170

Grab S, Knight J (eds) (2015) Landscapes and landforms of South Africa. Springer, World Geomorphological Landscapes

Grant S (2012) Botswana and its national heritage. Melrose Books

Gray M (2008) Geodiversity: developing the paradigm. Proc Geol Assoc 119(3–4):287–298

Gray M (2013) Geodiversity: valuing and conserving abiotic nature, 2nd edn. UK, Wiley Blackwell, Chichester

Gray M (2018) Geodiversity: the backbone of geoheritage and geoconservation. Chapter 1. In: Reynard E, Brilha J (eds) Geoheritage: assessment, protection and management. Elsevier, Amsterdam, pp 13–25

Gray M (2019) Geodiversity, geoheritage and geoconservation for society. Int J Geoheritage Parks 7(4):226–236

Hagos M, Nyssen J, Amare K, Poesen J (2019) Geosites, geoheritage, human-environment interactions, and sustainable geotourism in Dogu'a Tembien. In: Nyssen J et al (eds) Geo-trekking in Ethiopia's tropical mountains. GeoGuide. Springer Nature Switzerland AG

Henriques MH, Brilha J (2017) UNESCO global geoparks: a strategy towards global understanding and sustainability. Episodes 40:349–355

Henriques MH, dos Reis RP, Brilha J, Mota T (2011) Geoconservation as an emerging geoscience. Geoheritage 3:117–128

Henriques MH, Neto K (2015) Geoheritage at the Equator: Selected Geosites of São Tomé Island (Cameron Line, Central Africa). Sustainability 7:648–667

Henriques MH, Tavares AO, Bala ALM (2013) The geological heritage of Tundavala (Angola): an integrated approach to its characterization. J Afr Earth Sci 88:62–71

Isaacs SM, Manatsha BT (2016) Will the dreaded 'yellow monster' stop roaring again?: An appraisal of Botswana's 2015 Land Policy. Botswana Notes Rec 48(1):383–395

Knight J (2014) Shoshong: a short history. Kwangu, Gaborone, Botswana

Knight J, Grab S, Esterhuysen AB (2015) Geoheritage and Geotourism in South Africa. In: Grab S, Knight J (eds) Landscapes and landforms of South Africa. Springer

Larwood GJ, Prosser DC (2008) Geotourism, conservation and society. Geol Balcanica 28:97–100

Mbaiwa JE, Sakuze LK (2009) Cultural tourism and livelihood diversification: the case of Gcwihaba caves and XaiXai village in the Okavango Delta, Botswana. J Tour Cult Chang 7(1):61–75

McFarlane MJ, Gieske A, Wormald J (2002) The origin of the Loe hole of Matsieng. Botswana Notes Rec 34:11–24

Migoń P (2018) Geoheritage and world heritage sites. In: Reynard E, Brilha J (eds) Geoheritage: assessment, protection, and management. Elsevier, Amsterdam, pp 237–250

Mukwada G, Sekhele N (2017) The potential of community-based geotourism in rural development in South Africa: The case of Witsie cave project. J Asian Afr Stud 52:471–483

Ngwira PM (2015) Geotourism and geoparks: Africa's current prospects for sustainable rural development and poverty alleviation. In: Errami E, Brocx M, Semeniuk V (eds) From geoheritage to geoparks: case studies from Africa and beyond. Springer, pp 25–33

Ngwira PM (2019) A review of geotourism and geoparks: is Africa missing out on this new mechanism for the development of sustainable tourism? Geoconserv Res 2(1):26–39

Olson K, Dowling R (2018) Geotourism and cultural heritage. Geoconserv Res 1(1):37–41

Prosser CD (2013) Our rich and varied geoconservation portfolio: the foundation for the future. Proc Geol Assoc 124(4):568–580

Prosser CD, Díaz-Martínez E, Larwood JG (2018) The conservation of geosites: principles and practice. In: Reynard E, Brilha J (eds) Geoheritage: assessment, protection, and management. Elsevier, Amsterdam, pp 193–212

Raharimahefa T (2012) Geoconservation and geodiversity for sustainable development in Madagascar. Madagascar Conserv Dev 7 (3):126–134

Reimold WU (1999) Geoconservation—a southern African and African perspective. J Afr Earth Sci 29:469–483

Reimold WU, Whitfield G, Wallmach Th (2006) Geotourism potential of southern Africa. In: Dowling R, Newsome D (eds) Geotourism. Butterworth-Heinemann, Oxford, pp 42–62

Reynard E, Brilha J (2018) Geoheritage: a multidisciplinary and applied research topic. In: Reynard E, Brilha J (eds) Geoheritage: assessment, protection and management. Elsevier, Amsterdam, pp 3–9

Robbins LH, Murphy ML, Stevens NJ, Brook GA, Ivester AH, Haberyan KA, Klein RG, Milo R, Stewart KM, Matthiesen DG, Winkler AJ (1996) Paleoenvironment and archaeology of Drotsky's cave: western Kalahari Desert, Botswana. J Archaeol Sci 23:7–22

Schlüter T (2006) Geological atlas of Africa: with notes on stratigraphy, tectonics, economic geology. Springer, Geohazards and Geosites of Each Country

Schneider GIC (2003) The roadside geology of Namibia. Sammlung Geologischer Führer, Gebrueder Borntraeger, Stuttgart

Schneider GIC, Schneider MB (2005) Gondwanaland Geopark – A proposed Geopark for Namibia. Gondwanaland Geopark Project. Report to UNESCO. Available at: http://portal.unesco.org/en/files/47468/12665840421Gondwana_Park_sm.pdf/Gondwana%2BPark%2Bsm.pdf

Schumann A, Muwanga A, Lehto T, Staudt M, Schlüter T, Kato V, Namboyera A (2015) Ugandan geosites. Geol Today 31(2):59–67

Segadika P (2006) Managing intangible heritage at Tsodilo. Mus Int 58 (1/2):31–40

Sharples C (1995) Geoconservation in forest management-principles and procedures. Tasforests 7:37–50

Stephens M (2020) Geoconservation for sustainable development. In: Leal Filho W, Azul A, Brandli L, Lange Salvia A, Wall T (eds) Life on land. Encyclopedia of the UN sustainable development goals. Springer, Cham

Stephens, M. (this book) Gorges of Eastern and Southeastern Botswana. In: Eckardt, F. (ed.) Landscapes and Landforms of Botswana. Springer.

Stephens M, de Wit M, Isaacs SM (this book) Caves of Botswana. In: Eckardt F (ed) Landscapes and landforms of Botswana. Springer

Stephens M, Hodge S, Paquette J (2013) Geoconservation of Volivoli Cave, Fiji: a prehistoric heritage site of national significance. Geoheritage 5(2):123–136

Taru P, Chingombe W (2016) Geoheritage and the potential of geotourism in the Golden Gate Highlands National Park, South Africa. Afr J Hospitality Tourism Leisure 5(2):1–11

Tavares AO, Henriques MH, Domingos A, Bala A (2015) Community involvement in geoconservation: a conceptual approach based on the geoheritage of South Angola. Sustainability 7:4893–4918

Thomas D, Shaw P (2010) The Kalahari environment. Cambridge University Press, Cambridge

Thomas MF, Asrat A (2018) The potential contribution of geotourism in Africa. Chapter 12. In: Dowling R, Newsome D (eds) Handbook of geotourism. Edward Elgar Publishing, pp 168–189

Tlhapiso M, Stephens M (2020) Application of the Karst Disturbance Index (KDI) to Kobokwe Cave and Gorge, SE Botswana: implications for management of a nationally important geoheritage site. Geoheritage 12(39). https://doi.org/10.1007/s12371-020-00461-8

Toteu SF, Anderson JM, de Wit M (2010) 'Africa Alive Corridors': forging a new future for the people of Africa by the people of Africa. J Afr Earth Sci 58(4):692–715

Viljoen R, Viljoen M, Anhaeusser C (eds) (2016) Africa's top geological sites. Struik Nature

Williams F (2020) Safeguarding geoheritage in Ethiopia: challenges faced and the role of geotourism. Geoheritage 12:31. https://doi.org/10.1007/s12371-020-00436-9

Żaba J, Gaidzik K (2011) The Ngorongoro Crater as the biggest geotouristic attraction of the Gregory Rift (Northern Tanzania, Africa) - geographical setting. Geotourism 1–2(24–25):3–26

Senwelo M. Isaacs is Director Geospatial in the Directorate of Intelligence and Security, Ministry of Presidential Affairs and Public Administration and was Lecturer in Geographic Information Systems in Archaeology at the University of Botswana. She was Principal Land Use Planner in the Ministry of Lands and Housing, Department of Town and Regional Planning. She was also an Environmental Impact Assessment Practitioner and Private Archaeologist. She was a Senior Teacher in the Ministry of Education and served in various schools mainly in the Greater Gaborone area. She holds an MSc and MPhil in Environmental Science and is a PhD candidate Environmental Science (All from University of Botswana). She mainly investigated landforms of the hardveld in Botswana, including the geodiversity and geomorphology of caves and gorges and geoarchaeology. Her research interests include geoarchaeology, karst geomorphology, geoheritage, remote sensing and geographic information systems, land use planning and sustainable socio-economic development of local communities.

Mark Stephens is Associate Professor of Geography at the University of The Bahamas and was previously Senior Lecturer in Geomorphology at the University of Botswana. He holds a M.Sc. in Quaternary Science, and Ph.D. in Geography (both University of London). He mainly investigated landforms of the hardveld in Botswana, including the geodiversity and geoconservation of caves and gorges. His research interests include tropical environmental change, natural hazards, karst geomorphology, geoarchaeology, and geoheritage.

Zoogeomorphology of Botswana

22

Jeremy S. Perkins

Abstract

Zoogeomorphology is a relatively new discipline and is set for changes this century in Botswana that have not been seen for millennia. The co-evolution of landscapes and wildlife amidst a changing climate means that the precise role of each is difficult to determine. As southern Africa becomes hotter and drier due to anthropogenic climate change this century, past adaptation strategies such as wildlife movements along (Balinsky's 1962) 'drought corridor' will no longer be possible due to land use/land cover change and agriculture and human-related expansion. As the Sixth Great Extinction unfolds, it offers a unique opportunity to study just how significant different biota are in determining geomorphology, albeit as our climate changes. Fauna that has remained intact since the Miocene will largely disappear from African shores and the palaeo-dune fields of the Kalahari may well become reactivated. The real significance of micro-organisms and invertebrates will then be realised as the bioturbators remain to shape the landscapes around them without the distraction posed by humans and the unique mega-fauna that surrounded them.

Keywords

Zoogeomorphology • Kalahari • Drought corridor • Migration • Bioturbators

22.1 Introduction

'The study of the effects of animals as geomorphic agents', or zoogeomorphology (Butler 2019) has developed only recently (Butler 1995) and quite markedly over the last decade (Butler 2013; Butler et al. 2013, 2018). Humans have played a major role as geomorphic agents by altering landscapes, often at unprecedented scales and magnitude, and continue to do so, through land use land cover change (Venter et al. 2016), urbanisation (Chen et al. 2020), and climate change (Butler 2019). Broad theoretical aspects of zoogeomorphology at the species level have been well covered in the literature, although specific in-country examples are often lacking. This chapter focuses on three key aspects of zoogeomorphology:

(i) the way in which landscapes have impacted animals over time by influencing their key movements and migrations, and
(ii) finer-scale impacts of animals in the landscape, within the specific context of Botswana.
(iii) The question as to whether animals influence geomorphology at anything but at a local scale.

It is striking that the seminal work on the landforms of Botswana by Grove (1969) emphasised the role of tectonics and climate change in shaping what we see around us today but made only one reference to what is now considered to be zoogeomorphology—by mentioning the role hippos can play in influencing the route floodwaters take through the Okavango Delta (see also Stigand 1923). To use the terminology of Molnar (2003), tectonics set the initial conditions by creating the Kalahari Basin (nature) while climate change (nurture) infilled it with Kalahari Sand and shaped the dune system between the Late Pliocene and Early Pleistocene (Grove 1969). The Okavango system is formed due to downwarping and movement along the terminus of the African Rift system with wet episodes evidenced by

J. S. Perkins (✉)
Department of Environmental Science, University of Botswana, Private Bag UB, 0022 Gaborone, Botswana
e-mail: perkinsjs@ac.ub.bw

F. D. Eckardt (ed.), *Landscapes and Landforms of Botswana*, World Geomorphological Landscapes,
https://doi.org/10.1007/978-3-030-86102-5_22

concentric shorelines around the Okavango Delta and Makgadikgadi Pans (Jacobberger and Hooper 1991). Mega palaeo lake Makgadikgadi was recently linked to the beginning of human evolution, with migration a key strategy when the lake dried up due to climate change (Chan et al. 2019).

In many respects, the flora is the painting, but the canvas it plays out upon was determined predominantly by much bigger forces, such as the filling up of the Kalahari Basin with windblown sand and tectonics that re-directed the rivers, and climate change that led to the creation of mega lakes and dune fields. This chapter is primarily concerned with how animals can influence the shape of the canvas itself as well as how they must utilise it to adapt to climate change, much as they have done in the past. The painting in itself is fascinating, but it is the interactions of animals with the canvas that all too often gets lost in its viewing and that is the real focus here. The latter has been transformed in the last century by the actions of humans, as Grove (1977) puts it, *'Man's activities are subjecting life in many semiarid lands to stresses of a kind and degree which it may not have experienced in the past, and may be unable to meet effectively'* (p. 303). Climate change seems likely to expose this with drastic consequences in terms of species extinctions as range shifts and movements that enabled many ungulates to adapt in the past will simply not be possible today.

22.2 Landscapes Impacting Animals

22.2.1 Palaeo Pathways

The more one studies the ecosystems and landscapes of Africa, the greater one is struck by the importance of the African rift valley system which has been splitting the African continent since the late Eocene-Oligocene (Pastier et al. 2017). Its termini are the Olduvai Gorge in Kenya (Reynolds et al. 2016) and the Okavango Delta/palaeo lake Makgadikgadi (Fadel et al. 2020; Chan et al. 2019) in Botswana—both of which seemingly tussle for the coveted title of being associated with the beginning of humankind. The biodiversity associated with each terminus is considerable even to this day, with both thought to play essential roles as refugia when the climate changes, much as will happen this century.

At this scale, Balinsky (1962) emphasised the existence of a 'drought corridor' in Africa that ran from the Kalahari-Namib ecosystem along the Rift Valley, then through Ethiopia to the Sahara-Sahelian zone (Fig. 22.1). The movement of mammals along the 'drought corridor', as well as range shifts, enabled them to adapt to the changing climate (Coe and Skinner 1993; Berry 1997). In wet periods the rainforests of the Congo expanded to cut off the 'drought corridor' that was then reinstated during hotter and drier periods (Balinsky 1962). As the climate becomes hotter and drier in southern Africa this century (Engelbrecht et al. 2015), movement along the 'drought corridor' will again be integral to the survival of many key ungulate populations (Perkins 2019). Unfortunately, such 'palaeo pathways' are today blocked by land use and land cover changes and high human density around key geomorphological barriers such as Lake Malawi (Perkins 2020). Moritz and Agudo (2013) emphasise the need to '*manage and restore eco-evolutionary dynamics across heterogenous landscapes*' so as to ensure '*habitat connectivity with long-term climatic refugia*'.

The Kalahari sand plain is characterised by dunes and inter-dune valleys, shallow depressions or pans and dry river valleys, all of which impart significant geomorphic variation on the landscape. Dryland ecosystems are characteristically dis-equilibrial (Sietz et al. 2011) and the relict nature of these landforms today owes much to the extreme climatic excursions that formed them (Rinaldo et al. 1995). While rainfall, fire and the action of herbivores (Huntley and Walker 1982) modify the vegetation upon them, the dry river valleys, pans, palaeo lakes and dunes remain an enduring legacy of past climate change with which animals have co-evolved.

Mobility is central to understanding how many key ungulates survive in the savannas of Africa and it has also long been recognised that migratory systems are more productive than those in which the ungulates are sedentary (Fryxell et al. 1988). Unfortunately, migrations are under unprecedented pressure globally (van Oosterzee 2017) from land use land cover change driven by humans (Holechek and Valdez 2017), that includes fences (Jakes et al. 2018), with the wildebeest in the Serengeti increasingly isolated (Msoffe et al. 2019) as are large herbivore populations in southern Africa (Perkins and Ringrose 1996; Perkins and Naidoo 2016).

Biogeomorphic processes associated with range shifts can be separated from the dry season concentration around water points and wet season dispersal that water-dependent herbivores such as elephants, buffalo and zebra follow. The water that collects in shallow depressions or pans and dry river valleys provide the steppingstones for this dispersal and are linked by a complex network of game trails as animals fulfil their needs for water and salts such as sodium. Elephants are ecosystem engineers and powerful earthmovers (Haynes 2012) and while it is rainfall that enables the dispersal of water-dependent herbivores from the permanent waters of the Chobe-Linyanti-Kwando system it is geomorphology and the availability of surface water that directs it (Chaps. 6 and 7; Fig. 22.2).

Many key wildlife populations make movements and foraging decisions at the landscape level, with the Kalahari far from a homogenous plain in this regard. Zebra and wildebeest graze the Makgadikgadi Pan grasslands in the

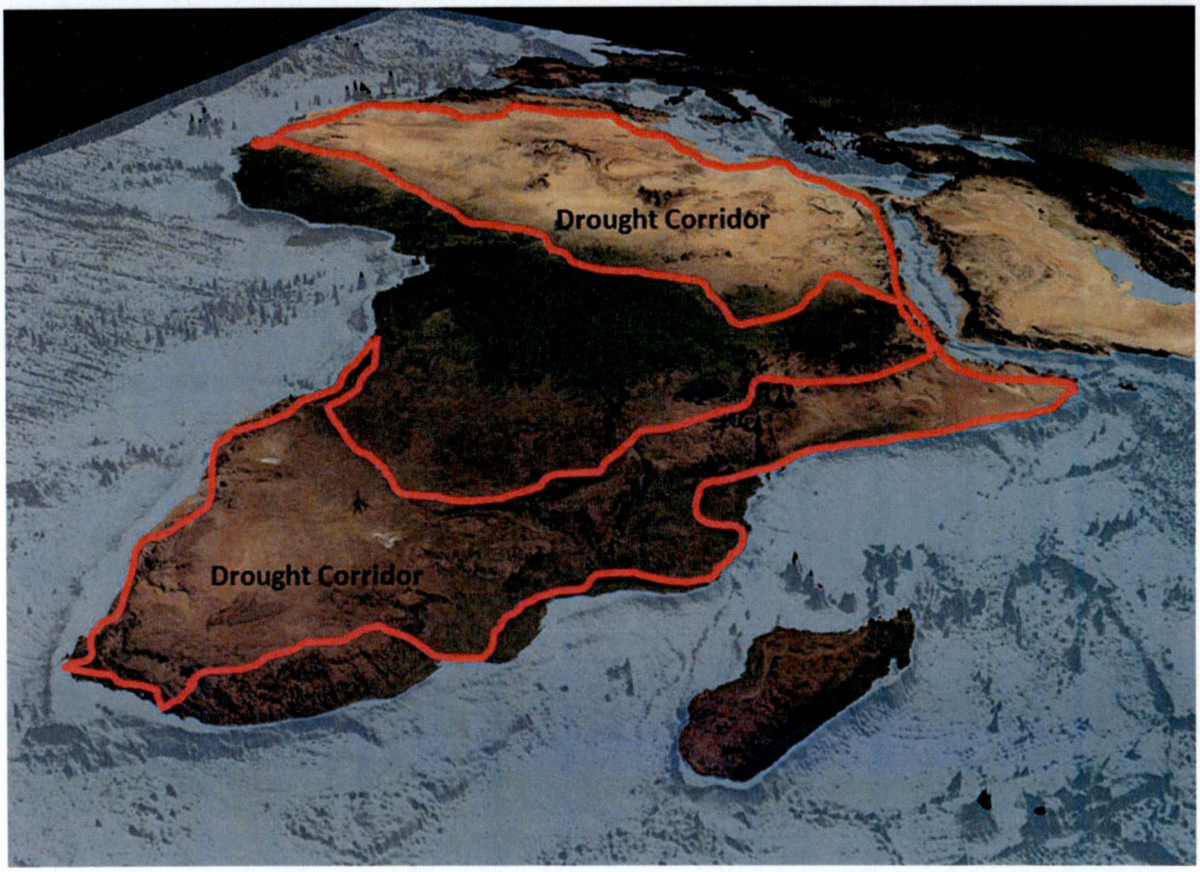

Fig. 22.1 Drought corridor—areas with marked dry seasons in which mean monthly rainfall is less than 10 mm in at least three consecutive months (Balinsky 1962). Location of drought corridor is illustrative only with more detailed spatial definition available from Bouvet et al. (2018)

wet season and migrate to the Boteti River in the dry, when the floodwaters from the Okavango Delta arrive. A great diversity of biogeomorphic processes are linked to these movements and the apex predators (lions) that follow them maintain the functional integrity of semi-arid savannas. Megaherbivores, such as elephants, have shown to be able to reverse decades of loss of soil carbon and nitrogen pools and affect the long-term functioning of ecosystems (Sitters et al. 2020).

The consequences of constraining such mobility have been all too clearly shown in the Kalahari where blue wildebeest (*Connochaetes taurinus*) display wet season concentration in the belt of mineral-rich pans that runs from Ncojane to Hukuntsi (the Schwelle) (Chap. 10; Fig. 22.3) and dry season dispersal from it along dry river valleys such as the Okwa (DHV 1980). The blue wildebeest calve in the Schwelle drinking the water that collects there as well as eating the sodium-rich soil found in the pans (Knight et al 1987). Like the eland and gemsbok they eat melons and survive by digging up roots and tubers (Knight 1995), but unlike them, the blue wildebeest must find surface water in a drought. Historically this meant moving to the Okavango Delta (Chaps. 2 and 3) and as this became blocked by fences to the Boteti River and Lake Xau, the most southerly distributary of the system (DHV 1980). It was the longest blue wildebeest migration in Africa, involving a trek of some 800 km (Bonifica 1992), four times that of the world-renown Serengeti—Mara migration, dispersing nutrients throughout the Kalahari (de Queiroz 1992) and subjecting the pastures to a rest and recovery grazing regime.

In the 1980s drought when the blue wildebeest dispersed from the Schwelle they found their path blocked by fences (Fig. 22.3), which combined with competition for water and grazing with people and their livestock at Lake Xau (Williamson and Williamson 1981; Williamson and Mbano 1988). This resulted in the blue wildebeest population crashing from over 350,000 to less than 20,000, from which it has never recovered (Perkins and Ringrose 1996). As the latter point out the veterinary fences for disease control imposed a geographic constraint on mobility that is unparalleled in nature and continues to this day to profoundly affect the functioning of the Kalahari system. Indeed, it has

Fig. 22.2 Elephants at a water hole in northern Botswana

heralded a new era of utilisation in which the spatial scale of utilisation has changed from regional movements in a wildlife dominated system, to more local movements dominated by cattle. It represents an intensification of land utilisation with a corresponding drastic change in biogeomorphic processes, in particular, mediated through fire effects and bush encroachment. Unfortunately, this sequence of increasingly isolated movements and decline seems likely to be repeated for many key wildlife populations throughout the drought corridor and will be greatly accentuated by climate change.

22.2.2 Changing the Nature and Scale of Zoogeomorphology

The great painting of the Kalahari, where ecosystem functioning and biogeomorphic processes were driven by fluctuations in rainfall, nutrients and surface water (nurture) at the landscape level (nature), has collapsed over the last century and appears set to do so further as the climate warms (Mayaud et al. 2017). It was the onset of deep borehole drilling that 'broke down' the 'age-old' protection of the Kalahari from permanent grazing by cattle (Cooke 1985) and also served to dislocate the ecology of the system from the geomorphology that sustained it. As a result, the Kalahari has shifted from a system dominated by wildlife and utilisation by hunter-gatherers (for the last 40,000 years) to one dominated by domestic stock. It is a trend that has characterised Africa (Hempson et al. 2017) and in the Kalahari has transformed the painting (vegetation) and impacted the canvas (landscape), primarily by altering key ecological processes and the spatial and temporal scale at which they operate. The so-called piosphere effect, or patterning which results from the concentration of large herbivores around a water point, provides a good example.

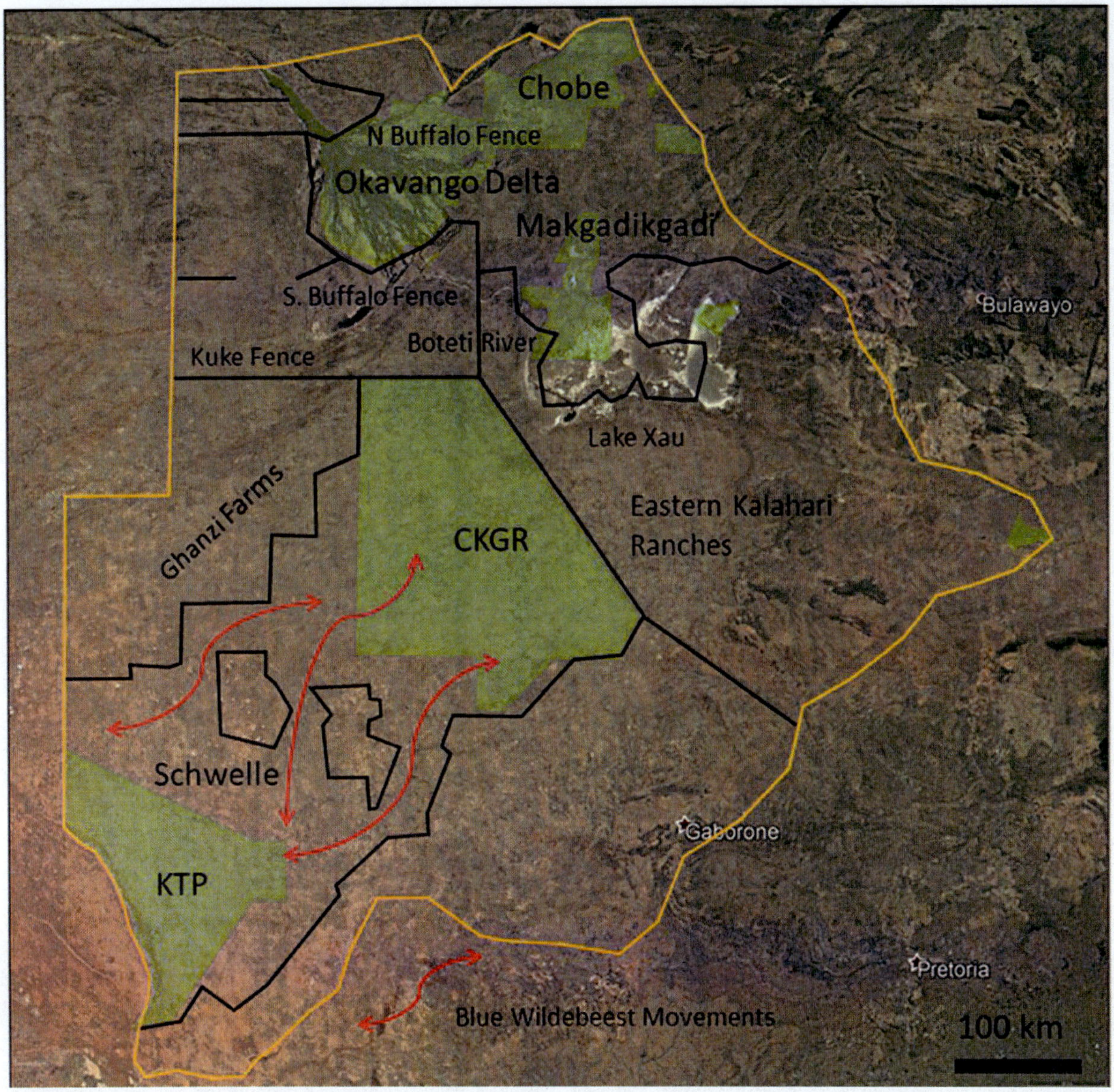

Fig. 22.3 Kalahari blue wildebeest movements constrained by fences. The full range would have extended from northern Botswana to the Schwelle but is now restricted to residual fenced corridors between the Central Kalahari Game Reserve (CKGR) and the Kalahari Trans-Frontier Park (TFP)

22.3 Impacts of Animals on the Landscape

22.3.1 Piospheres

The way in which livestock impacts on erosion processes (Trimble and Kendel 1995; Butler 2013), particularly by creating bare soil (Pages et al. 2019), is neatly illustrated in Botswana by the impact of livestock grazing in both the Kalahari and the eastern quarter of the country (hardveld)—where most people reside, and the soils and relief are relatively more diverse. Heavy stocking rates in the latter have accentuated soil erosion rates, while in the Kalahari there is little or no surface runoff, as infiltration rates are high due to the extensive sand cover. Wind erosion tends to be localised except on the finer soils of the dry and bare palaeo lakes where the dust erosion can be significant (Bryant et al. 2007).

In Botswana livestock grazing around waterpoints has led to distinct zones or piospheres (Lange 1969) (Fig. 22.4), involving a bare ground zone around the borehole (0–200 m), an extensive bush thickened zone (200–5000 m+)

Fig. 22.4 Piospheres in the eastern Kalahari (CKGR to the west)

and depending upon borehole density, an outlying grazing reserve to become established (Hess et al. 2020). The centripetal movement of nutrients (Tolsma et al. 1987) to the kraal-waterpoint axis pollutes both the groundwater (Schwiede et al. 2005) and the air (Butterbach-Bahl et al. 2020). Bush encroachment in the outlying areas has detrimentally altered the land capability of millions of hectares of rangeland in southern Africa (O'Connor et al. 2014), changing water and energy budgets in the process (Schreiner-McGraw et al. 2020).

In the hardveld, bush encroachment has been accompanied by pronounced soil erosion with rills and gullies opening up over extensive areas, with riparian zones especially vulnerable when grazed continuously. In fact, bush encroachment can 'mask' the occurrence of pronounced soil loss and areas that resemble badlands as the bushes stand up on pedestals as the surrounding soil is washed into the rivers, all under a canopy of thorn bushes.

In the southern Kalahari, it is too dry for bush encroachment to occur, except on the pans and dry river valleys, such that the palaeo linear dunes (Fig. 22.5) around the borehole itself become re-activated and form spectacular dune features (Figs. 22.6 and 22.7) (Perkins 2019). The dunes do not advance outwards but appear to be strictly confined to the zone immediately around the borehole, with dune crests active up to a distance of 1.2 km from the borehole—where the cover of *Stipagrostis amabilis* had been reduced by trampling (Perkins 2018). Wiggs et al. (1995) found that below about 14% mean vegetation cover the dunes became active, with Thomas et al. (2005) predicting enhanced dune activity in the southern dune field by 2039, regardless of the emission scenario used, unless appropriate land management practices are adopted (Mayaud et al. 2017).

The southern Kalahari illustrates how the actions of humans, by shifting the system from a wildlife to a livestock dominated one, can impact upon the geomorphology of the area. Water points are the key as they sedentarise the ungulates that use them with their provision for wildlife in and around protected areas, linked to an array of impacts that will prove detrimental to the resilience of the Kalahari (Perkins 2020). Geomorphology, particularly in the form of dry river valleys and pans, holds the key to understanding how the Kalahari ecosystem functions and how its wildlife populations can best be conserved and utilised. Unfortunately, the role of geomorphology in shaping the utilisation of the Kalahari system has been almost entirely overlooked.

22.3.2 Examples of the Impact of Animals on the Landscape

Elephants dig holes in dry riverbeds to access water in the dry season (Purdon et al. 2018), and along with other water-dependent herbivores such as buffalo (Bennitt et al. 2016) and zebra (e.g., Naidoo et al. 2014), trek back from their wet season range utilising the last remaining water in the pans and depressions along the way, as the dry season sets in (see Fig. 22.8). Elephants wallowing have long been thought to be behind the formation of pans (Weir 1972) along with a structural hydro-geological explanation. Geomorphology can interact with wildlife in very simplistic and

Fig. 22.5 Southern Kalahari without piospheres

Fig. 22.6 Southern Kalahari dunes activated by cattle

Fig. 22.7 Southern Kalahari piosphere—'sacrifice zone'

also complex ways. In 2018, hundreds of buffalo drowned in the Chobe River when they stampeded into the river and were unable to exit due to the steep banks (Reuters 2018).

Geophagy in the Kalahari blue wildebeest has already been emphasised with many ungulates attracted to salt licks, with elephants in northern Botswana a good example, where they can be seen actively excavating areas of salt accumulation. In a dystrophic system such as the Kalahari, relative differences in mineral content matter greatly and significant spatial variation in their distribution is expressed through the geomorphology of the landscape. Excavators such as ant bears create unique geomorphology when they excavate termite mounds (Butler 1995) and in doing so create habitats for other species that can even help them persist in the area when surface temperatures rise excessively (Pike and Mitchell 2013). Biogeomorphologically, Dandurand et al. (2018) link condensation corrosion from the occupation of over 60,000 bats as important in influencing the interior shape of Drotsky's Cave at Gwichaba Hills (Chap. 14).

Apex predators are known through trophic cascade effects to profoundly affect entire food webs and the functioning of ecosystems (Morris and Letnic 2017), often through indirect effects that are only realised after the key predator species have been lost from the system (e.g. Beschta and Ripple 2016). Cattle are a significant geomorphic agent (Trimble and Kendel 1995) with the onset of cattle keeping a key marker of the Anthropocene (Smith and Zeder 2013). Zerboni and Nicoll (2018) detail how the introduction of cattle and herding into North Africa between 6000 and 8000 years ago left a number of zoogeomorphological features that include tracks and trails, rock shelters and accentuated surface processes like erosion and dust mobilisation. Similar features exist throughout Botswana where livestock keeping can be traced back at least several thousand years (Hall 1986). Significantly livestock has profoundly changed the savannas of southern Africa, particularly in terms of their functioning, appearance, composition and dynamics.

Fig. 22.8 Piosphere utilised by elephants in northern Botswana

22.3.3 Termites

In many respects, termites are the hidden hand influencing the zoogeomorphology of savannas and a key driver of ecosystem functioning, in turn influencing the role of herbivory and fire. Regular patterning on the landscape (Pringle et al. 2010) and the regular spacing of mounds (Schuurman 2007; Levick et al. 2010) provide clues as to their importance, although most of their work occurs underground and under a protective covering of soil. In the Okavango Delta termite mounds are associated with tree islands and their own unique geomorphology by influencing the local flooding regime (McCarthy et al. 2012) just as hippos do with their trails (Figs. 22.9 and 22.10).

Termite fungiculture dates back to some 31 Ma (Roberts et al. 2016) which may have coincided with the African rift valley initiation in the Paleogene (O'Bryan et al. 2018). Termites are ecosystem engineers, sequestering carbon and recycling nutrients that drive primary and secondary production and buffer them against climate change (Tarnita et al. 2017). While production rates vary by three to four orders between wood, grass, soil or fungus-feeding termites, they are believed to be responsible for between 1 and 3% of global CH_4 emissions (Kirschke et al. 2013 from Chiri et al. 2020). Termite '*mounds are really the supermarkets of the savanna'* (Palmer from Pennisi 2015; p. 597) reprocessing more than 90% of the dry wood in some semi-arid savannas, in the case of the Macrotermitinae (O'Bryan et al. 2018).

The need to move away from our pre-occupation with large mammals in our attempts to understand ecosystem functioning and landscape appearance is taken one step further by Cavicchioli et al. (2019) who emphasise the fundamental role played by micro-organisms in carbon and nutrient cycling, and bestowing resilience. Nitrogen fixing soil crusts have long been felt to be important in the Kalahari (Skarpe and Henriksson 1987) and have been shown to be inversely proportional to the trampling pressure exerted by domestic stock. The direct and indirect feedback loops between termite activity and geomorphology are considerable as they influence a number of key factors such as soil

Fig. 22.9 Northern Botswana showing hippo trails

properties, slope, above-ground biomass, nutrient status, resistance to climate change (Bonachela et al. 2015) and soil erosion by wind and water.

22.3.4 Biotic Versus Abiotic

It is easy to get lost in the specifics of zoogeomorphology, such as the local impacts of whitefronted bee-eaters nesting in mud cliffs (Fig. 22.11), aardvarks digging burrows (Butler et al. 2013), and miss the bigger picture such as quite how burrowing impacts at the ecosystem level (Louw et al. 2019). Zoogeomorphology is now beginning to grapple with these issues that lie at the core of whether or not animals influence the landscape at anything but at the local scale. Land–ocean–atmosphere couplings through the hydrological cycle have long suggested that profound changes to ecosystem functioning can occur through human-induced land use land cover changes, as mediated through changes to albedo (Charney 1975). In this respect, it is the arid and adjoining semi-arid lands that have attracted the most attention and will be the first to undergo extensive degradation due to reduced vegetation cover and increased erosion (Kleidon et al. 2000).

Zoogeomorphology runs the risk of placing too much emphasis on prominent species of animals, including alien and invasive species, and how they influence the local terrain and not enough on the abiotic processes that ultimately drive the system. The loss of over half a million blue wildebeest and red hartebeest in the drought of the 1980s, due to reduced mobility to key resource areas due to veterinary cordon fencing for disease control (Perkins and Ringrose, 1996), has meant that much of the primary biomass in extensive wildlife areas of the Kalahari remains unutilised. As a result, widespread and severe catastrophic fires can sweep through these areas in the late dry season, accentuated further by the climate becoming hotter and drier as a result of anthropogenic climate change (Bowman et al. 2020). While triggered indirectly by human beings, the ecosystem level changes that occur to hydrological and mineral cycles through such fires far exceed those that can be expected from the interactions of the species with the landscape itself (Lewin 1986). Mining for minerals provides a good example as to how humans can directly and profoundly modify the

Fig. 22.10 The 'flood' moving towards Maun with termite mounds creating islands

landscape and create new habitats in the process (Fig. 22.12).

Open-cast mining for minerals, especially diamonds, copper, gold and coal, has created its own unique geomorphology that includes open pits, waste rock dumps, spoil heaps and tailing dams.

Indeed, the ability of human beings to influence the geomorphology around them in both direct and indirect ways should not be underestimated. Significantly, many of the waste rock dumps are not static and inert but are being actively transformed by bacteria and archaea. These 'rock-eating' acidophiles are inert when buried deep underground, but when exposed to moisture and oxygen on the surface erodes the landscape by breaking down sulphide-containing minerals in rocks and releasing acids into the environment (acid mine drainage) with damaging consequences (Johnson and Hallberg 2003). The harshest environments on the earth provide permanent habitat to micro-organisms and invertebrates, emphasising their resilience. The salt pans of Makgadikgadi have their own unique ecosystem and like their eastern African rift valley equivalents have very high microbial biomass (Buatois et al. 2020).

Cavicchioli et al. (2019) put humanity on notice with regard to our need to appreciate the role micro-organisms play in the biosphere, for at least 3.6 billion years and will play in future extinction events. Landscapes can be expected to change profoundly in their appearance this century as cultivated lands are *'likely to be much more vulnerable to climatic variability and much more likely to tip over into more permanently degraded landscapes'* (Palmer from Pennisi 2015; p. 597). Global human-made mass (of which concrete dominates) now exceeds all living biomass, with profound implications for ecosystems and biodiversity and the key climatic and biogeochemical cycles (Elhacham et al. 2020) that sustain us.

Dietrich and Perron (2006) ask the most critical question of all, as to whether or not biota makes much difference to the landforms, we see around us. They note the abundance of familiar-looking valley networks on images of Mars and that on Earth tiny invertebrates influence landforms at much larger scales. The global loss of the world's largest megabiota (Enquist et al. 2020), including herbivores (Ripple et al. 2015) and carnivores (Ripple et al. 2014), does of course define the Anthropocene (Dirzo et al. 2014) and while there is some debate as to when this period should begin (Diaz et al. 2020), there seems little evidence currently to suggest that without transformative change, the outcome for humanity will not be good.

Fig. 22.11 Whitefronted Bee-Eater—Okavango Panhandle

There can be little doubt that complex eco-evolutionary inter-relationships between animals and the landscape have developed over time and have been modified by climate change. Many adaptations that worked in the past, such as the use of the 'drought corridor', are no longer possible. As the Sixth Great Mass Extinction progresses (Dirzo et al. 2014), we are likely to realise how little we knew about zoogeomorphology and how integral it was to our way of life and well-being. Bats provide a good example with over 1,300 species providing an array of ecosystem services that are important also for conservation and disease (Nature Ed 2017). The latter and our failure to respect nature have been linked by some to the outbreak of Covid-19 (Tollefson 2020), which in turn has impacted heavily upon the future of Africa's dwindling wildlands and protected areas (Lindsey et al. 2020).

The Sixth Great Extinction represents a tragic global experiment which should reveal the relative role of biotic and abiotic factors in shaping the landscapes we see around us. Most definitely, however, the painting will change in ways in which we have not seen before, but more importantly, we will come to realise that we have destroyed the very canvas on which we depended.

Bioturbation was the focus of Darwin's last book (Meysman et al. 2006) and the latter authors emphasise that the profound ecological and evolutionary role that bioturbation has played in both the oceans (Gingras and Konhauser 2015; van de Velde et al. 2018) and on land (Benton 1988) would have pleasantly surprised Darwin, not least because his book on earthworms was poorly received. When it comes to the role of animals in landscape evolution it is clear that there is much that is yet to be understood and a

Fig. 22.12 Waste rock dump and rock dassies (inset)

unique topographic signature remains to be found (Dietrich and Perron 2006).

22.4 Conclusion

Regarding the Kalahari as an undifferentiated sand plain belies significant variation at the landscape level that has shaped animal movements and behaviour via eco-evolutionary feedbacks. At a continental scale when the climate changed in the past Balinsky's (1962) 'drought corridor' was the key to understanding how the key ungulates adapted and are distributed as they are today. Such macro-level movements are no longer possible and while intra-regional and more localised migrations occur today, they are increasingly under threat from land use land cover change. Finer scale impacts of animals on the landscape, such as digging and burrowing, not only influence what we see but provide habitats for a multitude of species, not least because of the relatively cooler micro-environment they provide.

It is poignant to reflect on our contribution to zoogeomorphology in the Anthropocene. Turner and Anton (2004) emphasise that '*while the biota of Africa has changed and its diversity has shrunk, the continent is still the home of the longest-enduring mammalian fauna, with a higher proportion of Pliocene species than remains on any other continent. The richness is of long-standing because it is the product of a biomic patterning whose origins can be traced back to the Miocene. Put most simply, the modern mammalian fauna of Africa is the oldest in the world. One of our fossil relatives from among the australopithecenes visiting the Serengeti today might well comment on the absence of this or that species but would still happily recognize virtually all the animals on the plain. We cannot be so careless as to let that situation change - can we?'* (Turner and Anton 2004; p. 242).

Humans walking around the Kalahari in 2100 seem unlikely to experience such diversity and will probably be struck by the simplicity and uniformity of the ecosystems they encounter, as well as how hot and dry it is. Indeed, when they see the huge pits surrounded by overburden like those at Jwaneng and Orapa diamond mines, in particular, they will probably wonder at what it was we were looking for or trying to create. It will be an enduring human imprint on the landscape throughout the world and one that attests to our seemingly insatiable desire to change and modify what is around us for nothing more than the simple pursuit of money. Sadly, it is not cave drawings like those at Tsodilo Hills that we will leave future generations to ponder over, but severely modified, exploited and polluted landscapes that attest to our complete disregard for sustainability and by extension, our own survival. As the Sixth Great Extinction unfolds this century, it is likely that we will only appreciate the profound co-evolution that has occurred between organisms and landscapes, via a process of species elimination. It can only be hoped that our ransacking of the planet at least leaves the bioturbators behind so that life on Earth can start over, long after humans have demised.

Acknowledgements This chapter benefited greatly from a review by Professor Mike Meadows.

References

Balinsky BI (1962) Patterns of animal distribution on the African continent (summing-up talk)

Bennitt E, Bonyongo MC, Harris S (2016) Effects of divergent migratory strategies on access to resources for Cape buffalo (*Syncerus caffer caffer*). J Mammal 97:1682–1698

Benton M (1988) Burrowing by vertebrates. Nature 331:17–18. https://doi.org/10.1038/331017a0

Berry HH (1997) Aspects of wildebeest Connochaetes taurinus ecology in the Etosha National Park - a synthesis for future management. Modoqua 20(1):137–148

Beschta RL, Ripple WJ (2016) Riparian vegetation recovery in yellowstone: the first two decades after wolf reintroduction. Biol Cons 198:93–103

Bonachela JA, Pringle RM, Sheffer E, Coverdale TC, Guyton JA, Caylor KK, Levin SA, Tarnita CE (2015) Termite mounds can increase the robustness of dryland ecosystems to climatic change. Science 347:651–655

Bonifica (1992) Technical assistance to the project 'initial measures for the conservation of the Kalahari ecosystem. Commission of the European Communities. Bonifica S.p.a. Final Report

Bouvet A, Mermoz S, Le Toan T, Villard L, Mathieu R, Naidoo L, Asner GP (2018) An above-ground biomass map of African savannahs and woodlands at 25 m resolution derived from ALOS PALSAR. Remote Sens Environ 206:156–173. https://doi.org/10.1016/j.rse.2017.12.030

Bowman DM, Kolden CA, Abatzoglou JT, Johnston FH, van der Werf GR, Flannigan M (2020) Vegetation fires in the Anthropocene. Nat Rev Earth Environ. https://doi.org/10.1038/s43017-020-0085-3

Buatois LA, Renaut RW, Owen RB, Behrensmeyer AK, Scott JJ (2020) Animal bioturbation preserved in Pleistocene magadiite at Lake Magadi, Kenya Rift Valley, and its implications for the depositional environment of bedded magadiite. Sci Rep 10:6794. https://doi.org/10.1038/s41598-020-63505-7

Bryant RG, Bigg GR, Mahowald N, Eckardt FD, Ross S (2007) Dust emission response to climate in southern Africa. J Geophys Res Atmos. 112. https://doi.org/10.1029/2005JD007025

Butler DR (1995) Zoogeomorphology – animals as geomorphic agents. Cambridge University Press, Cambridge and New York

Butler DR (2013) Grazing influences on geomorphic systems. In: Shroder J (Editor in Chief), James LA, Harden CP, Clague JJ (eds) Treatise on geomorphology. Academic Press, San Diego, CA, vol 13, Geomorphology of human disturbances, climate change, and natural hazards, pp 68–73

Butler DR, Whitesides CJ, Wamsley JM, Tsikalas SG (2013) The geomorphic impacts of animal burrowing and denning. In: Shroder J(Editor in Chief), Butler DR, Hupp CR (eds) Treatise on geomorphology. Academic Press, San Diego, CA, vol 12, Ecogeomorphology, pp 271–280

Butler DR, Anzah F, Goff PD, Villa J (2018) Zoogeomorphology and resilience theory. Geomorphology 305:154–162. https://doi.org/10.1016/j.geomorph.2017.08.0360169-555X

Butler DR (2019) *Zoogeomorphology.* Int Encycl Geogr, 1–6. https://doi.org/10.1002/9781118786352.wbieg1119.pub2

Butterbach-Bahl K, Gettel G, Kiese R, Fuchs K, Werner C, Rahimi J, Barthel M, Merbold L (2020) Livestock enclosures in drylands of Sub-Saharan Africa are overlooked hotspots of N_2O emissions. Nat Commun 11:4644. https://doi.org/10.1038/s41467-020-18359-y

Cavicchioli R, Ripple WJ, Timmis KN, Azam F, Bakken LR, Baylis M, Behrenfeld MJ, Boetius A, Boyd PW, Classen AT, Crowther TW, Danovaro R, Foreman CM, Huisman J, Hutchins DA, Jansson JK, Karl DM, Koskella B, Welch DBM, Martiny JBH, Moran MA, Orphan VJ, Reay DS, Remais JV, Rich VI, Singh BK, Stein LY, Stewart FJ, Sullivan MB, van Oppen MJH, Weaver SC, Webb EA, Webster NS (2019) Scientists' warning to humanity: microorganisms and climate change. Nat Rev Microbiol 17:569–686. https://doi.org/10.1038/s41579-019-0222-5

Chan EKF, Timmermann A, Baldi BF, Moore AE, Lyons RJ, Lee S-S, Kalsbeek AMF, Petersen DC, Rautenbach H, Förtsch HEA, Bornman MSR, Hayes VM (2019) Human origins in a southern African palaeo-wetland and first migrations. Nature 575:185–189. https://doi.org/10.1038/s41586-019-1714-1

Charney JG (1975) Dynamics of deserts and drought in Sahel. Q J Roy Meteorol Soc 101:193–202

Chen G, Li X, Liu X, Chen Y, Liang X, Leng J, Xu X, Lio W, Qiu Y, Wu Q, Huang K (2020) Global projections of future urban land expansion under shared socioeconomic pathways. Nat Commun 11, 537. https://doi.org/10.1038/s41467-020-14386-x

Chiri E, Greening C, Lappan R, Waite DW, Jirapanjawat T, Dong X, Arndt SK, Nauer PA (2020) Termite mounds contain soil-derived methanotroph communities kinetically adapted to elevated methane concentrations. ISME J. https://doi.org/10.1038/s41396-020-0722-3

Coe MJ, Skinner JD (1993) Connections, disjunctions and endemism in the eastern and southern African mammal faunas. Trans R Soc South Afr 48, 233–255. https://doi.org/10.1080/00359199309520273

Cooke HJ (1985) The Kalahari today: a case of conflict over resource use. Geogr J 151:75–85

Dandurand G, Duranthon F, Jarry M, Stratford DJ, Bruxelles L (2018) Biogenic corrosion caused by bats in Drotsky's Cave (the Gcwihaba

Hills, NW Botswana). Geomorphology. https://doi.org/10.1016/j.geomorph.2018.10.027
de Queiroz JS (1992) Range degradation in Botswana: myth Or Reality? - Pastoral Development Network Papers 35b. pp 19.
DHV (1980). Countrywide animal and range assessment project. 7 vols. European Development Fund and Ministry of Commerce and Industry. Amersfoort, The Netherlands
Díaz SJ, Settele J, Brondízio ES, Ngo HT, Agard J, Arneth A, Balvanera P, Brauman KA, Butchart SHM, Chan KMA, Garibaldi LA, Ichii K, Liu J, Subramanian SM, Midgley GF, Miloslavich P, Molnár Z, Obura D, Pfaff A, Polasky S, Purvis A, Razzaque J, Reyers B, Chowdhury RR, Shin Y-J, Visseren-Hamakers I, Willis KJ, Zayas CN, (2020) Pervasive human-driven decline of life on Earth points to the need for transformative change. Science 366, eaax3100
Dietrich WE, Perron JT (2006) The search for a topographic signature of life. Nature 439(7075):411–418. https://doi.org/10.1038/nature04452
Dirzo R, Young HS, Galetti M, Ceballos G, Isaac NJB, Collen B (2014) Defaunation in the anthropocene. Science 345(6195):401–406. https://doi.org/10.1126/science.1251817
Elhacham E, Ben-Uri L, Grozovski J, Bar-On YM, Milo R (2020) Global human-made mass exceeds all living biomass. Nature 588:442–444. https://doi.org/10.1038/s41586-020-3010-5
Engelbrecht F, Adegoke J, Bopape M-J, Naidoo M, Garland R, Thatcher M, McGregor J, Katzfey J, Werner M, Ichoku C, Gatebe C (2015) Projections of rapidly rising surface temperatures over Africa under low mitigation. Environ Res Lett. https://doi.org/10.1088/1748-9326/10/8/085004
Enquist BJ, Abraham AJ, Harfoot MBJ, Malhi Y, Doughty CE (2020) The megabiota are disproportionately important for biosphere functioning. Nat Commun 11:699. https://doi.org/10.1038/s41467-020-14369-y
Fadel I, Paulssen H, van der Meijde M, Kwadiba M, Ntibinyane O, Nyblade A, Durrheim R (2020) Crustal and upper mantle shear wave velocity structure of Botswana: the 3 April 2017 Central Botswana Earthquake Linked to the East African Rift System. Geophys Res Lett 47(4). https://doi.org/10.1029/2019GL085598
Fryxell JM, Greever J, Sinclair ARE (1988) Why are migratory ungulates so abundant? Am Nat 131(6):781–798. https://doi.org/10.1086/284822
Gingras M, Konhauser K (2015) Digging deeper. Nature Geosci 8:825–826. https://doi.org/10.1038/ngeo2548
Grove AT (1969) Landforms and climatic change in the Kalahari and Ngamiland. Geogr J 135(2):191. https://doi.org/10.2307/1796824
Grove AT (1977) Desertification. Prog Phys Geogr: Earth Environ 1 (2):296–310. https://doi.org/10.1177/030913337700100204
Hall M (1986) The role of cattle in Southern African agropastoral societies: more than bones alone can tell. Goodwin Ser 5:83. https://doi.org/10.2307/3858150
Haynes G (2012) Elephants (and extinct relatives) as earth-movers and ecosystem engineers. Geomorphology 157–158:99–107. https://doi.org/10.1016/j.geomorph.2011.04.045
Hempson GP, Archibald S, Bond WJ (2017) The consequences of replacing wildlife with livestock in Africa. Sci Rep 7:17196. https://doi.org/10.1038/s41598-017-17348-4
Hess B, Dreber N, Liu Y, Wiegand K, Ludwig M, Meyer H, Meyer KM (2020) PioLaG: a piosphere landscape generator for savanna rangeland modelling. Landscape Ecol 35:2061–2082. https://doi.org/10.1007/s10980-020-01066-w
Holechek J, Valdez R (2018) Wildlife Conservation on the rangelands of eastern and Southern Africa: past, present, and future. Rangel Ecol Manage 71(2):245–258. https://doi.org/10.1016/j.rama.2017.10.005
Huntley BJ, Walker BH (1982) Ecology of tropical savannas. https://doi.org/10.1007/978-3-642-68786-0
Jacobberger PA, Hooper DM (1991) Geomorphology and reflectance patterns of vegetation-covered dunes at the Tsodilo Hills, North-West Botswana. Int J Remote Sens 12(11):2321
Jakes AF, Jones PF, Paige C, Seidler RG, Huijser MP (2018) A fence runs through it: a call for greater attention to the influence of fences on wildlife and ecosystems. Biol Cons 227:310–318. https://doi.org/10.1016/j.biocon.2018.09.026
Johnson DB, Hallberg KB (2003) The microbiology of acidic mine waters. Res Microbiol 154(2003):466–473
Kirschke S, Bousquet P, Ciais P, Saunois M, Canadell JG, Dlugokencky EJ, Bergamaschi P, Bergmann D, Blake DR, Bruhwiler L, Cameron-Smith P, Castaldi S, Chevallier F, Feng L, Fraser A, Heimann M, Hodson EL, Houweling S, Josse B, Fraser PJ, Krummel PB, Lamarque J-F, Langenfelds RL, Le Quéré C, Naik V, O'Doherty S, Palmer PI, Pison I, Plummer D, Poulter B, Prinn RG, Rigby M, Ringeval B, Santini M, Schmidt M, Shindell DT, Simpson IJ, Spahni R, Steele LP, Strode SA, Sudo K, Szopa S, van der Werf GR, Voulgarakis A, van Weele M, Weiss RF, Williams JE, Zeng G (2013) Three decades of global methane sources and sinks. Nat Geosci 6:813–823. https://doi.org/10.1038/ngeo1955
Kleidon A, Fraedrich K, Heimann M (2000) A Green Planet versus a desert world: estimating the maximum effect of vegetation on the land surface climate. Clim Change 44(4):471–493. https://doi.org/10.1023/A:1005559518889
Knight MH (1995) Tsama melons, Citrullus lanatus, a supplementary water supply for wildlife in the southern Kalahari. Afr J Ecol 33:71–80. https://doi.org/10.1111/j.1365-2028.1995.tb00782.x
Knight MH, Knight-Eloff AK, Bornman JJ (1987) The importance of borehole water and lick sites to Kalahari ungulates. J Arid Environ 15:269–281
Lange RT (1969) The piosphere, sheep track and dung patterns. J Range Manage 22:396–400
Levick SR, Asner GP, Chadwick OA, Khomo LM, Rogers KH, Hartshorn AS, Kennedy-Bowdoin T, Knapp DE (2010) Regional insight into savanna hydrogeomorphology from termite mounds. Nat Commun 1(6):1–7. https://doi.org/10.1038/ncomms1066
Lewin R (1986) In Ecology, Change brings Stability. Science 234 (4780):1071–1073
Lindsey P, Allan J, Brehony P, Dickman A, Robson A, Begg C, Bhammar H, Blanken L, Breuer T, Fitzgerald K, Flyman M, Gandiwa P, Giva N, Dickson K, Nampindo S, Nyambe N, Steiner K, Parker A, Roe D, Thomson P, Trimble M, Caron A, Tyrrell P (2020) Conserving Africa's wildlife and wildlands through the COVID-19 crisis and beyond. Nat Ecol Evol. https://doi.org/10.1038/s41559-020-1275-6
Louw MA, Haussmann NS, le Roux PC (2019) Testing for consistency in the impacts of a burrowing ecosystem engineer on soil and vegetation characteristics across biomes. Sci Rep 9:19355. https://doi.org/10.1038/s41598-019-55917-x
Mayaud JR, Bailey RM, Wiggs GF (2017) Modelled responses of the Kalahari Desert to 21stcentury climate and land use change. Sci Rep 7:3887
McCarthy TS, Humphries MS, Mahomed I, Le Roux P, Verhagen BT (2012) Island forming processes in the Okavango Delta, Botswana. Geomorphology 179:249–257. https://doi.org/10.1016/j.geomorph.2012.08.016
Meysman F, Middelburg J, Heip C (2006) Bioturbation: a fresh look at Darwin's last idea. Trends Ecol Evol 21(12):688–695. https://doi.org/10.1016/j.tree.2006.08.002

Molnar P (2003) Nature, nurture and landscape. Nature 426:612–614

Morris T, Letnic M (2017) Removal of an apex predator initiates a trophic cascade that extends from herbivores to vegetation and the soil nutrient pool. Proc R Soc London B 284:20170111

Moritz C, Agudo R (2013) The future of species under climate change: resilience or decline? Science 341(6145):504–508. https://doi.org/10.1126/science.1237190

Msoffe FU, Ogutu JO, Said MY, Kifugo SC, de Leeuw J, Van Gardingen P, Reid RS, Stabach JA, Boone RB (2019) Wildebeest migration in East Africa: status, threats and conservation measures. https://doi.org/10.1101/546747

Naidoo R, Du Preez P, Stuart-Hill G, Beytell P, Taylor R (2014) Long-range migrations and dispersals of African buffalo (Syncerus caffer) in the Kavango-Zambezi Transfrontier Conservation area. Afr J Ecol 2014(52):581–584. https://doi.org/10.1111/aje.12163

Nature Ed (2017) In praise of bats. Nat Ecol Evol 1, 0071. https://doi.org/10.1038/s41559-017-0071

O'Bryan CJ, Braczkowski AR, Beyer HL, Carter NH, Watson JEM, McDonald-Madden E (2018) The contribution of predators and scavengers to human well-being. Rev Article Nat Ecol Evol. https://doi.org/10.1038/s41559-017-0421-2

O'Connor TG, Puttick JR, Hoffman MT (2014) Bush encroachment in southern Africa: changes and causes. J Afr J Range Forage Sci 31 (2):67–88. https://doi.org/10.2989/10220119.2014.939996

Pages JF, Jenkins SR, Bouma TJ, Sharps E, Skov MW (2019) Opposing indirect effects of domestic herbivores on saltmarsh erosion. Ecosystems 22:1055–1068. https://doi.org/10.1007/s10021-018-0322-5

Palmer TM (2003) Spatial habitat heterogeneity influences competition and coexistence in an African acacia ant guild. Ecology 84:2843–2855. https://doi.org/10.1890/02-0528

Pastier AM, Dauteuil O, Murray-Hudson M, Moreau F, Walpersdorf A, Makati K (2017) Is the Okavango Delta the terminus of the East African Rift System? Towards a new geodynamic model: geodetic study and geophysical review. Tectonophysics, 712–713:469–481. https://hal-insu.archives-ouvertes.fr/insu-01534694

Pennisi, E. (2015) Africa's soil engineers: termites. Nature 347 (6222)

Perkins JS (2018) Southern Kalahari piospheres: looking beyond the sacrifice zone. Land Degrad Dev 29(9):2778–2784. https://doi.org/10.1002/ldr.2968

Perkins JS (2019) 'Only connect': restoring resilience in the Kalahari ecosystem. J Environ Manage 249:109420

Perkins JS (2020) Take me to the River along the African drought corridor: Adapting to climate change. Botswana J Agric Appl Sci 14 (1):60–71. https://doi.org/10.37106/bojaas.2020.77

Perkins JS, Ringrose S (1996) Development co-operation objectives and the beef protocol: the case for Botswana: a study of livestock/wildlife/tourism/degradation linkages. Unpublished Report commissioned by Metroeconomica Limited on behalf of the European Union. Gaborone

Perkins JS, Naidoo R (2016) Phytochorology, connectivity and resilience within the KAZA TFCA: a cautionary case of the Kalahari System. State of KAZA Symposium 2016, "Where have we come from, where are we now and where are we going?" Proceedings, 31 October - 2 November 2016, Victoria Falls Safari Lodge, pp 48–49

Pike D, Mitchell JC (2013) Burrow-dwelling ecosystem engineers provide thermal refugia throughout the landscape. Animal Conserv 16(6). https://doi.org/10.1111/acv.12049

Pringle RM, Doak DF, Brody AK, Jocque R, Palmer TM (2010) Spatial pattern enhances ecosystem functioning in an African Savanna. PLoS Biol 8(5):e1000377. https://doi.org/10.1371/journal.pbio.1000377

Purdon A, Mole MA., Chase MJ, van Aarde RJ (2018) Partial migration in savanna elephant populations distributedacross southern Africa. Sci Rep 8: 11331. https://doi.org/10.1038/s41598-018-29724-9

Reuters (2018) 400 buffalo drown after stampede into river between Botswana and Namibia. https://globalnews.ca/news/4647703/400-buffalo-drown-after-stampede-into-river-between-botswana-and-namibia/

Reynolds SC, Marston CG, Hassani H, King GC, Bennett MR (2016) Environmental hydro-refugia demonstrated by vegetation vigour in the Okavango Delta, Botswana. Sci Rep 6:35951. https://doi.org/10.1038/srep35951

Rinaldo A, Dietrich W, Rigon R, Vogel GK, Rodrlguez-lturbe I (1995) Geomorphological signatures of varying climate. Nature 374:632–635. https://doi.org/10.1038/374632a0

Ripple WJ, Estes JA, Beschta RL, Wilmers CC, Ritchie EG, Hebblewhite M, Berger J, Elmhagen B, Letnic M, Nelson MP, Schmitz OJ, Smith DW, Wallach AD, Wirsing AJ (2014) Status and ecological effects of the world's largest carnivores. Science 343:1241484. https://doi.org/10.1126/science.1241484

Ripple WJ, et al (2015) Collapse of the world's largest herbivores. Sci Adv 1:e1400103. 10.1126/ sciadv.1400103

Roberts EM, Todd CN, Aanen DK, Nobre T, Hilbert-Wolf HL, O'Connor PM, Tapanila L, Mtelela C, Stevens NJ (2016) Oligocene termite nests with in situ fungus gardens from the Rukwa Rift Basin, Tanzania, Support a Paleogene African Origin for Insect Agriculture. PLoS One 11(6):e0156847. https://doi.org/10.1371/journal.pone.0156847

Schreiner-McGraw AP, Vivoni ER, Ajami H, Sala OE, Throop HL, Peters DPC (2020) Woody plant encroachment has a larger impact than climate change on dryland water budgets. Sci Rep 10:8112. https://doi.org/10.1038/s41598-020-65094-x

Schuurman GW (2007) Termite diets in dry habitats of the Okavango Delta region of northern Botswana: a stable carbon isotope analysis. Sociobiology 47(2):373–389

Schwiede M, Duijnisveld M, Bottcher J (2005) Investigation of processes leading to nitrate enrichment in soils in the Kalahari Region, Botswana. Phys Chem Earth 30:712–716

Sietz D, Ludeke MKB, Walther C (2011) Categorisation of typical vulnerability patterns in global drylands. Global Environ Change 21:431–440

Sitters J, Kimuyu DM, Young TP, Claeys P, Venterink HO (2020) Negative effects of cattle on soil carbon and nutrient pools reversed by megaherbivores. Nat Sustain 3:360–366. https://doi.org/10.1038/s41893-020-0490-0

Skarpe C, Henriksson E (1987) Research note—Nitrogen fixation by cyanobacterial crusts and by associative-symbiotic bacteria in Western Kalahari, Botswana. Arid Soil Res Rehabil 1(1):55–59. https://doi.org/10.1080/15324988709381128

Smith BD, Zeder MA (2013) The onset of the Anthropocene. Anthropocene. https://doi.org/10.1016/j.ancene.2013.05.001

Stigand AG (1923) Ngamiland. Geogr J. 62(6):401–419. http://www.jstor.org/stable/1781166

Tarnita CE, Bonachela JA, Sheffer E, Jennifer A. Guyton JA, Coverdale TC, Long RA, Pringle RM (2017) A theoretical foundation for multi-scale regular vegetation patterns. Nature 541:398–401

Thomas DSG, Knight M, Wiggs GFS (2005) Remobilization of southern African desert dune systems by twenty-first century global warming. Nature 435:1218–1221

Tollefson J (2020) Why deforestation and extinctions make pandemics more likely. Nature 584:175–176. https://doi.org/10.1038/d41586-020-02341-1

Tolsma DJ, Ernst WHO, Verwey RA (1987) Nutrients in soil and vegetation around two artificial waterpoints in Eastern Botswana. J Appl Ecol 24(3):991–1000

Trimble SW, Kendel AC (1995) The cow as a geomorphic agent — a critical review. Geomorphology 13:233–253. https://doi.org/10.1016/0169-555X(95)00028-4

Turner A, Antón M (2004) Evolving Eden: an Illustrated Guide to the Evolution of the African Large mammal Fauna. Columbia University Press, New York. ISBN 0-231-11944-5

van Oosterzee P (2017) Wildlife interrupted. New Scientist 236 (3155):9. Wildlife interrupted. 236(3155):32–35. https://doi.org/10.1016/S0262-4079(17)32408-9

van de Velde S, Mills BJW, Meysman FJR, Lenton TM, Poulton SW (2018) Early Palaeozoic ocean anoxia and global warming driven by the evolution of shallow burrowing. Nat Commun 9:2554. https://doi.org/10.1038/s41467-018-04973-4

Venter O, Sanderson E, Magrach A, Allan JR, Beher J, Jones KR, Possingham HP, Laurance WF, Wood P, Fekete BM, Levy MA, Watson JEM (2016) Sixteen years of change in the global terrestrial human footprint and implications for biodiversity conservation. Nat Commun 7:12558. https://doi.org/10.1038/ncomms12558

Weir JS (1972) Spatial distribution of elephants in an African National Park in relation to environmental sodium. Oikos 23(1):1–13. https://doi.org/10.2307/3543921

Wiggs GFS, Thomas DSG, Bullard JE, Livingstone I (1995) Dune mobility and vegetation cover in the southwest Kalahari Desert. Earth Surf Proc Land 20:515–529

Williamson D, Mbano B (1988) Wildebeest mortality during 1983 at Lake Xau, Bostwana. Afr J Ecol 26:341–344. https://doi.org/10.1111/j.1365-2028.1988.tb00987.x

Williamson D, Williamson J (1981) An assessment of fences on the large herbivore biomass in the Kalahari. Botswana Notes Rec 13:91–94

Zerboni A, Nicoll K (2018) Enhanced zoogeomorphological processes in North Africa in the human-impacted landscapes of the Anthropocene. Geomorphology. https://doi.org/10.1016/j.geomorph.2018.10.011

Jeremy S. Perkins is Associate Professor in Range Ecology at the Department of Environmental Science at the University of Botswana. He undertook his PhD research on Kalahari cattle posts in 1988 and returned to Botswana in 1992 and has lived and worked there ever since.

Index

F. D. Eckardt (ed.), *Landscapes and Landforms of Botswana*, World Geomorphological Landscapes,
https://doi.org/10.1007/978-3-030-86102-5

C

D

T

U

V

W